THE NEW SOLAR SYSTEM

FOURTH EDITION

The New Solar System

EDITED BY

J. Kelly Beatty
Carolyn Collins Petersen
Andrew Chaikin

 SKY PUBLISHING CORPORATION

 CAMBRIDGE UNIVERSITY PRESS

PUBLISHED BY SKY PUBLISHING CORPORATION
49 Bay State Road, Cambridge, Massachusetts 02138   http://www.skypub.com

AND BY THE PRESS SYNDICATE OF THE UNIVERSITY OF CAMBRIDGE
The Pitt Building, Trumpington Street, Cambridge United Kingdom

CAMBRIDGE UNIVERSITY PRESS
The Edinburgh Building, Cambridge CB2 2RU, UK   http://www.cup.cam.ac.uk
40 West 20th Street, New York, NY 10011–4211, USA http://www.cup.org
10 Stamford Road, Oakleigh, Melbourne 3166, Australia

First published 1981
Second edition 1982
Third edition 1990
Fourth edition 1999

Printed in Canada

Typeface ITC Galliard 9/12pt   *System* QuarkXPress®   [DS]

Library of Congress cataloging in publication data

The new solar system / edited by J. Kelly Beatty, Carolyn Collins
    Petersen, Andrew Chaikin. — 4th ed.
        p.   cm.
    Includes bibliographical references and index.
    ISBN: 0 933346 86 7 (alk. paper)
        1. Solar system.   I. Beatty, J. Kelly.   II. Petersen, Carolyn Collins.
    III. Chaikin, Andrew, 1956 –   .
QB501.N47   1999
523.2 — dc21   98-34472
                                                        CIP

A catalogue record for this book is available from the British Library

ISBN 0-933346-86-7 (Sky edition: softcover)
ISBN 0-521-64183-7 (Cambridge Univ. Press edition: hardcover)
ISBN 0-521-64587-5 (Cambridge Univ. Press edition: softcover)

Contents

Preface

J. Kelly Beatty
Carolyn Collins Petersen
& Andrew Chaikin

Jupiter appeared as a slender crescent as Voyager 1 departed the Jovian system on 24 March 1979.

THE STUDY OF our solar system is arguably the oldest branch of astronomy. Thousands of years before the invention of the telescope, human eyes looked into the night sky and perceived the steady, purposeful motion of certain "stars" among their neighbors. The wandering planets were seen as physical manifestations of prominent gods, whose Roman and Greek names remain with us to this day. Ancient astronomers became obsessed with deducing how and why the planets move as they do. Early telescopes, too crude to resolve dim galaxies, nonetheless revealed the Moon's stark craters, Jupiter's satellites, and the spectacle of Saturn's rings.

Given this legacy, it is hard to believe that 40 years ago — at the dawn of space exploration — planetary science was at its nadir. In the postwar 1940s and 1950s, world-class telescopes like the 5-m Hale reflector on Palomar Mountain were revealing the majesty of the distant universe. In those heady days, planetary research was considered second-rate science, and consequently very few professional astronomers observed Mars, Jupiter, or other planets on a regular basis. At many professional facilities, only a small fraction of the available observing time could be used for solar-system studies.

By stark contrast, the planets loomed large at the newly formed National Aeronautics and Space Administration, where mission planners had already set their sights on the Moon, Venus, and Mars. However, even though these worlds were within our technological reach, it soon became obvious that we had little concept of what spacecraft should do once they got there. Moreover, NASA managers were shocked to learn how few astronomers had any relevant experience in the planetary field.

Historian Joseph Tatarewicz describes the agency's predicament in his 1990 book *Space Technology and Planetary Astronomy:* "In order to develop lunar and planetary probes, calculate trajectories, and develop scientific instruments, NASA planners had hoped to enlist the expertise of a science that had spent literally thousands of years preoccupied with the planets. Yet on the very eve of planetary exploration astronomers showed little interest." Tatarewicz recalls that NASA officials even "stood before groups of astronomers and implored them to enter a field of study for which there were few inducements."

One of the few astronomers actively — and overtly — pursuing solar-system research at that time was Gerard P. Kuiper. Having already distinguished himself with research on double stars and stellar evolution, Kuiper had turned his considerable talent to the planets in the 1940s. Working with the limited tools of the era, utilizing his skills as an observer and his rigorous scientific judgment, Kuiper amassed a body of knowledge that formed the basis for modern planetary science.

In 1961, on the eve of the first interplanetary missions, Kuiper and Barbara Middlehurst published *Planets and Satellites,* the third installment in a four-volume series that surveyed the state of knowledge about the entire solar system. In its preface Kuiper argued that, even in the era of planetary exploration, telescopic observations from Earth would remain important. Today, even though our robotic emissaries have surveyed at close range every one of the known planets except Pluto, it is clear that he was right. Were it not for the ongoing scrutiny of the solar system by patient professional and amateur observers here on Earth, much of what you will read in these pages would still lie waiting to be discovered. We would know little about the potential danger posed by Earth-crossing asteroids and nothing about the planets of other stars, to cite but two examples.

When Gerard Kuiper died in 1973, he left a void not easily filled. No single person has yet matched his influence on and command of planetary science. In fairness, however, the present-day study of "a planet" can involve a host of scientific disciplines. Few scientists today possess a complete working knowledge of any one world, let alone the entire solar system. Over the years the ranks of traditional planetary astronomers have been fortified with geologists, physicists, chemists, mathematicians, fluid dynamicists, biologists, and others. We suspect Kuiper would be pleased to know that the study of our solar system now occupies the talents of roughly 1,500 researchers worldwide.

Given the tremendous growth of planetary science in both scope and detail, the task confronting this book's editors once again — to summarize the current state of what we know about the solar system — could only be accomplished by bringing together a wide variety of specialists. Each author endeavored not only to provide the most up-to-date information available, but also to identify the gaps in our understanding that beg further investigation. Their presentations, taken together, may not appear entirely self-consistent. Many topics are the subject of disagreement or even outright feuding. Others enjoy a consensus of theoretical opinion yet lack observational confirmation.

Since the third edition of *The New Solar System* was published in 1990, there have been so many new developments in planetary science that most chapters had to be entirely recast and several new ones added. Indeed, this fourth edition is nearly twice the length of the first. Advances in solar astronomy have transformed our understanding of the Sun. Robotic eyes have provided glimpses of several asteroids — one of which proved to have its own satellite. We have seen objects in the distant Kuiper belt, thus certifying that the Sun's dominion extends far beyond Pluto's orbit. And the final chapter provides a census of the rapidly growing number of known worlds around other stars.

In assembling this book, we have attempted, as before, to bring the fruits of recent planetary exploration to the widest possible audience. This is neither a textbook nor a "coffee-table" volume — it lies somewhere in between. By the same token, we have encouraged our authors to avoid both sweeping generalizations and incomprehensible details. An abiding theme has been that our solar system is no longer a collection of individual bodies that can be addressed in isolation. It is instead an interrelated whole, whose parts must be studied comparatively. Above all, we strove to make this enjoyable reading for those with either casual or professional interest.

We have drawn upon the talents of many individuals. Artist Don Davis and illustrator Sue Lee have imbued these pages with their colorful vitality and uncompromising attention to detail. We thank Richard Tresch Fienberg and Leif Robinson of Sky Publishing, and Simon Mitton of Cambridge University Press for editorial direction. Our thanks also go to designer David Seabourne and CUP production manager Tony Tomlinson, and to Mary Agner, Vanessa Thomas, Tal Mentall, Cheryl Beatty, Lynn Sternbergh, and Paul Williams for editorial and design support. Myche McAuley and Susan LaVoie of NASA's Planetary Photojournal ensured that we had the best possible spacecraft images at our disposal. And finally we are deeply indebted to SPC production manager Sally MacGillivray and photo researcher Imelda Joson, whose tireless perseverance and dedication were truly inspirational.

This edition has been two years in the making, an interval during which, sadly, we mourned the deaths of Carl Sagan and Gene Shoemaker, singular members of the planetary-science community who had served us as contributors and honored us with their close friendship. Also lost were Clyde Tombaugh, the discoverer of Pluto, and Jürgen Rahe, who helped manage NASA's planetary programs for many years. We acknowledge their accomplishments and unflagging spirit in furthering our understanding of the solar system.

✳ ✳ ✳ ✳

The final chapter of *Planets and Satellites,* written by Kuiper, is entitled "The Limits of Completeness." There he took stock of the known worlds, equating "completeness" with knowing how *many* worlds orbited the Sun. Today we think of completeness in a much broader sense, and it remains unattainable. Consider, for example, Voyager 1's stunning image of a crescent Jupiter *(page vii),* which adorned the cover of this book's first two editions. Unobtainable from Earth, this view reminds us that our perspective — then, as now — is ever changing. As we write this, the Galileo, Mars Global Surveyor, and Lunar Prospector spacecraft continue their respective orbital vigils around Jupiter, Mars, and the Moon. Cassini and its Huygens probe are en route to Saturn, the NEAR spacecraft is closing in on asteroid 433 Eros, and Stardust is being readied for its sample-gathering dash through the periodic comet 81P/Wild 2. These missions, together with ongoing observations by the Hubble Space Telescope and by ground-based telescopes around the world, virtually guarantee that a fifth edition of *The New Solar System* will soon be a welcome necessity.

Exploring the Solar System

David Morrison

NASA's 70-m-diameter Goldstone tracking antenna near Barstow, California.

THE **EXPLORATION OF** our solar system has stimulated one of the most important scientific revolutions of the last third of the 20th century, comparable in significance to deciphering the genetic code of life. Planetary exploration has been carried out by the astronauts who traveled to the Moon, by robotic spacecraft that extend our reach to other planets, and by thousands of scientists working in observatories and laboratories on Earth. This international effort has yielded an initial reconnaissance of our cosmic neighborhood — the planets and other objects that share our solar system with the Earth. It has transformed dozens of planets and satellites from mysterious dots of light into real worlds, each with its own unique environment and history.

Why do we explore? An urge to explore seems to be a fundamental human trait, a hallmark of the most successful human societies of the past millennium. Exploration is partly motivated by a desire to understand our environment and the way it works. For some, the satisfaction of exploration can be achieved from reading books or surfing the Internet or performing computer simulations. For many others, there is an added dimension of experience. In addition to purely intellectual knowledge, we want a more personal involvement. We want to travel to new places, either directly or vicariously. We feel an urge to cross that river, to climb that mountain, or to set foot on that new world.

Only a few fortunate individuals have had the opportunity to travel into space, and even fewer (12, to be exact) have left their footprints on another world. Modern communications, however, allowed more than a billion people to share via television the experience of astronauts walking on the Moon. More than 100 million "hits" were made in a single day on Mars Pathfinder's Internet site. In an era when most of the frontiers of Earth have been reached, millions of people have been engaged to some degree in exploring the wider frontiers of our planetary system.

Figure 1. The Voyager spacecraft. In many ways, Voyager represents the "typical" planetary-science mission. Laden with a variety of instruments to study its targets in different wavelengths, it was sent to the outer planets on a flyby trajectory. Other missions combine landers, orbiters, and probes.

The basic human urge to explore may have motivated our solar-system missions, but without other factors it is unlikely that the necessary political and funding priorities could have been achieved. Over most of its short history, space travel was a direct product of the Cold War between capitalism and communism. Although nationalistic and geopolitical motives generated the resources that made solar-system exploration possible, we are fortunate that scientists often provided the detailed leadership to focus that effort on specific space goals.

In the United States, the National Aeronautics and Space Administration (NASA) was established in 1958 as a civilian agency and given a charter that places highest priority on scientific exploration and the acquisition and dissemination of new knowledge. Although the organizational details have changed, NASA has always had a space-science office or division, usually led by administrators with strong scientific credentials. NASA's planetary mission centers — principally the Jet Propulsion Laboratory (JPL) with contributions from Ames Research Center, Goddard Space Flight Center, and Langley Research Center —

developed strong, science-driven cultures to advocate and manage planetary explorations like the highly successful Voyager missions (*Figure 1*).

For scientific and programmatic guidance, NASA has turned throughout its history to various committees and councils of the National Academy of Science's National Research Council. In addition, the broad scientific community has played an essential role through a process of open competition and peer review to select the scientific teams and instruments for planetary spacecraft. This unique government-academic partnership ensured a dominant role for science in American planetary missions. When the European Space Agency (ESA) inaugurated its own planetary program, it followed a similar partnership model. Even deep-space missions of the former Soviet Union drew on American studies, thus establishing a consistent scientific foundation for a truly international effort. Partly as a matter of luck, Soviet planetary exploration focused on Venus, while the NASA effort stressed Mars and, later, the outer planets. The two programs together accomplished much more than either might have alone.

At the dawn of the Space Age, planetary studies occupied the backwaters of astronomy. Only a handful of scientists worldwide actively studied the physical or chemical properties of the planets and their satellites. Indeed, progress was so slow that textbooks written in the 1920s still adequately described our level of knowledge 30 years later. Around the time of Sputnik 1's launch in 1957, people still speculated about global oceans of water on Venus. Belief in plant life on Mars was widespread, as was the theory that volcanoes created most of the Moon's craters. Speculations aside, we did not know the surface composition of any solid planet or satellite in the solar system — except Earth.

A PLANETARY TOOLKIT

The most obvious limitation to knowing more about solar-system objects was — and still is — distance (*Table 1*). Our Sun illuminates its planetary realm like a single, bare light bulb in a huge meeting hall. The strength of sunlight decreases by the square of the distance from the Sun; likewise, the area subtended by planet or moon in our nighttime sky shrinks as the square of its separation from Earth. This double penalty makes the study of distant solar-system objects very difficult. For example, Pluto is presently about 30 AU from the Sun, so sunlight there is a mere 1/900 the intensity of what we enjoy here on Earth. Even though the diminutive planet is fully two-thirds the size of our Moon, it remains an unresolved, 14th-magnitude pinpoint of light in ground-based telescopes. If we could somehow bring Pluto inward to a point 1 AU from the Sun and view its fully illuminated disk from 1 AU away, it would outshine every star (except Sirius) in the nighttime sky. Astronomers would have then little difficulty mapping its major surface features or tracking its moon, Charon. In reality, when Pluto reaches aphelion in 2113, it will be 49 AU from the Sun and a very dim 17th magnitude — 25,000 times fainter than the limit of human vision.

Another hindrance to planetary astronomy is the presence of Earth's atmosphere. Its turbulent motions distort the clarity of our telescopic views, and its gases prevent much of the electromagnetic spectrum from reaching the ground (*Figure 2*). Light reflected from the planets and other solar-system objects is a Rosetta stone of information about their surfaces, atmospheres, and positions. Since human eyes are sensitive only to a very small part of this light, astronomers have developed detectors to study solar-system objects in as many wavelengths of light as possible.

A primary tool of such remote sensing is *spectroscopy* – the science of separating light into its component wavelengths. Certain spectral regions have turned out to be quite useful for the detection and study of specific characteristics: ultraviolet wavelengths for atmospheres and magnetospheric ions, the near-infrared for understanding the mineralogical makeup of a solid surface, and radio (especially radar) for mapping gross surface properties. The spectra of atmospheric gases are generally diagnostic of composition and can even be used to infer the quantities of gas present. Solid surfaces are more problematic. Simple ices (those of water, methane, and carbon dioxide) can be identified easily, but the mineralogical interpretation of rocky surfaces is sometimes ambiguous even with good observational data.

There are other useful observational tools in the planetary scientist's "toolkit." *Photometry* characterizes the changes in the

Key Parameters for Planetary Exploration

Planet	Distance from Earth (AU)	Greatest apparent magnitude	Largest diameter (arcseconds)	Mean solar constant (Earth=1)	One-way flight time (years)
Mercury	0.594	+0.6	8.4	6.67	0.18
Venus	0.267	−4.1	62.5	1.19	0.29
Mars	0.563	+1.7	6.0	0.431	0.71
Jupiter	3.966	−2.9	49.7	0.037	2.73
Saturn	8.293	+0.1	20.0*	0.011	6.05
Uranus	18.85	+5.7	3.7	0.003	16.03
Neptune	29.12	+7.8	2.3	0.001	30.60
Pluto	29.07	+13.7	0.1	0.0006	45.46

Table 1. The vast distance to any of Earth's planetary neighbors poses obstacles for both telescopic observation and visiting spacecraft. Distances, magnitudes, and apparent angular diameters reflect the values at inferior conjunction for Mercury and Venus, and at oppositions for the outer planets. (*Saturn's rings span 45.5 arcseconds at opposition.)

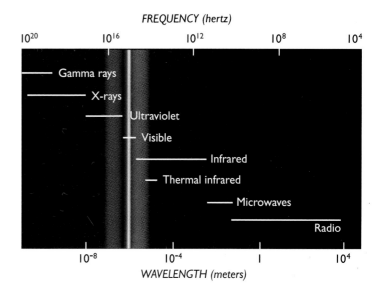

Figure 2. Only the narrow, visible-light region of the electromagnetic spectrum reaches Earth's surface relatively unimpeded by our atmosphere. Thus, to extend our spectral knowledge, instruments on interplanetary spacecraft study their targets in nearly every wavelength regime.

brightness of an object. For example, by monitoring variations in the light reflected from an asteroid over time (known as "obtaining a light curve"), we can learn how fast it spins, approximate its size and shape, and determine its color. *Polarimetry* takes advantage of another property of light: its polarization. When sunlight (or starlight) passes through an atmosphere, it is scattered and polarized by hazes and aerosols. Surfaces can polarize light as well, and polarimetry is particularly useful in the study of ring particles. *Astrometry,* usually thought of as a tool for tracking the precise positions of stars, is critically important in deducing the orbital characteristics of all solar-system bodies. Astrometric measurements of the exact positions of Pluto and Charon, for example, have given us reasonably precise estimates of their masses.

These remote-sensing techniques have extended our sensory range and given us views of solar-system objects that we could not otherwise obtain (*Figure 3*). The power of all these tech-

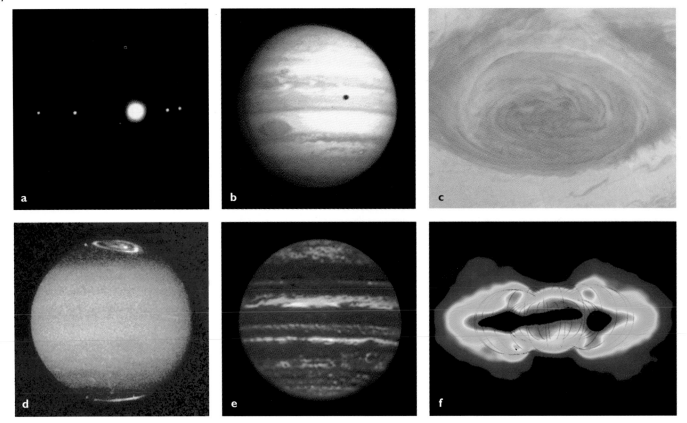

Figure 3. These images of Jupiter demonstrate the importance of looking at an object using as many vantage points and wavelengths as possible. Seen through small telescopes, Jupiter and its four largest satellites (*a*) reveal few distinguishing characteristics. In 1973, Pioneer 10 transmitted the first crude images (*b*) from the giant's vicinity. Most recently (*c*), the Galileo orbiter provided ongoing imagery of the constantly changing Jovian weather patterns, particularly the Great Red Spot. Astronomers back at Earth can study the planet at a wide range of wavelengths. The orbiting Hubble Space Telescope monitors the development of Jovian auroras in the ultraviolet (*d*). From the ground, at 4.8 microns in the infrared (*e*), holes in Jupiter's cloud structure are quite bright. Finally, a radio image (*f*) shows emission from charged particles cycling along field lines within the planet's magnetosphere.

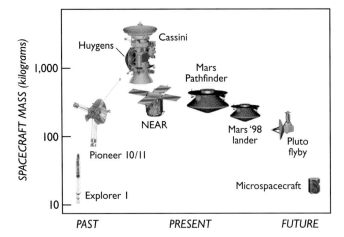

Figure 4. The capabilities of interplanetary spacecraft have seen remarkable advances over the past four decades — even though their size has waxed and waned. Miniaturized electronic components now make possible entire spacecraft no larger than the cameras on early lunar and planetary probes.

niques can be enhanced by placing detectors in orbit, thereby avoiding the degrading effects of Earth's atmosphere on resolution and spectral transmission. Moving these same sensors out to the planets themselves affords even better spatial and spectral resolution. Not surprisingly, therefore, interplanetary spacecraft have traditionally carried instruments covering a wide range of the electromagnetic spectrum.

THE EARLY YEARS: GETTING TO KNOW OUR NEIGHBORS

In the 1960s, when scientists first acquired the propulsive means to send instruments to the Moon and beyond, their immediate goal was a basic reconnaissance of the solar system. The first successful planetary mission, Mariner 2 (launched in 1962), had as its primary objective at Venus to determine the source of microwave emissions discovered by ground-based radio astronomers, and thus to answer the fundamental question of whether the planet's surface was hot (700° K) or temperate. Similarly, the first mission to carry a camera, Mariner 4 (launched in 1964), was designed to determine if the Martian surface was cratered and old or mountainous and geologically active, and also to measure the surface pressure of the atmosphere. These were truly basic questions, necessary to a first-order characterization of our nearest planetary neighbors.

In 1963, when the late Carl Sagan arrived at Harvard College Observatory as a young assistant professor, he gave a series of popular public lectures entitled "Planets are Places." At the time this was a radical idea, to think of the planets as other worlds to be compared with Earth. Scientists considering career paths in planetary science contemplated such questions as "What would it be like to stand on the surface of another planet? What does the ground look like? What is the temperature? What color is the sky?"

Planetary missions have changed a great deal since those early days. In response to budgetary changes and instrumentation advances, spacecraft have gotten first larger and then smaller (*Figure 4*). Some of the early planetary flights were focused on practical questions, serving as pathfinders for later scientific missions (*Table 2*). For example, the primary justification for the Surveyor lunar landers of the 1960s was to demonstrate that lunar dust had sufficient strength to bear the weight of the Apollo landers that would follow. The Pioneer 10 and 11 missions to Jupiter, the first spacecraft sent to the outer solar system, were built to determine whether a spacecraft could pass through the asteroid belt without being destroyed by collisions with small particles, and to assess the survivability of electronics in the intense radiation environment of the inner Jovian magnetos-phere. Without Pioneer, the later Voyager missions to the outer solar system would not have been possible.

The initial characterization of the planets was not all achieved by spacecraft. In the United States, NASA and the National Science Foundation provided funds to build new telescopes and equip laboratories, which created the foundation for a new multi-disciplinary field: planetary science. It was earthbound radio astronomers who discovered the high surface temperature of Venus and the magnetosphere of Jupiter; radar astronomers who established the rotation periods of Venus and Mercury and determined the distances between the planets with high accuracy; infrared and visible-light observers who measured the internal

Table 2. **Over the past four decades, spacecraft launched by the former Soviet Union (italic type) and United States have amassed an impressive list of milestones as space explorations have extended ever farther from Earth.**

Milestones in Solar-System Exploration

Spacecraft	Launch	Encounter	Object	Accomplishment
Explorer 1	1 Feb 1958	Feb 1958	Earth	detection of charged-particle belts
Luna 2	12 Sep 1959	15 Sep 1959	Moon	impact with surface
Luna 3	4 Oct 1959	7 Oct 1959	Moon	photograph of far side
Mariner 2	27 Aug 1962	14 Dec 1962	Venus	flyby
Ranger 7	28 Jul 1964	31 Jul 1964	Moon	photographs at close range
Mariner 4	28 Nov 1964	14 Jul 1965	Mars	flyby
Luna 9	31 Jan 1966	3 Feb 1966	Moon	photographs from surface
Venera 3	16 Nov 1965	1 Mar 1966	Venus	impact with surface
Luna 10	31 Mar 1966	3 Apr 1966	Moon	orbiter
Surveyor 1	30 May 1966	2 Jun 1966	Moon	controlled soft landing
Lunar Orbiter 1	10 Aug 1966	14 Aug 1966	Moon	photographic orbiter
Zond 5	15 Sep 1968	18 Sep 1968	Moon	round trip with life forms
Apollo 8	21 Dec 1968	24 Dec 1968	Moon	human crew (no landing)
Apollo 11	16 Jul 1969	20 Jul 1969	Moon	humans explore surface; samples returned to Earth
Luna 16	12 Sep 1970	20 Sep 1970	Moon	automated sample return
Luna 17	10 Nov 1970	17 Nov 1970	Moon	surface rover
Venera 7	17 Aug 1970	15 Dec 1970	Venus	soft landing
Mariner 9	30 May 1971	13 Nov 1971	Mars	long-life orbiter
Mars 3	28 May 1971	2 Dec 1971	Mars	soft landing
Pioneer 10	3 Mar 1972	3 Dec 1973	Jupiter	flyby
Mariner 10	3 Nov 1973	29 Mar 1974	Mercury	flyby (also on 21 Sep 1974 and 16 Mar 1975)
Venera 9	8 Jun 1975	22 Oct 1975	Venus	photographs from surface
Viking 1	20 Aug 1975	20 Jul 1976	Mars	photographs from surface; search for life forms
Pioneer Venus 1	20 May 1978	4 Dec 1978	Venus	long-life orbiter
Voyager 1	5 Sep 1977	5 Mar 1979	Jupiter	flyby
Pioneer 11	6 Apr 1973	1 Sep 1979	Saturn	flyby
Voyager 1		13 Nov 1980	Saturn	flyby
Vega 1	15 Dec 1984	11 Jun 1985	Venus	atmospheric balloon
ICE (ISEE 3)	12 Aug 1978	11 Sep 1985	comet	flyby through plasma tail of 21P/Giacobini-Zinner
Voyager 2	20 Aug 1977	24 Jan 1986	Uranus	flyby
Vega 1		6 Mar 1986	comet	photographs of 1P/Halley's nucleus
Voyager 2		25 Aug 1989	Neptune	flyby
Galileo	18 Oct 1989	29 Oct 1991	Gaspra	flyby of S-type asteroid
Ulysses	6 Oct 1990	13 Sep 1994	Sun	polar flyover at −80° latitude (European-built spacecraft)
Galileo		7 Dec 1995	Jupiter	orbiter, atmospheric probe
NEAR	17 Feb 1996	27 Jun 1997	Mathilde	flyby of C-type asteroid
Mars Pathfinder	4 Dec 1996	4 Jul 1997	Mars	automated surface rover

6

heat sources of the giant planets and discovered the rings of Uranus; and laboratory chemists studying meteorites and lunar samples who established the chronology and fundamental geochemistry of the solar system.

The first planetary missions were focused on answering a few specific questions, such as measuring the surface temperature of Venus or the bearing strength of the lunar soil. Influential scientists argued at the time that this was the proper way to carry out such an investigation: begin with a hypothesis, pose one or more specific questions to test the hypothesis, and fly a mission to make the critical measurements. Very quickly, however, it became apparent that spacecraft could do far more than answer a few predetermined questions (sometimes called "focused sci-

ence"). Traveling to other planets, and eventually orbiting them and landing on their surfaces, spacecraft had demonstrated a remarkable capacity for serendipitous discovery. All they had to do, in effect, was look around — and the result would be wonderful new discoveries. Besides, it was not cost-effective to send a spacecraft all the way to another world just to answer a few questions when so much more could be done with cameras and other broadly based investigations.

While nearly every spacecraft carried a payload customized for its particular target, a few basic types of measurements predominated. Only when cameras were added could the new era of exploration really begin. Remote-sensing instruments included, in addition to cameras, spectrometers to analyze the light for compositional information, as well as ultraviolet and infrared systems to extend spectral sensitivity. These devices were, in effect, small telescopes mounted together on a common platform that could be pointed toward specific regions of the target with high precision. Spacecraft could acquire color images and true spectra to deduce surface compositions. (Recent missions carry sophisticated instruments that combine photography and spectroscopy in a single device.)

A second major class of instruments measured electromagnetic fields and charged particles — not only the intrinsic magnetic field of a planet, but also the complex interactions of the electrons and ions that are trapped in the planet's magnetosphere. A third class of instruments, carried later on descent probes, made direct measurements of atmospheric composition, temperature, and pressure.

The data from all these measurements were digitally encoded and transmitted to Earth. The new multipurpose instruments required radio bandwidth to transmit all this information. High data rate became the key to planetary exploration, so more efficient spacecraft transmitters were designed, and the giant receiving antennas of the NASA Deep Space Network were built (*Figure 5*). As a result, our interplanetary data rates increased from 8 bits per second (bps) from Mars in 1965 to more than 116,000 bps from Jupiter in 1979.

The most successful mission of this era of initial reconnaissance was Voyager. Launched in 1977, the two Voyager spacecraft took advantage of a rare (once-in-176-years) alignment of the outer planets to achieve a "Grand Tour" of the outer solar system, flying at close range past Jupiter, Saturn, Uranus, and Neptune, each with an extensive system of satellites and rings (*Figure 6*). Each Voyager carried a dozen scientific instruments, and at Jupiter these spacecraft transmitted a detailed television image every 90 seconds. Even from Neptune, nearly 4 billion km from Earth, Voyager 2 sent us several images per hour (or an equivalent amount of other data). The twin spacecraft discovered rings, moons, and magnetospheres where none had been thought to exist. They vastly improved our understanding of the atmospheric dynamics on all four giant planets and sent back detailed images of 16 major satellites — several of them as large as planets themselves.

With each successive encounter, the Voyagers necessitated a rewrite of the texts that chronicle the advances in planetary science (like this one). Nothing we can ever do will equal this concentrated record of discovery and exploration. As Carl Sagan often pointed out, only one generation has the privilege of

Figure 5. Exploration of the solar system depends critically on the worldwide collection of tracking antennas known as the Deep Space Network. The largest DSN antennas, like this 70-m-wide dish located northwest of Barstow, California, can receive transmissions from distant spacecraft far weaker than one-billionth of a watt.

accomplishing the first scientific characterization of the solar system, and a great deal of that characterization was accomplished by Voyagers 1 and 2 (*Figure 7*).

A PROGRESSION OF EXPLORATORY MISSIONS

Voyager was a flyby mission. Launched on powerful Titan-Centaur rockets and accelerated by the gravity of each planet they passed, both Voyager spacecraft achieved escape velocity with respect to the Sun. At the turn of the 21st century, nearly 8 billion km from the Sun, they continue to transmit data from far beyond the classical realm of the planets.

A flyby is a great way to get an overview of a planetary target, but the time available for detailed studies is strictly limited because the spacecraft usually speeds by at 10 to 20 km per second. Typically, the Voyager cameras surpassed the resolution of the best Earth-based telescopes only a week or two before a given flyby, thus providing a useful encounter period of about a month's duration. The spacecraft took at most a few days to pass completely through each planet's magnetosphere. Opportunities for close-up views of satellites were much shorter, usually limited to a few hours. The result would be a few dozen good photographs of each object, some in color (by taking successive exposures through different filters), with the best resolution limited to just a handful of images.

Figure 6 (right). The long-lived Voyager spacecraft have completed virtually all of the ambitious "Grand Tour" of the outer solar system first envisioned by planetary scientists in the 1970s. The only world not visited was Pluto, which was cut out of the itinerary when the missions were downsized and launched later than expected.

There is no opportunity with a flyby to look a second time at an interesting feature. Flybys are best at providing an overview that is the necessary starting point of planetary exploration. Their strength, of course, is that the spacecraft can then move on to another target, as Voyager so beautifully demonstrated. It is also much easier (and less expensive) to fly past a target than to assume an orbit around it or to land on its surface. Thus, virtually all the early planetary missions were flybys.

The next evolutionary step is to orbit the target planet. With an orbiter, the time available for study increases from a few hours or days to the lifetime of the spacecraft (usually set by the exhaustion of either its fuel or its budget). In the case of the giant planets, with their many satellites, orbiters like Galileo and

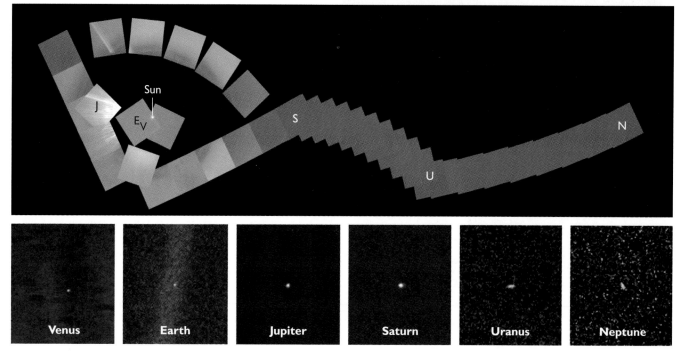

Figure 7. History will record that our robotic emissaries had traveled to the edge of the planetary realm by the close of the 20th century. In February 1990 the cameras aboard Voyager 1 pointed back toward the Sun and took the first-ever "portrait" of our solar-system. At the time the spacecraft was 6 billion km from Earth and situated 32° above the ecliptic plane. The 39-frame mosaic of wide-angle images (upper panel) captures the Sun and six planets. Voyager's narrow-angle camera then recorded telephoto views of the individual planets. Mercury was too near the Sun to be resolved, Mars was hidden amid scattered sunlight, and Pluto was simply too dim to register.

Cassini provide opportunities for multiple satellite flybys, allowing us to build up full coverage without orbiting each satellite individually.

Mariner 9 (launched in 1971) was the first successful planetary orbiter. The circumstances of its arrival at Mars illustrate the advantages of orbiting. At the time, Mars was shrouded in a global dust storm, making it impossible to photograph or measure the surface. If Mariner 9 had been a flyby craft, its mission would have been a failure. Instead, it simply waited in orbit for four months until the dust had settled, then began the planned mapping of the entire planet. Mariner 9 also demonstrated the power of a global examination. Three preceding Martian flybys, Mariners 4, 6, and 7, had made fascinating discoveries but, due to bad luck, had missed all of the younger geologic features. They spied none of Mars's great volcanoes, the rift valley Valles Marineris, or the evidence of ancient stream beds and water erosion. Indeed, there were suggestions made to terminate the exploration of the red planet after Mariners 6 and 7, on the

Figure 8. Jet Propulsion Laboratory director William Pickering, physicist James Van Allen, and rocket designer Wernher von Braun (left to right) hold a model of Explorer 1 and its integrated Sargeant rocket stage after the satellite's 1958 launch. Explorer 1 was the first American satellite and, by virtue of its discovery of the Van Allen radiation belts, the first planetary-science mission.

grounds that Mars was a dead planet. Fortunately, by then Mariner 9 had already been built, and it changed our view of Mars forever.

The most productive planetary orbiter, measured by the volume of scientific data obtained, was Magellan. Venus, the nearest planet to Earth, is perpetually shrouded in clouds, its surface invisible to both optical telescopes and spacecraft cameras. Microwave radiation, however, can penetrate the atmosphere, and radar operating at microwave frequencies can be used to map the surface, as first demonstrated by large radar telescopes on Earth. In 1978, the Pioneer Venus orbiter constructed a crude radar map of Venus, followed in 1983 by two Soviet radar mappers, Veneras 15 and 16. The resolution of the Pioneer global map was 50 km, and the Veneras obtained radar images of much of the northern hemisphere at about 2 km resolution.

While enticing, the results of the Pioneer and Venera missions were unable to answer first-order questions about the geology of Venus, such as the age of its surface or the possible presence of plate tectonism. Thus Magellan was built to obtain a global map at 100-m resolution using radar imaging. Orbiting from 1989 to 1992, it returned more data than all previous planetary spacecraft combined, yielding a global map of Venus more detailed than our knowledge of many of the submarine portions of Earth. Planetary geologists are still making discoveries about Venus as they sift through its treasure-trove of radar images.

After we have orbited a planet, mapped its surface, measured its magnetic and gravity fields, and observed its weather from above, the next step is usually an atmospheric entry probe or surface lander. On Mars or Venus, it is possible to combine the probe and lander, measuring detailed atmospheric properties during descent and deploying the lander on the surface. At the opposite extreme, Jupiter has no surface, and in late 1995 the Galileo atmospheric probe just kept descending until it was vaporized by high temperatures.

Early interplanetary spacecraft were actually derivatives of those used to study Earth's upper atmosphere (*Figure 8*). The first successful probe of another planet was the Soviet Union's Venera 4, which in 1967 entered the atmosphere of Venus, deployed a descent parachute, and transmitted measurements of density and temperature. At the time, Soviet scientists announced that Venera 4 had dropped through the entire atmosphere and crash-landed on the surface. However, American scientists immediately questioned this claim, since it indicated a surface pressure on Venus about five times lower than had been implied by radar measurements combined with data from the Mariner 5 flyby. After some detective work, the Soviet team found that the probe had ceased transmitting when the atmospheric pressure exceeded its design limit. Not expecting so massive an atmosphere, Venera 4's designers had not provided for pressures higher than 15 bars. Subsequent Soviet probes were given stronger hulls, and Venera 7 successfully reached the surface of Venus in 1970.

Technologically, the solar system's most demanding probe target is Jupiter. Any free-falling object enters a planet's upper atmosphere at a speed nearly equal to the escape velocity, and giant Jupiter has the highest escape velocity of any planet. All of this velocity and kinetic energy had to be carefully dissipated if the Galileo probe were to decelerate safely. The spacecraft slowed from 47 km per second to subsonic velocities in only 110 seconds, enduring an abrupt deceleration that was 228 times the gravitational acceleration on Earth. Then it deployed a parachute and descended for nearly an hour, making measurements and transmitting data to Earth. At a pressure of 24 bars and a temperature of 425° K, transmissions ceased, and about 9 hours later the aluminum-titanium probe had sunk to a level in the atmosphere where the components melted and evaporated. The Galileo probe thus became a part of Jupiter's atmosphere.

The first successful robotic landers were Luna 9 and Surveyor 1, both of which landed on the Moon in 1966. As mentioned earlier, Venera 7 successfully transmitted data from the surface of Venus in 1970, followed by many successful Soviet missions to Venus. Soviet space engineers also achieved the first controlled landing on Mars with their Mars 4 in 1970, but after an apparently flawless entry sequence the spacecraft ceased transmitting data after just 20 seconds.

The U.S. spacecraft Vikings 1 and 2 made the first scientifically successful Mars landings in 1976. Built and flown at a cost of approximately $2 billion (in 1998 dollars), the Vikings were also the most expensive planetary spacecraft ever (excluding Apollo missions). They undertook a daunting challenge: to land safely on a surface that had never been imaged at the resolution of the landers themselves (so that we could not know of rocks or other hazards in the landing area) and to do so autonomously (since the communications travel time between the spacecraft and Earth precluded intervention during the landing sequence). In the end, both spacecraft accomplished flawless landings, becoming highly capable, nuclear-powered, 1-ton laboratories on the Martian surface. Both also greatly exceeded their nominal 90-day lifetimes. In fact, Viking 1's lander survived three frigid Martian winters and continued to transmit data until November 1982, when it was silenced by an engineer's programming error. Together with the two Voyager spacecraft, the Viking mission represented the high point of what has been called the Golden Age of planetary exploration.

One thing the Viking landers lacked was mobility. No one could look at their beautiful panoramas of the Martian surface without wanting to see what lay beyond the nearby hills. NASA considered a plan to launch a Martian rover in 1984, but it — and many other proposed missions — fell victim to budget cuts in the 1980s. The first rover did not arrive on the red planet until 1997, and while Sojourner worked well, its 100-m range did not allow excursions beyond the horizon of its accompanying lander (Mars Pathfinder).

Ultimately we wish to move beyond landing and roving, and to return samples of planetary surfaces for study in terrestrial laboratories. Only in a modern lab will such rocks give up the secrets of their origin and age, reveal the internal chemistry and geologic history of their parent planet, and perhaps even provide evidence of fossil life. Samples of many asteroids arrive at Earth in the form of meteorites, though in only a few cases can we confidently relate the meteorites to their parent asteroids. A handful of lunar and Martian rocks have also been found on Earth, ejected into space by hypervelocity impacts and eventually colliding with our planet. The most impressive sampling effort took place as part of the Apollo program, when more than a half ton of Moon rocks were carefully selected by astronauts and brought back with them. In 1970, a small amount of lunar material was also returned to Earth robotically by Luna 16, followed by two other Soviet sample-return missions. Today sample-return missions are a prime focus of NASA's Mars-exploration program. Also planned are the return of material from a comet (a mission called Stardust) and from a near-Earth asteroid (Japan's MUSES C).

The first Golden Age ended following the 1978 launches of the two Pioneer Venus craft. In 1981 the administration of President Ronald Reagan seriously considered terminating the NASA planetary program entirely. Budget director David Stockman announced that he expected the United States to be out of the planetary-exploration business by 1984, and he even proposed switching off the Voyagers after their Saturn encounters and closing down the Deep Space Network. This tragedy was averted, but budgets continued to be extremely tight following the loss of the Space Shuttle *Challenger* in 1986. More than 10 years elapsed between the Pioneer Venus launches and those of Magellan and Galileo. By this time the Soviet Union was crumbling, and its planetary program soon fell victim to Russia's desperate financial troubles. The launch failure of the ambitious Mars '96 mission probably marked the end of an independent Russian planetary program. The more modest efforts of the European Space Agency and Japan, though successful, could not begin to fill the gaps left by the United States and Russia.

WHAT HAVE WE LEARNED?

By 1990, more than 30 successful planetary missions had been flown (excluding lunar missions), primarily by the United States and Soviet Union. Most of these were aimed at initial reconnaissance of the solar system, though a few (Viking, the later Venera missions, and Magellan) had moved beyond reconnaissance into extensive exploration.

To summarize the results of these missions, let us define initial reconnaissance as equivalent to obtaining images with more than 10,000 picture elements (pixels) on the surface, together with some characterization of the local environment including magnetic and gravity fields. An image of this size, equivalent to a square of 100 by 100 pixels, is comparable to those we see every day on television, in newspapers, and on most Internet sites. To place this imaging yardstick in perspective, note that only five objects in the solar system can be photographed at this resolution by the Hubble Space Telescope: Venus, Mars, Jupiter, Saturn, and Uranus. And of course the Hubble telescope cannot measure magnetic or gravity fields at all.

Using this definition, by 1998 we had achieved an initial reconnaissance of eight of the nine planets (all but Pluto), 16 large satellites, and six small bodies (three asteroids, the two moons of Mars, and Comet Halley). In the decade of the 1990s, long-term studies of the Sun by such spacecraft as the Japanese Yohkoh and the multinational Solar and Heliospheric Observatory have changed our understanding of solar physics. The contents of this book bear witness to the scientific productivity of all these space missions, together with observational, laboratory, and theoretical research carried out in parallel with them.

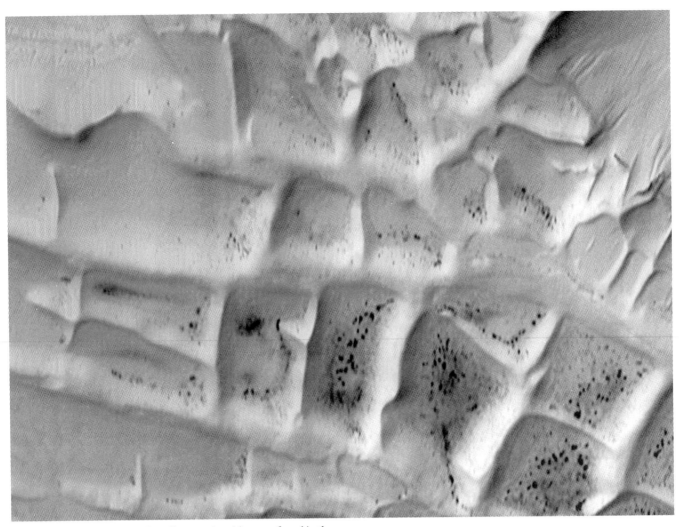

Figure 9. This enigmatic complex of intersecting ridges was found in the Martian south-polar region early in the mapping mission of NASA's Mars Global Surveyor. Each ridge is about 1 km wide, and the smallest discernible details are no more than 25 m across.

With high-resolution images it is possible to provide very basic characterization of a solar-system object, for example, to tell whether it is geologically active. Think of a typical "talking head" on the nightly television news: we can see all the main features but tend to miss most of the small wrinkles and blemishes. On small bodies such as asteroids it is possible to identify many individual craters with a 10,000-pixel image, since the largest craters are a fair fraction of the size of the object itself. But for a larger planet or the Galilean satellites of Jupiter, we must achieve a million-pixel image (1,000 by 1,000 elements) to resolve individual geologic features like impact craters, volcanoes, or lava flows. A million-pixel image is exemplified by the photographs we see printed in this book or on a high-resolution computer monitor. This is the sort of resolution obtained by Mariner 9 on Mars or by Voyager and Galileo in the Jovian system. With such images it is possible to distinguish different geologic units and to place these in a temporal sequence.

Until recently, a resolution of a kilometer or two was considered adequate for most geologic interpretations. However, the much higher-resolution images of the Jovian satellites obtained by Galileo and of the Martian surface by Mars Global Surveyor challenge this conclusion (*Figure 9*). The new photos, acquired with 5- to 50-m resolution, really look as if they were of different planets than those seen by the Voyagers and Vikings. They suggest geologic histories that could never have been guessed from the earlier data. As we analyze these higher-resolution images, we may have to rethink our ideas of what constitutes a general reconnaissance of a planetary surface.

Nonetheless, over the years a few common themes have emerged that cut across the planetary system. One of these is the ubiquity of impact cratering, testifying to a common intense bombardment of the inner planets in the first 500 million years of solar-system history as well as a continuing bombardment since. Every solid planet or satellite bears the scars of such impacts. It appears that the continuing flux of colliding objects is roughly the same for each of the inner planets (which are hit by both asteroids and comets) and a few times lower in the outer solar system (where comets are the only source of projectiles). Even a quick look at a spacecraft image reveals, from the number of visible craters, whether the object is geologically young or old (and hence geologically active or not). Thus the Moon, a world that has experienced little internal activity over the past 3 billion years, is heavily cratered. In contrast, Earth and Venus, both of which have typical surface ages of a few hundred million years, are rather sparsely cratered. At higher resolution, it is possible to count the numbers of craters of different sizes and their state of degradation, revealing the planet's general timetable of geologic evolution and allowing comparison of one object with another.

Volcanism provides another common element. Nearly every solid object more than a few hundred kilometers in diameter shows some evidence of internal melting and surface eruptions. For rocky planets and satellites, volcanism creates structures that are closely analogous to Earth's: shield volcanoes, calderas, rift zones, cascading flows, and even lava tubes. Surprisingly, many cold icy satellites also show evidence of fluidized eruptions, termed cryovolcanism. However, the "lava" in these cases could not have been molten silicate rock. More likely, at such low temperatures the working fluid is an exotic mixture of water and ammonia, icy slush, or simply warm ice.

Surface volcanism is an expression of the release of heat from the interior of a planet or satellite. This energy can remain from primordial times, a vestige of the object's accumulation of high-speed debris in the early solar system. It might also reflect the continuing decay of radioactive elements in the interior, or the dynamic heating created when a satellite interacts tidally with its parent planet. The greatest energy sources lie within the giant planets, three of which release interior heat in quantities comparable to the energy they absorb from the Sun. Among the giant planets only Uranus lacks such a heat source, for reasons that are not understood.

A consequence of this internal heating is that all of the larger solar-system objects are differentiated; that is, their interiors have sorted themselves into layers of different density, with the heavier metals at the center and the least-dense materials in the crust. Among the larger objects, only Callisto, the outer Galilean satellite of Jupiter, appears to have avoided a thorough differentiation.

One of the main reasons for making comparative studies of the solar system's members is to reveal the process by which the system itself formed and evolved. The general scheme has smaller, more oxidized, and metal-rich planets close to the Sun and larger, chemically reduced planets with their retinues of ice-rich satellites farther out. This arrangement is interpreted as the imprint of processes that were occurring within the disk of gas and dust from which the planets formed.

Beyond such broad generalizations, the compositions of the individual objects tell us a great deal about the details. In particular, the elemental and isotopic compositions of planetary atmospheres tell us how volatile materials were redistributed after the formation of planetary cores. Presumably the impacts of volatile-rich objects produced veneers of exotic materials, which today make up much of the atmosphere and hydrosphere of Earth and other terrestrial planets. These collisions also subtly modified the atmospheric composition of the outer planets. At this level of chemical detail, most of our data come from probes of the atmospheres of Venus, Mars, Jupiter, and, of course, Earth. Unique chemical data relevant to the formation and early history of the solar system are also obtained from laboratory analyses of meteorites and cosmic dust.

Finally, in any summary of the common aspects of the members of the solar system, the most general conclusion is that there are no valid generalizations, and that each world has experienced its own unique history. Time after time we have been surprised and thrilled as spacecraft revealed the unexpected. For example, scientists before Voyager had confidently predicted that the inner Galilean satellite Io would be heavily cratered and rather lunar in appearance, never anticipating its extraordinary level of volcanic activity. The little Uranian satellite Miranda was another complete surprise; we still do not understand the origin of its bizarre landforms. And we thought of asteroids as solid rock until the Near Earth Asteroid Rendezvous (NEAR) spacecraft showed that 253 Mathilde contains as much void space as rock in its interior. Veteran planetary scientist Laurence Soderblom once remarked that there are no dull satellites — once you look at them closely. Presumably we could generalize this sentiment to all the members of the planetary system.

STRATEGIES FOR FUTURE EXPLORATION

As we approach the 21st century, there is a resurgence of public and governmental interest in planetary exploration. However, the current level of financial support for NASA dictates that planetary missions be smaller and more efficient than their pre-

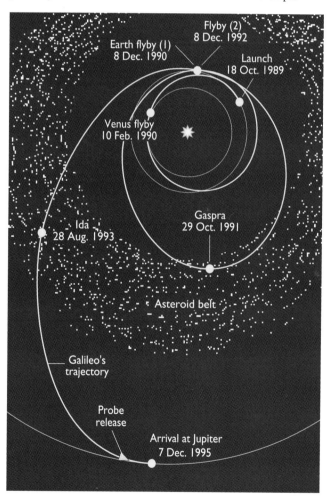

Figure 10. **The Galileo orbiter-probe spacecraft was so massive that no existing rocket had the power to launch it directly to Jupiter, its primary target. Instead, the spacecraft was placed on a looping trajectory that took it past Venus once and Earth twice — gaining enough velocity in the process to reach Jupiter. While the roundabout route took six years (versus just 21 months for the direct flights of Pioneers 10 and 11), it also afforded the opportunity for two flybys of asteroids along the way.**

decessors were. The Russians, Japanese, and Europeans are under similar pressure to reduce mission cost and complexity. Beginning in the early 1990s, NASA administrator Daniel Goldin demanded a new level of performance under the mantra "smaller, faster, cheaper." Cassini, with its mass of more than 3 tons and a cost in excess of $3 billion, for now stands as the last of the large missions in the tradition of Lunar Surveyor, Viking, Voyager, and Galileo (*Figure 10*). Largely because of Goldin's bold initiatives, the launch rate for NASA planetary missions has risen from two per decade in the 1980s to better than two per year in the late 1990s.

The reduction in cost and size of missions has been accompanied by changes in the way NASA missions are planned and executed. Individual "new starts," laboriously planned and marketed with delays of up to a decade between planning and launch, are being replaced with generic classes such as Mars missions or outer-planet missions, within which resources and priorities can be more easily allocated. In the case of Mars, NASA has committed to launching both an orbiter and a lander at each orbital opportunity (which occur about every 26 months) through the year 2010, under the general program name of Mars Surveyor.

Even more flexible are the Discovery missions, an ongoing series of low-cost initiatives selected from a competition held annually. The first Discovery mission, Mars Pathfinder, was launched in 1996 and landed on Mars in August 1997. Pathfinder operated for three months and included a small rover (Sojourner). Next came the NEAR mission, also launched in 1996, which flew past the main-belt asteroid 253 Mathilde in 1997 en route to reaching the near-Earth asteroid 433 Eros in 1999. Lunar Prospector, the lowest-cost mission of all at $63 million (including launch vehicle and operations), was launched in 1998 to map the lunar surface composition and gravity field. Next in the Discovery queue are Stardust, which is to return a sample from Comet Wild 2; Genesis, to sample the solar wind for clues to how our solar system formed; and Contour, a multiple-comet flyby mission.

The most important emerging themes in planetary exploration are the origin of the solar system and the search for evidence of life, past or present. Most of the Discovery missions to date address questions about our origins, which are often best answered by studying small primitive bodies such as comets and asteroids. Interest in extraterrestrial biology is even newer. Twenty years after the Vikings found a sterile Mars, discoveries in terrestrial biology are reinvigorating the field of exobiology and have created a broader umbrella discipline within NASA called astrobiology. Astrobiology is the scientific study of the origin, evolution, distribution, and future of life in the universe. To understand life's origin, we need to place terrestrial life in its cosmic context.

The search for evidence of past life is at the heart of the Mars Surveyor program. Although it is a frozen world today, Mars was not always so inhospitable. Large tracts of the surface were once washed by floods and drained by extensive river systems. Ancient lakes and hot springs have left their imprint as well. Significantly, Mars was its most Earthlike at the same time, between 3 and 4 billion years ago, that life first arose on our own planet. For the biological sciences, no discovery could be more exciting than the opportunity to study life forms having an independent, extraterrestrial origin. Surveyor's first priority is the search for evidence of fossil Martian life, but the possibility of extant life cannot be excluded. Over the next decade we will also identify the most promising landing sites on Mars, select biologically interesting rocks, and return those rocks to Earth.

The solar-system exploration program hopes to address many questions dealing with possible past or present life on Mars. Did conditions ever exist there that were conducive to the introduction of biology? Did Martian life, in fact, develop independently of that on Earth? Alternatively, did the exchange of impact-related debris between the two developing worlds seed one with microbes from the other? If life once existed on Mars, does it persist today in some protected environments? Could we detect and recognize it as such? If it has not survived, what went wrong? And what dangers might a surviving Martian ecosystem pose to the biological diversity on Earth?

Looking beyond Mars, we can examine the biological prospects for Europa. To many scientists, Galileo's images offer compelling evidence for a global ocean beneath the icy crust of this moon. But this is at best an indirect inference. The extension of Galileo's mission to allow more comprehensive studies of Europa is a first response to this interest. The next logical step would be to dispatch an orbiter equipped with radar to probe the Europan ice from afar. For Europa to now have a liquid-water mantle, substantial amounts of heat must be coming from the moon's interior. One can easily imagine the Europan equivalent of hydrothermal vents. On our planet, there is a thriving biota associated with these vents, independent of photosynthesis. By analogy, Europa could support similar lifeforms.

The search to understand our solar system's origins naturally raises questions about its distant fringes, where conditions may still resemble those in the solar nebula from which the planets formed. Such scientific curiosity, as well as a desire to complete the spacecraft tour of the major planets, is the motivation behind a mission to Pluto and perhaps other large icy objects in the Kuiper belt. NASA's Pluto Express, proposed for launch around 2004, is designed to reach the planet before its increasing distance from the Sun causes its tenuous atmosphere to collapse as frost onto the icy surface.

We face exciting opportunities in solar-system exploration. For NASA, lower-cost robotic missions provide the means for an expanded program, with special emphasis on Mars. The flexibility inherent in the competitively selected Discovery missions lets us respond rapidly to new conditions and scientific opportunities. Pluto Express will show us the last unexplored planet. Finally, a renewed interest in life's origins may set the stage for the kind of public and political support that could eventually propel humans — not just machines — to the surfaces of other worlds.

Origin of the Solar System

John A. Wood

FOR MORE THAN two centuries, scientific ideas about how the planets came to be were based almost entirely upon theory. There were few constraints on such speculation: astronomers knew that stars rotate and that the Sun's planets are nested in almost-circular orbits. But beyond these meager knowns, those who pondered the solar system's origin had nowhere to turn except to their own visions of how a huge amount of cosmic matter, obeying the laws of physics as then understood, might have organized itself into a Sun and planetary system. Theorists continued to offer variations on the same answer: the Sun and planets were born from a rotating disk of cosmic gas and dust. The flattened form of the disk constrained the planets that formed from it to have orbits lying in the same plane, or nearly so, all moving in the same direction the disk had turned. This hypothetical disk, the *solar nebula*, is where any discussion of the origin of our solar system must begin.

The idea of a solar nebula was first formulated in 1755 by the Prussian philosopher and physicist Immanuel Kant. Although his treatment of the problem was only qualitative, its precepts were remarkably similar to those considered fundamental today. Kant pictured an early universe evenly filled with thin gas. He thought such a configuration would have been gravitationally unstable, so it must have drawn itself together into many large dense clumps of gas. Kant correctly attached a great deal of importance to rotation: he assumed these primordial clumps of gas were rotating, and as they shrank the rotation spun them out into flattened disks. One of these disks became our solar system.

Kant's validation was long in coming, as generations of telescopes proved unable to resolve a disk associated with another star in the heavens. The first indirect evidence for disks came from studies of *T Tauri stars,* which are similar in mass to the Sun but very young — roughly a million years old. In the 1980s

The heart of the Orion Nebula, as recorded by the Hubble Space Telescope.

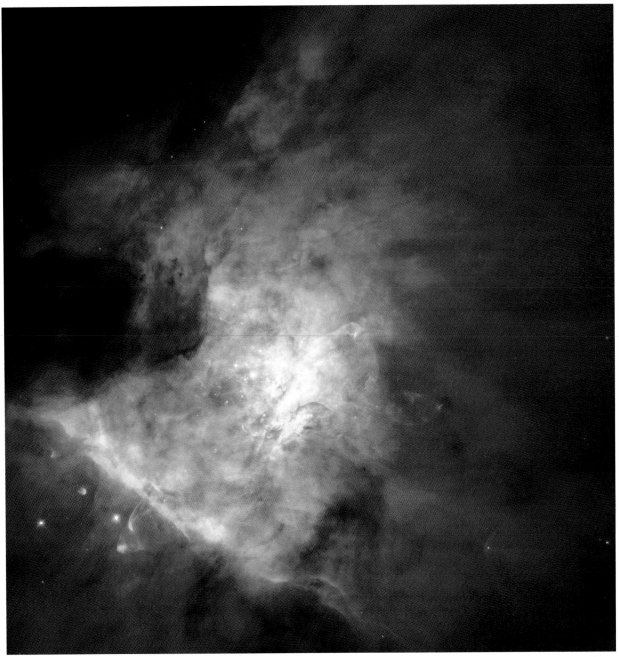

astronomers realized that about a third of T Tauri stars have "infrared excesses," that is, the amount of infrared radiation they emit is too great to be consistent with their output at visible wavelengths. This can be understood if the stars in question are surrounded by haloes of dust kept warm by short-wavelength radiation from the stars; the dust then reradiates the energy it receives at longer (infrared and radio) wavelengths. However, the strong infrared signatures implied the presence of enough dust, if distributed evenly in a sphere, to completely block our view of such a star at visible wavelengths. Only if the dust were arranged in a flattened disk, tilted somewhat to our line of sight, could we expect to see the star itself.

Our first direct views of these disks came a few years later, when observers used radio telescopes to see through the dark dusty clouds where stars are forming. Most recently the sharp eye of the Hubble Space Telescope (HST) has resolved several dozen disks at visible wavelengths in the Orion Nebula, a giant stellar nursery about 1,600 light-years away (*Figure 1*). They have been given the colorful name "proplyds," a contraction of the term *protoplanetary disks*. Some of these are visible as silhouettes against a background of hot, bright interstellar gas (*Figure 2*); others have been illuminated by nearby bright stars. The Orion disks are large compared with the solar system, and they contain more than enough gas and dust to provide the raw material for future planetary systems. The stars associated with these disks are very young, at most a few million years old.

FORMATION OF STARS AND DISKS

Today we realize that Kant, by and large, got it right. Stars and their disks form in much the way he pictured, by the gravitational collapse of huge volumes of thinly dispersed interstellar gas and dust onto appropriate nuclei. Many theoretical studies have attempted to model stellar collapse and its aftermath, but a completely realistic simulation remains an elusive goal. The setting for star formation is a huge cloud, a particularly dense concentration of interstellar matter with roughly 10,000 gas molecules per cubic centimeter — vanishingly small compared to Earth's atmosphere, yet much denser than most of interstellar space. The cloud is dark, cold (10° to 50° K), turbulent, and threaded by magnetic fields. By chance, it contains concentrations (cloud cores) that become the favored sites for gravitational collapse. The Orion Nebula provides just such an environment.

The magnetic fields in cloud cores tend to resist collapse, but sometimes gravity prevails, drawing cloud material through the field and concentrating it near the center of a core. As this nucleus grows, its attractive force overwhelms magnetic resistance, and core material begins to pour onto the nucleus or protostar at free-fall velocity. Pressure and temperature mount in the growing object, and it begins to radiate energy from its surface.

Rotation adds a complication to the picture. The original cloud core that produces a given star inevitably has some small amount of angular momentum, perhaps inherited from the slowly roiling turbulence of its parent cloud. Initially it may rotate only once every few million years. But as the core's substance collapses to the dimension of a star, the conservation of angular momentum dictates that it spin faster and faster.

Most of the collapsing core material has too much angular momentum to fall directly onto the protostar. Instead it attempts to follow an orbit (much like the orbit of a comet) around the protostar. However, the flows of material falling in from the northern and southern hemispheres of the rotating cloud core collide at its equatorial plane. The vertical motion of both flows cancels out, and their material is added to a disk that orbits about the protostar. This disk, not the protostar, carries most of the angular momentum inherited from the cloud core. But for the protostar to grow, matter must flow into it. Thus the disk's angular momentum must be redistributed outward, which allows the inner-disk material to join the protosun while outlying matter is spun to greater radii.

How disks accomplish this is not understood, but gravitational instability may once again be responsible. Once the accumulating disk achieves a mass roughly one-third that of the protosun, it becomes gravitationally unstable and changes from an elegant, symmetric item of cosmic dinnerware to something less regular, perhaps resembling a miniature galaxy. The dense lumps in such an asymmetric structure exert tidal forces on one another, which have the effect of moving angular momentum outward. This would feed some disk material into the protostar, decreasing the disk-protostar mass ratio and perhaps restoring gravitational stability. It may be that star/disk systems grow by a series of excursions into instability of this sort.

Real stars display phenomena, some very dramatic, that correlate with the processes pictured by theorists. In the earliest stages of collapse, protostars are deeply embedded in their dusty

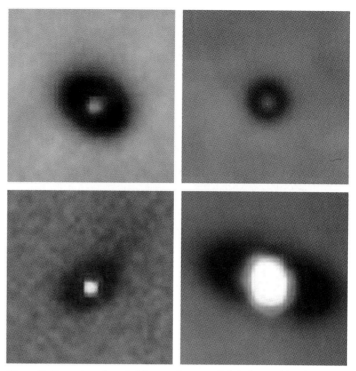

Figure 2. Examples of *proplyds* (protoplanetary disks) observed by the HST in the Orion Nebula. At the center of each is a T Tauri star. The disks surrounding them are two to eight times the diameter of our solar system.

parental clouds. Thus hidden, they remain observable only at radio and infrared wavelengths. Radio telescopes have detected powerful winds blowing away from such embedded objects in two opposing directions (*Figures 3,4*). These bipolar outflows are held responsible for turning back the inflow of dusty gas that feeds and conceals protostars. In time the outflows dominate, revealing the central masses as T Tauri stars. When seen in telescopes at visible wavelengths, such stars are girdled by disks (proplyds) and are still ejecting intense bipolar outflows. It takes about a million years to reach this stage of stellar evolution.

A solar-mass star remains in the T Tauri stage for another 10 million years (roughly) before it attains core temperatures high enough to initiate hydrogen burning and evolve further to an ordinary main-sequence star. During this time the associated dusty disk becomes less and less evident. Some of its substance probably continues to migrate into the central star; much of the residual gas may be heated so greatly by the star's ultraviolet radiation that it evaporates to interstellar space. A very small fraction of the dust and gas that passed through the disk may have collected into discrete lumps that remain in orbit around the star after the disk is gone, serving as the seeds of its planetary family.

THE SOLAR NEBULA

Before considering the formation of planets, let us take stock of the interstellar raw material from which the solar nebula emerged (*Figure 5*). The bulk of the matter in the solar system consists of atoms of hydrogen and helium created in the "Big Bang" some 11 to 16 billion years ago. Other chemical elements, which are less than one-thousandth as abundant as hydrogen, were formed by nuclear reactions in the interiors of earlier generations of stars

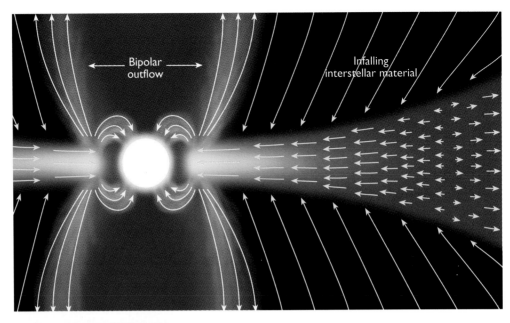

Figure 3. The solar nebula was not a pancake-shaped disk, but instead flared in vertical thickness farther from the protosun. The angular momentum of infalling interstellar matter was redistributed in a way that caused most of the mass to flow inward to the Sun (arrows), and the rest to be spun out to greater distances. Neither the means of this redistribution nor the origin of the nebula's bipolar outflows is well understood. In one model the young Sun remained coupled to the disk, across a gap between them, by magnetic field lines. Partially ionized disk gas followed field lines (yellow) that led part of it into the Sun, and slung another part of it centrifugally out of the solar system, forming a bipolar outflow.

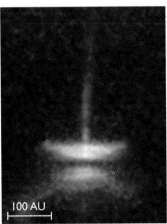
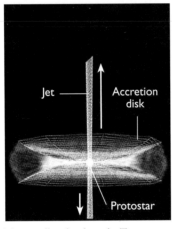

Figure 4. HH 30, a proplyd in the dark interstellar cloud on the Taurus-Auriga border, is another stellar birth site. This HST image actually shows the flaring shape of the protoplanetary disk, as well as bipolar outflows.

that existed between the time of the Big Bang and the origin of the solar system 4.55 billion years ago. These heavier, later-formed nuclides were released from their parent stars into the interstellar medium by stellar winds or stellar explosions.

Surprisingly little can be said about the physical state of these elements before and during the collapse process that formed the solar system. Extrapolating from our knowledge of the present interstellar medium, the most abundant element, hydrogen, existed chiefly as the diatomic gas molecule H_2 — thus justifying the often-used term "molecular cloud." Metallic elements (most notably magnesium, silicon, and iron, the principal ingredients of rocky planets) condense into solids at the highest temperatures and are termed *refractory* for this reason. They combined with oxygen and other elements while still in interstellar space to form tiny grains roughly 0.1 micrometer across — only about 1,000 atoms wide. The physical state of the elements lying between H (and He) and Mg, Si, Fe in atomic mass and abundance is very poorly known. In part they occurred as a variety of molecules in the gas phase, such as CO, N_2, NH_3, and free oxygen. In part they condensed into solid grains, as graphite (C) and

silicon carbide (SiC), for example. They also formed coatings of complex organic compounds and mantles of frozen ices on more refractory grains, though the ices would have evaporated when the grains were warmed as they fell toward the protosun.

Most of the matter present in the original solar nebula is gone now. It was drawn into the Sun, expelled into interstellar space, or incorporated into planets whose internal activity has reprocessed it into some new form. However, some of the primordial nebular material has survived and thus provides a crucial key to learning the details of how our solar system formed. The most abundant reservoir of unchanged nebular matter is in the form of comets. Because they remained small and far from the Sun, effectively in deep freeze for eons, these icy planetesimals retain most or all of the properties they had when they accreted in the outer nebular disk (see Chapter 24).

We do not yet have direct access to comets for study, but some of their ingredients are in our laboratories. Comets that approach the Sun are warmed by its heat, causing some cometary ice to evaporate and release embedded dust particles into space. Some of these are swept up by the Earth, along with other interplanetary dust particles, at which point they can be collected by research aircraft flying high in the stratosphere. Exactly which particles in these collections are cometary remains a puzzle. Viewed by electron microscope (*Figure 6*), many particles consist of clusters of tiny grains of minerals, organic compounds, and nondescript amorphous materials. Notably, these component grains tend to have roughly the same dimension, 0.1 micron, attributed to interstellar grains. Some fraction of them probably are just that, having fallen into the nebula 4.55 billion years ago. They then became embedded in the snowflakes that joined growing icy planetesimals.

Samples of more refractory primitive material, from the inner solar nebula, are preserved in the form of meteorites known as *chondrites*. These are fragments of asteroids, bodies that were not large and geologically active enough to completely reprocess the primitive nebular material. Although all chondrites were affected to some degree by thermal or hydrous metamorphism in their parent asteroids, the least-altered ones contain *bona fide* interstellar grains. We conclude such an origin based on the grains' anomalous isotopic compositions, which in each case records the particular nuclear reactions occurring in the unknown star that gave rise to it, long before our solar system formed. Collisions between asteroids release a shower of chondritic debris into space, some of which eventually reach the Earth's atmosphere as meteorites.

The solar nebula was hot near its center, tapering off to cold, then very cold, at its outermost margins. Of course, the environment near the infant Sun was warmed by its radiant energy. More important than this heat, however, was the Sun's mass and gravitational attraction. Close to the Sun, the nebula was its thickest and densest, and all the mechanical processes affecting the nebula — infall of molecular-cloud material, relative motions of nebular gas, turbulence, shocks — were stronger there and generated more heat than they did farther from the Sun.

Think of the nebula's falloff in temperature with heliocentric distance as defining three radial zones, like rings in a target. The innermost zone was too warm for water to condense as ice; objects forming there consisted entirely of silicate minerals and other refractory materials, ultimately becoming the terrestrial planets. The next zone was colder, water ice was stable, and a vast blizzard of snowflakes gave rise to the much larger Jovian planets. In the outermost and thus coldest zone, condensed matter was also icy. But it was too sparsely distributed to accrete into sizable planets; instead it remained dispersed in small icy planetesimals — comets — in what we now call the Kuiper belt. Remarkably, the planets assembled themselves very quickly. Although the process differed in detail from zone to zone, virtually everything was in place within 10 million years, by which time the solar nebula had largely dissipated.

ZONE 1: THE TERRESTRIAL PLANETS

Terrestrial planets were not made by simply sweeping up the refractory interstellar dust grains that fell into the nebula. We know this from the makeup of chondritic meteorites, which are samples of terrestrial planetary material as it first accreted. Chondrites consist in large part of small igneous spheroids called *chondrules* (*Figure 7*). These were once molten, which required temperatures in excess of 1,500° to 1,900° K. When the chondrules cooled and solidified they must have been dispersed in space, not aggregated as we see them now, in order to have maintained their droplike shapes. Over a century ago the English microscopist Henry Clifton Sorby argued that chondrules must have formed as "detached glassy globules, like drops of a fiery rain." Sorby further surmised that chondrules were "residual cosmic matter, not collected into planets, formed when conditions now met with only near the surface of the Sun extended much further out from the centre of the solar system" — a remarkably prescient assessment.

Efforts to simulate chondrules in laboratory furnaces have shown that a sample must be cooled relatively rapidly, in roughly an hour, to reproduce the properties of meteoritic chondrules. This means the chondrules could not have been melted by the ambient temperature of the innermost nebula, because there is no way to cool matter so quickly in that setting. Instead the workings of the inner nebula must have involved pervasive, local,

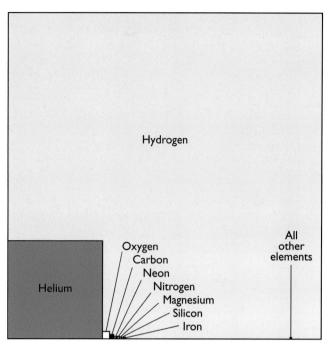

Figure 5. **One ton of a typical interstellar cloud, if cooled to well below 100° K, would yield 984 kg of hydrogen and helium gas, 11 kg of various ices, 4 kg of rock, and just under 1 kg of metal. The solar nebula was derived from such a cloud, and this diagram depicts the relative abundances of the nine principal elements now present in the solar system. These abundances have been derived from spectral studies of the solar atmosphere and laboratory analyses of chondritic meteorites.**

impulsive, high-energy events that drastically — but very briefly — affected its dispersed silicate material. We have little clue to the nature of these energetic pulses. Nor is it clear whether they occurred during the violent first million years of disk history, while interstellar material was still falling into the nebula, or in the 10 million years thereafter, when the nebula was thinner and more quiescent. Most researchers favor the latter period.

Severe thermal processing is capable of changing the chemical composition of planetary material, by selectively boiling off the more volatile chemical elements; these volatile elements are then free to recondense, perhaps somewhere else. Chemical fractionations of this sort must have occurred in the solar nebula; we see the evidence in individual chondrules, in their bulk aggregations (such as chondrites), and even in the planets themselves. For example, potassium, which is moderately volatile, exists in the Earth at only about one-fifth the abundance (relative to nonvolatile elements) that is present in average solar-system material.

Eventually, chondrules and dust began to stick together, creating larger and larger clods of chondritic material. This is a crucial moment in the history of planet formation, but like much else in the story it is very poorly understood. The gravitational attraction between such small objects is much too feeble to get the process of accretion started. Small particles that collided at less than about 1 meter per second might have stuck together because of Van der Waals attraction (weak, short-range forces caused by the uneven distribution of electrostatic charge). However, clumps bonded by Van der Waals forces alone would not be strong enough to survive mutual collisions in the nebu-

Figure 6. In this scanning electron microscope image, all the minerals, organic compounds, and amorphous materials in an interplanetary dust particle look the same. However, isotopic analysis reveals that some components of this dust actually solidified in interstellar space long before our Sun and its planets formed.

Figure 7. A 2-cm-wide section of the Mezö-Madaras chondrite, a tightly compacted mass of spheroidal and more irregular chondrules that fell near Harghita, Romania in 1852. Each chondrule formed as an independent igneous system. The section has been ground so thin, about 30 microns, that most of its minerals appear transparent when light shines through it.

la's turbulent zones. The onset of accretion would be easier to understand if some stronger "glue" were available.

There is no question that sticking did occur. Many chondrules in chondrites are coated with rims of dust particles gathered before they became grouped with other chondrules. Once the clusters exceeded a few centimeters across they became too heavy to be pushed around by turbulence in the gas, whereupon they settled and concentrated near the midplane of the nebula.

They also began spiraling in toward the Sun. An orbiting solid object, like a bit of chondrite, maintains a simple balance between centrifugal force (directed outward) and the Sun's gravity (inward). However, a parcel of gas in the nebula would have been pushed outward not only by centrifugal force but also by the gradient of gas pressure, which decreased outward in the disk. Consequently, gas required less centrifugal force and orbital velocity than solids did to remain in a given orbit, so any solid object in the same orbit traveled a little bit faster than the gas surrounding it. The resulting drag continually slowed the particle down, thus forcing it to spiral slowly inward through the nebula. As chondritic clusters crept inward they swept up and accumulated dust, loose chondrules, and smaller aggregations. Once such an object grew to a dimension of a kilometer or so, the gas-drag effect became relatively small, and the object's motion toward the Sun ceased. At about this size we dignify masses that were growing in the nebula with the name *planetesimal.*

Imagine being a passenger on the surface of one of these 1-km planetesimals. At the stage of evolution just described, the nebula teems with countless other planetesimals of similar size, but they are so widely dispersed that most of the time we can't even see any of them. Because of slight differences in their orbits, the planetesimals move relative to one another: a combination of up-and-down motion relative to the ecliptic plane, due to varying orbital inclinations, and in-plane motion with respect to one another due to orbital eccentricity. Because of these, every so often one of the other planetesimals zooms past us, then quickly disappears in

the distance. During the closest brushes, each planetesimal gives a tiny gravitational tug to the other, which changes the orbits of both slightly. And if we stay aboard our planetesimal long enough (maybe 1,000 years), eventually it will hit something of comparable size. If the collision velocity is slow, the two objects will merge into a single, larger mass; if fast, they knock each other apart. Either way, our ride is over!

High-speed computers allow us to study how planetesimals grow in such a system, and how fast it happens. Early on, planetesimal growth probably does not occur uniformly; instead, something like "the rich get richer and the poor get poorer" occurs. The largest objects experience runaway growth at the expense of their smaller cousins, and in some 20,000 years hundreds of bodies roughly the size of the Moon have been produced. Computer simulations have shown that such a population of bodies, through mutual orbital perturbations, coalescences, and catastrophic collisions, will eventually evolve into a family of bodies similar to the terrestrial planets. These simulations are limited in many ways, which makes them incapable of exactly reproducing Mercury, Venus, Earth, and Mars. But they do show that the terrestrial planets reached nearly their full size in about 10 million years, then continued to sweep up large planetesimals for another 100 million years.

The impacts of accreting planetesimals deposited huge amounts of kinetic energy in the planets being formed (*Figure 8*), enough to partly melt them. The earliest history of planetary surfaces was a chaos of solidifying crustal slabs, erupting lava, and giant explosions caused by the arrival of more planetesimals.

THE ASTEROID BELT

The border between Zones 1 and 2, the asteroid belt, is an untidy place. There is no object large enough to be called a planet in the asteroid belt; the aggregate amount of mass there is tiny, less than that in Earth's Moon. Yet we presume there was a smooth distribution of mass in the solar nebula before the plan-

Figure 8. As embryonic planets emerged from the chaos of the protoplanetary disk, they endured constant bombardment by a progression of ever-larger objects.

ets assembled, and the Titius-Bode law suggests that a planet should be in the belt's location (*Figure 9*). So what happened?

Planetary scientists believe the early appearance of Jupiter, with its mighty gravitational field, disrupted the orderly accretional assembly of a planet in this region. It may also have stunted the growth of Mars. Perhaps planetesimals in this transition zone never grew any larger than the asteroids are now, or it may be that for a time it was occupied by objects as big as the Moon or larger. If the latter, Jupiter eliminated them in the same way it removes material from the belt today. Those objects whose orbital periods are a simple integral fraction of the period of Jupiter's orbit (like 1:2, 2:5, 1:3) suffer *resonant perturbations* by Jupiter. The process might be likened to using a swing: small pushes, applied at exactly the right moment during each cycle of the swing, make it go higher and higher. Jupiter's resonant perturbations cause the orbital eccentricity of an asteroidal object to vary chaotically. In time it may collide with another asteroid, dive into the inner solar system (often hitting a planet or the Sun itself), or soar out to Jupiter's orbit or beyond. Ultimately, something almost certainly happens to remove it from the asteroid belt.

This may seem an ineffective way of removing things from the asteroid belt, because only a small percentage of its objects have orbital periods close to being integral fractions of Jupiter's period. However, asteroid orbits change periodically because of the mutual gravitational interactions that result when they pass close to one another; sometimes one of these encounters redirects an object into a resonant relationship with Jupiter. At that point it is destined to leave the asteroid belt.

ZONE 2: THE GIANT PLANETS

Farther out in the solar nebula, it was cold enough for its water to exist as ice. In this second zone, snowflakes were 10 times more abundant, by volume, than the silicate dust particles. This follows because oxygen — the main ingredient in water — is

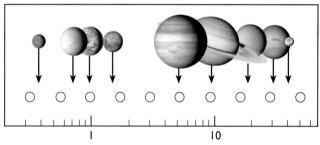

MEAN DISTANCE FROM SUN *(astronomical units)*

Figure 9. According to the Titius-Bode law, the spacing of planetary orbits follows a regular progression. In its traditional form, the law predicts the orbital radius of a given planet (in AU) by adding 4 to its corresponding value in the series 0, 3, 6, 12, 24, and so on, then dividing the sum by 10. Shown here is a simple variation on this empirical relation: each orbit is 75 percent larger than the one inside it. Points representing such a progression (anchored at 1 AU for Earth) appear as equally spaced yellow dots on this logarithmic plot. The planets conform to their predicted locations reasonably well; the most conspicuous discrepancy is the absence of a planet between Mars and Jupiter.

more abundant in the solar system than magnesium, silicon, and iron combined (*Figure 5*). Clearly the planetesimals that collected in the outer nebula, and ultimately the planets that formed from them, would have very different compositions than those of the terrestrial planets. However, the largest worlds, Jupiter and Saturn, do not have water as their main ingredient; instead they consist mostly of hydrogen and helium — a composition closer to the Sun's than to an icy planetesimal's. (Actually, in December 1995 the Galileo probe found the outermost Jupiter atmosphere contains much *less* oxygen than average solar-system matter, not more. But this superficial layer may not be representative of the planet as a whole; see Chapter 15.)

Jupiter and Saturn's compositions were not established by accreting snowflakes of pure hydrogen and helium, because temperatures in the solar nebula were not nearly cold enough to permit either of these gases to condense. Instead, these planets most likely gathered the bulk of their mass directly from the nebula, wholesale, without discriminating between solids and gases. Thus their compositions are essentially that of the nebula and, in turn, that of the Sun.

There are two ways in which they might have gathered in nebular material. Large cores or nuclei of ice and dust may have accreted first, much as terrestrial planetesimals did. When they became massive enough, their gravity began to attract and hold the nebular gas. The more gas these icy nuclei collected the heavier they became, and the greater their attraction for even more gas. The other possibility is that the very early solar nebula was massive enough to go through periods of gravitational instability, as described earlier. One form this instability might have taken, especially in the outer disk, was the separation and pulling together of gaseous, self-gravitating protoplanets massive enough to resist being dispersed by later tidal forces. In this case, nuclei of solid material would not have been required to get the process started.

Any good model for formation of the giant planets must explain why they differ in composition with radial distance from the Sun (*Figure 10*). Although hydrogen and helium dominate the compositions of Jupiter and Saturn, Uranus and Neptune consist mostly of the elements that form ices: oxygen, carbon, and nitrogen. In the icy-nuclei model, this trend could have occurred if the planetary nuclei grew so slowly out beyond Saturn that, by the time they were massive enough to attract gases, the nebula had largely dissipated. On the other hand, if all the giant planets started out as gaseous protoplanets, they would have needed to attract and absorb icy planetesimals — which enriched their proportions of O, C, and N to varying degrees (*Figure 11*).

The various planet-forming processes just described are interdependent and must have taken place in a particular sequence. However, the exact timing of these events is not well understood. *Figure 12* shows two possible sequences as they relate to the chronology of star and disk formation laid out earlier. Other scenarios are also possible.

According to the upper schedule, things happened — or began to happen — very early, while interstellar material was still actively falling into the nebular disk. This option is discounted by meteorite researchers, however, because the isotopic record in meteorites seems to argue against such an early beginning. Specifically, some inclusions present in chondrites once contained the short-lived radionuclide aluminum-26, whereas other inclusions in the same chondrites never did. If the latter inclusions did not form until the early solar system's ^{26}Al had decayed to indetectability, and if chondrite accretion had to wait until these inclusions became available, the sequence would have been delayed for several million years.

The lower schedule adheres to the ^{26}Al constraint, in that chondrites and terrestrial planets do not begin to form until well after interstellar collapse has ended and the Sun and protostellar disk are in place. Another constraint assumed here (one that cannot be proven with absolute certainty) is that the terrestrial planets were made of chondritic material, so their formation had to follow after that of the chondrites.

ROTATION AND SATELLITES

Somehow, the process of accretion imparted rotation to the planets. Planetesimals that struck a growing world on the right side, as viewed from the direction of their approach, increased the planet's spin in a prograde sense. However, an approximately equal number of planetesimals might be expected to strike the planets' left sides too, such that the spins imparted by the two families of accreting bodies would tend to cancel one another. If the planets grew from a very large number of small bodies, the averaging and cancellation of their many contributions should have left the planets with similar, rather slow rotation rates, all in the same direction and all about axes that were nearly perpendicular to the ecliptic plane.

However, this is not what we observe today. The planets spin at a wide variety of rates, two of them turn in a retrograde sense, and most of their rotation axes are tipped at substantial angles to the ecliptic perpendicular (*Figure 13*). This is consistent with accretion not from a large number of small bodies, but from a

Figure 10. **The outer planets have compositions that are distinct not only from the rocky worlds nearer the Sun but also from one another. In particular, the proportion of ice goes up — and that of gas (hydrogen and helium) goes down — farther from the Sun.**

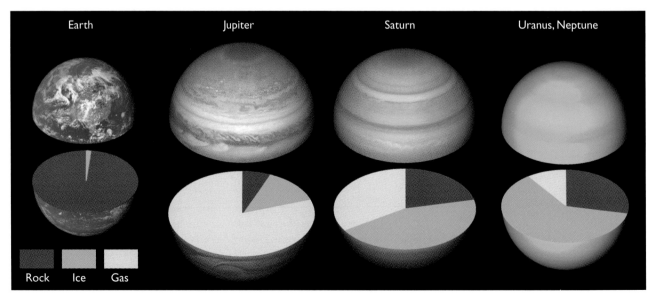

smaller number of large bodies, some of which were *quite* large. In this case the sum of all the planetesimals' contributions would have been unequal, leaving residual spins and tilts (obliquities) like those that characterize our planetary system. The evidence argues for such hierarchical growth, with the accretion of relatively large planetesimals occurring in the final stage.

We realize that Jupiter and Saturn did not accrete from solid planetesimals but instead mostly gathered gases directly from the nebula. Both planets spin quite rapidly, yet the manner in which they acquired their angular momenta is not well understood. But they too were subject to having their spins modified by the late addition of large planetesimals. The tilt of Saturn's rotation axis probably requires an oblique collision, near one of its poles, from something more massive than all of the terrestrial planets put together.

The giant planets were hot when they accreted, just as the terrestrial planets were, because as nebular material fell onto them its kinetic energy was converted to heat. This heat expanded the atmospheres of the giant planets to vastly larger dimensions than they have today. Thereafter they radiated heat, cooled, and shrank. In all likelihood, as each planet shrank a disk of gas, ice, and dust was left in orbit around it — a small analog of the solar nebula. From these disks emerged the regular satellites and ring systems, by means more or less analogous to the formation of planets and asteroids in the solar nebula. The irregular satellites, which have high eccentricities or inclinations (or both) relative to the equatorial plane of their primary planet, are thought to have been captured from the solar system at large. These include Phoebe, Triton, and many very small satellites, among them Phobos and Deimos, the satellites of Mars.

Earth's Moon is another story. As best we can surmise, when our planet had grown to nearly its present size, it was struck off-axis by a relatively large, fast-moving planetesimal. The energy of the impact heated the Earth and the impactor to very high temperatures, and it spalled off a plume of molten rock and vapor. Some of this debris fell back onto the Earth and some escaped to interplanetary space, but a portion of it settled into an incandescent disk orbiting the planet. We believe that such a disk would spread outward very quickly, and that once material was beyond the Roche limit, about 10,000 km above Earth's surface, gravitational instabilities would quickly gather it into a number of moonlets. (Inside the Roche limit individual objects cannot pull together because tidal forces induced by the central planet exert dispersing forces stronger than the objects' self-gravity.) Over a longer period of time, the hot moonlets coalesced into our Moon (see Chapter 10).

ZONE 3: COMETS

Where does the solar system end? The traditional answer has been at the orbit of Pluto (mean heliocentric distance: 39.4 AU). However, we now know the Sun's family extends to much greater distances. It was long suspected that large planets failed to form beyond Neptune not because there were no icy planetesimals to accrete them from, but only because the planetesimals were dispersed too thinly. Some 50 years ago Kenneth Edgeworth and Gerard Kuiper independently predicted that

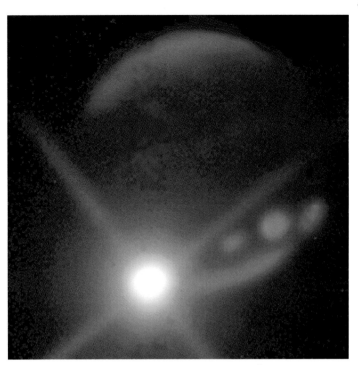

Figure 11. **Although their accretion effectively ended 4½ billion years ago, the planets continue to accumulate a trickle of matter even today. In this infared (2.3-micron) view of Jupiter, the bright "starburst" heralds the impact of fragment K from Comet Shoemaker-Levy 9 on 19 July 1994. Other infrared-bright spots to its right mark where earlier-arriving fragments struck. By accreting this ice-laden comet, Jupiter increased its content of oxygen, carbon, and nitrogen — albeit by extremely small amounts relative to the planet's total mass.**

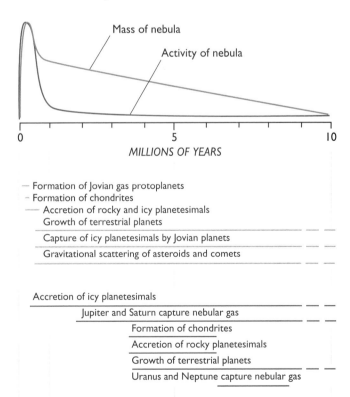

Figure 12. **By observing the formation of other stars, astronomers can estimate the rate of change in the mass and activity of a condensing nebular mass (top panel). But the chronology of planet formation is less certain, and the lower section outlines two possible timelines of planet-forming activity.**

such a belt of remnant building blocks still exists beyond Pluto. Recent astronomical observations, particularly by the HST, has confirmed the existence of these objects. Extrapolating from the sample observed, there must be more than 200 million objects, each a few kilometers in dimension, orbiting in the inner edge of what has come to be called the *Kuiper belt*. (Pluto itself, its satellite Charon, and the Neptunian moon Triton are probably the largest examples of Kuiper-belt objects; see Chapter 21.) The presence of all this solid matter extending the solar system's diameter beyond Pluto is consistent with the large disks found in association with young solar-type stars.

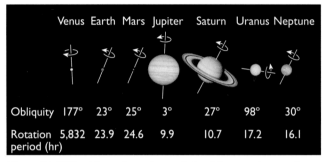

	Venus	Earth	Mars	Jupiter	Saturn	Uranus	Neptune
Obliquity	177°	23°	25°	3°	27°	98°	30°
Rotation period (hr)	5,832	23.9	24.6	9.9	10.7	17.2	16.1

Figure 13. Obliquities and rotation periods of the seven largest planets. *Obliquity* is the angle between a planet's equatorial plane and the plane of its orbit; values exceeding 90° (Venus, Uranus) mean the planet has retrograde rotation.

Figure 14. As evidence of the diversity of worlds now present in our solar system, consider this dramatic image acquired by Voyager 1 in February 1979. The swirling, hydrogen-dominated clouds of Jupiter form a colorful backdrop for two of its largest satellites. Seen projected against the Great Red Spot, Io (left) is indisputably the most volcanic object in the solar system. By contrast, Europa has an ice crust that may lie atop a deep global ocean of liquid water.

Short-period comets are very likely Kuiper-belt planetesimals that have been perturbed into the inner solar system by mutual interactions or by (as yet undiscovered) Pluto-size objects. Long-period comets are thought to have had a different origin, from within Zone 2. As they neared their final sizes, the giant planets gravitationally "stirred" the icy planetesimals around them into eccentric orbits. (NASA refers to such an interaction with a planet as a "gravity assist," and employs the technique to pump up the orbital energy of its spacecraft.) Most of the redirected planetesimals were ejected from the solar system altogether, doomed to wander "forever" in interstellar space. However, a great many did not quite reach the solar system's escape velocity; these hangers-on, still feebly bound to the Sun at vast distances from it, make up the *Oort cloud*, whose representatives occasionally revisit the planetary system in the form of long-period comets (see Chapter 5).

So where does the solar system effectively end? Outside Pluto's orbit, the Kuiper belt of icy planetesimals appears to extend to about 50 AU, then taper off. Much farther out lies the Oort cloud of comet nuclei, the last remote fringe of the solar system. The outer margin of the Oort cloud extends to roughly 1 light-year — and we should consider this the true limit of the Sun's realm.

Nearly four centuries of telescopic observation, combined with four decades of space exploration, have taught us this essential truth about the solar system: While the Sun and its planetary system surely arose from one grand spiral of gas and dust in a flurry of collective activity, the results are hardly a homogeneous set of characterless orbs. Instead, this grand scheme of formation has yielded amazing diversity (*Figure 14*) — as the following chapters will explore in much greater detail. It is humbling to realize that still other totally different kinds of planets, beyond our imagining, must be circling stars elsewhere in the galaxy.

The Sun

Kenneth R. Lang

The dynamic Sun, seen in X-rays near the peak of its 11-year activity cycle.

FROM AFAR, THE Sun does not look very complex. To the unaided eye it appears to be just a smooth, uniform ball of gas. This seeming simplicity once led astronomers to think of the Sun and other stars as uncomplicated, easily understood objects: gaseous spheres governed by simple laws of gravity, temperature, and pressure. However, each time astronomers scrutinize the Sun in greater detail, its behavior and structure turn out to be more elaborate than initially thought. We now know it is in constant turmoil, a churning, quivering body exhibiting unexpected features and behavior at every scale throughout its layered structure (*Figure 1*).

An unseen part of the Sun — its internal core — is nonetheless yielding its secrets as we develop better methods for understanding the nuclear reactions that occur there. Our model for answering the question, "How does the Sun shine?" now details how energy is liberated at its heart. While our predictions do not agree exactly with observations of subatomic particles (neutrinos) generated during the core's reactions, we expect future neutrino observations to provide important insights to the fundamental nature of the material universe.

The most familiar part of the Sun to us — its surface — is a seething, boiling place of constant change. Dark places of colossal magnetism spot the Sun's face, coming and going with cyclic regularity. Magnetic activity varies in an 11-year cycle, seemingly in tandem with the sunspot cycle. Without warning, the relatively calm solar atmosphere above sunspots can be torn apart by sudden explosions of unimagined intensity. These unpredictable events can adversely affect humans on Earth.

To fully understand our enigmatic star, scientists have extended their gaze past the visible solar disk deep into the Sun's interior and out again through its hot, tenuous gases and blowing winds. They have discovered unexpected flows and currents of material that churn the Sun's interior. Tantalizing clues hint at how the Sun generates its magnetic fields, which are responsible for most solar activity, how sunspots are held together, what heats its million-degree atmosphere, and where its powerful winds originate. These fascinating discoveries have become possible because the Sun is about 260,000 times closer than the next nearest star. This proximity gives us the opportunity to examine physical processes and phenomena that cannot be seen in detail on other stars.

ENERGIZING THE SUN

The Sun provides, directly or indirectly, almost all the energy on Earth, and it has apparently done so for a very long time. Without its light and heat, life would quickly vanish from the surface of our planet. In looking back at Earth's history, we find fossils of primitive creatures in rocks at least 3.5 billion years old, so the Sun was apparently hot enough to sustain life back then. Barring any unforeseen events, the Sun should continue to shine in much the same way for at least another 7 billion years.

Given this history, we are naturally curious about what keeps the Sun shining with such intensity. The only known fueling process that can sustain the solar fires for billions of years at their current rates involves nuclear reactions at the very center of the Sun (they are termed "nuclear" because it is the interaction of atomic nuclei that powers the Sun). The immense Sun — 109 times Earth's diameter and 333,000 times its mass — is very hot and dense in its core, too hot for whole atoms to exist. Innumerable collisions fragment the abundant hydrogen atoms into their constituent pieces. The nuclear protons and orbiting elec-

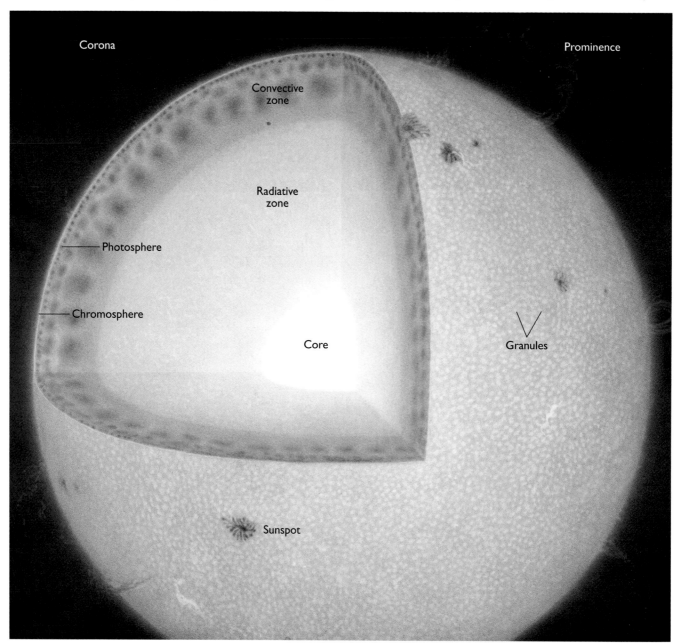

Figure 1. A cross-section of the Sun's interior. Energy produced through the fusion of hydrogen in the core is transported outward, first by countless absorptions and emissions within the radiative zone and then by convection. The wholesale motion of ionized matter within the interior generates magnetic fields that express themselves at the surface as sunspots, prominences, and active regions.

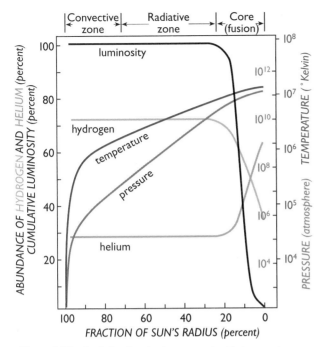

Figure 2. The Sun's luminosity (energy output), temperature, pressure, and composition all vary with depth in its interior. At the Sun's center, the pressure is 233 billion times that of the Earth's atmosphere at sea level.

Vital Statistics of the Sun

Radius		695,510 km (109 Earth radii)
Mass		1.989×10^{33} g (332,946 Earth masses)
Volume		1.412×10^{33} cm^3 (1.3 million Earths)
Density	(center)	151.3 g/cm^3
	(mean)	1.409 g/cm^3
Pressure	(center)	2.334×10^{11} bars
	(photosphere)	0.0001 bar
Temperature	(center)	15,557,000° K
	(photosphere)	5,780° K
	(corona)	2,000,000° to 3,000,000° K
Luminosity		3.854×10^{33} erg/s
Solar constant		1.368×10^6 erg/s/cm^2 = 1,368 W/m^2
Principal chemical constituents		
(by number of atoms):		
	Hydrogen	92.1 percent
	Helium	7.8 percent
	All others	0.1 percent

Table 1. Vital statistics of the Sun. Astronomers classify it as a *G2* star, a rather average type found abundantly in the galaxy. By coincidence, the Sun's nearest stellar neighbor, the Alpha Centauri system, contains a brilliant *G2* star as well.

trons are free to move throughout the solar interior. This ionized solar gas is a *plasma*, often termed the fourth state of matter (besides solids, liquids, and gases).

With their electrons gone, hydrogen nuclei (protons) can be packed together much more tightly than complete atoms. The nuclei are squeezed into a smaller volume by the pressure of material above, becoming hotter and more densely concentrated. At the Sun's center the temperature is nearly 15,600,000° K, and the density is 151 g/cm^3, more than 13 times that of solid lead (*Figure 2, Table 1*). This is hot and dense enough for *fusion* to occur: the very fastest-moving protons occasionally collide head-on and fuse to form deuterium. This starts a chain of events during which four protons are successively fused into one alpha particle (helium-4 nucleus) containing two protons and two neutrons.

However, an alpha particle is slightly less massive (by a mere 0.7 percent) than the four protons that combine to make it. This mass difference is converted into energy according to Einstein's famous equation $E = mc^2$. Because the velocity of light, *c*, is a large number, the annihilation of relatively small amounts of mass, *m*, provides large quantities of energy, *E*. Every second the Sun's central nuclear fire processes about 700 million tons of hydrogen into helium "ashes." In doing so, 5 million tons (0.7 percent) of this matter disappears as pure energy, and every second the Sun becomes that much lighter. Outside the solar core, where the overlying weight and compression are less, the gas is cooler and thinner, and fusion cannot occur.

Although the Sun is consuming itself at a prodigious rate, the loss of material is insignificant in comparison with its total mass. In 4.5 billion years the Sun has consumed only a few hundredths of 1 percent of its original mass. A more significant concern is the depletion of the Sun's hydrogen fuel within its nuclear furnace. About 37 percent of the hydrogen originally in the Sun's core has been converted into helium. Since nuclear reactions are limited to the hot, dense core, it will run out of hydrogen in about 7 billion years. As discussed in greater detail later, the Sun will then become a giant star, and life on Earth will end.

THE MYSTERY OF SOLAR NEUTRINOS

As confident as we are about understanding energy production in the Sun, there is at least one nagging problem marring our model. The nuclear reactions in the Sun's central furnace create prodigious quantities of neutrinos. These tiny, ghostlike particles travel at the speed of light almost unimpeded through the Sun, Earth, and nearly any amount of ordinary matter. Every second an estimated 70 billion solar neutrinos pass through each square centimeter of the Earth's surface facing the Sun.

It is possible to snag some of these elusive particles as they pass by. Massive tanks of fluid have been buried deep underground so that only neutrinos can reach them (*Figure 3*). Captured neutrinos have opened a window directly onto the Sun's energy-generating core. In addition, their existence offers new insights on the nature of the fundamental particles that make up matter in the universe. By finding solar neutrinos in roughly the predicted numbers, four pioneering detectors have now demonstrated that the Sun is indeed energized by hydrogen fusion. However, these experiments detect only one-third to one-half of the particles that theory predicts, a discrepancy known as the *solar-neutrino problem*. There are two possible explanations: either we don't really know how the Sun and stars create their energy, or we don't understand neutrinos.

Figure 3. The Super-Kamiokande neutrino detector has been built deep under a mountain in Japan. This huge stainless-steel vessel, 40 m tall and 40 m wide, has been filled with 50,000 tons of ultrapure water. Lining its walls are 13,000 photomultiplier tubes, which detect the telltale pattern of faint blue light made when a single neutrino from the heart of the Sun collides with an electron in a water molecule.

One solution would be to alter our model of the Sun's interior so that it produces the observed number of neutrinos. However, this would be inconsistent with other observations. The predicted rate of neutrino production depends on specific assumptions of temperature and composition, which also dictate the Sun's internal sound velocity (discussed later). The measured and predicted velocities do not differ by more than 0.2 percent down to very near the Sun's center. This plausibly pins down the core temperature very close to the calculated value of 15,600,000° K, and it strongly disfavors astrophysical explanations of the solar-neutrino problem.

Perhaps neutrinos have an identity crisis on their way to us from the center of the Sun. Neutrinos exist in three varieties, associated with the electron, the muon, and the tau particle. The first of these, the type generated in the Sun's core, is (for now) the only kind of solar neutrino that our detectors respond to. Could some of the electron neutrinos be switching to another type during their 8.3-minute journey from the center of the Sun to Earth, thereby escaping detection? This could neatly account for why we find only a fraction of the expected numbers. In one theoretical model, the switchover is modulated by the Sun's matter as the neutrinos pass through it.

Any such behavior would have profound implications for particle physics. In order to change states, a neutrino must have at least a tiny amount of mass. Recent measurements at the Super-Kamiokande detector in Japan indicate that atmospheric neutrinos generated by cosmic rays do, in fact, possess a very small mass whose exact value has yet to be determined. However, as presently understood, neutrinos are assumed to be completely massless. The Super-Kamiokande results may therefore require the development of new physics to explain these subatomic particles. If the neutrino changes are real, our observations may someday yield the actual neutrino mass and, in turn, the exact energy at which all the forces of nature become interchangeable.

The resolution of the neutrino problem rests with the refinement of our detection techniques. Experiments currently or soon to be under way should settle the question of whether solar neutrinos switch identities en route to Earth, providing stringent limits to, or measurements of, the neutrino mass. While the problem of the neutrinos complicates our model of the solar interior, we should recognize that this whole issue reveals something important about the particles that make up matter. Ultimately, understanding other stellar interiors rests upon our ability to make sense of the neutrino mystery that our Sun has handed us.

RADIATION AND CONVECTION

Because we cannot see inside the Sun, astronomers deduce its internal structure by combining theoretical equations, such as those for equilibrium and energy generation or transport, with observations such as the Sun's mass and luminosity (its total energy output per unit of time). The energy-generating core extends about one-quarter of the way outward to the visible surface. It thus accounts for only 1.6 percent of the Sun's volume — but about half of its mass. The rest of the Sun consists of two spherical shells that surround the core like nested Russian dolls. Gas within the inner shell is very calm and still, and energy is transported outward by radiation. This *radiative zone* reaches from the core boundary out to 71.3 percent of the Sun's radius. Above this lies the *convective zone*, within which churning, turbulent motion transports energy from the center to the surface.

Although light is the fastest thing around, radiation does not move quickly from the Sun's core to its visible surface. Instead it diffuses slowly outward in a haphazard, zigzag pattern, becoming absorbed, reradiated, and deflected repeatedly. Because of this continued ricocheting in the radiative zone, it takes about 170,000 years, on average, for radiation to work its way out from the Sun's core to the bottom of the convective zone.

The Sun's interior cools with increasing distance from the core as its heat migrates outward into an ever-larger volume. At the transition between the radiative and convective zones, the temperature has declined to about 2,200,000° K (*Figure 2*). At that point the solar interior becomes relatively opaque to the outward flowing radiation, causing the material below to become hotter than it would otherwise be. The pent-up energy must then be released through huge convection currents: hot material rises to the surface in roughly 10 days, cools by radiation to space, then sinks.

The convection zone is capped by a veneer of solar radiation, called the *photosphere* (from *photos*, the Greek word for light), at a temperature of 5,780° K. About 500 km thick, the photosphere is a zone where the gaseous material changes from being completely opaque to upwelling radiation to being transparent. It is the layer from which the light we actually see is emitted and where most of the Sun's energy escapes into space.

THE SOLAR ATMOSPHERE

We consider the photosphere to be the surface of the Sun, but it is not really a surface. The photosphere is surrounded by an overlying atmosphere. Its gases are so rarefied that we see right through them, just as we see through the Earth's atmosphere. The solar atmosphere includes, from its deepest part outward, the photosphere, chromosphere, and corona.

The Sun's temperature rises to about 10,000° K in the *chromosphere* (from *chromos*, the Greek word for color), a thin layer about 1,000 km thick. It becomes visible a few seconds before and after a total solar eclipse, creating a narrow pink,

Figure 4. **The Sun's corona, as photographed during the total solar eclipse on 26 February 1998. This event occurred near sunspot minimum, a time when the corona appears roughly symmetrical. To extract this much coronal detail, several individual images were combined and processed electronically. Note the numerous fine rays over the poles and the "helmet" streamers nearer the equator.**

rose, or ruby-colored band at the extreme limb of the Sun. We can also study the chromosphere by observing it in the red light of hydrogen (the alpha transition at a wavelength of 6563 angstroms) or the violet light of singly ionized calcium (at 3933 and 3967 angstroms).

Above the chromosphere lies the *corona*, from the Latin word for crown. Like the photosphere, the corona is also self-luminous, but its light is only about one-millionth as intense. Thus it is no surprise that the corona is only visible to the unaided eye during a total solar eclipse (*Figure 4*). However, the corona can be routinely observed at other times with special telescopes called coronagraphs, which use a small occulting disk to mask the Sun's face and block out the photosphere's light. Our best coronagraphic images come from high-flying satellites that are not hampered by the image-degrading effects of Earth's atmosphere.

The corona presents one of the most puzzling paradoxes of solar physics. At its base, in a *transition region* less than 100 km thick, the temperature jumps from 10,000° to 1,000,000° K (and there is a corresponding drop in gas density to keep the pressure constant). This abrupt rise in temperature is unexpected. Heat simply should not flow outward from a cooler to a hotter region — that would violate the second law of thermodynamics and common sense as well. As we'll soon see, solar physicists believe one of the mechanisms most likely responsible for heating the million-degree corona is the Sun's interacting magnetic fields.

The hot plasma of the corona emits most of its energy at ultraviolet, extreme ultraviolet, and X-ray wavelengths. Ultraviolet radiation also reaches us from the chromosphere and

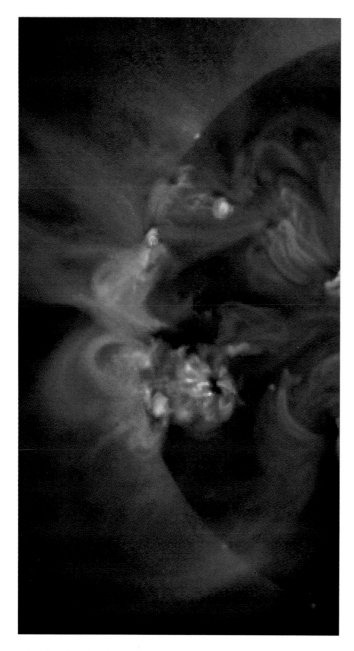

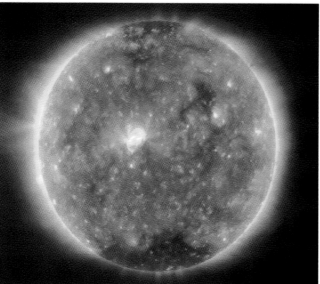

from the transition region. (The photosphere is too cool to emit strongly at these wavelengths, so it appears dark under the hot gas.) Unfortunately, these energetic wavelengths are partially or totally absorbed by the Earth's atmosphere, so they must be observed through telescopes in space. One very productive instrument has been the X-ray telescope aboard Japan's Yohkoh ("Sunbeam") satellite. Equally valuable are the ultraviolet and extreme-ultraviolet instruments aboard the Solar and Heliospheric Observatory, or SOHO.

Yohkoh and SOHO have revealed an unseen world of constant and violent change (*Figure 5*). They have shown that the corona is stitched together by thin, bright magnetized loops that constrain the hot, dense, X-ray emitting gas. These ubiquitous coronal loops can rise from inside the Sun, sink back down into it, or expand out into space. Indeed, the corona is in a continuous state of change, constantly varying in brightness and structure on all detectable spatial and temporal scales. Elsewhere X-ray and extreme-ultraviolet images delineate what are termed *coronal holes* (*Figure 6*). These have so little hot material in them that they appear as large dark areas seemingly devoid of radiation. Coronal holes are nearly always present at the Sun's poles.

The Sun's hot and stormy atmosphere expands in all directions, filling the solar system with a ceaseless flow of charged particles called the *solar wind*. This outflow of electrons and ions, threaded by magnetic-field lines, is powered by the million-degree corona, which creates an outward pressure that overcomes the Sun's gravitational attraction. The wind accelerates as it moves away from the Sun like water overflowing a dam. As the coronal plasma disperses into space, it is replaced by gases welling up from below.

Since the X-ray-emitting material in the low corona is constrained and molded by the Sun's magnetic fields, its form varies continually, adjusting to the shifting forces of magnetism. Related magnetic activity causes powerful gusts in the solar wind (see Chapter 4).

SUNSPOTS, THE SOLAR CYCLE, AND EXPLOSIVE ACTIVITY

The most intense solar magnetism is localized within *sunspots*, which are relatively dark regions of the photosphere as large as the Earth (*Figure 7*). The intense magnetic field within a sunspot acts as a kind of filter or valve, choking off the heat and

Figure 5 (top). Taken on 19 May 1992 with the Soft X-ray Telescope (SXT) aboard the Japanese-built Yohkoh satellite, this image shows a particularly dynamic portion of the Sun near the peak of its 11-year activity cycle. Numerous "active regions" (bright patches) in the corona indicate material heated to several million degrees. Above many of them are coronal loops, marking the location of hot plasma and concentrated magnetic fields arching high into the solar atmosphere.

Figure 6 (bottom). Small, bright hotspots pepper the solar disk in this far-ultraviolet view taken in the light of ionized iron (195 angstroms), which corresponds to a temperature of about 1,500,000° K. Other features include polar and equatorial coronal holes (dark regions), polar plumes, and filament channels. This picture was taken with the Extreme-ultraviolet Imaging Telescope aboard the SOHO spacecraft on 22 August 1996.

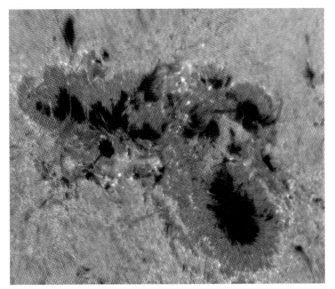

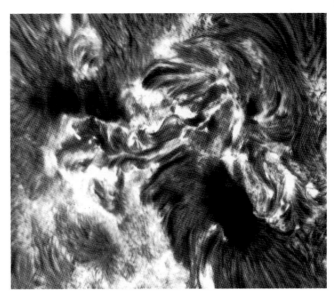

Figure 7 (above, left and right). Closeup views of a particularly large sunspot group, roughly 150,000 km across, taken on 28 June 1988. The left-hand photograph approximates a white-light view. The corresponding one at right, taken in the hydrogen-alpha emission at 6563 angstroms, shows how material in the Sun's lower atmosphere traces the magnetic field lines emerging from this active region.

Figure 8 (below). The number of sunspots (upper panel), their location (middle), and total area (bottom) have varied in an 11-year cycle for the past 100 years. Note the migration of spots from high latitudes, at the beginning of a cycle, to near the Sun's equator at its end; such historical plots are sometimes called "butterfly diagrams."

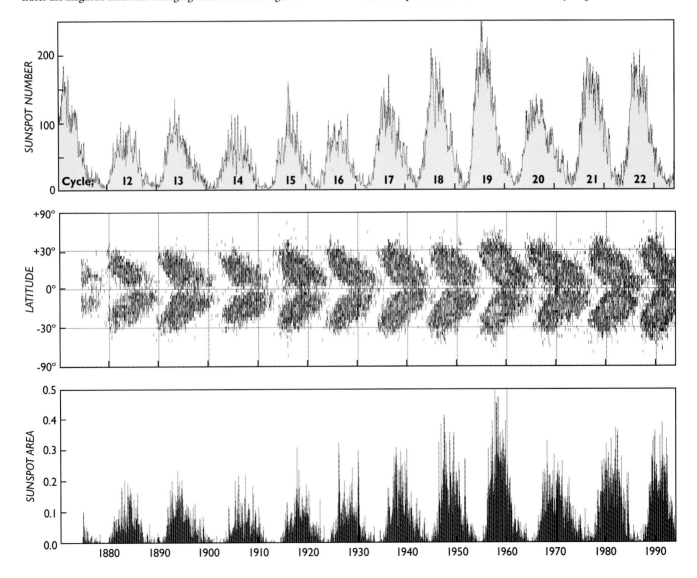

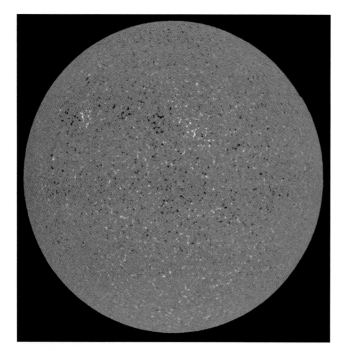

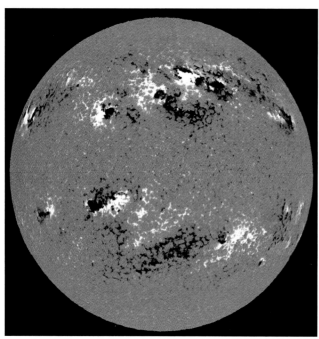

energy flowing outward from the solar interior. This keeps sunspots thousands of degrees cooler than the turbulent gas around them and makes them appear dark relative to their dazzlingly bright surroundings.

At the center of a sunspot, the magnetic field is a few thousand gauss in strength and oriented radially (perpendicular to the photosphere's surface). Sunspots normally occur in pairs of opposite magnetic polarity, joined by loops of magnetic field lines that rise into the overlying solar atmosphere. These loops are among the brightest coronal features detected in X-ray images of the Sun. The highly magnetized realm in, around, and above a bipolar sunspot group is called an *active region*.

The total number of sunspots visible on the Sun varies from a maximum to a minimum and back to a maximum, in a cycle that lasts about 11 years (*Figure 8*). The latitudinal positions of sunspots also vary during this cycle, moving from high solar latitudes toward the equator as the cycle progresses. The sunspot cycle describes a periodic variation in magnetic activity on the Sun (*Figure 9*). Since most forms of solar activity are magnetic in origin, they also follow an 11-year cycle. However, solar activity does not completely disappear at the minimum in the sunspot cycle.

The Sun's ever-changing surface magnetism produces unrest on an awesome scale. In a few short minutes, catastrophic *solar flares* can release stored magnetic energy equivalent to billions of thermonuclear explosions, raising the temperature of Earth-size regions by tens of millions of degrees. Although flares appear rather inconspicuous in visible light, they often produce enough X-ray and radio energy to outshine the entire Sun at these wavelengths.

Thanks to the detailed views afforded by arrays of radio telescopes on the ground and X-ray observatories in space, we know that solar flares are frequently triggered in compact structures at the tops of magnetized coronal loops (*Figure 10*). In just a few seconds, electrons within these loops can be accelerated to nearly the speed of light (the source of the powerful radio bursts). Ions are likewise accelerated, and the downward-moving beams strike the denser chromosphere below like bullets hitting a concrete wall. This interaction produces a cascade of high-energy X-rays and gamma rays.

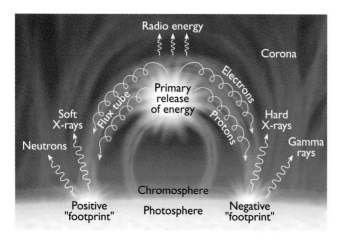

Figure 9 (top and middle). These maps show the strength and polarity of solar magnetic fields. Dark regions have south magnetic polarity, bright ones north polarity. During a solar-activity minimum (upper image, 27 December 1985), there are no large sunspots and tiny magnetic clusters pepper the entire photosphere. When the Sun is most active (middle, 12 February 1989), sunspots are most numerous and dominated by large bipolar sunspots within two parallel bands. The preceding, or rightmost, spots in the northern hemisphere usually have one magnetic polarity, while the following spots have the opposite polarity. In the southern hemisphere the polarities are exactly reversed.

Figure 10 (bottom). Solar flares tend to occur within concentrated magnetic loops extending from the photosphere into the corona. Energy released at the top of the loop creates a burst of radio energy. It also accelerates great numbers of electrons and protons. These are channeled down the loop and strike the chromosphere at relativistic speeds, which creates X-rays and gamma rays.

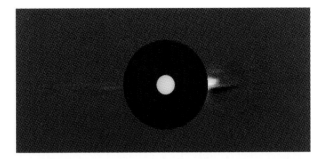

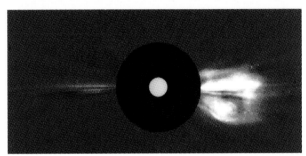

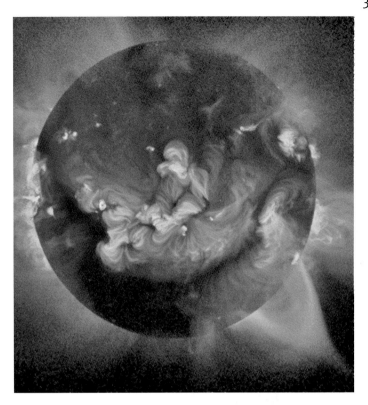

Figure 11 (left). **The Large Angle Spectroscopic Coronagraph aboard the SOHO spacecraft recorded an enormous coronal mass ejection (CME) over an 8-hour period on 15 January 1996. The expanding bubble contained roughly a billion tons of hot gas. Streaks on the opposite side of the Sun may have been magnetically linked to the CME. The black occulting disk blocks the glare of the Sun, whose visible edge is represented here by the white circle.**

Figure 12 (above). **A large helmet-type structure is seen in the south west (lower right) of this image obtained on 25 January 1992 by the Yohkoh satellite. The large cusp formed following a coronal mass ejection.**

TAKING THE SUN'S PULSE

To fully understand the Sun's cyclic activity, with its impressive array of coronal loops, sunspots, and explosive flares, we must look deep inside our star. This is where its magnetism is generated and sustained. One way to explore the Sun's unseen depths is by tracing the in-and-out, heavings of its visible surface using the powerful observational technique called *helioseismology*. During the 1960s solar physicists realized that the photosphere exhibited complex, periodic throbbing motions, which by 1968 were shown to result from pressure waves — literally sounds — that echo and resonate inside the Sun.

The sound waves are trapped inside the Sun and cannot propagate through the near vacuum of space. Even if they could reach the Earth, we could not hear them. The "loudest" ones are extremely low-pitched, as befits an object so large and massive, with frequencies that cluster around 0.003 hertz, or one vibration per 5 minutes. This is 12.5 octaves below the lowest note audible to humans, and many of the Sun's notes are even lower.

Nevertheless, when these sounds propagate upward to the photosphere they disturb the gases there and cause them to rise

Solar physicists believe that flares signal the release of tremendous amounts of magnetic energy, like a tightly twisted rubber band that suddenly snaps. The awesome explosions might be triggered when oppositely directed magnetic fields come together, merge, and annihilate, releasing their pent-up energy and then reconnecting to a stable magnetic configuration.

Another dramatic, magnetically energized phenomenon is termed a *coronal mass ejection*, or CME (*Figures 11,12*). These violent eruptions hurl billions of tons of material into space. They expand at high speed as they propagate outward from the Sun and quickly rival it in size. The associated shock waves accelerate and propel vast quantities of charged particles ahead of them. Like sunspots, CMEs and solar flares ebb and flow in step with the Sun's 11-year cycle of magnetic activity, and they may have a similar origin in the Sun's ever-changing magnetism.

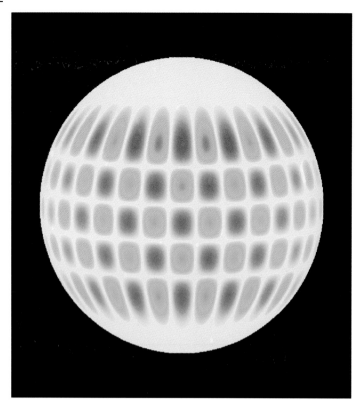

Figure 13. Sound waves resonating within the solar interior cause the photosphere to move in and out, rhythmically distorting the shape of the Sun. This heaving motion can be described as the superposition of literally millions of oscillations, including the one shown here for regions moving in (red spots) and out (blue spots).

and fall, slowly and rhythmically (*Figure 13*). These in-and-out motions can be tens of kilometers high and attain speeds of a few hundred meters per second. Such movements are imperceptible to the eye, but sensitive instruments on the ground and in space routinely pick them out. We detect them as tiny, periodic Doppler shifts in a well-defined spectral line or as minuscule variations in the Sun's total light output.

A major obstacle to helioseismic studies is the Earth's rotation, which keeps us from observing the Sun around the clock. Nightly gaps in the data create background noise that hides all but the strongest oscillations. The low frequencies that probe the solar interior to the greatest depths are especially difficult to detect. Fortunately, continuous observation is possible from space. Orbiting at the inner (L_1) Lagrangian point along the Sun-Earth line, the SOHO spacecraft has a continuous, uninterrupted view of our star. From this vantage point SOHO detected photospheric motions with remarkable precision — to better than 1 millimeter per second. Spacecraft have by no means rendered ground-based helioseismology obsolete, however. The Sun is being observed around the clock by a worldwide network of observatories known as the Global Oscillation Network Group (GONG). These electronically linked sentinels follow the Sun as the Earth rotates. In effect, the Sun never sets on the GONG network.

The Sun's surface oscillations are the combined effect of about 10 million separate notes, each of which has a unique path of propagation and traverses a well-defined section inside the Sun (*Figure 14*). Sound waves of different frequencies descend to different depths before being refracted back up to

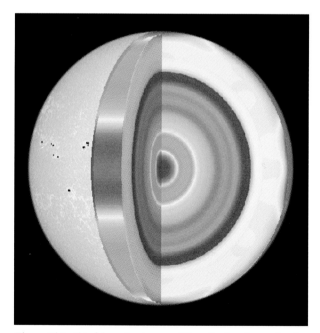

Figure 15. In this false-color image of the solar interior, red and blue correspond to faster and slower sound speeds, respectively, relative to a standard solar model (yellow). The velocity drop at the boundary of the Sun's core may result from some unstable burning process mixing the material there. Higher speeds just below the convection zone may reflect turbulence caused by adjacent parts of the interior rotating at different rates. Variations near the surface probably mark real temperature differences, with red areas hotter and blue ones colder.

Figure 14. Sound waves inside the Sun, like seismic waves in the Earth, do not travel in straight lines. As they travel toward the Sun's center, the waves are refracted back out. At the same time, the Sun's surface reflects waves traveling outward back in. How deep a wave penetrates and how far around the Sun it goes before it hits the surface depends on its wavelength.

the surface. Observations of many different oscillation *modes* (resonances set up by the waves) can be combined to probe various depths of the Sun. To trace our star's physical landscape all the way through — from its churning convection zone down into its radiative zone and core — we must determine the precise pitch (frequency) of all the notes.

Each sound wave has a corresponding propagation speed that depends on the temperature and composition of the solar regions through which it passes. Helioseismologists exploit this relationship to establish the Sun's radial variation in temperature, density, and composition. For example, a small but definite change in the observed sound speeds pinpoints the lower boundary of the convective zone at 71.3 percent of the radius of the photosphere. Theoretical sound velocities and those observed by SOHO are in close agreement but can differ by up to 0.2 percent. Two places where these discrepancies occur are significant. They suggest that turbulent material is moving in and out just below the convective zone, and that such mixing motions might also occur at the boundary of the energy-generating core (*Figure 15*).

INTERNAL MOTIONS: THE KEY TO MAGNETISM

Although vertical convection churns the Sun's outermost layers, rotation produces much faster, horizontal motions. Moreover, for more than three centuries, astronomers have known from watching sunspots that the photosphere rotates faster at the equator than it does at higher latitudes, decreasing in speed evenly toward each pole. The photosphere's sidereal rotation period ranges from 25.7 days at the Sun's equator to 33.4 days at 75° latitude.

The motion of the Sun's deep interior is likewise dominated by rotation. Sound waves moving with the direction of rotation appear to move faster, like birds flying with the wind, so their measured periods are shorter. Waves propagating against the rotation will be slowed and show longer periods. These opposite effects split a given oscillation into a pair of closely spaced frequencies. Since each oscillation probes a distinct region inside the Sun, this splitting reveals rotation at different depths and latitudes.

Armed with this sensitive technique, helioseismologists have found that the latitude-dependent rates exhibited by the photosphere persist throughout the convective region (*Figure 16*). However, the rotation speed becomes uniform from pole to pole nearly one-third of the way to the core, 220,000 km beneath the photosphere. Lower down the rotation rate remains independent of latitude, acting as if the Sun were a solid body.

Thus, the rotation velocity changes sharply at the top of the radiative zone. There it meets the overlying convective zone, which spins faster in its equatorial middle. We now suspect that the roughly 20,000-km-wide layer where these very different zones meet and shear against one another is the site of the solar dynamo, the source of the Sun's magnetism.

Our star's interior flows in other ways as well. In 1997, after a year of nearly continuous SOHO observations, researchers discovered that vast rivers of hot gas circulate within the Sun, describing a sort of solar meteorology (*Figure 17*). These flows are not the dominant, global rotational motions, but rather the ones found when rotation is removed from the data.

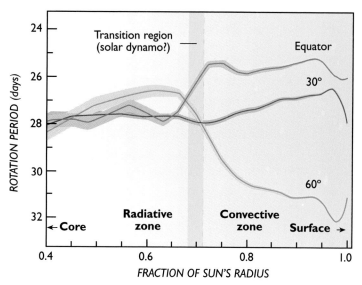

Figure 16. Internal rotation rates of the Sun at latitudes of 0°, 30°, and 60° have been inferred using data from the Michelson Doppler Imager (MDI) aboard the SOHO spacecraft. Just below the convection zone, the rotational speed changes markedly. Shearing motions along this interface may be the source of the Sun's magnetism.

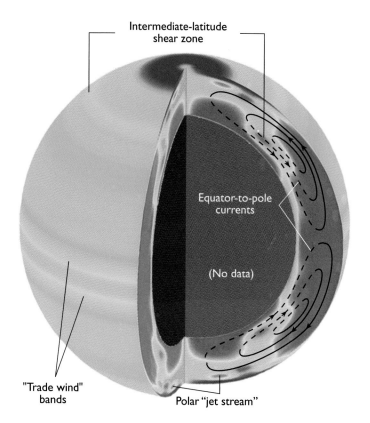

Figure 17. A surprising variety of large-scale flows occur in the Sun's interior. Red corresponds to faster-than-average flows, yellow to slower than average, and blue to slower yet. For example, on the left side, deeply rooted zones (yellow bands), analogous to the Earth's trade winds, travel slightly faster than their surroundings (blue regions). Sunspots may tend to form at the edges of these zones. Polar "jet streams" (dark blue ovals) move approximately 10 percent faster than their surroundings (light blue). There is slow movement poleward from the equator shown by the streamlines in the right-hand cutaway (the return flow below it is inferred).

34

Completely unexpected currents circle the polar regions and seem to resemble the jet streams in Earth's atmosphere. Ringing the Sun at about 75° latitude, the solar jet streams lie 40,000 km below the photosphere and thus cannot be seen at the visible surface. They move about 10 percent (130 km per hour) faster than the surrounding gas, and they are wide enough to engulf two Earths.

We also now realize that the outer layer of the Sun, to a depth of at least 24,000 km, is slowly flowing toward the polar regions at about 85 km per hour. At this rate an object would be transported from the equator to the pole in about three years. Of course, the Sun rotates much faster than that, about 7,100 km per hour at its equator. The combination of differential rotation and poleward flow may explain the stretched-out shapes of magnetic regions that migrate toward the poles.

Finally, researchers have identified internal rivers of gas moving in bands parallel to the equator at different speeds relative to each other. They are more than 64,000 km in width and move about 16 km per hour faster than the gases to either side. These broad, higher-velocity currents are reminiscent of the Earth's equatorial trade winds and the zonal jets in Jupiter's colorful atmosphere (see Chapter 15).

The first evidence for alternating fast and slow bands was detected on the photosphere in the mid-1980s, and their location migrates from higher latitudes toward the equator during the 11 years of the sunspot cycle. SOHO's observations show that these flows are not a superficial phenomenon but are instead deeply rooted, penetrating at least 19,000 km into the Sun. The full extent of the newfound solar meteorology could never have been deduced by looking solely at the visible layer of the solar exterior.

The boundaries between these high- and low-speed currents are located precisely where we expect sunspots to form. These are also the places where the gaseous material is slightly hotter than average. Other internal motions, when compared with a map of magnetic-field strength, reveal a mechanism for holding sunspots together. Strong magnetic concentrations apparently exist in regions where the subsurface gas flow converges. The churning gas probably forces magnetic fields together and concentrates them, thereby overcoming the magnetic resistance that ought to make such localized concentrations expand outward and disperse.

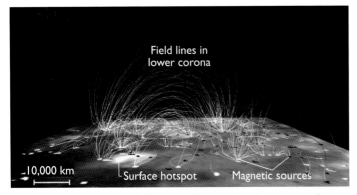

Figure 18. **Magnetic loops of all sizes rise far into the solar corona and connect regions of opposite polarity (black and white) in the photosphere. Energy released during reconnection events in this "magnetic carpet" is one likely cause for making the solar corona so hot.**

The currents of ionized gas (plasma) inside the Sun do more than merely confine the magnetic field locally, however. They are in part responsible for it. Any plasma conducts electricity, and when a conductor moves across a magnetic field, electric current is generated within it. This current creates more magnetism, which adds to and reshapes the original magnetic field. Conversely, a moving magnetic field will tend to drag a conductor along with it. The upshot is that gas and magnetic lines of force are intimately locked together in the Sun, as they are in plasmas throughout the universe.

POWERING THE CORONA AND SOLAR WIND

Writhing gases and shifting magnetic fields probably also explain why the solar corona has a temperature of at least 1,000,000° K even though it lies directly above the relatively cool 5,780° K photosphere. We have long known that visible photospheric radiation was not the reason behind this heating paradox. Sunlight passes right through the tenuous corona without depositing substantial quantities of energy in it. Some other mechanism must be transporting energy from the photosphere (or below) out to the corona.

After decades of trying to unravel the mystery of coronal heating, scientists think they have some plausible explanations. Ultraviolet observations by SOHO reveal that the Sun is a vigorous, violent place even when its 11-year activity cycle is in an apparent slump, and this fact may help explain why the corona is so hot. The whole Sun seems to sparkle in the ultraviolet light emitted by thousands of localized bright spots (*Figure 6*). These ubiquitous hot spots can reach 1,000,000° K and seem to originate in small magnetic loops of hot gas. Some of these spots appear to explode, hurling material outward at hundreds of kilometers per second. Apparently they are localized sites of magnetic reconnection, where oppositely directed magnetic fields merge. As such they mimic (on a much smaller scale) the releases of stored magnetic energy seen in solar flares.

SOHO has provided direct evidence for such a transfer of magnetic energy from the Sun's visible surface toward the corona above. The spacecraft's images of the photospheric magnetism reveal ubiquitous pairs of opposite polarity, with each pair joined by a magnetic arch that rises above them like a bridge connecting two islands (*Figure 18*). Energy flows from these magnetic loops when they interact, producing powerful electrical and magnetic "short circuits" that cause coronal plasma to heat up. Even closer scrutiny of such interactions became possible when NASA's Transition Region and Coronal Explorer (TRACE) reached orbit in April 1998. The spacecraft's three extreme-ultraviolet spectral bands were optimized for monitoring activity in the lower corona (*Figure 19*). Moreover, TRACE was placed in a special polar orbit above Earth's day-night terminator, which afforded continuous viewing of the Sun.

The corona is forever expanding into interplanetary space, filling the solar system with the solar wind's thin, constant outflow. This seemingly eternal gale of ionized matter has fast and slow components. The fast stream emanates mainly from higher latitudes, and measurements by the Ulysses spacecraft show it to be traveling at about 800 km per second. The slower wind, found nearer the solar equator, moves outward at about 450 km

CHAPTER THREE

Figure 19. Hot plasma in the Sun's lower corona follows strongly magnetized loops over an active region. The larger loops are about 200,000 km (15 Earths) across. Acquired at extreme-ultraviolet wavelengths, this snapshot from the TRACE spacecraft uses false colors to indicate temperature: blue corresponds to roughly 200,000° K, green to 900,000°, and red to 2,700,000°. The loops' thinness and differing temperatures suggest that coronal plasma is heated along small, very localized groups of field lines.

per second. Ulysses, which passed over both poles of the Sun during the mid-1990s, showed conclusively that the high-speed component pours forth from the polar coronal holes. There open magnetic fields allow charged particles to escape the Sun's gravitational and magnetic grasp (*Figure 20*).

Where does the slow solar wind come from, and what accelerates the fast component to such high speeds? Near a minimum in the solar-activity cycle, the corona takes on a simple appearance — it remains highly symmetrical and stable. During such magnetic lulls, the corona exhibits pronounced holes in the north and south (*Figure 6*). In contrast, the equatorial regions are ringed by straight, flat streamers of outflowing matter. The Sun's magnetic field shapes these streamers. At their base, coronal plasma is densely concentrated within magnetized loops rooted in the photosphere. Farther out, the streamers narrow into long stalks that stretch tens of millions of kilometers into space. These extensions confine material at temperatures of about 2,000,000° K within their elongated

magnetic boundaries, creating a torus of hot gas that extends around the Sun's midsection.

The streamers live up to their name. Material seems to flow along their open magnetic fields, thus creating the slow component of the solar wind. Occasionally the coronagraphs aboard SOHO record dense concentrations of material moving though an otherwise unchanging streamer — like seeing leaves floating on a moving stream. These much-larger coronal mass ejections punctuate the outward flow.

The slow-speed solar wind is an expected consequence of the corona's high temperature. The hot coronal plasma must expand into space, and it accelerates to supersonic speeds as it moves away from the Sun. Yet no one is certain about what gives the high-speed wind its additional push.

SOHO has provided clues to the wind acceleration mechanisms by observing hydrogen and oxygen ions in the regions where the corona is heated and the solar wind accelerates. In polar coronal holes, where the fast solar wind originates, the

Figure 20. A close-up of the south polar coronal hole (large dark region) acquired on 13 March 1996. The fast component of the solar wind emanates from these coronal holes. Plumes can be seen emerging from tiny bright spots. SOHO recorded this view at the far-ultraviolet wavelength of 171 angstroms, and it shows emission from highly ionized iron atoms at a temperature of 1,000,000° K.

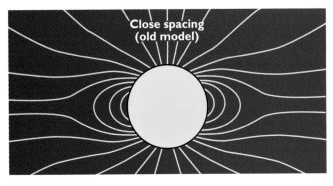

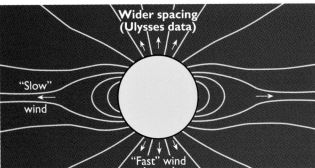

Figure 21. At one time researchers believed that the Sun's magnetic-field lines should be concentrated near its poles (upper panel). However, when it passed over the solar poles, the Ulysses spacecraft discovered that the field lines are spaced rather uniformly over all solar latitudes (lower panel). The high-speed solar wind escapes along the open magnetic field lines emanating from the polar regions, whereas the slow-speed wind probably flows along the stalk-like axes of equatorial coronal streamers.

more massive oxygen ions become far more agitated, acquiring a velocity approaching 500 km per second. Hydrogen, on the other hand, moves at only 250 km per second, suggesting that whatever heats the polar corona preferentially energizes the more massive ions. In contrast, within equatorial regions, where the slow-speed wind begins, the lighter hydrogen moves faster than the oxygen, as one would expect in a wind driven solely by thermal expansion.

Researchers are now trying to determine why the more massive oxygen ions move at greater speeds in coronal holes. One possibility is that the ions are whirling around magnetic field lines that stretch from the Sun (*Figure 21*). The heavier ions gyrate at frequencies where there is more ambient magnetic-wave energy, and they use that energy to attain higher velocities than lighter ions can. And because ions rarely collide with electrons in the low-density coronal holes, we have some hope that the outflowing ions retain some imprint of the process that caused their heating and acceleration. In contrast, frequent collisions in high-density equatorial streamers have likely erased any signature of the processes relevant to their origin.

THE INCONSTANT SUN AND ITS TERRESTRIAL CONSEQUENCES

Having examined the Sun in fine detail, we now step back and look at it from our viewpoint on Earth. Day after day our star rises and sets in an endless cycle, an apparently unchanging ball of fire whose heat and light make terrestrial life possible. A measurement of this life-sustaining energy has been called the *solar constant*. It is the average amount of radiant solar energy per second per unit area reaching the top of Earth's atmosphere at a mean distance of 1 AU, amounting to 1,368 watts of power per square meter.

We long considered the Sun's output to be truly constant because no variations could be reliably detected from the ground. However, exquisitely sensitive detectors aboard satellites above the Earth's atmosphere provide conclusive evidence for a slightly varying luminosity (*Figure 22*). The Sun's total radiative output is almost always changing, it seems, in amounts of up to a few tenths of a percent and on time scales from 1 second to 10 years and possibly centuries.

The Sun becomes brighter overall as the number of sunspots on its surface increases, and *vice versa*. This seems counterintuitive, since sunspots are cooler than their surroundings. However, the sunspot cycle is accompanied by variations in magnetic activity, which create an increase in luminous output that exceeds the cooling effects of sunspots. Indeed, the entire spectrum of the Sun's radiation varies in step with the 11-year cycle of solar magnetic activity.

The changes in visible solar radiation, as well as in the solar constant itself, are relatively modest. However, significant variations occur at the short, invisible wavelengths that contribute only a tiny fraction of the Sun's total radiation. Ultraviolet radiation is at least 10 times more variable than longer-wavelength visible radiation, and the Sun's X-ray emission varies by at least a factor of 100 (*Figure 23*).

Ultraviolet and X-ray energy is absorbed high in Earth's atmosphere, where the global mean temperature can double between the minimum and maximum of solar activity. Smaller changes in the Earth's global surface temperature, observed during the past century, correlate closely with the length of the sunspot cycle. For the time being, such cyclic variations seem to be greater than those produced by the greenhouse warming caused by human activity. However, our civilization's contribution to global warming will eventually surpass these natural

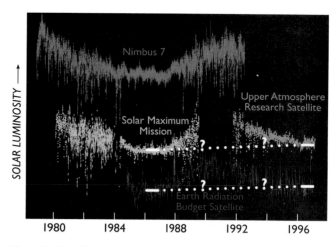

Figure 22. Complementary observations from several satellites show that the Sun's output fluctuates during each 11-year cycle (dips in curves), with a maximum coinciding with the peak in sunspot and magnetic activity. Some researchers believe the satellite data also suggest that the Sun was slightly brighter in 1996 than during the previous solar minimum (dashed lines). However, this interpretation is considered controversial, and a longer baseline of observations is needed.

Figure 23. X-ray images of the Sun, taken by the Yohkoh spacecraft, show dramatic changes in the corona from the time of the satellite's launch in August 1991, at the maximum phase of the 11-year sunspot cycle (left), to late 1995 when the active magnetic regions associated with sunspots had almost disappeared (right). During this interval the Sun's magnetic field changed from a complex tangle to a simpler structure. In response, the coronal gases became less agitated and cooled down, resulting in a hundredfold drop in overall X-ray brightness over four years.

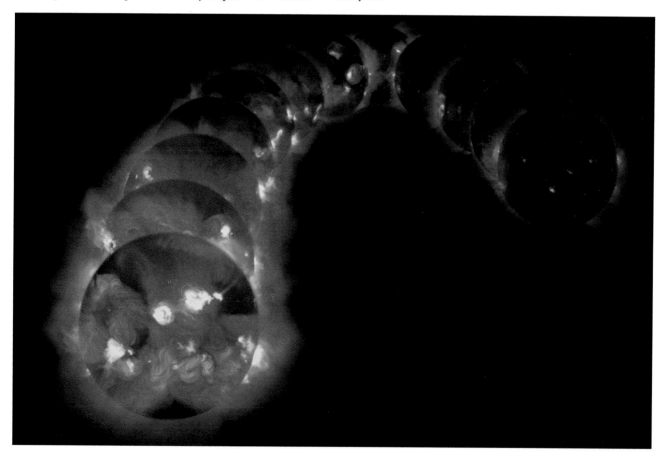

effects if we continue to alter the composition of the atmosphere without control (see Chapter 13).

The varying Sun has many other important consequences for life on Earth. For example, our protective high-altitude layer of ozone (O_3) is both produced by and protects us from ultraviolet sunlight. The ozone's abundance becomes enhanced, depleted, and enhanced again during the solar-activity cycle. Moreover, the Earth's magnetic cocoon, or magnetosphere, is constantly buffeted and reshaped by the solar wind, especially when explosive outbursts emanate from the Sun. Such events increase in intensity and frequency at the maximum of the solar-activity cycle, and these geomagnetic storms can affect us significantly (see Chapter 4). As our civilization becomes increasingly dependent on sophisticated systems in space, it becomes more vulnerable to this Sun-driven space weather.

THE REMOTE PAST AND DISTANT FUTURE

Nothing in the cosmos is fixed and unchanging. Everything moves and evolves, and that includes the seemingly constant and unchanging star of our solar system. The Sun began its life

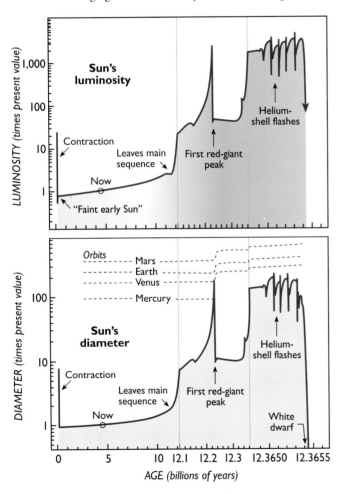

Figure 24. Billions of years from now the Sun will grow enormously in size and luminosity. Note the different time scales, expanded near the end of the Sun's life to show relatively rapid changes. The orbits of the planets enlarge due to mass loss from the Sun. By the time our star becomes a white dwarf, it will have only 0.51 to 0.58 of its present mass.

shining only 70 percent as brightly as it does now. It has slowly grown in luminous intensity with age, a steady, inexorable brightening that is a consequence of the increasing amount of helium accumulating in the Sun's core. Because fusion turns protons into helium nuclei at a four-to-one exchange rate, the sum total and number density of these subatomic particles has gradually decreased. Since gas pressure is proportional both to this number density and to the temperature, the only way to keep the core pressure in balance with the weight of overlying material has been for the core temperature to increase. This heating, in turn, has produced the gradual increase in the Sun's luminosity — enough to make its surface 300° K hotter and its radius 6 percent greater than when it first shone 4.5 billion years ago. Aside from this gradual brightening, we anticipate that the Sun will continue as it is now for several billion years more.

If the Sun was significantly dimmer when it formed, the young Earth's oceans should have been frozen solid at that time — even if we take into account the warming greenhouse effect of our atmosphere. Yet sedimentary rocks, which must have been deposited in liquid water, date from 3.8 billion years ago. There also is fossil evidence in those rocks for the emergence of life about that time. Thus, for billions of years the Earth's surface temperature was not very different from today.

The discrepancy between Earth's warm climatic record and an initially dimmer Sun has come to be known as the *faint-young-Sun paradox*. It can be resolved if the Earth's primitive atmosphere contained about a thousand times more carbon dioxide than it does now. The greater heating of the enhanced greenhouse effect could have kept the oceans from freezing. Over time the Sun grew brighter and hotter. The Earth could only maintain a temperate climate by turning down its greenhouse effect as the Sun turned up the heat. Our planet's atmosphere, rocks, and ocean apparently combined to decrease the amount of carbon dioxide over time. The recent increase in atmospheric carbon dioxide caused by the burning of fossil fuels is currently negligible compared to this long-term depletion, but it could produce noticeable global warming in the future if left unchecked.

In the end, our prospects are not all that great anyway. From both astrophysical theory and observations of other stars, we know that ultimately dramatic changes are in store for the Sun (*Figure 24*). Astronomers calculate that the Sun will be hot enough in 3 billion years to boil Earth's oceans away, leaving the planet a burned-out cinder, a dead and sterile place. Four billion years thereafter our star will balloon into a giant star, engulfing the planet Mercury and becoming 2,000 times brighter than it is now. Its light will be intense enough to melt the Earth's surface and to turn the icy moons of the giant planets into globes of liquid water. The only imaginable escape would then be interplanetary migration to distant planets with a warm, pleasant climate.

Such events are in the very distant future, of course. For now the Sun continues to provide us with an up-close laboratory of stellar astrophysics and evolution. What we learn from its seething surface and roiling interior, and how we ultimately solve the mystery of the neutrinos, will surely help us in understanding of the universe at large.

Planetary Magnetospheres and the Interplanetary Medium

James A. Van Allen
Frances Bagenal

The Van Allen belts, high-energy charged particles trapped in Earth's magnetopshere.

THE CONCEPT OF interplanetary space generally conjures up an image of a vast, tranquil void of complete emptiness. The environment between planets is indeed much nearer to a vacuum than that produced in any laboratory. However, in reality it is stormy with energetic particles and radiation flowing from the Sun. This "space weather" changes constantly in step with the variable and sometimes violent solar activity.

One of the excitements of the Space Age has been our unfolding of the complex story of the interplanetary medium and its interaction with the planets (magnetized or not), comets, and asteroids. Magnetospheres are the most important products of such interactions, but there are also other effects. For example, the direct irradiation of otherwise-unprotected surfaces alters their spectral properties and compositions. Atmospheres can be heated, ionized, and eroded by the impinging interplanetary medium. In some cases, magnetic fields shield surfaces from energetic particles very effectively, conceivably accommodating — or frustrating — the development of life.

Virtually all of this activity goes unseen at visible wavelengths. It is no coincidence that human eyes have evolved to be most sensitive over the narrow wavelength range at which the Sun's intensity peaks. However, "sunlight" includes a wide spectrum of wavelengths, all of which contribute to the *interplanetary medium*. There are far fewer solar ultraviolet and X-ray photons, but their shorter wavelengths make them more energetic than visible-light photons. Consequently, they create more pronounced effects on upper atmospheres and exposed surfaces. Our Sun is also a tremendous source of radio energy, and it frequently gives off tremendous bursts of X-rays that last from minutes to hours.

THE SOLAR WIND

One of the most engaging detective stories of modern science has been recognition of the existence and importance of solar "corpuscular radiation," that is, gaseous material shed by the Sun, distinct from light and other forms of electromagnetic energy. This outpouring from the Sun's atmosphere, a combination of ionized gas and the entrained solar magnetic field, is now usually termed *solar wind*.

In their great 1940 treatise, *Geomagnetism*, Sydney Chapman and Julius Bartels speculated that fluctuations in Earth's magnetic field (magnetic storms) are caused by solar corpuscular streams. A decade later, the German astronomer Ludwig Biermann showed that the streams were not simply intermittent bursts but were instead a continuous phenomenon (a conclusion he based on observations of comets' bluish gas tails). Over the past half century, instrumented spacecraft have confirmed some, but not all, of these early inferences concerning solar corpuscular streams and have provided a wealth of detailed knowledge about them.

The solar wind consists of a hot *plasma* — an electrically neutral mixture of electrons and ions (principally protons with some heavier atomic nuclei) at roughly 100,000° K. Its source is the Sun's atmosphere, or corona, and it is continuously present in interplanetary space. This gas flows radially outward at a typical speed of 450 km per second to at least 70 AU and probably much farther. The average speed of the flowing gas is remarkably independent of its distance from the Sun.

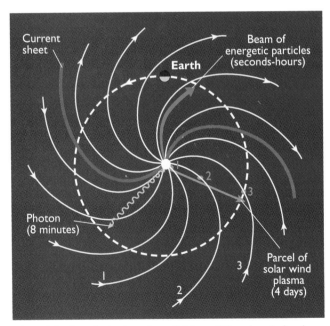

Figure 1. In the ecliptic plane the radial flow of the solar wind and the rotation of the Sun combine to wind the solar magnetic field into a spiral. A parcel of solar wind plasma (average speed: 450 km per second) takes about 4 days to travel from the Sun to Earth's orbit at 1 AU. The dots and magnetic field lines labeled *1, 2,* and *3* represent snapshots during this journey. Solar energetic particles travel much more rapidly. Because their trajectories are controlled by the Sun's magnetic field, these particles are scattered by irregularities in the field and thus spread out quickly as they move outward through the heliosphere. The heliospheric *current sheet* separates magnetic fields of opposite polarities.

Even though the solar corona ejects gas radially outward into interplanetary space, the continuous stream takes on, by virtue of the Sun's rotation, the approximate form of an Archimedean spiral (*Figure 1*). At the orbit of Earth, the spiral makes an angle of about 45° with a radial line from the Sun but becomes nearly perpendicular at the orbit of Saturn (9.5 AU) and beyond.

At Earth's orbit the number density of ions and electrons in interplanetary space is typically five particles per cm^3 under quiet conditions. This population density diminishes as the inverse square of the heliocentric distance, but sporadic order-of-magnitude fluctuations occur in response to varying solar activity. The solar wind's speed is also variable, and a variety of collisionless shock phenomena occur as fast streams overtake slow ones. Outright collisions are rare in this exceedingly dilute interplanetary medium. However, the particles and magnetic field are coupled by electric and magnetic interactions, and these wave-particle phenomena replace collisions as the agent of energy transfer. A wide variety of wave types are carried in the interplanetary medium.

The ionized, electrically conducting gas carries with it an entrained magnetic field that also originates in the solar corona (see Chapter 3). Since the dynamics of the interplanetary medium are dominated by the mass motion of the ionized gas it contains, the magnetic field lines become stretched out approximately parallel to the "wave fronts" of the plasma stream. The magnitude and direction of this field vary markedly from point to point, but it generally parallels the theoretical Archimedean spiral. At 1 AU the magnetic-field strength is typically 5×10^{-9} tesla (or 5 nanotesla), many orders of magnitude less than its levels in the solar chromosphere. Because the field lines assume a tightly wrapped spiral form far from the Sun, the magnetic field's strength does not decrease as quickly as the number density of the plasma does. Instead, it falls off as the inverse of heliocentric distance.

Until recently, direct observations of the solar wind were confined to a thin, pancake-shaped region near the plane of Earth's orbit (the ecliptic), which is approximately the equatorial plane of the Sun. However, studies of the scintillation of stellar radio sources have shown that solar-wind speed increases at locations well above and below the Sun's equator. A marked advance was made by the Ulysses mission, which extended our measurements to latitudes within 10° of the Sun's south and north poles in 1994–1995. At that time, near solar-activity minimum, the solar-wind speed increased abruptly at a latitude of about 20° from its low-latitude value of 400 km per second to 770 km per second (*Figure 2*). Although this finding validated our earlier, less-direct inferences, we do not yet understand why the solar wind is so much faster at high latitudes.

COSMIC RAYS

In 1912 Victor Hess found that a weak but mysterious radiation grew more intense with altitude in Earth's atmosphere. He surmised that it must come from outer space, a conjecture that was soon confirmed. We now realize that this *cosmic radiation* consists of atomic nuclei (principally protons) and a much smaller contribution of electrons. These are termed "primary" particles, and their energies range from tens of millions to many billions of electron volts. Because primary cosmic rays and their progeny are so energetic, they have had an important role in modern high-energy particle physics.

The compositional distribution of cosmic-ray nuclei resembles the relative abundance of nuclei in "universal" matter, as estimated from astrophysical evidence. This is not surprising, since it is widely believed that the more energetic cosmic-ray particles, sometimes called galactic cosmic rays, originate in violent astrophysical events such as supernovas and explosive stellar flares. The Sun also sporadically emits electrons, protons, and heavier ions from its active regions. These *solar energetic particles* range from a few thousand to tens of millions of electron volts in energy, and they travel outward through the interplanetary magnetic field (*Figure 1*). Bursts of such particles last for hours to days, and they pose a serious hazard for human interplanetary travel.

At ground level, instruments detect only cosmic-ray "secondaries," which result when primaries having energies of many billions of electron volts interact with molecules of gas in our atmosphere. As first shown in 1954, the count of secondaries varies by a few percent in step with the 11-year cycle of solar activity, with the intensity peak near the time of solar minimum and vice versa. At balloon altitudes this cyclic variation becomes considerably more pronounced, especially at high latitudes. In near-Earth space, the disparity is even greater, with the maximum cosmic-ray intensity more than twice that at its cyclic minimum. Thus, for lower-energy particles, whose secondaries are unable to reach the ground, the magnitude of the solar-cycle variation is much greater.

The mystery of this cyclic variation in galactic cosmic-ray intensity has intrigued many investigators. Undoubtedly, the effect is attributable to the Sun. The prevailing view is that the magnetic field entrained in the outflowing solar wind acts as a modulating agent. Cosmic rays enter the outer solar system and diffuse inward, but the turbulent magnetic field of the interplanetary medium tends to convect them outward and decelerate them. The process may be likened to a school of fish swimming upstream in a turbulent river. Thus, cosmic rays are relatively rare in the inner solar system — especially the low-energy particles, which are convected and decelerated most easily.

A further mystery, perhaps connected to the issue of solar-cycle modulation of cosmic rays, lies in the variation of cosmic-ray fluxes with solar latitude. At one time we expected that cosmic rays would have relatively free access to the inner solar system along conical spirals of field lines over the Sun's poles. While Ulysses did detect greater numbers of cosmic rays during its high-latitude passes, the increase was much less than expected, particularly at lower energies. An explanation may be rooted in Ulysses's finding that the polar magnetic field is more irregular and less streamlined than expected.

The challenge to space physicists is to determine how well the interplanetary magnetic field repels cosmic-ray particles of a given energy throughout the planetary system and, ultimately, to find the boundary (the *heliopause*) at which the Sun's modulating influence ends. We shall return to the subject of the search for the heliopause later in the chapter.

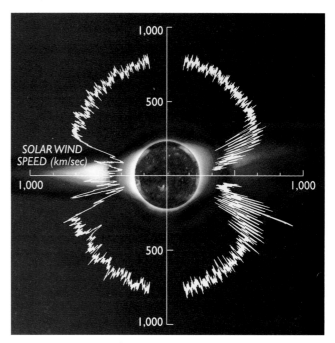

Figure 2. A composite of an ultraviolet image of the solar disk and white-light images of the solar corona form the backdrop for a radial plot of solar-wind speed versus latitude. These data, obtained by the Ulysses spacecraft between 1991 and 1996, show that the wind escapes into interplanetary space much faster from the Sun's polar regions than from near its equator.

PLANETARY MAGNETISM

The Sun's powerful magnetic field arises from the turbulent motion of electrically conductive matter, powered by the nuclear fusion going on in its core. The planets do not contain nuclear furnaces, yet somehow they still create and maintain intrinsic magnetic fields. One explanation is that they simply solidified in the presence of the Sun's extended field 4½ billion years ago. The ambient interplanetary field would have become "frozen" into the iron-bearing rocks of Earth and its neighbors. However, the interiors of the largest planets were — and remain — well above the *Curie point*, the temperature below which ferromagnetic materials retain an ambient magnetic field. Smaller bodies have cooled below the Curie point and thus may still have their primordial fields.

Alternatively, global magnetic fields in planetary objects can be generated by an internal *dynamo*. Theories of such a dynamo are complex, and their predictive value is quite limited. Nevertheless, there is general agreement that a dynamo has three basic requirements: (1) a rotating body (2) containing a fluid, electrically conducting region (3) within which convective motion occurs. The necessary rate of rotation is uncertain, but all the planets (except perhaps Venus) and major moons appear to spin sufficiently fast. Likewise, all the planets and many of the larger moons have electrically conducting interior regions. These can take the form of molten rock-iron mixtures (the terrestrial planets and major moons); hydrogen having metallic properties at high pressure and temperature (Jupiter and Saturn); or mixtures of water, ammonia, and methane (Uranus and Neptune).

Another requirement that differentiates the magnetic "haves" and "have nots" is the existence of a thermal gradient across their electrically conducting regions large enough to drive convective motions. In theory, the smaller, rocky objects of the solar system, including the terrestrial planets, should have stably stratified cores that are slowly cooling by conduction. Within rocky bodies having magnetic fields (Earth, Mercury, and possibly Ganymede), dynamo convection is powered by gravitational energy from primordial accretion, the decay of radioactive elements, and heat from chemical transformations. The giant planets, meanwhile, have retained much of their primordial heat, enough to drive vigorous internal convection and power their dynamos. In some cases, ongoing compositional differentiation may be providing an additional source of heat.

Although modern theories of self-sustained dynamos are complex, a plausible scenario was first provided by Eugene Parker in 1955. The fluid inside a planetary body tends to rotate differentially, with the innermost region spinning fastest. If an exterior (roughly north-south) magnetic field penetrates this conducting fluid, it will be wound up into coils of azimuthal or "toroidal" magnetic field. This coiling acts to concentrate and amplify the magnetic field strength in the planet's core. If the conducting fluid is heated and rises, the combination of convection and rotation carries the toroidal loops upward and twists them. The result is a field that rein-

forces the original (north-south) external field. At that point the process becomes self-sustaining, and the planet's magnetic field is regenerated continuously.

The simplest form of a magnetic field is a *dipole*, analogous to a bar magnet. The simplest general characterization of a planet's intrinsic magnetism is its equivalent dipole *magnetic moment*. The dipole moment divided by the cube of the planet's radius yields the average strength of the field along the magnetic equator (*Table 1*). For example, at Earth's magnetic equator this value is 0.305 gauss or 30,500 nanoteslas (nT). To generate a surface field this strong, thousands of kilometers from its core, Earth must have an impressively large magnetic moment, 7.91×10^{25} gauss cm^3. By comparison, laboratory electromagnets are typically of the order of 100,000 gauss cm^3. We cannot yet explain why Jupiter's magnetic moment is nearly 100 million times that of Mercury, though an obvious factor must be the absolute size of the conducting fluid region.

We also do not understand why dipole moments exhibit such a wide range of alignments with respect to the planets' rotation axes (*Figure 3*). Moreover, the dipoles of Mercury and Earth have a polarity opposite that of the other planets. Little significance is placed on this fact, however, since Earth's magnetic field has reversed its polarity many times over geologic time. Of

Characteristics of Planetary Magnetic Fields

	Rotation period (days)	Dipole moment (Earth=1)	Field at equator (gauss)	Polarity same as Earth's?	Angle between axes	Typical magnetopause distance (R_p)	Plasma sources
Mercury	58.65	0.0007	0.003	yes	14°	1.5	W
Venus	243.02(R)	< 0.0004	< 0.00003	—	—	—	W,A
Earth	1.00	1	0.305	yes	10.8°	10	W,A
Mars	1.03	< 0.0002	< 0.0003	—	—	—	—
Jupiter	0.41	20,000	4.28	no	9.6°	80	W,A,S
Saturn	0.44	600	0.22	no	< 1°	20	W,A,S
Uranus	0.72(R)	50	0.23	no	58.6°	20	W,A
Neptune	0.67	25	0.14	no	47°	25	S

Table 1 (above). The solar system's planets exhibit a wide range of magnetic properties. *(R)* indicates retrograde rotation. Earth's magnetic polarity is such that its south magnetic pole lies in the Northern Hemisphere. *Angle between axes* refers to the magnetic and rotational axes. *Typical magnetopause distance* is in the direction of the Sun and given in planetary radii. Under *plasma sources*, the letter *W* is for solar wind, *A* for atmosphere, and *S* for satellites or rings.

Figure 3 (below). No two planetary magnetic fields are alike, as demonstrated by the examples of Earth, Jupiter, and Saturn. Bar magnets represent the planets' dipole fields and magnetic polarities. Throughout a planet's orbit its obliquity and dipole tilt combine to produce a seasonal variation in the angle between the magnetic field axis and the radial distance from the Sun.

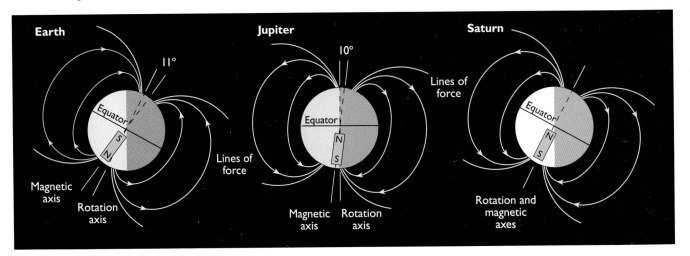

greater consequence is the combined effect of the dipole tilt and a planet's obliquity (the inclination of the equator to the orbital plane). This leads to diurnal and seasonal changes between the angle of the solar wind and the dipole moment, sometimes with significant consequences (discussed later).

A planet's external magnetic field can be represented most simply by a point magnetic dipole (a vector with defined magnitude and direction) located at its center. Saturn's field comes closest to the ideal case, in which the rotation and magnetic axes coincide. Point dipoles also provide reasonable approximations for Mercury, Earth, and Jupiter. Maps of the surface fields of Earth and Jupiter show stronger fields at the poles and a magnetic equator (*Figure 4*). By contrast, the magnetic fields of Uranus and Neptune are both offset from these planets' centers and highly inclined with respect to their rotational axes. This creates surface fields that are much stronger in one hemisphere and magnetic equators that weave over the surface. The dynamo motions in their interiors must be very different from those of other planets.

The situation with Mars is less certain. Early spacecraft equipped with magnetometers did not approach very close to the Martian surface. Weak fields were detected, but the planet's dipole moment is no more than 0.0002 that of Earth. Scientists argued for decades over whether the field is due to a dynamo or to currents in the ionosphere. Then, in September 1997, Mars Global Surveyor reached its destination. During its first few orbits the spacecraft came within 120 km of the Martian surface and found regions with surprisingly strong magnetic fields (about 400 nT). However, the field's patchy nature suggests that it is unlikely the result of an active internal dynamo. Instead, the magnetization is probably *remanent*, that is, frozen into expanses of solidifying lava. In time, we should learn whether this lingering field arose from an ancient Martian dynamo, or from ionospheric currents driven by the solar wind (which may have been strong if Mars's early atmosphere was denser than it is now).

TYPES OF MAGNETOSPHERES

Our knowledge of Earth provides the basis for understanding the electromagnetic properties of other planets. However, as our spacecraft emissaries have found, each world exhibits distinctive and unique magnetic properties. A strict definition of a planetary *magnetosphere* is the region surrounding a planet within which its own magnetic field dominates the behavior of electrically charged particles. The term does not imply a spherical shape but is used in a looser sense, as in the phrase "sphere of influence."

The solar wind has a negligible effect on the movements of large bodies. However, it has profound effects in their immediate vicinity, creating an amazing assortment of physical phenomena. Because the solar wind is a plasma, it behaves like an electrically conducting fluid. It is also magnetized; that is, it contains systems of electrical currents that survive from their origin in the solar corona. These two properties, when combined with its bulk flow, are essential to the creation of most magnetospheric phenomena (*Figure 5*).

Upstream of a planetary magnetosphere is a *bow shock*, a standing shock wave that results because the solar wind is supersonic — it moves faster than the waves that are propagating

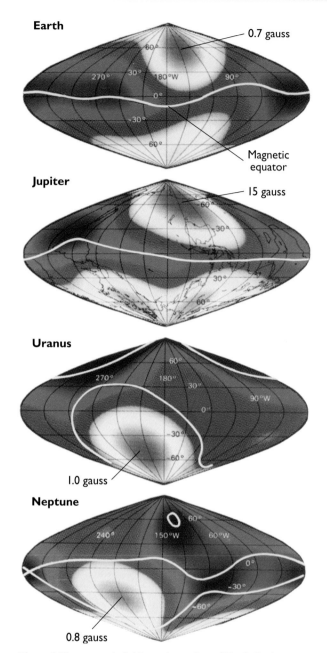

Figure 4. The magnetic fields on the surface of Earth, Jupiter, Uranus, and Neptune exhibit a wide range of strengths. Note especially how the dipole fields of Uranus and Neptune, which are markedly offset from the planets' centers, create distinctly asymmetric surface fields.

through it. Essentially, the solar wind cannot "sense" the magnetized obstacle in its path soon enough to move smoothly around it. The situation is comparable to the behavior of air around a supersonic aircraft. After passing through the bow shock, the solar wind encounters and flows around the *magnetopause*, the boundary between the solar-wind plasma and the magnetosphere. The magnetic field accompanying the solar wind merges with that of a planet and stretches it out to produce a long, turbulent *magnetotail*, or wake, on the downwind side of the planet. Earth's magnetotail, for example, is several million kilometers long. The *stagnation point*, on the planet's sunward side, is where the solar wind comes closest to the cen-

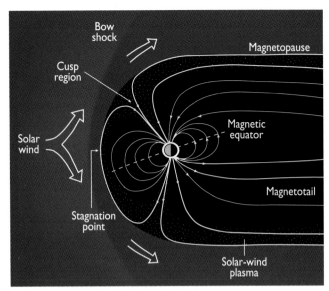

Figure 5. This portrayal of a generic planetary magnetosphere shows many of the features commonly found around magnetized planets.

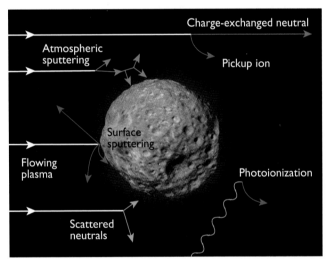

Figure 6. When plasma (ions and electrons) bombards a surface or atmosphere at high speed, a wide variety of processes can occur.

ter of the planet. This standoff distance depends upon the ever-changing solar-wind pressure. For Earth, it is usually about 64,000 km away — far above the atmosphere.

The stronger a planet's magnetic field, the larger its magnetosphere. In addition, the weak solar-wind pressures present in the outer solar system allow relatively weak planetary magnetic fields to carve out large cavities. Consequently, the sizes of planetary magnetospheres span a vast range, from the small magnetic bubble surrounding Mercury (which would fit entirely within Earth) to the giant magnetosphere of Jupiter (which is at least a thousand times the volume of the Sun).

The magnetopause is not completely "plasma-tight," however, particularly around the magnetic poles. Consequently, magnetospheres contain considerable amounts of plasma, the main source of which is usually the Sun. Solar-wind plasma has a characteristic composition: protons (H^+), about 4 percent alpha particles (He^{2+}), and traces of heavier nuclei, many of which are highly ionized. Ionospheric plasma has a composition that reflects the composition of the planet's upper atmosphere (for

example, O^+ for Earth and H^+ for the outer planets). Natural satellites or ring particles embedded in the magnetosphere can also generate significant quantities of plasma. The residence times of magnetically trapped particles vary widely — from hours to years. Their motion inside a magnetosphere depends on the relative strength of the coupling to the rotating planet compared with that to the solar wind.

As it races outward through the planetary system, the solar wind sometimes encounters a sizable object that lacks an intrinsic magnetic field. If such an object has a surface with a low electrical conductivity, no electrodynamic interaction occurs and the plasma runs directly into the surface. Downstream of the object is wake cavity largely devoid of plasma. This is the type of interaction created by Earth's Moon, and we assume that the same scenario applies to many small, rocky objects of the solar system.

If the nonmagnetic object possesses an atmosphere, the uppermost atoms and molecules are ionized by solar radiation or by impact with the flowing plasma in which they are imbedded (*Figure 6*). This *ionosphere* provides conducting paths for electrical currents to flow into the solar wind, where they create forces that slow and divert the incident flow. The barrier that separates the planetary plasma from the solar wind is referred to as the *ionopause* (analogous to the magnetopause), and the interactions are similar in many ways to those of a true magnetosphere. A bow shock forms upstream if the plasma flow is supersonic, and the solar-wind magnetic field drapes around the planet and forms a magnetotail. This type of magnetosphere is found at Venus and Mars, as well as at comets and satellites that have substantial atmospheres.

HOME TERRITORY: EARTH

Using simple radiation detectors aboard the Explorer 1 and 3 satellites in 1958, James Van Allen and his students discovered that a huge population of energetic charged particles surrounds Earth and remains durably trapped there by our planet's magnetic field. No one had predicted such an effect; indeed, those early instruments were designed to survey cosmic-ray intensities above the atmosphere. The trapped particles form two distinctively different *radiation belts*. (The term "radiation belts" has historical roots; it does not imply radioactivity.) Each population has the shape of a torus and encircles our planet in such a way that its central plane coincides roughly with Earth's magnetic equatorial plane (*Figure 7*).

In 1907, Carl Størmer showed theoretically that an energetic, electrically charged particle can be permanently trapped, or confined, within the magnetic field of a dipole. In any static magnetic field, the force on a moving charged particle is directed at right angles to both the direction of its motion and the magnetic vector. This *Lorentz force* causes the particles to move in spring-shaped paths (*Figure 8*), quickly moving back and forth in magnetic latitude while drifting slowly in longitude. The specific trajectory depends on the particle's momentum and electrical charge, and over time it sweeps out a toroidal volume encircling the dipole.

An important source of radiation-belt particles is neutrons produced when galactic cosmic rays and energetic solar particles

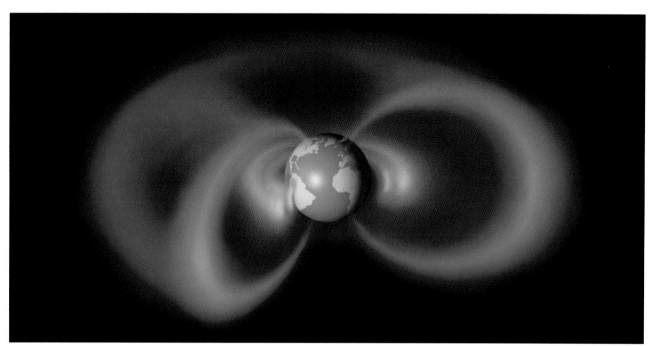

Figure 7 (above). The inner Van Allen belt contains a mixture of trapped charged particles: protons (yellow) with energies greater than 10 million electron volts and electrons (blue) exceeding 500,000 electron volts. The outer belt also contains protons and electrons, most of which have energies under 1.5 million electron volts. Earth's magnetic dipole is offset from its center by about 500 km. Consequently, one side of the inner Van Allen radiation belt comes closer to our planet's surface than the other side does. This region, termed the South Atlantic Anomaly, has affected satellites for four decades.

Figure 8 (middle right). Charged particles trapped by a planet's magnetic field follow complex trajectories. They gyrate around the lines of force (upper diagram), bouncing between the regions of stronger magnetic field at the ends of the field lines over intervals of seconds to minutes. They also drift around the planet on time scales of hours. The greater the particle's energy, the less time it takes for each motion.

Figure 9 (bottom right). The general shape and principal features of Earth's magnetosphere. Charged particles become trapped in a pair of radiation belts near the planet and in the magnetotail, which extends to the right (well outside this depiction).

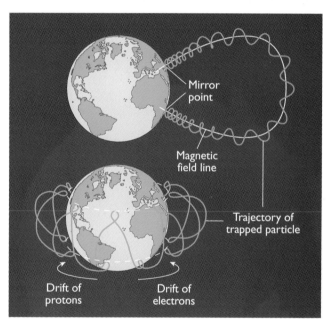

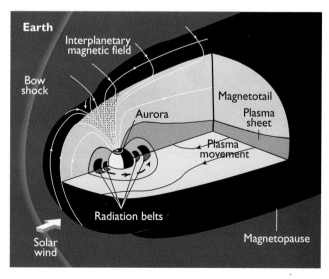

bombard Earth's atmosphere. As they fly off into space, a small fraction of these neutrons decay into protons and electrons, which are immediately snared by the magnetic field. The residence times of such particles in trapped orbits depend in part on the ambient field strength. In the strong field present close to Earth, residence times for protons having energies of tens of millions of electron volts are of the order of a decade. So, even though the decay of neutrons provides only a weak source of particles, they remain trapped long enough to accumulate in substantial numbers.

The magnetosphere, especially its outer reaches, is dynamic and constantly changing. "Low-energy" particles (with less than about 10,000 electron volts) have a much greater influence on gross physical phenomena than do their relatively rare "high-energy" counterparts. However, the latter pose hazards

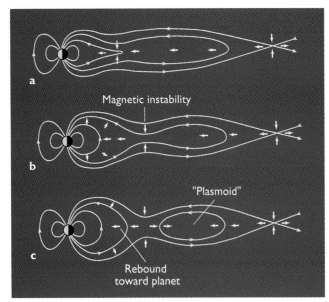

Figure 10. During a magnetic substorm, changes in solar-wind conditions cause the magnetotail to be pinched off close to Earth. Magnetospheric plasma is accelerated away from this disturbance; a blob of plasma is ejected down the magnetotail, while other particles cascade into Earth's polar regions, often causing auroras.

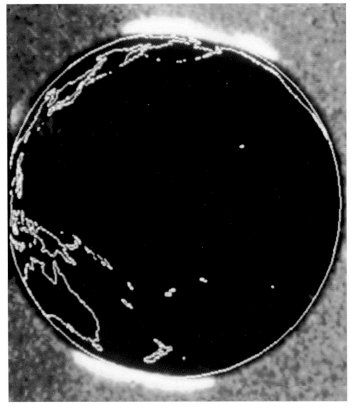

Figure 11. This extraordinary image from NASA's Dynamics Explorer 1 satellite shows both the aurora borealis ("northern lights") and the aurora australis ("southern lights"). These glowing ovals or rings are about 500 km wide, 4,500 km in diameter, and centered roughly on Earth's magnetic poles. Green lines show major land areas; Australia is at lower left and North America at upper right. Most auroral displays occur at altitudes of 110 to 240 km and appear green (dominated by 5577-angstrom emission from oxygen atoms). Above that, up to 400 km high (rarely to 1,000 km), the aurora has a ruby-red glow from oxygen ions emitting at 6300 and 6364 angstroms.

to electronic systems and living things, and their distribution places practical limits on the orbits around Earth where human crews and animals are safe from excessive radiation exposure. The most readily accessible region of safe flight lies at altitudes below 400 km. By contrast, the radiation dosage within the equatorial region of the inner radiation belt, at an altitude of about 2,500 km, is especially severe — even electronic instrumentation has a limited useful lifetime there.

Earth's magnetosphere comprises two very different regions, distinguished by the plasma sources and flows in each region (*Figure 9*). Close to Earth, the trapped oxygen ions, protons and electrons derived from the ionosphere corotate with the planet. This *plasmasphere* extends outward to between 25,000 and 40,000 km, and within it lies the terrestrial ring current that causes magnetic storms. The plasmasphere and radiation belts overlap in the inner magnetosphere, where the cold (2,000° K) ionospheric plasma limits the population of energetic radiation-belt particles through complex plasma-wave interactions.

Farther out, the magnetosphere is dominated by its interaction with the solar wind. The plasma in this outer region consists largely of protons and electrons that have leaked in across the magnetopause. Because convection cycles this plasma through the magnetosphere in a matter of hours, it does have a chance to build up to substantial densities. In the magnetotail, on the nightside of Earth's magnetosphere, the solar-wind–driven flow brings the plasma down to the equatorial plane, where it is concentrated in the *plasma sheet*. Changes in the solar wind's condition can disturb the convective flows and trigger an instability in the plasma sheet (*Figure 10*). This disruption, termed a *magnetic substorm*, occurs between 6 and 20 Earth radii down the tail. It causes plasma to flow away from the disturbance up and down the magnetotail, some of which follows magnetic-field lines back toward Earth. At such times the accelerated electrons bombard the upper atmosphere and can generate spectacular auroral displays (*Figure 11*).

Earth's magnetosphere is a natural laboratory for a variety of interactions of plasma with electric and magnetic fields. When the charged-particle population is disturbed or becomes unstable, it generates numerous electrostatic and electromagnetic waves. Taken together, these processes make Earth a strong radio source that radiates roughly 100 million watts into interplanetary space. However, despite its panoply of energetic processes, Earth's magnetic environment is dwarfed by the magnetospheres surrounding Jupiter and Saturn. Their strong planetary magnetic fields allow the magnetospheric plasma to tap the planet's rotational energy and accelerate particles to relativistic energies.

JUPITER: THE MAGNETIC GIANT

As early as 1955, Jupiter was recognized as a source of sporadic bursts of radio noise at a frequency of 22.2 megahertz (decametric wavelengths). These emissions are termed *nonthermal* because they arise from processes other than those associated with heat. Soon thereafter, another type of nonthermal radiation from Jupiter was discovered at much higher frequencies, from 300 to 3,000 megahertz (decimetric wavelengths). Unlike the bursty decametric emission, Jupiter's decimetric radiation has an intensity and spectral form that are nearly constant with time. It comes from a toroidal region whose central plane tilts about 10° to the planet's equatorial plane (*Figure 12*). In 1959, Frank Drake and Hein Hvatum interpreted the decimetric emission as synchrotron radiation emitted by electrons trapped in the Jovian magnetic field and moving at relativistic velocities.

These ground-based observations set Jupiter apart from all other planets and provided an impetus for sending spacecraft to probe its radiation belts directly. Pioneers 10 and 11 encountered Jupiter in late 1973 and 1974, respectively. They confirmed the existence of trapped, relativistic electrons packed several orders of magnitude more densely than in Earth's magnetosphere. Even during their brief encounters, each spacecraft absorbed a thousand times the dosage of high-energy electrons known to cause severe radiation sickness or death in humans. Both Pioneers suffered the failure of several transistor circuits and the darkening of exposed optics.

Jupiter's magnetic moment is tilted 9.6° to its rotational axis, in agreement with the evidence from ground-based radio observations. The general form of the Jovian magnetosphere resembles that of Earth, but its dimensions are at least 1,200 times greater (*Figure 13*). In fact, if we could see this enormous electromagnetic bubble in the nighttime sky, it would appear several times larger than the full Moon. Voyager observations later showed that the magnetotail of Jupiter extends behind it to the orbit of Saturn and beyond — at least 650 million km! These enormous dimensions result from a magnetic moment 20,000 times greater than Earth's and from the fact that the solar-wind pressure at 5.2 AU from the Sun is only about 4 percent of its value here at 1 AU.

Beyond these basic characteristics, the magnetosphere of Jupiter exhibits a rich variety of special and unique features. First, substantial amounts of trapped, energetic plasma exert pressure on the magnetic field, inflating it like an air-filled balloon. Because the field is weakest in its equatorial plane, the outward distention is most prominent there. Second, the magnetic field *corotates* with the planet's interior (once every 9 hours 56 minutes), and the plasma interacting with the field is

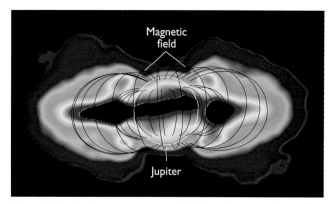

Figure 12. Jupiter is a strong source of both continuous and sporadic radio energy. This radio map shows the steady, decimetric-wavelength energy emitted by electrons with relativistic energies that are trapped in the Jovian magnetosphere. This type of synchrotron emission is not observed from any other planet but is a common feature of many astrophysical objects, including pulsars.

forced to circle Jupiter with this period as well. Therefore, centrifugal force pushes the plasma outward.

The combination of these two effects produces a distinctive disk of plasma, or *plasma sheet*, lying roughly in the planet's magnetic equatorial plane (*Figure 13*). As Pioneer 10 and the twin Voyagers flew through Jupiter's magnetosphere they measured particle fluxes that rose and fell as the tilted plasma sheet flopped north and then south over the spacecraft twice per 10-hour rotation period. The exact mechanism that accelerates so many particles to high energies, generating the hot plasma that inflates the magnetosphere, is not known. The source may well be the rapid rotation of Jupiter and the strong coupling of its magnetic field to the trapped magnetospheric plasma.

The first indication that Jupiter's moon Io plays a significant role in the magnetosphere came when E. K. Bigg noticed in 1964 that the bursts of Jovian decametric radio emission are strongly controlled by the moon's orbital longitude. Early clues of the "Io connection" came in the 1970s, when ground-based optical telescopes recorded an extended atmosphere of sodium atoms around Io itself and a cloud of S+ ions all the way around

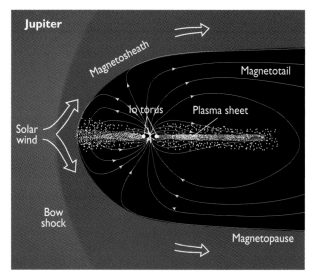

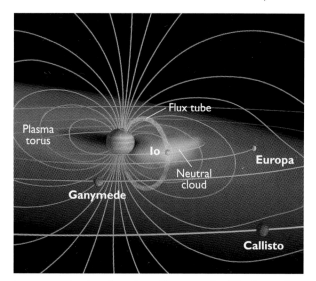

Figure 13. Jupiter's magnetosphere is an enormous envelope much larger than the Sun. Trapped plasma is concentrated in a disklike *plasma sheet* near the planet's magnetic equator. The

inner magnetosphere is portrayed in greater detail in the panel at right. Io is electrodynamically coupled to Jupiter and is the main source of the magnetosphere's plasma.

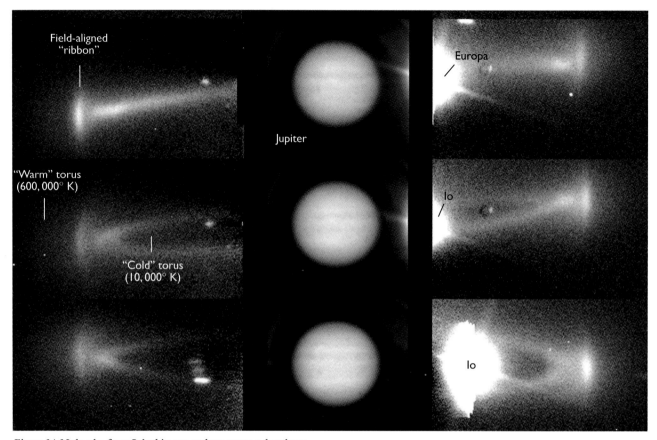

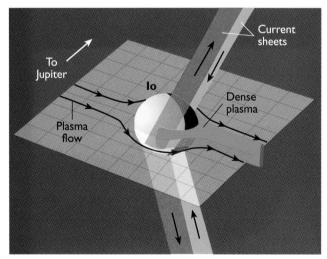

Figure 14. Molecules from Io's thin atmosphere create a doughnut-shaped ring of matter all along its orbit. Once ionized, these atoms follow the planet's wobbling magnetic field. Astronomers can see this torus thanks to the light emitted by sulfur ions at 6731 angstroms and by neutral sodium atoms at 5890 angstroms. This series shows S^+ emission recorded over 3 hours on 31 January 1991. Each "triptych" is a single image, but the left and right sides have been enhanced to bring out detail. The vertical "ribbons" of enhanced emission at far left and far right roughly coincide with Io's orbit.

Figure 15. The plasma interaction at Io with plasma in Jupiter's magnetosphere is complex and not completely understood. However, this innermost Galilean satellite serves as one end of a powerful *current sheet* (sometimes called a "flux tube") of electrons that connect it with the ionosphere of Jupiter. The electrical circuit created along this path is enormous, involving an estimated 400,000 volts and 2 trillion watts of power.

its orbit (*Figure 14*). As Voyager 1 approached Jupiter in early 1979 it began to detect strong ultraviolet emissions that suggested this toroidal cloud was dense with ions of sulfur and oxygen. Days later the spacecraft flew through Io's plasma torus and found thousands of ions and electrons per cm^3 (comparable to Earth's plasmasphere), temperatures of about 1,000,000° K, and a composition that suggested the dissociation products of sulfur dioxide (SO_2). The source of this gas soon became obvious: volcanic eruptions on Io (see Chapter 17). The ionization of SO_2 products from Io's atmosphere injects roughly a ton of material into the plasma torus every second. It quickly couples to the Jovian magnetic field and accelerates to the 10-hour corotation rate.

Since ground-based and Voyager observations suggested that the magnetospheric interaction with Io is complex, a close flyby of this moon became a major objective for the Galileo mission. Upon reaching Jupiter in December 1995, the spacecraft passed within 900 km of the surface of Io, ahead of the moon in its orbit and thus downstream of the satellite with respect to the rapidly circling magnetospheric plasma. The Galileo instruments detected a dense, cold (low-energy) plasma in the center of the wake and hot plasma on its flanks. The plasma appears to be deflected around Io. The spacecraft's magnetometer also showed a falloff in the magnetic-field strength in the moon's immediate vicinity.

Interpretation of these observations has proved tricky. There is no question that the interaction of the magnetospheric plasma with the atmosphere and ionosphere of Io generates large amounts of new plasma and drives electrical currents through the plasma (and maybe Io itself). Yet controversy rages over whether the observed magnetic signature requires Io to have an internal magnetic field, or whether the observations can be explained by electric currents created by the magnetosphere's interaction with the satellite and its atmosphere. Io most likely has an iron core. However, tidal flexing heats its outer layers, not the core, making a convection-driven dynamo problematic. Some scientists argue that the magnetic signature could be explained entirely by strong currents coursing through Io's ionosphere (and created by interactions with the surrounding magnetosphere). Others counter that the moon's ionosphere is too weakly conducting for these currents to arise.

The ambiguity may soon be resolved. In addition to ongoing detailed modeling of how the moon couples to its electromagnetic surroundings, the last two orbits of the Galileo mission are to include close flybys of Io, one pass upstream and the other over the pole. From these, we expect to get one set of magnetic-field and plasma measurements if Io has an internal magnetic field, and a very different signature if it does not.

Although many questions remain about the interplay between Io and the magnetospheric plasma, it is clear that Io is coupled electrodynamically to Jupiter itself (*Figure 15*). The bursts of decametric radio emission triggered by Io are generated close to Jupiter, presumably at the base of field lines that have been perturbed by Io. Centrifugal forces enable the corotating torus plasma to diffuse slowly outward against the confining forces of Jupiter's magnetic field, and over some tens of days the material fills the giant magnetosphere of Jupiter. However, instead of the ionized gas cooling as it expands, the plasma is somehow accelerated and heated, thereby inflating the middle and outer regions of the magnetosphere. The heating mechanism remains unknown, but the source of energy is generally thought to be the rotation of Jupiter, coupled to the plasma by the magnetic field. In any case, the particles then diffuse inward, gaining additional energy from the magnetic field as they are "recycled" back to the inner magnetosphere.

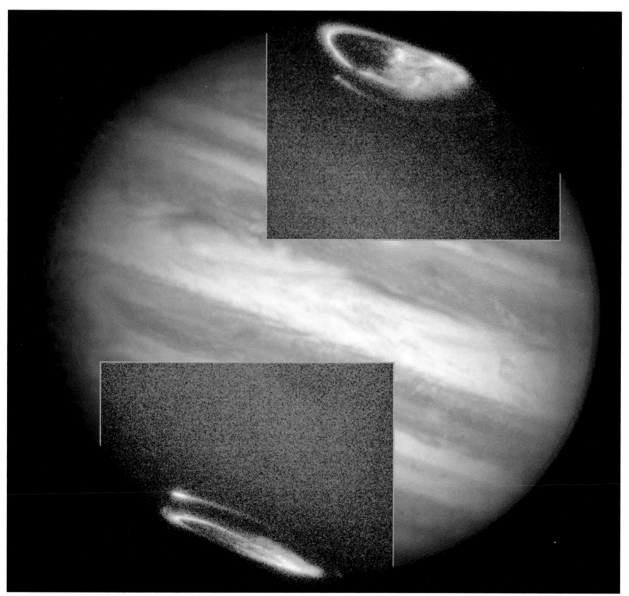

Figure 16. **High-energy electrons cascade into Jupiter's upper atmosphere and create bright auroral displays at ultraviolet wavelengths. This composite image, taken by the Hubble Space Telescope in September 1997, clearly shows auroral ovals in both the northern and southern polar regions. Elongated trails outside the ovals mark the locations where the powerful electrical current sheets from Io enter the Jovian atmosphere.**

On returning to the outer region of the torus, these energetic ions and electrons are redirected along the magnetic field into the atmosphere of Jupiter. The energy deposited by this process totals 10 to 100 trillion watts, and consequently it probably has an important influence on the temperature and dynamics of the polar regions of Jupiter's atmosphere. At the very least the process generates spectacular auroral emissions (*Figure 16*) whose surface brightness can exceed that of Earth.

The power for populating and maintaining the magnetosphere of Jupiter comes principally from the rotational energy of the planet and the orbital energy of Io, whereas the power source of Earth's magnetosphere is principally the solar wind. While rotation dominates the plasma flows throughout the day side of Jupiter's magnetosphere, it is clear that corotation cannot be maintained all the way down the tail. At some point the plasma begins to flow downstream, either as a steady wind or sporadically as *plasmoids* (similar to magnetic storms in Earth's magnetosphere). A copious amount of these energetic particles is discharged into interplanetary space. Galileo's excursions into the magnetotail should help us understand how this material is expelled. In a larger context, these data have gone far toward elucidating the origin of energetic charged particles in astrophysical settings like pulsars, which are often compared with Jupiter's giant rotating magnetosphere.

THE SYMMETRICAL SYSTEM OF SATURN

In contrast to the situation at Jupiter, prior to the arrival of spacecraft we knew nothing of the magnetic state of Saturn. No signature was apparent from Earth-based radio astronomy. Our speculations favored a strongly magnetized planet whose ring system prevented energetic electrons from being trapped close to the planet in synchrotron-emitting radiation belts. We suspected that an intense radiation belt existed outside of the outer edge of the main ring system.

Figure 17. Saturn's auroras, like Earth's, are powered largely by ions and electrons captured from the solar wind. This ultraviolet image was recorded by the Hubble Space Telescope in October 1997.

The discovery of Saturn's magnetosphere came in September 1979 as Pioneer 11 detected the presence of a bow shock 24 Saturnian radii (1.44 million km) from the planet's center on its sunward side. Soon thereafter, the spacecraft entered an intense, fully developed magnetosphere. As Pioneer 11 passed beneath the outer edge of the A ring, the instruments on board recorded a guillotinelike cutoff of the charged-particle population, as expected. The twin Voyager spacecraft followed soon thereafter, passing through the Saturnian system in November 1980 and August 1981.

These explorations revealed that Saturn's magnetic field is 600 times stronger than the magnetic moment of Earth but still considerably weaker than that of Jupiter. At first glance, the magnetosphere of Saturn can be thought of as a smaller version of the Jovian case, scaled down to 10 to 20 percent of its size. Both structures are dominated by rotation, satellites are the major sources of plasma, and centrifugal forces confine plasma to an equatorial disk. The major differences at Saturn, including the lack of synchrotron-emission belts, are due to the presence of large quantities of icy material in the rings, which efficiently absorb ions and electrons that strike them.

Low-energy ionized material permeates the Saturnian magnetosphere but at densities much lower than in Io's plasma torus. In addition to protons, the Voyagers found oxygen ions created by the dissociation and ionization of water molecules from the planet's rings and icy satellites. In 1993, Donald Shemansky and his colleagues used the Hubble Space Telescope to detect ultraviolet emission from OH molecules, which exist in a dense cloud that extends 30,000 km above and below the ring plane. To maintain this cloud, about 170 kg of water must be supplied by the rings every second. Neutral atoms and molecules outnumber charged particles by a factor of five to 10 throughout the magnetosphere. Saturn's is the only magnetosphere where neutral species are more abundant than ionized ones.

Another source of plasma is Titan, Saturn's largest satellite. It moves through the outer fringes of the magnetosphere at an orbital radius of 1.2 million km, or 20 Saturn radii. As discussed more fully later, Titan apparently loses nitrogen from its upper atmosphere to the fast-moving (corotating) magnetosphere and produces substantial plasma and magnetic effects in its wake.

In the inner magnetosphere, reactions between the plasma and the dense neutral cloud are the dominant source and loss processes. Outside about 10 Saturn radii, centrifugal force takes over and transports the plasma radially outward. The smaller scale of Saturn's magnetosphere and perhaps collisions with the dense neutral cloud seem to limit the generation of hot plasma, so the magnetosphere is not as inflated as that of Jupiter. Nevertheless, enough energetic particles precipitate into Saturn's atmosphere to produce bright auroral emissions (*Figure 17*).

Like their counterparts around Jupiter, the Saturnian satellites Rhea, Dione, Tethys, Enceladus, and Mimas are excellent absorbers of inwardly diffusing energetic electrons and protons. Yet, remarkably, the satellites selectively permit the inward migration of electrons at specific energies. Those electrons drifting in longitude at the same rate as a moving satellite are able to diffuse across its orbit as though the satellite were not present. Electrons with other energies, and thus drift rates, will most likely strike the satellite and be absorbed. Inside the orbit of Mimas, for example, nearly all of the surviving electrons have an energy of about 1.6 million electron volts. This selective effect is analogous to white light passing through a succession of colored filters. No similar phenomenon occurs in the magnetospheres of either Earth or Jupiter.

Inward-diffusing protons are less fortunate. They are strongly absorbed by Dione, Tethys, Enceladus, and perhaps dust in the tenuous E ring. Consequently, there is a near-total absence of low- and intermediate-energy protons within 180,000 km of the planet. Within 90,000 km of Saturn are protons with energies greater than 80 million electron volts and electrons having roughly 100 million electron volts. These potent charged particles could not have survived the gauntlet of inward diffusion from the outer magnetosphere; instead, they arise when cosmic rays interact with Saturn's upper atmosphere and ring material.

The count of all trapped particles drops dramatically at the outer edge of ring A, because they are absorbed by the ring material encountered there. In addition, Saturn's general magnetic field deflects most cosmic rays before they can reach the inner magnetosphere. As a result, the region interior to the A ring cutoff is nearly free of high-energy particles — it is the most completely shielded region of space within the solar system.

The imaging instrument on Pioneer 11 recorded an inner satellite first glimpsed from Earth in 1966 (since named Epimetheus) and discovered a previously unknown ring (F). Notably, the spacecraft identified both of these independently by the gaps they created in the magnetosphere's electron population. Pioneer 11 picked up other distinctive charged-particle absorption features, which were designated 1979 S3, S4, and S5. Fourteen months later Voyager 1 found another new ring, G, at the location matched by that of 1979 S3. The other two may be the signatures of as yet unconfirmed rings or small satellites.

ASYMMETRIC MAGNETOSPHERES

As Voyager 2 approached Uranus in January 1986, we wondered if our experiences with the symmetric magnetic environments of Earth, Jupiter, and Saturn would hold true for a planet that is quite literally spinning on its side. Pioneer 10 had already established that the solar wind extended beyond Uranus's orbit. An empirical relationship that relates the angular momenta and magnetic moments, the "Bode's law" of planetary magnetism, suggested that the magnetic moment of Uranus would be about one-tenth that of Saturn. Anticipation heightened in the early 1980s, when astronomers using the International Ultraviolet Explorer satellite observed "aurora-like" emissions from hydrogen in Uranus's upper atmosphere.

We knew that the rotational axis of Uranus would lie, in early 1986, within 8° of the planet-Sun line. If Uranus's magnetic and rotational axes were nearly parallel, as is the case for other magnetized planets, one magnetic pole would be pointed almost directly at the Sun and a very unusual magnetospheric shape would be expected. However, if the axes were inclined markedly to one another, each magnetic pole would move in a sweeping, conical path as the planet spun — creating an even more exotic magnetosphere.

Voyager 2 ended our many years of conjecture and sent magnetospheric theorists back to basics. The planet's magnetic moment is nearly the same strength as that predicted, but its orientation is *very* different from our expectations. Uranus's magnetic axis is tilted a huge 59° from the rotational axis and offset from the planet's center (*Figure 18*). As at Saturn, the presence of sizable satellites and a ring system controls the populations of charged particles in the inner magnetosphere. However, unlike the Saturnian situation, the diurnal wobble of Uranus's tilted magnetic equator creates a very complex relationship between the orbiting satellites and the magnetic field.

Uranus's magnetotail shows strong similarities to Earth's: two lobes with opposite polarities are separated by a cross-tail current and a flat layer of concentrated plasma. The plasma sheet lies in the magnetic equatorial plane near Uranus but bends parallel to the solar-wind flow about 250,000 km downstream. Unlike the situation at Earth, the whole tail structure rotates in space approximately about the Uranus–Sun line. This results from solar wind flowing in more or less along the Uranian spin axis.

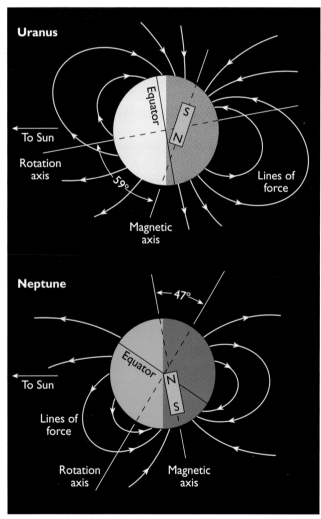

Figure 18. The magnetic fields of Uranus and Neptune are remarkably — and unexpectedly — alike. The large offsets from center means that the field strength at each planet's cloud tops varies widely from place to place. It also suggests that the field's source regions cannot lie in the cores but rather in a turbulent liquid mantle where dynamo-driving convection can be sustained.

Essentially, the solar wind is now approaching the planet from "over the pole." Thus, although the trapped plasma corotates with the planet once every 17 hours, over time it is circulated through the magnetosphere by solar-wind–driven convection. In some circumstances the particles trace out helical trajectories, spiraling sunward at the magnetic equator and antisunward at high latitudes. Within about 100,000 km of the planet, solar-wind convection is ineffective, allowing appreciable amounts of cold plasma to accumulate. Farther out, convection stirs up a plasma with higher energies, in the range of 1,000 electron volts. Voyager 2 found that this hotter population is almost completely absent from the inner, cold-plasma region. However, the Voyager data also recorded pulses of activity, akin to Earth's magnetic storms, that appear to propagate inward from the magnetotail.

Two factors make Uranus's magnetosphere rather empty: there are only a few, small icy satellites, and solar-wind convection circulates material through the magnetosphere in a few days. Voyager 2 found the plasma to consist almost entirely of protons. The dearth of heavier ions (specifically oxygen) means that water molecules are not sputtering off the icy satellites in profusion. The cold (low-energy) plasma probably arises from hydrogen atoms in Uranus's upper atmosphere and protons flowing out of the ionosphere. The hot plasma may derive from hydrogen atoms that become ionized farther out; they attain higher energies because corotational speeds increase farther from the planet.

Unfortunately, Voyager 2's pass through the complex and variable Uranian magnetosphere was too brief to pin down even the basic properties such as the dominant source of its plasma. Nevertheless, we offer one simple prediction: the magnetosphere's configuration will change radically during the next phase of the planet's 84-year orbit as the rotational pole swings away from the solar-wind direction. For example, early next century, the solar wind will impinge at latitudes close to the rotational equator. Tilted from the spin axis by 59°, the magnetic pole will gyrate toward and away from the solar wind every 17 hours (Uranus's spin period), creating a magnetospheric configuration similar to that found at Neptune.

In its fourth and final planetary flyby, Voyager 2 reached Neptune in August 1989. The spacecraft found a magnetic field substantial in strength, inclined 47° to the rotation axis, and having a large offset from center (*Figure 18, Table 1*). At the time of the encounter, Neptune's northern hemisphere was in mid-winter, tipped 23° away from the Sun. The large tilt of the magnetic dipole means that the angle between the solar wind and the dipole axis was vacillating between 20° and 114° during each 16.1-hour Neptunian day (*Figure 19*). When the angle is near 90° the configuration is, momentarily, symmetrical — like that of Earth, Jupiter, and Saturn. When the angle is small, however, the magnetic axis points "pole-on" into the solar wind (the configuration that we expected for Uranus before Voyager 2 discovered otherwise). Consequently, the magnetic field and magnetosphere appear to gyrate wildly as seen from outside the system.

The dramatic changes in geometry make it difficult to visualize the behavior of trapped plasma. Corotation again plays the dominant role, though convection driven by the solar wind has a cumulative effect over several planetary rotations.

The maximum coupling between the solar wind and the magnetosphere occurs for the "Earthlike" configuration. According to one model, at such times plasma situated on the day side at local noon will drift away from Neptune, while plasma on the midnight line will drift (more slowly) toward the planet. Thus, the plasma spirals inward or outward depending on its location at the time of Earthlike configuration. A second model envisions a four-cell convection pattern that corotates with the planet.

The dramatic changes that occur in the configuration of the magnetotail during every planetary rotation must further complicate the dynamics of Neptune's magnetosphere. During the Earthlike magnetic configuration, the magnetotail mimics Earth's tail, with lobes of opposite polarity separated by a current sheet. When the magnetosphere is pole-on to the solar wind, the magnetotail has a cylindrical shape. At such times, the magnetic-field lines directed toward the planet are on the outside of the cylinder. Field lines leaving the planet are on the inside, and a cylindrical current sheet separates them.

Neptune's large satellite Triton orbits well inside the magnetopause, and every second it supplies an estimated 10^{25} ions (200 g) of nitrogen into the surrounding magnetosphere. The N$^+$ ions detected by Voyager 2 are likely produced when magnetospheric plasma collides with the satellite's atmosphere, while the protons come from a large, diffuse hydrogen cloud that extends from Triton's orbit inward for more than 150,000 km. Closer in, much of the plasma is absent, apparently redirected by magnetospheric waves along the magnetic field and into the atmosphere of Neptune. However, this simple picture is not without its problems, and we know that considerable further study is required in order to understand the plasma configuration and dynamics of Neptune's magnetosphere.

SMALL MAGNETIZED WORLDS

Before 1974, it was thought that only large planets would have magnetic fields and that convective motions in smaller objects would have halted as the interiors cooled. The discovery of weak but significant magnetic fields of smaller bodies, first Mercury and recently Ganymede, has challenged our view of both the dynamo process and the nature of small-scale magnetospheres.

Mercury. During 1974 and 1975, the Mariner 10 spacecraft made three successive flybys of Mercury, providing the first and thus far only close-up observations of the innermost planet (see Chapter 7). Mercury takes 59 days to rotate, which might seem too slow to sustain dynamo activity. Nonetheless, the planet has a distinct though weak global magnetic field with a magnetic moment $1/1400$ that of Earth. Before Mariner 10, geophysicists suspected that Mercury's core should be completely solid. However, the presence of a magnetic field shows that at least a shell of molten material must still exist (perhaps iron mixed with sulfur to lower its melting point), and that it must be in convective motion.

The magnetosphere of Mercury contains closed field lines (that is, connected to the planet at both ends), but the field is too weak to maintain a belt of trapped particles. Low-energy plasma was detected by the Mariner spacecraft's traversals through the nightside magnetosphere, but we are uncertain

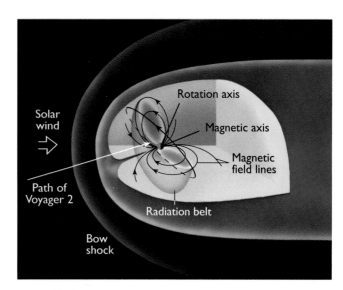

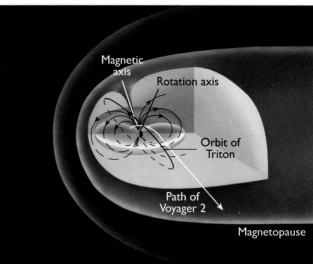

Figure 19. All of Neptune's rings and satellites (except Nereid) lie deep within the planet's largely empty magnetosphere. When Voyager 2 crossed the magnetopause on 24 August 1989 (upper panel), one pole of Neptune's highly inclined magnetic field was pointing toward the Sun. When the spacecraft made its exit 38 hours later (lower panel), the field's orientation was very different and rather Earthlike.

whether this plasma originates in the solar wind or from the ionization of Mercury's tenuous atmosphere (see Chapter 13). More notable, however, are the intense bursts of energetic particles detected in the magnetotail. These have been compared with magnetic substorms on Earth and suggest that Mercury's magnetosphere is dynamic. The Mercurian storms last about 1 minute (compared with hours at Earth), consistent with the much smaller scale of the planet's magnetosphere.

Ganymede. One of the greatest surprises of the Galileo mission was the discovery of a magnetic field intrinsic to the Jovian moon Ganymede. This creates a magnetosphere within a magnetosphere (*Figure 20*), the first of its kind found in a planetary system.

The Jovian plasma, in which Ganymede is embedded, flows at speeds much slower than the solar wind and, in particular, slower than the local speeds of waves that propagate in the magnetospheric plasma. It is a subsonic flow, hence no bow shock occurs upstream of Ganymede. We suspect that there is a small region close to the moon where magnetic field lines are closed. However, the magnetosphere of Ganymede is too small to have trapped plasma or a magnetotail. There is thought to be a pattern of convection, similar to solar-wind-driven convection at Earth, but in this case driven by the interaction of Ganymede's magnetic field with the magnetospheric plasma sweeping past it.

Orbiting Jupiter at a distance of 19 Jovian radii, Ganymede is located in the middle magnetosphere region. Consequently, the moon experiences a change in the surrounding plasma conditions and magnetic-field orientation as the tilted plasma sheet passes over the satellite twice per 10-hour rotation.

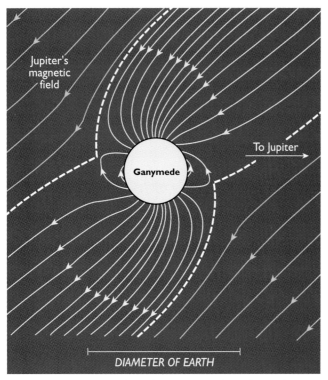

Figure 20. The magnetosphere of Ganymede is shown for conditions at the time of Galileo's first two flybys in 1996, from a perspective in front of the moon along its orbit. Gray lines represent the magnetic field of Jupiter, and yellow lines those attached to Ganymede.

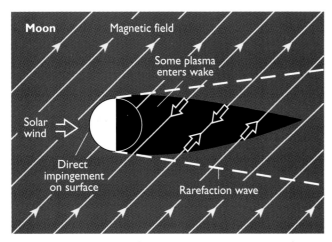

Figure 21. The solar wind collides directly with the Moon's surface. This creates a cavity downstream that is progressively filled in by plasma flowing along the magnetic field.

On Galileo's passes over Ganymede's magnetic poles, plasma detectors measured a substantial outflow of protons. These may result from the photodissociation of water vapor that has sublimated from the moon's icy surface (or has been sputtered off it by charged particles). The fact that no oxygen ions were detected suggests that the oxygen must be left behind on the surface.

PLASMA INTERACTION WITH ROCKS

Earth's Moon. The first extraterrestrial body to be investigated firsthand was the Moon, an object studied intensively by many American and Soviet flybys, orbiters, and landers. Yet the Moon has no global magnetic field — its magnetic moment is at least 10 million times weaker than Earth's. The Moon must therefore lack the convecting molten core necessary for an internal dynamo. Models of the lunar interior suggest that the core is small and at least partially solidified. Curiously, experimenters have discovered *localized* regions on the Moon with surface magnetic fields of 5 to 300 nT, but these appear to be geologic anomalies exhibiting remanent magnetization that has survived from long ago. How the primordial Moon could have developed such magnetized patches remains a nagging problem.

Our satellite passes through Earth's magnetotail a few days each month near the time of full Moon; for the remainder of the month it is outside the magnetosphere and immersed in the solar wind. Because the Moon lacks both a global magnetic field and a significant atmosphere, the solar wind strikes the lunar surface directly and some of its particles become infused into the rock and dust there. For example, the Apollo lunar samples contained significant amounts of trapped helium-3, an isotope that holds considerable promise as a fuel for fusion-powered energy systems. Extending from the Moon's antisolar side is a long plasma void (or plasma umbra) shaped like an ice-cream cone, with its apex downwind and with the Moon itself as the scoop of ice cream (*Figure 21*). Although this void gradually fills in farther downstream because of the lateral motion of solar-wind particles, it still may be one of the most nearly perfect vacuums in the solar system.

The existence of such a plasma void implies a system of weak electrical currents along its boundary; these currents were observed during the late 1960s by the lunar-orbiting spacecraft Explorer 35. In addition, the varying magnetic field entrained in the solar wind induces a system of transient electrical currents within the Moon itself, as was observed by the long-lived Apollo magnetometers placed on its surface.

Europa and Callisto. Compared with the strong interaction at Io and the magnetosphere of Ganymede, results from the initial Galileo flybys of Europa and Callisto were rather disappointing. Neither moon shows strong magnetic signatures, putting weaker upper limits on their surface fields of 30 nT for Callisto and 240 nT for Europa. These results pose interesting problems for planetary scientists. Why do the similar-sized Ganymede and Callisto have such different interiors? Furthermore, should Io's magnetic field be confirmed, why doesn't Europa have a dynamo? After all, Io and Europa have very similar interiors (see Chapters 17–19).

The Galileo particle and field data from the first few flybys of Europa and Callisto are puzzling. They show complex structure that does not fit the simple picture exemplified by the solar-wind interaction with the Moon. The signatures imply that magnetic fields are being induced in a conducting layer within each body by Jupiter's magnetosphere. Since both moons are largely ice, the most plausible conductors are sub-surface oceans of water (see Chapter 18 and 19).

Asteroids. The small sizes of asteroids imply that they should have very little interaction with the solar wind beyond unimpeded bombardment by charged particles. However, the significant perturbations of the interplanetary magnetic field measured during Galileo's flyby of 951 Gaspra in October 1991 changed this view. Margaret Kivelson and her colleagues estimate that Gaspra could have a level of magnetization, presumably remnant, comparable in strength to that found in chrondritic meteorites. A similar, though rather weaker signature was observed when Galileo flew past 243 Ida in August 1992. Because an asteroid is smaller than the diameter of the gyrations that solar-wind protons make in the interplanetary magnetic field, the waves generated by the asteroid's interaction propagate faster than the solar wind itself. This means that no bow shock forms upstream of these asteroids. Their interaction regions are wedge shaped, similar to a wake of a ship.

PLASMA INTERACTION WITH ATMOSPHERES

The scale of the interaction between a flowing plasma and the atmosphere of a nonmagnetic planet depends on the atmosphere's thickness. At one extreme is Mars, whose tenuous ionosphere is barely able to hold off the solar wind. At the other extreme is the case of a comet near perihelion. Its "atmosphere" extends for hundreds of thousands of kilometers. The circumstances for Venus and Titan lie near the middle of these extremes. Pluto's situation could range anywhere from a weak Marslike interaction to that of a comet. Among all these, we have the most information about Venus.

Venus. American and Soviet spacecraft have observed the magnetic properties of Venus for nearly four decades. The planet's magnetic moment is undeniably weak, at least 25,000 times less than Earth's, despite our sister planet's comparable size and probable molten interior. The lack of a comparable magnetic field suggests Venus lacks the temperature gradient and vigorous internal convection required for a dynamo. Nonetheless, the solar wind is prevented from reaching the surface by Venus's dense atmosphere and by electrical currents induced in its conducting ionosphere. The barrier that separates the planetary plasma from solar-wind plasma is referred to as the *ionopause.* The planet has a well-developed bow shock, but it possesses no population of trapped particles.

When it reaches Venus, the solar wind slows down and is deflected around the ionopause (*Figure 22*) and the interplanetary magnetic field is draped back to form a magnetotail. Sometimes the magnetic field lines become concentrated into a twisted bundle like a rope, and these "tubes" of magnetic flux are dragged through Venus's ionosphere. Oxygen and hydrogen atoms in the planet's extended atmosphere become ionized and picked up by the solar wind. Many of these ions come bom-

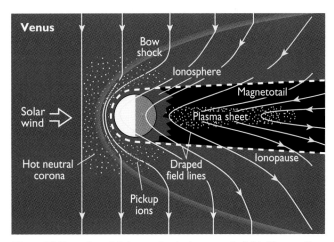

Figure 22. Even though it has no intrinsic magnetic field, Venus still interacts with and diverts the solar wind by virtue of its ionosphere and extended corona of hot gas derived from the upper atmosphere.

barding back into the atmosphere, further heating its uppermost layers. The rest are carried away by the solar wind. The present removal rate (10^{26} O^+ ions per second) is too low for this process to be a significant factor in the lack of water in Venus's atmosphere.

The solar wind's interaction with Venus changes on time scales of hours, depending on the wind's dynamic pressure and the orientation of the interplanetary magnetic field. Over the course of an 11-year solar cycle, variations in the solar wind and in the Sun's ultraviolet and X-ray output dramatically change the heating and ionization of Venus's upper atmosphere and hence the size of the interaction region.

Mars. Until the recent Mars Global Surveyor mission, sparse data from Mariner 4 (1965) and the Soviet orbiters Mars 2, 3, and 5 (1971–1974) and Phobos 2 (1989) represented our only observations of the magnetic properties of Mars. The upper limit on an interior dipole magnetic field is about 30 nT at the planet's equator. This magnetic moment is at least 5,000 times weaker than that of Earth. Nonetheless, the planet's ionosphere deflects the solar wind and the interplanetary magnetic field is draped around the ionopause, forming a magnetotail downstream. There are a weak bow shock and plasma phenomena similar to the interaction at Venus. The tenuous atmosphere of Mars implies that the ionosphere is tenuous too and perhaps insufficient to stand off the solar wind alone. In that case a very weak internal field might prove necessary. Alternatively, if the ionosphere of Mars were only weakly conducting, the interplanetary magnetic field would be able to penetrate into the ionosphere region, where it would pile up (since the accompanying plasma flow is slowed down) and be able to provide additional pressure to hold off the solar wind.

Mars Global Surveyor detected patchy crustal magnetization of up to 400 nT, which is comparable to the remnant magnetization measured in some of the meteorites that are believed to have been ejected from Mars. Conceivably, if enough of the surface exhibits similar levels, the total magnetic field may have a significant effect on the solar-wind interaction.

Phobos 2 detected abundant O^+ that had been ionized in Mars's upper atmosphere and picked up and carried away by the solar wind. The observed rates of escape (10^{23} oxygen ions and

10^{24} protons per second) suggest that this scavenging could have played a significant role in the evolution of the Martian atmosphere over the age of the planet — or at least since the decay of any internal dynamo.

The global magnetic field of Mars is too weak and the atmosphere too thin to protect the planet's surface from cosmic rays and bursts of solar-flare protons. Even if Mars had an active dynamo 1.3 billion years ago (when most of the known Martian meteorites crystallized), the field that may have induced the remnant magnetization now found in the meteorites was also weak, no more than 1,000 nT and thus comparable to Mercury's current field. Consequently, any biogenic materials present on the planet's surface must have been exposed to high doses of radiation.

Titan. The single flyby of Titan by Voyager 1 in November 1990 recorded an intriguing plasma-atmosphere interaction, the further study of which is a major objective of the Cassini/Huygens mission. The spacecraft's measurements suggest that the background magnetic field of Saturn is draped over Titan and forms a magnetotail downstream. The upper limit on the internal field of Titan is about 4 nT at the surface, implying a magnetic moment no greater than 0.00001 of Earth's.

Saturn's largest moon is similar in size to the Galilean satellites, but its dense nitrogen-methane atmosphere sets it apart from all other satellites (see Chapter 20). The combination of high surface pressure (about 1.5 bars) and low gravity (about 15 percent of that on Earth) results in an atmosphere that extends far above the surface. Its exobase is 1,400 km up — one-fourth of the moon's diameter. Solar ultraviolet light and electrons from Saturn's magnetosphere heat this extended upper atmosphere, driving off about 10 kg of neutral nitrogen and hydrogen atoms per second. This is comparable to the gas lost by an active comet.

Solar ultraviolet light and magnetospheric electrons also ionize Titan's upper atmosphere (particularly on the nightside) and produce a complex mix of ions. The resulting dense ionosphere is probably sufficient to deflect the tenuous plasma present at

Titan's location in the outer Saturnian magnetosphere. Thus, in many ways Titan's situation mimics how solar wind interacts with Venus, except that the uppermost layers of Titan's neutral atmosphere can be scavenged directly by the plasma flowing by. About 30 g per second is lost in this way.

Comets. The icy nuclei of comets are presumably not magnetized. However, as they pass through the inner solar system these objects create huge envelopes of escaping gases, which become ionized and interact with the solar wind (*Figure 23*). Indeed, as mentioned earlier, Ludwig Biermann's observations of the ionized-gas tails of comets provided the first evidence for the continuous nature of the solar wind.

The Sun causes both solar-wind pressure and radiation pressure (the latter from the "impact" of photons of light) on all objects in the solar system. For something the size of an asteroid or planet, these pressures have a negligible effect compared to gravitational forces. For cometary dust grains, which have vastly greater area-to-mass ratios, radiation pressure is quite important. Conversely, the solar wind has a much greater effect on ionized cometary gas than radiation pressure does. Comets' ionized tails are thus distinguished from their more familiar dust and neutral-gas tails by their form, optical spectra, and other detailed features (see Chapter 24).

The interaction process starts where the solar wind first picks up cometary ions, millions of kilometers from the nucleus. These ions not only slow down the solar wind but also create waves in the plasma that are ultimately responsible for the surprisingly abundant energetic particles detected at comets. After passing through a bow shock, the solar-wind plasma continues to slow down as more and more cometary ions are picked up from the increasingly dense "atmosphere" (coma). The deceleration of the solar wind causes the interplanetary magnetic field to pile up and drape around the comet's ionosphere, forming an extended magnetotail behind it. At Comet Halley, the Giotto spacecraft detected a well-defined boundary that separated the mixture of solar wind, cometary plasma, and interplanetary magnetic field from the field-free, nearly pure cometary ionosphere. This ionopause can be up to 20,000 km across, providing an area some 10 times larger than Venus's ionopause. Therefore, it is not surprising that a small comet can provide thousands of times more ions to interact with the solar wind than Venus can.

Pluto. While it is improbable that Pluto has an internal magnetic dynamo, even a little remanent magnetization could produce a significant magnetosphere in the weak solar wind 30 to 50 AU from the Sun. A surface field of 1 nT would be enough to stand off the solar wind from the surface, while 3,700 nT would produce a magnetosphere large enough to encompass Charon's orbit. Such surface fields could easily exist if Pluto retains a magnetization comparable to that of chondritic meteorites.

Pluto's atmosphere is only weakly bound to the planet, and an estimated 10^{27} to 10^{28} molecules per second may be escaping to space. For typical solar wind conditions at 30 AU, escape rates significantly greater than 1.5×10^{28} molecules per second slow the solar wind and create a cometlike interaction that might extend beyond the orbit of Charon. If the escape rate is less, then the estimated density of electrons should be sufficient to create an ionopause about 600 km above Pluto's sunlit

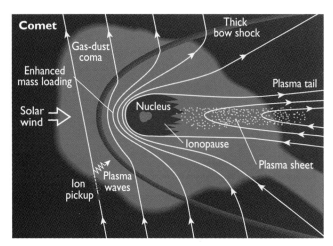

Figure 23. The gas lost from a comet's nucleus is quickly ionized by sunlight, creating a planet-scale obstacle to the solar wind. This portrayal of the complex interaction is not to scale: the nucleus is typically just a few kilometers across, whereas the cometopause is about 10,000 km to its sunward side and the bow shock about 1 million km away.

hemisphere. In this case, the interaction would be like that of Mars. The large variations in solar-wind flux that have been observed in the outer heliosphere suggest that the stand-off distance would vary from about 4 to 24 Pluto radii on time scales of days.

However, the nature of the solar-wind interaction would change dramatically if Pluto's atmosphere were to freeze out as the planet recedes from the Sun after perihelion. Around aphelion, both Pluto and Charon would simply absorb the solar wind on their dayside hemispheres (if Pluto is unmagnetized), similar to the situation with Earth's Moon.

THE HELIOPAUSE AND BEYOND

We now believe that the solar wind merges with the nearby interstellar medium and loses its identity roughly 100 AU from the Sun (*Figure 24*). The boundary at which this merging begins is called the *heliopause* and the region inside it the *heliosphere*. Instruments on Pioneer 11 made significant contributions to this subject until tracking of the spacecraft ended in January 1995 at a heliocentric distance of 42 AU. Pioneer 10 continues to provide valuable cosmic-ray observations from its location more than 70 AU from the Sun. Voyagers 1 and 2, respectively 70 and 55 AU from the Sun in mid-1998, continue

to function well as they head out of the solar system at more than 3 AU per year.

Moving outward from the Sun, the average increase in total cosmic-ray intensity measured by these spacecraft is 1.3 percent per AU, and the outer boundary is now known to lie beyond 70 AU (Pioneer 10). This cosmic-ray modulation boundary is doubtless related to the termination shock and heliopause but may not be identical to either.

In May and June 1991 a series of extraordinarily intense blast waves originated at the Sun. About 400 days later correspondingly strong bursts of radio waves were received by both Voyagers. Donald Gurnett has interpreted these events as evidence for the terminal shock being about 110 AU away and the heliosphere about 145 AU. Both distances probably fluctuate by tens of AU as the dynamic pressure of the solar wind varies with the level of solar activity. Direct observations of these transition regions by Voyager 1 are eagerly awaited. This craft's passage out into the interstellar medium will be revealed by at least three effects. First, the solar wind, flowing radially outward, will be replaced by an interstellar wind of different speed, ionic composition, temperature, and direction of flow. Second, the intensity and spectrum of the cosmic radiation will become constant, no longer modulated by the magnetic irregularities of the solar wind. (This expectation assumes that the interstellar medium

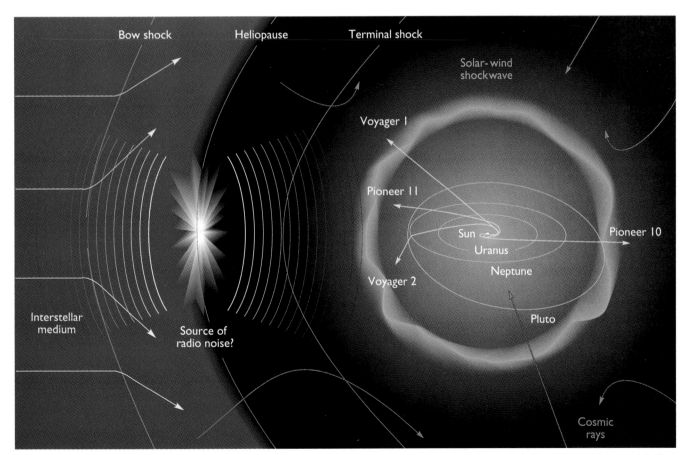

Figure 24. Space physicists believe the Sun's electromagnetic influence extends to a distance of roughly 150 AU, far beyond the orbit of Pluto. There, at what they term the heliopause, the outward-flowing solar wind meets the interstellar medium. Neither the termination shock, the heliopause, nor the bow shock has yet been directly

observed. However, space scientists believe a dense, high-speed shell of plasma ejected by the Sun in mid-1991 reached the heliopause in late 1992 and triggered a massive release of radio energy that was detected by both Voyager spacecraft.

contains a smoothed average of contributions for many extremely remote sources.) Third, the total cosmic-ray intensity will be greater than that at any point within the heliosphere. In addition, the local signature of the transition region may be detectable by the spacecraft's magnetometer and plasma-wave instrument.

Voyager 1's instruments are already near their limits of sensitivity for making such measurements. On the other hand, since relatively low-energy cosmic rays become more abundant farther away from the Sun, observations of such particles may offer the best potential for revealing the location of the heliopause. Even so, the transition to the interstellar medium may occur not at an exact distance but over perhaps many AU and at a mean distance that fluctuates with the 11-year cycle of solar activity.

The solar system does not end at the heliopause, however. In recent years dozens of small bodies have been discovered on the outer fringes of the heliosphere. These objects are thought to be relatively primitive, remnant icy planetesimals that condensed on the very fringes of the solar nebula or were kicked out by the giant planets. While their interiors may have remained unaltered, their surfaces have been bombarded by the solar wind or the interstellar medium for billions of years. Over such long periods even low-level radiation exposure can substantially change the chemical and optical properties of the surface materials. Unfortunately, close-up exploration of even the nearest Kuiper-belt objects will be decades in the future. In the meantime, studying the interaction of the solar wind or magnetospheric plasmas with satellites, asteroids, and comets closer to Earth will help us to understand the different surface processes at work and to interpret the spectroscopic signatures from more distant and presumably more primitive surfaces.

The application of knowledge derived from magnetospheric physics to the study of the radio, X-ray, and gamma-ray emissions from distant objects is already an active field that will likely play an increasingly important role in modern astrophysics.

Cometary Reservoirs

Paul R. Weissman

The Oort cloud, which envelops our solar system with trillions of distant comets.

WE COMMONLY THINK of the solar system as ending at the orbit of distant Pluto. However, the Sun's gravitational sphere of influence actually extends some 2,500 times farther out, halfway to the nearest stars. Moreover, all this extra space is not empty — it is populated with two vast reservoirs of comets, material left over from the formation of the Sun and planets. These reservoirs are the ultimate source of all comets that pass through the inner solar system. Yet we have recognized their existence only within the past few decades, in part because the nature of comets themselves is only now becoming clear despite many centuries of visual and telescopic observation.

Comets were once mistakenly thought to be clouds of luminous gas high in the Earth's atmosphere. If that were really true, observers at different points on the Earth's surface would see comets projected against different stars in the sky. But such is not the case, and early astronomers eventually realized that comets were moving through interplanetary space. However, they also recognized that cometary orbits were very different from those of the known planets. Why was this so?

In the 17th century Isaac Newton used his then-new law of gravitation, coupled with Kepler's laws of planetary motion, to show that the Sun-grazing comet of 1680 was moving through space along a path that looked very much like a parabola. (A parabolic orbit is not bound to the Sun but instead extends into interstellar space.) By 1705 Edmond Halley had used the equations of Newton and Johannes Kepler to compile the first catalog of 24 cometary orbits (*Figure 1*). Although the observations were too crude to permit anything but parabolic orbital solutions, Halley argued that the comets could instead be in very long-period ellipses and thus bound to the Sun. As he correctly pointed out, "...the Space between the Sun and the fix'd Stars is so immense that there is Room

Comet. An	Nodus Ascend.	Inclin. Orbitæ.	Perihelion.	Distan. Perihel. à Sole.	Log. Dist. Perihelix à Sole.	Temp. equat. Perihelii.	Perihelion à Nodo.	
	gr. ' "	gr. ' "	gr. ' "			d. h. '	gr. ' "	
1337	♊ 24.21.	32.11.	♉ 7.59. 0	.40666	9.609236	June 2. 6.25	46.22. 0	Retrog.
1472	♑ 11 46.20	5.20. 0	♉ 15.33.30	54273	9.734584	Feb. 28 22.23	123.47.10	Retrog
1531	♉ 19.25.	17.56.	♒ 1.39. 0	56700	9.753583	Aug. 24.21.18½	107.46. 0	Retrog
1532	♊ 20.27.	32.36.	♋ 21. 7. 0	50910	9.706803	Oct. 19 22.12	30.40. 0	Direct
1556	♍ 25.42.	32. 6.30	♑ 8.50. 0	46390	9.666424	Apr. 21.20. 3	103. 8. 0	Direct
1577	♈ 25.52.	74.32.45	♌ 9.22. 0	18342	9.263447	Oct. 26 18.45	103.30. 0	Retrog
1580	♈ 18.57.20	64.40. 0	♋ 19. 5.50	59628	9.775450	Nov. 28 15.00	90. 8.30	Direct
1585	♉ 7.42 30	6 4. 0	♈ 8.51. 0	109358	0.038850	Sept. 27.19.20	28.51.30	Direct
1590	♍ 15.30.40	29.40.40	♏ 6.54.30	57661	9.760882	Jan. 29. 3.45	51.23.50	Retrog
1596	♒ 12.12.30	55.12.	♏ 18.16. 0	51293	9.710058	Juli 31.19.55	83.56.30	Retrog
1607	♉ 20.21.	17. 2.	♒ 2.16. 0	58680	9.768490	Oct. 16. 3.50	108.05. 0	Retrog
1618	♊ 16. 1.	37 34. 0	♈ 2.14. 0	37975	9.579498	Oct. 29.12 23	73.47. 0	Direct
1652	♊ 28.10.	79.28. 0	♈ 28.18 40	84750	9.928140	Nov. 2.15.40	59.51.20	Direct
1661	♊ 22.30.30	32.35.50	♋ 25.58.40	44851	9.651772	Jan. 16.23 41	33.28.10	Direct
1664	♊ 21.14.	21.18.30	♌ 10.41.25	102575½	0.011044	Nov. 24.11.52	49.27 25	Retrog
1665	♏ 18.02. 0	76.05.	♊ 11.54.30	10649	9.027309	Apr. 14. 5.15¼	156 7.30	Retrog
1672	♑ 27.30.30	83.22.10	♉ 16.59 30	69739	9.843476	Feb. 20. 8 37	109.29. 0	Direct
1677	♏ 26.49.10	79 03 15	♌ 17.37. 5	28059	9.448072	Apr. 26.00.37½	99.12. 5	Retrog
1680	♑ 2. 2.	60.56. 0	♐ 22 39.30	0061 2½	7.787106	Dec. 8.00. 6	9.22 30	Direct
1682	♉ 21.16.30	17.56. 0	♒ 2.52.45	58328	9.765877	Sept. 4.07 39	108.23 45	Retrog.
1683	♍ 23.23.	83.11. 0	♊ 25.29.30	56020	9.748343	Juli 3. 2 50	87.53.30	Retrog.
1684	♐ 28.15.	55.48.40	♏ 28.52. 0	96015	9.982339	Mai 29.10.16	29 23 00	Direct.
1686	♓ 20.34 40	31.21.40	♊ 17.00.30	32500	9.511883	Sept. 6.14.33	86.25.50	Direct.
1698	♐ 27.44.1	11 46.	♑ 00.51.15	69129	9.839660	Oct. 8 16.57	3. 7. 0	Retrog.

Figure 1. The catalog page from Edmond Halley's 1705 paper, "A Synopsis of the Astronomy of Comets." Orbits for the 24 entries are all listed as parabolic because the quality of the observations were not good enough to derive more precise, elliptical orbits. Halley correctly recognized that the comets of 1531, 1607, and 1682 were in fact the same comet returning at 76-year intervals.

enough for a Comet to revolve, tho' the Period of its Revolution be vastly long."

Halley noticed the comets of 1531, 1607, and 1682 listed in his catalog had remarkably similar orbits, and their appearances were spaced roughly 76 years apart. He suggested this was the same comet returning each time and predicted its reappearance in 1758. It was spotted on Christmas day of that year by a German amateur astronomer and named "Halley's Comet" in honor of the successful prediction. This celebrated object returned to perihelion most recently in 1986, and records of it have been found as far back as 239 BC in ancient Chinese texts.

COMETARY GROUPS

As more comets were discovered and tracked, astronomers started dividing them into two groups, a tradition that continues today. *Long-period* comets are those taking more than 200 years to orbit the Sun. The recent bright comets Hyakutake and Hale-Bopp (*Figure 2*) belong to this group. *Short-period* comets complete an orbit in less than 200 years and often less than 20 years; comets Encke and Halley are good examples. The distinction is somewhat arbitrary and based on the fact that orbits accurate enough to predict cometary returns have only generally been available for about 200 years.

Another key difference between the two designations is that long-period comets tend to enter the inner solar system randomly from all directions. In contrast, most short-period comets have orbits inclined no more than about 40° to the ecliptic plane, the plane of Earth's orbit. As a result, short-period comets travel around the Sun in the same direction that the Earth and other planets do, while many long-period comets move in the opposite sense (their inclinations exceed 90°). This difference in the inclination distributions of the long and short-period comets provides an important clue to their different source regions.

In the 1800s, as observations and orbital calculations became more precise, it became obvious that the periods of many long-period comets were *exceedingly* long, up to a million years or more. (By comparison the orbital period of the most distant planet, Pluto, is a mere 248 years). Such orbits must extend more than 20,000 AU from the Sun. In fact, it appeared that about one-third of the long-period comets were coming into the planetary region on hyperbolic trajectories, which are not gravitationally bound to the solar system. This led to the idea

Figure 2. Two recent visitors from the Oort cloud were comets Hyakutake (C/1996 B2; top), as photographed on 26 March 1996, and Hale-Bopp (C/1995 O1) on 17 March 1997. Long-period comets like these can appear at any time and approach the planetary region from any direction; bright ones typically arrive every 5 to 10 years or so. Comet Hyakutake was average in size, but its appearance was spectacular because it approached especially close to Earth, only 0.10 AU (about 15,000,000 km). By contrast, Comet Hale-Bopp was an unusually large and dynamic comet — at least 10 times more active than Comet Halley at comparable distances from the Sun. This made Hale-Bopp appear quite bright, even though when closest to Earth it was a distant 1.32 AU (197,000,000 km) away.

that comets somehow formed in interstellar space, then became gravitationally captured by the Sun during close encounters with the solar system.

However, careful calculations proved that these were not true interstellar interlopers. By extrapolating the comets' locations backward in time, astronomers showed that each orbit had been changed by gravitational perturbations as the comet entered the planetary region. Long-period comets are so weakly bound to the Sun that even distant perturbations from the giant planets, primarily Jupiter, are enough to change their orbital periods substantially. Furthermore, astronomers realized that their cal-

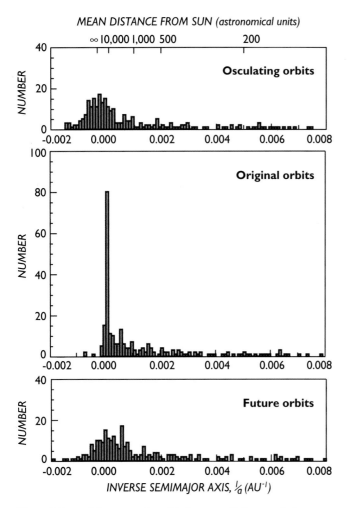

Figure 3. Dynamicists gauge the orbital energy of a long-period comet using the parameter $1/a$, where a is the semimajor axis of the orbit. As seen in the upper panel, about one-third of these objects' osculating orbits (their apparent orbits while passing through the planetary system) have negative values, implying that they are arriving from interstellar space. But note the distribution of "original" orbits for the same group of comets (middle panel), which have been integrated backward in time and referenced to the solar system's center of mass. The spike of comets coming from the Oort cloud is easily visible at very small positive values of $1/a$. These comets are all members of the solar system. A few apparently interstellar comets (negative values) are likely the result of small errors in observation or computation. "Future" orbits (lower panel) track comets as they leave the planetary region, again referenced to the solar system's center of mass. The gravitational pull of the planets alters these trajectories, and very few dynamically "new" comets return to the Oort cloud. Comets with negative values of $1/a$ are ejected into interstellar space and will not return.

culations should not be referenced only to the Sun (yielding what are termed *osculating heliocentric orbits*), but rather to the center of mass of the entire solar system *(barycentric orbits)*.

When these corrections were performed, almost all of the hyperbolic long-period comet orbits became elliptical, that is, bound to the solar system. The few comets that still appeared to be hyperbolic were only weakly so, and small errors in the observations could probably account for most of these cases. In addition, Harvard astronomer Fred Whipple was the first to realize that the jetting of gas and dust from a comet's surface could act like a small rocket engine as it approaches the Sun, thus altering the trajectory slightly. Such *nongravitational forces* can make an orbit appear hyperbolic when it is actually elliptical.

Eventually enough corrected orbits, called *original orbits,* were compiled to begin a study of their statistical distribution. A very interesting fact emerged. About one-third of the comets occupy an extremely narrow range of orbital energy (*Figure 3*); these have distant orbits just barely bound to the Sun with periods exceeding 1,000,000 years. The remaining two-thirds have orbital periods spread rather evenly down to 200 years.

In the late 1940s Dutch astronomer Jan Oort became interested in the problem of the origin of the long-period comets. Oort's principal expertise was in galactic studies, and he had gained fame in the 1920s for solving the problem of the rotation of the Milky Way galaxy. A colleague at Leiden Observatory, Adianus van Woerkom, had shown that the broad, flat distribution of cometary orbital periods could be explained by planetary perturbations, which tend to scatter the comets randomly to both larger and smaller orbits. But what then caused the pileup at near-zero energy? Oort recognized that this spike had to represent the source of the long-period comets, a vast cloud of objects lying far beyond the planets and extending to the edge of the Sun's gravitational influence. As these "Oort cloud" comets enter the planetary system for the first time, their courses are altered by the planets' attraction to them. Those gaining orbital energy escape the solar system and become interstellar wanderers. Comets losing energy become more tightly bound to the Sun and thus fall among the flat distribution of orbital energies calculated by van Woerkom.

Oort showed that the orbits of the comets in the cloud were so weakly bound to the Sun that random passing stars could perturb them. Although stars rarely come very close to one another, on average about 10 to 12 stars must pass within 200,000 AU of the Sun every million years. Over the lifetime of the solar system this is close enough to slowly stir the cometary orbits, in effect randomizing their orientations and pumping them up to longer-period orbits. This explains why comets appear to enter the planetary system from all directions. The perturbing influence of nearby stars also robs many comets of orbital angular momentum, causing them to "fall" toward the Sun and into the planetary region from all directions.

THE ORIGIN OF LONG-PERIOD COMETS

Oort calculated that the "dynamically new" comets entering the planetary region for the first time would have orbits that extended out to between 100,000 and 300,000 AU from the Sun. He also estimated that the cometary cloud must total

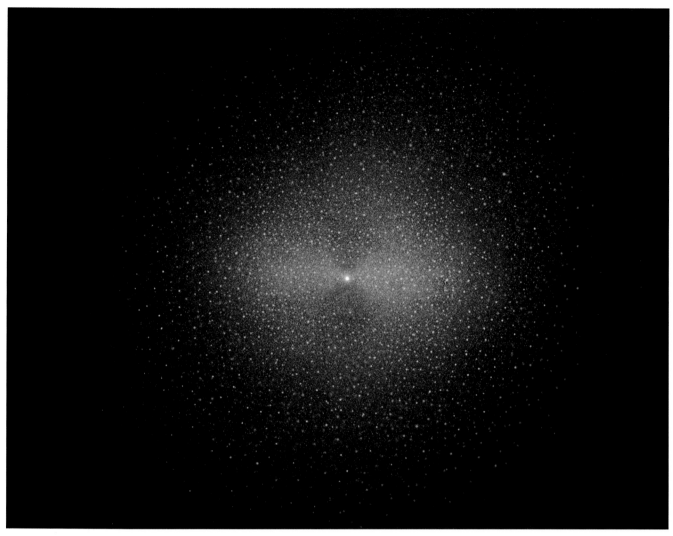

about 190 billion comets in order to supply the number of long-period comets seen passing through the planetary region. Oort's deductions about the orbital energy distribution are even more impressive when one considers that he did so with only 19 well-measured cometary orbits. This was a truly remarkable achievement.

Today we have at least 15 times as many accurate orbits for long-period comets to help us understand the dynamics of the Oort cloud. For example, we now know that the average orbital aphelion of a dynamically "new" long-period comet (one only recently perturbed toward the Sun) is about 44,000 AU, somewhat closer than Oort suggested. The reason for this smaller distance is that, in addition to random passing stars, the Oort cloud is also perturbed by gravitational tides generated by stars both in the Milky Way's disk and, to a lesser extent, in the galactic core. The tides arise because the Sun and a comet in the cloud typically differ in distance from these massive concentrations of matter and thus feel slightly different gravitational tugs from them. It turns out that the tide due to the galactic disk is somewhat stronger than the perturbations calculated by Oort from random passing stars. Consequently, comets with orbital aphelia beyond 200,000 AU are easily lost to interstellar space, and the Oort cloud thus does not extend out as far as originally thought (*Figure 4*).

Figure 4. **The Oort cloud, which envelops our solar system with perhaps trillions of icy objects, has never been seen. But its existence was postulated a half century ago to explain the trajectories of long-period comets. The cometary population occupying the inner, more massive portion of the cloud preferentially lies near the ecliptic plane. But in the cloud's outer reaches the orbits have been randomized due to the attraction of passing stars and other external forces. The Oort cloud's outermost edge is perhaps 30 trillion km (200,000 AU or 3 light-years) from the Sun! This tenuous boundary is probably ellipsoidal in shape because of gravitational effects of mass in the nucleus and disk of our galaxy.**

The cometary cloud was described by Oort as being similar to "a garden, gently raked by stellar perturbations." We now know that such effects are not always gentle. Occasionally, random passing stars come so close to the Sun that they pass right through the Oort cloud, violently perturbing the cometary orbits along its path. A star is expected to pass within 10,000 AU of the Sun every 35 million years and within 3,000 AU every 400 million years. The closest stellar approach over the entire 4.5-billion-year history of the solar system has probably been about 900 AU, still far beyond the planetary region.

In the late 1970s astronomers recognized that violent perturbations on the Oort cloud could also be caused by giant molecular clouds (GMCs) in the galaxy. These massive accumulations of cold hydrogen are the birthplace of stars and solar systems, and they range from 100,000 to 1,000,000 times the mass of the Sun. If the solar system came close to a GMC, gravitational perturbations would be so strong that many comets would be ripped away and flung into interstellar space; many others would cascade into the planetary realm.

GMC encounters are expected to occur only once every 300 to 500 million years — so, while violent, they are also very infrequent. Still, some astronomers have proposed that the Oort cloud could not have survived these incursions over the solar-system history, and that it has to be replenished by capturing interstellar comets. Unfortunately, the capture of comets from interstellar space is too inefficient a process to keep the Oort cloud populated. Some other solution was needed. At least part of the problem was solved with the realization that GMC perturbations are not quite as severe as earlier thought. Their total effect over the history of the solar system has probably been about the same as that for the summed effect of all passing stars.

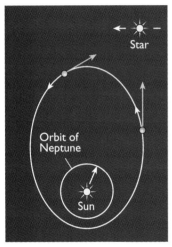 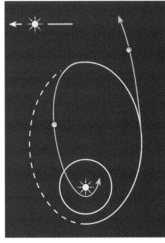

Figure 5. When a star passes close to the solar system, its gravitational attraction causes some comets in the outer Oort cloud to lose orbital angular momentum and "fall" into the planetary region (red); others gain energy and angular momentum and escape to interstellar space (blue). Meanwhile, many comets in the denser inner Oort cloud (not shown) are perturbed as well, and some of these migrate outward to replace those comets lost during the stellar encounter.

Figure 6. Early in solar-system history comets moved in the ecliptic plane among the outer planets *(a)*. Gravitational interactions with these planets pumped the comets into ever larger orbits *(b)*, after which the gravitational attraction of random passing stars, giant molecular clouds, and the galactic tide randomized their orbital inclinations and made the Oort cloud more spherical *(c,d)*. The circles are 20,000 AU from the Sun — the distance beyond which Oort-cloud comets can be thrown back into the planetary system by stellar and GMC perturbations and become visible as long-period comets.

In 1981 theorist Jack Hills suggested that close stellar passages could prove really violent if the Oort cloud included a dense inner core of comets, one too close to the Sun to be easily perturbed by the more gentle external perturbations. A close-passing star could send "showers" of the core's comets into the planetary region (*Figure 5*). Computer models show that the rate of comets coming near the Sun would jump substantially, up to 300 times the steady-state infall rate, before returning to normal levels in 2 to 3 million years. What would a comet shower look like to people on Earth? Nowadays we see a bright naked-eye comet about every 5 years on average. But during the peak of such a shower, a new comet might appear every 5 days or so! A dozen bright comets would be visible in the night sky at any time. It would truly be a cosmic spectacular!

Hills' suggestion prompted others to investigate the possibility of a dense inner Oort cloud, in the hope that it could solve the problem of long-term depletion. Celestial dynamicists Martin Duncan, Thomas Quinn, and Scott Tremaine showed that some of the icy planetesimals ejected out of the planetary zones by the growing protoplanets early in the solar system's history would naturally be captured into Oort-cloud orbits stretching between 6,000 and 200,000 AU from the Sun. Objects at the outer edge of the cloud would be lost rapidly due to stellar and GMC perturbations (but not due to galactic tides, which are much less effective). However, these same forces would also draw comets outward from the dense core to replace the outer cloud's lost members (*Figure 6*).

This evolution, and several other factors, have led to a substantial increase in the estimated population of the Oort cloud to about 6 trillion comets. Only one-sixth of these reside in the outer, dynamically active region first described by Oort; the majority are more tightly bound in the cloud's inner core. If we use our best estimate for the average mass of a cometary nucleus, then the entire Oort cloud represents about 2.4×10^{29} g of matter — 38 times the mass of the Earth!

Where did all those comets come from? Oort speculated in 1950 that the comets had been ejected from the asteroid belt by the giant planets during the formation of the solar system. However, that same year Fred Whipple suggested that comets were icy conglomerates ("dirty snowballs"). This meant that comets must have formed much further from the Sun, in locations cold enough for water ice to condense. Later dynamical studies suggested that the Oort cloud comets probably came from the Uranus-Neptune zone. Because Jupiter and Saturn are so massive, they would have ejected any icy bodies in their zones beyond the Oort cloud and into interstellar space. Uranus and Neptune, with smaller masses, could not easily throw so many

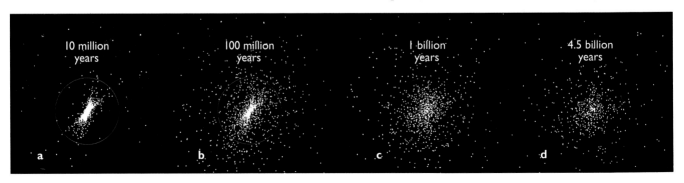

comets onto escape trajectories, and a larger fraction of the comets in their zones ended up in the Oort cloud.

However, some doubt has been cast on this scenario. Although Jupiter and Saturn may have placed only a small fraction of the comets from their zones into the Oort cloud, the amount of primordial material orbiting near them was much greater than that near Uranus and Neptune. The net amount likely ejected by Jupiter and Saturn still adds up to a sizable fraction of the Oort cloud's total mass. Many asteroids are probably out there as well. Dynamical studies by Harold Levison and me suggest that asteroids ejected from the inner-planet region should comprise about 2 percent of the total Oort cloud population.

THE ORIGIN OF SHORT-PERIOD COMETS

With the Oort cloud now reasonably well characterized, astronomers have turned their attention to the origin of the short-period comets. Traditionally, short-period comets were simply thought to be long-period comets that had been repeatedly perturbed by the planets onto smaller orbits. This seemed a good explanation, but it was difficult to confirm because computer simulations in the 1970s and early 1980s were not good enough to model the dynamics accurately over so many returns. Also the question arose as to whether a comet might exhaust all its volatile ices — burn itself out — before it could evolve to a short-period orbit.

In 1980 astronomer Julio Fernandez resurrected an idea (first proposed by Kenneth Edgeworth in 1949 and Gerard Kuiper in 1951) that a vast belt of comets in near-circular orbits could exist beyond Pluto. Because orbital periods become quite long far from the Sun, and because icy planetesimals accrete into planets at rates inversely proportional to their orbital periods, Edgeworth and Kuiper each speculated that huge numbers of unaccreted comets should still exist on the fringes of the planetary realm (*Figure 7*).

Fernandez realized that such a belt of comets could provide a far more efficient source for the short-period comets than the distant Oort cloud, for two reasons. First, the required change in orbital energy would be far less than that for comets coming from the Oort cloud. Second, and more important, the comets would begin their evolution in orbits already close to the ecliptic plane, where perturbations by the planets and the frequency of planetary encounters are their greatest.

About this time, a completely different avenue of research hinted that comet belts might exist around nearby stars. The Infrared Astronomical Satellite (IRAS) discovered that several nearby stars emit significantly more infrared radiation than expected. The most logical explanation was that they are surrounded by warm, dusty material, and IRAS actually resolved extended disks of warm dust around several of the stars. In the case of Beta Pictoris (the second brightest star in the constellation Pictor), the evidence suggested that the dust lies in a narrow disk, seen edge-on. This was confirmed by remarkable visible-light photographs taken in the mid-1980s (*Figure 8*). Using a coronagraph to block the light from Beta Pictoris itself, astronomers recorded an edge-on dust disk that extends 800 to 900 AU from the central star.

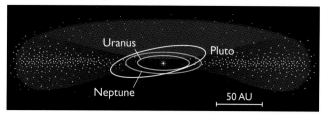

Figure 7. In 1949 and 1951 two astronomers independently proposed the existence of a band of comets closer to the Sun, objects left over from the solar system's formation. Known today as the Kuiper belt, this distant reservoir is the dominant source of short-period comets, whose orbits preferentially lie near the ecliptic plane.

Figure 8. The edge-on disk of material around the star Beta Pictoris extends out about 900 AU to either side. (The star itself is blocked out by a small occulting disk placed in the telescope's optical path.) With an estimated mass of tens or even hundreds times that of Earth, this disk of matter is similar to what astronomers think the Kuiper belt might look like if viewed from far outside our solar system.

It seemed clear to me and others that the dust could result from collisions between comets in each disk and from the sputtering of material off the comets' surfaces by cosmic radiation. Moreover, because dust is very effective at scattering light, the dust would be much easier to detect than the comets themselves. Meanwhile, back in our own solar system, astronomers thought that the comet belt beyond the planets might blend smoothly into the distribution of comets in the inner Oort cloud. Because most of the stars with infrared excesses are much younger than the Sun, their disks might represent comets still in the process of being ejected from their respective planetary systems.

A major breakthrough in understanding the dynamics of cometary reservoirs in our own solar system came in 1988. Duncan, Quinn, and Tremaine had used powerful computers to track the orbital evolution of comets moving inward from the Oort cloud and from a hypothetical comet belt beyond Neptune. Although both sources could produce short-period comets, those from the Oort cloud tended to have orbits with fairly large inclinations and thus appeared to be scattered over much of the sky. On the other hand, comets that initially orbited in the ecliptic beyond Neptune tended to *remain* in low-inclination orbits as they approached the Sun — a characteristic very similar to that observed for short-period comets. Thus, not only did the trans-Neptunian belt produce comets with the right kind of orbits, but the Oort cloud also yielded comets with the *wrong* orbital characteristics, particularly inclination. Duncan, Quinn, and Tremaine suggested that this new cometary reservoir beyond Neptune be called the Kuiper belt. (While Edgeworth's earlier paper was missed or not remembered by most astronomers, he certainly shares credit for proposing the belt's existence.)

The Oort cloud still produces some of what we classify as short-period comets. About 20 known comets, like Halley and Swift-Tuttle, have orbits with high inclinations and periods between 20 and 200 years. Dynamical simulations show that these bodies likely did originate in the Oort cloud. So the short-period comets we observe actually come from both cometary reservoirs, which has led to a revised classification scheme. Short-period comets with modest inclinations and periods, typically less than about 20 years, are termed *Jupiter-family* or *ecliptic* comets. Those with high inclinations and periods of 20 to 200 years are called *Halley-type* or *random* comets. Astronomers differentiate between the two classes using a number called the *Tisserand parameter*, which is calculated from a comet's orbital elements (geometric characteristics). French dynamicist François Tisserand derived this formula a century

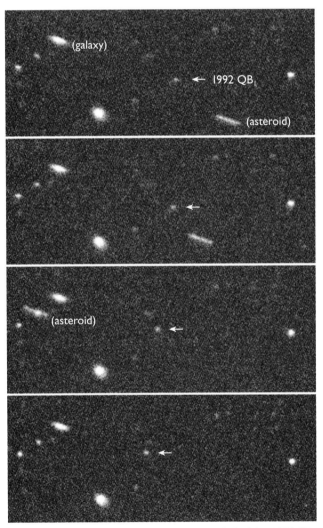

Figure 9. At 23rd magnitude, the distant Kuiper-belt object 1992 QB₁ (arrows) is not much to look at. These discovery images, the fruit of a 5-year search by observers David Jewitt and Jane Luu, show the object's slow motion against the background stars of the constellation Pisces, very near the ecliptic plane. Also captured in the 40-arcsecond-wide fields are two asteroids and a distant galaxy.

ago as a means of identifying returning periodic comets, and this parameter remains nearly the same for a given comet even if its orbit is changed by a close approach to Jupiter.

FROM THEORY TO REALITY

Unlike the Oort cloud, the Kuiper belt was still a purely theoretical construct in the late 1980s. Although the Oort cloud cannot be "seen" in the usual sense, we do observe comets coming directly from it. We can also show by dynamical modeling that a distant cloud is the only logical explanation for these comets. However, many astronomers were far more reluctant to accept the Kuiper belt on the basis of theory alone, since the comet orbits had to evolve considerably for them to reach the inner solar system and become visible.

So the search began for objects in orbits beyond Neptune. Past searches, such as the one that led to the discovery of Pluto, had already constrained the size and number of large bodies that might exist in this region. In addition, studies of the motion of Halley's Comet (whose aphelion lies beyond Neptune's orbit) set an upper limit on the total mass that could be in such a comet belt: it was somewhere between 0.8 and 1.3 Earth masses. Despite these limits, a lot of comets could still lie waiting to be discovered in that region.

In August 1992 astronomers David Jewitt and Jane Luu found the first of these trans-Neptunian objects, a 23rd-magnitude speck of light moving very slowly in Pisces (*Figure 9*). The object was designated 1992 QB₁, using the naming convention for asteroid discoveries (it could not officially be classified as a comet because it did not display a coma). The new object was estimated to be about 320 km in diameter, quite large as comets go, and reddish in color. By comparison, the potato-shaped nucleus of Comet Halley is only 16 km long and 8 km wide.

The job of determining the orbit of 1992 QB₁ fell to astronomer Brian Marsden. Using the discoverers' observations and those from other observatories, Marsden calculated that the object has an orbit inclined about 2.2° to the ecliptic. It takes 292 years to complete one orbit, and its mean distance from the Sun is 44.0 AU. This is 14 AU beyond Neptune and 4.5 AU beyond the mean distance of Pluto. But the orbit of 1992 QB₁ is not as eccentric as Pluto's, and in fact at aphelion Pluto is farther from the Sun.

Other Kuiper-belt discoveries followed, and by early 1998 the total count stood at 60, not including Pluto and its satellite Charon, with about 10 new ones being found each year. Dynamicists believe that the orbits of 1992 QB₁ and other Kuiper-belt objects are probably stable for the age of the solar system. Many are located far enough beyond Neptune's orbit that the planet's gravity cannot easily perturb them out of their current orbits. Others are in *mean-motion resonances* with Neptune. This means they orbit the Sun with a period that is a simple multiple of Neptune's period. Most crowded (at least so far) is the 2:3 resonance, in which an object circles the Sun twice in the time Neptune takes to go around three times. Interestingly, Pluto is in this resonance. Mean-motion resonances prolong the dynamical lifetime of Pluto and the Kuiper objects, despite the fact that their orbits approach or even cross that of Neptune.

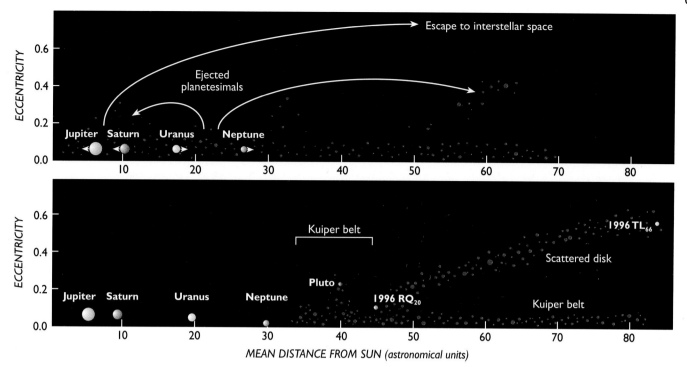

Figure 10. **Early in the solar system's history, the giant planets deflected countless smaller icy bodies out of their formation zones *(upper panel).* The sheer number of these gravitational interactions caused Jupiter and Saturn to migrate closer to the Sun, while Uranus and Neptune were forced outward — the latter by an estimated 5 AU. A great many of the ejected planetesimals escaped to interstellar space. Others, with not quite as much velocity, ended up in the Oort cloud. The recent discovery of the distant object 1996 TL_{66} *(lower panel)* suggests that it resides within the Kuiper belt but actually represents a dynamically distinct "scattered disk" of objects with modest eccentricities.**

A particularly unusual Kuiper-belt object (KBO) is 1996 TL_{66}, which comes as close as 35 AU to the Sun and is the most distant known KBO at 133 AU. This highly eccentric orbit has a period of 772 years. Perhaps 1996 TL_{66} was thrown into such an orbit after a recent close encounter with Neptune. But this object was more likely among a great many bodies initially in the Uranus-Neptune zone or the Kuiper belt that were forced into eccentric orbits early in the solar system's history (*Figure 10*). Because of their very long orbital periods, objects relegated to this *scattered disk* had fewer opportunities to encounter a planet. While most eventually did and are now gone, about 1 percent of them should have survived to the present day.

Where do Pluto and Charon fit into this picture? Although it is generally labeled as a planet (see Chapter 21), many solar-system astronomers think of Pluto as the largest icy planetesimal to grow in the region beyond Neptune. While Pluto has a satellite and a thin atmosphere, traditionally considered proof of planetary status, we now know of asteroids with satellites, and satellites with atmospheres. Moreover, Pluto is smaller than Titan, Ganymede, Callisto, and Triton — all satellites of the Jovian planets. Consequently, Pluto's classification as a planet has increasingly been questioned, especially given its dynamical similarities to other objects in the Kuiper belt.

COUNTING THE KUIPER-BELT OBJECTS

How many comets reside in the Kuiper belt? Telescopic searches, which to date have covered only a few square degrees of sky, had found 60 objects by early 1998. These are between 100 and 760 km in diameter, assuming that their surfaces reflect only 4 percent of the sunlight striking them (a value typical for comets), and all are between 30 and 50 AU from the Sun. By extrapolating the discovery statistics to a 60°-wide band all around the ecliptic, we would expect to find 70,000 such objects. Assuming a density of 1 g/cm³, their combined mass

would be a few percent of an Earth mass.

But the objects discovered with ground-based telescopic searches are only the brightest (and thus largest) objects in the Kuiper belt. Fainter objects have been glimpsed, however. Using the Hubble Space Telescope to examine a tiny area of sky, Anita Cochran and her colleagues found evidence for about 30 Kuiper belt objects roughly 10 to 20 km across, about the same size as the nucleus of Comet Halley. They estimate that roughly 200 million comets of similar size should reside in the Kuiper belt between 30 and 50 AU from the Sun.

Another estimate comes from the number of objects necessary to provide the observed population of short-period comets. As deduced in dynamical simulations, the Kuiper belt must contain 6.7 billion comets with a diameter of at least 2 km. Yet another way to "see" the belt is through the gravitational effect it has on other bodies in the solar system. The Voyager and Pioneer spacecraft are now passing through the Kuiper belt, and the analysis of their trajectories has revealed the first hints of what may be perturbations from the gravitational pull of billions of comets.

By putting all these numbers together, we can derive an approximate mass distribution of the objects in the Kuiper belt (*Figure 11*). This mass distribution is known as a *power law,* because the number of comets in any size interval (N) is proportional to that interval's radius R raised to a specific power (q).

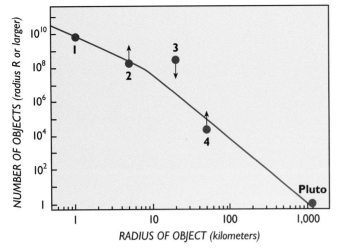

Figure 11. Estimating the total number of comets of a given size involves several observational inputs. For example, *1* is the number of objects required to provide all the short-period comets observed in the planetary region. *2* is the total number of comets implied by faint Kuiper-belt detections made with the Hubble Space Telescope. *3* is the number of comets needed to maintain the slight eccentricity in the orbit of Charon, Pluto's satellite. Point *4* is the estimated number of Kuiper-belt objects at least 100 km across, based on discovery statistics to date. The resulting size distribution is called a *broken power law*, in that the slope is shallower for small comets (radii less than 20 km) and steeper for large ones.

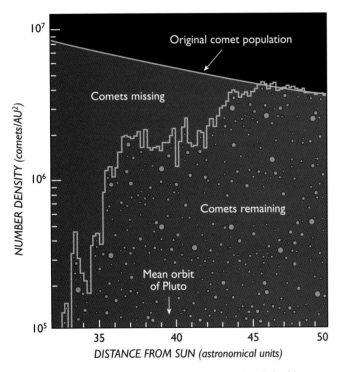

Figure 12. Despite the fact that only a few dozen Kuiper-belt objects have been found to date, theorists believe that a great many more await discovery outside the orbit of Pluto. The smooth curve at top represents the minimum population of comets believed to be present when the solar system formed. The jagged curve represents the minimum number that must be left in order to supply the short-period comets. In the 4½ billion years since the Kuiper belt formed, a combination of processes (such as collisions and long-term perturbations by Neptune and Uranus) have removed most of the objects in the Kuiper belt's inner regions.

The mathematical function is $\delta N \sim aR^q \, \delta R$. However, Kuiper-belt objects appear to follow a "broken" power law, in which the exponent q has two different values: 3 for comets smaller than 10 to 20 km in radius and 4½ for larger comets. This size distribution is surprisingly similar to one that had been proposed for the nuclei of long-period comets, as based on hundreds of observations and on the distributions found in computer simulations of how small bodies accreted in the early solar system.

If we combine all the bodies in Figure 11 and assume they have a density of 1 g/cm³, then the total mass of the Kuiper belt between 30 and 50 AU is slightly more than that of Mercury. This is actually quite a bit less than we would expect if all the material left over from the origin of the solar system was still there (*Figure 12*). Most likely, the objects that formed beyond Neptune have experienced a violent collisional history and have literally ground each other down to dust through repeated impacts. In addition, many objects have likely been ejected from the belt due to orbital encounters with the outer planets.

Farther out, perhaps around 100 AU, collisions are much less frequent and many more Kuiper belt objects should have survived. If our solar system is similar to what we see around Beta Pictoris and other stars, then the Kuiper belt may extend out to about 1,000 AU and contain tens of Earth masses of primordial comets.

If stars like Beta Pictoris have Kuiper belts, might they also have Oort clouds? Perhaps. Oort clouds may be common if other solar systems have giant planets that can eject the primitive comets to distant orbits. If each star has its own comet cloud, then during a close encounter of two stars their Oort clouds will pass through each other. Collisions will be rare because the typical space between any two comets should be 1 AU or more. Nonetheless, each star system will expel many comets into space. These interstellar comets should be easily recognizable if they were to pass close to the Sun, because they would approach the planetary system at much higher velocities than the comets from our own Oort cloud do.

To date, no such interstellar comets have been detected. This is not too distressing because the planetary system is a very small target when compared to the vastness of interstellar space. There is about a 50:50 probability that we should have seen one interstellar comet by now, and even those odds may be somewhat optimistic. Still, one day we may see one of these interstellar wanderers, offering us tantalizing evidence that other comet clouds exist around distant, unseen solar systems.

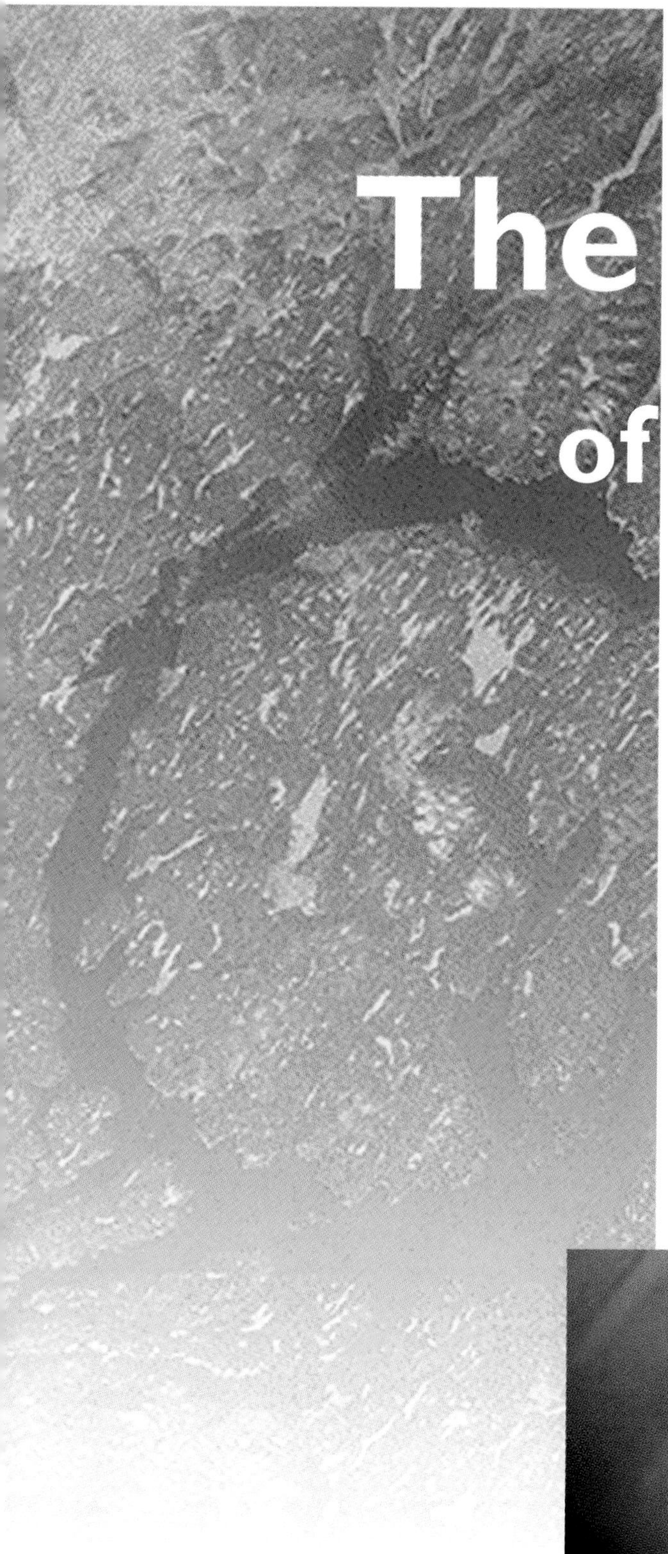

The Role of Collisions

Eugene M. Shoemaker
&
Carolyn S. Shoemaker

I N JULY 1994, the inhabitants of Earth witnessed a serendipitous experiment in planetary-scale impacts, as fragments of a comet slammed into the upper atmosphere of Jupiter. The series of collisions created huge dark stains on the planet *(Figure 1)*, which remained recognizable for months thereafter — even in backyard telescopes. Never before in the history of modern astronomy had so many of the world's telescopes been trained on a single point in the sky. At the end of "impact week," our most important conclusion was a newfound appreciation that such impacts have happened on Earth in the past and could happen again, at any time, with devastating affects to civilization.

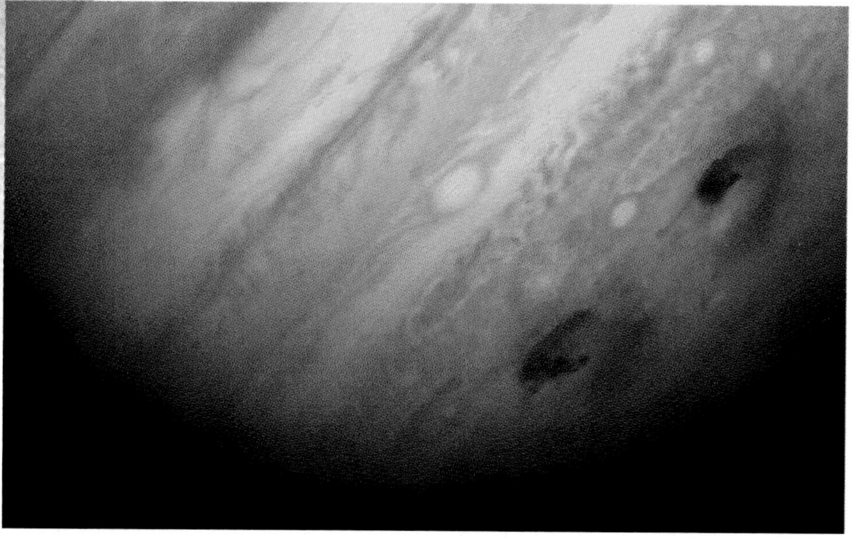

69

The 70-km-wide Manicouagan impact structure in Quebec, Canada.

Figure 1. This image of Jupiter shows the impact sites of fragments Q (along bottom), G and D (overlapping), and L (upper right) from Comet Shoemaker-Levy 9. The comet fragments struck Jupiter between 16 and 22 July 1994, entering its atmosphere from the south at a 45° angle. The G site has concentric rings around it, with a central dark spot 2,500 km across. The outermost ring's inner edge has a diameter about the size of the Earth.

The comet's crash underscores the role played by collisions in shaping planetary surfaces. One of the most striking discoveries from four decades of space exploration is that most of the solid surfaces of planets and satellites — from Mercury to the satellites of Neptune — are heavily cratered. Indeed, impact seems to have been the most fundamental process on the terrestrial planets. Without impact, Mercury, Venus, Earth, Mars, and the Moon — a group of bodies whose very formation probably depended on the collision and accretion of smaller objects — would not exist.

Impact craters are by far the dominant landforms observed on most of the rocky and icy bodies surveyed in the solar system to date. Remarkably, however, we have been slow to come to this realization, in part because only within the last half century have we learned to recognize an impact crater for what it truly is.

IMPACT BASICS

Think of the cratering process as a very efficient delivery system for a vast amount of kinetic energy. When a sizable solid body strikes the ground at high speed *(Figure 2)*, shock waves propagate into the target rocks and into the impacting body. At collision speeds of tens of kilometers per second, the initial pressure on the material engulfed by the expanding shock waves is millions of times the Earth's normal atmospheric pressure. This can squeeze even dense rock into one-third of its usual volume. Stress so overwhelms the target material that the rock initially flows almost like a fluid.

A rarefaction (decompression) wave follows the advancing shock front into the compressed rock, allowing the material to move sideways. As more and more of the target rock becomes engulfed by the shock wave, which expands more or less radially away from the point of impact, the flow of target material behind the shock front is diverted out along the wall of a rapidly expanding cavity created by the rarefaction wave. The impacting body, now melted or vaporized, moves outward with this divergent flow and lines the cavity walls. Decompressed material sprays out of the cavity as an expanding conical sheet. Rocky material continues to flow outward until stresses in the shock wave drop below the strength of the target rocks. In the case of small craters, the ground motion is arrested at this time.

In large impact structures *(Figure 3)*, the rock walls slump inward soon after excavation of the initial, or *transient*, cavity. On Earth, this has occurred in all craters larger than 3 km in diameter that formed in soft sedimentary rocks, and in most craters larger than about 4 km across in strong crystalline rocks. As the slumping material converges inward it produces a pronounced central hill or peak in most cases. A few large craters exhibit a ring-shaped inner ridge or more complex central structure. Evidence of uplifted rocks at the center and of subsidence and inward flow from the sides are important clues for the recognition of the deeply eroded impact structures on Earth.

When a large asteroid or comet impacts a solid body, more happens than just the formation of a single crater. The energy released from the collision of a 10-km-wide object with Earth exceeds by five orders of magnitude that of our most powerful terrestrial earthquakes. The list of possible impact-related effects is sobering: large earthquakes, volcanic activity, major tsunami waves (if the impact is at sea), acid rain, sunlight-blocking clouds of dust, cessation of photosynthesis, and collapse of the food chain. Fortunately, as discussed later, such cataclysms

Figure 2. **During the formation of a simple impact crater, a compression wave spreads outward from the center into the target material. A wave of rarefaction (decompression) moving behind the shock front allows material mobilized in the event to be ejected in a conical sheet. Engulfed by the shock wave, the colliding meteoroid melts and partly vaporizes. Simulations indicate that for all but very low angles of impact, less than about 20°, the crater produced is circular.**

have become very rare as the solar system has matured. However, in a sense the last stage of planetary accretion is still going on — albeit at a very slow rate.

THE CRATERING RECORD OF EARTH

If a projectile is large enough, it can survive passage through the Earth's atmosphere more or less intact and strike the ground or the ocean at high velocity. The threshold size for survival depends on the material strength and density of the body and on its velocity at the time of encounter. For a stony body, this size appears to be about 100 m. Most smaller objects are sheared apart and then slowed dramatically by aerodynamic drag (see Chapter 26). A historically recent example is the meteoroid that produced the great Tunguska fireball over western Siberia on 30 June 1908. We are not yet certain whether the object was an asteroid, a comet, or something in between. However, dynamical simulations and the tentative identification of cosmic particles in the resin of surviving trees favor something asteroidal. In any case, the impactor was probably about 50 m across. It did not strike the ground but instead shattered abruptly in the atmosphere about 6 km up. The ensuing blast felled trees over 2,150 km² of tundra but failed to produce a crater. On the other hand, fragments from disrupted, relatively small iron meteoroids sometimes reach the ground with a substantial fraction of their infall velocity and produce swarms of small impact craters. Half a dozen such swarms are known, including a cluster of craters produced by a fall observed in the Sikhote-Alin region of eastern Siberia in 1947.

Sometimes a large iron mass or a very compact cluster of fragments strikes the ground with sufficient energy to produce a single crater. An iron asteroid 40 to 50 m in diameter survived atmospheric entry 50,000 years ago to form Meteor Crater (known officially as Barringer Crater) in northern Arizona. Meteor Crater is 1.2 km in diameter and 200 m deep *(Figure 4)*. The surrounding raised rim consists partly of rocky fragments ejected from the crater and partly of bedrock that has been lifted up and shoved radially outward. The rim has a gently rolling, hummocky character that reflects the irregular distribution of large lumps in the ejected debris. Roughly half of the volume of the crater stems from material thrown out, and half from the outward displacement of rocks in the crater walls. Ejected debris lies stacked on the crater rim in inverse stratigraphic order; that is, the original sequence of layers in the target bedrock is preserved but upside down. The assortment of relatively unshocked meteorites found in the crater's vicinity suggests that numerous fragments were detached and decelerated during atmospheric passage of the principal body, which may have broken apart just before impact.

Beneath the crater floor is a lens-shaped body of *breccia*, rock that has been smashed by the shock wave. Abundant

Figure 3. Craters excavated by objects at least several kilometers across are more complex than the simple bowl form. Typically the sides of the transient cavity slump inward, enlarging the crater's final width and often forming terraces. The inward-slumping mass and upward rebound of lower-lying rocks creates a central peak or plateau. The largest impact basins sometimes exhibit multiple, concentric rings.

glass, produced by shock melting of the rocks, occurs near the base of the breccia. It contains microscopic spheres of meteoritic iron — the metamorphosed remains of the impacting body. The shock initially excavated a cavity to a depth of nearly 400 m. The breccia, which was at first smeared along the cavity walls, collapsed toward the center to produce a much shallower final crater. Relatively fine-grained ejecta, arrested in their flight through the atmosphere, then showered down and left a layer of mixed debris about 10 m thick on the crater floor. Subsequently, the rim and upper walls of the crater were eroded, and lake-bed sediments 30 m thick now lie in the center.

Figure 4. Meteor Crater, near Flagstaff, Arizona, is one of the youngest impact craters on Earth. It was excavated about 50,000 years ago when an iron mass (or perhaps several) struck flat-lying sedimentary rocks at more than 11 km per second. Between 15 and 20 megatons of kinetic energy were released during the impact, which left a bowl-shaped crater 1.2 km in diameter and 200 m deep, surrounded by an extensive blanket of ejecta. The crater now has a somewhat squarish outline, lake sediments blanket the floor, and erosion has removed 15 to 25 percent of the ejecta.

Because it formed recently (geologically speaking) in a relatively dry environment, Meteor Crater is the best-preserved sizable impact feature on Earth. Geologists have come to recognize that the altered rocks found there provide important clues for the identification of impact craters and structures elsewhere on Earth. In addition to the presence of shock-melted glass and distinctive macroscopic and microscopic deformations of unmelted rocks, two high-pressure forms of crystalline silica were discovered at the site. These shock-formed minerals, coesite and stishovite, occur at many other impact localities around the world.

Crater-hunters identify other terrestrial impact sites primarily by their general structure and by evidence of shock metamorphism in the rocks. The craters themselves are ephemeral; they tend to fill in or erode away very quickly. If erosion has not

encroached too deeply, some of the breccia may be preserved. Impact structures larger than about 30 km in diameter commonly have a fairly thick layer of congealed, shocked-melted material, collected into a pool on the crater floor *(Figure 5)*. The contamination of this melt by material from the impacting body creates enhanced amounts of trace elements like the noble metals (including platinum, iridium, and gold), which are relatively abundant in meteorites but greatly depleted in the crust of the Earth.

Except for small, very young excavations associated with iron or stony-iron meteorites (Meteor Crater being the archetypical example), the impact record of Earth generally consists of craters or eroded circular structures at least 3 km in diameter. Because kinetic energy scales as the square of velocity, a crater of that size can be formed by a stony body roughly 150 m across — $\frac{1}{20}$ the crater's size — if it arrives at a velocity typical of Earth-crossing asteroids, about $17\frac{1}{2}$ km per second. This 20:1 ratio is a good "rule of thumb" for gauging the impactor size for a crater on our planet.

Approximately 150 impact structures have been recognized on Earth by the two basic criteria of structural form and shock-induced metamorphism. Another 30 exhibit suggestive but inconclusive evidence for impact origin. Three to five new impact sites are recognized each year, frequently as circular features seen from orbit or as subsurface features discovered in the course of drilling for oil or water. Earth's known impact craters are generally younger than 500 million years (because the geologic record is more complete for recent times), but some of the largest are more ancient. The largest impact structures, about 300 km in diameter or a little less, are Vredefort in South Africa (2.02 billion years old) and Sudbury in Canada (1.85 billion years old).

Most of the recognized impact structures occur on the continental shields, where the Earth's oldest rocks are exposed, stable, and well preserved. Geologists have studied the shields of North America, Europe, and Australia most thoroughly, and not surprisingly most of the well-documented impact sites occur on these *(Figure 6)*. An analysis by Richard Grieve suggests that on the American and European shields, an average of five craters larger than 20 km in diameter have been produced per million square kilometers per billion years. Where the geologic record is most complete in North America, the production of craters down to 10 km diameter has been about four

Figure 5 (right). **The Manicouagan impact structure in Quebec, Canada, is about 210 million years old and 70 km in diameter. A central peak of shock-metamorphosed rock is surrounded by a thick layer of frozen impact melt that pooled on the original crater floor. The ring-shaped reservoir fills a valley carved by glaciers out of soft sedimentary rocks that had slumped inward along the original walls of the crater.**

Figure 6 (below). **Over time geologists have identified and confirmed more than 150 impact craters on Earth. The counts are higher in North America and Europe because surveys are more complete in these regions. However, in recent years many impact sites have been located in remote sections of Australia.**

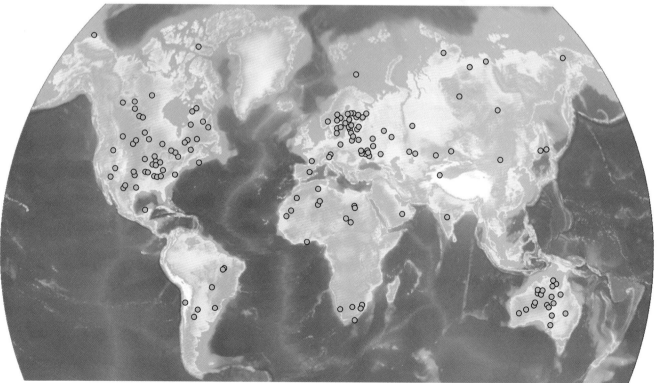

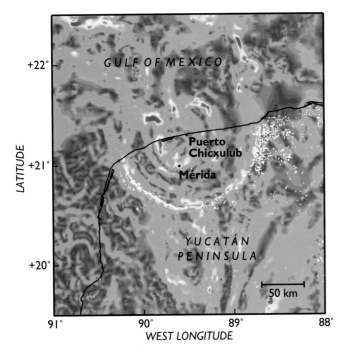

Figure 7. The 180-km-wide impact structure known as Chicxulub straddles the coastline of the Yucatán Peninsula, Mexico. Even though it is completely buried, the crater can be discerned in gravity data processed to accentuate shallow subsurface geologic structure. (Note that the data are incomplete, particularly in the portion that lies under water.) A striking series of concentric features reveals the hidden crater's location and internal structures. White dots represent locations of water-filled sinkholes called *cenotes* after the Maya word "dzonot." Formed in much younger limestones that overlie the crater, the cenote ring marks a zone where groundwater can most easily flow to the sea creating freshwater springs at the east and west sides of the crater. Somehow the crater is able to reach up through several hundred metres of sediment and tens of millions of years of time to influence groundwater flow.

times greater. This is in agreement with the size distribution of young craters observed on the Moon. These estimates pertain to the last several hundred million years of Earth's history.

Of course, the number of craters seen on Earth should match our estimates based on the number of objects destined to collide with it. For Earth-crossing asteroids, 3.5 craters greater than 20 km in diameter should be produced per million square kilometers per billion years. If we include the contribution expected from collisions with comets, the total predicted cratering rate is in excellent agreement with the geologic record. The correspondence is good at smaller sizes, too. On average, asteroid impact should create about three craters at least 10 km across on Earth's land areas every million years. Geologists have indeed found two 10-km impact craters no older than 1.1 million years. One is in Russia, and the other is occupied by Lake Bosumtwi, the sacred lake of Ghana's Ashanti tribe. In addition, one or perhaps two great strewn fields of impact-glass *tektites* (beads of once-molten mineral glass that solidified in midair) have been formed in the last million years. This glass was probably thrown from a crater or craters much larger than 10 km. Within the last few years, the newly recognized Chesapeake Bay crater has been identified as the source of the vast North American tektite strewn field.

Evidence indicates that the well-preserved Chicxulub structure on the north coast of the Yucatán is the long-sought impact that coincides with the Cretaceous-Tertiary (K/T) boundary 65 million years ago (*Figure 7*). Roughly 180 km across, Chicxulub is buried under about 1 km of marine sediment; there is no rim or central depression on the surface to betray its existence. Indeed, much of the early evidence for Chicxulub's presence was circumstantial: ringlike arcs in subsurface gravity data, deposits of tektites in nearby Haiti, and chaotically disturbed sediment beds around the perimeter of the Gulf of Mexico. The crater's location was confirmed through wells drilled by Petróleos Mexicanos during searches for new oil fields.

Prior to 1980, no one suspected that a major impact could have triggered the mass extinction of species that had been recognized at the K/T boundary. However, in that year Luis and Walter Alvarez and their colleagues Frank Asaro and Helen Michel announced that they had found a large excess of the element iridium in a thin clay layer laid down in sediments precisely at the boundary. Iridium is extremely rare in Earth's crust but considerably more abundant in meteorites. The Alvarez team calculated that the iridium excess worldwide corresponded to the impact of an asteroid about 10 km across — an estimate that remains valid today. Since 1980, iridium anomalies have been found in the K/T boundary claystone at more than 80 localities on the continents and in the ocean basins. The relative proportions of other metals, including platinum, osmium, and gold, have been shown to be close to the relative proportions in primitive stony meteorites. Of special significance was the discovery of shocked mineral grains and rock fragments precisely at the Cretaceous-Tertiary boundary.

In the past two decades researchers have come to realize that large impacts can have global climatic consequences for Earth (see Chapter 13). For example, a strike in the ocean would throw enough water into the stratosphere to initiate a period of greenhouse warming. The evaporation of more ocean water and release of carbon dioxide would ensue, possibly raising temperatures in the lower atmosphere and the uppermost oceans by more than 10° K. Such a pulse of heat could explain why certain species were exterminated at the end of the Cretaceous, while others living in protected environments (deep in the ocean or at high latitudes) survived.

Enormous quantities of atmospheric nitrogen would have been consumed by the impact fireball (*Figure 8*), creating nitrogen oxides and strong nitric acid rain that greatly increased the acidity of soils, lakes, and shallow ocean waters. Sulfur in the Yucatán target beds would also have added a strong sulfuric acid component to the fallout. Carbon soot found in the boundary clay suggests that the continents were engulfed in widespread forest fires. Vegetation killed by a combination of darkness and cold — or by a subsequent combination of heat, acid rain, and possibly other causes — may have provided an abnormal supply of tinder that fueled this conflagration.

Paleontological evidence shows that flowering plants largely disappeared, at least locally, for several thousand years. At the same time, conditions in the surface layer of the ocean also were inimical to many forms of life. Temperatures fluctuated widely and biological productivity plummeted, probably in response to

dramatically changing climatic conditions and acid rain. Kenneth Hsü has referred to this situation as the "Strangelove ocean." Various species of marine organisms gradually died off over perhaps several tens of thousands of years. The altered marine ecology, with various organisms disappearing forever, may have affected the survivors as significantly as did the physical changes in their environment.

Another episode of mass biological extinction occurred about 35 million years ago, late in the Eocene epoch, which also appears to be linked to large impacts at that time. Paleontologist Gerta Keller and her colleagues have found tiny glass spherules of impact melt (tektites) in layered deep-sea sediments that suggest at least three large impacts occurred over an interval of about 0.5 to 1.0 million years. A strong iridium anomaly is associated with the middle spherule layer. Besides the layers with impact glass, geologists are aware of a number of impact craters that were formed about 35 million years ago. Apparently the Earth was subjected at that time to a pulse of bombardment that is most readily explained as a mild comet shower.

There are hints of other closely timed impacts in the geologic record. Craters younger than 5 million years form the largest peak in the distribution of ages for large impact structures. This apparent surge may be due, in part, to the fact that it is easier to recognize and date young, fresh-looking craters than old, eroded ones. However, two known large craters and several strewn fields of impact glass were all formed between 700,000 and 1.1 million years ago. One of the latter, the Australasian tektite field, is among the largest known. It may have been produced by two large impacts separated in time by about 100,000

Figure 8. One day, 65 million years ago, a large asteroid (or possibly a comet) about 10 km across approached the Earth. Striking just north of the Yucatán peninsula, it formed a 180-km-wide crater known as Chicxulub. One likely consequence of this catastrophic event was the eradication of much of the flora and fauna then existing on Earth — probably including the dinosaurs.

years. The close spacing in time of these impact events, if not a statistical fluke, suggests that a short but significant increase in the number of Earth-crossing bodies occurred at the dawn of human existence.

Noble-metal enrichments have now been discovered in stratigraphic association with each of the four global mass extinctions of the last 100 million years. These took place about 11, 35, 65, and 91 million years ago. As noted, we have solid evidence that large impacts coincided with the Cretaceous-Tertiary and late Eocene events. The odds are good that at least three and possibly all four of these mass extinctions are somehow linked to single or multiple impacts.

THE MOON: WINDOW TO THE PAST

Unlike the Earth, where erosion, deformation, and renewal of the crust tend to obscure the effects of impact, the surface of the Moon preserves a pristine record of bombardment by solid bodies extending back several billion years. Because of the absence of an atmosphere, even microscopic particles strike the Moon at high speed, producing countless tiny craters on exposed rocks *(Figure 9)*. Larger particles pound and churn the surface into a layer of ground-up rocky debris, the *regolith*, that

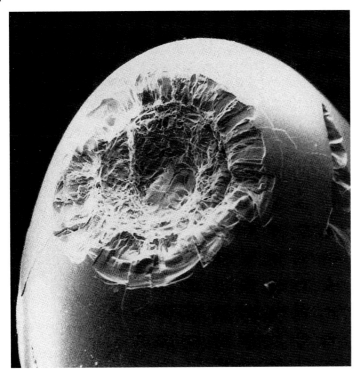

Figure 9. Less than 1 mm wide, a spherule of glass from a lunar sample bears a tiny crater on its tip. Note the pattern of radial and concentric fractures, a miniature of that found in craters many kilometers across.

Figure 10. The ray pattern of the crater Tycho, 85 km across and situated in the Moon's southern hemisphere, is strikingly distinct.

Figure 11 (right). Lunar Orbiter 5 photographed Tycho under oblique lighting that emphasized its complex topographic detail. Clearly visible are hummocky debris on the rim, terraced walls, a prominent central peak, and the frozen pool of impact melt on the crater floor.

blankets nearly all of the Moon. This layer averages about 3 m thick where it covers lava flows some 3 billion years old. Most of the Moon's surface has been darkened by accumulation in the regolith of black, impact-produced glass.

Asteroids and comet nuclei striking the Moon have produced conspicuous craters. The youngest of these are surrounded by bright deposits of freshly excavated rock that extend radially outward in discontinuous bright streaks called *rays*. One of the youngest and perhaps the most spectacular of the Moon's large ray craters is Tycho, 85 km in diameter *(Figures 10, 11)*. A continuous deposit of ejecta surrounds Tycho beyond its rim crest for another 85 km on average. The rays form a great asymmetric splash whose fingers can be traced to distances of up to about one-fourth of the lunar circumference.

Within the rays and the outer part of the continuous ejecta deposit are abundant small *secondary craters*, formed by chunks of rock and clots of debris flung out of Tycho. Just outside the crater rim's crest, which rises 1 km or so above the average level of the surrounding terrain, ejected debris flowed down portions of the outer slopes in viscous, lobate masses. Some of the most fluid, shock-melted material (termed *impact melt*) filled small depressions with smooth, dark deposits resembling frozen ponds or lakes. The terraced crater walls arose when great slabs of rock slumped inward during collapse of the initial cavity. Shock-melted rock pooled on the terraces and created frozen rivulets several kilometers long down to the main crater floor, itself filled with a deep layer of once-molten material. A prominent central peak, formed by the inward and upward flow of material as the crater walls slumped, rises about 2 km from this now-solid "lake bed."

All of the events that led to Tycho's present appearance took place rather quickly. Some, like the formation of impact melt and the excavation of the initial cavity, happened within a few minutes. Slumping of the crater walls and formation of the central peak may have taken an hour or so. The rain of ejecta, formation of secondary craters, and emplacement of impact melt

probably occurred within a few hours after the impact. The times involved for some of these events vary at other impact sites, taking proportionally longer as the crater size increases.

Other large ray craters on the Moon closely resemble Tycho. However, craters smaller than about 20 km in diameter generally have smooth walls devoid of terraces. Evidently, up to this size the bedrock walls are strong enough to prevent collapse in the weak lunar gravity (which is about one-sixth that of the Earth). Where terraces are missing, central peaks are absent as well. Craters 7 to 20 km across generally have fairly smooth, level floors, which probably consist of impact breccia that settled from the walls of the initial cavity, as at Meteor Crater. Pools of frozen impact melt are not easily recognizable in or around these smaller craters. Below diameters of about 7 km, lunar ray craters have a simple bowl shape.

Certain features are characteristic of all ray craters. The exterior slopes of each rim are marked by a pattern of small hills or hummocks (again resembling Meteor Crater) and by lower-lying ridges that are roughly radial in orientation. Abundant small secondary craters invariably occur near the outer limit of the ejecta blanket and in the rays. As small meteoroids continued to pound the lunar surface, the rim deposits and floors of old craters gradually darkened and the rays disappeared. About two-thirds of the large craters found on the Moon's great lava plains (the maria) lack rays, but hummocky rim deposits and surrounding swarms of secondary craters confirm their impact origin.

The 1994 mission of the Clementine spacecraft acquired just under two million digital images of the Moon at visible and infrared wavelengths. This bonanza makes possible the global mapping of rock types and has enabled the first thorough geologic investigation of the far side and of the poles (see Chapter 10). Clementine's imaging and altitude measurements confirmed that the far-side basin called South Pole-Aitken is about 2,500 km in diameter and averages 12 km in depth — the biggest and deepest basin in the solar system. Evidence suggests that the transient cavity from the huge basin-forming impact must have been at least 1,000 km across. Such an impact would have excavated the entire crust and incorporated large amounts of the lunar mantle from depths of 120 km with rocks that make up the floor of the basin. The south pole itself is in a deep irregular topographic depression, much of which is permanently shadowed and the location of recently discovered deposits of water ice. Near the south pole lies Schrödinger (320 km in diameter), one of the two youngest and least modified of the great multiring impact basins on the Moon.

Our inventory of lunar samples comes from only a small number of landing sites. However, thanks to isotopic dating techniques, the ages of these samples can be determined very accurately. Meanwhile, geologic units at the landing sites can be correlated in relative age with similar units in other areas using orbital photography and the relationships of superposition of the mapped units. Age correlations can also be made by matching different units' areal densities of small craters (the older a given surface, the more superposed small craters it will have of any specific size).

Using these methods, Don E. Wilhelms has mapped all craters at least 10 km across that formed on the lunar nearside in the last 3.3 billion years, after many of the Moon's maria had

appeared. He concludes that, since then, an average of 10 craters in this size range have been produced per million square kilometers per billion years. The corresponding rate for craters at least 20 km in diameter is 2.4. Any extrapolation of these rates to Earth must take into account the two bodies' differences in gravity, which affects sizes of the initial cavities and how readily their walls slump (thus enlarging the crater diameters). Additionally, Earth's stronger gravitation draws in the bodies in its vicinity better and increases their collision speed. When all these corrections have been made, Wilhelms's lunar values yield rates on Earth of 12.5 and 2.3 craters per million square kilometers per billion years (for craters at least 10 km and 20 km across, respectively). These are about half what we find either from the present flux of asteroids and comets or from the Earth's geologic record of impact over the last several hundred million years.

Is this discrepancy real, or is it due to uncertainties either in our calculations or in the statistics of Earth's cratering record? Quite possibly, the difference *is* real — collisions on the Earth may well have occurred twice as often during the last several hundred million years as during the previous 3 eons. We believe the long-term flux of asteroids derived from the main belt has remained steady to within about 10 percent, so variations in their arrival rate are probably not the cause. Rather, we must look to the comets to explain a doubling of the cratering rate in

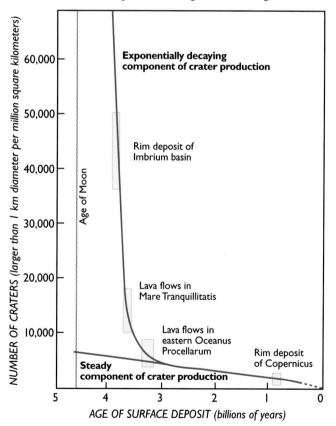

Figure 12. Lunar surfaces of different ages exhibit different crater densities. The rapid cratering rate during the late heavy bombardment fell off dramatically between 3.9 and 3.3 billion years ago, giving way to a slower, steady rate of crater production. This dramatic falloff is reflected in the varying ages and crater densities determined at Apollo landing sites (small rectangles, whose dimensions correspond to uncertainties).

recent geologic time. Perhaps a few relatively strong comet showers caused the average number of comets passing near the Earth to increase, or maybe the "background" rate of Earth-crossing comets has risen. Possible sources of these near-Earth comets will be discussed later.

In tracing the geologic record of the Moon, we find that collisions occurred very frequently prior to the eruption of the mare lavas but dropped off rapidly between 3.9 and 3.3 billion years ago *(Figure 12)*. For example, the crater density found on a great sheet of ejecta surrounding the Imbrium basin, formed 3.85 billion years ago, is six times greater than on lava plains that erupted only 550 million years later. The abrupt decrease in the intervening impact rate can be mimicked mathematically by combining one cratering rate that decayed exponentially (a 50-percent decrease every 100 million years) with another that remained steady with time. According to this model, the rate of formation of craters dropped off by a factor of 35 between 3.85 and 3.3 billion years ago. An even steeper decline apparently occurred between the formation of the Nectaris basin, estimated at 3.92 billion years ago, and the time of the Imbrium basin. Evidently, 11 of the Moon's largest impact basins formed in this one 70-million-year interval. Moreover, about twice as many smaller craters appeared in that brief span than in the billions of years that followed it.

From these observations we can infer the existence of two general populations of Earth-crossing bodies. One group became depleted with time as they collided with the planets or were ejected from the solar system. The second group corresponds to the Earth-crossing objects we observe today — a population that is apparently renewed from other regions of the solar system at approximately the same rate that it is lost. Without this renewal, the "steady" population would quickly dwindle away.

The first, rapidly decaying population may have originated in at least three ways: (1) the impacting bodies were a remnant of the principal batch of small *planetesimals* from which most of the Earth accreted; (2) they were injected into Earth-crossing orbits from the asteroid belt, perhaps following major collisions there; (3) a large planetesimal of Uranus or Neptune was perturbed into an Earth-crossing orbit and became tidally disrupted (fragmented) during a close approach to one of the terrestrial planets. The first two sources might suffice to account for the population's 100-million-year half-life. Only the sudden injection of objects on collision-prone orbits, as in the third case, would satisfactorily account for the apparently abrupt drop in the cratering rate just after 3.92 billion years ago.

The episode of rapid cratering near 3.9 billion years ago has been referred to as the *late heavy bombardment* (LHB). One view holds that the heaviest pummeling occurred as the Moon formed about 4.5 billion years ago. The bombardment and cratering declined thereafter in a generally steady way for about 1¼ billion years. From this standpoint, the LHB simply represents the final vestiges of planetary accretion. All evidence of the Moon's earlier, more intense bombardment has been lost or obscured by emplacement of the final large basins and their regional deposits of ejecta.

However, the rapid falloff of the cratering rate between 3.92 and 3.85 billion years ago argues instead that a discrete pulse occurred, as first suggested by Fouad Tera, Dimitri Papanastassiou, and Gerald Wasserburg. In our view, it seems most likely that multiple pulses were superimposed on a general decline in cratering rate and that the LHB was the last of these pulses. The best explanation for pulses is the breakup of large objects on Earth-crossing orbits, either by collision or tidal disruption.

CRATERING ON MERCURY AND VENUS

Comets and asteroids appear to excavate craters on Mercury at a rate roughly comparable with that on Earth *(Table 1)*, though the proportions due to various classes of impactors differ. Fewer than one-fifth of the known asteroids traversing Earth's orbit pass inside Mercury's as well. However, the confluence of their orbits occurs in the relatively small volume of space between Mercury and the Sun, and this increases their likelihood of impact. Moreover, about 10 percent of the asteroids that can hit Mercury should occupy orbits that lie entirely inside the Earth's. Finally, an object nearing Mercury travels faster, because of the Sun's gravitational acceleration. When it strikes the planet, the impact energy is greater and a proportionately larger crater is formed (Mercury's relatively weak surface gravity compared with Earth's is also a factor). Taken together, these effects partly offset the markedly fewer asteroids that pass in the vicinity of Mercury and yield a cratering rate per unit area of surface about half that on the Earth.

Roughly one-third of the comets passing inside the Earth's orbit also penetrate inside Mercury's average aphelion (the point in its orbit farthest from the Sun). The same orbital congestion that increases the chance of asteroidal impact also holds for comets, resulting in a cratering rate from cometary objects about 30 percent higher for Mercury than for the Earth.

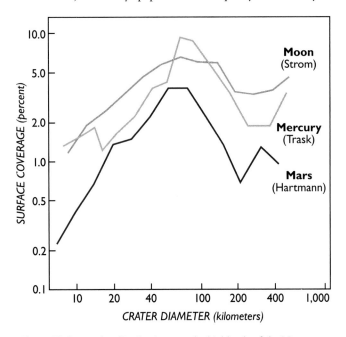

Figure 13. Crater size distributions on the highlands of the Moon, Mercury, and Mars (authors whose data have been used are named by each curve). The coverage on all three bodies peaks for craters with diameters of 40 to 100 km, and the distributions' similar shapes suggest that the same population of impacting projectiles produced most of the craters on the ancient highlands of all three planets.

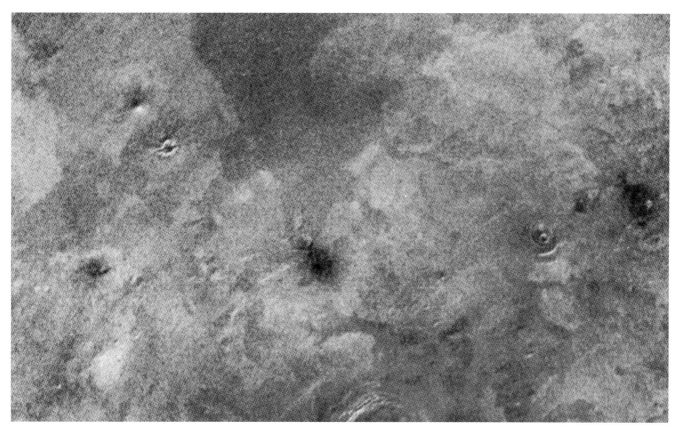

Figure 14. **The atmosphere of Venus shields the planet from direct strikes by small incoming projectiles. Instead, the would-be impactors disintegrate before reaching the ground, sometimes leaving radar-dark "powder burns" on the surface.**

Since the same basic classes of objects strike the Moon, Earth, and Mercury, some idea of Mercury's cratering history can be deduced from the impact records of the other two. For example, one would expect that, if the flux of crater-forming objects near the Earth changed, at least in recent eras, a corresponding change should have occurred for Mercury. We calculate that the present production of craters greater than 20 km diameter on this innermost planet is about two-thirds of that on the Moon. This proportion may have remained nearly the same for several billion years. It was thus no surprise when Mariner 10 revealed in 1974 that Mercury has a heavily cratered surface very similar in appearance to the Moon's (see Chapter 7). Fresh craters on Mercury look remarkably like their counterparts on the lunar maria: the youngest ones have rays, but older ones do not. The most distinctive difference exists in the patterns of secondary craters. These lie closer to the rims of their primaries on Mercury than on the Moon, because Mercury's higher surface gravity confines crater ejecta to shorter ballistic ranges.

Mercury's surface, like the Moon's, bears a cratering record that probably extends far back in time. Craters 20 km across are about six times more abundant on the smooth plains of Mercury than on the 3.3-billion-year-old lunar maria. This suggests that the smooth Mercurian plains were in place about 3.9 billion years ago, somewhat earlier than when the Imbrium basin formed on the Moon. In terms of packing density and size distribution, Mercury's highland craters match those of the lunar highlands fairly closely *(Figure 13)*, which implies that the same population of bodies was involved. If the orbital distribution of the impactors resembled what we now observe among the Earth-crossing asteroids, the cratering rates on Mercury may have been about half the rate on the Moon during the late

heavy bombardment. The enormous Caloris basin, probably the last giant impact structure created on Mercury, exhibits about three-fourths the density of craters as the Nectaris basin on the Moon and probably predates it slightly.

The impact record of cloud-veiled Venus remained a mystery until very recently. During the 1970s, low-resolution images from Earth-based radar telescopes displayed many circular features on the planet that were thought to be large craters and impact basins. From these observations, geologists concluded that the density of large craters was somewhat similar to that of the heavily cratered regions of the Moon and Mercury. It was a puzzling result. Venus is similar to the Earth in size, mass, and presumably its heat budget, and the planet's surface is hundreds of degrees hotter than Earth's. Consequently, if Venus's geology were truly Earthlike, all that heat should have thinned its lithosphere (the brittle upper surface layer) and thus made it susceptible to deformation during the convective stirrings of the interior. How, then, could ancient craters be preserved?

The mystery was resolved by the Soviet spacecraft Veneras 15 and 16, and later by the U.S. orbiter Magellan. Their radar images of Venus reveal roughly 1,000 impact craters. The smallest of these is 1.4 km in diameter (there are none as small as Meteor Crater); the largest, Mead, is 285 km across. Craters appear to be spread rather randomly over the planet. Their spacing is far sparser than in the lunar highlands but remarkably similar to the density found on the American and European continental shields.

Inner-planet Cratering Rates

Impacting object	Mercury	Venus	Earth	Moon	Mars
"Venus-crossing asteroids"	0.4	0.4			
Earth-crossing asteroids	1.5	3.1	3.5	3.5	1.2
Mars-crossing asteroids					6?
Comets	1.6	(1.3)	1.2	1.5	0.6
All objects	3.5	3.5	4.7	5.0	8

Table 1. Estimates of the current cratering rates on the inner planets caused by collisions with asteroids and comets. Each entry is the average number of craters larger than 20 km in diameter produced on a surface area of 1 million square kilometers every billion years. These rates are uncertain by a factor of about two. *Venus-crossing asteroids* cross the orbit of Venus (and in some cases Mercury's) but not the orbit of Earth. Only one such candidate has been discovered: 1998 DK_{36}. *Mars-crossing asteroids* do not cross the orbit of any other terrestrial planet. The entry in parentheses refers to the theoretical crater production due to comets if Venus had no atmosphere.

Magellan data revealed other patterns called "splotches." These do not resemble impact craters but may instead be a surface manifestation of the breakup (in Tunguskalike manner) of large asteroids and comets just before they strike the surface of Venus. There are more than 400 such structures, none of which surround actual craters *(Figure 14)*. Some exhibit small dark halos with bright surface markings in the center, while others have no central marking, and a few are bright with dark centers. Also evident in Magellan radar images are 57 craters surrounded by parabolic-shaped "hoods." These parabolas always open to the west, with a crater just inside or west of the apex. It appears that they form when fine material ejected from the craters interacts with the planet's prevailing east-to-west winds at 50 km in altitude (see Chapter 8).

Almost precisely half of the known Earth-crossing asteroids are Venus-crossing as well. However, the cratering rate must be adjusted for Venus's relative proximity to the Sun, its slightly weaker surface gravity (relative to Earth's), and the likely existence of as-yet undiscovered objects that remain entirely inside the orbit of Earth. When these factors have been accounted for, the calculated asteroidal cratering rates for Venus and Earth are nearly the same *(Table 1)*. Similar calculations show that the cratering rate by cometary impact would be about 14 percent higher on Venus than here — *if* that planet did not have its dense atmosphere.

In fact, the degree to which Venus's atmosphere shields its surface is the single largest uncertainty in estimating the planet's impact history. Craters less than 20 km in diameter are distinctly underrepresented. Above this threshold, the size-frequency counts resemble those for young craters on the Moon. Since comets, rather than asteroids, probably create most of the young craters on the Earth and Moon larger than 60 km in diameter, we suspect that the largest craters on Venus have been produced by comet hits. As explained more fully in Chapter 8, the paucity and apparently random placement of craters on Venus means that none of the planet's surface is geologically ancient. Instead, all evidence suggests that virtually the entire planet was completely resurfaced by volcanic flows about 500 million years ago.

THE CRATERING RECORD OF MARS

Unlike Venus, Mars has a surface that abounds with impact craters. However, many have been significantly modified by erosion, sediment deposition, or volcanism; those that have not resemble, in most respects, craters on the Moon and Mercury. One rather remarkable difference is that the material thrown from most large, fresh Martian craters appears to have hugged the ground as it flowed radially outward to form lobes, rather than being deposited by ballistic ejection (see Chapter 11). This fluidlike behavior may indicate that the target material contained ice, which melted or vaporized when struck, or perhaps atmospheric gases became trapped beneath the outward-moving ejecta blanket.

Mars is a geologically diverse world. Many of the Martian plains are only sparsely cratered, while others are pitted about as heavily as the lunar maria. This diversity implies that the Martian surface varies greatly in age from place to place. If the population of Mars-crossing asteroids has been steady with time, some of the most ancient plains should be about 3 billion years old. However, we suspect that the actual number of Mars crossers has dwindled steadily over the eons. Therefore, the Martian cratering rate was probably higher in the past, and some of the more ancient plains may actually be no more than about 1 or 2 billion years old.

The craters in the heavily pummeled Martian highlands, found almost entirely in the southern hemisphere, resemble in both abundance and size distribution the highlands of the Moon and Mercury. As with Mercury, the lunarlike size distribution suggests that one population of impacting bodies was responsible for both the Moon's late heavy bombardment and most of the highland craters on Mars. If these bodies had orbits like those of modern-day Earth-crossing asteroids, their cratering rate on Mars would have been only 40 percent of their rate on the Moon during the late heavy bombardment. On the other hand, if enough of them originated in the outer solar system, Mars and the Moon could have been subjected to about equally intense battering at that time. Either way, the Martian highlands bear a record of impact scars that may span nearly 4 billion years.

IMPACTS IN THE OUTER SOLAR SYSTEM: LESSONS LEARNED FROM COMET SHOEMAKER-LEVY 9

The Voyager missions have greatly extended our firsthand knowledge of the outer solar system, where the record of collisions is written in ice. It is the history of icy planetesimals and comet nuclei, of swarms of icy shards striking the frozen crusts of outer-planet satellites at high speed. In geologically recent times, the bombardment has been dominated almost entirely by comets. An important subset of these is the so-called Jupiter family of short-period comets. They were once long-period objects that became captured by Jupiter after a succession of close encounters with it and perhaps other outer planets. An orbital period of less than 20 years, typical for the Jupiter family, arises only after an unlikely series of momentum-robbing passes ahead of Jupiter or an even less likely single pass at extremely close range.

The paths of comets near the massive outer planets, and especially near Jupiter, are strongly focused by the planet's gravitational field. Such objects are thus accelerated as they "fall" inward, and their velocities increase significantly. Consequently, the cratering rate on Io, the innermost Galilean satellite, is about four times that of Callisto, the outermost one. Such gravitational focusing can endow an incoming projectile with enough kinetic energy to wreak total havoc on smaller satellites. At Saturn, for example, the moons Mimas, Enceladus, Tethys, and Dione probably were all "destroyed" by large impacts several times during the late heavy bombardment, only to reassemble themselves from the rings of debris that would have been left circling the planet. Indeed, the beautiful Saturnian ring system could well be the remains of one or more shattered satellites that were prevented from reaccumulating by the planet's strong gravity (see Chapter 16).

In recent years we have come to realize that asteroids and comets are unlikely to be solid, completely cohesive chunks of rock or ice. Collisions with one another may have left their interiors heavily fractured, or they may have accumulated from lumps of matter that never fused into a single firm mass in the first place. This notion of a solar system filled with flying "rubble piles" has interesting consequences for objects that pass very near a sizable planet (including Earth). Computer simulations show that they can become distorted or even torn apart during such an encounter.

We do not yet know how often these tidally induced breakups occur, or how they might skew our cratering statistics. However, the evidence that such fragmentation does happen is mounting. For example, Earth has several large double craters. A cleanly separated, side-by-side crater pair cannot have occurred after a single object broke in two during its atmospheric passage. Instead, the object must have first become fragmented when it swung by our planet at close range, then collided with Earth on a subsequent pass. Another line of evidence is that comets sometimes split as they pass near the Sun.

The reality of tidal disruption was driven home by the discovery of Comet Shoemaker-Levy 9 (D/1993 F2), which had been captured on photographic plates taken on 23 March 1993. From the outset, this object looked strange: it had a bar-shaped nucleus surrounded by a series of tails and two dusty "wings" that extended from both ends of the bar. In the weeks that followed, powerful telescopes revealed a perfectly straight train of nuclei *(Figure 15)* that eventually numbered 21. It soon became apparent that the comet was in orbit around Jupiter. Eventually, calculations showed that the comet had come apart when it passed 21,000 km from Jupiter's cloud tops on 8 July 1992 *(Figure 16)*.

A few previous comets had been known to orbit the planet temporarily; one has even slipped in and out of Jovian orbit more than once. Nor was a passing comet's breakup by Jupiter unprecedented. What set Comet Shoemaker-Levy 9 apart was that its many pieces were doomed to strike the planet at 60 km per second. Strike they did, in July 1994, in a succession of dramatic collisions that created titanic fireballs (visible from Earth in the infrared) and stained Jupiter's upper atmosphere with huge, dark clouds that lingered for many months afterward *(Figures 1,17)*. The unprecedented event was watched and recorded by observatories worldwide, the Hubble Space Telescope, and even the Galileo spacecraft (then en route to Jupiter).

Since then, observers and theorists have struggled to understand what, exactly, happened. Unfortunately, all the impact sites were behind the limb of Jupiter as seen from Earth, and by

Figure 15. Within days of its discovery on 23 March 1993, Comet Shoemaker-Levy 9 became the telescopic target of astronomers worldwide. Its "string of pearls" appearance is evident in this false-color image acquired four days after discovery. Fifteen months later the comet's fragments struck Jupiter.

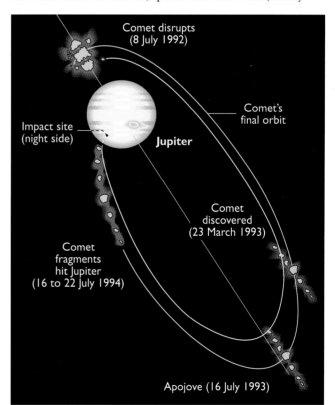

Figure 16. When the comet's nucleus skirted past Jupiter in July 1992, tidal stress (unequal pulls on its near and far sides) tore the object apart. These fragments were just about their farthest from Jupiter when discovered.

the time they rotated into view the fireballs had died away. Moreover, the event created a range of complex phenomena that were not anticipated. For example, it was hoped that the nucleus fragments would release their enormous kinetic energy deep enough in the Jovian clouds to dredge up normally unobservable atmospheric constituents. However, disentangling the spectroscopic signature of material from the comet and from Jupiter's deep atmosphere has proved difficult.

Scientists continue to debate what we learned from the series of impacts. The depth of penetration into Jupiter's atmosphere,

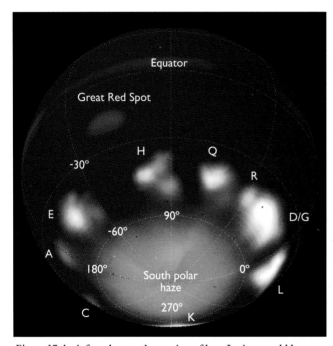

Figure 17. An infrared, comet's-eye view of how Jupiter would have looked after all fragments from Comet Shoemaker-Levy 9 reached their fateful end. At a wavelength of 2.3 microns, Jupiter itself appears dark; the impact sites are bright because fine particles deposited high in the atmosphere are reflecting sunlight. Infrared-bright aerosol haze is also present continuously over the planet's polar regions.

Figure 18. The Jovian moon Callisto displays a remarkably straight, 620-km-long chain of impact craters collectively named Gipul Catena. Thirteen such chains have been identified on Callisto (12 on its Jupiter-facing hemisphere) and three on neighboring Ganymede.

plume and fireball formation, electromagnetic effects, and chemical synthesis are all being studied. Even the size and cohesiveness of the original comet and of the individual fragments remain uncertain. Zdenek Sekanina suggests that the original nucleus was about 10 km across before its breakup. However, tidal physics indicates a diameter between 1 and 2 km. If the precursor object was $1\frac{1}{2}$ km across, then each fragment was perhaps 300 to 500 m in size. Assuming it had a density near 0.5 g/cm^3 (a value by no means certain), then a typical fragment delivered 10^{27} ergs of kinetic energy to Jupiter — the explosive equivalent of nearly 25,000 megatons of TNT.

Notably, a number of crater chains have been found on the Jovian moons Ganymede and Callisto that almost certainly were formed under similar circumstances *(Figure 18)*. These chains mark where ancient comets ceased to exist soon after being pulled apart by Jupiter's gravity. Given the cratering rate and the number of crater chains observed on Ganymede and Callisto, a $1\frac{1}{2}$-km-wide comet probably strikes Jupiter about once per century. Moreover, if the frequency of strikes increases as the inverse square of diameter, then 500-m-wide comets must collide more often, almost once per decade. If this is true, how have the atmospheric scars of other impacts escaped detection throughout four centuries of telescopic observation? Perhaps the comet and its fragments were larger (and thus rare) after all. Some orbital analyses imply that the precursor was up to 15 km across. Alternatively, small comets may not be as numerous as we suspect. This paradox is just one of the many mysteries left in the wake of Comet Shoemaker-Levy 9.

CURRENT SOURCES OF IMPACTORS

The events of July 1994 verified an important tenet of planetary science: impacts happen. The challenge is to find out how often they happen and what kinds of objects are involved. While scientifically we are curious about the collision rates throughout the solar system, the information with the most practical long-term application is of course the likelihood of a major impact here on Earth.

Of the small bodies that have been found crossing the orbits of the planets, objects of asteroidal appearance are most common in the vicinity of the Earth and Mars, while comet nuclei

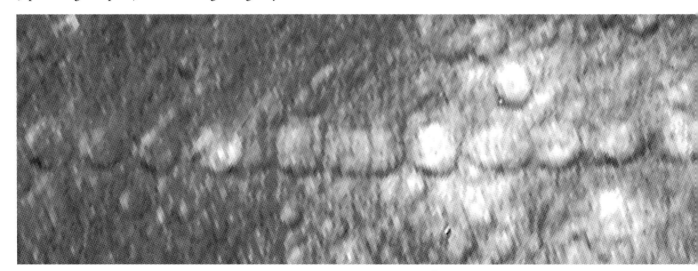

predominate at and beyond the orbit of Jupiter. Traditionally, we designate a small body a "comet" if it exhibits an envelope, or coma, of gas and dust. However, the dividing line between comets and asteroids has become blurred (see Chapter 23). For example, Comet 133P/Elst-Pizarro is active but has an asteroidal orbit; the asteroid 1996 PW is inactive but occupies a nearly parabolic comet-like orbit. Thus, a number of coma-lacking "asteroids" that cross the orbits of giant planets are very probably inactive comets.

Roughly 500 of the numbered asteroids, about 5 percent of the ones with well-determined orbits, are known planet crossers. One of these, 944 Hidalgo, crosses the orbits of both Jupiter and Saturn. It travels on a path similar to the orbits of certain short-period comets, which characteristically are ejected from the solar system after close passes by Jupiter on time scales of less than a million years. About a half dozen other Jupiter crossers have been discovered, and these also cross Mars's orbit. The ones whose colors have been measured, including Hidalgo, are similar in color as well as in orbital characteristics to weakly active comets. Another "asteroid," 2060 Chiron, crosses the orbits of Saturn and Uranus. Discovery in recent years of a coma implies it is a large comet nucleus.

Almost all other known planet-crossing asteroids make close approaches to one or more of the terrestrial planets. About half of these objects do not overlap any planet's orbit at present, but over time their orbits and Mars's orbit evolve in ways that lead to slight overlap and possible collisions. Most of these "shallow Mars crossers" can be distinguished from other main-belt asteroids only by elaborate theoretical investigation of their motion. Combining these studies and the results of photographic surveys, we calculate that there are about 20,000 Mars crossers brighter than absolute visual magnitude 18. (*Absolute magnitude* is how bright something would appear if placed 1 AU from both the observer and the Sun.) Each of these objects has about a 50 percent chance of ultimately hitting a planet or being thrown entirely out of the solar system. Most are probably fragments of bodies stranded in very shallow Mars-crossing orbits after the planets formed. At present, Mars is being bombarded mostly by a group of asteroids that do not cross the Earth's orbit (Table 1).

By mid-1998, astronomers had discovered more than 500 asteroids that approach or cross the orbit of Earth (*Table 2*). Objects that currently come within 1.3 AU of the Sun but do not pass inside 1.017 AU (the maximum outward excursion of Earth's orbit) have been named *Amor* asteroids. Those with orbits larger than Earth's but with perihelia inside 1.017 AU are *Apollo* asteroids. In addition, *Aten* asteroids have been found whose orbits have semimajor axes smaller than 1 AU. Most of the Aten orbits lie inside the orbit of Earth and overlap Earth's orbit at aphelion. Small and intrinsically faint, most known members of these three asteroid classes are observable only when relatively close to the Earth. The majority are less than 2 km in diameter. The closest known passage of any such object to Earth occurred in December 1994, when the previously unknown asteroid 1994 XM₁ passed a scant 104,700 km away — less than one-third the Moon's distance (*Table 3*).

Based on their discovery rate during systematic photographic surveys, the estimated total of Earth-crossing asteroids brighter

Populations of Near-Earth Asteroids

Asteroid type	Number discovered	Percent discovered	Estimated population
Atens	30	(40)	80 ± 50
Apollos	263	(40)	700 ± 300
Amors	209	(70)	300 ± 150
All near-Earth asteroids	502	47	1,080 ± 500

Table 2. **By mid-1998, astronomers had discovered several dozen Earth-crossing asteroids brighter than absolute visual magnitude 18. The diameter of an object at this limiting magnitude generally ranges from 0.9 to 1.7 km, depending on the object's reflectivity.** *Percent discovered* **values are based on the rate of discovery in systematic sky surveys and are uncertain.**

The Closest Asteroid Encounters

Designation	Miss distance (AU)	(km)	Flyby date (UT)	Diameter (m)
1994 XM₁	0.0007	104,700	9.8 Dec. 1994	9
1993 KA₂	0.0010	149,600	20.9 May 1993	6
1994 ES₁	0.0011	164,600	15.7 Mar. 1994	7
1991 BA	0.0011	164,600	18.7 Jan. 1991	7
1995 FF	0.0029	433,800	27.2 Mar. 1995	18
1996 JA₁	0.0030	448,800	19.7 May 1996	220
1991 VG*	0.0031	463,800	5.4 Dec. 1991	6
1989 FC**	0.0046	688,200	22.9 Mar. 1989	280
1994 WR₁₂	0.0048	718,100	24.8 Nov. 1994	140
1937 UB***	0.0049	733,000	30.7 Oct. 1937	900

Table 3. **Diameter estimates assume a reflectivity (albedo) of 14 percent, but these objects' actual albedoes — and thus their true diameters — are very uncertain. *This object may have been a returning piece of artificial space debris. **Now named 4581 Asclepius. ***Also known as Hermes.**

than absolute visual magnitude 18 is about 1,100 (*Table 2*). Thus, the Earth resides within an asteroid swarm, though one about a thousand times smaller than the collection of main-belt asteroids of comparable size. If the orbits of the known Earth crossers are representative of the entire population's, about five objects brighter than absolute magnitude 18 strike the Earth every million years, on average. A comparable number collide with Venus, and some collide with Mars, Mercury, or the Moon. The typical dynamic lifetime of these maverick bodies is only a few million years, though some probably last much longer; over time, about a third are ejected from the solar system.

Clearly, a large majority of the Earth-crossing asteroids cannot have survived in their present orbits over most of geologic time. Yet the cratering records of the Earth and Moon suggest that the population has been in approximate equilibrium over the last 3 billion years. Apparently these objects arrive from other regions of interplanetary space at about the same rate at which they are swept up or ejected. A fraction were once shallow Mars crossers that were deflected inward during close encounters with Mars. Most have come chiefly from the main asteroid belt, where catastrophic collisions among asteroids create large numbers of fragments (see Chapter 25).

Within the last few years, four asteroids have been observed to have moons of their own; one of these is Dionysus, an Apollo object. This implies that such asteroids may be collisionally evolved rubble piles. Some of the fragments then become subject to strong resonant perturbations, such as when their orbital periods are simple fractions of the orbital period of Jupiter. Over time, perturbations and encounters with Mars combine to redirect some of these toward Earth's vicinity. In addition, some Earth-crossing asteroids may be extinct comets that have been driven into very short-period orbits by close encounters with Jupiter and by the rocketlike thrust provided by their own escaping gases. Comet 2P/Encke, for example, has been propelled into an orbit like those of some Earth-crossing asteroids.

Active comets also play an important role in impact cratering on the Earth and the other terrestrial planets. Each year, on average, astronomers find about three comets with nearly parabolic orbits that pass within 1 AU of the Sun. Most of these are from the Oort cloud, a gigantic, diffuse cloud of comets surrounding the Sun at an average distance of about 40,000 AU (see Chapter 5). A more important impactor source, at least for the terrestrial planets, is the Jupiter family of short-period comets.

Combining the discovery rate of bright comets with the brightness of dormant comet nuclei observed far from the Sun, we estimate that about 30 comets with solid nuclei brighter than absolute magnitude 18 (in blue light) pass inside the orbit of the Earth each year. These appear to be coated with dark material left behind when the ices sublimate. A typical dark comet nucleus of magnitude 18 has a diameter of 2½ km. Given the present flux, comet nuclei of this size and larger should collide with the Earth about once every 10 million years, on average. On the same basis, we expect comet nuclei at least 10 km in diameter to hit the Earth about once every 100 million years, and to pass within 4.7 million km of the Earth once every 100 million years. (This is the miss distance of Comet IRAS-Araki-Alcock, C/1983 H1, which radar observations showed to have a nucleus with a mean diameter of 9.3 km.)

Today small comet nuclei pass Earth's vicinity about one-tenth as often as asteroids of the same size do, but this rate cannot have remained constant over time. During intervals of tens to hundreds of millions of years, stars passing near the solar neighborhood have probably produced comet "showers" of varying intensity. The frequency of comet passages through the inner solar system probably changes distinctly over intervals of 10 million years. We cannot determine from comet observations alone how near the present value is to the long-term average. Surveys of large ray craters on the Moon suggest that comets are now coming in at about twice the mean background rate, and that the present flux roughly equals the average (including showers) over the last billion years. There is even a hint that we are now experiencing the tail of a weak comet shower that reached a peak about a million years ago.

TARGET EARTH

It seems probable that, over the past several hundred million years, several comet nuclei at least 10 km across have collided with the Earth. Projectiles of this size produce craters more than 150 km in diameter when they hit, or many large craters if they break up, and eject enormous amounts of material into the atmosphere. The odds are that about two-thirds of such impacts occurred in the ocean, creating giant tsunamis ("tidal waves") and driving huge volumes of water vapor into the upper atmosphere. As previously noted, the collision of large comets or asteroids with Earth may be responsible for a number of mass extinctions recognized in the paleontological record.

A lively debate has arisen as to whether mass extinctions and impact events are periodically distributed in time. From their analysis of mass extinctions in the last 250 million years, paleontologists David Raup and John Sepkoski suggested in 1984 that such events recur regularly, about once per 26 million years. Their conclusion sparked a flurry of papers in which various astrophysical mechanisms that might produce periodic comet showers were explored. Two studies suggested that the Sun has a faint undiscovered companion star that revolves on a highly eccentric orbit with a period of 26 million years. Another hypothesis assumed that a fairly massive undiscovered planet orbits beyond Pluto and periodically perturbs an unseen disk of comets in its neighborhood. A third theory involved oscillations of the Sun up and down through the massive, central plane of the Milky Way.

Much discussion followed the publication of these intriguing hypotheses. One problem common to all of them is establishing accurate ages for the extinction events, which are generally poorly known. Consequently, the apparent periodicity found by Raup and Sepkoski may not be statistically significant. Similarly, an apparent periodicity found in Earth's impact record, which seems to peak at roughly 30-million-year intervals, may not be significant either. Moreover, there are serious dynamical difficulties with each of the proposed astronomical "clocks."

However, excellent evidence does exist for a correlation between large impacts and some specific mass extinctions. We believe that as many as four or five weak-to-mild comet showers may have occurred in the last 100 million years. (By our reckoning, a "shower" corresponds to a surge in the comet flux at least three times above average.) This is roughly the frequency of mild comet showers expected from the random passage of stars through the Sun's close neighborhood. Within this 100-million-year interval, the times between apparent peaks in the cratering rate, or between the corresponding mass extinctions, appear moderately uniform but no more so than might be expected by chance.

The essential point is that evidence accumulated during the 1980s suggests that the collisions of large objects with the Earth have played a major role in the destruction (and evolution) of life here. When an impact triggers the loss of species and even whole families of organisms, ecological or environmental space opens up for new ones. Various species of mammals, for example, multiplied rapidly after the Cretaceous-Tertiary extinction. It can be argued that the presence of the human race on Earth may be due, in no small part, to chains of events initiated by large impacts about 65 and 35 million years ago.

David Morrison and Clark Chapman have compared the risk of the impact hazard with the risks of other terrestrial calamities (*Figure 19*). Mortality rates due to various natural disasters are clearly higher in an immediate sense for a small area of our world. However, a large impact, even though there may be thousands or millions of years between occurrences, would have the potential for global consequences from which civilization and, indeed, humankind might not survive. Scientists continue to debate at what energy threshold, and under what impact scenarios, a global-scale catastrophe would ensue *(Figure 20)*. Although there is no consensus yet, estimates tend to cluster around impactor diameters near 1 km.

The issue then becomes how best to find them. Virtually all known bodies with the potential to strike Earth have been discovered photographically and more than half have been found by dedicated programs. Photographic searches are usually carried out by looking into the night sky more or less directly opposite the Sun. Asteroids and comets thus seen at opposition are nearly fully illuminated by sunlight and most likely are nearly at their closest to Earth. Discoveries to date have been biased toward the Northern Hemisphere because more searchers are located there, and there is a second bias toward the part of the month surrounding new Moon.

Search programs include an improved Spacewatch program in southern Arizona, the Near-Earth Asteroid Tracking (NEAT) program carried out on the Hawaiian island of Maui in conjunction with the U.S. Air Force, the LONEOS program of Lowell Observatory in Arizona, and Lincoln Laboratory's LINEAR program in New Mexico. These programs promise to surpass the productivity of the old photographic programs in finding what are collectively called "near-Earth objects" (NEOs). Even so, all existing search strategies rely on finding asteroids and inactive comets only when they are passing relatively close to Earth, because they are too faint to be spotted when farther away. Even if a system of large telescopes was built to conduct a dedicated NEO search, we could only expect to discover and catalog about 90 percent of all 1-km-wide Earth-crossing asteroids within a decade. Comets — particularly dark, extinct ones in short-period orbits — will be harder to survey with any measure of completeness and thus potentially pose the greater threat to Earth.

In March 1998, astronomers announced that 1997 XF_{11}, a 3-km-wide asteroid, would come very close to Earth — and perhaps hit us — in the year 2028. Although later calculations showed that 1997 XF_{11} actually poses no such threat, the initial announcement still caused great consternation throughout the world. There was the sudden realization that the nightmare of collision could come true within the lifetimes of many living today, and that humankind is not yet ready to defend Earth against asteroidal or cometary impact.

The lesson from all this is plain: our capabilities to discover and track NEOs need improvement; the ways in which we should respond to the discovery of a possible impactor should be clear; and the methods to mitigate its danger must undergo continuous study and change. Obviously, we very much need to know the extent of the hazard we face, but changing technology, economics, and world politics will play critical roles in how we will meet the challenge of an imminent collision. At present we know of no body that threatens Earth in foreseeable time. However, perhaps 90 percent of all sizable Earth-crossing asteroids remain to be discovered, and long-period comets may appear at any time from any direction. Exploratory research and constant surveillance of the skies must be our answer to the threat of impact.

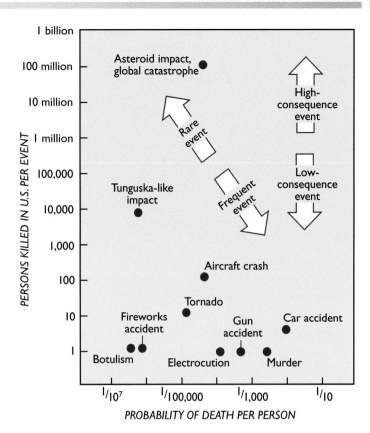

Figure 19. **Averaged over a human lifetime, the chance of being killed by an asteroid or comet impact is about the same as dying in a plane crash. Impacts are much rarer than aircraft accidents, but orders of magnitude more persons would be killed if even a small asteroid stuck Earth.**

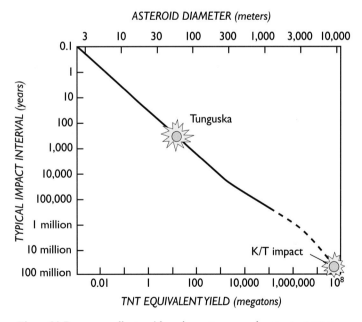

Figure 20. **Because small asteroids and comets are much more numerous than large ones, the chance of Earth receiving a local-scale impact is proportionately greater than the likelihood of a globally catastrophic event. However, since we remain uncertain of the absolute numbers of such objects, a given probability derived from this curve could be uncertain by a factor of three (at lower right) to 10 (at upper left).**

Mercury

Faith Vilas

AS THE CLOSEST planet to the Sun, Mercury guards the secrets of its planetary properties very tightly. Its mean heliocentric distance is only 0.3871 AU, and it completes one orbit in just 88 days. From Earth's perspective, Mercury's tight orbit means the planet can never appear more than 28° from the Sun in the sky. Egyptian, Greek, Roman, and Mayan civilizations all envisioned this elusive, quickly moving "star" as a fleet-footed messenger in their myths and religions (*Figure 1*).

Detailed telescopic observations of Mercury began in the early 1800s. Between 1881 and 1889, Giovanni Schiaparelli conducted the first serious attempt to map the surface features of Mercury. In 1934, Eugenios Antoniadi published a book of observations of the planet, which included a map as well. These cartographers and their contemporaries derived a rotational period of 88 days, equal to Mercury's orbital period — a mistake that stood unchallenged for decades. Astronomers began to suspect that something was awry in 1962, when microwave studies showed that the presumably dark half of Mercury was actually quite hot. Two years later, radar observations by Gordon H. Pettengill and Richard B. Dyce showed that the spin period is actually 58.6 days. This is exactly two-thirds of Mercury's orbital

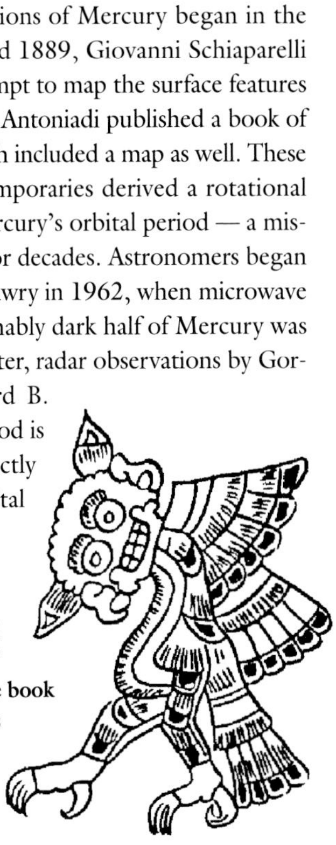

A portion of Mercury's Caloris basin, a multiring impact more than 1,300 km in diameter.

Figure 1. This pre-Hispanic drawing possibly represents Skull Owl from *Popul Vuh,* the book of creation of the Quiche Maya. The book describes four owls who served as messengers for underworld lords and are believed to represent the planet Mercury.

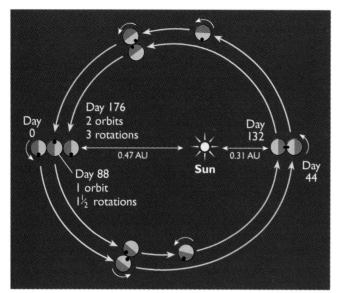

Figure 2. Mercury's 3:2 spin-orbit coupling means that after two orbits, the planet has rotated three times. This peculiar arrangement means that for a hypothetical astronaut on Mercury (black dot) sunrises occur only every 176 days.

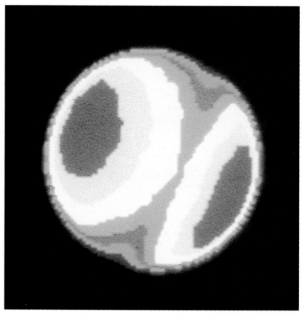

Figure 3. Mercury's "hot poles," depicted in red, appear clearly on both the night (left) and day (right) hemispheres in this map of radio emission at 3.6 cm in wavelength. This microwave energy is escaping to space from about 70 cm below the surface. It is a consequence of the much-stronger sunlight that these locations receive when the planet passes through perihelion of its rather eccentric orbit.

period, and it means that the planet's spin and orbit are locked into a repetitive 3:2 cycle (*Figure 2*).

In hindsight, it is easy to see how early telescopic observers were misled. Mercury's orbital motion is also coupled with Earth's: the planet circles the Sun 54 times in 13.00600 Earth years. From Earth we see the planet well separated from the Sun about six times each year. Three of these elongations occur when Mercury is located at northern declinations and three at southern declinations. Because of the 3:2 spin-orbit coupling, during any 13-year interval observers always see hemispheres centered at longitudes of 90° and 270° during the planet's best northern excursions. Other longitudes are turned our way only when Mercury has an elongation at southern declinations. Since all the early astronomers observed from the Northern Hemisphere, they were unknowingly seeing the same two faces of Mercury over and over. If any of them had been observing the planet from the Southern Hemisphere, the true rotational period might have been detected well before the radar observations of the 1960s.

The spin-orbit coupling, combined with Mercury's pronounced orbital eccentricity of 0.21, produces another curious effect. At successive perihelion passages, the planet first presents one particular hemisphere, then the one opposite it, toward the Sun (*Figure 2*). In fact, long before astronomers knew anything for certain about Mercury's surface, they defined its prime meridian (0° longitude) as the longitude facing the Sun during the planet's first perihelion passage of 1950. By this definition, the central meridians of 0° and 180° always face the Sun at perihelion.

When Mercury is closest to the Sun, its surface temperature reaches about 740° K — one of the highest in the solar system. (The surface of Venus is hotter, due to the greenhouse effect caused by the dense atmosphere.) The 0° and 180° longitudes are the locations of what have been called "hot poles," a phenomenon that can be observed from Earth (*Figure 3*). Con-

versely, the central meridians of 90° and 270° face the Sun only at aphelion, so their maximum temperature is significantly lower, about 525° K. On Mercury's night side, the Sun does not shine for months at a time due to the planet's slow rotation. Surface temperatures there can drop to about 90° K, ranking among the coldest spots in the solar system.

Mercury likely began its existence spinning much faster than it does today. Astronomers generally believe that, over time, the Sun's proximity raised large tides within Mercury's molten core that gradually robbed the planet of angular momentum. However, the details of this simple explanation remain murky. For example, such a "spin down" might logically have ended when the rotation period had slowed to 44 days (a 2:1 spin-orbit coupling), rather than 58.6 days. Perhaps the planet started out with a long rotation period, but that seems unlikely. Alternatively, Mercury's core might not have melted until the spin down was complete (also considered unlikely). Whatever the reason, most dynamicists suspect that Mercury had slowed to the present rotation rate within 500 million years after its formation.

THROUGH THE EYES OF MARINER 10

Our first good looks at Mercury occurred when the Mariner 10 spacecraft flew past the planet not once, but three times in 1974–75. Despite three flybys, orbital geometry limited the imaging coverage to 45 percent of Mercury's surface (*Figures 4,5*). The surface features recorded by Mariner 10 bore no resemblance to those mapped by ground-based astronomers. At first glance, Mercury looks very similar to the Moon: an airless surface dominated by craters ranging in size from 100 m (the camera's resolution limit) to basin-sized impact scars. On closer inspection, Mercury's surface can be divided into two types of terrain: *highlands* and *lowland plains.*

The highlands contain heavily cratered areas mixed with rolling clearings called *intercrater plains* (*Figure 6*). The heavily cratered areas on Mercury bear proportionately fewer impacts than the lunar highlands do, especially for craters having diameters of 50 km or less. However, the abundance of craters as a function of size matches the size-frequency distributions found on the other terrestrial planets, suggesting that the same general population of primordial objects was colliding with all the terrestrial planets early in solar-system history.

This relative paucity of craters on Mercury as compared to the Moon came as a surprise. Evidence exists for an early, major resurfacing event covering all of Mercury's exterior, obliterating a large fraction of the early cratering record. Because of their shallow depths, the smaller craters were preferentially covered. At larger scales, 15 ancient basins were apparently buried by the

material from this early event. We could be seeing material extruded from the interior of the planet. Alternatively, the material could be ejecta from giant, basin-forming impacts.

Key to solving the missing-crater mystery are the intercrater plains. This type of level or gently rolling terrain covers and embays many of the heavily cratered areas, so its emplacement probably caused the obliteration of the smaller-diameter craters. The intercrater plains likely formed 4.2 to 4.0 billion years ago, during the first part of what is known as the "late heavy bombardment" (see Chapter 6). At that time, planetesimals left over from planetary formation rained down on the terrestrial planets and created most of the basins and craters we observe today. The plains also bear many craters no more than 15 km across. These often appear in chains or clusters, have elongated shapes, and are shallow or open at one end. Such characteristics suggest that they are *secondary craters*, formed by large chunks of debris ejected from a larger impact somewhere on Mercury's surface.

Figure 4. During three flybys in 1974–75, the Mariner 10 spacecraft recorded 3,500 useful images of the planet. However, due to the repetitive orbital geometry of these encounters, Mercury presented much the same sunlit face each time. Planetary geologist Mark Robinson carefully reconstructed this mosaic and that in Figure 5 from individual frames acquired by the approaching spacecraft. From this perspective, Mercury looks generally similar to the Moon.

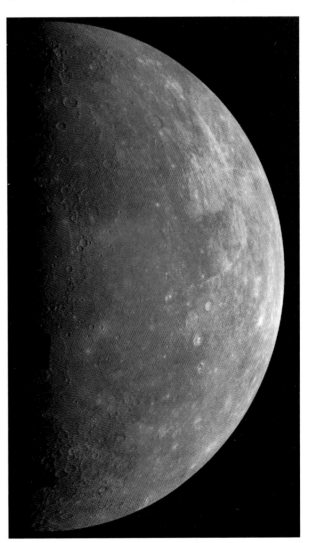

Figure 5. This aspect of Mercury, recorded as Mariner 10 receded from the planet, shows large tracts of *smooth plains,* a terrain type that may be due to extensive volcanism or the splashed-out debris from large impacts. Partially visible along the day-night terminator north above center is Caloris basin, a gigantic multiringed impact scar. The two faces of Mercury seen in figures 4 and 5 do not join up along the terminator but rather along the bright limb at far left and far right.

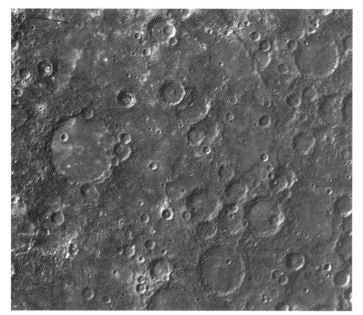

Figure 6. The *highlands* are believed to be Mercury's oldest surviving surface type. These areas exhibit clusters of overlapping craters with smoother *intercrater plains* lying between them. The largest crater in this view is about 175 km across.

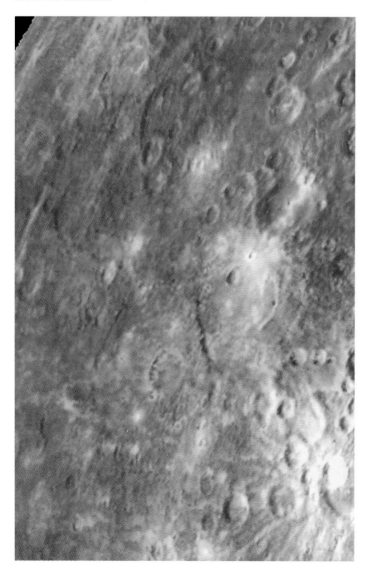

Mercurian craters all show the same morphological features as their lunar counterparts. The smaller ones are all bowl-shaped. With increasing size, they develop flatter interior surfaces, scalloped rims, central peaks, and terraces on their inner walls. Fresher craters all have halos (dark or bright) and crater rays. However, the proportions and scales differ from those of lunar craters due largely to Mercury's stronger gravity, which is nearly 2½ times that of the Moon. On Mercury, an impact's continuous sheet of ejecta and secondaries falls 65 percent as far from the crater rim as occurs around lunar craters with the same diameter. Mercury's craters and their surrounding ejecta blankets degrade more quickly than lunar craters do, because secondary material falls closer to the crater rim and strikes with greater fallback velocities. This crater degradation allows scientists to establish relative ages for craters and other geologic features. It helps us to sort out which surface features overlie or cut across older markings, thus establishing the overall stratigraphy of the planet's surface.

The *lowland plains*, sometimes called smooth plains, occur within and around the enormous Caloris basin, on the floors of other large basins, and in the north polar region (*Figure 5*). Craters are considerably sparser on the lowland plains than on the intercrater plains, suggesting that the former are somewhat

Figure 7 (left). A false-color composite of ultraviolet-, orange-, and clear-filtered images from Mariner 10 reveals distinct geologic units on the surface of Mercury. Reddish areas contain fewer opaque minerals, while green indicates increasing iron "maturity" (longer exposure to the solar wind and micrometeoritic bombardment). The yellow splash at lower right marks the location of Kuiper, a relatively fresh, 60-km-wide ray crater believed to have formed about 1 billion years ago.

Figure 8. An enlarged portion of the false-color composite shows evidence supporting the theory that the plains on Mercury are volcanic flows. The floor of 120-km-wide Rudaki crater and the smooth areas to its west and south have distinct boundaries and a different color than that of the surrounding terrain. By contrast, the diffusely scattered material near Homer basin appears bluer (more opaque) and is probably a layer of impact debris.

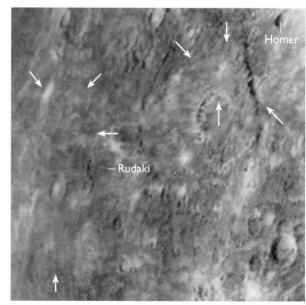

Figure 9. Caloris basin, centered at 30° north, 195° west, is 1,340 km in diameter. The basin's rim is marked by rough blocks 1 to 2 km high, and its floor is covered with smooth plains crisscrossed by wrinkle ridges and fractures. The Van Eyck formation, a collection of radially spreading hills and ridges, is apparent to the basin's northeast (upper right). Hummocky plains (the Odin formation) lie east of the basin rim and grade into Mercurian smooth plains.

younger. They probably formed near the end of the period of heavy bombardment 3.8 billion years ago. But what created them? The age of the lowland plains and their intimate proximity to the larger basins suggests that molten rock extruded onto the surface following the giant basin-forming impacts. This theory, while favored, is not necessarily correct. No volcanic features such as flow lobes, domes, or cones — commonly recognized as volcanic sources — have yet been found. Alternately, the smooth plains could be huge sheets of molten material splashed onto the surface during the impacts themselves. This uncertain origin poses one of the great mysteries of the planet's history.

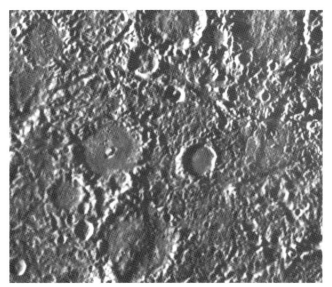

Figure 10. The disrupted landscape at the antipode of Caloris basin exibits a hilly and lineated (fractured) terrain that is unique to Mercury. The hills are 5 to 10 km wide and up to 1,800 m high.

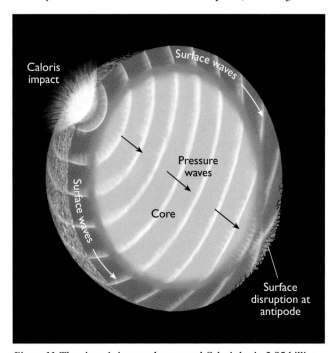

Figure 11. The gigantic impact that created Caloris basin 3.85 billion years ago sent intense seismic waves around and throughout the planet. These came to a focus at the antipodal point, where the ground shook and heaved violently.

One clue to the source of the smooth plains may come from the fact that there is little variation in the albedo (reflectivity) of Mercury's various terrain types. On the Moon, we see a distinct difference between the relatively dark maria (mean albedo: 0.06 to 0.07) and the ancient lunar highlands (0.11 to 0.18). Rock and soil samples collected during the Apollo program suggest a history of volcanism. Laboratory spectra of these specimens correlate well with the spectra obtained telescopically for different areas on the Moon. All of this evidence points to the maria being young basaltic magma extruded following the large basin-forming impacts on the Moon. By comparison, the difference in albedos between Mercury's smooth plains (0.12 to 0.13) and intercrater plains (0.16 to 0.18) is somewhat less. This range is similar to that found on the Moon between bright, impact-emplaced plains and the highlands.

Are Mercury's lowland plains huge volcanic eruptions from the interior or giant sheets of impact melt? Recent reanalyses of the decades-old Mariner 10 data might answer this question. Mark Robinson and Paul Lucey have found that a surface's reflectivity at ultraviolet and orange wavelengths can be used to determine the abundance and maturity of iron and the concentration of opaque minerals it contains. By good fortune, Mariner 10 acquired many of its images through ultraviolet and orange filters (*Figure 7*), and these are now being used to identify differences in the minerals lying on Mercury's surface. In particular, the plains west of Rudaki crater (*Figure 8*) show distinct color boundaries that correspond to their geologic boundaries. This material also fills the crater Rudaki, and there appears to have been a flow of material in the area. In contrast, material along the margin of the crater Homer appears to be much more opaque, yet it has a diffuse boundary not shaped by the local topography. It could well be a sheet of impact melt. On the other hand, the Rudaki plains — and, by extension, other tracts of the smooth plains — originated as a volcanic outflow from the interior.

CALORIS BASIN

The most prominent of the giant impacts on Mercury is Caloris basin, about half of which was in sunlight during Mariner 10's flybys (*Figure 9*). This huge excavation spans 1,340 km, as defined by the Caloris mountain terrain — an annulus of smooth-surfaced blocks rising 1 to 2 km. Its individual massifs are up to 50 km long and create a ring 100 to 150 km wide. The patchwork of rough-surfaced plains nestled among these mountains is probably a deposit of fallback material and impact melt created by the impact.

Two other units are considered to be ejecta from the basin. The Caloris lineated terrain (known as the Van Eyck formation), extends roughly 1,000 km beyond the Caloris mountains. It consists of long hilly ridges and grooves oriented more or less radially to the basin's center. About 800 km farther out is a broad ring of low hills some hundreds of meters high. Grading into the (younger) smooth plains in this area, these hummocky plains are named the Odin formation.

The Caloris impact ranks as the second largest such event recorded on any of the terrestrial planets. (The South

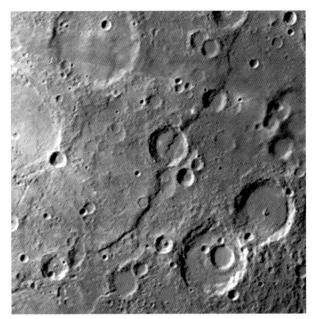

Figure 12. Discovery Rupes, one of the longest lobate scarps on Mercury, extends 500 km and is up to 2 km high in some places. It transects two craters, and the foreshortening seen in their floors argues that Discovery is a thrust fault caused by compressional stress in the crust.

Pole–Aitken basin, on the Moon's far side, is 2,600 km across.) The cataclysmic Caloris impact occurred about 3.85 billion years ago when an object roughly 150 km across struck Mercury with the energy equivalent of a trillion 1-megaton hydrogen bombs. It was an event with global consequences for the young planet, and its manifestations are not limited to the region around the basin.

For example, at the basin's *antipode* (the spot on Mercury 180° from the impact site) is an unusual tract of terrain unlike any other in the solar system. Roughly the size of France and Germany combined, it consists of numerous hills and depressions that crosscut the preexisting landforms (*Figure 10*). The Caloris impact generated strong seismic waves that radiated outward from the site. Surface waves rolled through the crust, while compression waves passed through the deep interior (*Figure 11*). Minutes later these seismic shocks came to a focus at the antipode, lurching the ground more than 1 km upward and fracturing the crust to great depth. Once the violent shaking ended, the antipodal area had been reduced to a jumble of large blocks — the ones preserved in the hilly and lineated terrain.

THE IRON PLANET

Astronomers first estimated Mercury's mass by observing the planet's interaction with Venus and with comets passing in its vicinity. Careful tracking of how Mercury affected Mariner 10's trajectory yielded a much more precise mass, 3.3×10^{26} g, and a mean density of 5.43 g/cm^3. This is close to the bulk densities of Earth (5.5) and Venus (5.2) but much greater than that of Mars (3.9). The density values for large planets reflect, in part, the compression of their deep interiors by all the overlying mass. However, this is less of a factor with smallish Mercury, and consequently its high density means that its

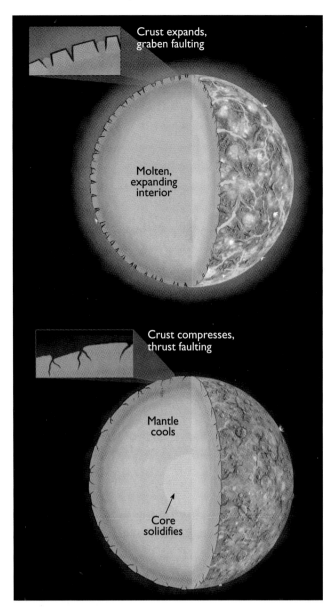

Figure 13. Global expansion (top), which may have occurred early in Mercury's history, creates extensional stress that leads to normal faults in the surface. Conversely, as Mercury cooled its entire globe shrank slightly (bottom), which compressed the crust and triggered the formation of numerous thrust faults like Discovery Rupes.

interior has a greater proportion of heavy elements than do interiors of the other terrestrial planets. Mercury's composition is roughly 70 percent iron (with some nickel) and 30 percent silicate material.

Most of this iron is believed to be concentrated in a large core that extends to 75 percent of Mercury's radius of 2,440 km. With so much iron present, it is not surprising that Mercury has its own magnetic field. Such a field was detected by Mariner 10 during its first and third flybys, at which time it passed through a small magnetosphere on Mercury's night side. However, the existence of a planetwide magnetic field is problematic. First, Mercury rotates so slowly that it is difficult to imagine how a circulation-driven dynamo can be sustained in the core. Second, this small planet has cooled considerably over time and its core now is likely almost completely solid. Any remaining liquid

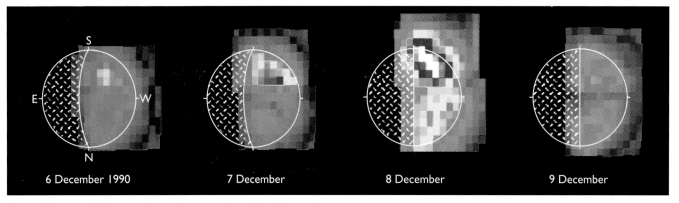

6 December 1990   7 December   8 December   9 December

Figure 14. Images of Mercury taken on four successive nights in December 1990 at the yellow-light wavelength of emission from metallic sodium vapor. The amount of sodium in Mercury's atmosphere varies both with location and over time.

probably exists as a thin shell enriched in sulfur — too little fluid to provide for dynamo circulation. Some researchers have proposed that the field could be due instead to remanent magnetism, that is, a magnetic field frozen into the solid core. In any case, the large amount of iron (solid or liquid) thought to reside in Mercury's core must be involved somehow in creating the observed weak magnetic field. We simply do not yet have enough information to deduce its source.

Some hints about the state of Mercury's core can be gleaned from long, sinuous features on the surface known as *lobate scarps.* Unique to Mercury, these curved features range in length from 20 to 500 km, and in height from a few hundred meters to 2 km (*Figure 12*). They are thrust faults, which result from compressional stress in the crust. This compression was probably caused by a 1- to 2-km decrease in Mercury's radius that occurred when the planet's mantle cooled, and its core cooled and partly solidified (*Figure 13*). We can estimate when this cooling and contraction occurred from the scarps' locations. None of them are embayed by intercrater plains material, suggesting that they are younger than those plains. We see lobate scarps on the smooth plains surrounding Caloris basin, so apparently the scarps postdate both the impact and the volcanism that formed the smooth plains. Thus, the scarps — and by implication the planet's contraction — date from roughly the time that Caloris basin was created, near the end of the period of heavy bombardment.

In theory, mantle cooling and core solidification should have shrunk Mercury's radius by 6 to 10 km — not just the 1 or 2 km evidenced in the lobate scarps. This disparity can be explained if much of the contraction took place before the existing crustal topography was in place. It is also possible that some of the core has not yet solidified, a scenario quite consistent with the existence of a magnetic field.

MERCURY SINCE MARINER 10

What can be said about the 55 percent of the Mercurian surface not imaged by Mariner 10? No spacecraft missions are planned to visit Mercury at this time, and the planet is situated too near the Sun to be observed by orbiting observatories such as the Hubble Space Telescope. Ground-based instrumentation has

advanced significantly over the last quarter century, however, and scientists have again turned their attention to studies of the innermost planet.

Although Mariner 10 had found traces of hydrogen, helium, and oxygen surrounding Mercury, Andrew Potter and Thomas Morgan discovered two additional "gases" — neutral sodium and potassium atoms — in the mid-1980s. Oxygen is the most abundant element in the atmosphere, followed by sodium and hydrogen (see Chapter 13). The hydrogen and helium derive mainly from solar-wind ions. However, the oxygen, potassium, and sodium likely come from the vaporized rock created momentarily when micrometeoroids strike Mercury's surface. (The infalling meteoroids also contain up to 10 percent water, which dissociates into H and O.) Alternatively, these freed atoms may result from sputtering of the surface regolith by ions trapped in Mercury's magnetosphere.

Because sodium and potassium scatter solar photons very efficiently, their emissions at visible wavelengths are easily observed from the Earth. Perhaps most interesting is that the abundances of these elements vary both in time and location across Mercury's surface (*Figure 14*). The sodium and potassium are apparently being injected into the atmosphere from discrete locations on the surface. One obvious explanation is that the surface varies in composition from place to place. The sodium emission is sometimes strongest over paired locations north and south of the Mercurian equator. These are roughly located where the auroral cusps of a dipole magnetic field would be located, suggesting some kind of interaction of the surface with the magnetosphere. Perhaps heavy ions rain down in greater numbers near the Mercurian poles, increasing the rate of sputtering there or reverting to neutral atoms that accumulate onto the surface after striking it.

The sodium and potassium are probably telling us something about the compositions of rocks on the surface. For many years, astronomers have tried to glean information about the surface mineralogy from the spectrum of the sunlight reflected off Mercury's surface, but the Sun's interference makes acquiring these spectra very difficult. As mentioned, Mercury's high density implies that there is a significant amount of iron or iron-nickel in the planet's interior. Understandably, astronomers have searched for evidence of iron-bearing silicates in Mercury's surface material, but the results have been inconclusive.

If anything, the available evidence points us in the opposite direction. Spectra of the whole disk of Mercury hint of an absorption feature near 0.9 micron that corresponds to an iron-poor form of the mineral pyroxene. More intriguing data have come at mid-infrared wavelengths (7.5 to 13.5 microns), where Mercury's surface shows spectral features that have been attributed to the iron-free silicate mineral plagioclase feldspar $((Ca,Na)(Al,Si)AlSi_2O_8)$, in agreement with the implied low iron content. *Figure 15* shows an especially interesting matchup of an infrared spectrum of Mercury and a laboratory spectrum of an Apollo sample from the iron-depleted lunar highlands. These results beg the question of why Mercury appears to have so much iron in its interior and so little in its surface rocks.

We can also probe the gross physical structure and soil properties of the unimaged areas of Mercury using the powerful radar systems at Arecibo, Puerto Rico, and Goldstone, California. The planet's radar reflectivity is near 6 percent, quite similar to the low reflectivities of 4 to 7 percent measured for the Moon. This result implies (not unexpectedly) that Mercury is covered with a loose regolith of pulverized rock produced by the constant bombardment of the surface. Little bedrock lies exposed on the planet.

At radio wavelengths, radiation from Mercury is dominated by heat coming from the topmost layer of the surface. The regolith is significantly more transparent to the escaping radio (microwave) energy than the lunar regolith is. This can be explained if Mercury's regolith has very low abundances of iron and titanium (less than 6 percent by weight). Such low abundances at the surface would imply a lack of extrusive basaltic volcanism, which is at odds with some of the other results discussed earlier and in the next paragraph.

Astronomers have also used radar to derive the heights of portions of the Mercurian surface. These topographic profiles of Mercurian smooth plains show a distinctive down-bowing that is characteristic of a rigid surface layer sagging under its own weight. Similar deformation has occurred under the lava-flooded lunar maria. The down-bowing occurs at varying distances from Caloris basin, suggesting that outpourings of volcanic magma (not a blanket of impact ejecta) could be the cause of these smooth plains.

While our radar images are understandably crude compared to the Mariner 10 images, they have given astronomers a first look at the side not photographed (*Figure 16*). The unseen hemisphere bears a large, circular, radar-bright feature surrounding a small, dark center. Although unlike any other feature imaged by radar on Mercury, this feature matches the appearance in radar images of shield volcanoes on Venus and Mars. The Mercurian spot would be consistent with a shield volcano 500 km across topped by a central caldera 70 km in diameter, sizes comparable to the great shield volcanoes of both Venus and Mars.

Probably the most stunning observations of Mercury during the past decade have been the detection of radar-bright spots near the planet's north and south poles (*Figures 16,17*). These features' high radar albedos and strong polarizations mimic the appearance of ice deposits on the Galilean satellites and the south polar ice cap on Mars.

Could ice really exist at the poles of the planet closest to the Sun? Apparently, the answer is yes! Mercury's rotation axis is very nearly perpendicular to the plane of its orbit, and theoretical modeling confirms the floors of some craters within about 6.5° of the poles have regions that are never exposed to sunlight. It is therefore entirely possible to maintain water ice in these permanently shadowed niches for geologically long times. The radar features have been carefully mapped and compared with Mariner 10 images of the polar regions. Remarkably, the most pronounced radar features in the imaged hemisphere can be matched with specific craters, and all these craters lie within the latitude range where permanent shadowing should exist. Not yet understood, however, is the mechanism by which some combination of volatiles might migrate to the Mercurian poles and become "cold trapped" onto these shadowed crater floors.

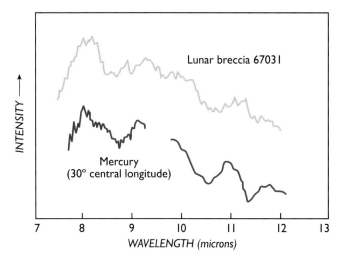

Figure 15. A mid-infrared spectrum of Mercury, centered near 30° in longitude, is a good match to a laboratory spectrum of a sample from the iron-poor lunar highlands. The broad maximum at 8.1 microns strongly implies a plagioclase-dominated composition, because the mineral's sodium- and calcium-rich endmembers have strong emissions at 7.8 and 8.1 microns, respectively. The lunar sample contains about 90 percent plagioclase and 10 percent pyroxene (an iron-bearing silicate).

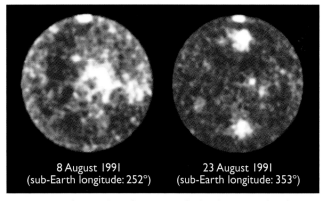

Figure 16. In these 1991 radar images, the bright spot at the planet's north pole (top) is most readily explained as a deposit of ice — a seemingly impossible conclusion given Mercury's very high surface temperatures. The view at left shows virtually the entire hemisphere of Mercury that went unseen by the Mariner 10 spacecraft. In the view at right, the two broad brightenings astride the central meridian appear to be areas of very rough terrain, perhaps marking the locations of huge impact basins. In both images Mercury's north pole is tipped about 10° toward Earth.

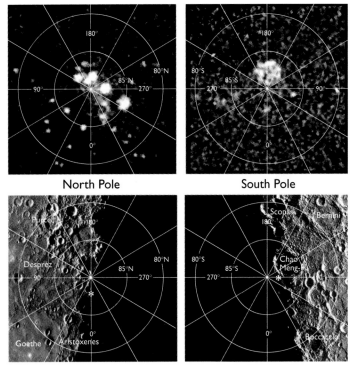

North Pole   South Pole

Figure 17. **The upper pair of images shows radar-bright deposits, caused by what is most likely ice, near the poles of Mercury. First spotted in 1991, the circular patches coincide with the shadowed floors of large, fresh craters seen in photographs taken by Mariner 10 (bottom pair). The match becomes nearly perfect if the poles' positions are allowed to shift from their Mariner-based locations (grid centers) to new ones (asterisks).**

MERCURY'S ORIGIN: STILL A MYSTERY

As the planet closest to the Sun, Mercury is often viewed as an end-member of planetary formation. We generally accept the idea that it formed from the condensation of material in the solar nebula. Presumably the growing planet collected planetesimals from throughout the terrestrial-planet region, but with an emphasis on those near Mercury's present orbit. The fact remains, however, that the high density of Mercury cannot be explained by our current models of condensation and accretion in the early solar system as a function of heliocentric distance.

Perhaps Mercury formed with a larger proportion of rocky matter, but then some mechanism removed much of its exterior silicate shell and left little more than the largely iron core and some mantle material. If Mercury formed in its present orbit, an early energetic phase of the Sun could have volatilized the planet's outer crust and mantle. Another possibility is that one or more major impacts could have removed the crust and mantle. Such devastation might have been the rule, not the exception, in the early solar system. Indeed, a cataclysmic impact of a Mars-size object with Earth is presently the most likely explanation for the Moon. An interesting possibility along these speculative lines is that Mercury could have formed anywhere among the terrestrial planets, then a major impact not only removed its outer shell but also knocked it into its current orbital niche.

Mercury remains an enigmatic planet, a world of extreme contrasts and unexpected surprises. Just when scientists believe they understand what is happening on and within it, innovative observations probe a little deeper into Mercury's secrets. And each new revelation seems more improbable than the last.

Venus

R. Stephen Saunders

IN THE PAST few years, planetary scientists have studied in detail a world that can perhaps teach us a great deal about Earth. Venus was expected to be like our planet in many ways. It is nearly as large as Earth (*Figure 1*), and everything we know about Venus suggests that it is made of the same stuff as Earth and in about the same proportions — except for one key ingredient. Venus is dry. It has almost no water. This lack of water may explain why Venus is, in fact, very *different* from Earth and apparently has always been so.

Adivar crater on Venus, as seen in a radar image from the Magellan orbiter.

Figure 1. Even before Magellan's arrival, scientists realized that the landscape of Venus (left) was dominated by vast plains and lowlands, with few landmasses comparable to Earth's continents (right).

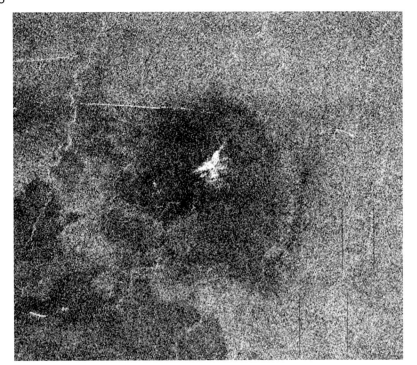

Figure 2. This image covers an area of approximately 50 km by 60 km and is located in the Lakshmi region of Venus at 47° north latitude and 334° east longitude. The dark circular region and associated central bright feature are thought to be the remnants of a meteoroid smaller than the size necessary to create an impact crater.

The virtual absence of water explains, for example, why Venus has a dense atmosphere made up mostly of carbon dioxide (CO_2). If Venus had ever had oceans, that carbon dioxide would have formed limestone and other carbon-bearing rocks. As much carbon dioxide has come out of Earth's interior, mostly in volcanic eruptions, as we find in the entire Venus atmosphere. However, our oceans remove CO_2 from the atmosphere and keep Earth habitable. Because carbon dioxide is an excellent greenhouse gas, sunlight that penetrates into the Venusian atmosphere stays as heat, and today the planet's surface is about 750° K.

We have known for decades that Venus was a dry planet. However, the full significance of that fact, especially with regard to surface processes, could not be imagined because the clouds and haze in the planet's dense atmosphere shield its landscape from our view. Fortunately, we can see through to the surface using radar imaging. Our first crude identifications of Venusian surface features came in the 1960s, when powerful radio telescopes near Goldstone, California, and Arecibo, Puerto Rico, bounced radar pulses off the planet and analyzed the returning echo. In time this Earth-based technique matured and revealed many discrete features. However, it never did resolve the surface of Venus with enough detail to be geologically definitive.

Global-scale radar mapping of Venus began with NASA's Pioneer Venus mission in the late 1970s. This was followed in the mid-1980s by Veneras 15 and 16, twin spacecraft from the former Soviet Union that mapped much of the planet's northern hemisphere. The results from these missions established basic surface units distinguished on the basis of variations in radar properties, surface textures, topography, and morphology. True revelation, however, did not come until the arrival of the Magellan orbiter, which systematically mapped Venus from Sep-

tember 1992 to October 1994, when it plunged into the atmosphere. Magellan was in a polar orbit. Consequently, as Venus slowly rotated through its 243-day period beneath the spacecraft's radar gaze, the entire globe was eventually revealed.

Magellan imaged 98 percent of the planet's surface at a resolution of 120 to 300 m. The spacecraft carried a side-looking radar system that "illuminated" a narrow strip of surface with radio energy as the spacecraft moved along its orbit, then it recorded the reflected signal. Magellan's radar looked off to one side at an angle of 35°, compared to the 10° used by Veneras 15 and 16. This allowed us to distinguish terrains with rough textures fairly easily. Also on board was a second radar system that sent radar pulses straight down to continuously gauge the height of the topography below.

Regions that are bright in the reconstructed radar images have higher radar reflectivity. Such areas are often rougher than their surroundings on the scale of the radar's wavelength, a few centimeters to a few meters. Strong radar echoes come from rough-textured lava flows or the blocky surfaces surrounding impact craters. Sometimes terrain looks bright if tilted so it faces toward the spacecraft, like the side of a hill. In either case, more radio energy than average is reflected back to the spacecraft receiver. In some cases, particularly at the highest mountain elevations, strong radar reflectivity seems to indicate a kind of "shiny" terrain thought to contain a metal component with a high dielectric constant, such as pyrite (FeS) or iron oxides like magnetite (Fe_3O_4).

The Magellan radar images and altimetry readings have been painstakingly compiled into global maps that allow us to identify and map different terrain types and geologic units. The geologic characterizations extracted from Magellan data build upon the basic criteria established by the Pioneer Venus and Venera results. These, along with the application of such stratigraphic interpretations as overlapping and crosscutting relations, have helped us to obtain as much information on the chronology of rock sequences as possible.

SURPRISES FROM IMPACT CRATERS

One important clue to the age of a planetary surface is the number, size, and condition of the impact craters it bears. Impact craters are seen on all the solid surfaces of the solar system (see Chapter 6), formed by the infall of asteroids and comets at 20 to 30 km per second or more. Impact events are more or less random in time and space when viewed over millions of years. Thus the older the surface, the more craters it should have. Unfortunately, such crater counts only provide relative ages of the surfaces involved. For the Moon we have been able to put absolute numbers on the lunar time scale by dating the samples returned by the Apollo astronauts. On other worlds, however, there is great uncertainty in the absolute ages. In spite of such uncertainties, craters have been extremely useful in unraveling the sequence of geologic events on the Moon, Mars, and outer-planet satellites.

September 1992 was an exciting time for those awaiting the first global view of Venus. We already knew that the planet was cratered, thanks to the low-resolution radar images from the Arecibo and Goldstone radio telescopes on Earth and from the Veneras. As the first days of mapping progressed, the Magellan science team eagerly pored over the image "noodles," each one representing a piece of Venus 20 km wide and 15,000 km long. The first few images scrolled across several large impact craters.

After a few weeks of mapping, with about 10 percent of Venus revealed, a profound puzzle began to emerge. Volcanism dominated the surface. Lava flows and various kinds of volcanoes were everywhere. Faults and fractures crisscrossed the landscape. Wind streaks were abundant. All of these things were fascinating but not unexpected. The craters provided the big surprise. On the Earth, Moon and Mars, craters range from very fresh scars to barely visible, eroded circular ghosts. However, apparently little erosion had affected the Venusian craters. They all looked as if they had formed recently.

Further, as mapping proceeded, the craters appeared to be scattered randomly over the surface. Unlike the Moon or Mars, there were no highly cratered (ancient) regions intermixed with lightly cratered (younger) regions. All the surface appeared to be about the same age and geologically young. If we use the lunar chronology and crater densities as a guide, the average surface age of Venus could be no more than about 500 million years. We were seeing only the last 10 percent of Venus's geologic history recorded in its surface.

As it turned out, the impacts seen during the first month of mapping proved to be representative of the entire planet's crater population. No one has been able to demonstrate convincingly that the approximately 1,000 Venusian craters are distributed in any arrangement other than randomly scattered. They must truly represent an average time of accumulation of less than 500 million years, and as such they provide important constraints on the evolution of Venus — but in very unexpected ways.

The finding of a sparse, pristine, and randomly distributed crater population immediately led to an interesting scientific debate. One explanation is that volcanic eruptions destroy craters about as fast as they form, so that we will always see about the same number of craters on Venus. This is the "equilibrium hypothesis," or what some geologists call the uniformitarian model. Another interpretation is the "global-catastrophe hypothesis," according to which massive volcanic eruptions approximately 500 million years ago resurfaced the entire globe and thus erased all the craters that had formed earlier. Having been wiped clean, the surface again began to collect craters, and nothing much else happened geologically except for some faulting and a few large volcanoes.

Curiously, this debate about Venus echoes a debate among early geologists a couple of hundred years ago. The catastrophists believed that most of Earth's rocks formed over a short period of time in the Noachian floods, while the uniformitarianists held that "the present is the key to the past." In other words, geologic time was vast, and enormous surface modifications result from slow changes that we can see operating today. In that early historic debate, the uniformitarianists ultimately won out. For Venus, in the opinion of this writer, the catastrophe model will in time be judged more nearly correct than the equilibrium model.

Aside from their number and distribution, the craters themselves have proved very interesting. We anticipated that the dense atmosphere of Venus would shield the planet from small asteroidal and cometary debris, allowing only larger bodies to reach the surface. Magellan images show no intact craters less than about 3 km across. This means that no objects larger than about 30 m in diameter make it to the surface with enough of their initial velocity to produce a crater. Small objects either are destroyed by the atmosphere or are slowed so much that they simply fall onto the surface at low speed.

However, the craters' appearances were difficult to predict, and they show several intriguing characteristics. Even though the smallest meteoroids never reach the ground, the energy they dissipate during atmospheric passage apparently creates strong winds and shock waves that disrupt the surface, producing diffuse, radar-dark "splotches" (*Figure 2*). Sometimes the splotches have a bright center, apparently marking a cluster of small impacts, ejecta, and debris from the broken-up meteoroid. What caused these features? Since smooth surfaces generally appear

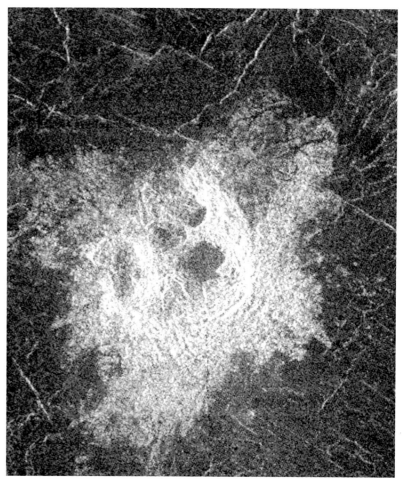

Figure 3. **The multiple crater Lillian, a 13.5-km-diameter crater in the Guinevere region of Venus (25.6 °N, 336.0°). This crater is actually a cluster of four separate craters whose rims touch or overlap.**

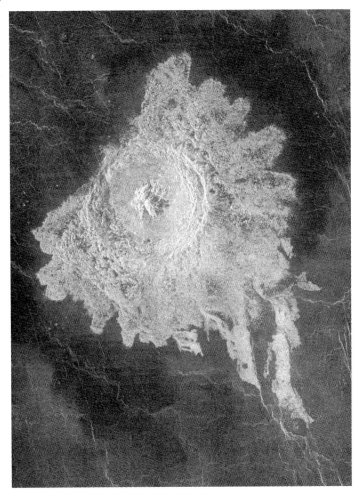

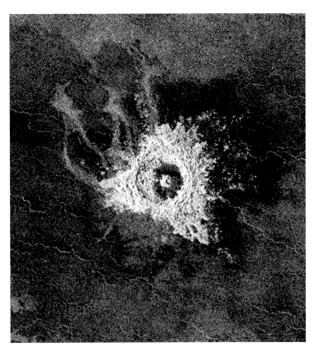

Figure 5. Jeanne is a 19.5-km-wide impact crater located at 40.0° north latitude and 331.4° east longitude. The distinctive triangular shape of the ejecta indicates that the impacting body probably hit obliquely, traveling from southwest to northeast.

Figure 4. The crater Aurelia is centered at 20.3° north latitude and 331.8° east longitude. With a diameter of 32 km, it displays a circular rim, terraced walls, and central peaks. Geologists are intrigued by how the radar-bright (rough) ejecta surrounding the crater creates an asymmetric pattern that indicates the impactor's incoming direction (from upper left).

dark in a radar image, it could be that the meteoroid's atmospheric shock wave was energetic enough to pulverize and flatten the surface below. Another explanation is that the surface could be blanketed by a fine material that fell from the sky — the dusty remains of the original meteoroid. More than half of the impact craters on Venus have associated dark margins, and most of these are prominently located to the west of the crater's center. This drifting is another effect caused by the dense atmosphere, which rotates east to west around Venus (see Chapter 13).

As their size increases, infalling objects reach a point where they punch through the atmosphere and strike the surface. However, craters less than about 30 km in diameter are usually irregular or multiple excavations (*Figure 3*). An irregular outline and hummocky floors tell us that a meteoroid probably broke apart during passage through the dense Venusian atmosphere. After breaking up, the meteoroid fragments struck nearly simultaneously, creating the crater cluster.

Most of the nearly 1,000 craters scattered over the surface of Venus are gemlike works of art. Almost all appear to have formed recently — their rims are sharp, and the surrounding blankets of ejected debris look undisturbed. However, such appearances are deceiving, as Venus's craters have been accumu-

lating over the past few hundred million years at an average rate of only one or two per million years. The excavations appear fresh because the processes of erosion on Venus are slow. Despite the dense atmosphere, wind speeds are too low to wear down the terrain, and as mentioned water is not a factor.

A good example of one of the more typical Venusian craters is Aurelia (*Figure 4*). Named for the mother of Julius Caesar, 32-km-wide Aurelia is a complex crater with a circular rim, terraced walls, and central peaks — all rather common features. The pattern of surrounding ejecta spreads out like a butterfly's wings, with relatively little in the direction from which the meteoroid descended. This asymmetry is seen frequently around craters on other planets, and it results from an oblique (not vertical) impact. The collision's energy melted a large amount of the target rock, which can be seen flowing out into the ejecta blanket. Much of this apron appears bright in the radar image, as does the crater floor, which implies that it is very rough.

Another crater with asymmetric ejecta and bright flows is Jeanne (*Figure 5*). It is surrounded by dark material of two types. On the southwest side is an area of smooth (radar-dark) lava flows that run out into the surrounding brighter flows like long fingers. The very dark area on the northeast side is probably covered by smooth material such as fine-grained sediment. This dark halo mimics the shape of the ejecta blanket. It may have been caused by an atmospheric shock or pressure wave produced by the incoming body. Jeanne also displays several distinct lobes to its northwest. These features may have formed by fine-grained ejecta transported by a hot, turbulent wind immediately after the impact. Alternatively, they may denote flows of impact melt.

The Magellan scientists were able to combine topography obtained by the radar system's altimeter with the images and reproject them in the computer to produce perspective views. The example seen in *Figure 6* includes the midsized impact craters Howe, Danilova, and Aglaonice, which are situated relatively near one another in a region south of the equator known as Lavinia Planitia. All three craters have floors covered with smooth, radar-dark lava. Oblique, three-dimensional portrayals like this one proved very useful in determining the often-complex geologic relationships present on the surface.

Many of the largest Venusian craters have another very interesting feature, which is illustrated by the crater Adivar (*Figure 7*). Some 50 km above the surface, the atmosphere of Venus maintains a strong, continuous flow, analogous to a jet stream, that blows more than 150 km per hour from east to west. When a large crater forms, its ejecta are thrown up into this flow and carried downwind. Surrounding Adivar's rim is ejected material that appears bright in the radar image due to the presence of rough fractured rock. However, a much broader area has also been affected by the impact, particularly to the west of the crater. Radar-bright materials, including a jetlike "tail" due west of the crater, extend for over 500 km across the surrounding plains. A darker apron surrounds the bright area in a horseshoe shape. These radar-dark (smooth) parabolic hoods surround many Venusian craters. However, bright streaks are rarer. Seen only on Venus, they result from the interaction of crater materials (the colliding meteoroid, ejected target rock, or both) and high-speed winds in the upper atmosphere. The precise mechanism that produces the streaks is poorly understood, but it is clear that the dense atmosphere of Venus plays an important role in the cratering process.

Figure 6. This simulated three-dimensional view is from a vantage point southwest of Howe, the 37-km-wide crater centered in the lower portion of the image. To its upper left is Danilova, with a diameter of 48 km, and at upper right is Aglaonice, 63 km across. The orangish hues are based on color images recorded by the Soviet Union's Venera 13 and 14 spacecraft, which viewed the surface as seen by sunlight filtered through the planet's dense clouds.

PLANETWIDE VOLCANISM

Volcanoes are among the most common geologic features in the solar system, and they are ubiquitous on Venus. Long before Magellan's arrival, evidence for a basaltic (volcanic) surface composition had been recorded by the seven Venera landers. Some of these spacecraft acquired images of the Venusian surface (*Figure 8*), which revealed flat-looking rock plates separated by small amounts of soil. The surface looks like some basalt flows in Hawaii and on the Snake River Plains of Idaho. The Veneras' chemical analyses determined that the rocks are indeed basaltic in composition. Basalts generally form from the melting of mantle materials, and they are abundant on each of the terrestrial planets and the Moon.

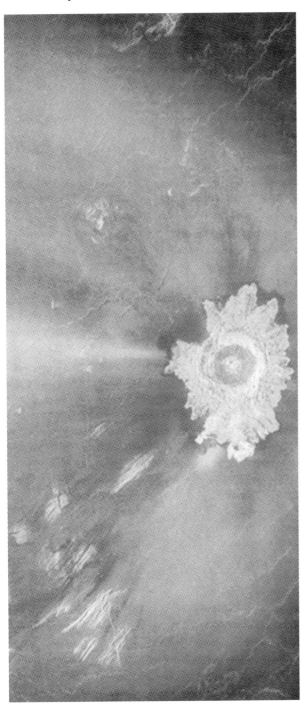

The global topographic mapping done by the Pioneer Venus orbiter and Magellan show that relatively flat plains cover about 85 percent of the surface of Venus. They tend to occupy the planet's lowest elevations (*Figure 9*). Magellan's radar images reveal that these plains are essentially all volcanic, though the source vents for all these flows are generally not apparent. However, the plains do contain meandering "rivers" that appear similar to lava channels on Earth and the Moon. These must have transported molten rock of some kind, since water is not presently stable on the Venusian surface. Although the channels generally lack tributaries, they often display levees, deltas, and abandoned bends. One particular channel, named Baltis Vallis, is 6,800 km in length — the longest sinuous rille yet identified in the solar system (*Figure 10*).

Approximately 1,100 volcanic constructs have been identified on Venus, which have been classified into groups according to the diameters of the bases. The most common edifices are small shield volcanoes with roughly circular outlines, diameters less than 20 km, commonly with summit pits (*Figure 11*). Such small shields are abundant on Venus. They are often found clustered together in shield fields that include identifiable lava flows. While the volcanic deposits associated with these shields are small in terms of volume, their sheer numbers and global distribution on Venus suggest that they may have contributed significantly to the formation of the crust.

Volcanic constructs of intermediate size, 20 to 100 km across, may have circular shapes like small shields or distinctive radial deposits emanating from a circular or elongated central vent. Flat-topped, steep-sided domes also occur in this size range (*Figure 12*). The shape of these latter domes implies that they formed from sluggish, viscous lava. They look similar to chemically differentiated lava domes on Earth. Although they are much larger, these domes are the best evidence for evolved (silicon-rich) magmas on Venus. Magma viscosity can also be increased if the melt is partly crystallized or contains abundant gas bubbles. The latter situation may result from the high atmospheric pressure at the Venusian surface (about 90 bars). In this high-pressure environment, gases that would ordinarily escape freely from the magma remain trapped within it, leading to explosive eruptions.

Some of the largest constructs on Venus, those at least 100 km in diameter, share many characteristics with large volcanoes on Earth. These often display lava flows emanating radially outward from a region of current or former high relief. Sapas Mons (*Figure 13*) is typical of the planet's large, shield-type volcanoes. Measuring 400 km across its base and 1.5 km high, Sapas has a collapse caldera at its summit. It's lava flows extend for hundreds of kilometers across the surrounding, fractured plains.

Coronae form a distinctive class of large volcanoes that were first identified on Venus in Venera radar images. Coronae are characterized by large, concentric rings of fractures, within which large volcanic outpourings have occurred repeatedly

Figure 7. **This 30-km-wide crater, named for the Turkish educator and author Halide Adivar (1883–1964), is located just north of the western Aphrodite highland at 9° north latitude, 76° east longitude. Its bright, streamlined hood and tail resulted from the interaction of ejected debris with high-altitude winds blowing from the east (right).**

Figure 8. On 5 March 1982 the Venera 14 lander touched down on Venus at 13° south latitude and 310° east longitude, where it survived for 60 minutes before succumbing to the planet's heat. In that time it radioed to Earth these images of the Venusian surface, which include parts of the lander at bottom (a mechanical arm can be seen in the upper image, a lens cover in the lower one). The landscape appears distorted because Venera 14's wide-angle camera scanned in a tilted, sweeping arc. The horizon appears in the upper left and right corners of each scene, and the views are remarkably free of atmospheric haze. Note the dominance of slabby or platy rocks, separated by minor amounts of soil. The composition and texture of these rocks is similar to terrestrial basalts.

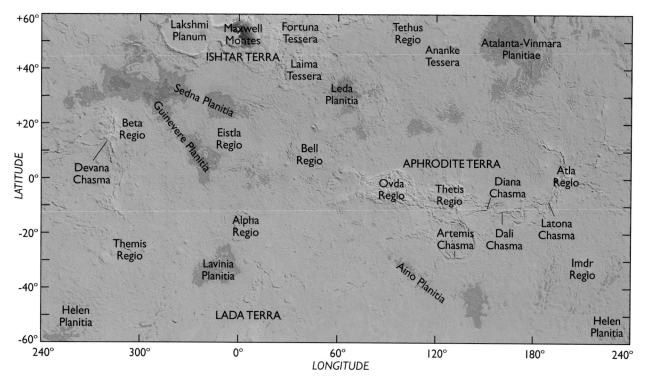

Figure 9. It took Magellan's radar altimeter 24 months to map 98 percent of Venus. In this Mercator-projected view, red corresponds to the highest elevations, blue to the lowest. Maxwell Montes, the planet's highest mountains, rise 12 km above the mean elevation. Even though Venus exhibits a range of elevations comparable to that of Earth, the two planets have distinct topographies. Earth has many high-standing continents and low-lying ocean floors, whereas about 60 percent of Venus's terrain lies within 500 m of the mean planetary radius (its equivilent of sea level). The scorpion-shaped feature extending along the equator between 70° and 210° east longitude is Aphrodite Terra, a continentlike highland that contains several spectacular volcanoes at its eastern end: Maat Mons, Ozza Mons, and Sapas Mons.

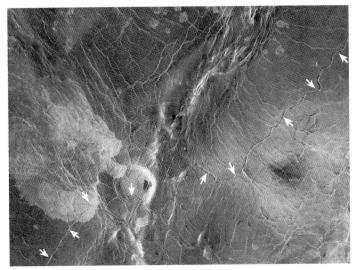

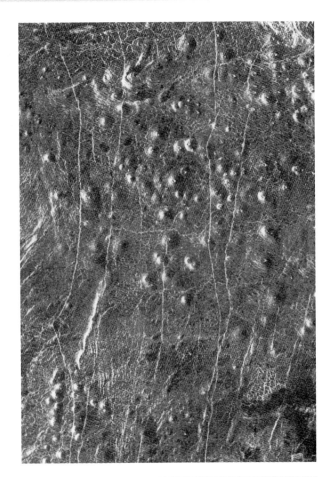

Figure 10. Baltis Vallis (indicated by arrows) snakes across the Venusian landscape for an amazing 6,800 km. Its great length and nearly constant width imply that it was fed by lava at a vigorous rate. The lava must also have been a very fluid material, such as komatite (a high-temperature basalt), carbonatite (an igneous rock made of calcium carbonate), or sulfur. Whatever the material, channels like this one record the last stages of widespread plains emplacement on Venus.

Figure 11 (right). Scores of small volcanic domes, all less than 15 km across, pepper the southern flank of a highland area known as Tethus Regio. Their shape is more reminiscent of Hawaiian-type shield volcanoes than of cinder cones. These pocklike edifices often appear in dense swarms and may total in the hundreds of thousands.

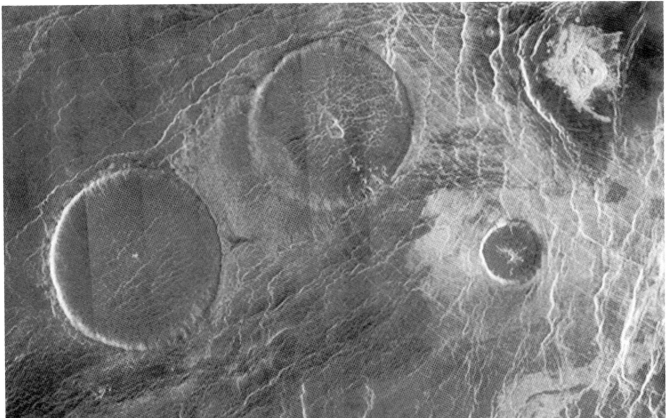

Figure 12. These "pancake" domes in the Eistla region of Venus are 65 km in diameter with broad, flat tops less than 1 km high. The cracks and pits commonly found in these features result from cooling and the withdrawal of lava. Note the more-fluid flow emitted from the dome above center that moved toward the other large dome at lower left. An irregularly shaped, radar-bright impact feature is at upper right.

(*Figure 14*). Such features are often surrounded by an annulus of ridges and troughs, which cut (in places) across fractures aligned more or less radially. The centers of the features also contain radial fractures as well as volcanic domes and flows. Coronae are thought to form due to the upwelling of hot material from within the mantle of Venus. The rising mantle plume causes the overlying crust to bulge and fracture, which leads to frequent episodes of volcanism. When the upwelling subsides, the bulge deflates, producing the fractured annulus.

Both large volcanoes and coronae on Venus are preferentially associated with rift zones, volcanic rises, and chasms concentrated at the equatorial latitudes. Indeed, the spatial distribution of volcanic centers on Venus is concentrated in an area bounded by Atla, Beta, and Themis Regiones which are large topographic swells thousands of kilometers across. Beta, in particular, displays all the characteristics of a dynamic rift zone (*Figure 15*). These volcanic swells probably sit directly over localized mantle upwellings, and they may be actively erupting now. However, despite their association with rift zones, volcanic features on Venus are not concentrated along linear boundaries or in chains as on Earth. This is the primary evidence against a global system of crustal spreading and subduction on Venus. Instead, the planet probably has developed a different mechanism for crustal recycling (see Chapter 12).

Stratigraphic relationships clearly show that the large volcanoes have erupted onto the widespread plains in geologically recent times. This is borne out by the fact that there is a relative paucity of craters on large volcanoes compared to the global average. It is also true of the most recent coronae deposits. However, structures associated with corona formation are seen throughout the planet's stratigraphic record. From this we conclude that coronae have a long evolutionary history. The sequence of feature emplacement paints a picture of change over time. Initially the volcanoes spewed forth abundant lavas thick enough to bury preexisting craters and form sinuous channels. Then the activity became localized and concentrated at volcanic rises and individual edifices.

EXPRESSIONS OF TECTONISM

Fifteen percent of the Venus surface comprises highland plateaus and mountain belts. Many of the terrains typical of highlands are found within Aphrodite Terra, which straddles the equator from 45° to 210°. Western Aphrodite is composed of regions called Ovda and Thetis, which stand about 3 to 4 km above the planet's mean radius (6,052 km). Ovda and Thetis are dominated by highly deformed crust called *tessera* terrain. First identified in the Venera data, tessera terrain is characterized by a very rough surface, topography that is elevated relative to its surroundings, and a complex deformation history involving at least two sets of intersecting structures (*Figure 16*).

Figure 13. Magellan scientists combined radar-reflectivity data with topographic altitudes to create this computer-generated perspective view of the large volcano Sapas Mons. Another volcano, Maat Mons, appears on the horizon.

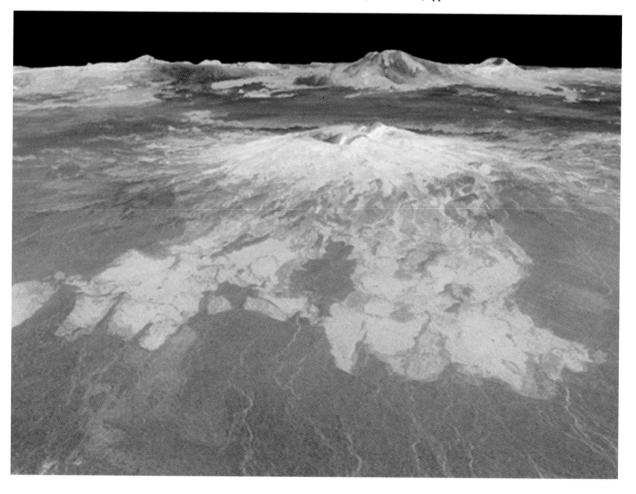

Tessera terrain occupies about 8 percent of the surface area of Venus and may occur as small (tens of kilometers across) "islands" within the plains. The tesserae are embayed by the volcanic flows that cover the widespread plains and thus are commonly the oldest preserved unit in any given area. This observation is supported by the work of Mikhail Ivanov and Alexander Basilevsky, who found that crater counts on tesserae are up to 40 percent higher per unit area than the global average. However, the tesserae are not ancient highlands like those on the Moon. Instead, they record complex deformation that occurred just prior to the eruption of the global plains.

Ridges, fractures, and graben are ubiquitous within tesserae and are accompanied by minor amounts of volcanism. Large

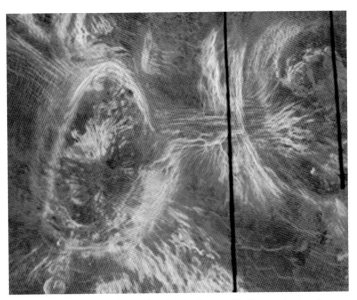

Figure 14. **This mosaic of Magellan data in Fortuna Regio contains two large coronae: Bahet (at left, 230 km across) and Onatah (350 km across).**

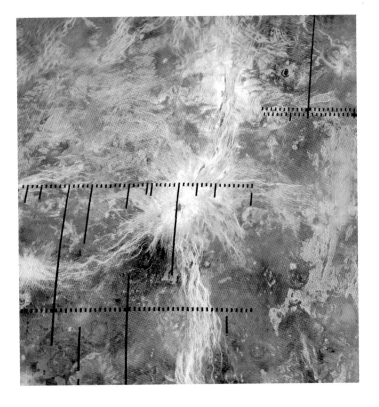

tessera plateaus, such as Ovda and Thetis, typically have relatively steep-sided margins that stand higher than the plateaus' interiors (*Figure 16*). At the time of this writing, the sequence of tessera evolution is under debate. Do the extensional structures we find there predate those involving contraction, or vice versa? This sequence is critical to understanding whether the plateaus formed as regions of crustal thickening due to magmatism or due to horizontal compression. The local gravity signature of large tessera plateaus shows them to be *isostatically compensated* at present. This means they are floating in equilibrium atop the mantle, like icebergs, and not supported from below by deeper upwellings. Just the opposite is true for prominent volcanic rises like Atla Regio and Beta Regio, which have attained significant elevation largely because they are being pushed upward and supported from below by hot, rising mantle plumes.

The eastern portions of Aphrodite are cut by a series of a deep, narrow canyons that extend eastward toward Atla Regio, where they are joined by large volcanoes such as Maat Mons and Sapas Mons. As they cross Atla and Beta, these *chasmata* consist of abundant faults and graben, which form rift zones with up to 6 km of vertical relief (*Figure 17*). The shape and scale of these zones are similar to those of terrestrial rifts such as the East African Rift. Despite this resemblance, in the Venusian situation volcanism is confined to individual edifices along the rifts and does not directly emanate from the rift itself, as occurs at mid-ocean ridges on Earth. This argues against these rifts being sites of plate spreading.

Perhaps the most recognizable tectonic feature on Venus is a high-standing "continent" at far northern latitudes called Ishtar Terra. Mountain belts feature prominently within western Ishtar (*Figure 18*), while the eastern half consists largely of Fortuna Tessera. In between is an impressive mountain chain called Maxwell Montes. (Named after famed physicist James Clerk Maxwell, it is the only Venusian feature that honors a man; all the rest recall important women in history or mythology.) Maxwell forms the eastern boundary of an elevated plateau, Lakshmi Planum, which is bounded on its north, west, and south by the mountain belts Danu, Akna, and Freya Montes, respectively. Lakshmi sits 4 km above the planet's mean radius and is dominated by volcanic plains and two major shield volcanoes.

The steep slopes (up to 35°) and unique elevation (12 km) of the Maxwell Montes mean that something unusual is taking place underneath. Either this chain formed relatively recently in Venusian history, or the strength of the underlying crust is quite high, or the mountains consist of low-density rock. Although most investigators agree that Ishtar was somehow created under compression, the mechanisms of formation of this enigmatic structure remain equivocal (see Chapter 12).

Figure 15. **The dramatic complex known as Beta Regio is considered one of the youngest and most dynamic features on Venus. Its two radar-bright peaks, Rhea Mons (top) and Theia Mons, are sliced open by a huge north-south rift known as Devana Chasma. Many faults radiate outward from the two tectonic centers, about 1,100 km apart. The rough surface of Beta Regio makes it very reflective at radio wavelengths, and it was among the first features seen on Venus in the 1960s during the infancy of ground-based radar experimentation.**

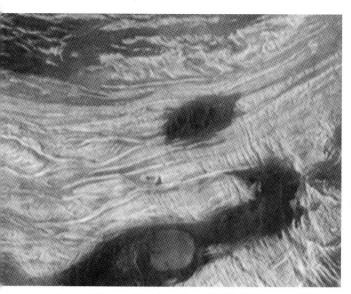

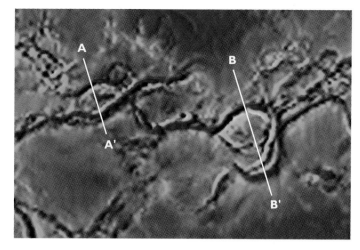

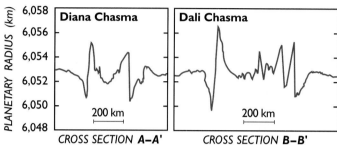

Figure 16. The northern boundary of Ovda Regio, one of the large highlands ringing the equator of Venus. Arcing ridges mark a boundary where the elevation drops 3 km from Ovda to the surrounding plains. Some of the ridges have been cut at right angles by extension fractures. Dark material, interpreted to be lava, fills the region between the ridges. The curvilinear, banded nature of these ridges suggests that compression contributed to their formation. The image is about 300 km across.

Figure 17. Diana Chasma and Dali Chasma are two major canyons near the east end of Aphrodite Terra. Each has a raised rim on one side that abruptly drops several kilometers to the floor below (heights in the graphs are exaggerated by a factor of 100).

One of Magellan's discoveries is that tectonic features are rather ubiquitous throughout the volcanic plains. Unlike the situation on Earth, where such features tend to be concentrated in localized areas, on Venus they are distributed across large regions. Wrinkle ridges are the most common feature on the plains. These long (10 to 50 km), narrow (about 1 km), sinuous features often occur in evenly spaced, parallel sets. Most likely they form under compressive stress. Wrinkle ridges are superimposed upon, and thus postdate, the widespread regional plains. However, they are not seen on the lobate flows associated with large volcanoes. Thus wrinkle ridges must be associated with the planetwide plains-forming event. Perhaps they reflect the stress fields that accompanied the formation of topographic features in many discrete, local situations. Interestingly, however, a global map of wrinkle-ridge orientations shows them to align circumferentially to the Aphrodite Terra highland. This suggests that the ridges may have been influenced by the stress field associated with the highland.

The plains also include regions of more concentrated tectonic deformation, termed *ridge belts,* which were identified in the planet's northern hemisphere in the Venera 15 and 16 data. The belts typically rise a few hundred meters above the surrounding plains, and they can be tens of kilometers wide and hundreds long. They may consist of a single broad arch or a complex assemblage of smaller ridges. Ridge belts are concentrated within or near the major lowlands, including Atalanta Planitia and Lavinia Planitia. In Lavinia, the ridge belts are joined by fracture belts of similar dimensions, though oriented orthogonally to the ridges (*Figure 19*).

Lavinia's ridge belts are examples of localized deformation in the broad, lowland plains. The ridge-fracture systems there and elsewhere on Venus are embayed by lava flows that themselves bear wrinkle ridges typical of distribution plains deformation. This suggests that both features may have formed under a similar stress field. Additional evidence for broad crustal deformation comes from 6,800-km-long Baltis Vallis, which shows variations in elevation along its length of up to 2 km. Assuming that Baltis formed on an initially flat or gently sloping surface, these topographic variations reflect vertical movements of the crust that have occurred since the plains were emplaced. The tilting of plains at tessera boundaries also reflects broad scale topographic adjustments of the crust.

All these stratigraphic relations give us details on the sequence in which Venus's features came to be. The tessera and lineated plains are the oldest units, and they indicate early episodes of faulting, folding, and perhaps mountain building. These units are typically embayed by the volcanic plains. Due to the absence of identifiable individual lava flows, we suspect that the plains resulted from widespread lava flooding. Plains emplaced more recently include bright, digitate, dark, and mottled plains, all of which are linked to volcanism in some form (individual vents, fractures, and coronae). Other features associated with recent volcanism are large individual edifices like Sapas Mons and the clusters of shields and domes. In a number of areas, ridge belts and fracture belts deform older plains, and these must therefore correspond to some of the most recent tectonic activity that has occurred on Venus.

The distribution of these geologic units is not uniform. The region bound by Atla, Themis, and Beta is dominated by fracture belts (corona chains), mottled plains, and volcanic edifices.

Tessera terrains are relatively rare in this area. In comparison, the area bound by Ishtar Terra to the north and Aphrodite to the south contains large concentrations of tesserae and regional plains, with few large volcanic edifices present. Many ridge belts are clustered in the north polar region between longitudes 180° to 240°, and a second major ridge-belt province is located near the south pole in Helen Planitia and Lavinia Planitia. In addition, this part of the planet contains the greatest concentration of large, visually distinct lava flow fields. We suspect that this southern region, along with the features Beta, Atla, and Themis, contains some of the planet's most recent lava deposits. Also probably quite young are the chains of coronae found along fracture belts in these regions.

REMINDERS OF EARTH

Venus has extremely slow rates of erosion compared to the spectacular effects caused by water and wind on Earth or even Mars. The atmosphere of Venus, though dense and moving quickly at higher altitudes, is actually rather sluggish right at the surface. There typical wind speeds are about as fast as a human walks. Although this is fast enough to move sand and dust, the slow speed makes the particles ineffective as cutting tools and agents of erosion. In Magellan's images, we see wind-related streaks in the lee of obstacles (*Figures 20,21*). These are not deposits of dust; instead, the wind has changed the surface in ways that make it rougher (and thus bright in the radar images) or

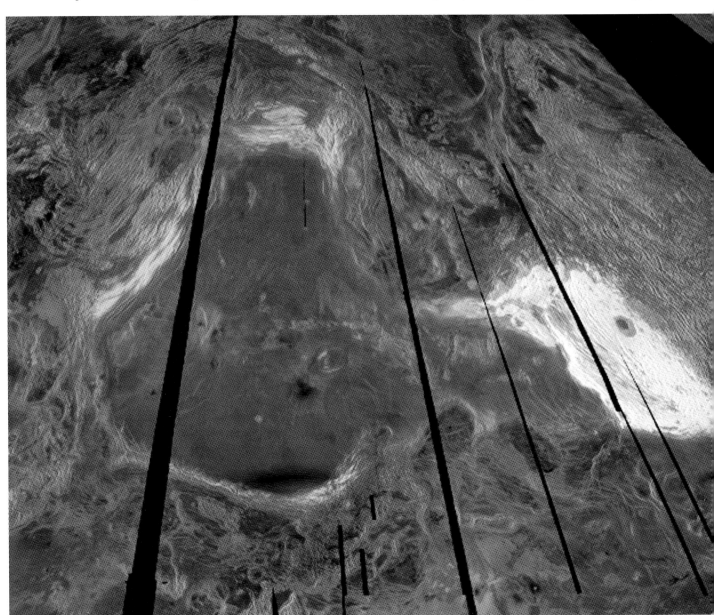

Figure 18. Western Ishtar Terra includes pear-shaped Lakshmi Planum, a volcanic plateau surrounded by mountain belts, and the radar-bright Maxwell Montes at right. Maxwell is the highest mountain chain on Venus, rising almost 12 km above the planet's mean radius. Its western (left) slopes are very steep, whereas the eastern slopes descend gradually into Fortuna Tessera. Broad ridges and valleys that make up much of Ishtar and Fortuna resulted from lateral compression within the crust.

Most of Maxwell Montes reflects radar strongly, a common trait on Venus at high altitudes. This phenomenon is thought to result not from a rough surface but from the presence of an unusual radar-reflective mineral similar to pyrite. The prominent circular feature in eastern Maxwell is Cleopatra, a double-ring impact basin about 100 km across. Black wedges and rectangles indicate areas where the Magellan orbiter acquired no data.

smoother (darker). In *Figure 20*, notice how the wind changes direction from a southeast-northwest flow at the right of the image to an east-west flow at the eastern edge of the outflow channel. It is possible that many of the wind features on Venus were formed during short-lived "storms" created in the atmosphere by impact events.

Another type of surface process is landslides. Although not very common on Venus, landslides have produced some spectacular features (*Figure 22*). Some of these exhibit characteristics typical of terrestrial slumped blocks (masses of rock that slide and rotate down a slope instead of breaking apart and tumbling). In some cases the heavily scalloped hillside suggests that much more material has slid away than we see at the foot of the slope. Possible explanations for the missing debris are that it may have been covered by lava flows, the debris may have weathered, or that the radar may not be recognizing it because the individual blocks are too small.

These landslides and dust streaks — like the volcanism, faulting, and tectonism discussed earlier — are familiar features on planetary surfaces. In fact, except for the lack of water, most of the geologic processes that occur on Earth can also be seen in one form or another on Venus. The absence of water is actually a boon for planetary geologists. Because so little erosion has taken place, we have been able to study craters, volcanoes, tectonic features (faults), and the effects of wind and landslides in their pristine, original state.

Nonetheless, there are some profound differences between Earth and Venus. Most obvious (so far) is that each loses its heat in a different way. On Earth the crust is recycled laterally, through plate growth, motion, and subduction. By contrast, the volcanic and tectonic structures we've seen on Venus suggest that the recycling occurs vertically, with mantle upwellings triggering volcanism and mantle downwellings resulting in compression. One theory about why Venus does not exhibit lateral plate movement is its lack of water. On Earth, water is important in the formation of chemically differentiated magmas that comprise continental crust, and these in turn help maintain the cycle of subduction.

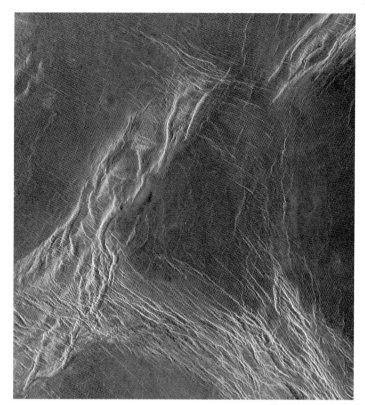

Figure 19. This 400-km-wide mosaic of Magellan images is centered in Lavinia Planitia at 38° south latitude and 348° east longitude. A broad belt of tectonically formed ridges runs from upper right to lower left. Radar-dark areas between the ridges are relatively smooth and probably filled with lava flows. Note the set of thinner fractures running from upper left to lower right. This intersection pattern is seen throughout much of Lavinia and suggests that compressional and extensional forces have affected a very large region.

Figure 20. This section of Navka Planitia covers 180 km in width and 78 km in height. The two radar-bright deposits at center outline a channel that flowed from a 60-km-wide crater outside the frame to the south. Within the channel, outlined by "bathtub ring" deposits, are small cones most likely of volcanic origin. At the end of the outflow channel are bright features that may be sand dunes.

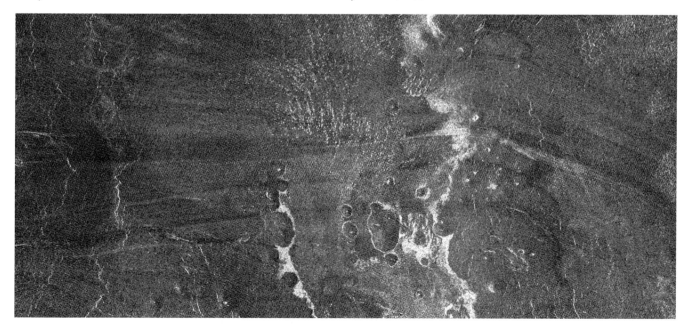

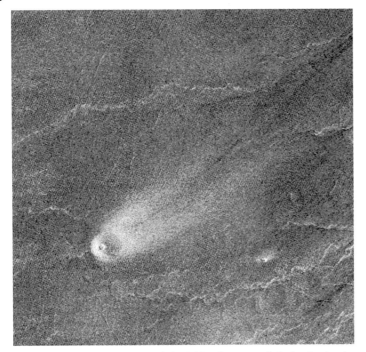

Figure 21. A cometlike tail extends for 35 km from a small, isolated volcanic structure. Although only 5 km across, the volcano has slowed the region's northeast-trending winds enough to cause deposition of this radar-bright material. This scene is located at the western end of Parga Chasma at 9.4° south latitude and 247.5° east longitude.

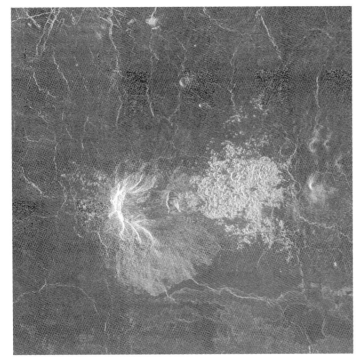

Figure 22. The bright feature seen slightly south of center is a volcano, 15 to 20 km in diameter, with a large apron of blocky debris to its right and some smaller aprons to its left. Several large landslides dropped down steep slopes and were carried by their momentum out into the smooth, radar-dark lava plains. At the base of the east-facing or largest scallop on the volcano is what appears to be a large block of rock, 8 to 10 km in length. This image is about 120 km wide.

What is amazing about Venus is that volcanic and tectonic resurfacing is not an ongoing process. Instead, it seems to have run rampant over a short period of time, obscuring all traces of what had occurred during the first 90 percent or so of the planet's history. This obliteration has robbed us of the opportunity to learn how and when the evolutionary paths of Venus and Earth diverged. The geologic rule of uniformitarianism — "the present is key to the past" — does not apply to Venus in the long-term sense. And so we wonder whether the ubiquitous volcanism seen on Venus today mimics the very early Earth or perhaps predicts an Earth yet to be. Will our planet's dynamic tectonism grind to a halt once its water disappears? Such speculations remain unanswered. For now, we are satisfied in having finally lifted the cloudy veil that has kept Venus's surface hidden for so long.

Planet Earth

Don L. Anderson

A PLANET'S SURFACE provides geologists with clues as to what is happening inside. But most of these clues are ambiguous because so many other processes contribute to surface characteristics. The record of the origin of our planet has been erased many times, due in part to erosion by wind and water, and in part to the continual recycling of material back into the interior and the repaving of the ocean basins by seafloor spreading. As a consequence, much of the Earth's exterior is less than 100 million years old, and even its oldest rocks are less than 4 billion years old.

The solar system's other solid planets and smaller worlds generally have more ancient surfaces, and this tells us two things that we cannot learn just from the Earth itself. One is that, in the early days of the solar system, violent and destructive impacts were common as larger bodies swept up and devoured smaller ones. The other is that most other worlds preserved evidence of these early happenings, while ours did not.

Geophysicists try to sidestep the mixed signals they encounter at ground level by studying the interiors of planets directly. On this planet, the effort has been rather successful: Earth is the only body for which we have detailed information, including three-dimensional images of the internal structure, from the surface to the center. And while we realize that other planets do not all share Earth's present behavior, they may at least have been put together in similar ways. So, intensive study of Earth's interior may yield knowledge that is applicable elsewhere — just as our study of other worlds will shape our understanding of Earth.

Early views of Earth's origin envisioned a gentle rain of dust and small particles that slowly accumulated layer by layer. A planet growing this way would remain relatively cool, building up heat mainly by the slow decay of radioactive elements. According to this scenario, an initially cold, homogeneous

The full Earth (showing Africa), as photographed by the crew of Apollo 17.

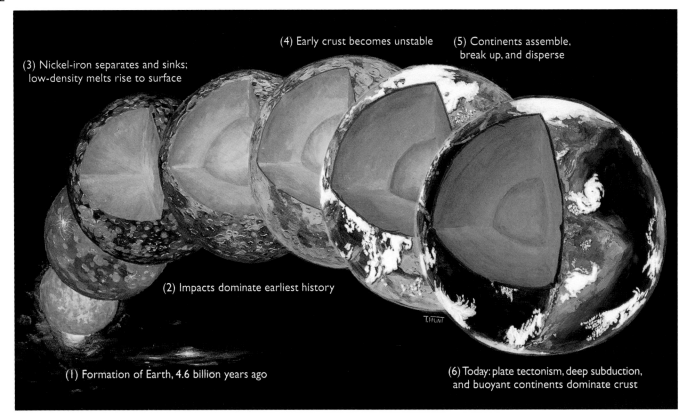

(3) Nickel-iron separates and sinks; low-density melts rise to surface

(4) Early crust becomes unstable

(5) Continents assemble, break up, and disperse

(2) Impacts dominate earliest history

(1) Formation of Earth, 4.6 billion years ago

(6) Today: plate tectonism, deep subduction, and buoyant continents dominate crust

Figure 1. Geophysicists do not yet know the exact circumstances of Earth's formation, but our planet's exterior must have been completely molten at least early in its history. Much of the energy needed to melt its outer layers came from innumerable collisions with interplanetary material left over from planetary formation.

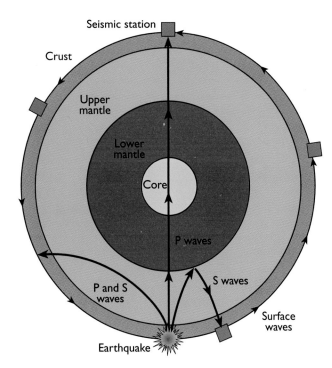

Figure 2. Earthquakes trigger different kinds of diagnostic seismic waves that travel around Earth and through its interior at 3 to 15 km per second. Compression *(P)* waves move almost twice as fast as shear *(S)* waves; they can also pass through the liquid outer core, which the S waves cannot.

Earth eventually heated up, started to melt, and formed a buoyant crust and a dense core in a way that somehow left behind (in most versions of this story) a homogeneous mantle.

However, planetary evolution has not been so simple. The energy associated with a single large impact is enough to melt, or even vaporize, much of the impactor and the planet it strikes. If the Moon really came into being when a Mars-size object struck the Earth (see Chapter 10), the energy from that collision would have melted much of the Earth itself. Even smaller hits, of which there were many more, would have caused widespread melting where they penetrated and generated shock waves.

So did the Earth start out cold or hot? The answer depends on whether it accreted slowly (100 million to 1 billion years) or rapidly (100,000 to 10 million years). As described in Chapter 2, theorists now lean toward the latter view. If that was indeed the case, kinetic energy would have been delivered faster than the growing Earth's ability to conduct and radiate it away as heat, so our planet would have remained molten, at least in its outer parts, as it accreted (*Figure 1*). However, every giant impact was essentially an instantaneous accretion event, and planets that grew by gathering up relatively large objects experienced widespread melting over and over again.

We assume, therefore, that rocky planets were molten as they grew — at least partially, and at least once. At such times their component materials had the opportunity to separate according to melting points and densities. The "heavy" materials sank toward the interior, creating cores, and the "light" ones rose to the surface, creating crusts. This process of gravitational separation, called *differentiation*, played a key role in the early histories of Earth and the other terrestrial planets. Differentiation is akin to what takes place in a blast furnace or fat-rendering plant.

By heating and boiling, the original material is reduced to frothy scum, dense dregs, and a "purified" liquid in between. However, over time these worlds may have acquired more internal stratification than simply a light crust and a dense core; as we shall see, the mantle situated between them can itself become layered according to chemistry and density.

DIVIDING EARTH'S INTERIOR

Seismology is the geophysicist's principal tool for probing planetary interiors. In a sense, the Earth is a huge spherical bell that is periodically "struck" by earthquakes. We learn about the interior by listening to how the Earth "rings" — that is, by noting how seismic waves move away from the source point, or focus, of an earthquake (*Figure 2*). Of the four types of seismic waves, two travel around the Earth's surface like rolling swells on an ocean. A third type, called primary or *P* waves, alternately compress and dilate the rock or liquid they travel through, just as sound travels. Secondary or *S* waves propagate through rock (but not liquids) by creating a momentary sideways displacement or shear, like the movement along a rope that is flicked at one end. Both P and S waves slow down when moving through hotter material, and they are refracted or reflected at the boundary between two layers with distinct physical properties.

Rocks' physical properties vary with depth due to increasing temperature and pressure and, in places, changes in chemistry or physical state. For example, the most common minerals in the crust and upper mantle are all unstable farther down. As pressure increases, the atoms in their crystals become more tightly packed, and their density increases. These changes are gradual except at phase transitions (such as when carbon transforms from graphite to diamond under pressure). Phase transitions cause a rapid or abrupt change in physical properties, which can often be measured as a velocity change by seismic techniques.

In the early 1980s, several groups of researchers discovered that seismic waves could be used to produce three-dimensional maps of the Earth's interior, a technique known as *seismic tomography*. The word "tomography" derives from the Greek word for a cutting or section, and in effect geophysicists create a series of cross-sections of the interior at various depths (*Figure 3*). Seismic tomography is a very powerful technique that has revolutionized our study of Earth's interior.

In fact, geophysicists have long relied on physical properties such as density and seismic velocity, rather than chemistry or composition, to distinguish the principal divisions of the Earth's interior. In 1906, the British geologist Richard D. Oldham found that at a certain depth, compression or P waves slow sharply and S waves cannot penetrate further. It was the first evidence that Earth has a liquid core. Only three years after Oldham's revelation, the Yugoslavian seismologist Andrija Mohorovičić discovered that the velocity of seismic waves takes a large jump about 60 km down. This Mohorovičić or "Moho" seismic discontinuity marks the crust-mantle boundary, where changes in rock chemistry and crystal structure occur. At the core-mantle boundary, averaging 2,890 km in depth, the composition of rock changes from silicate to metal-

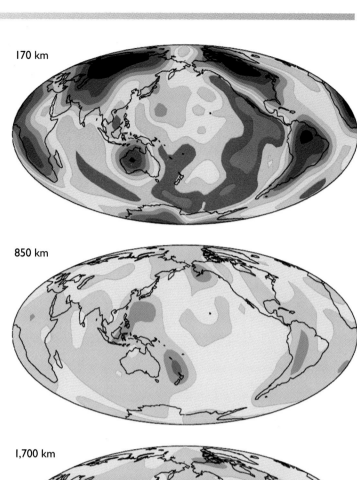

170 km

850 km

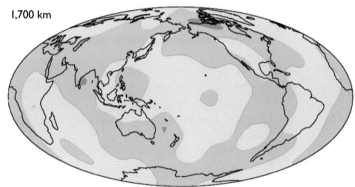

1,700 km

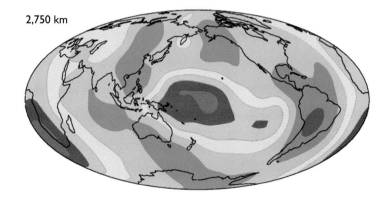

2,750 km

Figure 3. **This series of maps shows the state of Earth's interior at different depths as determined by seismic tomography using shear (S) waves. The red regions have slower-than-average seismic velocities and are therefore realtively hot. The green regions have faster velocities and are therefore colder. In the 170-km-deep map, notice the association of hot upper mantle with oceanic ridges and continental tectonic regions. A large hot region is apparent beneath the central Pacific near the core-mantle boundary in the 2,750-km map.**

Earth's Interior Structure

Region	Depth (km)	Percent of Earth's mass	Percent of mantle-crust mass
Continental crust	0–50	0.374	0.554
Oceanic crust	0–10	0.099	0.147
Upper mantle	10–400	10.3	15.3
Transition region	400–650	7.5	11.1
Lower mantle	650–2,890	49.2	72.9
Outer core	2,890–5,150	30.8	
Inner core	5,150–6,370	1.7	

Table 1. **A summary of Earth's internal structure, as deduced from decades of probings with seismic techniques.**

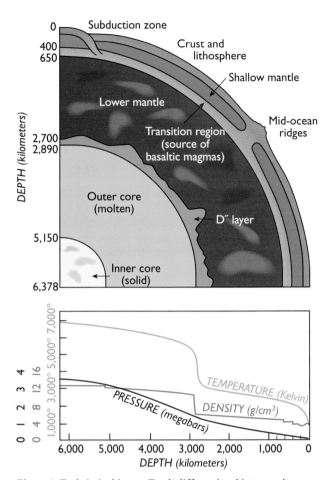

Figure 4. **Early in its history, Earth differentiated into a series of layers with distinct physical and perhaps compositional properties. Lateral variations of seismic velocity exist at all depths in the mantle and inner core.**

lic and its physical state changes from solid to liquid. This boundary is also known as the Gutenberg discontinuity, after Beno Gutenberg, who made the first accurate determination of its depth.

The *crust, mantle,* and *core* account for 0.4, 67.1, and 32.5 percent of our planet's mass, respectively. Seismic discontinuities allow a further division of the Earth into an inner core, outer core, lower mantle, transition region, upper mantle, and crust (*Table 1, Figure 4*). These regions are not necessarily all chemically distinct, nor can we assume that each of them is chemically homogeneous.

The inner core was discovered by the Danish seismologist Inge Lehmann, and its boundary is now known as the Lehmann discontinuity. It is solid, primarily the result of "pressure-freezing" (most liquids will solidify if the temperature is decreased or the pressure is increased). Probably the entire core was once molten, but over time it has lost enough heat for the inner portion to solidify. This now "floats" in the center of the outer core, which essentially decouples it from the mantle. Notably, seismic monitoring over the last several decades has permitted seismologists Xiaodong Song and Paul Richards to detect a small displacement of the inner core relative to the mantle. They have found that the inner core actually rotates with respect to the mantle (*Figure 5*).

The outer core remains liquid due to its high temperature and the fact that iron alloys melt at lower temperatures than do common rocks. The viscosity of the outer core is very low, probably not much greater than water. We expect it to behave in general like other fluid parts of the Earth. Rapid motions of molten iron in the core are responsible for Earth's magnetic field (see Chapter 4) and for some of the subtle jerkiness in our planet's rotation.

Just above the core is a 200- or 300-km-thick layer, called *D″*, that may differ chemically from the rest of the lower mantle lying above it. It may represent material that was once dissolved in the core, or dense material that sank through the mantle but was unable to sink into the core. The D″ layer comprises about 3 percent of Earth's mass, or about 4 percent of the mantle. Above that, the lower mantle has relatively uniform seismic properties compared to the upper mantle, but subtle variations have been detected that may indicate a seismic discontinuity at an average depth of 1,000 km. There are broad high-velocity

features in the lower mantle beneath North and South America and across the southern part of Eurasia. These may represent cold downwellings.

Smaller seismic discontinuities occur at several depths in the mantle and halfway through the core. These are often attributed to phase transitions, but they may signify changes in composition. The two largest ones in the mantle, 400 and 650 km down, represent abrupt rearrangements of the atoms in the major mantle minerals. Large variations in seismic velocity have also been found from place to place. Thanks to seismic tomography, we now realize that the Earth's upper mantle exhibits as much variation horizontally as it does vertically (*Figure 6*).

EARTH'S COMPOSITION

Ours is the only planet for which we can speak with some confidence about its bulk composition or chemical makeup. By combining Earth's mass with seismic determinations of the radius and density of the core, we have deduced that the Earth is about one-third iron and that this iron is concentrated toward the center of the planet. In fact, our planet's solid inner core, which is smaller than the Moon but three times denser, may be pure iron and nickel. The outer core is slightly less dense than molten iron, a characteristic that requires about 10 percent of some lighter elements such as sulfur, oxygen, or both. These elements are considered likely because they are cosmically abundant and would readily dissolve in the hot metallic soup.

Earth is the largest terrestrial planet and contains slightly more than 50 percent of the mass in the inner solar system, excluding the Sun. Compared to Earth, the dense planet Mercury contains proportionately more iron; Mars and the Moon contain substantially less iron, even though they may have small cores. On the basis of its similarity to our planet in size and density, Venus probably has an Earthlike core. But a solid inner core may be absent, because we expect the interior of Venus to have slightly lower pressures and possibly higher temperatures.

The bulk of the Earth is in its mantle, the region between the core and the thin crust. We can sample the top of the mantle in several ways. Fragments of it are exposed in eroded mountain belts and brought to the surface by volcanic eruptions. The major mantle minerals excavated in these ways are olivine $(Mg,Fe)_2SiO_4$ and pyroxene $(Mg,Fe)SiO_3$; thus, iron is present but only as a minor constituent.

The most abundant material we see erupting from the mantle is *basalt*, and it must exist there in vast quantities. Basaltic magma is rich in the elements calcium and aluminum and is less dense than upper-mantle material, which allows it to erupt into or onto the crust. The ocean floor is covered with basalt. Iceland and Hawaii (*Figure 7*) are two examples of thick basalt piles that have accumulated on the ocean floor. Hidden from our view under seawater is a 40,000-km-long network of volcanoes — the oceanic ridge system — which generates new oceanic crust at the rate of 17 km³ per year. In fact, the majority of the Earth's crust was made in this way.

However, at depths below 60 km in the mantle, cold (solid) basaltic material converts under pressure to a form of rock called *eclogite*, which is much denser than shallow-mantle rocks because it contains garnet, a complex, aluminum-bearing silicate mineral. Large bodies of eclogite can sink through the upper mantle, which probably explains why the crust on Earth never gets thicker than about 60 km. Inside smaller terrestrial planets, like Mars, the pressures at a given depth are lower, so their crusts can extend farther down without converting to dense eclogite. Within a hotter planet, like Venus, a thick basaltic crust would melt at its base rather than convert to eclogite.

Although our direct samples of Earth's interior are limited to the crust and shallow mantle, we know from seismic tomography that broad regions with low seismic velocities extend to depths of at least 400 km under oceanic ridges and other volcanic terrains. Magmas and rock-magma mixtures have low densities and low seismic velocities, so it seems reasonable that the basalt source region lies below about 400 km. When a hot

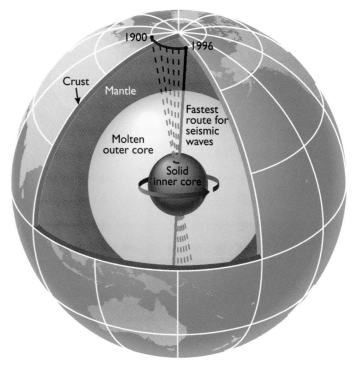

Figure 5. Geophysicists have discovered that the route of the fastest seismic waves through Earth's interior is gradually shifting eastward because the inner core is rotating slightly faster than the rest of the planet. This finding may help explain how Earth's magnetic field reverses polarity.

Figure 6. Earth's interior exhibits considerable variation with both depth and location. In these plots, red indicates the slowest seismic velocities and the hottest material, blue the fastest and coldest. The panel at left shows continents and plate boundaries (yellow lines) surrounding the Atlantic Ocean. The middle panel shows a slice into the underlying mantle to a depth of 550 km (with depth exaggerated by five times). The cutaway at right goes to a depth of 2,890 km, the mantle-core boundary; it also shows the solid inner core suspended within the outer core (red). The inner core and mantle also show inhomogeneities.

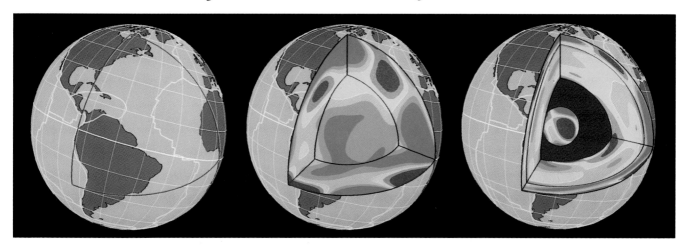

silicate rock or low-density magmatic mush ascends from that great depth, it eventually separates into molten liquids (which erupt at volcanoes) and crystals (which stay behind in the mantle or form new crustal material).

Thus, geophysicists have been able to identify three compositional layers in the outer Earth: (1) the buoyant crust, containing low-density minerals dominated by quartz (SiO_2) and metal-poor silicates called feldspars; (2) the uppermost mantle, con-taining olivine, pyroxene, and other refractory minerals, which crystallize at high temperatures, and thus settle out of rising magma mushes; and (3) a "fertile" layer, below 200 to 400 km, that contains a large basaltic component and therefore abundant calcium and aluminum. This third layer is dense when cold, due to the garnet it contains, and buoyant when hot, because garnet and related minerals melt easily to form basalt. Much of the mantle at depths between 100 and 400 km is at its melting point.

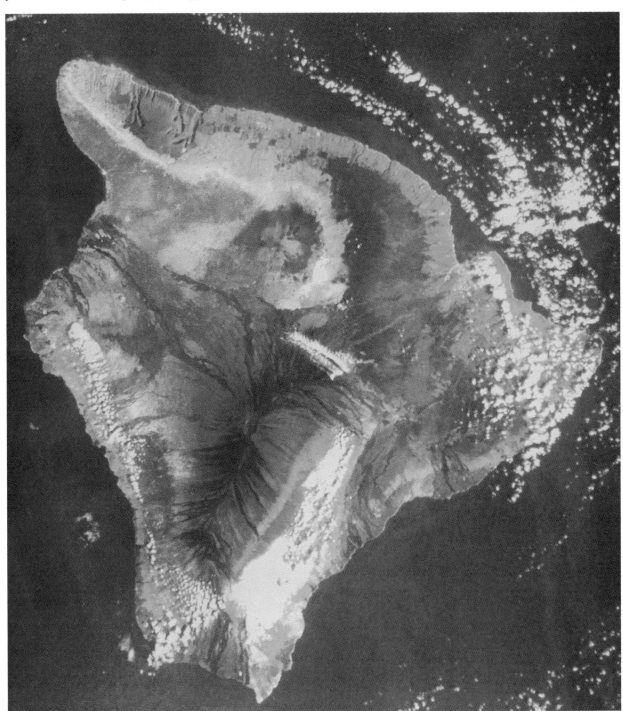

Figure 7. The island of Hawaii consists entirely of outpourings from Earth's mantle. The now-dormant volcano Mauna Kea, in the island's northern half, has become the site of numerous astronomical observatories. But Mauna Loa, to its south, is still quite active, especially along its southeast flank. The Hawaiian Islands are part of a long chain of peaks that formed either as the Pacific lithospheric plate slowly moved northwest over a plume of upwelling mantle material, or as a crustal crack propagated toward the southeast due to volcanic loading and other stresses. Hawaii, at the southeast end of the island chain, and Loihi, to its south, are the regions now active. This photograph is a composite of images from Landsat spacecraft.

Underneath all of this is the lower mantle. If Earth's elemental abundances match those in the Sun and primitive meteorites, then the massive lower mantle must be mainly silicon, magnesium, and oxygen. It probably also contains some iron, calcium, and aluminum. Although Ca and Al are well represented in Earth's crust, the crust is too thin to yield Ca:Si or Al:Si ratios for the whole Earth as high as those found in the Sun, meteorites, and, by inference, the planets. Moreover, little calcium or aluminum exists in upper-mantle rocks (otherwise basalts could not rise through them en route to the surface), nor are they present in the core.

The lower mantle may be richer in silicon than the layer above it. The reasoning behind this assumption is as follows: Primitive meteorites and the Sun have about one magnesium atom for every silicon atom. In the mantle, this 1:1 ratio would favor the formation of the mineral enstatite ($MgSiO_3$, a pyroxene) over forsterite (Mg_2SiO_4, an olivine). However, we know from its surface exposures that the upper mantle is olivine-rich and has a Mg:Si ratio of about two. Farther down, at the high pressures present in the lower mantle, Mg_2SiO_4 decomposes to two new minerals. One is periclase (MgO), which has the crystal structure of ordinary table salt, $NaCl$. The other is an ultrahigh-pressure form of enstatite (which, incidentally, has the same crystal structure as many of the new high-temperature superconductors). This enstatite variant propagates seismic waves at much higher velocities than periclase does and matches the seismic velocities we have observed for the lower mantle. Therefore, at great depths $MgSiO_3$ would appear to be the most abundant mineral. Some seismic evidence also indicates that the lower mantle has more iron (as FeO) than the upper mantle does; it may be similar to the mantles of the Moon and Mars, which we also suspect to be rich in FeO.

WHERE ON EARTH IS THE CRUST?

Planets grow by colliding with other objects, an energetic process that results in melting or even vaporization. Most of the energy is deposited in the outer layers, except for the small number of truly giant impacts that are as likely to destroy the target object as add to its bulk; these may melt a large fraction of a planet. A global ocean of magma can segregate incoming material into solid and liquid fractions that float and refractory crystals and iron-rich melts that sink.

Planetary geologists have invoked such a global magma ocean to explain the Moon's anorthositic highlands (calcium- and aluminum-rich silicates that floated to the surface) and its "KREEP" basalts, which cooled from the final liquid dregs that were highly enriched in potassium (K), rare-earth elements (REE), and phosphorus (P). A similar process probably occurred inside Earth, except that its much higher pressures caused dense garnet-bearing eclogite to form instead of buoyant anorthosite. In fact, high-grade anorthosite is fairly rare on Earth. Therefore, one key product of Earth's magma ocean did not float to the surface but sank from view. Had we not obtained actual anorthositic samples of the Moon, the magma-ocean concept might never have occurred to terrestrial geologists.

The crust of the Earth would be about 200 km thick if most of the low-density and easily melted material in the interior had

Elemental Fractions in the Crust

Rubidium	(Rb)	68	Sodium	(Na)	13
Cesium	(Cs)	67	Aluminum	(Al)	2.4
Thorium	(Th)	55	Calcium	(Ca)	0.9
Barium	(Ba)	49	Silicon	(Si)	0.7
Uranium	(U)	47	Iron	(Fe)	0.07
Lanthanum	(La)	27	Magnesium	(Mg)	0.06
Strontium	(Sr)	21			

Table 2. **The abundance of various elements in Earth's crust, as a percentage of their estimated abundance in the whole Earth.**

separated out during Earth's formation. Yet the average terrestrial crust (20 km) is considerably thinner than the lunar crust (100 km) — even though the Moon has only 2 percent of Earth's volume. Does this mean that our planet did not have a magma ocean? Or has the crustal material mostly remained in or returned to the mantle?

The lunar crust is so thick and contains so much of the Moon's calcium and aluminum that it must have formed very efficiently, with its light crustal minerals rising directly to the top of a deep magma ocean. However, on a larger body like Earth, the pressures far down in a magma ocean are so great that buoyant minerals never form. Instead, dense crystals such as garnet and pyroxene soak up the calcium and aluminum. These, by and large, stay in the mantle and may even sink to the base of a magma ocean, thus limiting the crust's thickness.

Even so, the high concentrations of some elements in the Earth's crust (*Table 2*) tell us that most of the mantle *must* have differentiated either during accretion or shortly thereafter. These elements happen to be ones that are not easily incorporated into the high-pressure minerals that form at depth in a magma ocean. Therefore, Earth probably made its crust quite efficiently, but very likely the missing crust resides somewhere in the mantle. That is, the amount of crust now at Earth's surface is much less than the *potential* crustal material and probably represents only a small fraction of the total volume of the crust that has been generated in 4½ billion years.

There is also a good reason why Earth cannot have a thick "secondary" crust — that is, one formed by continental collision, mountain building, or the accumulation of volcanic materials. Wherever these processes cause our planet's crust to thicken to more than about 60 km, the low-density crustal minerals convert to denser ones, causing the bottom of the crust to "fall off" or, technically, to delaminate. But even if delamination did not occur, the great pressure present below 60 km makes the seismic velocities there so high that a seismologist would call this deep-lying material part of the mantle, not the crust. In fact, "crust" is inherently a physical concept, and its properties and thickness are derived from seismology. However, since erosion and volcanism supply us with many samples of the lower crust and shallow mantle, we know that the crust truly is compositionally distinct. As mentioned, it is calcium-, aluminum-, and silicon-rich compared to the shallow mantle, so changes in physical properties at the crust-mantle boundary are accompanied by changes in chemistry as well.

Figure 8. With the oceans emptied and the continents obscured, this map reveals a seafloor shaped by ceaseless geologic activity. It uses a combination of shipboard bathymetry and satellite-derived radar altimetry of the ocean's surface, which reflects the underlying topography.

Figure 9. Earth's major lithospheric plates are in motion with respect to one another. At divergent boundaries (such as midocean ridges) the plates move apart, only to collide and overlap at convergent boundaries (subduction zones). Plates slide past each other along transform faults, the most famous of which is the San Andreas fault that runs the length of California.

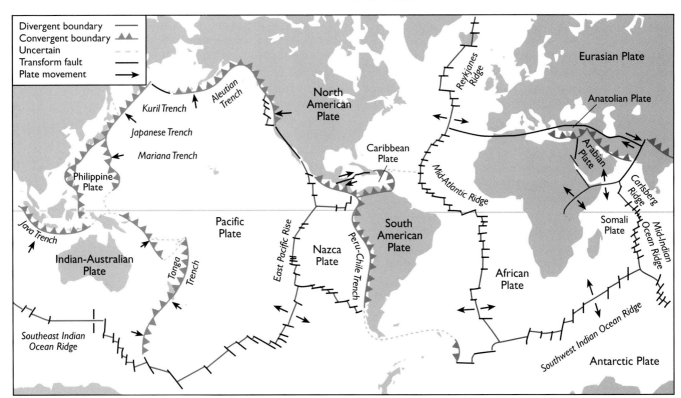

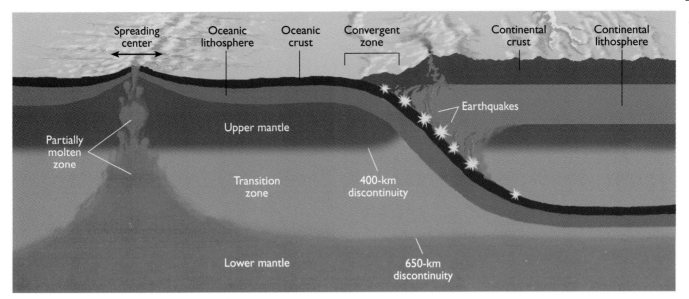

Figure 10. The basalt that forms oceanic crust does not come from immediately below the lithosphere but from a much-deeper transition zone in the mantle. As it rises, this material decompresses and may become partially molten; finally it erupts at a spreading center. As the new oceanic crust moves away from its formation site, it cools and thickens, eventually becoming dense enough to plunge back into the mantle. This subduction occurs dramatically along zones of convergence marked by deep trenches, frequent earthquakes, and active volcanism. In some places the slab may sink below the 650-km discontinuity.

THE LITHOSPHERE

Although most of the crust and mantle are solid, we know from seismic velocities, the abundance of volcanoes, and the rise of temperature with depth in wells and mines that much of the outer part of the Earth is near or above its melting point. In fact, the coldest part of our planet is its surface. Since cold rocks deform slowly, we refer to this rigid outer shell as the *lithosphere* (the "rocky" or "strong" layer). On the Earth the lithosphere is not a single seamless shell, but rather a patchwork of rigid, snugly fitting *plates* that ride atop the mantle (*Figures 8,9*). These plates — eight large ones and about two dozen smaller ones — are moving with respect to one another, and their interactions are collectively called *plate tectonism*, a subject to be discussed later.

At depths between about 50 and 100 km, lithospheric rocks become hot and structurally weak enough to behave as fluids — at least over geologic time. This portion of the upper mantle is called the *asthenosphere* (or "weak" layer), and it may be partially molten. Seismic velocities under young seafloor and tectonic regions are so low that some partial melting is required to depths as great as 400 km. Below the asthenosphere, the temperature continues to climb, but the solidifying effects of high pressure become dominant. So at still-greater depths the Earth again becomes strong and harder to deform. The region between the 400- and 650-km seismic discontinuities is called the *transition region* or *mesosphere* (for middle mantle), and the basalts that make up midocean ridges and new oceanic crust may be derived from this region.

Within the Earth's lithosphere, rocks are so cold and their viscosity so high that they support large loads and fail by brittle fracture rather than by deforming smoothly. New lithosphere forms at midocean ridges and thickens with time as it cools and moves away from the ridge, a process termed *seafloor spreading* (*Figure 10*). It also becomes denser and loses the high-temperature buoyancy it had initially; eventually it tries to sink back into the interior. From the way it deflects under large submarine volcanoes and enters deep-sea trenches, the oceanic lithosphere acts as an elastic plate whose thickness varies from zero at the midocean ridges to about 40 km under older seafloor.

Let's examine the formation and evolution of oceanic lithosphere a bit more closely. Portions of the Earth's mantle, especially the asthenosphere, behave like hot plastic and are in continuous, slow convective motion that brings heat from the interior to the surface. Upwelling mantle material partially melts in the asthenosphere, where a segregation takes place; denser refractory crystals are left behind when the lighter, easily melted material erupts upward. This buoyant, erupting melt creates the lithosphere's topmost layer, the oceanic crust, which is basaltic and averages about 6 km in thickness.

The lithospheric layer beneath this is essentially normal mantle material that has lost its basaltic component. The basalt is missing either because it rose as a melt to the crust or, at depths of roughly 60 km, it reverted to dense eclogite and sank as large blobs through the upper mantle and into the mesosphere. However, melting these blobs restores their buoyancy, and the resulting magma mush will rise toward the surface. If it has a clear upward path, as occurs along the midocean ridges, it will erupt as basalt onto the seafloor. The magma does not always ascend vertically, however, and may first migrate laterally through the mantle for great distances. Elsewhere its path may be blocked completely, so it pools on the underside of previously formed lithosphere. We are ignorant of the composition of the lower oceanic lithosphere, but it is probably a mixture of basalt and refractory crystals. Substantial amounts of eclogite in the older lithosphere would help explain why it eventually sinks back into the interior.

120

In contrast, the continental lithosphere is about 150 km thick, and its crustal and upper-mantle components are both buoyant relative to the normal mantle below. Continents therefore float around like icebergs and do not directly participate in the deeper circulation currents of mantle convection. But lateral movement in the mantle can and does move these lithospheric icebergs around, and once they come to rest they can insulate the underlying mantle and cause it to warm. In the course of this *continental drift*, continents can override the thinner oceanic lithosphere along *subduction zones* — linear or arcuate features characterized by deep oceanic trenches and large volcanic cones. If the oceanic lithosphere is still young and thus hot, it tends to slide under the continent at a shallow angle; older, thicker lithosphere is denser and tends to dive steeply into the mantle.

On Earth and elsewhere, the lithosphere is an important element in planetary dynamics (see Chapter 12). If it gets too cold or too thick, it can shut off the access of hot magmas to the surface or become too hard to break and descend (subduct). If its proportion of light minerals is too great, it will stay buoyant and will not sink back into the mantle. If there are too many plates or if they are moving rapidly, they likewise may not become dense enough to subduct. Thus, there are a variety of ways to "choke up" the surface. In the extreme, a lithosphere may get too thick to break anywhere, creating one uninterrupted plate

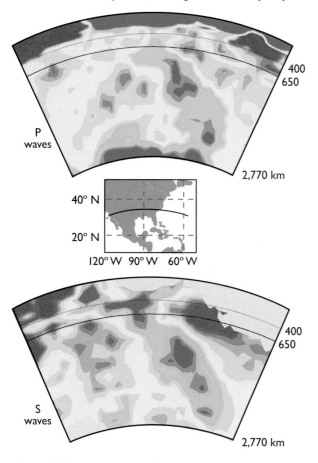

Figure 11. Seismic tomography has been used to create a cross-section of a descending slab of oceanic crust beneath North America. Blue indicates faster wave velocities (colder material), while reds are slower (warmer) ones. The large blue slab has probably subducted into the deep lower mantle over the last 100 million years.

that can slide around as a unit on the underlying mantle. On such a "one-plate planet," a huge meteoritic impact or the mass of a large new volcano could alter the planet's moment of inertia enough to make the whole outer shell rotate with respect to the spin axis.

Several mechanisms can fragment a lithosphere. Hot mantle upwellings can both heat and deform it. Diverging mantle currents below can create extensional stresses on its base. A lithosphere moving over an ellipsoid-shaped (rotating) planet will experience large stresses due to the changing contour of the surface. Tidal despinning, caused by a massive object in the planet's immediate neighborhood, can also generate large stresses in the surface layer and a global fracture pattern. If the lithosphere becomes too dense it may sag and break. Several subduction zones currently exist entirely under Earth's oceans. But it is not clear if they are the result of an instability of the oceanic lithosphere, or if subduction started at the edge of a continent and later migrated toward the ocean. We do know that both oceanic ridges and island arcs can migrate relative to the underlying mantle and Earth's spin axis.

The spin axis of a planet is controlled by the distribution of masses on the surface and in the interior. By analogy, the rotation of a spinning top is controlled by its shape, and its spin axis will change if bits of clay are attached to the surface. The physics of planetary reorientation is the same. If a large impact or a new volcano redistributes the mass, the planet will reorient itself relative to the spin axis so that the mass excesses lie closer to the equator. This shift is termed *true polar wander*. Both Mars and the Moon have apparently reoriented themselves to accommodate the effects of impacts or volcanoes. By contrast, *apparent polar wander* occurs whenever the continents drift relative to the magnetic pole. Geologists find evidence for such drift in magnetically aligned crystals in ancient rocks.

On the present-day Earth, true polar wander is very slight and results mainly from the rearrangement of mass due to melting glaciers. But the situation was apparently quite different in the past. Major shifts of the Earth's lithospheric shell relative to its spin axis might have followed convective rearrangement of mass in the interior, plate subduction, or the build-up of heat beneath large continents. These mass adjustments might be responsible for some major events in the geologic record, such as the breakup of supercontinents discussed in the next section.

Earth's rotation axis has apparently moved about 8° in the past 60 million years and 20° in the past 200 million — a period of time when the configuration of continents and subduction zones was also changing dramatically. There is some evidence that large areas shifted by about 90° near the time of the "Cambrian explosion," a dramatic change in the evolution of biota on our planet some 540 to 570 million years ago. This may have been caused by an episode of true polar wander, triggered by a change in the configuration of major subduction zones.

PLATE TECTONISM

Planets have various options for relieving themselves of their internal heat. Earth chooses the plate-tectonism option, whereby heat rises and dissipates at the midocean ridges, and giant slabs of lithosphere sink and cool the mantle at plate boundaries. Most of our planet's interior heat is removed by this cyclic, repetitive mechanism. Arthur Holmes suggested that the oceans were a source of new crustal material as long ago as the 1920s, but it was not until the early 1960s that Harry Hess (and later Robert Dietz) refined the scenario of a dynamic, self-renewing seafloor and focused attention on the midocean ridges and deep-sea troughs. The associated volcanism occurs mostly under water, but the ridges can be traced around the world by their bathymetry and their seismic activity.

The process begins at regions of extension, usually at boundaries of tectonic or geologic provinces where the lithosphere is weak or where stress can be concentrated. This is where continental rifts and oceanic ridges first form. Volcanoes act as an early warning system for the onset of extension, because surface magmatism requires both buoyant melts and the absence of overlying compression. Rifting, drifting, and midocean ridges are all indicators of long-lived extension.

Newly formed lithosphere appears at midocean ridges and other spreading centers, then cools and contracts as it moves away. Consequently, the ocean depth above it increases in a smooth and characteristic way as a function of distance from the ridge and, therefore, of age. The oceanic lithosphere also thickens with age and eventually becomes denser than the mantle material below; in response, it sinks back into the mantle at subduction zones.

Most of the ocean floor is less than 90 million years old, and nowhere is it older than 200 million years. It takes about 200 million years for the oceanic lithosphere and shallow mantle to cool to a depth of about 100 km, and when this is inserted back into the hot mantle it becomes, in effect, an ice cube in a warm drink. Subduction is the main mechanism by which mantle deeper than 100 km cools. Earthquakes have been recorded at depths as great as 670 km, and seismic tomography has shown that the cold oceanic plates, or slabs, sink at least this far into the mantle (*Figure 11*).

Subduction is also the mechanism for returning volatiles to the Earth's upper mantle. This keeps the melting temperature low and provides the type of "recycling chemistry" (sediments, seawater altered oceanic crust) that is seen in ocean island basalts and other "midplate" volcanoes. In fact, plate tectonism may require the presence of water to weaken the lithosphere and to make mantle melting easier. Venus and Mars may lack this process because they are drier inside. Once a planet has lost its volatiles, it may "freeze-up" — and subduction may be hard to reestablish later, particularly if the planet has also cooled (yielding a thick lithosphere).

Earth is apparently unique among the known worlds in its use of deep subduction as a cooling mechanism, and it occurs here because the lithosphere gets cold enough to become unstably dense and sink. On some other planet with a thicker crust, a hotter surface, or a colder interior, the lithosphere may be permanently buoyant. (In fact, on Earth the continents *are* permanently buoyant, a combination of thick low-density crust capping a buoyant upper-mantle "root" extending down to about 150 km.)

Smaller planets cool more rapidly than large ones, have lower gravity, and experience less vigorous internal convection. Therefore, a lithosphere of a given thickness would be harder to break

up and subduct on something smaller than Earth. The Moon (with 1 percent of Earth's mass) and probably Mars (11 percent) are single-plate planets. Their interiors are never exposed to the cooling effect of subducting lithosphere because their outer layers behave as more-or-less rigid shells. Except for isolated volcanoes, they must lose their internal heat by conduction. Mantle upwellings can focus heat on one portion of the shell, weaken and thin it, and permit magmas to erupt onto the surface. This occurs on Earth too: midplate volcanism accounts for about 10 percent of the heat flowing from the terrestrial interior (*Figure 11*).

The Earth actually exhibits at least three tectonic styles. The oceanic lithosphere recycles itself. The continents are buoyant; they may break up and reassemble, but they remain at the surface. A third characteristic is the way continents affect and are affected by the underlying mantle and adjacent plates. They are maintained against erosion — rejuvenated, in a sense — by compression and uplifting (mountain building) at their boundaries with other plates, by the sweeping up of island arcs at their leading edges, and by eruptions of basalt onto, into, or under the continental mass.

Since material flows from hot to cool parts of a convecting system, continents will tend to drift away from hot mantle zones and come to rest over cool ones. When viewed from Africa, the continents are drifting away from each other at rates of some 5 to 10 cm per year. When this motion is traced back in time, we find that about 180 million years ago the continents were assembled into a supercontinent called Pangea (*Figure 12*). Moreover, for at least several hundred million years prior to that, the southern continents (Africa, South America, Australia, and Antarctica) plus India were a single assemblage, Gondwana. About 360 million years ago, Gondwana was centered on the South Pole, but it moved toward the equator just prior to its breakup. Initially, the continents' separation was rapid, but it slowed as the distances between them increased.

As the continents moved apart, the Atlantic Ocean opened up along the Mid-Atlantic Ridge and the Pacific Ocean shrank. Part of the Pacific lithospheric plate disappeared beneath the continental plates surrounding it. If the Mid-Atlantic Ridge remains "midway", then its drift rate, as seen from Africa, is only one-half as fast as North America's. Most "hotspots" are near midocean ridges, probably because of extensional stresses and thin lithosphere, and they do seem to drift less slowly than continents. This may be purely a geometric, rather than deep-seated, effect. Ridges themselves are relatively slowly moving compared to the centers of plates.

Most of the continents are now sitting on or moving toward cold parts of the mantle. The exception is Africa, which was the core of Pangea. As they move around, the continents encounter oceanic lithosphere and force it to subduct into the mantle. Many active subduction zones are currently at the leading edges of continents. Perhaps all such zones formed along continental margins, after which some of them migrated to their present midocean locations.

There is another conceivable type of plate tectonism. If a large temperature difference does not exist between the surface and the interior, or if plate generation is very rapid, or if the crust-lithosphere system is completely buoyant — then deep

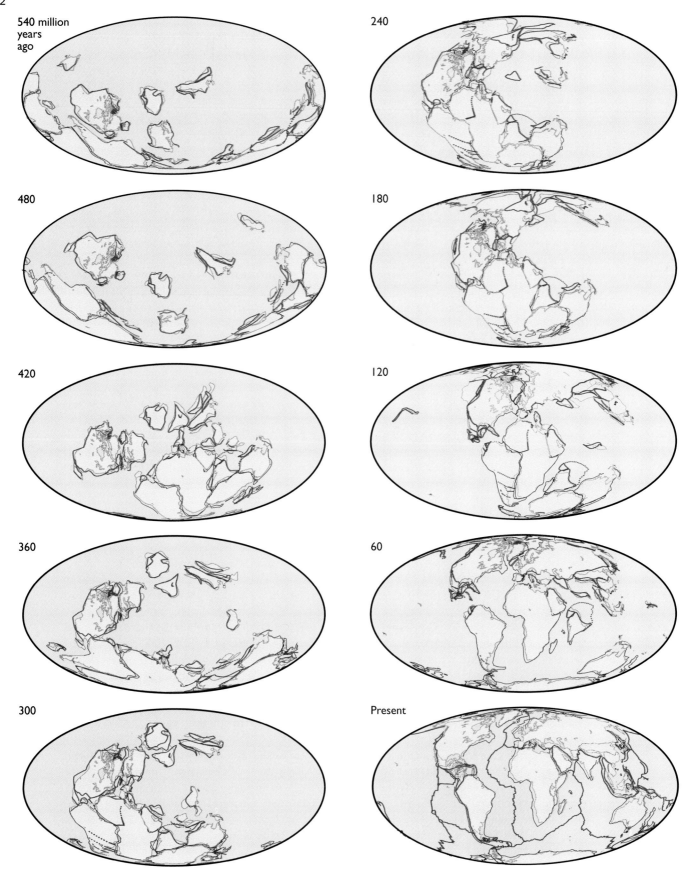

540 million years ago

480

420

360

300

240

180

120

60

Present

Figure 12. The Earth's face has changed dramatically in the last half billion years, as shown here in 60-million-year intervals. Note the assembly of Gondwana at the south pole prior to its incorporation into the supercontinent Pangea. Pangea moved northward across the equator over 150 million years; its eventual breakup created the Atlantic Ocean and greatly diminished the extent of the Pacific Ocean. There may have been roughly 90° of true polar wander near the onset of the Cambrian period, about the time portrayed in the oldest map.

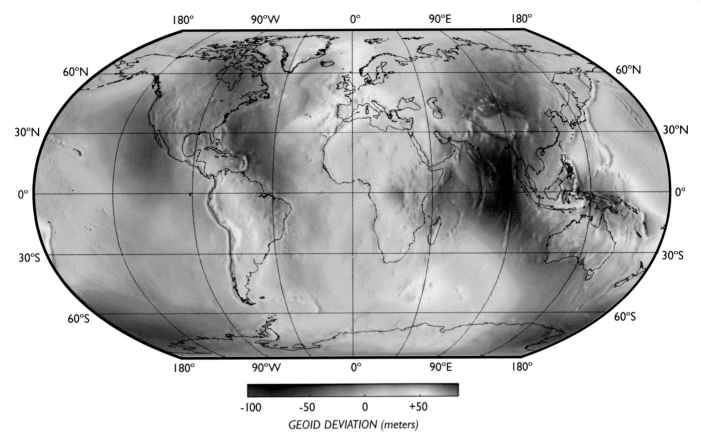

Figure 13. After allowing for the Earth's polar flattening (about 22 km, or 0.34 percent), geophysicists have analyzed satellite-tracking data to reveal that our planet displays residual geoid highs and lows, given here in meters with respect to the ideal spheroid. The highs correspond to subsurface concentrations of dense rocks, the lows to accumulations of less-dense rocks probably within the mantle.

subduction cannot happen. Consequently, the plates must remain near the surface, and their interactions will result in "pack-ice" underthrusting (much the way ice floes behave in the polar oceans). The convergence zones will be diffuse, elevated jumbles characterized by deformation, plate thickening, shallow underthrusting, and lithospheric doubling. Venus and the early Earth may have experienced this tectonic style, for we still see evidence of it in western North America, parts of western South America, and Tibet.

EARTH'S GEOID

On an entirely fluid planet the shape of its surface — the *geoid* — is not controlled solely by rotation. Concentrations of mass in the interior (actually, pockets of anomalously high density) attract the fluid, cause it to pool above them, and make the regional surface stand high. The geoid is usually defined with respect to the perfect ellipsoid that the planet would assume if its interior were completely fluid, with density changing only with depth. The result on a real planet is a global pattern of broad undulations, with heights of some hundreds of meters and a variety of wavelengths.

On Earth, the surface of the ocean approximates the geoid, but a more accurate figure for the entire planet has been obtained by tracking the motion of low-altitude satellites (*Figure 13*). While these geoid, or gravity, data cannot identify subsurface structures unambiguously, they can be used to calculate the contribution from isostatically compensated continents, slabs, and density variations in the lower mantle. (A continent is considered to be in a state of *isostasy* if equilibrium exists between

gravity's downward pull on the mass sitting above sea level and the upward push of the mantle on the continent's low-density "root." Icebergs, in a sense, float isostatically in sea water.)

At very long wavelengths, there are equatorial geoid highs centered on the Pacific Ocean and Africa. Geoid lows occur in a polar band extending through North America, Brazil, Antarctica, Australia, and Asia. Brad Hager, Robert Clayton, and Adam Dziewonski have shown that this pattern correlates with the seismic velocity distribution in the lower mantle, as expected. The long-wavelength geoid highs arise from upwellings of hot mantle material that deform the core-mantle boundary and the Earth's surface upward. At the same time, the hot upwelling mantle is expected to be buoyant and thus relatively low in density, and seismic waves travel through it more slowly than in cold material elsewhere.

Except for Africa, the continents are in or near geoid lows. We think they migrated into these regions as they moved away from Africa after the destruction of Pangea. Tomography has shown that the deep mantle under Africa exhibits very slow seismic velocities. This is probably a result of the absence of subduction under the central part of Pangea, plus continental insulation.

The major geoid highs of moderate wavelength are associated with subduction zones stretching from New Guinea to Tonga and along the Peru-Chile coastline. These highs, centered on the equator, undoubtedly contribute to the moment of inertia that controls the orientation of the Earth's spin axis relative to its mantle. At shorter wavelengths, subduction zones show up as geoid highs, or mass excesses. This is expected as long as the descending slabs are cold, dense, and supported from below by a strong or dense lower mantle.

From *Figure 13*, it is apparent that Earth's present-day expressions of tectonism correlate poorly with its geoid. However, there is *good* correlation with the continental and subduction-zone configurations of the past. For example, the geoid high centered over Africa has about the shape and size of Pangea, and geoid lows correspond roughly with where regions of subduction should have existed prior to extensive opening of the Atlantic Ocean. This is an excellent demonstration of the time-scales on which planetary processes operate — the heat trapped under the supercontinent of Pangea more than 100 million years ago continues to escape from the mantle today. The still-hot mantle has thus elevated the continent of Africa; it represents a geoid high.

Such long-enduring processes are hard to fathom on human time-scales. Indeed, only within the past several decades have we come to appreciate the internal turmoil that continuously shapes the landscapes around us. We were learning about the roles of continental drift and plate tectonism on the Earth at the same time we realized that every other world has a unique style of operation. As far as we know Earth is the only planet that has active plate tectonism, oceans, and life. One wonders if these facts are interrelated.

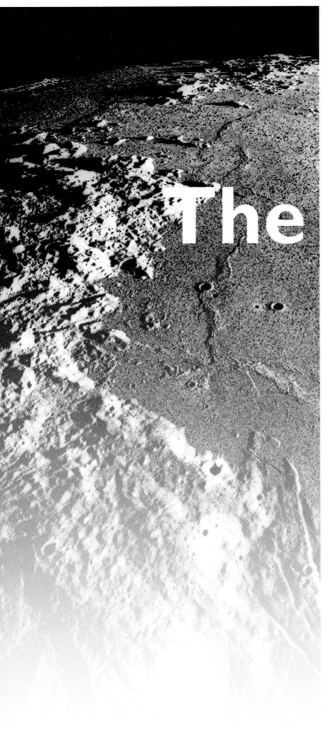

The Moon

Paul D. Spudis

THE **MOON OF** Earth has been a source of inspiration and curiosity throughout history. For millennia, people have gazed at its changing shape and wondered about its nature and origin. The first real answers have come within our lifetimes: we have witnessed the transformation of the Moon from a remote, passive mirror of the Sun to a planetary body with its own complex history. Twelve humans have walked on the lunar surface to gather samples, take photographs, and make other scientific measurements. An even greater number of robotic explorers have scrutinized the Moon from close range.

Thanks to these and other remarkable achievements, we have begun to unravel the lunar story. Today our knowledge of the Moon is deeper and broader than for any other solar-system object save Earth. By studying the processes and evolution of this nearest planetary body, we achieve not only a deeper understanding of geologic processes in general, but a fuller appreciation of the still more complex histories of the terrestrial planets.

LUNAR BASICS

It makes sense that we should know the Moon so well — after all, it is near enough to Earth that crude surface features can be distinguished with the naked eye. By simply looking up at the Moon (*Figure 1*), you can discern that its surface consists of two major types of terrain: relatively bright highlands (or *terrae* in Latin) and darker plains sometimes called the lunar "seas" (or *maria*).

The Moon orbits the Earth every 27.3 days, which is also the time it takes to rotate once on its axis. In other words, the Moon's "day" is equal to its "year." In consequence, we see only one hemisphere of the Moon, called the *near side*. The unseen

The western edge of Mare Serenitatis, as photographed by the crew of Apollo 17.

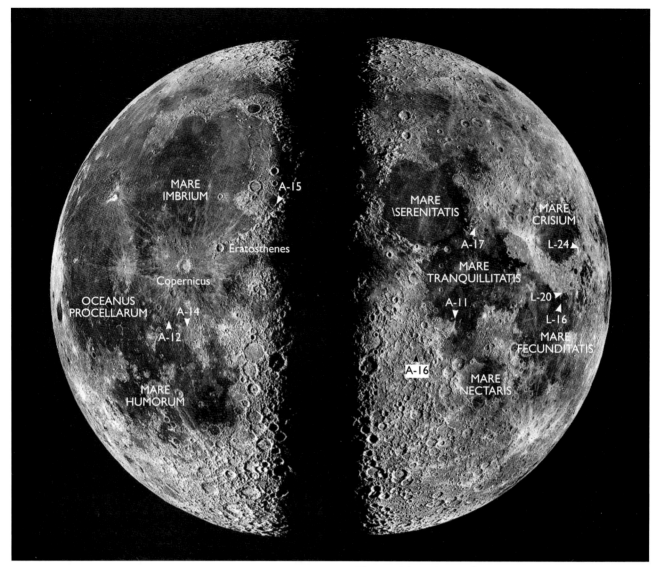

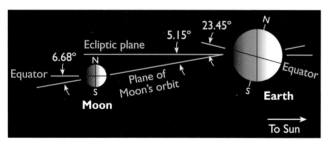

Figure 1 (above). A pair of Lick Observatory photographs have been labeled to show selected lunar features and the location of the nine Apollo (*A*) and Luna (*L*) sample-return sites.

Figure 2 (left). The near-vertical orientation of the Moon's spin axis to the ecliptic plane creates pockets of permanent sunlight and nighttime at the lunar poles. Note that the plane of the lunar orbit falls in neither the Earth's equatorial plane nor the ecliptic, suggesting that some dynamic process early in the history of the Earth-Moon system may have perturbed these orbital relations.

hemisphere (the *far side*) is sometimes termed the "dark side," but each hemisphere receives the same amount of sunlight. The Moon revolves around Earth in an orbit inclined about 7° to the ecliptic plane (the plane in which the Earth orbits the Sun), but the lunar spin axis is nearly perpendicular (1.5°) to the ecliptic (*Figure 2*). This relation has some important consequences: at the lunar poles, the Sun always appears at or close to the horizon. Thus, certain areas near the pole may lie in either permanent sunlight or permanent darkness (*Figure 3*). The existence of such regions may be of great importance to future lunar habitation, as water ice delivered over time by comets should be stable in the permanently dark areas, where the temperature is only about 50° K.

Seen close up or through a telescope (*Figure 4*), the terrae resolve into an apparently endless sequence of overlapping craters, ranging in size from small pits at the limit of resolution on even the best photographs to large multiringed basins — one of which exceeds 2,600 km in diameter. All of the basins and nearly all of the craters are the consequence of meteoritic impact (see Chapter 6). Indeed, the great number of impact scars in the lunar highlands serves to remind us that the Moon's early history was exceedingly violent. At least the top few kilometers of the crust have been broken up, crushed, and repeatedly mixed by the force of these collisions.

The dark maria cover about 16 percent of the lunar surface and are concentrated on the hemisphere facing Earth. While the

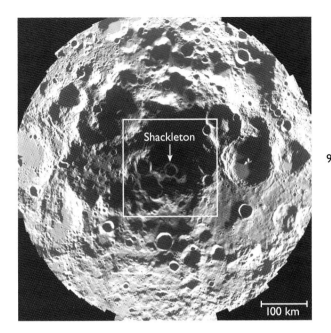

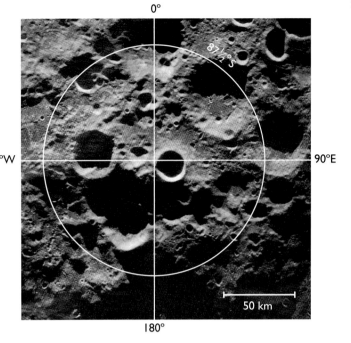

Figure 3 (above, left and right). The Moon's south pole, as seen (at left) in a mosaic of visible-light images by the Clementine spacecraft and (at right) in a radar image from Arecibo Observatory in Puerto Rico. Earth is toward the top in both views. Clementine's view has revealed what appears to be a major depression near the south pole (center), evident by the presence of extensive shadows around the pole. This depression either was formed by impact of an asteroid or comet or is part of the enormous South Pole-Aitken basin on the far side. A significant fraction of the dark area near the pole may be permanent shadow and sufficiently cold to trap water of cometary origin in the form of ice. Both views reveal a 20-km-wide crater, named Shackleton, whose floor never sees the Sun.

maria occur almost everywhere within impact basins, they are geologically distinct. Thus, it is important to distinguish between such features as the Imbrium basin (a large, ancient impact structure) and Mare Imbrium (the dark, smooth volcanic plains that later filled the basin). The maria are significantly younger than the highlands and thus have accumulated fewer craters. This difference in crater density is quite pronounced and easily seen through even a small telescope. Long before Apollo astronauts hopped across the lunar surface, geologists recognized that a substantial amount of time had elapsed between the heavy bombardment of the highlands and the final emplacement of the visible maria.

In the very best telescopic photographs, raised lobes can be seen in some mare regions, which led to the idea that the maria consist of volcanic lava flows. Photographs taken by spacecraft in lunar orbit show confirming evidence for such an origin, including sinuous lava channels (called *rilles*), domes, cones, and collapse pits. Chemical analyses made in 1967 by automated Surveyor landers — and later the study on Earth of actual lunar samples — showed that the maria are indeed volcanic outflows. They appear darker than the terrae because of their higher iron content; the lunar soil becomes momentarily molten where a meteorite hits it, and the heat produces glasses that are iron-rich and thus dark in color.

Geologists can go beyond the scrutiny of the Moon's impacts and volcanic landforms. They can assess the lunar surface in a fourth dimension, time, by determining the relative ages of geologically discrete surface units. According to the geologic law of superposition (*Figure 5*), younger materials overlie, embay, or intrude older ones. This simple but powerful methodology has allowed us to make geologic maps of the

Figure 4. Obtained in 1972 by the crew of Apollo 16, this photograph is predominantly of the lunar far side — the hemisphere never seen before the space age. The large dark circle at upper left is Mare Crisium, which is on the eastern limb of the Moon as seen from Earth; below it are Mare Smythii and Mare Marginis. Innumerable craters scar the ancient, light-colored highlands, which have an albedo (reflectivity) of 11 to 18 percent. The darker, smoother maria (albedo: 7 to 10 percent) are younger regions flooded long ago by volcanic outpourings from the interior. These two basic terrains are distinctly visible even to the naked eye.

entire Moon and to produce a formal stratigraphic sequence for events throughout its history (*Table 1*). However, stratigraphic analysis cannot by itself determine the *absolute* ages of surface units. Our understanding of those ages, as well as compositions and rock types, had to await the return of samples from the lunar "field trips" undertaken by the Apollo and Luna missions.

UNDERSTANDING THE LUNAR SAMPLES

From 1969 to 1972, six Apollo expeditions set down on the Moon, allowing a dozen American astronauts to explore the lunar landscape and return with pieces of its surface (*Figure 6*).

The initial landing sites were chosen primarily on the basis of safety. Apollo 11 landed on the smooth plains of Mare Tranquillitatis, Apollo 12 on a mare site near the east edge of the vast Oceanus Procellarum. These first missions confirmed the volcanic nature of the maria and established their antiquity (older than 3 billion years). Later missions visited sites of increasing geologic complexity. Apollo 14 landed in highland terrain near the crater Fra Mauro, an area thought to be covered with debris thrown out by the impact that formed the Imbrium basin. Apollo 15 was the first mission to employ a roving vehicle and the first sent to a site containing both mare and highland units (the Hadley-Apennine region). Apollo 16 landed on a highland site near the rim of the Nectaris basin. The final lunar mission in

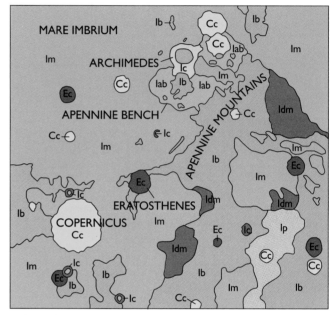

Figure 5 (above, left and right). At left is a photograph of the Apennine Mountains region and at right its corresponding geologic map. Geologists have used this area to define the lunar stratigraphic system *(see Table 1)*. Debris thrown out during the formation of the crater Copernicus (yellow) lies atop all other units and is therefore the youngest. For example, faint, distant rays of Copernicus' ejecta overlie the crater Eratosthenes (green). Eratosthenes was created on top of mare basalts (pink), which in turn fill the floor of the crater Archimedes (purple). Note that Archimedes lies both on the smooth deposits of the Apennine Bench Formation (light blue) and on the

Imbrium basin (dark blue); however, the Apennine Bench embays mountains rimming the Imbrium basin mountains and is thus younger. Thus the scene's relative ages increase as follows: Copernicus, Eratosthenes, mare deposits, Archimedes, Apennine Bench Formation, Imbrium basin.

Table 1 (below). The basic system of lunar stratigraphy has evolved, in a relative sense, from thorough scrutiny of the Moon with telescopes and orbiting spacecraft and, in an absolute sense, from the isotopic dating of lunar rocks and soils.

System of Lunar Stratigraphy

System	Age (10⁹ years)	Remarks
pre-Nectarian	began: 4.6	Includes crater and basin deposits and many other units formed before the Nectaris basin.
	ended: 3.92	Impact; includes formation of lunar crust and its most heavily cratered surfaces.
Nectarian	began: 3.92	Defined by deposits of the Nectaris basin (a large multiring basin on the lunar near side);
	ended: 3.85	includes almost four times as many large craters and basins as the Imbrian system; may also contain some volcanic deposits.
Imbrian	began: 3.85	Defined by deposits of the Imbrium basin; includes the striking Orientale basin on the
	ended: 3.15	Moon's extreme western limb, most visible mare deposits, and numerous large impact craters.
Eratosthenian	began: 3.15	Includes those craters that are slightly more degraded and have lost visible rays; also
	ended: about 1.0	includes most of the youngest mare deposits.
Copernican	began: about 1.0 (to present)	Youngest segment in the Moon's stratigraphic hierarchy; encompasses the freshest lunar craters, most of which have preserved rays.

Figure 6 (above). Apollo 17 astronaut-geologist Harrison ("Jack") Schmitt uses a special rake to collect small rock chips from the Moon's Taurus-Littrow Valley in December 1972.

Lunar-sample Missions

Mission	Arrival date	Landing site	Latitude	Longitude	Sample return
Apollo 11	20 July 1969	Mare Tranquillitatis	0° 67' N	23° 49' E	21.6 kg
Apollo 12	19 Nov. 1969	Oceanus Procellarum	3° 12' S	23° 23' W	34.3 kg
Apollo 14	31 Jan. 1971	Fra Mauro	3° 40' S	17° 28' E	42.6 kg
Apollo 15	30 July 1971	Hadley-Apennine	26° 6' N	3° 39' E	77.3 kg
Apollo 16	21 Apr. 1972	Descartes	9° 00' N	15° 31' E	95.7 kg
Apollo 17	11 Dec. 1972	Taurus-Littrow	20° 10' N	30° 46' E	110.5 kg
Luna 16	20 Sep. 1970	Mare Fecunditatis	0° 41' S	56° 18' E	100 g
Luna 20	21 Feb. 1972	Apollonius highlands	3° 32' N	56° 33' E	30 g
Luna 24	18 Aug. 1976	Mare Crisium	12° 45' N	60° 12' E	170 g

Table 2 (left). A total of nine "field trips" to the Moon have returned with samples of the lunar surface. In addition to these missions, the United States successfully soft-landed five automated Surveyors and the Soviet Union four other Lunas (including two Lunakhod rovers) on the Moon between 1966 and 1973.

the series, Apollo 17, was sent to a combination mare-highland site on the east edge of the Serenitatis basin.

The Soviet Union has acquired a small but important set of lunar samples of its own, thanks to three automated spacecraft that landed near the eastern limb of the Moon's near side. Luna 16 visited Mare Fecunditatis in 1970, and Luna 24 went to Mare Crisium in 1976. A third site, in the highlands surrounding the Crisium basin, was visited by Luna 20 in 1972.

Altogether, these nine missions returned 382 kg of rocks and soil (*Table 2*), the ground truth that provides most of our detailed knowledge of the Moon. In addition, we have now recognized distinct samples of the Moon that have landed on Earth as meteorites. These serendipitous acquisitions were probably thrown to Earth during impacts on the Moon and provide us with additional, random samples of the lunar surface. Although we do not know their exact source regions, their compositions provide us with additional information on the fine details of lunar chemistry and petrology.

While the most exhilarating discoveries came from studies of lunar material completed years ago, today scientists around the world continue to examine these samples, establishing their geologic contexts and making inferences about the regional events that shaped their histories. What we've learned about the Moon's three major surface materials — maria, terrae, and the soil-like regolith that covers both — is summarized in the following paragraphs.

Regolith. Over the history of the Moon, meteoritic bombardment has thoroughly pulverized the surface rocks into a fine-grained, chaotic mass of material called the regolith (also informally called "lunar soil," though it contains no organic matter). The regolith consists of single mineral grains, rock fragments, and combinations of these that have been cemented by impact-melted glass. Because the Moon has no atmosphere, its soil is directly exposed to the high-speed solar wind (see Chapter 4), gases flowing out from the Sun that become implanted directly onto small surface grains. The regolith's thickness depends on the age of the bedrock that underlies it and thus how long the surface has been exposed to meteoritic bombardment. Regolith in the maria is 2 to 8 meters deep, whereas in highland regions its thickness may exceed 15 meters.

Not surprisingly, the composition of the regolith closely resembles that of the local underlying bedrock. Some exotic components are always present, perhaps having arrived as debris flung from a large distant impact. However, this is the exception rather than the rule. The contacts between mare and highland units appear sharp from lunar orbit, which suggests that relatively little material has been transported laterally. Thus, while mare regoliths may contain numerous terrae fragments, in general these derive not from faraway highland plateaus but are instead material excavated locally from beneath the thin lava flows of the maria.

Figure 7. This mare basalt (top), sample 15016 from the landing site of Apollo 15, crystallized 3.3 billion years ago. The hand specimen's numerous vesicles (bubbles) were formed by gas that had been dissolved in the basaltic magma before it erupted. By shining polarized light through paper-thin slices of a lunar rock (above), geologists can learn much about its crystal structure and composition. Sample 15016 exhibits the minerals plagioclase (lath-shaped black and white crystals), pyroxene (lath-shaped colored crystals), olivine (roundish, brightly colored grains), and ilmenite (opaque).

Impacts energetic enough to form meter-size craters in the regolith sometimes compact and weld the loose soil into a type of rock called *breccia*. Once fused into a coherent mass, a regolith breccia no longer undergoes the fine-scale mixing and "gardening" taking place in the unconsolidated soil around it. Thus, regolith breccias are "fossilized soils" that retain not only their ancient composition but also the chemical and isotopic properties of the solar wind from the era in which they formed.

Maria. Thanks to our lunar samples, there is no longer any doubt that the maria are volcanic in nature. The mare rocks are basalts (*Figure 7*), which have a fine-grained crystalline or even glassy structure (indicating that they cooled rapidly) and are rich in the elements iron and magnesium. Basalts are a widespread volcanic rock on Earth, consisting mostly of the common silicate minerals pyroxene and plagioclase, oxide minerals such as ilmenite, and sometimes olivine (an iron-magnesium silicate). The lunar basalts display some interesting departures from this basic formulation. For example, they are completely devoid of water — or indeed any form of hydrated mineral — and contain few volatile elements in general. Basalts from Mare Tranquillitatis and Mare Serenitatis are remarkably abundant in titanium, sometimes containing roughly 10 times more than is typically found in their terrestrial counterparts.

The mare basalts originated hundreds of kilometers deep within the Moon in the total absence of water and the near-absence of free oxygen. There the heat from decaying radioactive isotopes partially melted the mantle, creating magma that ultimately forced its way to the surface. The occurrence of mare outpourings within impact basins is no chance coincidence. The crust beneath these basins must have been fractured to great depth and thinned by excavation by the cataclysmic impacts that formed them. Much later, magmas rose to the surface through these fractures and erupted onto the basin floors.

Figure 8. The western edge of Mare Serenitatis, looking north, as photographed from Apollo 17. The mare's surface exhibits numerous deep rilles (bottom center) and wrinkle ridges that resulted from strain and deformation within the massive basalt sheet. Mare Imbrium is on the horizon at upper left.

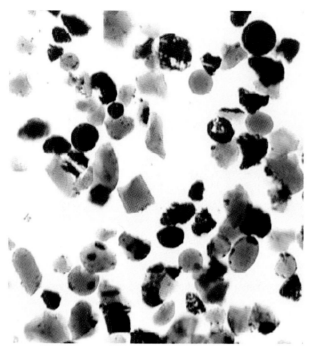

Figure 9. These flecks and spherules of Apollo 17 orange glass are roughly 0.03 mm across. The lunar equivalent to terrestrial ash deposits, they were sprayed onto the Moon's surface about 3.7 billion years ago in erupting fountains of basaltic magmas. The black particles are pieces of orange glass that have crystallized over time.

Although they may appear otherwise, the maria are typically only a few hundred meters or less in thickness. These volcanic veneers tend to be thinner near the rims that confine them and thicker over the basins' centers (as much as 2 to 4 km in some places). What the maria may lack in thickness they make up for in sheer mass, which frequently is great enough to deform the crust underneath them (*Figure 8*). This has stretched the outer edges of the maria (creating fault-like depressions called grabens) and compressed their interiors (creating raised "wrinkle" ridges).

Basalts returned from the mare plains range in age from 3.8 to 3.1 billion years, a substantial interval of time. Small fragments of mare basalt found in highland breccias solidified even earlier — as long ago as 4.3 billion years. We do not have samples of the youngest mare basalts on the Moon, but stratigraphic evidence from high-resolution photographs suggests that some mare flows actually embay (and therefore postdate) young, rayed craters and thus may have erupted as recently as 1 billion years ago.

A variety of volcanic glasses — distinct from the ubiquitous, impact-generated glass beads in the regolith — were found in the soils at virtually all the Apollo landing sites. They even were scattered about the terrae sites, far from the nearest mare. Some of these volcanic materials are similar in chemical composition, but not identical, to the mare basalts and were apparently formed at roughly the same time.

One such sample, tiny beads of orange glass, came from the Apollo 17 site (*Figure 9*). They are akin to the small airborne droplets accompanying volcanic "fire fountains" on Earth, like those in Hawaii. The force of eruption throws bits of lava high into the air, which solidify into tiny spherules before hitting the

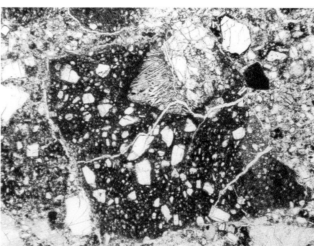

Figure 10. A breccia from the lunar highlands, sample 67015, collected at the Apollo 16 landing site near Descartes crater. This rock is termed polymict because it consists of numerous fragments of pre-existing rocks, some of which are themselves breccias. Fused into a coherent mass about 4.0 billion years ago, this breccia demonstrates dramatically how impacts have altered rocks on the lunar surface.

ground. The Moon's volcanic glass beads have had a similar origin. The orange ones from the Apollo 17 site get their color from a high titanium content (greater than 9 percent) and some of them are coated with amorphous mounds of volatile elements like zinc, lead, sulfur, and chlorine. Their existence suggests that the Moon does have volatile elements deep within its interior.

Terrae. One could easily imagine the lunar highlands to contain outcrops of the original lunar crust — much as we find in Earth's continents. But what really awaited the astronauts was a landscape so totally pulverized that no traces of the original outer crust survived intact. Instead, most of the rocks collected from the terrae were breccias (*Figure 10*), usually containing fragments from a wide variety of rock types that have been broken apart, mixed, and fused together by impact processes. Most of these consist of still-older breccia fragments, attesting to a long and protracted bombardment history.

The highland samples also include several fine-grained crystalline rocks with a wide range of compositions. They are not breccias, but they *were* created during an impact. During large impacts, the shock and pressure were so overwhelming in part of the crust that the "target" melted completely, creating in effect entirely new rocks from whatever ended up in the molten melee. Of course, the colliding "bullets" become part of this mixture, and these impact-melt rocks contain distinct elemental signatures of meteoritic material.

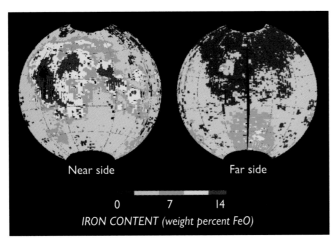

Near side   Far side

0   7   14
IRON CONTENT (weight percent FeO)

Figure 11 (above). A new technique developed by Paul Lucey and others uses Clementine color data to map the distribution of iron in the lunar surface. As these near-side and far-side maps show, areas with very low iron content (dominated by anorthosite) make up large regions of the highlands, particularly on the far side. Conversely, the dark, near-side maria are indeed iron-rich lavas. Note also the "anomaly" of elevated iron concentration associated with the floor of the South Pole-Aitken basin on the far side; this huge impact may have stripped off most of the lunar crust, partially exposing the upper mantle.

Virtually all of the highlands' breccias and impact melts among our lunar samples formed between about 4.0 and 3.8 billion years ago. The relative brevity of this interval surprised researchers — why were all the highland rocks so similar in age? Perhaps the rate of meteoritic bombardment on the Moon increased dramatically during that time, resulting in a violent period of cataclysm. Alternatively, the narrow age range may merely mark the conclusion of an intense and continuous bombardment that began much earlier, around 4.5 billion years ago, the estimated time of lunar origin. We cannot distinguish between these two models with the data in hand. To resolve the enigma, we must return to the Moon and sample its surface at carefully selected sites.

A substantial number of small, whitish rock fragments found in the mare soils returned by Apollo 11 and 12 astronauts had a composition totally unlike basalts and virtually unmatched on Earth. They consisted almost entirely of plagioclase feldspar, a silicate mineral rich in calcium and aluminum but depleted in heavier metals like iron. At the time a few prescient researchers postulated that these rocks came from the lunar highlands. The last four Apollo missions, sent to highland landing sites, confirmed that plagioclase feldspar dominates the lunar crust. More recent global data showed that vast regions of the highland surface are made up of this aluminum-rich, iron-poor component

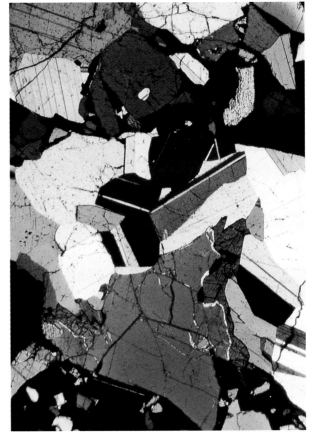

Figure 12. Sample 76535 *(left)*, collected from the lunar highlands by Apollo 17 astronauts, is a coarse-grained igneous rock containing plagioclase feldspar (white) and olivine (dark). Seen microscopically *(above)*, the large crystals indicate that this rock cooled slowly, well below the lunar surface. It is a single rock type, unmixed with other material by impact. Such rocks are relatively rare in the lunar sample collection.

(*Figure 11*). The resulting implication was broad and profound: at some point in the distant past much of the Moon's exterior — and perhaps its entire globe — had been molten.

The detailed nature of this waterless "magma ocean" is only dimly perceived at present; for example, the lunar surface may not have been everywhere completely molten. The consequences seem clear, however. In a deep, slowly cooling layer of lunar magma, crystals of low-density plagioclase feldspar would have risen upward after forming, while higher-density minerals would have accumulated at lower levels. This segregation process, termed *differentiation*, left the young Moon with a crust that was, in effect, a low-density rock "froth" tens of kilometers thick consisting mostly of plagioclase feldspar. At the same time, denser minerals (particularly olivine and pyroxene) became concentrated in the mantle below — the future source region of mare basalts.

It is unclear to what depth the magma ocean extended, but the volatile-element coatings discovered on some mare glasses provide an important clue. If the Moon's exterior really was once molten, the most volatile components in the melt would have vaporized and escaped into space. However, the volatile-coated glasses sprayed onto the lunar surface long after the magma ocean solidified. If the glasses' compositions did not change in their upward migration from the lunar interior, they imply that volatile-rich pockets remained (and perhaps still exist) in the upper mantle. The implication, therefore, is that the magma ocean was at most only a few hundred kilometers deep.

The highland samples returned by the last four Apollo crews provided other surprises. Unlike glasses and basalts, which quench quickly after erupting onto the surface, some of the clasts in the highland breccias contained large, well-formed crystals, indicating that they had cooled and solidified slowly, deep inside the Moon. These igneous rocks sometimes occur as discrete specimens (*Figure 12*). At least two distinct magmas were involved in their formation. Rocks composed almost completely of plagioclase feldspar, with just a hint of iron-rich silicates, are called *ferroan anorthosites*. These appear to be widespread in the highlands, and they are extremely ancient, having crystallized very soon after the Moon formed (at least one anorthosite having formed 4.50 to 4.52 billion years ago).

The highlands' other dominant rock type is also abundant in plagioclase feldspar, but it contains substantial amounts of olivine and a variety of pyroxene low in calcium. This second class of rocks is collectively termed the *Mg suite*, so called because they contain considerable amounts of the element magnesium (Mg). These rocks appear to have undergone the same intense impact processing as the anorthosites, but their crystallization ages vary more widely — from almost the age of the Moon to about 4.3 billion years ago.

The anorthosite and Mg-suite rocks could not have crystallized from the same parent magma, so at least two (and probably many more) deep-seated sources contributed to the formation of the early lunar crust. Conceivably, both magmas might have existed simultaneously during the first 300 million years of lunar history. This would contradict our notion of the Moon as a geologically simple world and greatly complicate our picture of the formation and early evolution of its crust.

During early study of the Apollo samples, an unusual chemical component was identified that is enriched in incompatible trace elements — those that do not fit well into the crystal structures of the common lunar minerals plagioclase, pyroxene, and olivine as molten rock solidifies. This element group includes potassium (K), rare-earth elements (REE) like samarium, and phosphorus (P); geochemists refer to this element combination as *KREEP*. It is a component of many highland soils, breccias, and impact melts, yet the *relative* abundances of these trace-elements remain remarkably constant wherever it is found. Moreover, its estimated age is consistently 4.35 billion years. These characteristics have led to the consensus that KREEP represents the final product of the crystallization of a global magma system that solidified eons ago, so this date reflects the "age" of the lunar crust and mantle as a whole.

Evidence for chemically distinct, widespread volcanic rocks in the highlands — KREEP-rich or otherwise — remains tenuous. Some highland rocks are compositionally similar to mare basalts yet exhibit KREEP's trace-element concentrations. For example, the Apollo 15 astronauts returned with true basalts that probably derive from the nearby Apennine Bench Formation, a large volcanic outflow situated along the Imbrium basin's rim. These

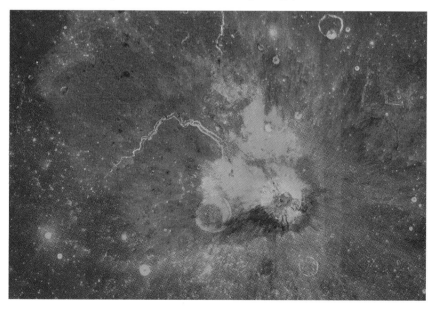

Figure 13. The Aristarchus region is one of the most diverse and interesting areas on the Moon. In this false-color Clementine mosaic, the colors represent ratios of brightness values from blue, red, and near-infrared images; these ratios serve to cancel out the albedo variations and topographic shading, thus isolating the color differences related to composition or mineralogy. Consequently, fresh highland materials appear blue, fresh mare materials are yellowish, mature mare soils are purplish, and the dark pyroclastic (volcanic) glasses on the plateau are deep red. The compositions of materials buried beneath surface debris are revealed by crater excavations and by steep slopes such as those along the walls of the rilles. The rilles formed primarily in lavas, except for the cobra-head crater of Vallis Schröteri, which formed in highland materials.

"KREEP basalts" have a well-determined age of 3.85 billion years, so the Imbrium impact must have occurred before this date and probably just before the Apennine Bench Formation extruded onto the surface. Thus, although the extent and importance of highland volcanism remains unknown, it apparently took place early in lunar history and contributed at least some of the KREEP component observed in highland breccias and impact melts.

Apart from actual lunar samples, our knowledge of the distribution of rock types on the Moon derives from a wide variety of remote-sensing data. Some of this information is obtained from telescopic observations of the near side, but most has come from the global survey conducted by the Clementine orbiter in 1994. Clementine carried several imaging cameras sensitive to the visible and near-infrared parts of the spectrum, where absorption bands characteristic of the common lunar rock-forming minerals are found. From Clementine multiwavelength images (*Figure 13*), we can correlate certain colors with specific geologic units, thus creating the first-ever global maps of the regional rock units that make up the lunar crust.

These remote observations indicate several important facts about the Moon's crust. First, global mapping of iron content confirms that anorthosite is the dominant rock type of the highlands (*Figure 11*), providing powerful support for the hypothesis of the magma ocean. Another mapping technique using Clementine images has shown that basalts that are highly enriched in titanium, while abundant in the returned sample collection, constitute only a small fraction of the surface lava flows globally. Earth-based spectra indicate that a wide variety

of volcanic flows cover the maria, only about a third of which are similar to our sampled basalts. Additionally, these data show that pure anorthosite deposits occur in the inner rings of several basins, including Nectaris, Humorum, and Orientale, and in the central peaks of some craters, including Aristarchus (*Figure 13*). Both of these latter observations are confirmed by the Clementine data, which also indicate many new rock occurrences that we are just beginning to inventory and catalog. By using the large craters and basins of the Moon as natural "drill holes," we will soon be able to complete a three-dimensional reconstruction of the crust that should offer great insight into the origin and evolution of the Moon.

THE LUNAR INTERIOR

What little we know about the internal structure of the Moon comes from both landed experiments and the tracking of orbital spacecraft. Seismic measurements were made at the Apollo 12, 14, 15, and 16 landing sites. Mild moonquakes shake the lunar interior from time to time and were recorded by the four seismic stations. Some of these seemed to emanate from the upper mantle, while others came from deeper within. Most important, the seismometers were able to record occasional impacts on the Moon — from both natural sources and impacting spacecraft. We derive our knowledge of the interior by measuring how the resulting shock waves of differing frequency and type propagate through and around the lunar globe.

On the basis of these data, we believe that, early in lunar history, an intense meteoritic bombardment shattered and frac-

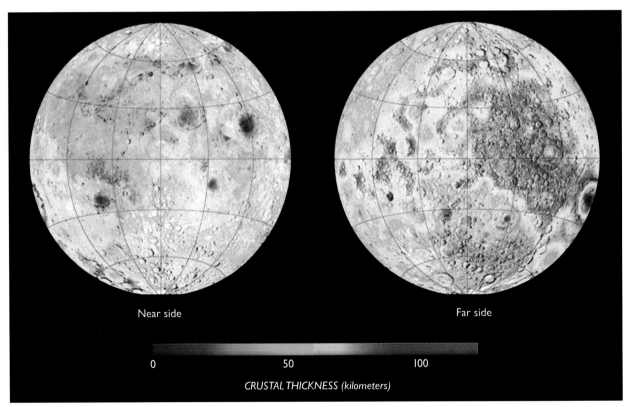

Near side   Far side

CRUSTAL THICKNESS (kilometers)

Figure 14. By combining gravity data with plausible assumptions about crustal density, geologists have deduced that the thickness of the lunar crust varies widely. It is relatively thin (blue) beneath basin floors, where the large impacts have stripped off much of the Moon's outermost layers. The crust appears thickest (red) on the far side, especially around the periphery of the South Pole-Aitken basin – a situation probably due to the pileup of material ejected during the basin's formation.

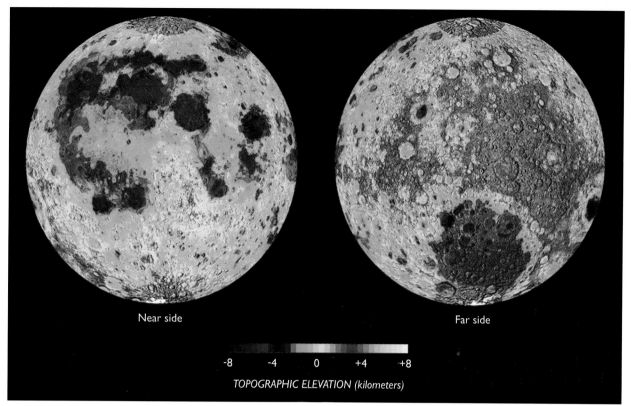

Near side

Far side

-8   -4   0   +4   +8

TOPOGRAPHIC ELEVATION (kilometers)

Figure 15. The laser altimeter on Clementine gave us our first comprehensive look at the topography of the Moon. Surprisingly, the Moon shows nearly the same range of elevation exhibited by the Earth: at least 16 km from the lowest to the highest points. Earth's wide range is caused by the complex dynamics of plate tectonism, while the Moon's stems from the preservation of ancient impact basins. Note that while the near side is relatively smooth, the far side shows extreme topographic variation. The large circular feature centered on the southern far side is the South Pole-Aitken basin, 2,600 km in diameter and over 12 km deep.

tured the crust to a depth of a few tens of kilometers. In the eons since, impacts have continued to pound the uppermost crust and mix it to depths of at least 2 km but perhaps down to 10 or 20 km. The seismic data also suggest that below about 25 km fractures in the crust are self-annealing, and that rocks deeper still may be largely intact (except those under large impact basins, such as Imbrium).

Gravity mapping obtained from orbiting spacecraft such as Clementine allows us to look at the structure and thickness of the crust globally (*Figure 14*). On average, the lunar crust is about 70 km thick, but it varies from a few tens of kilometers beneath the mare basins to over 100 km in some highland areas. Under some of the largest basins, the crust was weakened (and indeed partially removed) so much that the mantle has bulged upward. One manifestation of this movement is that basin floors are frequently raised and fractured. Moreover, the intrusion of dense mantle material into the crust changes the local gravity field. An orbiting spacecraft that passes over these mass concentrations, or *mascons*, experiences slight changes in velocity that can be used to map the mascons' locations.

The lunar crust varies from region to region, but does it contain stratified layers as well? The sparse seismic results do not require the presence of different rock types in the lower crust. However, we know that mafic (iron- and magnesium-rich) rocks exist in abundance on the rims of the large impact basins Imbrium and Serenitatis — precisely where material blasted out from great depths in the crust should have come to rest. These basaltic rocks have some peculiar properties. They were formed 3.9 to 3.8 billion years ago (the age of the last basin-forming impacts) but cannot be made by melting any combination of the known highland rock types. They also contain rock and mineral clasts of relatively deep-seated origin and have no soil or regolith-breccia fragments within them. If these rocks were thrown out as molten ejecta during the cataclysmic blasts that formed the basins, they provide direct evidence for a lower crust that is more mafic than the average upper crust.

The mantle constitutes about 90 percent of the volume of the Moon and is thought to consist of an olivine-pyroxene mixture that varies both regionally and with depth in complex ways that are not fully understood. Source regions for the mare basalts apparently were situated 200 to 400 km below the surface, so our mare samples provide tracers of the upper-mantle compositions. For example, at least some zones in the mantle must contain large concentrations of ilmenite (an iron-titanium oxide), because they spawned the titanium-rich basalts found in Mare Tranquillitatis and Mare Serenitatis. However, Clementine's finding that these basalts have minor areal extent makes this compositional anomaly much less acute.

Until the Clementine mission, we had no global map of the topography of the Moon. The spacecraft carried a laser altimeter that repeatedly measured the distance to the surface, and over time these data were combined into a global map of lunar topography (*Figure 15*). This new global map has given us several great insights into the nature and structure of the Moon. First, the dynamic range of topography is much greater than we had previously thought, over 16 km, comparable to that of the Earth! Second, the near side appears relatively smooth, with typical relief of only 5 to 6 km, whereas the far side exhibits the full 16-km range of relief. This difference is caused by widespread infilling of the near-side basins by mare basalts and the relative paucity of maria on the far side (*Figure 4*). Moreover, the wide range of relief on the far side is caused mostly by the presence of the huge South Pole-Aitken basin *(Figure 15)*, the largest excavation on the Moon, which is surprisingly deep — over 12 km, on average.

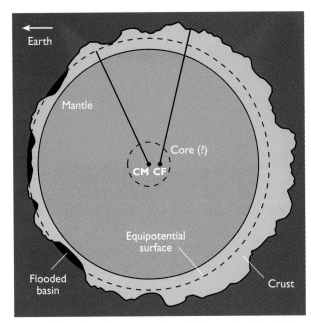

Figure 16. A schematic cross section of the lunar interior, which may or may not include a small metallic-iron core. The Moon's center of mass (*CM*) is offset by 2 km from its center of figure (*CF*), so an *equipotential surface* (which experiences an equal gravitational force at all points) lies closer to the lunar surface on the hemisphere facing Earth. Therefore, magmas originating at equipotential depths will have greater difficulty reaching the surface on the far side, accounting for the paucity of mare deposits there.

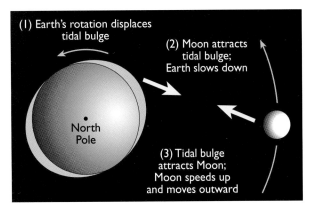

Figure 17. Ocean tides do not actually lie along the Earth-Moon line. Over time, this misalignment slows the Earth's rotation and causes the Moon to move farther away.

The Moon's center of mass is offset from its geometric center by about 2 km in the direction of the Earth, probably because the crust is generally thicker on the far side (*Figure 14*). This may not seem like much of an offset, but it may explain why so few maria exist on the far side of the Moon. Imagine a subsurface boundary akin to a global water table, attracted toward the center of mass with equal gravitational force at every point (*Figure 16*). Because of the 2-km offset, this *equipotential surface* lies farther from the top of the crust on the far side. It is possible, therefore, that basalt magmas rising from the interior reached the surface easily on the near side, but encountered difficulty on the far side.

The Moon currently has no global magnetic field. Yet many of our lunar samples cooled in the presence of a surprisingly strong magnetic environment that was most intense 3.6 to 3.8 billion years ago (an estimate considered crude because of our sparse sampling of the crust). The "paleomagnetism" found in certain lunar samples has led some researchers to postulate that the Moon once possessed a significant global magnetic field produced by dynamo motion within a metallic-iron core.

However, the size — and even the existence — of this metallic core remains unresolved. First, the low uncompressed bulk density of the Moon (3.3 g/cm^3) means that it is depleted in iron relative to other terrestrial planets and particularly with respect to the Earth (4.5 g/cm^3). Second, the best estimate of the lunar moment of inertia implies that the Moon's interior has a nearly uniform density throughout and that an iron-rich core can be no larger than about 400 km in radius. Third, the Moon's weak interaction with the Sun's magnetosphere argues that a highly conducting lunar core can be no greater than 350 to 450 km in radius. Such a core would constitute some 2 to 4 percent of the total lunar mass.

HYPOTHESES OF LUNAR ORIGIN

In their surveys of the solar system, astronomers have discovered dozens of satellites around other planets. Yet, of the four inner planets, only the Earth and Mars have moons (and the latter's are probably captured asteroids). Our Moon is large as satellites go, particularly when compared to the modest size of Earth itself. The creation of the Moon was thus an unusual event in terms of general planetary evolution, and our knowledge of the solar system — however detailed — would be profoundly incomplete without determining how our enigmatic satellite came to exist.

Traditionally, scientists have investigated three models of lunar origin. In the simplest hypothesis, termed *co-accretion*, the Earth and Moon formed together from gas and dust in the primordial solar nebula and have existed as a pair from the outset. A second concept, called the *capture* scenario, envisions the Moon as a maverick world that strayed too near the Earth and became trapped in orbit — either intact or as fragments torn apart by our planet's strong gravity. According to the third model, termed *fission*, the Earth initially had no satellite but somehow began to spin so fast that a large fraction of its mass tore away to create the Moon.

We had hoped that our astronauts would return with results that would allow us to choose decisively from among these three models. Instead, the Apollo samples have persuaded us that none of these models are completely satisfactory. First, the

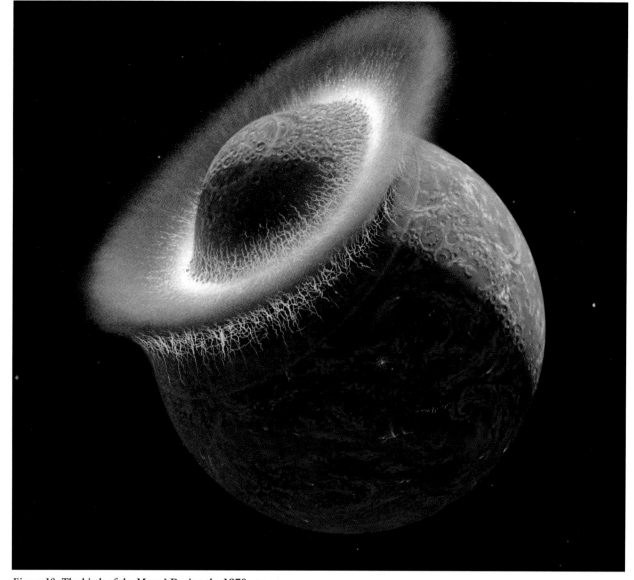

Figure 18. **The birth of the Moon? During the 1970s, two teams of scientists independently proposed that an object perhaps the size of Mars could have collided with Earth and thrown enough matter into orbit to create the Moon.**

Moon's bulk composition appears to be similar, but not identical, to the composition of the Earth's upper mantle. Both are dominated by the iron- and magnesium-rich silicates pyroxene and olivine. One important distinction is that, unlike Earth, the Moon generally lacks volatile elements. Another involves the relative dearth in lunar material of what are termed siderophile ("metal-loving") elements such as cobalt and nickel, which tend to occur in mineral assemblages containing metallic iron. Thus origin by co-accretion would appear doubtful.

A second key constraint comes from oxygen's three natural isotopes: ^{16}O, ^{17}O, and ^{18}O. Ratios of the abundances of these isotopes found in lunar and terrestrial materials exhibit a single trend, indicating that the Moon and Earth originated in the same part of the solar system (see Chapter 26). By contrast, these same ratios are different in major meteorite groups derived from the asteroids or Mars. So the Earth and Moon *must* share some common genetic link.

Beyond this geochemical evidence, the mystery of lunar origin has several physical clues. For example, given the dynamics of close encounters, it would have been all but impossible for our young planet to have captured of a body of lunar size. Moreover, the Earth-Moon system possesses a great deal of angular momentum, but far less than that needed for fission. Also, the Moon's orbit does not lie within in the plane either of the Earth's equator or the ecliptic plane (*Figure 2*). Finally, the Moon is gradually receding from the Earth at roughly 3 cm per year — a curious effect caused by the gravitational coupling of the Moon and our oceans. Tidal bulges raised in the ocean do not lie directly along the Earth-Moon line but actually precede it (*Figure 17*) because Earth's rotation drags them along for some distance before they can adjust to the Moon's changing location in the sky. This misalignment slows the Earth's rotation slightly (0.00001 second per century). The Moon in turn is pulled forward in its orbit, speeds up, and inches farther away (3 cm per year). The two worlds must surely have been closer in the distant past. However, we may never know just how close because the orbital recession going on now cannot be extrapolated back to the time of lunar origin.

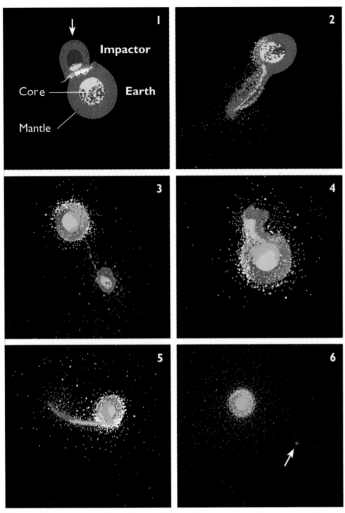

Figure 19. A computer has simulated the effects of a giant impact involving Earth 4.5 billion years ago. The mantles of both Earth and the impactor are vaporized; some of this material ends up in orbit and forms a circumterrestrial disk from which the Moon (arrow) coalesces shortly afterward.

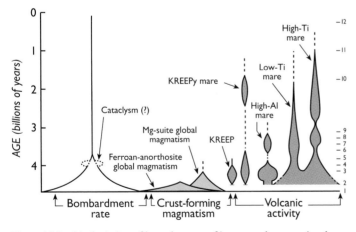

Figure 20. In this depiction of how the rates of impact and magmatism have varied on the Moon over time, the width of each envelope corresponds to the intensity of activity. Key dates derived or inferred from lunar exploration are shown at right: *1,* Moon forms; *2,* magma ocean solidifies; *3,* age of "average" highland surface; *4,* Orientale and Imbrium basins form; *5,* Apollo 11 and 17 maria; *6,* Luna 16 mare; *7,* Luna 24 mare; *8,* Apollo 15 mare; *9,* Apollo 12 mare; *10,* late Imbrium mare flows; *11,* Copernicus impact; *12,* Tycho impact. Most geologic activity is confined to the Moon's first 1½ billion years of existence.

Recently, a fourth idea for the Moon's birth — the *giant-impact* hypothesis — has gained popularity and even something of a consensus among planetary scientists. In 1975 two research teams (William Hartmann and Donald Davis, and Alastair Cameron and William Ward) independently proposed that a giant object hit the infant Earth some 4½ billion years ago (*Figure 18*). The off-center blow ejected a mixture of terrestrial and impactor material into orbit around Earth that soon coalesced to form the Moon. Since most of this material would have been a white-hot vapor, this scenario can explain both the Moon's dearth of volatiles and its enrichment in refractory elements (those that remain solid at high temperature). To create a proper Moon, the impactor's iron and siderophiles must have already been concentrated in its core before the collision; that core then became incorporated into the mantle of the Earth. Theorists' calculations show that at least half to nearly all of the lunar mass was derived from the outer layers of the colliding body (*Figure 19*). Remarkably, they also suggest that the ejected matter collapsed into a disk almost immediately and coalesced into nearly final form within just a few years.

The giant-impact hypothesis neatly explains (or allows for) the orientation and evolution of the Moon's orbit, as well as the Earth's relatively fast spin rate. Moreover, it makes the uniqueness of the Earth-Moon system seem more plausible. That is, impacts of such cataclysmic magnitude might have occurred only rarely, rather than being a requirement for planetary formation. Part of the reason for this model's current popularity is doubtless because we know too little to rule it out: key factors such as the impactor's composition, the collision geometry, and the Moon's initial orbit are all undetermined.

The advent of the giant-impact hypothesis has not solved the problem of lunar origin. For example, the close genetic relation of Earth and Moon (inferred from the oxygen-isotope ratios) is not an obvious consequence of a giant impact, especially if most of the lunar mass derived from the projectile. Also, the colliding object must have been at least the size of Mars to throw a Moon's worth of mass into orbit. Consequently, research into the effects of such cataclysmic impacts in early planetary history continues at a brisk pace. But this model for lunar origin appears to explain the most salient features of the Moon with the minimum amount of special pleading.

THE ONCE AND FUTURE MOON

Much has happened to the Moon since its formation, and three decades of intensive study by spacecraft now enable us to devise an outline of lunar history (*Figures 20,21*). However, the following scenario should be regarded only as a progress report. Many chapters in this history are still obscure, and some of the speculations here could easily be disproved by further research or exploration.

Assuming that a giant impact did create the Moon, the assembly period was quite brief by planetary standards. In fact, chunks of debris cascaded together so rapidly that the growing sphere became very hot and melted almost completely to a depth of at least a few hundred kilometers. As this magma ocean gradually cooled and crystallized, meteorites continued to bombard the Moon at a very high rate, fragmenting and mixing the upper-

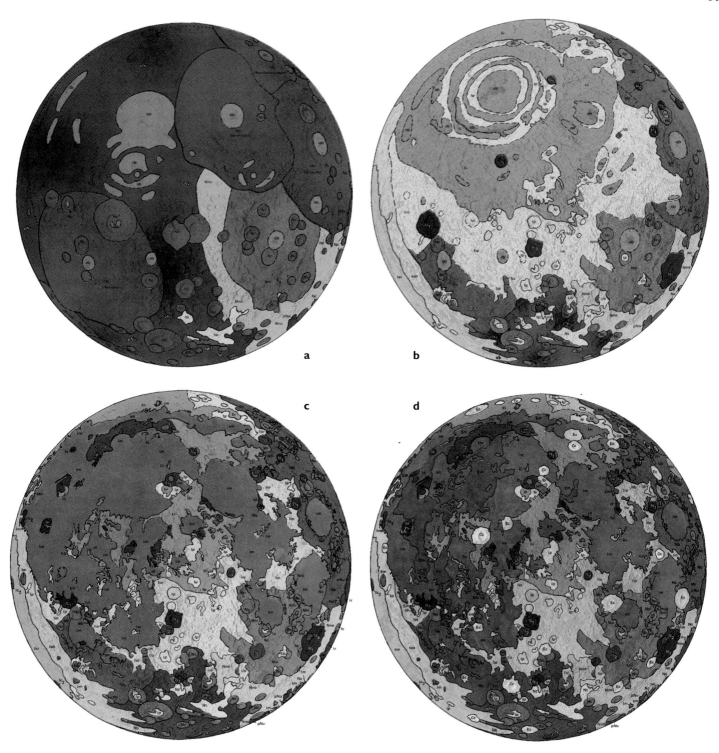

a b c d

most portions of the primordial crust. The Moon's molten outer shell solidified by about 4.3 billion years ago, when the last residues of the original magma system crystallized as the KREEP source region. This was not the end of the Moon's magmatic life, however. Deep within the lunar mantle, radioactive heat created zones of magma that were forced upward and onto the surface as eruptions of volcanic lavas (the maria).

Meanwhile, violent collisions continued to overturn and mix the upper crustal materials thoroughly, destroying most of the original geologic formations within the primordial crust and

Figure 21. Geologic maps of the evolution of the lunar near side at four key dates: *(a)* Just before the mammoth impact that formed the Imbrium basin, about 3.85 billion years ago. Brown represents pre-Nectarian and Nectarian deposits; pink is ancient mare basalts (now obliterated). *(b)* Just after the Orientale basin impact (beyond the limb at left), about 3.8 billion years ago. Deposits from the Imbrium basin (blue) dominate much of the lunar near side; purple signifies post-Imbrium deposits. *(c)* At the end of the Imbrian Period, about 3.2 billion years ago; widespread mare basalts (red and pink) have largely covered the Imbrium basin deposits. *(d)* At present; greens and yellows represent craters and the ejecta that surround them. The face of the Moon has remained largely unchanged for 3 billion years.

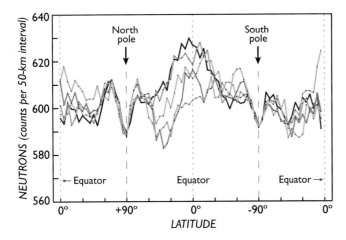

Figure 22. The number of medium-energy, or "epithermal," neutrons detected by Lunar Prospector early in its mission, plotted as a function of latitude. Each line represents one orbit's data. The dips observed at the north and south poles are due to neutron collisions with hydrogen, which mission scientists believe is evidence of water ice at the poles.

surface outflows of volcanic rock. Some of the larger impacts created multiring basins that penetrated below the broken, intermixed debris layer and threw deep-seated, pristine samples of the Moon's interior onto the surface for our collection and inspection several eons later.

The Imbrium and Orientale basins represent the last major impacts on the Moon. The Imbrium impact took place an estimated 3.85 billion years ago, and the Orientale impact probably occurred within a few tens of millions of years thereafter. At about this time the cratering rate was declining very rapidly, and more volcanic flows were being preserved from destruction. Mare volcanism may well have been more extensive before the Imbrium basin was formed, but just how much more is not known.

After about 3.0 billion years ago, the cratering rate apparently became relatively constant. The flooding of impact basins by molten basalts also began to fall off rapidly about then. Conceivably, some very small amounts of basalt surfaced onto the maria until the crater Copernicus appeared (roughly 1 billion years ago). But the dominant geologic activity on the Moon ever since has been the ongoing peppering of the surface by meteorites, punctuated by the occasional formation of a large crater. For all practical purposes, the Moon is now geologically dead.

Despite its violent beginnings, the Moon became quiescent long ago and now affords us the opportunity to examine a "fossil" from the early solar system, a planetary body frozen in time. Its most active geologic period, from 4.5 to about 3 billion years ago, perfectly complements the observable geologic record of the Earth, for which rocks older than 3 billion years have been almost completely destroyed (see Chapter 9). Thus, the Moon holds secrets of planetary processes that we could barely imagine before the Space Age, and we see in its battered surface many of the processes ubiquitous throughout the solar system.

Although we have explored the Moon extensively with spacecraft, much of it remains mysterious. Clementine has gone a

long way toward satisfying our need for global remote-sensing reconnaissance of the surface. More recently the Lunar Prospector spacecraft arrived on the scene to map the global abundance of uranium and thorium (which track KREEP) and measure the lunar magnetic field. Another of its instruments detected gas emitted from the lunar interior, refining our estimate of the Moon's bulk composition. Finally, we have obtained a high-fidelity map of the near-side gravity field — our best method for probing the lunar interior, in the absence of a global surface network.

Whereas Clementine data gave us hope that patches of water ice may lie hidden in the permanent darkness near the Moon's south pole, Lunar Prospector has all but proven the existence of water ice near the south pole and also discovered what appears to be additional ice near the north pole *(Figure 22)*. The spacecraft's neutron spectrometer detects neutrons ejected when cosmic rays collide with atoms in the Moon's crust. Many of the initial "fast" neutrons created by these collisions escape into space. Others bounce off other atoms before flying away, and some are slowed ("cooled") significantly when they strike something similar in mass, such as the nucleus of a hydrogen atom. During its passes over the poles, Lunar Prospector detected increases in these slower neutrons and decreases in the number of medium-speed neutrons that bounce off atoms other than hydrogen. This coincidence indicates that the regolith contains substantial hydrogen, which must exist predominantly in molecules of ice. (Some hydrogen is implanted by the solar wind.) Conservatively, at least 10 million tons of ice is mixed with the uppermost regolith in the shadowed polar regions.

Orbiters can only do so much, however, and lunar scientists hope someday to place a new generation of instruments directly on the lunar surface. For example, a global network of geophysical stations would help elucidate the Moon's mantle and core structure, variations in its crustal thickness, and its enigmatic paleomagnetism. Measuring the heat flowing from the interior would constrain the Moon's enrichment in radioactive uranium and thorium and, by association, other refractory elements. Of course, we would welcome the return of additional lunar samples to answer nagging unknowns, such as the age of the youngest lunar lava flows on the Moon. Such a flow could be identified from orbital remote-sensing data, whereupon a small probe is dispatched to collect a single "grab sample" of regolith. Isotopic dating of rock chips within such a sample would yield an absolute age for the lava flow and thus add a key datum to lunar thermal and geologic history.

Right now, there are no plans for future human missions to the Moon, but if launch costs can ever be lowered significantly, we may eventually return there. The establishment of a permanent presence on the Moon opens up scientific vistas that are difficult to foresee clearly. Each Apollo mission provided a surprise, and undoubtedly far more rock types and geologic processes await discovery. From a permanent outpost or base on the Moon, we could begin a detailed exploration of our complex and fascinating satellite that could last for centuries — uncovering not only its secrets, but the early history of our home planet as well.

Mars

Michael H. Carr

The robotic rover Sojourner ambles on the ruddy plains of Mars in July 1997.

WE HAVE LONG been fascinated by Mars. Its red color, periodic brightenings, and slow looping movements across the starry background make it particularly distinctive in the night sky. Over the past two centuries our fascination with Mars has been stimulated largely by the prospect that life may exist there and the certainty that it will be the first planet beyond Earth to be visited by humans. In the early years of the 20th century, before the advent of more powerful telescopes, remote-sensing techniques, and spacecraft missions, it was widely believed that advanced civilizations had developed on Mars. Many observers saw long, linear markings on the planet and theorized the existence of a canal system built to transport water from the poles to the parched equatorial deserts. As the 20th century progressed, belief in intelligent life on Mars dimmed considerably. Even so, as late as the early 1960s some maps of the planet prominently portrayed oases and canals stretching across the dusty plains. In addition, many astronomers continued to believe that seasonal changes in the surface markings (*Figure 1*) could be due to changing vegetation patterns.

These provocative speculations were laid to rest rather abruptly with the advent of interplanetary spacecraft in the late 20th century. With each flyby and lander, old misconceptions evaporated in the face of scientific fact, and our picture of Mars changed dramatically. The early perception of Mars as a place hospitable to life began to unravel in the 1960s, when Mariners 4, 6, and 7 flew by at close range. Their pictures revealed a cratered and apparently lifeless landscape somewhat resembling the Moon's. However, our perception changed again after Mariner 9 arrived at Mars in 1971. This spacecraft was an orbiter, not a flyby craft like its predecessors, and this proved fortunate. When it arrived the entire planet was shrouded by a dust storm. By early 1972, however, the dust had cleared, and

141

Mariner 9 began a systematic exploration using a variety of instruments. Unlike earlier explorations, this one revealed the complex, very *un*-Moonlike planet that we know today. Soviet spacecraft sent to Mars during the 1970s returned data confirming the planet's geologic diversity.

With these first survey missions completed, researchers turned their attention to the question of life on Mars. In 1976 NASA placed two spacecraft in orbit around the planet, each of which dispatched a lander to the Martian surface. All four craft worked flawlessly and returned data to Earth for more than four years. Although the landers' complex biology experiments did not find any life (see Chapter 27), they revealed much about the reactivity and physical properties of the surface. Other lander experiments observed the landscape as it changed through the seasons, recorded the local meteorology for more than two Martian years, determined the surface chemistry, and listened for marsquakes. Meanwhile, the two orbiters photographed the entire planet, measured its thermal properties, and monitored the water content of the atmosphere. By the mission's end, all of Mars had been photographed at a resolution of a few hundred meters — and much of it at resolutions down to 10 m.

The Vikings were followed by three missions that failed either in part (Phobos) or totally (Mars 96 and Mars Observer). But success returned with the highly popular Mars Pathfinder, which bounced its way to a soft landing in the Chryse Planitia region on 4 July 1997 and deployed a rover to inspect and analyze the various rocks at the landing site. Later the same year Mars Global Surveyor reached the planet and began to study the Martian surface geology and mineralogy, global topography, magnetism, and gravity.

Meanwhile, an unexpected source of new information about Mars came to light in the early 1980s. Several researchers suggested that a small group of meteorites found on Earth might come from Mars. These specimens all consisted of igneous rock, with most ranging in age from 150 million to 1.3 billion years — in sharp contrast to the 4.5-billion-year age of most other meteorites. The rocks all have similar oxygen-isotope ratios, which are unlike those found in other meteorites and in Earth and Moon rocks. Researchers concluded that these meteorites *had* to come from a planet that was volcanically active fairly recently. Their only plausible source is Mars. This origin was subsequently confirmed by analysis of gases trapped in some of the meteorites, gases with chemical and isotopic compositions almost identical to those identified in the Martian atmosphere by the Viking landers. By 1997 we had recognized 12 such meteorites (see Chapter 26). They were ejected from the surface of Mars during large impacts and subsequently fell on the Earth after spending several million years in space. One of these, designated ALH 84001, differed from the others in being very ancient. In 1996 a research

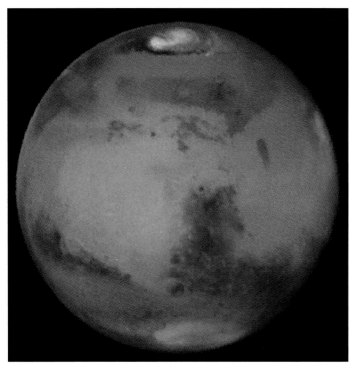

Figure 1 (above). The sharpest view of Mars ever taken from Earth was obtained by the Hubble Space Telescope on 10 March 1997 — the last day of spring in the Martian northern hemisphere. The image shows familiar bright and dark markings known to astronomers for more than a century. The annual deposit of carbon dioxide frost is rapidly subliming at the north pole, revealing the much smaller permanent water ice cap. The receding frost veneer also reveals the dark, circular sea of sand dunes that surrounds the north pole (Olympia Planitia). Other prominent features in this hemisphere include Syrtis Major Planitia, the large dark area just below center. Clouds of water ice shroud the giant impact basin Hellas (near the bottom of the disk) and several great volcanoes in the Elysium region near the eastern (right) limb.

Figure 2 (right). Two weeks after touching down on Mars on 20 July 1976, the Viking 1 lander returned this spectacular panorama from its site in western Chryse Planitia. The pair of large rocks at left (collectively nicknamed "Big Joe" by mission scientists) is about 2 m across and 8 m from the camera. Drifts of fine-grained sediment cover much of the scene; in some places the drifts have been obviously scoured and sculpted by wind.

team announced that this 4.5-billion-year-old meteorite, found in Antarctica, contains organic molecules and microscopic, rod-shaped structures that could be of biological origin. While such supposition is under debate by scientists, the presence of carbonates and organic molecules in the rock *do* suggest that conditions on early Mars were more Earthlike than they are today.

The results of our explorations of Mars and its meteorites paint a picture of a dry, cold world that is, in the larger sense, very unlike the wet, warm Earth. We know that the planet has had a long and varied geologic history. At the broadest scale Mars exhibits a striking, enigmatic global asymmetry, with most of its younger surfaces in one hemisphere and most of the older surfaces in the other. Volcanism appears to have occurred throughout the past and possibly up to the present day. Tectonic deformation has taken place both on local and regional scales. The surface has also preserved a long record of impacts, despite having been extensively modified by wind, water, and ice.

While many of the processes that have shaped the planet are familiar to us on Earth, the results on Mars are spectacularly different. Huge volcanoes have accumulated atop broad regional bulges. Extensive fault systems disrupt the surface. Vast canyons have formed. Several areas have been subjected to episodic floods of enormous magnitude, and closer to the poles ice has apparently caused pervasive modification of the surface.

Why do familiar geologic processes have such large-scale effects on Mars? In what ways are these processes similar to, and different from, those at work on Earth? What was the sequence of events that led to the configuration of the Martian surface we see today? This chapter will try to answer these questions as they apply to the surface; the Martian interior is discussed in Chapter 12 and its atmosphere in Chapter 13. However, the histories of the surface, atmosphere, and interior are inextricably intertwined, and any discussion of the Martian landscape leads inevitably to consideration of climate and volatile inventories. Indeed, one of the most fascinating aspects of Mars's surface is the set of clues it provides about former climatic conditions that may have been more hospitable to life.

Martian geology differs from Earth's mainly because of the planet's smaller size. Mars has a radius of 3,396 km, compared with 6,378 km for Earth. As a result its interior must have cooled relatively quickly after forming, and consequently Mars now has a lower heat flow and less volcanic activity than Earth does. Mars also differs from Earth in that it lacks plate tectonism; that is, its crust is now rigid and static. Consequently, the planet lacks analogs to many of the Earth's most prominent features that result from plate motions: mountain chains, oceanic troughs, mid-ocean ridges, and lines of volcanoes.

Another distinction is that the Martian surface is very cold. Temperatures there perhaps reach the freezing point of water (273° K) in summer, but the average daily temperature is some 50° colder. In winter, polar temperatures fall to 150° K. The planet is so cold partly because of its distance from the Sun (1.5 times the Earth's) and partly because its atmosphere is too thin to provide any greenhouse warming. Under present conditions, the ground is permanently frozen down to depths of 1 km or more, liquid water cannot exist anywhere on the surface, and exposed ice is stable only at the poles. Thus, liquid water, which is such an effective erosive and weathering agent here on Earth, is not available on present-day Mars. (As we will see later, this was not always true.)

THE VIEW FROM THE SURFACE

To truly understand Martian geology, global observations from orbiting spacecraft must be augmented by the "ground truth" of detailed surface studies. Currently, our only close-up views of the Martian surface come from the places where two Viking landers and Mars Pathfinder touched down. The Viking 1 landing site, in Chryse Planitia at 22.3° north, 48.0° west, is on a plain that looks featureless from orbit except for impact craters and wrinkled ridges like those on the lunar maria. As viewed from the lander (*Figure 2*), the surface appears to be a gently rolling region strewn with centimeter- to meter-sized blocks. These rocks exhibit a wide range of color, shape, and texture, probably reflecting variations in origin and the length of their exposure on the surface. Many are angular with coarsely pitted surfaces.

The blocks are thought to be pieces of the local volcanic rocks excavated by impact. Why there are so many of them,

Figure 3. A view of the rock-strewn Ares Vallis floodplain, as recorded by Mars Pathfinder in July 1997. The two hills on the horizon, near the left edge, are 1 km from the lander. Between and partly covering the rocks is dust, the result of chemical weathering of exposed rock surfaces here and elsewhere. Near the lander in the foreground the dust is partly cemented, probably by sulfates and other soluble salts. At right, the rover Sojourner is analyzing the elemental composition of a large boulder nicknamed "Yogi."

compared to those found on the lunar maria, is unclear. Perhaps the presence of an atmosphere protects the surface from microscopic erosion by meteoritic particles. It is also possible that the Martian wind winnows out small particles and carries them elsewhere, so that a Moonlike regolith of intermixed coarse and fine material does not develop everywhere on Mars.

Fine-grain debris is seen interspersed among the blocks and has accumulated into dunelike features that appear prominently in several parts of the scene. The lander set down on an extensive patch of this dusty material, which turned out to be all the spacecraft's 3-meter-long sampling arm could obtain for analysis. (The arm picked up what appeared to be rock fragments, but these all turned out to be clods of dirt.) The individual dust particles are generally very small, well under 10 microns across. From its chemical composition, and from simulations of the results of the biology experiments, the debris is thought to be a mixture of iron-rich clays and a poorly crystallized, hydrated alteration product of volcanic rocks called palagonite. This surface material thus differs from the lunar regolith because it is not merely pulverized rock — it is a product of chemical weathering processes such as oxidation and hydration.

Near the surface, tiny particles are cemented together to form a friable hardpan. This crust probably results from cementation of the clays by soluble salts such as sulfates and nitrates. The surface layer has three other peculiar chemical characteristics. First, if water is added to it, oxygen is given off. This sug-

gests the presence of some reactive oxidant, probably a peroxide. Second, the "soil" tested by the Vikings contains no organic compounds despite the continual rain of small amounts of such material in meteoritic debris. The implication is that organics are destroyed by a combination of the oxidants and ultraviolet radiation at the surface (see Chapter 27). Finally, the dust contains several percent of some magnetic mineral. The most likely candidate is maghemite, an iron oxide created when water leaches Fe^{2+} ions from bedrock. Alternatively, the magnetic phase could be a titanium-rich form of magnetite derived directly from basaltic rocks without the intervention of water.

Mars Pathfinder also landed in Chryse Planitia at 19.3° north, 33.6° west, 800 km to the east-southeast of the Viking 1 lander (*Figure 3*). The site was chosen because it lies at the mouth of a large outflow channel, called Ares Vallis, thought to have been carved by a catastrophic flood in the distant past. The channel's source is situated deep in the highlands to the south, but the landing site is in a place where the flood spilled onto the adjacent plains. As the water spread out, it would have deposited entrained rocks and boulders eroded from the highlands upstream. It seemed likely therefore, that a variety of different kinds of rocks from different regions would be available for analysis in the area where the spacecraft landed.

Those predictions were apparently borne out, as Pathfinder found a region more rocky than the Viking 1 site. Some of the rocks form stacks leaning in the direction of the flood (as

inferred from pictures taken from orbit). Horizontal lines on hills in the distance could be the former shorelines of the flood, or they might indicate the existence of sedimentary deposits.

Pathfinder's rover, Sojourner, examined a number of rocks and fine-grain drifts near the lander. We expected that some rocks at the site would be basalts, which forms by partial melting of the mantle. Basalt is the most common igneous rock type on Earth, the Moon, and probably Mars (all Martian meteorites, for example, are basalts of some type). However, Sojourner's analyses of several rocks revealed a surprisingly wide range in silicon content (*Figure 4*, *Table 1*). In fact, silicon's abundance in at least two of these rocks is much too high for

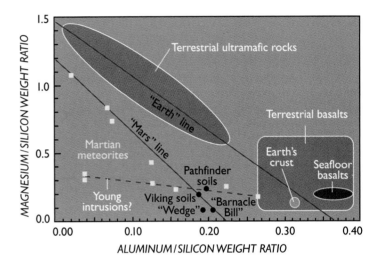

Figure 4. Key elemental ratios found at the Viking and Pathfinder landing sites are compared with those of Martian meteorites and terrestrial materials. (Hawaiian lava is a typical basalt; ultramafic rocks, rich in iron and magnesium, are usually found deep in Earth's crust.) Martian rocks are chemically distinct from their terrestrial counterparts, probably a consequence of the two planets having formed at different distances from the Sun.

Composition of the Martian Surface

Elemental oxide	Average Mars "soil"	"Barnacle Bill"	"Wedge"	Terrestrial oceanic basalt
Na_2O	2.0	3.2	3.1	0.2
MgO	7.1	3.0	4.9	11.0
Al_2O_3	9.1	10.8	10.0	15.0
SiO_2	51.6	58.6	52.2	50.7
SO_3	5.3	2.2	2.8	0.2
K_2O	0.5	0.7	0.7	0.1
CaO	7.3	5.3	7.4	12.0
TiO_2	1.1	0.8	1.0	0.8
Cl	0.7	0.5	0.5	0.2
FeO	13.4	12.9	15.4	8.4

Table 1. The compositions of fine-grained "soils" and two rocks from the Mars Pathfinder landing site differ noticeably from a typical basalt from the Earth's seafloor. The Martian rocks appear much more enriched in volatile elements (sodium, potassium, and sulfur) than their terrestrial equivalent, suggesting that they were partially coated with dust when analyzed. However, the Martian rocks do exhibit a higher iron-to-magnesium ratio than terrestrial basalts do. "Barnacle Bill" also contains more silicon than is typical for basalts, which suggests that it originated in a magma chamber within the crust rather than being derived from melting of the mantle.

Figure 5. Patches of frost surround the Viking 2 lander in this view taken in 1977 during the spacecraft's first Martian winter. Here the icy layer is perhaps no more than a few hundredths of a millimeter thick. The frost probably arose from water vapor transported to the site during dust storms. Dust-water aggregates precipitated onto the surface when carbon dioxide froze out of the atmosphere and adhered to them.

Figure 6 (below). The major geologic features of Mars. Heavily cratered terrain (reddish brown) formed at the end of heavy bombardment around 3.8 billion years ago. White areas are younger, more sparsely cratered terrains. Flood channels occupy three main areas: around the Chryse basin, northwest of the volcanic region of Elysium, and on the northeast rim of the large impact basin, Hellas. Valley networks are confined mostly to the oldest terrain.

basalts, and they may have been formed by some process other than partial melting. These rocks, nicknamed "Shark" and "Barnacle Bill," are more analogous to a type of fine-grained quartz-rich rock found on Earth called andesite. The high silicon content may have some implication for the amount of differentiation in the Martian crust.

Meanwhile, Pathfinder and Sojourner found that the drifts of dust within view of the lander have compositions unlike those of the rocks but similar to those of the fine-grained material measured by the Viking landers, with a few differences. The samples at Ares Vallis tend to have less sulfur and more titanium than the Viking samples. They are probably a mixture of weathering products derived locally and those redistributed by the winds from elsewhere on the planet.

The Viking 2 lander set down in Utopia Planitia at 47.6° north, 225.7° west, part of a vast plain extending over much of the high northern latitudes. As seen from orbit, the region is characterized by complex bright and dark markings, polygonal fractures, and a variety of textures attributed to repeated deposition and removal of material by the wind. All around the lander is a level, boulder-strewn, but otherwise featureless plain stretching out to the horizon (*Figure 5*). The blocks are remarkably uniform in size and appearance; most are angular and deeply pitted. No dunes can be seen, though fine-grain debris is interspersed among the blocks, commonly in their lee. Geologists think the site got its blocky appearance when it was covered by a lobe of ejecta from Mie, a 90-km-diameter crater situated 200 km east of the landing site. The compositions of the dusty material at the two Viking sites are almost identical. This suggests that the fine-grain debris is chemically homogenized over the whole planet as a result of mixing by the annual dust storms.

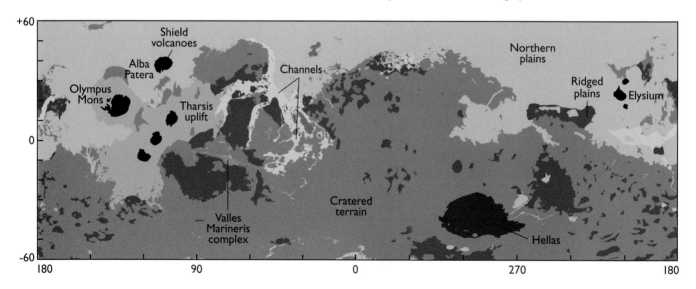

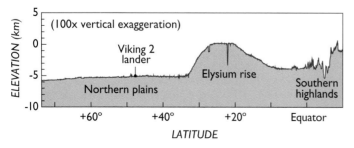

Figure 7 (left). Using a laser altimeter from orbit, Mars Global Surveyor measured the heights of features along this 4,800-km-long stretch of the Martian surface. The low elevation of the northern plains was known from Viking data, and they are seen here to be remarkably flat. To the south (right) is Elysium, a broad volcanic province centered at 210° longitude (near the right edge of Figure 6). Elysium's plateau is cut by a chasm roughly 3½ km deep.

HIGHLANDS AND PLAINS: A DICHOTOMY

Mars is decidedly not uniform with regard to the global distribution of its geologic features (*Figure 6*). It almost seems as if halves of two dissimilar planets have been fused together. Most of the southern hemisphere and part of the northern hemisphere (especially surrounding the 330° longitude meridian) are covered with heavily cratered highlands. These lie 1 to 4 km above the Martian equivalent of sea level. In contrast, much of the northern hemisphere is covered with sparsely cratered plains. The plains are mostly well below the mean surface level, except in the volcanic regions of Tharsis and Elysium (*Figure 7*). The cause of the hemispherical disparity is unknown; conceivably it is the result of a giant impact very early in the planet's history.

The abundance of large craters in the highlands suggests that the terrain has survived with relatively little modification for eons. We know the Moon experienced a prolonged period of heavy meteoritic bombardment ending around 3.8 billion years ago, after which the cratering rate became very low. Mars has probably had a similar history. Its heavily cratered highlands formed just prior to 3.8 billion years ago, and all the more sparsely cratered surfaces were emplaced afterward. Although the highlands of the Moon and Mars resemble one another, there are differences. First, Martian craters tend to be more degraded, perhaps reflecting faster erosion early in the planet's history. Second, branching valleys that resemble terrestrial river valleys are common throughout the highlands. Third, extensive smooth areas lie between the Martian highlands' craters. These areas probably result from relatively high rates of volcanism, and other forms of deposition, during and immediately after the heavy bombardment.

A fourth difference between the Martian and lunar highlands involves the blankets of debris thrown out during the craters' for-mation. Large lunar craters typically have coarse, hummocky emplacements close to their rims. Farther out, the ejecta have a radial "fabric," which merges outward with lines of secondary craters. In contrast, the ejecta around most Martian craters form thin sheets, each with a lobate outer margin that is clearly defined by a low ridge (*Figure 8*). The pattern resembles the result when pebbles are dropped into mud — an analogy that may be quite close to reality. Most geologists think that the distinctive patterns around Martian craters are a consequence of water or ice being present in the ground at the time of impact. The ejected slurry of material continued to flow outward after being deposited around the crater. In support of this idea, we see numerous places where the ejecta sheet flowed around obstacles in its path. One puzzle is the existence of such flow patterns high on the flanks of large volcanoes, where water-saturated ground would not be expected. An alternate explanation is that the patterns around Martian craters result from the ejecta's interaction with the atmosphere.

Under present climatic conditions, ice is unstable at low latitudes both at and below the surface (*Figure 9*). Over time it sublimes into the atmosphere, then freezes out at the poles. However, at latitudes higher than about 30° to 40°, ice is stable just below the surface, and we should expect ground ice at depths greater than about a meter. In these regions we do indeed find evidence for ground ice. At latitudes poleward of 30° the terrain has a softened appearance, as though the surface materials had oozed and so caused a rounding of all the landforms (*Figure 10*). In addition, at these same latitudes, every cliff or steep slope is accompanied by flowlike features extending up to 20 km away from its base. Both characteristics may be telling us that the cratered highlands at these latitudes contain abundant buried ice, which apparently acts as a lubricant and causes the surface materials to creep downhill. This movement is especially evident in the talus that accumulates on steep slopes. In contrast, at low lati-

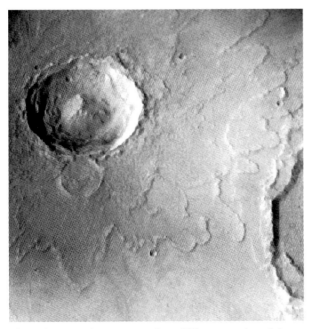

Figure 8. **Yuty, an impact crater about 20 km across, has a lobate, layered ejecta blanket. This distinctive pattern probably formed because the airborne debris was saturated with groundwater and thus tended to flow across the ground as a fluidlike mass after being ejected outward.**

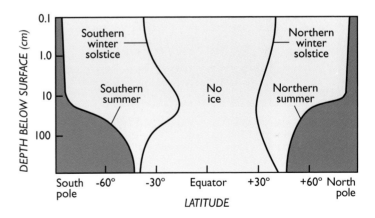

Figure 9. **Today liquid water does not exist anywhere on the Martian surface, and the stability of water ice depends largely on its subsurface depth and latitude. For example, ice is unstable within about 30° of the equator (tan); over time, near-surface materials at these latitudes have probably lost any water that might have been originally present down to depths of more than 100 m. Farther to the north and south (light blue), ice could exist within 1 m of the surface during local winter but would be driven into the atmosphere as vapor at other times. In the polar regions (dark blue), water ice is always stable down to kilometer depths. This model assumes that the Martian atmosphere is well mixed and carries only enough water vapor to create a layer on the surface 12 microns thick.**

Figure 10. Compare these Viking images of Martian craters. Those in the top image (the largest is 24 km across) are typical of the craters found near the equator. They have sharp rim crests, ridges and escarpments between them are well defined, and small craters are distinctly visible. Those below, however, are typical of the heavily cratered highlands found on Mars at latitudes poleward of 35°. Crater rims are rounded, and other features are indistinct and poorly preserved. The differences seen in this comparison have been ascribed to the presence of ground ice at high latitudes, which permits the gradual creep of material near the surface.

tudes we see no rounding of the topography and no flows at the bases of slopes. Talus simply accumulates on steep slopes and protects them from further erosion.

The extensive plains dominating Mars's northern hemisphere differ from the highlands mainly in their generally lower elevation and in being quite sparsely cratered. They clearly postdate the period of heavy bombardment but have a wide range of ages. The most heavily cratered plains, such as those of Lunae Planum, probably formed more than 3.5 billion years ago — shortly after the end of heavy bombardment. By contrast, very sparsely cratered plains in Tharsis are probably no older than 500 million years. Some of the equatorial plains, particularly those in the volcanic regions of Tharsis and Elysium, are formed of numerous flows superimposed on one another (*Figure 11*), as on the lunar maria. In other areas flows are not visible, but numerous ridges are (again resembling the maria). These plains appear to be volcanic as well, though other origins are possible. On all these plains, impact craters have the distinctive flowed-ejecta patterns described earlier.

Many low-lying plains areas, particularly those at high latitudes, do not fall into the two categories just described. Some have a distinctly mottled appearance because the ejecta around their impact craters are much brighter than the intervening areas. Others exhibit strange polygonal fracture patterns. Still others appear to be etched or partly covered with easily eroded debris. These plains are probably of diverse origin, with volcanism, wind, and ice all participating in their formation. Some researchers have speculated that large seas were created downstream of the outflow channels following the huge floods. Many peculiar textures and long linear features, which suggest shorelines, could be explained by long-ago oceans. Others believe only modest bodies of water remained after the floods. These froze rapidly, creating massive ice deposits that are responsible for the unusual textures seen at the surface. High-resolution mapping of these plains will provide evidence to support (or disprove) the idea of standing water on the Martian surface.

VOLCANOES

Among Mars's most impressive features are its large volcanoes. Several occur in the Tharsis region (centered on the equator at 105° west), a huge uplift in the Martian crust about 4,000 km across and 10 km high at its center. On the northwest flank of Tharsis are the large volcanoes Arsia Mons, Pavonis Mons, and Ascraeus Mons. Just beyond its northwest edge is Olympus Mons, the tallest volcano on the planet. All four constructs are enormous by terrestrial standards. The main edifice of Olympus Mons, 550 to 600 km across, rises more than 24 km above its surroundings and is rimmed by a cliff 6 km high in some spots (*Figures 12,13*). North of Tharsis lies Alba Patera, more than 1,500 km wide. For comparison, the largest volcano on Earth — Hawaii's Mauna Loa — is 120 km across at its base and has a summit 9 km above the ocean floor. The Elysium province (centered at 25° north, 210° west) is also at the center of a crustal bulge. Some 2,500 km across, Elysium is smaller than Tharsis and so are its volcanoes.

Olympus Mons and its like-size siblings resemble large terrestrial shield volcanoes like those of the Hawaiian Island chain.

Figure 11. **Overlapping lava flows cascade down the flank of Alba Patera, a huge shield volcano in Mars's northern hemisphere. The longest flows seen in these Viking orbiter images have traveled for at least 225 km, making them many times larger than eruptions on the Earth.**

Each has a summit pit or caldera, with numerous long flows and channels cascading down its flanks. The main difference between volcanoes on Mars and Earth is size. The biggest Martian volcanoes are at least 100 times more massive than their terrestrial equivalents, their summit calderas are 10 to 100 times larger, and the flows on their flanks 10 to 100 times longer. Calderas of such size indicate large magma chambers within the volcano, and the long lava flows indicate more copious eruptions than are typical on Earth.

The impressive sizes of the volcanoes themselves probably result mainly from the lack of plate tectonism on Mars. Hawaii's shield volcanoes have relatively brief active lifetimes because the Pacific plate on which they stand constantly moves northwest. Once separated from its magma source in the mantle below the plate, a Hawaiian volcano falls silent and a new one forms to its southeast. In contrast, volcanoes on Mars remained over their magma sources and continued to grow as long as magma was available. Compared to elsewhere on Mars, few impact craters are superimposed on the large shield volcanoes in Tharsis and Elysium. This suggests that the mountains' outermost veneers of lava are relatively young. However, evidence from surrounding flows indicates that the volcanoes have been accumulating for much of Martian history. Thus the planet's large shield volcanoes may actually be quite old, despite their youthful exteriors.

Not all Martian volcanoes created towering shields. In fact, some (like Alba Patera) show relatively little vertical relief. Tyrrhena Patera, situated northeast of the large impact basin Hellas, is unlike any of the volcanoes discussed so far. It is surrounded by deeply eroded, stratified deposits (*Figure 14*). Most probably, the eruptions of Tyrrhena Patera consisted primarily of ash rather than lava. In this respect the volcano more resembles Mount St. Helens than Mauna Loa. Its more explosive eruption style could arise from magma with a higher proportion of silicon than that which formed the Tharsis volcanoes. Alternatively, the magma could have encountered water or ice en route to the surface.

The Tharsis bulge has clearly played a major role in the evolution of Mars. On it are the planet's largest volcanoes and most of its youngest features. A vast system of fractures surrounds Tharsis and affects almost a third of the planet's surface. These fractures are an apparent response to the huge loads on the underlying crust caused by the bulge's immense mass. On the eastern flank is a huge equatorial canyon system, and numerous flood features scar its eastern periphery. What caused the Tharsis bulge is not known. Perhaps it formed by an upward doming of the crust, possibly in response to buoyant convection in the underlying mantle. Another theory holds that Tharsis is simply a thick accumulation of volcanic materials. Whatever its origin, the Tharsis bulge clearly arose very early in the planet's history, because all its visible lava flows (including very old ones) have moved downhill in directions consistent with the present topography.

The Martian meteorites tell us that volcanoes have been active on the red planet within the last 150 million years. Geologically speaking, this is very recent; it implies that Mars is volcanically active today. However the rates of volcanism are much lower than on Earth, making it unlikely that we will someday witness an eruption as it occurs.

CANYON SYSTEMS

To the east of Tharsis, just south of the equator, is a vast system of canyons collectively known as Valles Marineris (*Figures 6,15*). The system begins in the west with a complex fracture zone called Noctis Labyrinthus, at the summit of the Tharsis bulge. It then extends about 4,000 km eastward until the canyons merge with areas of what is termed *chaotic terrain* (to be described later). Depths range from 2 km at the east and west ends to over 7 km in the central section, where three parallel canyons merge to create a chasm over 600 km wide. The canyons appear to have formed largely by faulting. Their walls typically have long, straight sections, and linear offsets are common at the base of the walls, as are triangular-faceted spurs — all common attributes of fault scarps on Earth.

However, the canyons have been shaped by more than just faulting. In places, particularly on the south rim of Ius Chasma, deep branching side valleys suggest fluvial (water) erosion, caused not by rainfall but rather by the seepage of groundwater. Elsewhere the canyons have been widened by enormous landslides. In the west, at least, fluvial erosion appears to have played

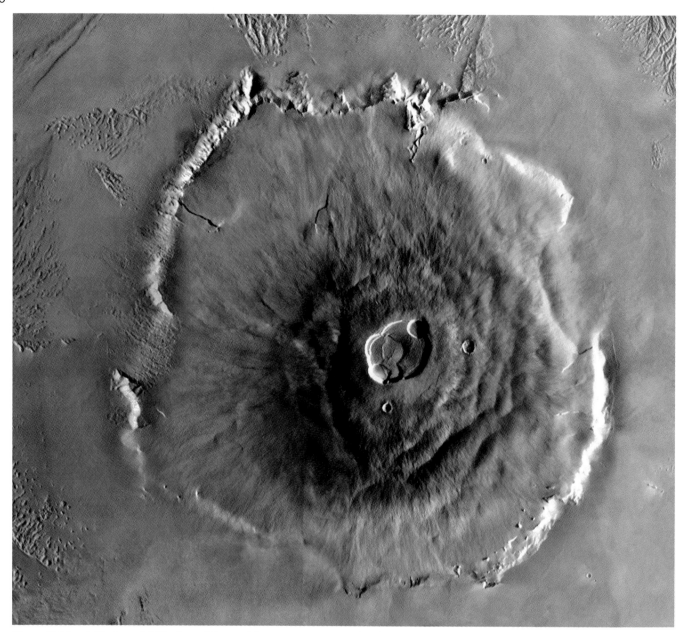

Figure 12 (above). The huge shield volcano Olympus Mons covers an area the size of Arizona. Its peak is roughly 24 km above the broad plateau on which the mountain sits and more than 26 km above Martian "sea level." The volcano's undulating flanks slope at an average angle of 4°. Somewhat north of its summit is a conspicuous caldera, a complex nest of collapse craters 90 km across. Long before spacecraft discovered its true nature, Olympus Mons and the clouds that frequently attend it were recognized by telescopic observers as a bright spot on Mars named Nix Olympica (Snows of Olympus).

Figure 13 (below). Olympus Mons dwarfs the principal peaks of the Hawaiian Island chain (here depicted as if they formed in a straight line). Many parallels can nonetheless be drawn between Martian and terrestrial shield volcanoes. Heights are exaggerated by five times.

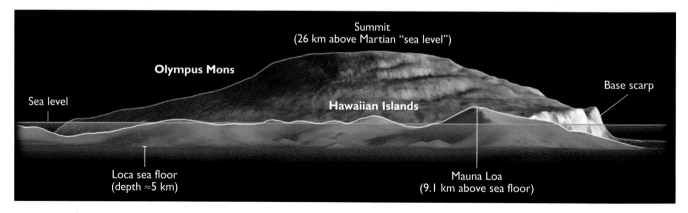

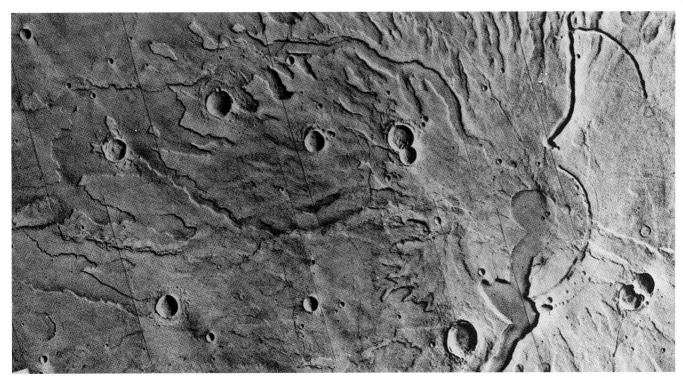

Figure 14. Tyrrhena Patera, in the Martian southern highlands, is one of several ancient, degraded volcanic centers found on the planet. Partly eroded deposits (probably volcanic ash), with superimposed impact craters, surround an indistinct central crater about 45 km wide.

a minor role in the formation of the canyons. There the system consists largely of interconnected closed depressions, with no throughgoing drainage. Indeed, Hebes Chasma is a large, closed canyon entirely isolated from the others. Farther east the canyons are more continuous, and at the extreme eastern end are many water-formed features such as teardrop-shaped islands and streamlined walls. Thus the Martian canyonlands had a complex origin, with faulting and subsidence common in the west and fluvial effects common in the east.

One of the most intriguing aspects of the canyons is the presence, in places, of thick stacks of layered sediments. They are particularly common in the Coprates, Hebes, Ophir, and Candor sections (*Figure 16*). The only satisfactory explanation is that these sediments were deposited under water. Accordingly, many geologists think standing bodies of water partly filled the canyons in times past. These lakes were probably fed by water seeping from the canyon walls. It has also been suggested that the lakes within Candor and Ophir were released catastrophically as the ridge dividing Candor Chasma from the main canyon to its south failed. This would explain the pattern of erosion of the canyon-floor sediments and the previously mentioned fluvial features found abundantly at the downstream (eastern) end.

CHANNELS AND VALLEY NETWORKS

Most of Mars's large flood channels occur to the north and east of the canyons and converge on Chryse Planitia (*Figure 6*). Many of the channels emerge from *chaotic terrain*. These are areas of jostled blocks situated 1 to 2 km below the surrounding landscape and surrounded by an inward facing cliff. They appear to have formed when the surface collapsed. Some tracts of chaotic terrain merge westward with the canyons; others are completely isolated. Large channels emerge out of these chaotic areas and extend thousands of kilometers northward across

Chryse Planitia until they merge with the low-lying plains at high northern latitudes (*Figure 17*). Other large channels originate from box canyons to the north of the Valles Marineris complex. They all emerge full-size from areas of chaos and have few, if any, tributaries. They tend to be narrow and deeply incised where they cross cratered highlands, but broad and shallow on the volcanic plains. Their paths are easily recognized by scour marks and numerous teardrop-shaped islands.

The abrupt beginnings, lack of tributaries, abundance of sculpted landforms, and strong resemblance to terrestrial flood features all suggest that these channels formed by catastrophic outpourings of water. One of the largest flood zones known on Earth is the Channeled Scablands of eastern Washington, which was carved about 10,000 years ago. Numerous branching channels, tens of kilometers wide, are believed to have formed within a few days when a large ice dam collapsed, thereby releasing water from a large lake in western Montana. The Scablands event discharged an estimated 10,000,000 m^3 per second at its peak, over 100 times the maximum flow of the Mississippi River. In contrast, the discharges of the Martian floods are estimated to have been 1,000 to 10,000 times those of the Mississippi.

The Martian floods appear to have had at least two causes. Catastrophic release of water pooled in the canyons has already been mentioned. Another likely cause is the sudden eruption of groundwater under high pressure from areas of chaotic terrain (*Figure 18*). If enough water were trapped beneath a thick expanse of permafrost, pressures could mount to very high levels. Then if the overlying frozen seal were breached, such as by faulting or by formation of an impact crater, the water would erupt onto the surface under high pressure. The release of water

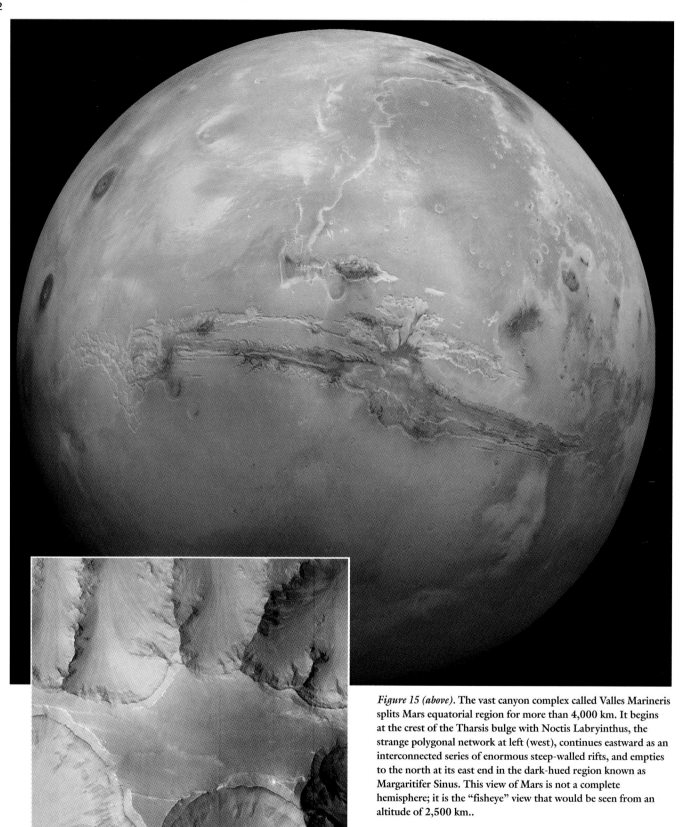

Figure 15 (above). The vast canyon complex called Valles Marineris splits Mars equatorial region for more than 4,000 km. It begins at the crest of the Tharsis bulge with Noctis Labryinthus, the strange polygonal network at left (west), continues eastward as an interconnected series of enormous steep-walled rifts, and empties to the north at its east end in the dark-hued region known as Margaritifer Sinus. This view of Mars is not a complete hemisphere; it is the "fisheye" view that would be seen from an altitude of 2,500 km..

Figure 16 (left). An image from Mars Global Surveyor resolves features as small as 6 m in a 10-km-wide portion of Coprates Chasma, which is located in the middle of Mars's vast Valles Marineris canyon complex. Steep, gully-cut slopes descend to the north and south from the high-standing plateau at center. The plateau is underlain by multiple rock layers that vary in thickness from a few meters to tens of meters. The layers may be sedimentary or volcanic in nature.

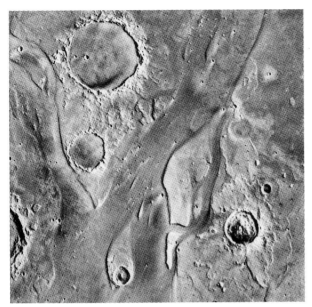

Figure 17. Ares Vallis downstream of the Mars Pathfinder landing site. The channel has streamlined banks and encloses numerous teardrop shaped islands. The largest crater is 62 km across. The channel was carved by a large flood with a peak discharge rate more than 1,000 times that of the Mississippi River. The flood flowed down the regional slope from the south (bottom) and probably created a large but temporary lake well to the north of this picture.

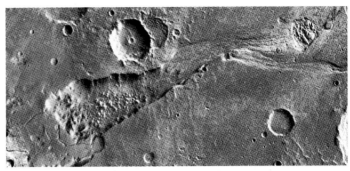

Figure 18. A large outflow channel, 20 km wide, emerges from the depression at left and continues eastward toward an even larger channel called Simud Vallis. Inside the depressed area is what geologists term *chaotic terrain* — an irregular jumble that may have resulted after groundwater erupted and flowed away, causing the ground above it to collapse.

Figure 19. Branching, dendritic channel networks, like these in Mars's southern highlands, are common. The largest neighboring craters are roughly 35 km in diameter.

could be so rapid that it would erode the underground aquifer's host rock and carry it away. The overlying surface layer would then collapse, creating chaotic terrain. This appears to have been the mechanism responsible for most of the large floods around the Chryse Basin. On the basis of age estimates for the eroded surfaces (deduced from the number of impact craters they bear), it appears that these breakouts have repeated episodically throughout much of Martian history.

Valley networks provide evidence for a different kind of fluvial erosion. In many ways the networks resemble typical terrestrial river valleys rather than the flood features just described. They are smaller than the flood channels, have tributaries, increase in size downstream, and lack sculpted bedforms (*Figure 19*). Valley networks occur almost everywhere in the cratered highlands, but only in a few places are they seen on younger terrains. Thus whatever process created them was more effective very early in the planet's history.

The branching patterns and small sizes of these valley networks suggest they, like most terrestrial river valleys, formed by slow erosion of running water rather than by large floods (*Figures 20,21*). Under present climatic conditions, small rivers would freeze too quickly to cut the valleys that we see. The valleys have, accordingly, been viewed as strong evidence that Mars was once significantly warmer. (In contrast, the floods involve such huge discharges that freezing would be insignificant even under present climatic conditions.)

If a warmer climate did exist in the past it did not last long, because the amount of fluvial erosion is limited and localized. The preservation of impact craters indicates that only a trivial amount of erosion has occurred in the last 3.5 billion years. Moreover, the valley patterns themselves suggest that the fluvial

action was brief. Individual trunk channels did not have time to extend themselves, capture streams from adjacent basins, and control the drainage over large areas. Such territorial consolidation happens relatively quickly on the Earth, so the fluvial action on Mars was either short-lived, very intermittent, or not truly analogous to what occurs here.

The valley networks are among the most puzzling of Martian surface features, and geologists are hard pressed to explain them. Perhaps they formed as terrestrial river valleys do, by the slow erosion of rock by running water. Presumably a warm climate persisted early in Martian history, allowing ample time for the older and more common valley networks to cut into the highlands. The far fewer, younger valley networks could have formed during large, short-lived, infrequent climatic swings late in Mars's history. Other researchers believe, however, that streams under an ice cover and fed by groundwater springs could have cut the valleys under present climatic conditions. Some modelers even theorize that the valleys were not cut by liquid water but by ice. They point to the difficulty of having

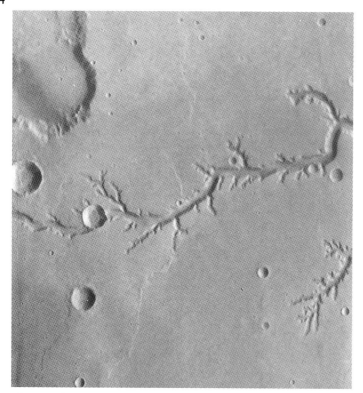

Figure 20. **All the tributaries of Nirgal Vallis, a branching valley network, have blunt ends, and there is no indication of any erosion between them. These characteristics suggest erosion by groundwater-fed streams rather than surface runoff. The picture is 60 km across.**

 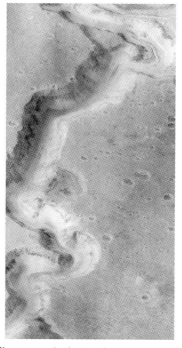

Figure 21. **Nanedi Vallis, a long, winding canyon in the Xanthe Terra region of Mars, was imaged by the Viking 1 orbiter during the 1970s (left panel) and at very high resolution by Mars Global Surveyor in January 1998 (right panel). Rocky outcrops jut from the canyon's walls, and weathered debris litters its floor. The origin of this 2½-km-wide channel is enigmatic; features like terraces and the small stream in its floor (seen near the top at right) suggest continual flow and downcutting. However, the lack of tributaries and tight meanders suggest formation by collapse. Both processes likely played a role in Nanedi's formation.**

warm conditions on early Mars because of the lower luminosity of the young Sun (see Chapter 3). Finding the networks' true cause has important consequences for the climatic history of Mars and the possibility that life could have evolved there.

Nonetheless, the geologic evidence for abundant water on the Martian surface in the past is compelling. From observations of fluvial erosion alone, geologists estimate that the planet once possessed enough water to cover it entirely with a layer up to 500 m deep. (For comparison, all of Earth's surface water, if spread in a uniform layer, would be 3 km deep.) Yet present-day Mars is very dry. The atmosphere contains only minute amounts of water, and only at the north pole has water ice been detected with certainty.

So where did the water go? Only a small amount can be stored at the poles, either in the residual cap or in the layered terrains described below. Water should be present in the ground as hydrated minerals, but again the amount involved is probably small. Some atmospheric water vapor must have been lost to space, after being split by sunlight into its component hydrogen and oxygen. At present, very little hydrogen is escaping from the upper atmosphere. Even if this rate were maintained over billions of years, the amount of lost water would still be small compared to huge volumes required to cut the channels and valleys. However, if past climatic conditions maintained more water in the atmosphere, then loss to space could have been significantly greater. We are forced to conclude that if Mars had abundant water near the surface in the past, as appears likely, then much of it is probably still there — either as ice buried near the surface or as a liquid at greater depths.

POLAR REGIONS

As mentioned above, Mars's polar regions could provide a modest-sized reservoir for water. These areas look distinctly different from the rest of the planet. Thick layers of sediment occur at both poles and extend outward to about the 80° latitude circle. In the south, they lie on the cratered highlands; in the north, they overlie plains. Cut into the deposits are valleys that curl outward from the poles, appearing as a distinct swirl in spring when the CO_2 frost sublimes from the sunward-facing slopes (*Figures 22,23*). Layering is visible on these frost-free slopes down to the resolution of the available Viking photographs, about 20 m. In the north, the layered deposits are surrounded by a vast "sea" of sand dunes (*Figure 24*), which form an almost complete dark collar around the polar regions. Dune fields of comparable scale do not occur in the south, though small ones are common within craters at high latitudes.

The polar deposits are believed to be mixtures of ice and dust. In the north, when winter's deposit of CO_2 frost completely sublimes, a residual cap of water ice is exposed. (A water-ice cap has not been observed in the south possibly because a residual cap of carbon dioxide has always been present during our observations of the south pole.) The relative proportions of water and dust in the layered terrains, and their respective accumulation rates, are probably being modulated by climatic variations. For example, dust storms observed in 1977 by the Viking spacecraft (*Figure 25*) deposited an estimated 0.4 mm of dust at the north pole. At this rate it would take 6 to 10 million years to accumu-

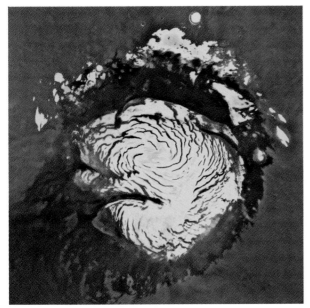

Figure 22. **A composite of Viking-orbiter images reveals the involved spiral of Mars's north polar cap. During local winter, atmospheric carbon dioxide freezes out onto the polar terrain and creates a thin white veneer covering a much greater area. But in summer the CO_2 sublimes and returns to the atmosphere, exposing a residual cap of water ice about 600 km across.**

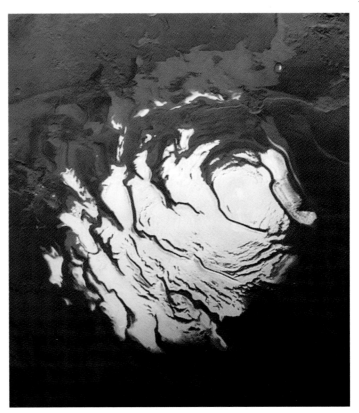

Figure 23. **Mars's south polar cap, as it appears near its minimum size (about 400 km across) during local summer. In contrast to the residual cap at the north pole, this one is believed to consist mainly of frozen CO_2. The true south pole of Mars lies just at cap's edge at lower right. The false blue shading arises from the dim lighting on the cap when the Viking 2 orbiter acquired these images in September 1977.**

late a layer 2 to 3 km deep, its apparent thickness. However, the precession of Mars's orbit and of the planet's rotation axis causes deposition to alternate between the poles in a 51,000-year cycle. These cyclic variations must surely alter the frequency and intensity of dust storms, but in ways not yet known.

Impact craters are rather sparse on the polar deposits, implying that they are young. Perhaps the deposits have eroded away and been redeposited throughout Martian history in response to long-term changes in illumination. The poles are particularly affected by changes in obliquity (the tilt of the spin axis with respect to the orbit plane), and Mars's obliquity may have changed quite dramatically in the past (see Chapter 13). At times one pole would be directly facing the Sun throughout the summer, and any water ice in the sediments would tend to sublime into the atmosphere. Having thus lost their "cement," the polar layered deposits may have been blown away by the wind, only to reform when lower obliquities returned.

SUMMARY

We have seen that Mars resembles Earth in numerous ways. The red planet has an atmosphere (albeit a thin one) and has been both volcanically and tectonically active. Its landscape is covered with rocks and minerals that are familiar to us. Water has eroded parts of the Martian surface and reacted with materials there to produce weathering products. Ice and wind have modified the surface as well.

Yet, in several very profound ways, Earth and Mars remain very different. The Earth's rigid outer rind, its lithosphere, is divided into large plates that move laterally with respect to one another. New lithosphere forms at mid-ocean ridges, and old lithosphere is consumed at subduction zones. But the Martian

Figure 24. **An enormous dune field surrounds the ice of Mars's north polar cap. The dunes here align roughly north-south, and the vague circular forms are probably buried craters.**

Figure 25. Dust storms, such as this 300-km-wide one observed within the Argyre basin in early 1977, carry micron-size particles high into Mars's atmosphere. Dust storms often grow so large that they cloak the entire planet — as happened twice during 1977 as the four Viking spacecraft looked on.

lithosphere, despite all its volcanism, is very stable. Anything erupted onto the surface simply remains in place. In fact, it is the *lack* of plate tectonism that allows the Martian volcanoes to attain their enormous sizes. Another major distinction is climate — and particularly the role of water in weathering, erosion, transport, and deposition. On Earth these processes work to reduce surface relief, constantly competing with the volcanism and deformation that combine to create relief. Throughout most of Martian history, liquid water could not exist on the surface. While fluvial erosion has occurred there, its cumulative effect has been trivial. Consequently, wherever relief has been created remains largely intact. The result is a spectacular planet on which geologic features of enormous scale and a wide variety of origins and ages are preserved.

In the eyes of many, the most crucial difference between Mars and Earth remains the widespread development of life. The prospects for finding evidence of past (and even present) Martian organisms now seem much better than in the 1970s, when the negative results from Viking seemed to dash the hopes of biologists. Now we realize that the most primitive organisms on Earth, those least evolved from the common ancestor of all terrestrial life, live in hydrothermal environments. Such places must have been common on early Mars as well: evidence suggests that liquid water flowed abundantly when the planet was rife with volcanic activity. However, despite all this suggestive evidence, we simply do not yet know whether life *did* start on Mars. In view of the current plans of NASA and other nations for sustained Mars exploration, including sample returns in the not-too-distant future, a definitive answer to this question — and many others — lies well within our reach.

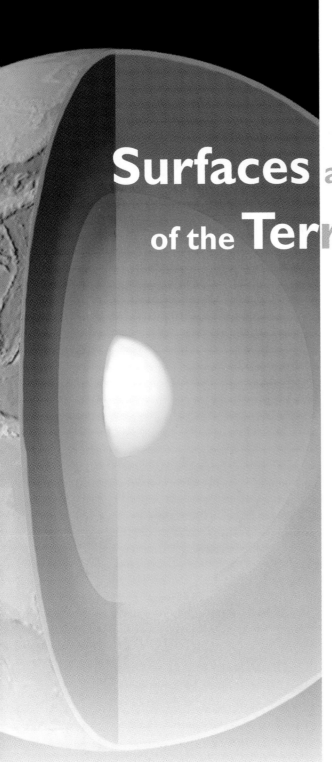

Surfaces and Interiors of the Terrestrial Planets

James W. Head III

A schematic cross-section through Earth reveals its immense iron core.

N THE LAST 35 years, two parallel revolutions in the Earth sciences have radically altered our perceptions of planets and how they work. The first was the development of the theory of terrestrial plate tectonism. During the 1960s we came to realize that Earth's geologic expressions are not an assortment of isolated puzzles to be solved individually, but rather a record of the movement and interaction of a small number of large lithospheric plates operating in an integrated, global manner. The second revolution resulted from the unfolding view, thanks to spacecraft exploration, of planetary surface features. Almost overnight, it seems, the Moon and planets changed from astronomical objects to geologic ones. In doing so, they began

157

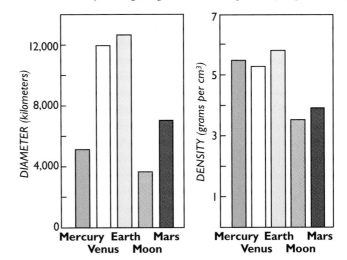

Figure 1. Mercury, Venus, Earth, Mars, and the Moon are crudely comparable in size, and all five have relatively high densities. Collectively known as terrestrial planetary bodies, their silicate-dominated compositions are distinct from those of the outer planets and their satellites, which are rich in ices and other volatile elements.

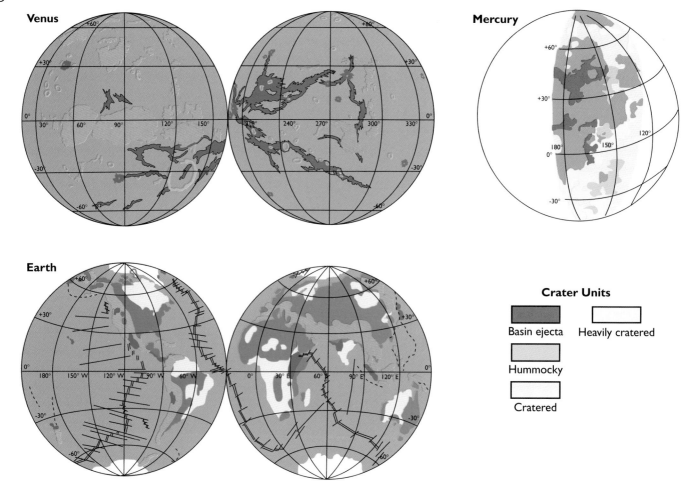

Venus

Mercury

Earth

Crater Units

Basin ejecta   Heavily cratered

Hummocky

Cratered

Figure 2. These generalized geologic maps of terrestrial planetary bodies bring together data from numerous spacecraft investigations and a variety of other sources. Several geologic surface types are common to all five worlds, including cratered terrain and volcanic plains. Others are limited to one or a few of them, such as polar deposits (Earth, Mars), tesserae and coronae (Venus), and folded mountains (Earth, Venus). In general, the surfaces of the Moon and Mercury are more primitive and less complex than those of Earth, Venus, and Mars. Of the five worlds, only the Earth and Venus are believed to have undergone extensive surface modification in the last 2 billion years. One puzzling feature common to Earth, Mars, and the Moon is a dichotomy of terrains that results in a hemispherical asymmetry. On Earth, the dichotomy is between the continents and the volcanic plains of its ocean basins; on the Moon it is between ancient, heavily cratered highlands and mare plains (which are concentrated on the lunar near side); on Mars it is represented by northern lowlands and southern cratered uplands. The origins of these hemispherical asymmetries are not fully understood.

to provide a framework within which Earth could be viewed not as a single data point, but as one of a family of planets.

The point is that, prior to the 1960s, geologic thought focused on Earth and in particular on specific areas of its surface. In a rather passive sense, this mindset was analogous to the pre-Copernican, Earth-centered view of the solar system. Now we have abandoned such geologic chauvinism, working and thinking instead in terms of a new solar system in which Earth's history is inextricably linked to those of the Moon and the other terrestrial planets. During the 1990s, Earth scientists began to apply their understanding of recent geologic history to the first 4 billion or so years of our planet's existence. Simultaneously, others consolidated their understanding of the formative years of planetary evolution (the first 1½ billion years) and began applying it to Earth. The study of Venus will provide a driving force for this convergence. It is the most Earth-like of the terrestrial planets in terms of gross physical properties (*Figure 1*), and we must learn to what extent it does — or does not — mimic Earth in detail.

THE INGREDIENTS FOR PLANETARY EVOLUTION

Three fundamental processes — impact cratering, volcanism, and tectonism (crustal movement) — give planetary surfaces most of their characteristic features (*Figure 2*). In addition, as later sections will show, each terrestrial planet expresses these processes with a unique signature. Our studies of the surfaces of these worlds reveal an amazing diversity of characteristics, a few emerging themes, and a host of basic new questions.

However, we cannot address questions of the planets' evolution only by studying their surfaces, any more than we can tell the life story of a group of human beings just by looking at them. Like humans, planets are complex systems in which most of the driving forces and regulating processes are hidden below the surface. To study these forces and processes, indirect measurements are required. Indeed, the surface features must be studied in detail to infer the nature of the interior and how it might have changed with time.

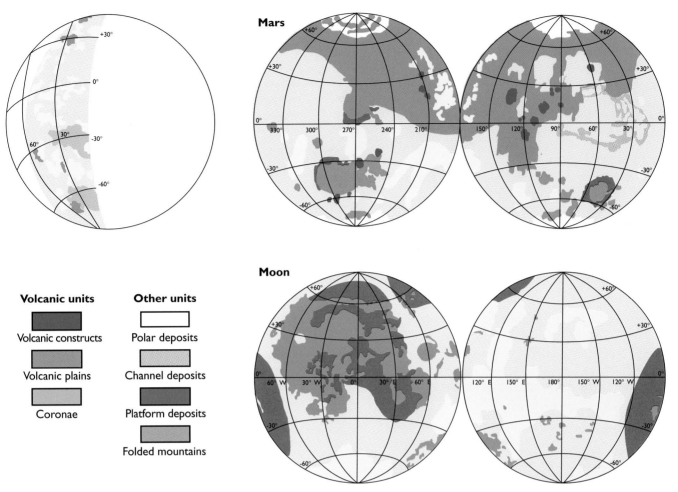

Mars

Moon

Volcanic units

Volcanic constructs

Volcanic plains

Coronae

Other units

Polar deposits

Channel deposits

Platform deposits

Folded mountains

Once we know the answers to basic questions — whether a planet's interior is homogeneous or layered, whether heat in the interior has been high enough to cause volcanic activity, and so on — we can ask more sophisticated questions, such as: How do planets differ in their basic internal structure, and how does this determine their evolution? What role do a planet's size, and its position in the solar nebula during its formation, play in its further development? Questions like these are basic tools of comparative planetology, which views the solar system as a single "experiment" in which differences among planets can be explained by variations in initial composition, distance from the Sun, and so on.

The most basic process that shapes the early evolution of a planet is *differentiation*, the segregation of a relatively homogeneous interior into layers or pockets of different composition (*Figures 3,4*). There is ample evidence that each of the terrestrial planets, as well as Earth's Moon, has undergone this segregation, producing a crust, mantle, and core. Differentiation can occur rather rapidly and be rather catastrophic. The process of core formation, in which denser, iron-rich material sinks to the deepest interior, is thought to occur in the first few percent of a planet's history, when the interior is hottest. The amount of gravitational potential energy released by the process of core formation is incredibly high, enough to melt the entire surface of a body the size of Earth.

Volcanism and magmatic activity are essentially a second phase of differentiation that takes place over longer periods.

Heat in the interior, caused by the decay of radioactive elements, causes partial melting of the mantle. The resulting magmas, which are hotter and less dense than their surroundings, rise toward the surface to form crustal intrusions and extrusions such as lava flows and volcanoes. The amounts and rates of lava production have varied from planet to planet (and over time on each individual planet). However, volcanism's influence on the surfaces of all five worlds has been widespread. For example, the dark lunar maria first appeared about 4 billion years ago and were almost completely emplaced about 1½ billion years later. Volcanic activity on Mars and the formation of its great shield volcanoes continued well into the most recent half of the solar-system history, and of course volcanic activity on Earth (and possibly Venus) occurs frequently even now.

In the early history of the solar system, impacts were another cause of differentiation. A meteoritic projectile strikes a planetary surface at velocities measured in kilometers per second. The kinetic energy concentrated at that point can easily equal the total annual heat flow of Earth! Consequently, the cataclysmic bombardment associated with the final stages of planetary formation was energetic enough to melt the outer parts of the Moon and terrestrial planets. In these global magma oceans, lighter mineral species floated to the top and formed the planet's primary crusts. Later internal heating and melting then produced materials for a secondary crust, and reworking of these earlier crusts has yielded tertiary crusts (*Figure 5*).

It is obvious that heat, which drives differentiation and subsequent volcanism, is the single most significant variable in shaping a planet's evolution. Among the most basic questions we can ask about a planet are how much heat it had initially, and how it has gotten rid of this heat over time (*Figure 6*). These simple questions are the keys to understanding planetary evolution. At a more detailed level, we can ask how much heat derives from specific sources: the energy acquired at the outset during the accretion of new material, the planet's position relative to the Sun, electromagnetic heating, gravitational energy released by core formation and other density instabilities, large impacts, the decay of radionuclides, and tidal interactions. We can also ask how heat is distributed within the interior over time, what is known as a planet's thermal history. We can estimate the rates of change of

temperature as a function of depth and use this information to help predict the physical state of material (liquid, partially molten, or solid). And we can explore how heat is transferred from one part of the interior to another — by conduction, convection, or advection, the direct transfer of heat by movement of molten material from the interior to the surface. We will return to these questions after considering the internal structure and geologic history of the Moon and the terrestrial planets in detail.

THE MOON

As our closest neighbor in space, the Moon was first to occupy the attention of geologists studying other planetary surfaces (see Chapter 10). The Moon is the largest satellite, relative to its parent body, in the solar system, and it has long been studied as if it were a planet in its own right. Apollo and Luna samples, remote sensing, and surface seismic data show that the Moon has been internally differentiated into a crust, mantle, and possibly a small core (*Figure 7*). Seismic data and geologic mapping show that the lunar maria are relatively thin (a few km maximum thickness) and perched on a globally continuous feldspar-rich crust. This crust is thinner on the central near side, about 55 km, but may be up to 100 km thick on the far side.

Following its formation, believed to have resulted from the impact of a Mars-sized body with Earth, the Moon was subjected to a period of heavy bombardment that resulted in widespread melting. Opinions differ on whether the melting was globally extensive (a magma "ocean") or regional (magma "lakes"). Whatever the exact details, melting was accompanied by fractional crystallization and separation of plagioclase feldspar, a low-density silicate mineral. This floated to the surface to create the primary crust, which survives today as the lunar highlands (*Figure 8*) and the residual upper-mantle layers. The residual layers below the crust were denser than the underlying mantle, which probably caused them to sink toward the interior and, perhaps, to form a core. During this latter period, partial melting of the mantle created magmas that flooded the lunar surface and

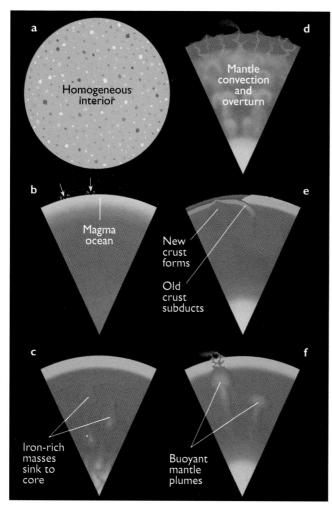

Figure 3. Differentiation is the process by which a body's initially homogeneous primitive materials (*a*) become segregated in planetary interiors. A number of different mechanisms contribute to differentiation. For example, the kinetic energy delivered by impact bombardment (*b*) can cause widespread — even global — melting of near-surface layers. Denser material collects and sinks to form a core (*c*), and the heat released from this event can trigger further differentiation. Deep-seated compositional and thermal instabilities (*d*) can cause materials to become less dense than their surroundings, resulting in rising plumes that undergo mixing and further differentiation. On Earth's ocean floors, differentiation also occurs as slabs of oceanic crust (*e*), mixed with water and sediment, are forced down into the crust and remelted. On Venus (*f*), vertical crustal accretion may continue until density instabilities cause the crust to founder, allowing planetwide resurfacing by volcanism.

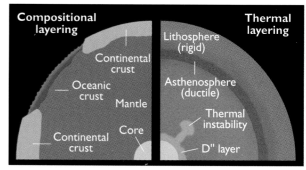

Figure 4. Planetary bodies are commonly divided into compositionally discrete layers (left half), such as a dense, iron-dominated core, a mantle rich in iron- and magnesium-silicates, and a crust that is of primary, secondary, or tertiary origin (or a combination of these). Thermal layering (right half) is caused by changes in temperature that affect the properties of planetary materials. For example, within Earth's crust the boundary between the rigid lithosphere and the more plastic aesthenosphere occurs at a depth of about 100 km. Deeper down, the D" layer at the core-mantle boundary may trigger rising mantle plumes that produce surface hotspots.

formed a secondary crust. By about 2½ to 3 billion years ago, basaltic lavas had covered approximately 17 percent of the lunar surface, preferentially filling the interiors of the giant impact basins on the near side to form the lunar maria (*Figure 9*).

Tectonic activity on the Moon stands in stark contrast to that of our own planet. The Moon lacks any evidence of the major lateral crustal movement so typical of Earth. Instead, lunar tectonism is characterized by downward movement, due to loading by the thousands of cubic kilometers of lava that have poured out onto the Moon's surface. The limited array of lunar tectonic features occurs predominantly in and near the maria. Linear rilles and graben, formed by crustal extension, were followed by sinuous (wrinkle) ridges formed by contraction.

Instruments left by the Apollo astronauts found that the amount of heat now flowing from the lunar interior is far less than that of Earth, consistent with a body that is losing heat by conduction. It appears that at present the outer 800 to 1,000 km of the Moon acts as a relatively rigid shell, or *lithosphere*. However, the presence of the highlands crust, mare basalts, and related tectonic features show that earlier in its history the Moon had a hotter interior. How was this heat lost?

The Moon's thermal evolution was shaped most of all by the formation of the globally continuous primary crust (*Figure 10*). Because the Moon has a large surface area relative to its volume, cooling by conduction was very efficient and the lithosphere cooled and thickened rapidly. In contrast to Earth, whose rigid lithosphere became divided into a number of moving, interlocking plates, the Moon quickly became a "one-plate planet." Plate tectonism never developed, and the Moon has lost heat primarily by conduction ever since. The small fraction of the surface covered by the volcanic maria indicates that advective cooling played a minor role. Evidence from tectonic features suggests that the Moon underwent a change from a net global expansion prior to about 3.6 billion years ago to a net contraction that continues today.

Gravity data show that there are large concentrations of mass, or *mascons*, associated with the youngest mare-filled basins. The spacing and type of tectonic features around the mare margins provide evidence that the surface was flexing and subsiding due to the load of the lavas on the lithosphere. The mascons plausibly represent the last outpourings of mare lavas, emplaced on a lithosphere that was so thick that it was able to support this load. The lack of a significant dipole magnetic field at present, combined with evidence for a fossil magnetic field in some lunar samples, is consistent with this internal thermal evolution.

Virtually no major internally generated geologic activity has manifested itself on the lunar surface for the last 2½ billion years. The Moon preserves a record of the impact bombardment and volcanism that dominated the early solar system, and thus it serves as a Rosetta stone for interpreting the surface geology of other terrestrial planets.

MERCURY

Mercury is one of the most poorly studied and enigmatic planets (see Chapter 7). Clues to its unusual nature come from the fact that it is about one-third the diameter, but about the same density, as Earth (*Figure 1*). These characteristics offer an

opportunity to study the influence of size and internal structure on a planet's geologic history and thermal evolution. Mercury also raises the question of whether a planet's initial starting conditions govern its evolution.

The Mariner 10 spacecraft returned images of about 35 percent of the planet's surface and at first glance revealed a lunarlike terrain (*Figure 11*). However, Mercury differs from the Moon in several important respects. Large areas of relatively ancient intercrater plains may indicate that early volcanism (during the period of heavy cratering) was more extensive on Mercury than on the Moon. Large, extended scarps on Mercury (*Figure 12*)

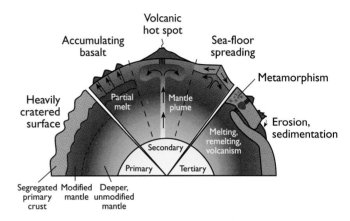

Figure 5. Various mechanisms lead to the formation of a planet's primary, secondary, and tertiary crusts. Primary crust, created early in planetary history, is still preserved in places like the lunar highlands. Partial melting of the mantle and volcanic activity lead to the formation of secondary crusts, largely of basaltic composition. Tertiary crust results from the recycling of primary and secondary crust, as is typified by Earth's continents and (perhaps) the tessera terrain of Venus.

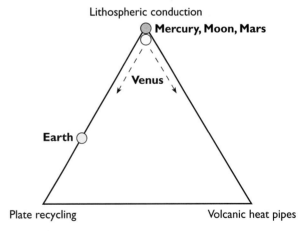

Figure 6. Methods of heat transfer within the terrestrial planetary bodies. At present, the Moon, Mercury and Mars lose their heat almost completely by the passive conduction through the lithosphere. Earth loses most of its heat through formation of the oceanic lithosphere at mid-ocean ridges, and through the spreading and cooling of this lithosphere. On Venus, conduction is the primary heat-loss mechanism today, though episodes of plate tectonism (left arrow) or widespread volcanic resurfacing (right arrow) may have occurred in the past. Advection, the direct transfer of heat by molten material to the surface through volcanic conduits, does not appear to be a significant process on any of these bodies. However, it almost completely dominates heat loss on the Jovian satellite Io.

attest to episodes of regional compression and perhaps even global contraction that would have resulted from a modest decrease in the planet's circumference during solidification.

Although Mercury's surface resembles the Moon's, its high density (5.4 g/cm³) suggests an interior more like Earth's. However, the high albedo of the surface, and the similarity of its spectrum to that of the lunar regolith, show that Mercury must be internally layered, with iron likely comprising about 60 to 70 percent of the interior by mass (*Figure 7*). The planet also exhibits a dipolar magnetic field thought by many to be internally generated. All these characteristics imply strongly that Mercury has a large, perhaps partially molten, iron core approximately the size of Earth's Moon.

Why is there such a large core relative to those of other planetary bodies? Two competing ideas have emerged. Early studies of solar-system formation emphasized the strong falloff in temperature and pressure as the distance increased from the center of the collapsing solar nebula. To a first order, this trend explains the primary distinction between the inner and outer planets. The solid, silicate-rich terrestrial worlds formed in a realm of higher temperatures and pressures, while the giant planets condensed at colder temperatures and lower pressures typical of the outer reaches of the nebula. It was viewed as logical that the planet closest to the Sun would have garnered an unusually high proportion of refractory materials such as iron.

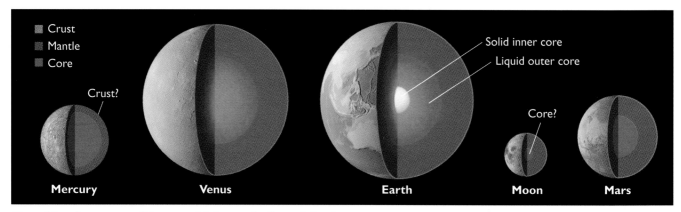

Figure 7. Interior structure of the terrestrial planetary bodies, as deduced from a wide range of observations. The Earth's inner and outer core take up about half its radius, while that of Mars is smaller. Mercury's core is believed to extend to more than two-thirds of the planet's radius. It is not clear whether the Moon has a discrete core.

Figure 8. The lunar highlands offer the best-preserved example of primary crust among the terrestrial planetary bodies. Impact craters of all sizes, including giant impact basins measuring hundreds of kilometers across, dominate this Apollo 16 view.

Figure 9. Secondary crust is represented by volcanic plains that were derived from partial melting of the mantle. On the Moon, secondary crust exists as mare deposits that filled the giant impact basins over a period of several hundred million years, ending around 3.2 billion years ago.

The idea that Earth's Moon formed as a result of a large impact raised questions about similar events elsewhere in the solar system. Could Mercury also have been struck by a large object, a collision big enough to strip away much of the planet's outer crust and leave it abnormally rich in iron? These two hypotheses can be tested by further exploration and ground-based observation. Like many scientific controversies, the truth may lie somewhere in between these two extremes.

How has Mercury evolved with time? One key is the existence of unusual lobate scarps like the one pictured in *Figure 12*. Some researchers have hypothesized that these features are part of a system caused by evolution into the resonance that now exists between the planet's rotation and its orbital period. Tidal despinning, which probably produced this resonance, is also thought to have caused contraction around the equator and extension at high latitudes. However, incomplete photographic coverage of Mercury makes testing this hypothesis difficult. Alternatively, the scarps may be clues to the planet's thermal evolution. As the planet cooled, global contraction could have shrunk the planet's circumference and caused the large lobate scarps seen by Mariner 10.

The existence of these scarps tells us something about the sequence of events in Mercury's thermal evolution. Core formation must have occurred prior to the end of heavy bombardment, because the intense release of gravitationally induced energy associated with this event would have caused large-scale melting and planetary expansion. Cooling and contraction of the interior and surface followed, including

solidification of part of the immense core. This should have led to a decrease in planetary radius of several kilometers, somewhat more than the roughly 1 to 2 km determined from the geometry and distribution of the large scarps. Mercury provides an excellent example of how modest changes in planetary radius can induce large-scale changes in surface tectonism — an idea often invoked for Earth before the advent of the plate-tectonism perspective.

The geology of Mercury suggests that this planet, like the Moon, has maintained a globally continuous crust and lithosphere throughout the history recorded in its surface. Mercury has lost heat primarily by conduction (also like the Moon), with its lithosphere increasing in thickness over time. In addition, if

Figure 11. Another possible example of secondary crust is the smooth-plain terrain of Mercury (upper right), which appears to embay older cratered terrain (lower left) and may be of volcanic origin. The large crater above center, Gauguin, is 70 km across.

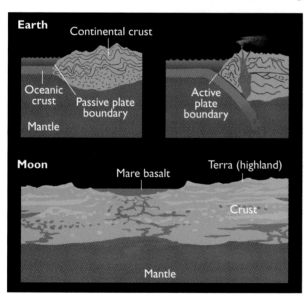

Figure 10. The Moon and Earth have distinctly different crusts. Energy from impact cratering melted the Moon's exterior to considerable depth, which allowed crystals of low-density silicates to float to the surface and create a global primary crust, the lunar highlands. Subsequent partial melting of the mantle caused basaltic magma to fill in low-lying surface areas, such as the major impact basins. These maria represent a secondary crust sitting passively on the primary (highland) crust. On Earth, crustal formation continues today. A thin oceanic (secondary) crust is created at divergent plate boundaries and returns to the mantle at subduction zones. Along plate boundaries, oceanic crust deforms, melts, and becomes incorporated with sediments to form the continents (tertiary crust).

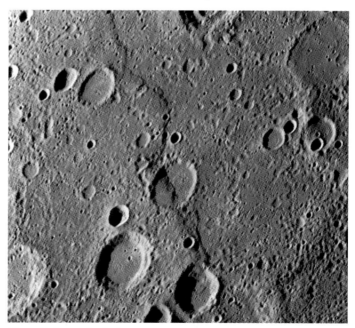

Figure 12. Impressive fault scarps appear on the third of Mercury seen well by Mariner 10. Santa Maria Rupes, seen here, is about 300 km long. These features were formed by compressional forces and overthrusting, probably because Mercury shrank slightly early in its history as it cooled. Crustal stress, caused by changes in the planet's rotation rate, may have played a role as well.

cooling caused enough contraction to create the scarps, then the entire lithosphere would have been subjected to contractional stresses from that point on. This may explain why secondary crustal formation (in the form of marelike lava plains) may be even more limited than on the Moon. With the crust in a continual state of compression, few pathways would exist for magmas to reach the surface. Although the outer core of Mercury is probably still molten, this relatively shallow internal dynamism is not reflected in the third of the surface observed in detail to date.

In summary, Mercury is a one-plate planet. Its lunarlike surface has been modified by global tectonic features linked to the large-scale evolution of its interior and (probably) its orbit. Despite its high levels of initial heat, an Earthlike interior, and probable present-day molten outer core, there is no evidence of internal activity over the last several billion years. To understand this quiescence, and to unravel the formation of observed tectonic features, we will need to assess the roles played by tidal despinning and changes in radius. Further insights will come with global geochemical and geologic mapping of Mercury and refined knowledge of its internal structure.

MARS

The red planet has a diameter about one-half that of Earth, but its density is closer to that of the Moon (*Figure 1*). Mariner and Viking spacecraft images revealed that Mars is more geologically diverse and complex than the Moon and Mercury. Like the Moon, it shows a distinct hemispheric asymmetry in the distribution of geologic units (*Figure 2*; see Chapter 11).

Mars has differentiated into a crust, mantle, and core. Gravity data (obtained from the careful tracking of orbiting spacecraft) suggest a relatively low-density crust. However, there are regional variations in thickness and density that must have formed in earliest Martian history. This early crustal formation must have influenced subsequent thermal evolution. The planet's distinct, dense core is between 1,300 and 2,000 km in radius. Spacecraft observations show little evidence for an intrinsic magnetic field now, but paleomagnetic signatures in Martian meteorites and recent Mars Global Surveyor measurements suggest that Mars once had one.

In many ways, Mars appears similar to the Moon. A one-plate planet, it has an ancient, heavily cratered highland (primary) crust and a globally continuous lithosphere. Volcanic flows have resurfaced low-lying areas and produced numerous edifices — some quite massive. At least two lines of evidence indicate that the Martian lithosphere has thickened with time: the presence of mass concentrations similar to the lunar mascons, and the fact that the giant shield volcanoes, which formed intermediate to late in Martian history, are supported by the crust. Although we see hints that the lithosphere's thickness has changed over time, the general global cooling and thickening trend is clear.

Fundamental differences from the Moon also exist, however. The extremely large topographic rises of Tharsis and Elysium, and their associated tectonic and volcanic features, have no parallel on the Moon. If the Moon can be characterized as a one-plate planet with vertical tectonism linked to loading, flexure,

and downward subsidence in the mare basins, then Mars has all this plus interior upwelling, vertical uplift, and voluminous associated volcanism. (Another distinction from the Moon is Mars's extensive reservoir of volatiles.)

A major departure from Mars's general surface geology is the Tharsis rise, a broad topographic bulge exhibiting ancient heavily cratered units, fracture systems, and young shield volcanoes (*Figure 13*). This 8,000-km-wide region, centered at 14° south latitude and 101° west longitude, stands about 10 km above the surrounding terrain. Some of its volcanic shields extend another 16 km in altitude. The vast majority of the linear rilles and fractures seen on Mars surround the Tharsis region. Valles Marineris, an enormous equatorial canyon system, extends radially away from Tharsis and is probably related to faulting that accompanied the bulge's evolution.

Mars has numerous tectonic features comparable to those seen on the Moon. However, the Tharsis region, with its concentration of tectonic and volcanic features, is a distinctive area unlike virtually any other observed on the terrestrial planets. The origin of Tharsis is uncertain. Some geophysicists believe it is predominantly a massive uplift of the crust caused by some dynamic process deep within the planet's mantle. Others propose that the underlying lithosphere is thin, rendering it more susceptible than elsewhere to volcanic outpouring, topographic buildup, and related stresses. Whatever the origin of Tharsis, the nature of Martian tectonism is still vertical, rather than horizontal.

The existence of Tharsis has puzzled planetary scientists for decades. They wonder how interior processes could produce so much uplift, and how convective upwelling in the interior could have consistently lasted over what appears to be several billion years. In addition, they must explain how the large amounts of melting necessary to produce the tremendous shield volcanoes could continue for billions of years. Why do the chemical and thermal anomalies needed to drive this activity occur only on limited places on Mars, and not on other planets?

Recent studies of the interior of Mars have provided new insight into the formation of Tharsis. In most planets, convection in the mantle is thought to assume relatively regular patterns of upwelling and downwelling. The number of mantle plumes, and their dimensions, are linked to the thickness of the layer they originate in, and whether the layer is being heated from below or within. In some cases, variations in temperature, composition, and density in and below the mantle can cause instabilities and deviations from simple patterns of convective heat transfer. For example, the subduction of lithospheric plates on Earth is thought to modify mantle convective patterns. In addition, instabilities at Earth's core-mantle boundary often produce mantle plumes and surface hotspots on a variety of scales. Even so, measured over the history of the planet, these perturbations are thought to be regional and relatively transient.

How could a gargantuan, Tharsis-building anomaly occur and be sustained over such a long time on Mars? Recent studies suggest that the situation might be very different than previously thought in some very important ways. The new insight comes from an understanding of the importance of changes in the phases of minerals as a function of changing temperature

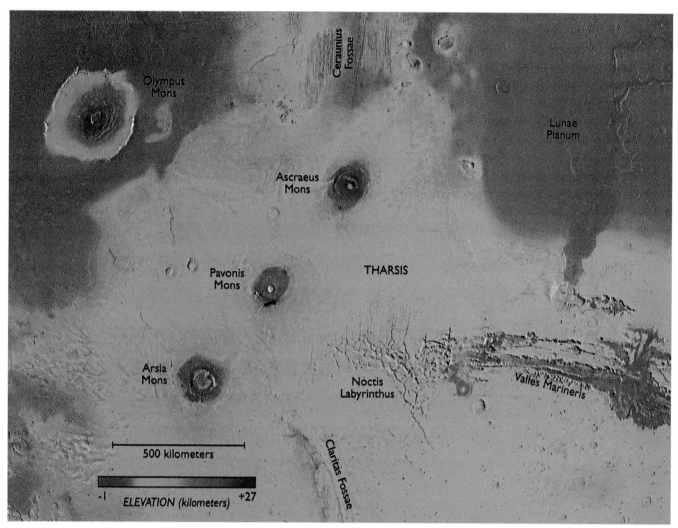

Figure 13. **A key result of the Mariner 9 mission was the discovery of a spectacular concentration of tectonic uplift and large shield volcanoes on Mars. Commonly called the Tharsis bulge, this region enompasses more than 6,500,000 km^2 – nearly the area of the 48 contiguous United States. Along the crest of Tharsis, which itself stands some 10 km above the planet's mean radius, are three large shield volcanoes: Arsia Mons (20 km in total elevation), Pavonis Mons (18 km), and Ascraeus Mons (26 km). To their northwest is towering Olympus Mons (26 km), the largest volcano in the solar system. Extensive fracture systems, including the enigmatic complex called Noctis Labyrinthus, cut the uplift's margins more or less radially to its center. The origin of Tharsis remains uncertain. It could be an enormous collection of volcanic outflows piled atop a stiff, inflexible lithosphere. More likely, the entire region has been elevated by a strong and sustained upwelling within the Martian mantle.**

and pressure with depth (*Figure 4*). Also important is whether these phase changes release heat (and are thus exothermic) or absorb it (endothermic). Within Earth's interior, for example, a phase change of the mineral olivine has been shown to be endothermic. This heat-absorbing effect may be significant, shutting off the transfer of material and heat across the phase-change boundary.

Olivine's transition should occur much deeper in the interior of Mars than in Earth. Computer modeling predicts that it should lead to only one or two major upwellings, and eventually only one survives. This scenario provides a very close approximation to the observed history of volcanism on Mars. The planet exhibits two major loci of volcanism and tectonism (Elysium and Tharsis), and activity on larger Tharsis continued well after Elysium fell quiet. If this explanation is correct, it shows that variations in the properties of planetary interiors under different conditions can have important effects on the overall evolution of planetary surfaces and mechanisms of heat loss.

One of the most exciting aspects of Mars is the abundant evidence pointing to a warmer, wetter climate long ago (see Chapter 13). Near its boundary with the northern plains, the older, higher terrain of the southern hemisphere bears numerous channels tens of kilometers wide and hundreds of kilometers

long. These are reminiscent of the channels that formed on Earth during catastrophic flooding in ancient times. Today, there are indications that the Martian regolith is the major reservoir for all this water (mostly as ice). But where did the water come from?

Mars's greater distance from the Sun suggests that this planet would have accreted from more volatile-rich materials than Earth did. The final phases of accretion and the initial outgassing of the interior would have concentrated water at or near the surface. However, much of it was probably lost to space in the high-temperature environment that followed differentiation and core formation. Opinions differ as to how much water remained and where it would have resided. Perhaps some was left in the mantle, to be released during the planet's earliest history when more heat was flowing from the interior. Alternately, water could have been delivered to the surface by comets late in accretion, and the planet's stable lithosphere would have prevented it from becoming mixed into the interior.

It is possible that Mars was dry on the inside and wet on the outside in its early history. Alternatively, perhaps both interior and exterior volatile reservoirs existed but were kept relatively isolated from each other by the globally continuous lithosphere. Further exploration by Surveyor spacecraft will provide important data for understanding the history of volatiles on Mars. Scientists hope these missions will shed light on the nature of the atmosphere, the interaction of the atmosphere and the surface, the nature of climate change, and the possible existence, origin, and evolution of life on Mars.

In summary, Mars is an example of a one-plate planet with underlying similarities to the surfaces of the Moon and Mercury, but with an atmosphere, hydrosphere, and cryosphere. The surface units thus reveal an early geologic chronology much like that of the Moon and Mercury, though volcanism (particularly in the Tharsis region) extended well into the last half of solar-system history, perhaps even up to the present. However, the absolute chronology is not known because of the lack of documented samples from the surface units on Mars. Another aspect that sets Mars apart from the Moon and Mercury is the unusual concentration of volcanic and tectonic activity (Tharsis and Elysium). The origin of these two centers may lie in mineral phase changes that altered circulation patterns in the interior.

VENUS

Because Venus has long been thought of as a near twin of Earth, not only in size but in position in the solar system, its geology has been of extreme interest. Decades of probing its dense, cloudy atmosphere with Earth-based radar, robotic landers, and radar-equipped orbiters have revealed that Venus's surface is anything but Earthlike (see Chapter 8). This revelation has only increased the fascination of scientists, who wonder why the two planets appear to have taken very different evolutionary paths.

The first clue that Venus might be different, geologically speaking, came from the nearly global altimetry data gathered by the Pioneer Venus orbiter in the late 1970s. About 60 percent of the surface was found to lie within 500 m of the planet's mean radius, and only 5 percent lies more than 2 km above it. This arrangement is in contrast to the distinctly bimodal distribution of Earth's topography, which arises from the density contrast between the lighter continents and the heavier ocean basins, as well as the thicker continental crust. However, despite the strongly unimodal distribution of altitudes on Venus, the total range of elevations is comparable to that of Earth.

High-resolution radar imaging by Soviet Venera 15 and 16 orbiters and NASA's Magellan orbiter made it possible to study the planet's geologic character in detail. Surprisingly, Venus turned out to be quite different not only from Earth but from the other terrestrial planets as well. Instead of major terrains with two different ages (the equivalent of lunar highlands and maria, Mars's heavily cratered terrain and northern lowlands, or

Earth's continents and seafloor), the entire surface of Venus appears to be no older than a few hundred million years.

Even more surprisingly, the distribution of impact craters was initially found to be indistinguishable from a random one. To some researchers, it appears that Venus was subjected to a planetwide episode of resurfacing, after which impact craters accumulated again. Furthermore, the resurfacing had to be rapid compared to the impact rate, in order to maintain the apparent randomness of the crater distribution. (Another clue to the swiftness of the resurfacing was that very few craters seen on Venus today have been embayed by lava flows.) This evidence for a massive resurfacing episode has important implications for Venus's geophysics and, at the very least, has sparked a lively debate about its cause that continues today.

In any case, it appears that our geologic mapping and stratigraphic analyses can only address the last 10 or 20 percent of Venus's history. At the beginning of this record, we see intensive tectonic deformation over nearly the entire globe. This has given rise to regions of intensely fractured crust, called *tesserae*, which comprise less than 10 percent of the planet's exposed surface (*Figure 14*). Some geophysicists believe tessera terrain comes about through the normal evolution of crustal structure formed and modified by mantle convection patterns. In one model, tesserae initially formed above large areas of mantle upwelling (hotspots or plumes) as regions of enhanced volcanism and crustal thickening; subsequent cooling and gravitational collapse then converted these volcanic plateaus into the highly deformed tessera terrain. Alternatively, tessera terrain may form over zones of mantle downwelling when mantle flow patterns and crustal deformation coincided.

Another hypothesis builds on the planet's high surface temperature and on the exponential relationship between temperature and rate of strain. In this scenario, mantle convection is closely linked to the overlying lithosphere. Over most of solar-system history, Venus has a weak lower crust that deforms readily, resulting in very high levels of surface strain. Therefore, tessera deformation occurs globally. At some point late in the planet's history, the heat flux declines, the lower crust becomes more ductile, and rates of surface deformation decrease rapidly. This latter model illustrates how important surface and near-surface temperatures can be. Is there evidence for major changes in surfaces temperatures late in the history of Venus? Some recent calculations suggest that the release of gases associated with volcanism could enhance the planet's greenhouse effect and thus significantly increase the surface temperatures. In this case, deformation patterns and styles would be more variable.

Following tessera formation, several stages of extensive volcanism occurred, burying vast areas of tesserae and forming what we see now as regional plains (*Figure 2*). Several types of tectonic deformation have subdivided these plains. From oldest to youngest, they are (1) the formation of the tesserae, (2) dense fracturing, (3) the formation of broad ridges, and (4) the formation of wrinkle ridges like those seen on the lunar maria. Some scientists have interpreted these tectonic features as representing successive, planetwide episodes of contraction, then extension, then contraction again, and finally extension. The last of these episodes, widespread wrinkle-ridging, occurred very close in time to the emplacement of the most ubiquitous

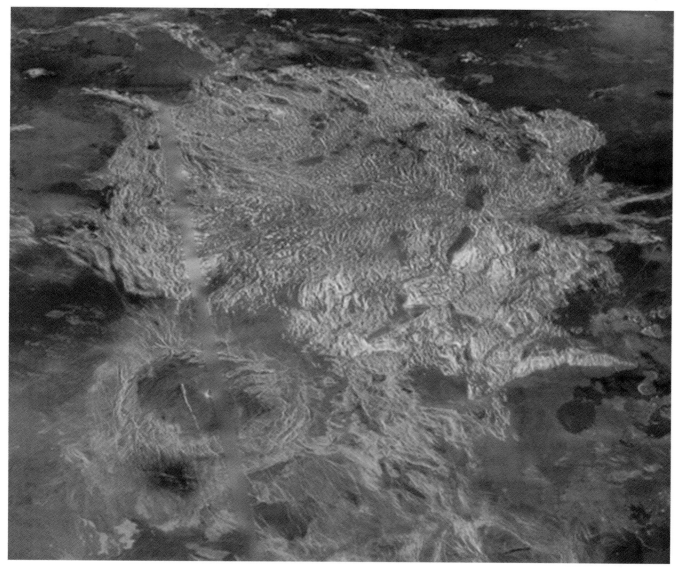

Figure 14. Alpha Regio, a radar-bright "upland" some 1,300 km across, was the first feature to be discerned on Venus using Earth-based radar. Alpha Regio is a good example of complexly deformed, high-standing terrains called tesserae, which cover about 8 percent of Venus's surface. Although tesserae are the oldest visible landforms on Venus, they are young by planetary standards, having formed within the last billion years.

plains unit. This period marked the transition to the present stage of Venus's history, in which the dominant geologic activity is regional rifting and related volcanism. It is this most recent stage that appears to have lasted longest, even though the resulting tectonic and volcanic features cover no more than 20 percent of the surface of Venus. Apparently the general intensity of tectonism and volcanism has been much lower recently than in earlier times.

Before Magellan's reconnaissance, some scientists had predicted that Venus would display evidence for Earthlike plate tectonism. This would not have been surprising for a planet so much like Earth in size, mass, and (presumably) its allotment of radioactive elements. If the two planets had comparable heat budgets, it seemed reasonable to predict that both would have developed a plate-tectonic "engine," driven by mantle convection, to get rid of excess heat. Indeed, the period of time we see preserved on Venus's surface is equivalent to the period on Earth when global geodynamic processes were dominated by plate tectonics.

At present, however, Venus shows no sign of plate tectonism. Instead, global activity appears to have moved back and forth between periods of contraction and extension, with deformed structures appearing at a gradually declining rate. In addition, although Venus's volcanic plains formed at a rate comparable to the Earth's midocean ridges, they were emplaced in an entirely different style, as extensive floods of lava (*Figure 15*). Then the style changed; for the last few hundred million years, Venus has been dominated primarily by rift-related volcanism occurring at rates comparable to or even lower than present intraplate volcanism on Earth. At present, Venus appears to be a one-plate planet losing heat largely by conduction (*Figure 6*). Scientists want to understand how two planets that are similar in so many ways could be so different in others. They would also like to know what implications Venus's story has for the early history of Earth.

What is known about the nature of Venus's crust and interior? The Soviet Venera landers measured the compositions of several plains units. Most of these were consistent with basalts and with the morphologies observed in Magellan images of

Venus's volcanic plains. Venera 8 measurements suggest that some areas may be more silica-rich, thus tending toward granitic in composition. Although no measurements were made in the highlands (tesserae), some researchers have suggested that these areas, like the continents on Earth, may be more completely differentiated. In any case, the surface compositions show that the interior of Venus is differentiated and that a basaltic crust has been extracted from the mantle. From spacecraft data, scientists estimate that crustal thickness averages 25 to 40 km, with local variations — particularly in tessera regions — that probably exceed 50 to 60 km. How the crust attained these thicknesses, and how they have changed with time, are not well understood.

Another means of investigating the Venusian interior has come from precise tracking of orbiting spacecraft like Magellan. By noting minute accelerations and decelerations, scientists have calculated the planet's gravity field, which varies with the density of the crust and mantle. On Earth, large-scale variations in gravity are generally not correlated with topography and crustal density anomalies, being instead related to broad convective motions in the mantle. Venus appears markedly different from Earth in this respect; its large-scale gravity anomalies are remarkably well correlated with topography.

Figure 15. **Radar studies prior to Magellan had shown that volcanic flows cover about 80 percent of Venus. This particularly stunning example occurs in the southern highland area called Lada Terra. Molten rock first pooled against a long, complexly folded ridge (far left). After breaching the ridge, the lava cascaded toward the right and flooded an area of approximately 100,000 km². The radar-dark flows have relatively smooth surfaces, while the radar-bright ones are rough. This scene covers 630 by 550 km.**

This may be telling us something important about Venus's interior, namely, that it lacks an equivalent of Earth's *asthenosphere*, a partially molten layer between the lithosphere and the mantle (*Figure 4*). On Earth, the asthenosphere has the effect of decoupling convection patterns in the deep mantle from lithospheric plate motions. However, Earth's asthenosphere exists because water, subducted along with descending oceanic plates, lowers the melting point of lithospheric rocks and makes them more plastic. Venus's water is thought to have escaped into space during a so-called runaway greenhouse episode. Consequently, some researchers say, it has no asthenosphere.

On the other hand, the strong correlation between Venus's gravity and topography may be saying simply that the topography is supported by the strength of the underlying rocks. This would also have considerable significance. Because of the high temperatures on Venus, some geophysicists believe its rocks are so weakened that major topographic relief cannot exist for more than a few tens of millions of years. With this in mind, they have suggested that features such as the towering Maxwell Montes (whose slopes are steeper than those of any mountain range on Earth) must be geologically very young, and that they are supported dynamically, by ongoing mantle convection, rather than structurally. Here again, however, the lack of water may play an important role by giving rocks added structural strength. Venus's dry lithosphere may be able to support far greater loads, and for longer times, than Earth's.

No direct information exists about the deep interior of Venus, but because of the high, Earthlike mean density, virtually all researchers assume that Venus underwent core formation. Some investigators believe that the core is now completely solid, while others suggest that core solidification is under way now or has not yet commenced. Theoretical models predict that layering within the mantle of Venus is even more likely than on Earth.

Venus has essentially no intrinsic magnetic field. An immediately appealing idea is that this may be due to Venus's very slow rotation rate, but theory suggests that these factors are not related. Of course, Venus could have had a magnetic field in the past, but unfortunately the surface temperatures are currently above the Curie point, the maximum temperature at which the record of past magnetic fields would be preserved in rocks.

In order to understand the present internal thermal state without direct seismic and heat-flow readings, information must be gleaned by other, indirect means. By comparing the planet's topographic profiles to models of lithospheric behavior, the thickness of the lithosphere has been estimated at 10 to 40 km. Recent laboratory tests on the behavior of very dry rocks suggest that the crust of Venus may be almost as strong as the mantle. The implication is that at present Venus loses less heat than Earth does. Yet, if we assume that Venus and Earth incorporated similar abundances of radioactive elements, then more heat is being generated in Venus's interior than we now see escaping through its surface. Alternatively, Earth and Venus may have been endowed with fundamentally different abundances of heat-producing radionuclides.

Unlike Earth, Venus is thought to possess a stagnant lithospheric "lid," in which heat is lost primarily by conduction. However, has that always been the case? Has Venus undergone changes in the way it gets rid of internal heat? Some models

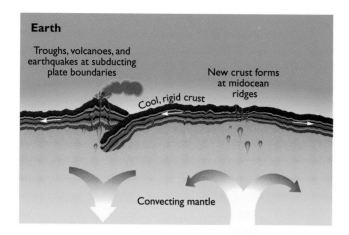

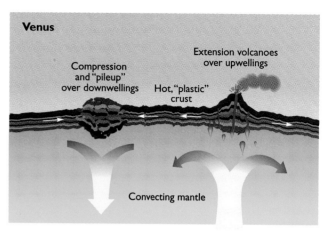

Figure 16. In contrast to plate tectonism on Earth (left panel) Venus may pile up old crust over convective downwellings in the mantle and produce new crust over upwellings (right). At present Venus is a one-plate planet whose internal heat escapes to space largely by conduction.

assume a typical thermal evolution but predict evolutionary changes in mantle convection patterns (*Figure 16*). These scenarios imply very high rates of deformation early in Venus's history, due to a highly convective interior and a highly deformable lithosphere that becomes incorporated into the convecting mantle. Surface recycling into the mantle continues over a prolonged period. All the while, however, the interior gradually cools, convective vigor wanes, and the crust and lithosphere become more rigid. A transition occurs about 500 million years ago to a one-plate planet that is dominated by hot-spot volcanism (*Figure 17*).

Other models involve one or more episodes of catastrophic resurfacing in the history of Venus — an idea that has been attractive to theorists wishing to explain Venus's observed crater distribution. However, many have questioned the plausibility of such global resurfacing. One possible explanation is a scenario called episodic plate tectonics. This view holds that at several times in its history Venus is a one-plate planet with a stable lithosphere that loses heat by conduction and thickens over time. The lithospheric "lid" causes the interior to heat up, which in turn enhances mantle convection. More vigorous convection causes the lithosphere to break apart and become recycled rapidly. This overturn is accompanied by planetwide volcanism that resurfaces Venus almost completely. However, the episode also allows the interior to cool, causing a return to stabilization, and the beginning of another cycle.

Another possible explanation is related to the style of crustal formation on Venus. The Moon's secondary crust, the maria, comprises only a small percentage of its surface and total crustal volume. However, what happens when, unlike the Moon, a planet is large enough to produce vast quantities of secondary crust over geologic time — but does not recycle it as Earth does? On Earth, sea-floor spreading causes the oceanic crust to accrete laterally. On Venus, the lack of plate-tectonic features suggests that secondary crust accumulates vertically, by adding layers of solidified mantle rock to the underside of the lithosphere. Models of this latter scenario show that, over time, the buoyancy of the silica-rich crust is countered by the cold, dense layers of mantle rock that accrete along its underside. The crustal and upper-mantle rocks eventually deform and separate, like layers of plywood, in a process called *delamination*, and the dense slabs descend into the interior. At the same time, hot mantle material from the deep interior ascends and melts as its

overburden of pressure is lifted. The accumulation of hot magma breaks through the thinned lithosphere and causes a phase of widespread surface volcanism. Following this overturn event, vertical crustal accretion resumes, though at much reduced rates. The process is predicted to repeat itself at intervals of 300 to 750 million years.

This vertical-accretion model predicts many of the features we see in Venus's geologic record. Resurfacing events would erase the first 75 to 85 percent of the planet's history, as observed. Following overturn, the last vestiges of the intense crustal deformation would correspond to the oldest preserved unit in Venus's stratigraphic record (the tesserae?). It could also explain the observed major changes in the style and intensity of deformation, following tessera formation, to more focused rifting like the long, sinuous belts of ridges seen in Magellan images (*Figure 15*). Finally, associated with this change, there is evidence for a substantial decrease in volcanic eruptions and a change from large-scale regional plains emplacement to focused local sources.

At present, there is no firm consensus among geophysicists on this wide range of models. Nonetheless, we now have a much clearer picture of the evolution of Venus's surface over the last few hundred million years, and of how it relates to Earth and the other terrestrial planetary bodies. Such comparisons offer a very promising way to understand some of the basic principles and trends in the evolution of terrestrial planetary bodies. During the part of Venusian history recorded in its rocks, Earth has been dominated by plate tectonics, with the most significant activity concentrated along the boundaries of lithospheric plates. The emerging picture of Venus reveals a very different situation dominated by vertical processes, not lateral movement. The general stratigraphic relations suggest initial intense deformation (as evidenced by the tesserae), a brief but intense period of volcanic flooding, then relative quiescence. No evidence of plate recycling is seen at present on Venus. Yet there are important clues that Venus has changed its style of heat loss from the past to what we see today. Some models, in fact, predict a style of heat loss that is both catastrophic and episodic. Our future exploration of Venus will include surface stations to make long-term seismic and heat-flow measurements, balloons and landers to assess the atmosphere and global surface composition, and eventually spacecraft capable of returning samples for laboratory analysis.

EARTH

Our own planet has many geologic and geophysical features that set it apart from the other terrestrial planets. Of course, the vast majority of Earth's surface is concealed by liquid water and vegetation. Although the atmosphere, hydrosphere, and biosphere are not considered terrain units, they are nonetheless significant agents in the modification of the solid crust beneath us. If they could be stripped away, as in *Figure 2*, we would see that, compared to the other terrestrial planets, Earth possesses some units that are different (platform deposits), and some units that are either much less widespread (cratered terrain) or much more so (volcanic plains).

The exploration of the ocean basins revealed their basic geologic and morphologic differences from the continents and paved the way for the development of the theory of plate tectonism (see Chapter 9). From this theory has come the realization that Earth is divided up into a series of rigid lithospheric plates, and that the formation, lateral movement, interaction, and destruction of these plates is responsible for most of Earth's large-scale geologic features. Plates collide to produce folded mountain belts, strings of volcanoes, and deep trenches (where the plates flex, subduct into the mantle and are destroyed). Continental rift valleys and vast plateaus of basalts accompany plate breakup, while new crust forms at the gradually separating midocean ridges. Unfortunately, this dynamic activity has erased a very significant fraction of the geologic record, and most of that record remaining from the first half of solar-system history has been destroyed (*Figure 18*).

Despite the apparent uniqueness of many aspects of Earth, several terrain units appear comparable to some seen on the other planets. The distribution of meteorite craters suggests that the ancient rocks of Earth's surface (the 10 percent comprising the continents' Precambrian shields) are the nearest terrestrial analog to cratered terrain. However, craters in this unit on Earth are much sparser, and their ages much younger, than on the most ancient portions of other planets. The vast, basaltic plains of the ocean floor are among the youngest of Earth's rocks, having formed within the last 200 million years. Separate phases of volcanism have built broad basaltic plateaus (Ethiopian and Indian flood basalts, for example) and conical mountains and craters (Mount St. Helens and Hawaii) over the last 65 million years.

Thanks to a global network of seismometers, and to Earth's dynamic geologic activity, the present nature and structure of Earth's interior is very well known — at least relative to that of other terrestrial planetary bodies. Samples from the crust and interior provide important information on differentiation and fractionation. Earth's core consists of iron with some nickel and sulfur mixed in. It spans about half the radius of the planet, with

Figure 17. The surface of Venus is dominated by extensive volcanic plains (comprising over 80 percent) and shield volcanoes are often superposed on these plains. This oblique perspective, synthesized from Magellan radar images, shows the volcano Maat Mons, which towers 8 km above the surrounding plains. (Vertical elevations have been exaggerated 10 times.)

an inner solid portion and an outer liquid layer (the source of Earth's magnetic dynamo). The silicate-rich mantle is subdivided into an outer layer that extends down to about 700 km, and an inner layer dominated by higher-density phases of mantle minerals. Two-thirds of the crustal surface area is made up of young, thin basaltic (secondary) oceanic crust. The remaining third is the much thicker and older tertiary continental crust. Although superficially similar to the lunar highlands and maria, Earth's crust has a strikingly different structure (*Figure 10*), in large part because of plate-tectonic processes.

Plate tectonism also determines the unique way in which Earth gets rid of its internal heat. Although the lithosphere (which averages about 100 km in thickness) is relatively rigid, it is in continuous lateral motion. New lithosphere is created by volcanism at diverging plate boundaries such as the mid-ocean ridges. The lithosphere thickens progressively with age, and flexes downward and dives into the mantle along subduction zones. Thus, the plate-tectonic process on Earth provides a kind of conveyor belt for heat loss. This is in marked contrast to the smaller terrestrial planetary bodies, whose lithospheres thicken over time by conduction. Earth's lithosphere overlies the partially molten asthenosphere, which forms in the narrow zone (about 100 to 125 km down) where the pressure and temperature are right for upper-mantle materials to melt. This layer is partly responsible for the ability of lithospheric plates to move rapidly, and it is a significant source of magma for surface volcanism.

Manifestations of plate-tectonic activity dominate Earth's surface. But we also find evidence for thermal anomalies in the mantle, namely, the volcanoes that are produced over mantle plumes ("hotspots"). The resulting edifices grow; loading and flexing the lithosphere much as the maria do on the Moon. On Earth, however, the rapid lateral migration of the lithosphere moves the volcano away from the hotspot, and a new volcano forms where it once was. The result is a chain of volcanic peaks, the Hawaiian islands being one well-known example. These mantle plumes appear to result from instabilities that develop in the thermal boundary layer at the core-mantle boundary. Indeed, occasionally in Earth history, unusually large instabilities have produced "megaplumes" that triggered vast outpourings of lava. These enormous eruptions have caused major modifications to the atmosphere, plate-recycling patterns, oceanic chemistry, the biota, and the magnetic field.

Differentiation of Earth's interior into a core and mantle began very early, possibly coincident with the late stages of accretion, and continues today. We have no record of primary crust or of any secondary crust produced prior to about 3.8 billion years ago. The earliest evidence of tertiary crust (the reworked primary and secondary crust seen in continents) dates to about this time, and it suggests that no stable continents had emerged until more than 500 million years after Earth's formation. Continental crust has grown continuously in areal extent since that time. The rate of this growth is controversial, but it may have peaked about 3.0 billion years ago. Notably, despite billions of years of internal mixing by convection and plate recycling, isotopically distinct volcanic source regions still persist in the mantle. Still unknown is whether these represent chemical layering or smaller-scale heterogeneities.

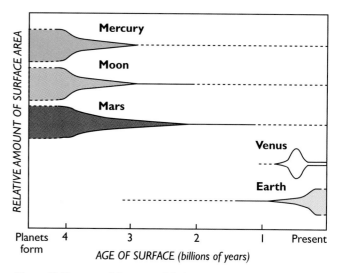

Figure 18. The ages of the terrestrial planets' current surfaces, with colored shading representing the total surface area of each body. Most regions on the Moon, Mars, and Mercury are several billions of years old, though volcanic activity continued for some time on Mars. Two-thirds of Earth's surface (its ocean basins) formed within the last 200 million years. As on Earth, most of Venus's surface formed relatively recently, apparently due to processes very different from terrestrial plate tectonism.

Crustal spreading and lithospheric recycling are amazingly efficient. If the present rates have been maintained over the last 4 billion years, the oceanic crust has been renewed at least 20 times. Indeed, detailed studies of the eroded cores of ancient mountain belts show that they mark the location of earlier collision and subduction zones, a process repeated many times as rifting opened new oceans, leading to spreading, subduction, continental collision, and closure. It is tempting to believe that plate-tectonic processes dominated the entire geologic history of Earth, but have they really? What can we determine about the way in which Earth has handled its heat budget over time?

The early Earth received heat from accretional impacts and core formation. Together, these sources probably gave the early Earth a vigorously convecting mantle and a thin lithosphere. How did our planet deal with so much heat in this early period? One idea is that Earth was completely covered by very rapidly moving oceanic crust that quickly and continuously recycled itself. The heavy bombardment typical of this era may have played a role in crustal recycling, perhaps producing an early primary crust like the Moon's. Alternatively, there may have been some sort of catastrophic event that caused foundering of the crust during the first billion years, setting the stage for plate tectonics. None of these suggestions is completely satisfactory, however, and the history of Earth in its infancy remains elusive.

Today, the elegant simplicity of plate tectonism suggests a steady-state regularity in Earth's geologic activity. However, if the early Earth was similar to our emerging picture of the surface and interior of Venus, it may have been characterized by catastrophism. Indeed, there is evidence in the early geologic record for major phases and events that were not repeated later. Some examples are the amassing of early continents, the emplacement of iron-rich volcanic rocks known as komatites, and the emplacement of feldspar-rich rocks known as anorthosites.

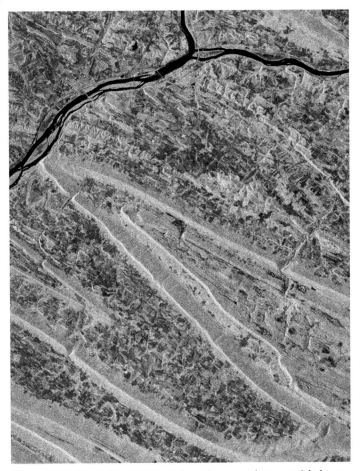

Figure 19. Earth's surface is the most dynamic among the terrestrial planets. Shown here is a false-colored radar image of the folded and eroded Appalachian Mountains near Sudbury, Pennsylvania (the view is 30 km wide, and the Susquehanna River runs across the top). Only small portions of our planet are covered with rock units whose ages are measured in eons. Instead, most of Earth is covered by oceanic crust only tens of millions of years old. The continents are largely tertiary crust, like that seen here, derived from reworking of earlier secondary crust.

However early Earth rid itself of excess heat, the lithosphere certainly thickened with time as the planet cooled. After the first billion years, this thickening was accompanied by growing accumulations of stable continental crust. Although plate tectonism is understood in principle, the process as we see it today requires a relatively rigid lithosphere several tens of kilometers thick — a condition likely to have characterized Earth for the past several billion years. There is no consensus, however, on how or when plate tectonism might have been initiated. In any event, these processes set Earth on an evolutionary course distinct from that of any planetary body observed to date. Lithospheric plate recycling, localization of geologic activity along plate boundaries, efficient mantle mixing, the formation of vast quantities of secondary crust and of tertiary continental crust (*Figure 19*) are some of Earth's unique characteristics.

SYNTHESIS AND PROGNOSIS

By studying terrestrial planetary bodies and the processes responsible for their origin and evolution, planetary scientists seek to understand the basic themes seen in the terrestrial plan-

ets. We want to identify the major factors that cause the differences in the interiors and geologic histories of the planets. Like social scientists who debate the significance of "nature" and "nurture" in human development, we want to know how much is predetermined by starting conditions and how much by subsequent events. We have already made significant progress toward these goals, and clear themes are beginning to emerge. However, substantive and satisfying answers to these questions will not be known until well into the next millennium. By then, we hope, sophisticated geophysical networks will have determined the internal structure and will be monitoring the conditions on all the terrestrial planetary bodies. Landers will routinely make sophisticated geochemical assays on planetary surfaces and return samples to laboratories on Earth.

Initial compositional variations are clearly an important part of the "genetic" makeup of a planet. We recognize that the relative proportions of volatile elements and refractory elements varied with distance from the center of the collapsing solar nebula. Yet there was very likely a significant amount of mixing of planetesimals from different parts of the solar system during the rather chaotic final period of planetary accretion. Moreover, we must allow for the late-stage "singular" events, such as the titanic impacts that formed the Moon and, perhaps, stripped away much of Mercury's low-density crust. A challenge for the future lies in sorting out these several factors that clearly determine planetary starting conditions, then tracking their influence on subsequent planetary evolution. For example, one unanswered and rarely addressed issue is what role the Moon-forming impact played in creating Earth's seemingly unique characteristics — oceans, plate tectonism, and even the development of life.

Once the planets formed, their temperatures, compositional, and density variations led to differentiation into internal layers of significant proportions (core, mantle, and crust), a process that continued throughout their history. The timing and duration of the formation of these layers differed considerably, however. Core formation appears to have occurred rapidly and in earliest history. The amount of gravitational potential energy released likely caused significant surface melting on the larger planets, but the geologic record of this melting has been obscured by impact cratering. Although the mantle material that remained after core formation became a distinct layer, it has continued to evolve throughout geologic time. Heat in the interior causes convective mixing and partial melting of mantle materials, and the hotter, less-dense portions of these magmas migrate toward the surface to form crust. Over time, the rates of heat production and patterns of heat loss have changed, which help modulate mantle differentiation.

The terrestrial planets' outermost layers, their crusts, exhibit a very wide range of ages, rates, and styles of formation (*Figure 20*). Primary crust, as seen on the ancient highlands of Earth's Moon, can result from melting associated with the impact energy of late heavy bombardment. The fate of the primary crusts on Earth and Venus is currently unknown. Secondary crust, formed by partial melting of the mantle and subsequent volcanic eruptions, can assume several forms. The maria, for example, cover a small fraction of the lunar surface and were emplaced in the first half of solar-system history. Martian volcanic plains had similar origins, but their emplacement continued for billions of years and (unlike the Moon) may represent the vast majority of the thickness of Mars's crust. In addition, Mars's relatively small size apparently caused mantle upwellings to become localized into a very small number of focused centers. Among these were Tharsis and Elysium, where a significant part of Mars's secondary crustal formation was concentrated. Earth's seafloor represents secondary crust that is very young, thin, and actively forming today. In contrast to the vertical crustal accretion typical of the Moon and Mars, Earth's oceanic crust moves laterally, much as a conveyor belt, and is thrust back into the mantle at subduction zones. Venus is covered with secondary crust roughly 500 million years old — a small percent of the planet's age. Unlike Earth's, it appears to have formed by vertical crustal accretion late in Venus's history, perhaps by catastrophic processes. Tertiary crust, formed by the reworking of primary and secondary crust, is best represented by Earth's continental crust. It is the product of billions of years by erosion, intrusion, and tectonic reworking. The tesserae of Venus and their associated deposits are thought by some scientists to be another example of tertiary crust.

Layering in planetary interiors is caused not only by compositional variations but also by thermal structure. We need to know what the sources of internal heat were and are, as well as how thermal structure influences the state of materials. A major challenge for the future is to determine the relative importance of the heat sources present during each planet's earliest history. These include such diverse factors as solar energy, accretional impacts, electromagnetic heating, core formation, radioactive decay, and tidal interactions. By the time they were a few hundred million years old, planetary bodies had acquired internal thermal layering and convection patterns to process excess heat from these sources. A distinct thermal gradient evolved within each world that affected compositional layering and the behavior of materials under different temperature-pressure conditions. Thermal boundaries formed and evolved, causing local and large-scale convection patterns to change.

The lithosphere is a planet's outer thermal boundary layer. It represents the interface between the heat in the interior and loss of that heat to space. An important emerging theme is that a lithosphere's initial configuration is crucial to its planet's subsequent evolution. The relatively small Moon, Mercury, and Mars formed an unbroken lithosphere in their earliest history and became "one-plate" planets. For eons, they have shed heat to space largely through conduction. Conductive heat loss caused a thickening of these bodies' lithospheres and, as a sort of bonus, the preservation of much of the record of early solar-system history. Increasing lithospheric thickness, together with the efficient cooling of the mantle, made the formation of magmas more difficult and their transport to the surface increasingly unlikely.

Earth, however, adopted a different mode of heat transfer: seafloor spreading and plate tectonics. New crust and lithosphere constantly forms at divergent boundaries, moves laterally away, cools and thickens, then is swept back into the mantle at subduction zones. This very efficient mechanism appears to have been operating on Earth for at least a few billion years, but how it originated is currently unknown. We recognize that the lithosphere's thinness, relative to the total radius of Earth, favors crustal recycling. However, we cannot yet explain how plate tectonism was initiated.

Venus, approximately the same size and density as Earth, and nearby in the solar system, has a surface that is comparable in average age to that of Earth. However, the planet's unusual geology, the uniqueness of its cratering record, and the presently globally continuous lithosphere (no active plate tectonism) attest to a geologic and thermal history very different from that of Earth. For example, the surface appears to have been catastrophically renewed by tectonism and volcanism in its recent geologic past. Venus, then, may be the exception both to the examples of slowly waning heat loss represented by the one-plate planets, and to the "equilibrium" heat loss represented by Earth. Venus may undergo periods of globally catastrophic heat loss, separated by intervals of geologic quiescence. Such an episodic history is now thought more plausible, though the implications of such events are just now being considered. How would they influence the Venusian atmosphere? Would higher surface temperatures affect the deformation of rocks? Have such episodic and catastrophic changes occurred in the past history of Earth? Could such an event have been responsible for the initiation of subduction and plate tectonism early in Earth history?

And what about the future of Earth itself? Earth's lithosphere has thickened overall with time, and some scientists predict that plate tectonism will cease — creating another one-plate planet — a few hundred million years from now. What then? Will Earth's surface become like those of the Moon, Mars, and Mercury, accumulating ever more craters over time? Or will heat buildup in the interior lead to breakup and overturn of the crust and catastrophic resurfacing, as is thought to have happened on Venus?

Solar-system exploration in the last decades of the 20th century has taught us more about the surfaces and interiors of Earth's neighboring planetary bodies than had been gleaned in all the preceding years of telescopic observation. This new perspective, whether related to the importance of impact events in Earth's geologic and biological history, or to the origin and fate of plate tectonics, has provided us with a comparative basis on which to understand our own home planet. We have no doubt that our understanding will grow in the decades ahead, and we are excited by that prospect.

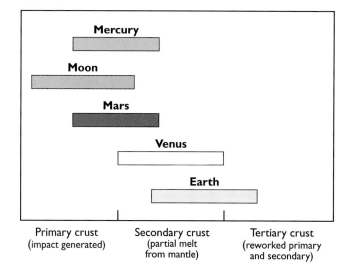

Figure 20. The types of crust now exposed on the terrestrial planetary bodies vary widely, from the largely primary crusts of the Moon to the heavily modified tertiary crust represented by Earth's continents.

Atmospheres of the Terrestrial Planets

Bruce M. Jakosky

A PLANET'S ATMOSPHERE is arguably one of its most important features. We live at the bottom of a sea of air and breathe it every day in order to stay alive. This very fact underscores its significance to us and to most (if not all) of the life on Earth. Through its control of the climate of a planet, an atmosphere may determine whether life can exist at all. By interacting with a planet's surface, and through geologic and geophysical processes in its interior, atmospheric gases can even affect the evolution of the planet as a whole.

To understand all these causes and effects, we must first understand the nature of the atmosphere itself — its origin, evolution, and current state — as well as the processes that affect its motions and behavior on all time scales. Only then can we begin to see the connections to other planetary processes and the nature of each planet as an integrated system. Fortunately, we have three examples of relatively dense atmospheres in the inner solar system: those of Venus, Earth, and Mars. Each planet has taken a different evolutionary path, and their atmospheres are a manifestation of those differences (*Figure 1*). The other terrestrial worlds are Mercury and the Moon (for our purposes, we can consider the Moon to be a planet). Their lower gravity and much less extensive geologic activity will help us to understand the nature of atmospheres because theirs are so thin as to be almost nonexistent.

It should not be too surprising to find that planets have atmospheres; they are a natural consequence of planetary formation and evolution. Gases such as water, carbon dioxide, nitrogen, and the noble gases, for example, are all present in the materials out of which the Sun and planets formed. If a planet is big enough, its interior will be heated both by its initial formation and by the subsequent decay of natural radioactive elements in its interior. This heating releases many gases from

Delicate clouds in the predawn Martian sky, as seen by Mars Pathfinder in 1997.

Figure 1. Mercury, Venus, Earth, and Mars have atmospheres that span the entire range from negligible traces to thick, dense air masses. The relative visibility of their surfaces reflects these differences. Mercury's is always in full view, while the Martian landscape is occasionally hidden by clouds or airborne dust. Typically about 40 percent of Earth is shrouded by clouds at any given time, and the surface of Venus is never seen in visible light from space.

within the rock, which can then migrate to the surface. There they accumulate, unless some process either removes them to space or forces them back into the interior.

THE ATMOSPHERES' CURRENT STATES

Since all the terrestrial planets formed from similar materials and underwent similar interior processes, the atmospheres of Venus, Earth, and Mars consist of the same gases — though in very different amounts and proportions (*Table 1*). The most abundant ones in Earth's atmosphere are nitrogen (N_2) and oxygen (O_2), together accounting for more than 98 percent of its molecules. Both of these gases are present in the atmospheres of Venus and Mars, but they are not the predominant species. The most common thread in composition appears to be carbon dioxide (CO_2). Not only does it dominate the gas mixture on Venus, but it plays an important role in determin-

Atmospheric Compositions Compared

Planet	Molecule	Abundance (bars)	Fraction of total
Venus	CO_2	86.4	0.96
	N_2	3.2	0.035
	Ar	0.0063	0.000070
	H_2O	0.009	0.000100
Earth	N_2	0.78	0.77
	O_2	0.21	0.21
	H_2O	0.01	0.01
	Ar	0.94	0.0093
	CO_2	0.000355	0.00035
Mars	CO_2	0.0062	0.95
	N_2	0.00018	0.027
	Ar	0.00010	0.016
	H_2O	3.9×10^{-7}	0.00006

Table 1. The atmospheres of Earth, Venus, and Mars contain many of the same gases — but in very different absolute and relative abundances. Some values are lower limits only, reflecting the past escape of gas to space and other factors.

ing the atmospheric and surface temperatures on all three planets. Water, one of the most abundant volatile species in the solar system, is present in each atmosphere but is not a major constituent of any of them.

The Venusian atmosphere contains about 100 times as much gas as does Earth's, which in turn contains about 100 times as much as Mars's. Thus, the three planets span a tremendous range in atmospheric abundance. These differences in composition and abundance appear to outweigh any of their similarities. However, the differences between Venus and Earth are much less important when we also consider the gases *not* in their atmospheres. For example, CO_2 can be removed very efficiently from the Earth's atmosphere by dissolving into the oceans, where it precipitates onto the ocean floor as deposits of carbon-containing minerals such as limestone. If we count the terrestrial CO_2 contained in the global deposits of limestone (the Rock of Gibraltar, the white cliffs of Dover, and so on), the total amounts of CO_2 on Venus and Earth are actually quite similar. So are their total amounts of nitrogen (*Table 2*). Mars's volatile inventory is much lower for two reasons: it is only about a tenth as massive as Earth or Venus and it may have lost a substantial fraction of its atmospheric gases to space throughout time.

Thus an atmosphere represents only one portion of a planet's volatile inventory. On Earth, and possibly on Venus and Mars, gases do not stay in the atmosphere permanently but continually cycle into and out of the crust. The composition and mass of the atmosphere today, then, represent a steady-state balance. Therefore, one goal of our studies is to understand what the history of supply and loss of gas has been, and what the atmosphere might have been like throughout time.

In today's atmospheres, the gases help to determine the surface temperature of each planet. In their absence, the temperature is determined by a balance between sunlight absorbed at the surface and infrared (heat) energy emitted by the surface back to space. The more solar energy that is absorbed, or the closer the planet is to the Sun, the hotter the surface will be. If Earth did not have an atmosphere, this energy balance would maintain its surface at about 255° K. This is about 30° K less than its actual average surface temperature and 18° K below the freezing point of water.

Certain gases, however, absorb infrared energy emitted by the surface very efficiently and thereby trap it in the atmosphere. The energy eventually is lost back to space, but not as efficiently. As a result, the atmospheric temperature increases. This process is somewhat similar to what happens in a greenhouse — sunlight gets in very easily, but heat has a difficult time getting out, so the temperature rises. The planetary-scale version of this process is termed the *greenhouse effect*. Carbon dioxide and water vapor cause Earth's atmosphere to retain heat, which raises its temperature. The atmosphere of Venus contains a total of about 92 bars of CO_2 — 260,000 times more than that in Earth's atmosphere. Consequently, CO_2 (and other gases) raise the surface temperature of Venus to about 750° K, three times what it would otherwise be. By contrast, the Martian atmosphere contains only 6 millibars of CO_2 and very small amounts of water vapor. As a result, greenhouse warming amounts to only 6° K and raises the average surface temperature to about 220° K (*Figure 2*).

Inventories of Volatile Compounds

Planet	CO_2 (g/g)	H_2O (g/g)	N_2 (g/g)	Ar (10^{-10} cm³/g)
Venus	9.6×10^{-5}	$> 2 \times 10^{-5}$	2×10^{-6}	20,000
Earth	16×10^{-5}	2.8×10^{-4}	2.4×10^{-6}	210
Mars	$> 3.5 \times 10^{-8}$	$> 5 \times 10^{-6}$	4×10^{-8}	1.6

Table 2. **Expressed as a fraction of each planet's mass (except argon), the total volatile inventories of Venus, Earth, and Mars are rather small — even when the volatiles trapped in their surfaces and interiors are included.**

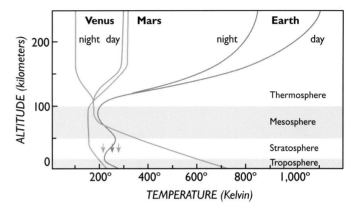

Figure 2. **A key diagnostic of an atmosphere's overall character is its temperature as a function of altitude. For example, based on its temperature profile, Earth's atmosphere divides into several distinct layers (horizontal bands) that the other planets do not share. In addition, the upper atmospheres of Venus and Earth exhibit a striking diurnal cycle (day-night pairs of curves). Arrows indicate the cooler surface temperatures that would occur in the absence of any greenhouse warming.**

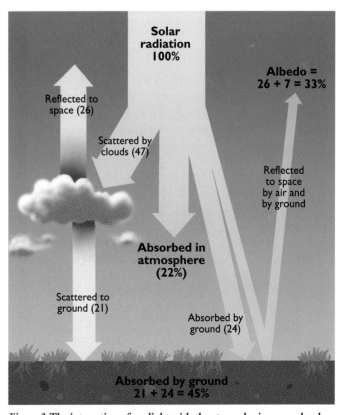

Figure 3. **The interaction of sunlight with the atmospheric gases, clouds, and surface of Earth has many avenues. Ultimately, about one-third of the solar energy reaching our globe is reflected back to space.**

Not all of the absorption of sunlight by these planets occurs at their surfaces. For example, Earth's atmosphere contains ozone (O_3), which absorbs the ultraviolet portion of the incoming sunlight at altitudes 20 km above the surface. Since the ultraviolet wavelengths carry a significant amount of solar energy, the middle part of the atmosphere becomes hotter. This increase in temperature serves to divide our atmosphere into distinct layers. The *troposphere* is the lowest portion; its temperatures are determined by the absorption of sunlight at the surface and its radiation of energy back to space. In the *stratosphere*, temperature increases with altitude due to the absorption of ultraviolet light by ozone (*Figure 2*). The Martian atmosphere contains too little ozone to provide much heating. However, airborne dust can absorb sunlight and heat up the middle part of the Martian atmosphere, and the red planet's frequent dust storms provide a continual source of this dust.

In each atmosphere, one or more of the important gases can *saturate* — reach the maximum concentration of vapor that the atmosphere can hold — and then condense. The result of this condensation is water clouds in the atmosphere of Earth, water or carbon-dioxide clouds at Mars, and clouds of sulfuric acid (H_2SO_4) in the Venusian atmosphere. The presence of these

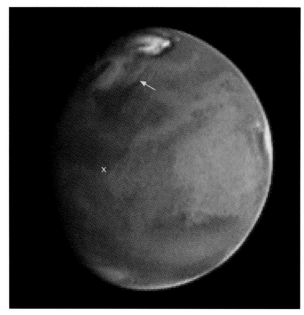

Figure 4. The Martian atmosphere can be a dynamic place. On 27 June 1997 (upper image) the Hubble Space Telescope recorded a large cloud mass to the north of the Mars Pathfinder landing site, which is marked with an *X*. A localized dust storm (orange arrow) is also evident in the Valles Marineris canyon complex. By 9 July (lower image), the clouds were less obvious and the dust storm gone. However, a strong climatic front (blue arrow) had developed near the north polar cap.

Figure 5. Delicate water-ice clouds hang in the predawn sky as seen from the Mars Pathfinder landing site in August 1997. The clouds appear to be 10 to 15 km in altitude.

clouds affects the energy balance of the planet by reflecting sunlight back into space and by reflecting heat from the surface back downward and keeping it from escaping (*Figure 3*).

The abundant clouds above us result from the presence of a global-scale ocean and the evaporation of its water. Winds can move parcels of atmosphere around; in the process, some will cool off and become saturated. Their excess water condenses as clouds of either liquid droplets or ice crystals, depending on temperature. When cloud particles get large enough, they can fall to the surface as rain or snow. Water runs off the continents and back into the ocean, completing the global water cycle.

In the Martian atmosphere, winter temperatures in the polar regions get low enough that CO_2 condenses as ice onto the surface. Since CO_2 is the dominant gas, as much as one-third of the atmosphere condenses onto the polar regions each winter, forming frost or ice deposits up to 1 or 2 m thick. These seasonal deposits *sublimate* back into the atmosphere (going from solid directly to vapor, without becoming liquid) in the spring. Mars does not have a global ocean that can supply water vapor, but its north polar cap consists of water ice. During summer, small amounts of water ice sublimate into the atmosphere, and the winds will distribute it globally. At various times and places, this water vapor becomes saturated and creates clouds (*Figures 4,5*). Water frost can be deposited onto the surface at night or during winter, but it disappears again when temperatures rise.

The opaque sulfuric-acid clouds of Venus are created by interactions between sulfur dioxide (SO_2) gas and water vapor. Chemical reactions with the surface should be removing sulfur quickly from the atmosphere, so the presence of SO_2 requires a continuous source of new sulfur into the atmosphere. This source is most likely surface volcanism. The total mass of the clouds is relatively small, amounting to only about 0.012 g in a vertical column 1 cm square. This is enough, though, to make clouds that have a dramatic effect on the planet's visible appearance from space, its energy balance, and the thermal structure of its atmosphere (*Figure 6*).

ATMOSPHERIC CIRCULATION

Actual atmospheric temperatures are determined not only by the balance of energy received from the Sun and lost to space but also by movement of the gas as well. If a parcel of gas expands, it cools off; it heats up if compressed. Thus, rising parcels experience lower pressure and expand, while cool, sinking parcels become compressed and warmer. Regional temperature variations, which result from differences in heating by the Sun, cause parts of the atmosphere to have different densities. As denser parcels of air try to sink and lighter ones try to rise, atmospheric motions occur, accompanied by changes in temperature. Thus, these motions tend to redistribute heat energy throughout the atmosphere and, of course, create local and global patterns of wind. We all are familiar with the results of this atmospheric circulation — afternoon thunderstorms on warm days, the trade winds used by sailing ships to cross the oceans, and the passage of weather fronts during the winter seasons.

At each altitude, the pressure is determined by the weight of the overlying air mass. This means the lower parts of the atmosphere are more compressed and thus have higher pressures. If

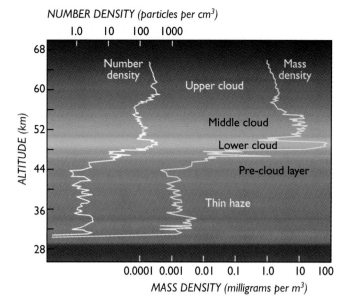

Figure 6. Clouds in the atmosphere of Venus form discrete layers that are fairly uniform from place to place. These are not condensations of water vapor, a gas that exists only in trace amounts on Venus. Instead, they consist almost entirely of droplets of sulfuric acid.

the atmosphere is locally *isothermal* (the same temperature at all altitudes), we can describe the vertical structure with what is known as the *scale height* of the atmosphere. This is the change in height needed to allow the density or pressure to drop by a factor of *e*, about 2.7. Typical scale heights on the terrestrial planets are about 10 km. As a result, for example, the air pressure in Denver (the "Mile-High City") is about 16 percent less than at sea level.

Suppose we take a parcel of air in this isothermal atmosphere and push it upward a bit. Because it moves into a less dense region, it expands and cools off slightly. Similarly, if we push a parcel of air downward, it compresses and heats up. If we could completely stir up the atmosphere, these changes in temperature would make the lower altitudes warmer than the higher ones even in the absence of sunlight. However, during the day, sunlight heats the surface and the surface in turn heats the lowermost part of the atmosphere. This extra warming causes air parcels there to rise, and as a result the atmosphere becomes well mixed. In this well-mixed gas, the resulting variation of temperature with altitude is called the *adiabatic* temperature profile. "Adiabatic" means simply that no energy is gained or lost from parcels of air as they move (even though the temperature changes).

An adiabatic profile does not actually describe the temperature gradient in Earth's troposphere, however, because of the presence of water vapor. As a parcel of air rises and cools off, water vapor becomes saturated and begins to condense. This condensation gives off latent heat, much as the condensation of water onto the outside of a cold glass provides heat that warms up the drink inside. Latent heat keeps the atmosphere from cooling off as much as it otherwise would. As a result, the temperature does not drop as quickly with altitude when water vapor is present. The resulting trend is known as a "wet adia-

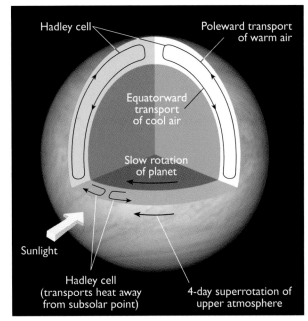

Figure 7. Rising motions occur wherever the strongest atmospheric heating occurs — near the subsolar latitude of Mars and Earth, and on the day side of Venus (as shown here). This movement is balanced by sinking parcels of air elsewhere, creating a net flow that keeps gases and heat from building up at any one location.

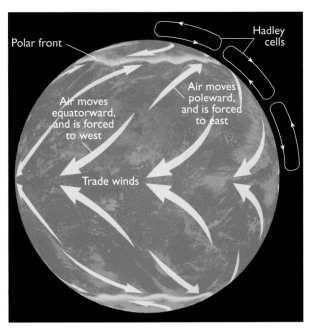

Figure 8. Zonal winds (along lines of constant latitude) result on Earth from the Coriolis force. Air is redirected eastward when it moves toward either pole and westward when it moves toward the equator. These east-west motions are analogous to the belts and zones in outer-planet atmospheres.

batic" temperature profile. The Earth's troposphere actually follows a wet adiabat quite closely on average, cooling off by 6° K for every kilometer in height up to an altitude of about 15 km (*Figure 2*). However, substantial variations occur from place to place, with regions of upwelling air, for example, having slightly different temperatures than regions of downwelling air. As a result, for example, over any given location we see clouds form at discrete altitudes rather than being present as a uniform haze.

The Martian atmosphere does not decrease in temperature with altitude quite as fast as predicted from its adiabatic value. In this case, water vapor is not the reason. The atmosphere is heated when sunlight is absorbed by airborne dust, and the relatively limited vertical mixing that occurs is insufficient to produce an adiabatic profile.

In contrast, the atmospheric temperatures of Venus are quite close to their adiabatic values all the way from the surface up to about 60 km. Some altitudes show temperatures that are consistent with constant overturning; others, however, show a smaller decrease than expected, indicating that the atmosphere is not being mixed at that location and at that time. Although clouds of liquid sulfuric acid are present in Venus's atmosphere, the temperatures do not follow a wet-adiabatic profile for sulfuric acid. The amount of liquid is too small for condensation to provide much heat.

In addition to the vertical variations in temperature, higher latitudes tend to be cooler because they receive less sunlight. As a result, polar air is denser than equatorial air. The warmer equatorial air has a tendency to rise and move toward higher latitudes, while the denser polar air sinks and moves equatorward to take its place. This general overturning of the atmosphere between low and high latitudes does not occur in a single glob-

al cell. Rather, the motions may be broken up into several smaller cells, the most prominent of which is the near-equatorial *Hadley cell* (*Figure 7*).

Multiple cells and zonal winds (which occur along lines of constant latitude) both result from the rotation of the Earth. As air moves from lower to higher latitudes, the rotation of the Earth forces an additional motion from west to east. This motion arises from the *Coriolis effect* — each parcel of air must conserve its angular momentum as it moves poleward. Physically, the situation is analogous to a spinning ice skater who pulls his arms in toward his body and thus spins faster. As a parcel of air moves toward higher latitude, it moves closer to Earth's rotation axis and therefore will "spin" faster by moving more quickly, that is, eastward, around the pole. This eastward movement breaks up the meridional (equator-to-pole) cell into multiple cells, and it leads to belts of air moving toward the east at high altitude and toward the west at low altitude. In fact, the overall motions of the atmosphere are even more complex, with the air at some latitudes moving eastward and at others westward (*Figure 8*).

Similar movements occur on Mars. A global-scale Hadley cell cycles the atmosphere between low and high latitudes and sets up a pattern of east-west motions. The global circulation on Mars is similar to Earth's because the Martian day is much like ours, 24.6 hours long. The biggest differences between the two planets, in terms of atmospheric dynamics, are the absence of a Martian ocean (no condensation/evaporation effects) and the presence of large-scale topography on Mars. The largest Martian mountains rise about 25 km into its atmosphere, compared to about 10 km for the largest mountains on Earth. At their peaks, the atmospheric pressure is about one-tenth that at the surface. Apart from its global motions, the Martian atmosphere generally rises as it passes over these mountains and sinks over the low-altitude regions. This motion is somewhat similar to the rising motions that occur over terrestrial islands during the day as the islands heat up more than the oceans.

One important effect of meridional overturning is that sunlight absorbed near the equator is converted to heat within parcels of air, and those parcels are moved to higher latitudes. In effect, this redistributes the Sun's heat from low to high latitudes and makes the global temperatures much more uniform than they otherwise would be. Without this redistribution, for example, Earth's polar regions would be frozen over completely year-round. The oceans also contribute to this heat transfer toward the poles.

The atmosphere of Venus also undergoes similar global-scale motions. They differ from those on Earth and Mars because of Venus's slow rotation (once every 243 days in the direction opposite that of Earth). Because Venus goes around the Sun every 225 days, the two motions combine to make the Venusian day effectively 117 days long (the time between successive sunrises). Because this day is so long, atmospheric motions are very different on Venus. The Hadley-cell circulation goes all the way from the equator to the poles without interruption. You might think that air would also flow from the planet's warm, sunlit side to its cooler night side. However, the atmosphere is so massive that the temperature changes very little between day and night. In addition, the heat transported by winds is able to remove almost all of the temperature variations at the surface. As a result, the equatorial and polar temperatures are almost exactly the same.

However, a complication arises high up in the Venusian atmosphere. We can track the planet's high-altitude winds by watching the patterns seen in its cloud tops at ultraviolet wavelengths (*Figure 9*). Remarkably, the zonal winds (along a line of constant latitude) move in the same direction as Venus's rotation, but at about 100 m per second — fast enough to circle the planet in only 4 days! The winds move this way at all altitudes, so there is no "return" motion in the opposite direction somewhere else. Rather, this *superrotation* seems to result from the mixing of momentum in the atmosphere due to turbulence, combined with the uniform, slow heating of the atmosphere by the Sun. Superimposed on this fast motion is a daily expansion of the dayside atmosphere due to solar heating and a contraction on the night side. These daily tides exert a force on the surface that may be affecting the planet's rotation rate. In fact, they might explain why Venus spins so slowly.

Another type of motion on Earth results from the west-east motions at high latitudes in winter. The polar regions get very cold, and as a result atmospheric temperatures change dramatically with latitude. This temperature gradient drives a strong jet of winds at mid-latitudes, where the gradient is greatest. Such rapid winds are not stable, though, and a *baroclinic instability* develops. That is, tongues of cold polar air move toward the equator, and tongues of warm mid-latitude air move poleward, even as the whole atmosphere continues to move from west to east. As a result, if you were situated in Colorado or England during winter, you would first feel warmer air passing by, then colder air several days later, then another burst of warm air. This is the origin of winter-hemisphere weather fronts. At the boundary between the warm and cold tongues of air, clouds condense and precipitation follows. The tongues of air generally tend to form at the same physical sizes, so the passage of these weather fronts occurs somewhat regularly.

The same types of weather fronts occur on Mars, driven by the same strong latitudinal gradient in atmospheric temperature in the winter hemisphere. On Mars, however, the absence of an ocean-and-continent structure and the relative thinness of the air mass makes the atmosphere much more uniform in its response to the temperature variations. Martian weather fronts pass overhead much more regularly than on Earth, with a remarkably uniform period of about 3 days.

The winds of Mars are also well known for their ability to raise surface dust into the atmosphere on local and global scales. Martian dust is extremely fine, with individual particles typically 1 micron across (comparable to those in cigarette smoke). Once the dust is raised, it can remain aloft for weeks before settling back onto the surface. Local dust storms, perhaps 100 to 1,000 km across, are relatively common. Occasionally, these can grow to global proportions, shrouding the entire planet in a veil of dust (*Figure 10*). The redistribution of dust by these global storms plays an important role in the centimeter- and meter-scale geology of the surface and in the evolution of the polar deposits.

The dust also plays an important role in Martian atmospheric dynamics. When substantial amounts of dust become airborne, sunlight is absorbed in the atmosphere rather than at the surface. This causes the middle of the atmosphere to become much warmer. This direct heating is much more efficient than the heating from the surface that occurs when the atmosphere is clear. As a result, circulation in the Hadley cell can intensify dramatically, and the winds can carry it to higher latitudes. At the same time, atmospheric heating becomes much more uniform

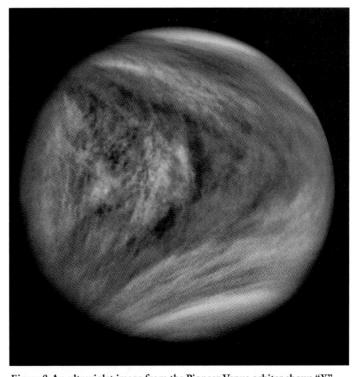

Figure 9. An ultraviolet image from the Pioneer Venus orbiter shows "Y"-shaped patterns in the Venusian cloud tops. These patterns, due to varying amounts of SO_2 and sulfur in the upper atmosphere, reflect the circulation of the atmosphere in the region of the cloud tops.

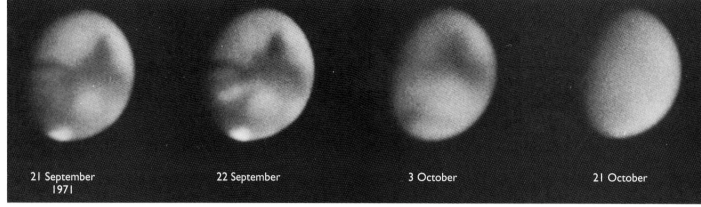

| 21 September 1971 | 22 September | 3 October | 21 October |

Figure 10. Because the orbit of Mars is distinctly elliptical, its perihelion (which occurs near midsummer in the southern hemisphere) tends to be the season of dust storms. The largest events redistribute fine-grained dust planetwide and typically last a month or two. A particularly vigorous dust storm in mid-1971 began in the Hellas region and quickly enveloped the entire planet in an orange shroud.

Two Tenuous Atmospheres

Species		Mercury (atoms/cm³)	Moon (atoms/cm³)
Hydrogen	(H)	200	< 17
Helium	(He)	6,000	2,000–4,000
Oxygen	(O)	< 40,000	< 500
Sodium	(Na)	20,000	70
Potassium	(K)	500	16
Argon	(Ar)	$< 3 \times 10^7$	4×10^4

Table 3. The various atoms in the atmospheres of the Moon and Mercury are distributed so thinly (expressed here as number density at the surface) that they rarely collide with one another.

with altitude, diminishing the tendency of the air mass to mix vertically. The reduced vertical mixing may act as a shut-off valve for the global dust storms, by lowering the tendency to raise new dust from the surface and allowing the dust already in the atmosphere to settle back onto the surface.

The heating of the Martian atmosphere when dust is present results in cooling of the surface, because less sunlight makes it to the ground. A similar effect has been predicted for Earth in the event of a large asteroid impact or a global nuclear war. Airborne dust and soot would absorb sunlight and produce a near-surface cooling, causing surface temperatures to drop below freezing and affecting essentially all life on Earth. The same thing is observed to a lesser extent from the largest volcanic eruptions, with stratospheric dust producing a cooling of the surface by 1° or 2° that can last for a year.

TENUOUS ATMOSPHERES OF THE MOON AND MERCURY

It is perhaps an exaggeration to call the traces of gas around the Moon and Mercury "atmospheres." Their density is so low that collisions between atoms are extremely rare; atoms and molecules bounce around the surface on ballistic trajectories. Consequently there are no atmospheric dynamics in the usual sense. Gases remain above the surface for a short time, so the atmospheres reflect a balance between supply of molecules to the atmosphere and very rapid removal of them from it. The steady-state composition may reflect whatever is being outgassed from the interior, so the atmospheres may tell us about recent geologic and interior processes. And the tenuous atmosphere acts as a source of volatiles that can create deposits of ice in the cold polar regions.

Gases detected in the lunar and Mercurian atmospheres (*Table 3*) include hydrogen, helium, oxygen, sodium, potassium, and argon. We normally do not think of sodium or potassium as being gases. On Mercury and the Moon, however, all atoms will behave similarly once they are released at the surface, and an atom of sodium, for example, differs from an atom of helium primarily by its mass. Curiously, of all of the different atoms present

in surface materials, only some are present in any abundance in the atmosphere. Atoms that have not been found yet — despite searches for them — include common rock-forming elements such as calcium, lithium, silicon, aluminum, iron, and titanium.

There can be several sources of atoms in these atmospheres. They can be supplied directly by the impact of comets and micrometeorites onto the surface. They can be released from surface materials due to the "impact" of sunlight by a process termed photodesorption or, at the higher temperatures on Mercury, directly by evaporation. Most atoms in these atmospheres will be ionized by solar ultraviolet light. Once ionized, the electric and magnetic fields of the outflowing solar wind pick them up and either push them back onto the surface at high velocity or remove them from the planet entirely. The lifetime of atoms in the lunar and Mercurian atmospheres depends on how efficiently they are ionized. Lifetimes on the Moon vary from about an hour for sodium up to a couple hundred days for hydrogen (*Figure 11*). The atoms typically can survive long enough to hop around the surface on ballistic trajectories for several tens of bounces. With enough bounces, an atom can "random walk" either to the unilluminated side of the planet or to the polar regions. When they strike the cold surface in either place, they will tend to stick. On the night side, the atoms that stick will stay on the surface until the next sunrise.

Molecules that make it into the polar regions may be deposited inside craters or depressions. At these high latitudes, such regions may never see direct sunlight and are permanently shadowed. Temperatures there can be below 50° K, even on Mercury, and volatiles deposited on the surface can survive for long periods. In fact, radar measurements of the Mercurian polar regions show the presence of what are thought to be substantial deposits of water ice (see Chapter 7). The Moon also appears to have water frozen at its poles, and hydrogen derived from this ice has been detected by the Lunar Prospector spacecraft (see Chapter 10). These polar deposits are not ice "caps," as occur on Earth and Mars, but probably exist as mixtures with dirt on the crater floors. Conceivably, the ice would not be obvious even to future lunar and Mercurian astronauts standing directly on it. Nonetheless, such polar ice deposits are a sensitive indicator of the integrated effect of supply and loss processes.

SOURCES OF VOLATILES

If we are to understand the composition and nature of planetary atmospheres, then we need to look at both their origin and subsequent evolution. Terrestrial atmospheres actually are somewhat ephemeral, with gases continually being added and removed. Atmospheric sources will be considered in this section, and loss processes ("sinks") in the next.

The terrestrial-planet atmospheres obtained their initial inventories while each planet was still forming. Gas probably was not accreted directly from the protoplanetary nebula, however, because the inner planets had too little mass to attract it. Any hydrogen or helium that accumulated must have been of very low abundance or been lost to space. Instead, the present-day atmospheres are thought to be "secondary," having come from other sources. The most obvious sources of volatiles were the comets and the planetesimals from which the inner planets were assembled.

Because they formed in the outer parts of the solar system, far from the Sun, comets are very rich in ices, H_2O and CO_2 in particular. Over time they can be sent in close to the Sun by the gravitational effects of the giant planets or of passing stars. As they collided with the forming planets, they would have released their volatile gases.

Even the solid planetesimals that provided most of the mass of the planets contained volatiles. They would have been present, for example, as water that is chemically bound to minerals or as elemental carbon. In fact, some of the meteorites that come to us from the asteroid belt contain up to 20 percent water by mass. The water exists in hydrated minerals and clays. If the forming planets took in even a tiny fraction of materials like these, then the amounts of water supplied would have been enormous.

Other important gases, such as the noble gases or nitrogen, might have come from either comets or planetesimals. We would like to know which of these were the major source of volatiles to the terrestrial planets. If the planet-forming planetesimals dominated, then in all likelihood a relatively small portion of volatiles has been added to the planets during the last several billion years. Primordial planetesimals no longer exist in the inner solar system, though escapees from the asteroid belt can still collide with Earth and its neighbor worlds. On the other hand, if comets and other icy objects from the outer solar system have dominated the supply of volatiles, then these objects must represent a continuing supply of gas now and into the future.

Can we identify the portion of the atmospheres that may have been supplied by comets? Fortunately, there are some "fin-

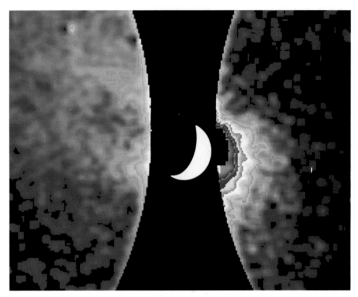

Figure 11. **A very tenuous lunar atmosphere forms when meteoroids strike the Moon and release gases trapped in rocks and soil. The freed gas is then blown back by solar radiation. Recorded in February 1991 using a filter that isolates emission from sodium atoms, this false-colored image shows a "coma and tail" of sodium emanating from the Moon. The lunar crescent, which was blocked during the exposure, is included for scale.**

gerprints" that can answer this question. One of them is the ratio of deuterium (D) to hydrogen (H). A deuterium atom consists of an atom of hydrogen with an additional neutron. As such, it is twice as heavy as hydrogen and behaves slightly differently in chemical reactions. All matter contains both D and H atoms, with an original ratio of 1.5×10^{-5} having been set at the time of the Big Bang. Various processes have modified this ratio through time. Earth's oceans, for example, are much richer in deuterium; they have a D:H ratio of 1.6×10^{-4}. Similarly, all meteorites (and thus the asteroids) have values within about a factor of two of Earth's.

The difference between the terrestrial D:H ratio and the primordial value undoubtedly reflects the processes by which hydrogen was incorporated into the objects from which the planets formed. It should not be too surprising to find such differences from one place in the solar system to another. We know there was considerable structure in the protoplanetary nebula and disk, and those variations are reflected in many elemental and isotopic ratios. Most processes that affect planets after their formation act to increase the D:H ratio (for example, the preferential loss of hydrogen to space, described later).

The ratio of D:H now has been measured in three different comets, and the average value is about twice the terrestrial value. If these examples are representative, then Earth's water cannot have been supplied entirely — or even mostly — from comets, because the D:H ratio in our oceans would be higher than it is. Theoretical estimates argue that if comets were the predominant source of Earth's water, then many more of them must exist in the outer reaches of the solar system than we believe is possible. This means that the bulk of Earth's water must have been supplied during its formation, rather than steadily throughout geologic time. By analogy, we think that supply of the other gases also occurred early.

We cannot easily apply similar arguments to the source(s) of water for Mars and Venus, because their D:H ratios have been complicated by subsequent evolution. Specifically, unlike the Earth, both Mars and Venus have probably lost substantial amounts of hydrogen (derived from water), which has created an enhanced D:H ratio that makes a quantitative interpretation difficult.

Assuming that most volatiles accreted with the planetesimals, we turn to their subsequent release to the atmosphere. This occurs in one of three ways: (1) during accretion itself, due to the intense heating that occurs when a planetesimal strikes the surface; (2) at the time of core formation, when Earth (and possibly Venus and Mars) may have been entirely molten; and (3) steadily throughout geologic time, during the partial melting of the interior that accompanies volcanic activity.

As the planets grew larger during accretion, their increasing size and mass caused the late-arriving objects to strike their surfaces at velocities no less than 11 km per second (for Earth), and possibly as high as 20 to 40 km per second. The force of the impact would have turned the kinetic energy of motion largely into heat. The energy supplied from Earth's accretion may have sustained a thick, hot, steam atmosphere containing much of its water. This hot envelope would have persisted until the impact rate declined enough for the water to condense. As the steam atmosphere collapsed, substantial quantities of gas may have become incorporated into the solid portion of the planet. A similar scenario likely unfolded at Venus. However, Mars probably did not reach the size required to support a steam atmosphere until very late in its formation, if at all.

Volatiles also may have been released during core formation, which probably occurred soon in the evolution of Earth (and, by analogy, Venus) judging from calculations of the energy of accretion. Late-accreting planetesimals would have heated these planets' surfaces enough to melt their outer crusts. Any free iron in this accreting material would be heavy enough to sink through the molten rock, converting its gravitational potential energy to heat by friction. Further melting ensued, which would have allowed more iron to sink, resulting in more heating. Once *some* melting occurred, then, any additional release of energy probably led to catastrophic core formation and global melting.

Rapid core formation would have released a substantial portion of the volatiles contained within the planet. We know that gases were released early, as deduced from the ratio of argon-40 to argon-36 now present in Earth's atmosphere. Argon is a noble gas — it does not react chemically. Therefore, once released into the atmosphere, argon cannot readily be removed

Figure 12. The 18 May 1980 eruption of Mount St. Helens, a typical subduction-zone volcano. Such volcanoes tend to erupt explosively, powered by volatiles liberated during the subduction of sea-floor sediments.

from it; it is thus an excellent tracer of volatile activity. The isotope ^{36}Ar was incorporated into the planetesimals out of which our solar system formed. By contrast, ^{40}Ar (which contains four additional neutrons in its nucleus) is created only by the radioactive decay of potassium-40. Both the radioactive and stable isotopes of potassium are incorporated into forming minerals. When an atom of ^{40}K decays to ^{40}Ar, it remains stuck inside the mineral until released by melting. The half-life of ^{40}K, the time during which half of it will decay to ^{40}Ar, is 1.25 billion years. Thus, if core formation (and global melting) occurred early, internal gases would have been released before much ^{40}Ar had accumulated; if core formation occurred late, there would have been much more ^{40}Ar available. Given how much potassium exists in the Earth and how much ^{36}Ar and ^{40}Ar are present in Earth's atmosphere, it appears that core formation and the release of argon to the atmosphere occurred very soon after Earth formed, taking no more than a few tens of millions of years. Undoubtedly, additional argon has been released throughout time, but the bulk of it must have been outgassed early.

The final mechanism for releasing gases into the atmosphere is volcanism. On the Earth, there are two predominant types of volcanic eruptions: those at mid-ocean ridges, created by the melting of material upwelling from the deep mantle, and those at subduction zones, where volatile-rich sediments from the ocean floor are pulled beneath the continents and melted. Each of these releases volatiles to the atmosphere. However, the eruptions at mid-ocean ridges involve mantle material that most likely has never been near the surface, thereby releasing "juvenile" gases to the surface. Since many of the interior's volatiles were released during core formation, the volatile content of mid-ocean-ridge basalts is relatively low. By contrast, volcanism at subduction zones recycles volatiles from the sea floor (mostly carbonates) back into the atmosphere, along with a small component of juvenile gases (*Figure 12*).

Have these same processes acted on Mars? There is some isotopic evidence in Martian meteorites that a global-scale "event" occurred very soon after the planet formed. This event most likely was core formation, probably accompanied by the release of some volatiles that were soon lost to space. We suspect that much of the present Martian atmosphere probably arose from the steady release of gas during volcanic events throughout geologic time (*Figure 13*). There is no convincing evidence for global-scale plate tectonics or subduction zones on Mars and thus no means for the wholesale recycling of volatiles to the atmosphere. Some recycling may have occurred, however, if volcanic eruptions overrode older sediments containing carbonate minerals. In that case, the lava's heat may have broken down the carbonates and released some CO_2 back into the atmosphere. It is also possible that Mars's catastrophic floods allowed the release of small amounts of CO_2 dissolved in the water, much as carbon dioxide in a bottle of soda escapes when the bottle is opened. The amounts of CO_2 released in these scenarios are not likely to be substantial, however.

Unfortunately, we still know little about the history of outgassing on Venus. The argon isotopes in its atmosphere appear to have been "contaminated" by argon from the solar wind, making interpretations pertaining to the earliest outgassing dif-

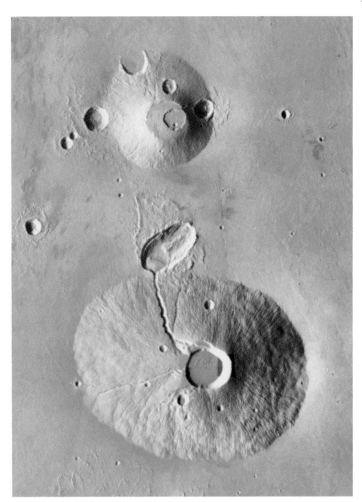

Figure 13. The Martian shield volcanoes Uranius Tholus (top) and Ceraunius Tholus are similar in size and appearance to their Hawaiian counterparts. This type of volcano is much less explosive because the magma comes from the mantle and has a low volatile content. Note the elongated 30-km-long crater on the north margin of Ceraunius; it has been partially filled by — and perhaps triggered — what appears to be a small volcanic flow down the mountain's side.

ficult. In addition, the surface of Venus is probably too hot for carbonate minerals to be stable; certainly they are not stable below even very shallow depths. Therefore, recycling from the atmosphere back onto the surface or into the subsurface probably is not a major process as it is for the Earth. Some outgassing or recycling of gases is likely to be occurring at present, given the presence of chemically unstable sulfur dioxide in the atmosphere, as discussed earlier.

Finally, and intriguingly, we can speculate about whether a barrage of "small comets" is continually supplying gas to terrestrial-planet atmospheres. The purported existence of these objects is based on dark spots seen in ultraviolet images of Earth. The spots are ostensibly caused by high-altitude water vapor, which absorbs sunlight at these wavelengths (*Figure 14*). On the basis of the size of the spots and the frequency at which they were observed to occur, investigators Louis Frank and John Sigwarth suggested that Earth's atmosphere is being bombarded thousands of times every day by small comets. These bodies would be perhaps 10 m in size and consist almost entirely of water ice with a thin, dark, carbon-rich coating. If

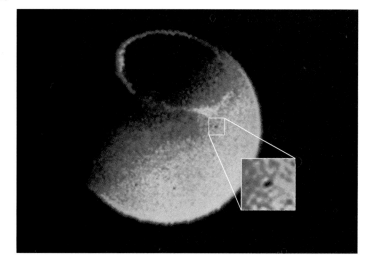

Figure 14. This ultraviolet "hole" may have resulted from absorption of ultraviolet sunlight by a cloud of water vapor high atop Earth's atmosphere. Some scientists speculate that the dark spots are caused by the impact of icy 10-m objects into the upper atmosphere, though there is no universal agreement that such objects actually exist. These spots may instead be an instrumental artifact, or they may result from an altogether different physical mechanism. The bright ring is Earth's northern auroral oval.

Figure 15. This 3-cm-wide piece of metamorphic stone from west Greenland is one of the oldest rocks ever found on Earth. It crystallized at least 3.85 billion years ago when a sedimentary deposit was heated and compressed. Because sedimentary rocks form from waterborne particles, by implication liquid water has probably existed on Earth's surface for at least 3.9 billion years.

these objects are real, then they could have supplied much of Earth's water over geologic time. Of course, the arrival frequency may not have remained constant, so it is not possible to determine what fraction of Earth's volatiles they might have brought. Similarly, it is not possible to extrapolate their impact rate on Mars or Venus. There have been no observations that demonstrate the existence of these small comets, however, and few scientists are convinced that they are real.

THE EVOLUTION OF ATMOSPHERES

Scientists take several approaches in trying to understand the evolution of planetary atmospheres throughout geologic time. For example, theoretical arguments can suggest how atmospheres might change under certain conditions. Observations of surface geology can help us infer processes that have acted in the past and their implications for future climate. Finally, isotopic measurements help us trace planetary evolution. Each of these approaches provides unique and independent information on the evolution of atmospheres. In the end, the different approaches must reach the same conclusions. After all, each planet and its atmosphere have had a unique and detailed history — only our ability to understand that history is uncertain.

In looking at the Earth's geologic record, we see strong indications that the composition of our atmosphere has changed through time. The amount of atmospheric CO_2 is thought to have dropped tremendously, and the amount of O_2 to have increased substantially, over the last 4 billion years. This evidence is largely circumstantial. It implies that the early Earth had liquid water on its surface. For example, the oldest known rocks, dating back 3.5 to 3.9 billion years, are metamorphic rocks (*Figure 15*) that were formed by the actions of pressure and heat on buried sediments. The existence of sedimentary rocks, in turn, requires that liquid water was present when they formed. We don't yet know whether sedimentary rocks were

localized or widespread at that ancient time. Nonetheless, some liquid water must have been present.

However, based on our understanding of stellar evolution, we think the Sun was 30 percent dimmer 4 billion years ago than it is today. As such, most climatologists expect that the entire Earth should have been frozen over. For example, the amount of CO_2 in the atmosphere today would not have provided sufficient greenhouse warming to allow liquid water to exist at the surface during these earlier epochs. Thus, there must have been more greenhouse gas in the atmosphere at that time, enough to raise the temperature above the melting point of ice. Although some greenhouse warming could have been provided by methane or ammonia gas, these species would not have been stable in the Earth's early atmosphere. Carbon dioxide is the most readily available greenhouse gas and, therefore, probably provided the warming. By some estimates, the amount of CO_2 in the early terrestrial atmosphere might have been as much as 1,000 times greater than it is now.

Meanwhile, we know of geologic features roughly 2 billion years old that were created by glacial erosion. Therefore, the atmospheric temperature at that time must have allowed surface ice to accumulate at least occasionally, even as the Sun was growing more luminous. This means that the efficiency of the atmospheric greenhouse — and by implication the amount of atmospheric CO_2 — must have declined over time.

There is a feedback mechanism between carbon dioxide and Earth's global climate that may regulate the amount of CO_2 in the atmosphere. In the modern atmosphere, CO_2 is readily removed and deposited onto the ocean floor in the form of carbon-containing minerals. This occurs when CO_2 dissolves in rainwater and the oceans, where it reacts with minerals washed off the continents to form carbonate minerals such as limestone. The carbonates precipitate onto the ocean floor, which eventually is recycled into the interior through subduction (see Chapter 9). As the sea-floor sediments and oceanic crust are pushed to great depth, the higher temperatures there release the CO_2 locked away in the carbonates. The gas makes its way back to the surface in explosive, subduction-zone volcanism (such as at Mount St. Helens in the Pacific Northwest or Mount Pinatubo in the Philippines), thus completing the global cycle of CO_2. The amount of CO_2 now present in the atmosphere represents a balance between these competing processes of supply and loss (*Figure 16*).

Since carbonates form most efficiently at high temperatures, the system is naturally self-regulating. Carbon dioxide is removed from the atmosphere if the air is warm and allowed to accumulate if it is cool. Although the formation of sea-floor sediments is helped today by organisms in the oceans (which make shells out of the CO_2 and help to precipitate it as minerals), carbonate sediments will form even in the absence of life simply by this geochemical regulation. Thus, Earth was endowed with a mechanism that drove up the CO_2 abundance in its early, cold atmosphere and provided sufficient greenhouse warming to allow liquid water to be stable. As the Sun heated up, the removal of CO_2 from the atmosphere became more efficient and the abundance of this gas declined to its present value. The net result is a change in the amount of CO_2 that just matches the changes required to keep liquid water at the surface.

The amount of O_2 in the Earth's atmosphere has increased with time, even as the CO_2 abundance was decreasing. Sediments older than about 2 billion years formed in an environment that was, for the most part, oxygen poor. Any free oxygen would have reacted with reduced minerals at the Earth's surface. Only after these "sinks" for oxygen were used up did O_2 begin to accumulate in the atmosphere. Of course, the source for the increasing oxygen was photosynthesis. Photosynthetic organisms have existed for at least 3.5 billion years, even though oxygen began to accumulate much later. There is very little evidence, however, to tell us whether the abundance of oxygen has remained constant (or nearly so) for the past 2 billion years. We suspect that it has changed substantially over that time.

The atmosphere of Venus has probably evolved substantially through time. One rationale for this assertion is theoretical. If the Earth were put in Venus's orbit, its atmosphere would get hotter and thus able to hold more water evaporating from the ocean. The additional water vapor would trap more heat from the Sun and raise the temperature further, thereby allowing even more water vapor to evaporate. This positive-feedback loop, termed a *runaway greenhouse*, would eventually result in the complete evaporation of the oceans. All Earth's water would reside in the atmosphere, whose temperature would soar to values at least as high as those on Venus today.

With sunlight during Venus's earliest history being 30 percent weaker than at present, however, its atmosphere was probably much cooler than it is today. A runaway greenhouse might not have occurred. Today, the distance from the Sun inside of which runaway conditions occur, and outside of which an atmosphere would stabilize at lower temperatures, is located between the orbits of Earth and Venus. Four billion years ago, when the Sun was dimmer, this boundary may have been inside Venus's orbit. Conceivably, the early climate on Venus was much more clement. Temperatures might have

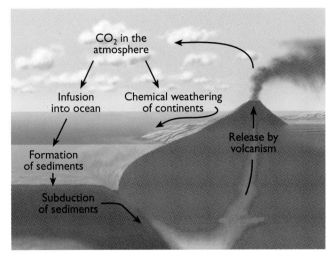

Figure 16. On Earth, carbon dioxide cycles between land and sky. Gaseous CO_2 is removed from the atmosphere to form sea-floor sediments; plate tectonism causes these sediments to be subducted into the upper mantle, and subduction-zone volcanism later releases the CO_2 back into the atmosphere. A balance thus exists between the loss of atmospheric CO_2 and its resupply, though the gas's actual abundance in the atmosphere may vary over time.

been warm, but not so warm as to evaporate an entire ocean. Over time, as the Sun heated up, the critical boundary would have moved outward and crossed Venus's orbit. At that point the planet's climate would have catastrophically "jumped" to a runaway-greenhouse state, with all of the water and CO_2 being pushed into the atmosphere.

Once in the atmosphere, the water molecules would have been dissociated by sunlight with relative ease, allowing hydrogen to escape to space. In fact, the present D:H ratio in Venus's atmosphere indicates that just such a loss occurred. The ratio is strikingly high, about 130 times that in the Earth's oceans or atmosphere. Assuming that Venus started with the D:H ratio seen elsewhere in the inner solar system, the current value implies that Venus has loss an amount of water equivalent to a global layer 3 m thick. This result is not unique, however. Impacts of water-rich comets and asteroids through time undoubtedly have added water to the atmosphere, new water whose much-lower D:H ratio would partially "reset" the value in Venus's atmosphere. It would take only a few impactors the size of Halley's comet, some 10 km across, to supply as much water as is presently in the Venus atmosphere. Therefore, the only thing we can say with certainty is that Venus has lost a considerable fraction of its total water inventory. Indeed, it may have even started out with an Earthlike ocean.

Mars, on the other hand, shows considerable evidence for climate change, both in its surface geology and in its atmospheric composition. On the basis of cratering statistics, Martian surfaces range in age from very young to as old as about 4 billion years. The oldest surfaces, concentrated in the southern hemisphere, have two features that suggest that the planet's early climate may have been very different from the present one. First, there are dendritic networks of valleys that look as if they had been carved by water flowing over the surface (*Figure 17*). Whether the water fell as precipitation or was released from underground is uncertain. However, the dendritic nature of the valleys suggests a slow,

Figure 17. Branching river valleys like these occur on the ancient portions of the Martian surface. Typically 1 km wide and several hundred meters deep, they may have been formed by the runoff of precipitation from the atmosphere or by the "sapping" of water from the subsurface. Either way, the valleys' dendritic nature suggests that they formed gradually, and that water must have been more stable at the Martian surface than it is today.

Figure 18. On Mars many ancient impact craters are very heavily degraded, like this trio of 45-km-wide examples, and the Martian landscape bears very few craters smaller than about 15 km across. These characteristics suggest erosion by water — yet another indication that the Martian climate was very different between about 4.0 and 3.5 billion years ago.

erosional process, rather than a sudden, catastrophic one. This implies that water must have been more stable during the earliest periods of Martian history than it is now.

Second, the impact craters on older surfaces have been very heavily eroded and degraded (*Figure 18*). This degradation has proceeded far enough that craters smaller than about 15 km in diameter are almost entirely absent and many larger ones are barely recognizable. Some of the less-degraded craters show gullies on their interior walls that indicate that erosion by water may have played a role. In any case, the erosion of the craters in the past occurred faster than it did later. The most likely explanation is that the early Martian atmosphere was more substantial and, again, that liquid water was more stable, than at present.

The transition to a colder, drier climate occurred about 3.5 billion years ago, as implied by the number of impact craters on these surfaces. What caused this cooldown, especially when the total heat supplied from the Sun was increasing? Three obvious processes may have played a role.

First, impacts of large asteroids during the tail end of Mars's formation may have ejected atmospheric gas to space. If we count the theoretical amount of atmospheric erosion due to impacts since the planet formed, some 99 percent of the atmosphere may have been lost. However, we are especially interested in the loss of gas beginning 4.0 billion years ago, corresponding to the oldest surfaces that we see. In this case, it is likely that only 50 to 80 percent of the atmosphere has been removed.

The second process involves stripping of the atmosphere by the force of the impinging solar wind. Specifically, atoms in the upper atmosphere become ionized by sunlight. Now having a net charge, they are picked up by the magnetic field of the solar wind streaming by at 400 km per second. Some of the ions slam into the upper atmosphere at very high velocity. When

they hit, they can knock upper-atmospheric atoms into space. Theoretical models suggest that this process can cause the loss of perhaps 1 bar of gas — about 100 times as much gas as is present in the Martian atmosphere. The loss of so much gas would have triggered a significant change in the Martian climate. If such loss actually did occur, it would preferentially remove the lighter isotopes of a given volatile and leave the remaining atmosphere enriched in its heavier isotopes. In fact, this type of enrichment has been observed. The ^{38}Ar:^{36}Ar ratio in Mars's atmosphere is about 30 percent larger than anywhere else in the solar system. Similar enrichments are seen in the ratios of the stable isotopes of carbon, oxygen, and nitrogen. Therefore, solar-wind stripping of the Martian atmosphere has probably occurred. Photochemical processes have probably contributed to the removal of some gas to space, though not as much as the solar-wind-related loss.

The third process for Martian atmospheric evolution may be the simplest — though it also is the most controversial. Mars's atmospheric CO_2 could have dissolved in the liquid water that existed early in the planet's history and reacted with the surface to form new minerals such as carbonates. In this way, carbon dioxide could have been removed from the atmosphere and sequestered into the regolith. We know that this process occurred to some extent, because some carbonate minerals are found in Martian meteorites. However, whether the regolith or crust actually contains a substantial CO_2 reservoir is uncertain. Each bar of CO_2 removed from the atmosphere would form enough carbonates to create a global layer several tens of meters thick. However, we have no evidence to suggest the presence of such large amounts of carbonates.

Both Earth and Mars appear to have undergone an additional loss process very early in their history, presuming that their initial atmospheres consisted primarily of hydrogen. (Large quantities of hydrogen may have been released when water from accreting planetesimals reacted with iron near the surface of the forming planet.) The escape of hydrogen from the top of the atmosphere would have mimicked the solar wind's escape from the Sun, and this wholesale outflow of hydrogen would have dragged other molecules along. The evidence that this process occurred is seen in the relative abundances of atoms heavy enough to have been only partly removed from the atmosphere. For example, the atmospheres of both Earth and Mars show that lighter isotopes of the noble gas xenon have been depleted relative to the heavier ones. Venus probably experienced similarly selective losses, but there are no measurements yet that would tell us the extent to which such loss occurred.

RECENT CLIMATIC CHANGES

Not all atmospheric evolution took place billions of years ago. Both Mars and Earth have undergone climatic changes during the last 1 to 10 million years. In both cases the changes were induced by the variations in the shape of the planets' orbits and, in particular, by the changing tilt of their polar axes.

A rapidly rotating planet undergoes shifts in the orientation of its polar axis for the same reason that the tilt of a spinning top changes due to the influence of Earth's gravity. For the top, the axis of rotation *precesses*, or "walks", around the point on which it is balanced. For Mars and Earth, the external force is the gravitational pull of the other planets, Jupiter in particular. Mars takes about 173,000 years to complete its precessional cycle, while Earth takes 25,800 years.

The Earth also undergoes oscillations in its *obliquity*, the tilt of its polar axis with respect to a line perpendicular to its orbit. Currently, Earth's axial obliquity is 23.5°, but this value changes by about ± 1.5° with a period of oscillation of 41,000 years. Remarkably, these slight variations are responsible for Earth's ice ages. During these ice ages our planet is cooler overall, which causes large amounts of water that has evaporated from the oceans to be deposited on land as snow and ice. During the last major glacial epoch, for example, the entire northern part of the United States was covered with ice, and sea level dropped by more than 100 m. We have established the link between the onset of ice ages and Earth's obliquity by noting the ratios of ^{18}O to ^{16}O in deep cores from Greenland and Antarctica that sample ice laid down several hundred thousand years ago. Because the evaporation rates of water containing ^{18}O and ^{16}O are different, and because this difference depends on temperature, the ^{18}O:^{16}O ratios tell us what the globally averaged temperature was when the ice formed. The changes in temperature, we find, correspond to changes in the tilt of the polar axis (*Figure 19*). This forcing of the climate by the changing orientation of the polar axis is known as the Milankovitch theory.

Mars feels the gravitational pull of Jupiter more strongly than Earth does, and it does not have a nearby, large Moon to dampen out its axial oscillations. As a result, the axial obliquity of Mars varies quite a bit from its current value of 25°, ranging between 15° and 35° (*Figure 20*). This oscillation takes about 100,000 years, with the most extreme excursions occurring

every million years. When the obliquity is low, the Sun does not rise as high above the horizon at the poles during summer. As a result, polar temperatures are colder, and more CO_2 and H_2O ice than average freeze out there. When the obliquity is high, the summertime Sun rises high in the polar sky, temperatures there are warmer, and the polar caps lose much of their frost.

One effect of Mars's changing obliquity is seen in the behavior of polar water ice. When water sublimates from the summertime polar cap, it is redistributed globally by the winds, and an annual exchange of water occurs between the northern and southern caps. At present, several tenths of a millimeter of ice

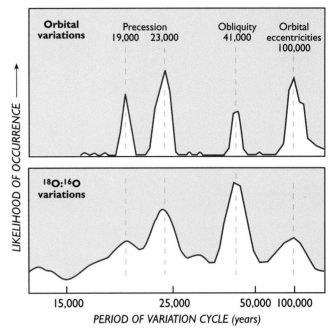

Figure 19. A comparison of Earth's axial obliquity and the ratio of ^{18}O:^{16}O in cores through ice sheets over recent geologic history. The fact that oscillations in the obliquity and the variations in the ^{18}O:^{16}O ratio occur on the same time scales suggests strongly that they are related, most likely by a climatic response to the tilt of Earth's polar axis.

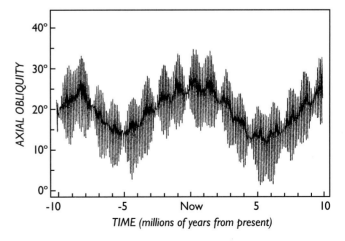

Figure 20. Calculations show that the axial obliquity of Mars oscillates with a period of about 100,000 years, with even wider excursions occurring every million years or so. These swings change the energy balance at the planet's poles, causing water and CO_2 to move into the polar regions and back out again.

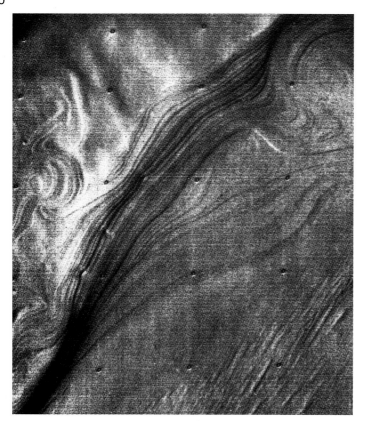

Figure 21. **Much of the Martian polar regions appears to consist of alternating layers of dust and water ice (or dust and ice mixed together in different proportions). These layers probably formed as a result of the changing tilt of the polar axis, and they reflect the behavior of the climate for the last 100 million to 1 billion years.**

may sublimate from the north polar cap during summer. During periods of high obliquity, however, as much as several tens of centimeters may be removed each year. At such times the atmosphere would hold much more water vapor, and the exchange between the caps would be much greater.

Interestingly, layered deposits of ice and dust at the Martian poles appear to show evidence for this quasiperiodic climatic change (*Figure 21*). The individual layers are too thick to be formed in a single year. They probably result instead from the net exchange of water between the two polar caps over 10,000 years or more. Calculations show that the Martian obliquity changes chaotically, meaning that its exact history cannot be determined and that the obliquity may have held any value between about 0° and 60° during the last 10 million years. As a result, our estimates of water and dust migration to and from the polar regions over these periods are uncertain. However, the small number of impact craters seen on these deposits proves that they are relatively young. This implies that substantial amounts of material must have moved into and out of the polar regions relatively recently.

Possible Milankovitch cycles on Venus are of much less interest than on the Earth or Mars. Despite its retrograde rotation, Venus has an axial tilt of less than 3°. More importantly, the Venusian atmosphere is so efficient at spreading solar energy uniformly that the temperatures are very nearly the same everywhere. Thus, changing the tilt of its polar axis should not affect the planetwide distribution of energy.

ATMOSPHERES AND LIFE: PAST, PRESENT, AND FUTURE

One of the significant aspects of the atmospheres of the terrestrial planets is the role that they can play in the origin and subsequent evolution of life. Certainly, Earth's atmosphere is key to understanding the nature of much of the terrestrial biosphere. We assume that the environmental conditions necessary for the origin and continued existence of life include the presence of liquid water (to act as a medium for transport of nutrients and waste products), access to the biogenic elements (especially carbon, hydrogen, oxygen, and nitrogen), and a source of energy. We might imagine that biological systems can plausibly and independently arise on any planet where these conditions are met (see Chapter 27).

Clearly, the early Earth had both liquid water and the biogenic elements. The dominant source of biologically useful energy at that time probably was not sunlight, as is the case today. Instead, it may have been organic molecules created by lightning in the early atmosphere or brought in from space by impacts and accretion. Other possibilities are energy from chemical weathering of surface materials or early hydrothermal systems. This last source may be particularly important, in that it allows for a straightforward conversion of geothermal energy to chemical energy as water circulates through the crust in the vicinity of recently emplaced volcanic rocks.

Our immediate planetary neighbors may well have harbored life sometime in their pasts. Mars appears to have had (or have) all of the necessary ingredients at one time or another. Liquid water certainly has been present at the surface and in the subsurface throughout geologic time. The surface (and presumably the subsurface) has access to all of the biogenic elements. Furthermore, with volcanism and possible chemical weathering of surface minerals, there are abundant sources of energy that have been available throughout time. Venus may have had the necessary environmental conditions for life early in its history, during a clement epoch that may have existed before the hot, thick greenhouse that we see today. Clearly, life on Venus is not likely at this time.

Earth has not always had conditions conducive to the existence of life, however. The violent impact environment that existed during the accretion process precluded its existence. In particular, the impacts of very large protoplanets near the end of planetary formation would have affected the climate in ways that would have annihilated any living organisms. In particular, much of the heat from an impact would have been transferred to the atmosphere, raising its temperature to perhaps several thousand degrees and triggering the wholesale evaporation of the world's oceans. The collision of an object about 450 km across with Earth would provide enough energy to vaporize the entire ocean and raise the global temperatures to such high values. After such an impact, the surface would cool only slowly, taking several thousand years before the water recondensed.

Although the declining collision rate at the end of accretion allowed life to form and to exist continuously, impacts have had a major effect on Earth's atmosphere and biology throughout history. For example, the collision of a 10- to 15-km-wide asteroid with Earth 65 million years ago is widely believed to have led to the extinction of the dinosaurs (along with most other species) and thus had a dramatic effect on the evolution of life. Other impacts throughout time are also thought to have affected life's development on Earth.

It is important to recognize that large impacts continue to occur in the solar system. Sizable asteroids and comets occupy orbits that thread through the inner solar system, and many of these will eventually collide with one of the terrestrial planets. As much as the past history of life on Earth has been influenced by the role of impacts, the future of life here is uncertain in the sense that such events occur in a random fashion that we cannot predict. There is no doubt, however, that they will happen.

Other factors are also affecting the evolution of Earth's atmosphere. Except for the effects of large impacts, our planet's surface layer is now undergoing the most rapid changes in its nature of which we are aware. Humans are adding greenhouse gases to the atmosphere, primarily CO_2. The chlorine derived from our chlorofluorocarbons is depleting the upper atmosphere of its ozone, thereby changing the nature of the sunlight that reaches the surface (*Figure 22*). We are removing forests at a tremendous rate, thereby altering not only the cycling of CO_2 into and out of the atmosphere, but also the amount of sunlight reflected from the ground and the amount of water that enters the atmosphere via transpiration.

There is no scientific consensus yet that our contributions to the atmosphere's greenhouse gases have produced a measurable effect on the surface temperatures. However, scientists generally agree that such an effect will occur. If our production of CO_2 remains unchecked, the atmosphere's abundance of this gas is expected to double by the middle of the next century. Atmospheric models imply that this increase in CO_2 will cause the average surface temperatures worldwide to rise between 2° and 5° K. Obviously, such a sudden warming would have dramatic effects on the nature of the climate, the distribution of rainfall over the surface, and possibly even the circulation of the oceans (*Figure 23*).

As we learn more about the nature of Earth's climate, we are discovering that variations occur on essentially all time scales. Presumably, other ongoing atmospheric changes will in time become pronounced enough to measure. The question will remain, though, whether the changes we observe are natural variations or human-induced effects. As a society, we face substantial moral, ethical, and practical issues about the latter category. Working through these issues successfully during the next century will be one of the major hurdles that will determine whether we as a civilization can endure.

A related issue is whether humans can intentionally alter the climate of the Martian or Venusian atmospheres, a process termed *terraforming*. For example, it may be possible to trigger a significant release of CO_2 into the atmosphere of Mars. If enough CO_2 were available there, the additional greenhouse warming could conceivably be sufficient to raise the surface temperature above the melting point of ice. Of course, even if liquid water could be made to exist, carbonate minerals would start forming again and thus remove CO_2 from the atmosphere relatively quickly. The climatic uncertainties are large enough that we cannot rely on making Mars a "lifeboat" to which humans

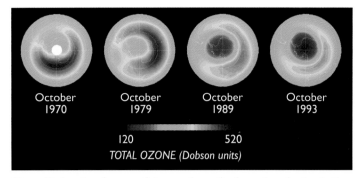

Figure 22. A series of spacecraft images shows the dramatic reduction in the amount of ozone (dark-blue "holes") over the Southern Hemisphere in recent years. Ozone depletion is an example of the subtle ways in which humans can and have influenced the climate around them, often in ways that were not anticipated.

Figure 23. An icon of its era, this photograph by the crew of Apollo 17 serves to remind us that Earth's entire biosphere exists in the thin veneer where land, ocean, and atmosphere meet.

can migrate if the terrestrial climate becomes too harsh to support life. Of course, there are substantial ethical and moral issues connected to terraforming Mars in the first place. Some of these concern whether Mars harbors living organisms today.

Interestingly, the discovery that we can affect the Martian atmosphere comes at the same time as the realization that we are having a substantial effect on the terrestrial atmosphere. Clearly, we cannot cease doing things that affect Earth's atmosphere. Rather, we need to understand what changes we are causing and what their ramifications will be, then modify our behavior in order to keep the Earth livable. Looking back to our goal of understanding Earth and the planets as a whole, then, it is clear that achieving these goals becomes possible only by understanding the present-day atmosphere, its interactions with the terrestrial geology, and where humans and the rest of the biosphere fit in that global system.

Interiors of the Giant Planets

William B. Hubbard

THE GAS-GIANT planets differ greatly from the terrestrial worlds in their mass, size, and chemical makeup. Taken together, Mercury, Venus, Earth, Mars, and the Moon amount to only 2 Earth masses of material — a total that pales when compared to the four giant outer planets. Jupiter alone amounts to 318 Earth masses; Saturn adds another 95, Uranus about 14.5, and Neptune 17.2. The four giant planets likewise greatly surpass Earth in size (*Figure 1*). They can be logically divided into two subclasses, with Jupiter and Saturn forming one like-sized pair, and Uranus and Neptune a second, smaller pair. Our key objective in studying these planets' interiors is to understand how sharp chemical divisions arose between the two types of giant planets as they formed, and how these divisions are expressed through the great differences in mass and size.

A schematic cross-section through a gas-giant planet, from its upper atmosphere to its dark, hot core.

Figure 1. A size comparison of Jupiter, Saturn, Uranus, Neptune, and Earth. The rotational flattening of these planets is also shown to scale. Note how nearly identical Uranus and Neptune are in size.

To understand the nature of these differences, consider the average composition of the Sun — not as it is today, but as it was in the solar system's infancy, before nuclear reactions had begun in the Sun's core. In the outer solar system where the giant planets formed, temperatures of this protosolar matter would have been quite low, between 100° to 200° K.

Under these early circumstances, pairs of hydrogen atoms combined to form molecules of gas. (In the present-day Sun's interior, hydrogen exists as fully ionized protons and electrons at temperatures of many millions of degrees.) After hydrogen, the second most abundant constituent was helium, present in atomic form as a noble gas. Other noble gases such as neon and krypton were also present, but their very low abundances contribute insignificantly to the total mass. By number, the cool nebular gas contained about five hydrogen molecules for every helium atom. By mass, these two elements contributed about 71 and 27 percent, respectively. Thus, hydrogen and helium together account for a little more than 98 percent of the mass of this primordial gas (*Figure 2*).

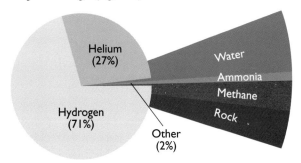

Figure 2. The distribution of molecules, by mass, in the cool gas of the primordial solar nebula. Hydrogen and helium, the most dominant and volatile components, account for 98 percent of the nebula's mass. Yet these gases were never in a condensed state in the primordial solar system. Condensable solid materials account for only 1½ to 2 percent of the solar system's mass.

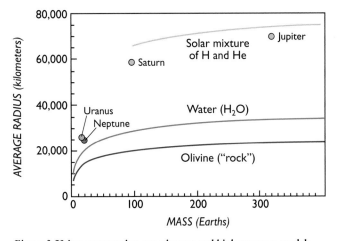

Figure 3. Using compression experiments and high-pressure models, theorists can calculate the size of hypothetical Jovian planets having various masses and compositions. The red curve is the radius-mass relation predicted for a liquid body of solar composition with its interior in equilibrium. Somewhat below the mass of Saturn, this curve actually turns upward, back to *higher* radii, with the details depending on the planet's age. At the bottom are curves for liquid water and olivine (a silicate mineral that is appropriate for "terrestrial" planets). Dots show actual values for the four giant planets.

Other elements present included carbon, nitrogen, and oxygen. Under the temperature and pressure conditions in the outer regions of the primordial disk where the giant planets formed, these three atoms bonded with the readily available hydrogen to form methane (CH_4), ammonia (NH_3), and water (H_2O), respectively. In the language of planet-modeling theorists, these compounds are termed "ices" because within a cold nebula they condense into solids. Together they dominate the 2 percent of the condensable matter from which giant planets could initially form. Some of these molecules probably existed as solids and others as gases in the cold giant-planet zone. Initially, some carbon might even have been found in carbon monoxide (CO) and carbon dioxide (CO_2) rather than methane. The remainder of the planet-building matter, the metal-and-silicate "rock" out of which the terrestrial planets formed, represents only about a quarter of the condensable material — a mere 0.5 percent of the entire nebula.

As we move outward from the Sun, the giant planets prove to be progressively less complete samples of this primordial nebular matter. Jupiter resembles a large aggregation of material close to solar composition, highly compressed but at much lower temperatures than in the Sun. Saturn is similar, except that more of the primordial hydrogen and helium are missing. Uranus and Neptune appear to be primitive samples of the nebula's ice and rock, but they possess only a small fraction of the primordial allotment of hydrogen and helium. In contrast, the planets of the inner solar system retain only the rocky component of the initial solar composition; the gas and ice are almost entirely missing.

We can verify such assertions about bulk composition most directly by comparing the giant planets' radii with their masses. *Figure 3*'s lowest curve corresponds to hypothetical worlds composed of pure olivine (a magnesium-iron silicate representing a planet's "rock" component). This is the curve that would apply to bodies with Earthlike compositions but masses comparable to the Jovian planets'. Such giant Earths, which do not exist in our solar system, would be much smaller than Jupiter and its siblings. The wide separation between the curves for solar composition and those for other abundant materials shows that Jupiter and Saturn are indeed composed mostly of hydrogen and helium. Conversely, Uranus and Neptune must contain little hydrogen or helium, for their radii are much too small to be consistent with solar composition. Instead, they lie only slightly above the curve for water, and it is not unreasonable to suppose that water contributes substantially to their content.

Intuition suggests that as planets grow more massive, they should also get substantially larger. The increase in size with mass is actually quite gradual for huge objects like the giant planets. Specifically, although Jupiter has more than three times the mass of Saturn, it is only slightly bigger. This behavior is due to the very great compression at work deep in these planets' interiors. Pressures at their cores, in excess of 10 million bars, cause the molecular and atomic structure of matter to break down, and the compression curve "softens." Thus, as more matter is added to a large planet, its interior begins to collapse somewhat at the atomic level and its growth in radius slows. As we shall see, eventually size actually *decreases* as the mass increases, such that there exists a maximum radius for a cold body of a given composition. For an object with solar abundances this maximum is

Dimensions of the Giant Planets

	Equatorial radius, r_e (km)	Polar radius, r_p (km)	Oblateness $(r_e - r_p)/r_e$
Jupiter	71,492	66,854	0.0649
Saturn	60,268	54,364	0.0980
Uranus	25,559	24,973	0.0229
Neptune	24,766	24,342	0.0171

Table 1. Since the giant planets have no solid surfaces, their radii are given here at the 1-bar pressure level. Jupiter and Saturn are sufficiently oblate that their disks appear distinctly "flattened" when viewed through a telescope (see Figure 1).

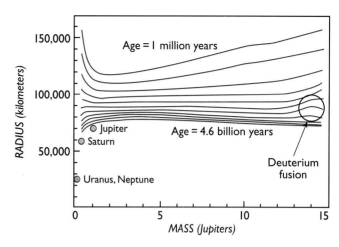

Figure 4. The evolution of giant planets and brown dwarfs of up to 15 Jupiter masses. As a brown dwarf ages, it briefly flares in radius (circle at right) as its deuterium fuses with protons to yield atoms of helium-3. However, lower-mass objects like Jupiter and Saturn retain their deuterium because the pressures inside them are not high enough to trigger the fusion reaction.

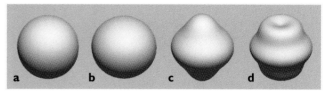

Figure 5. The ways that a planet changes shape when it rotates provide information about its interior structure. These responses can be quantified by numerical values called harmonic coefficients. When not rotating *(a)*, a planet will be perfectly spherical. Saturn spins very rapidly, and its distortion due to the harmonic coefficient J_2 is shown approximately to scale in *(b)*. Saturn's distortion corresponding to J_4 is exaggerated by about 10 times in *(c)*, and its distortion corresponding to J_6 is exaggerated by about 100 times in *(d)*.

about 80,000 km, reached at a mass about four to five times that of Jupiter. (The Sun exceeds this maximum radius by nine times and contains 1,000 times the mass of Jupiter. However, the radius-limiting relationship does not apply to the Sun, which is inflated primarily by its very hot interior.)

In theory, therefore, Jupiter is not the largest possible planet. Astronomers have long looked beyond our solar system for *brown dwarfs*, which are something akin to "super-Jupiters." Collapsing protostellar clouds of less than about 80 Jupiter masses never get hot enough in their centers to initiate hydrogen fusion — they never become stars. Instead, we suspect that these massive, substellar objects have evolutionary histories similar to those of Jupiter and Saturn. At present, several brown dwarfs have been detected in binary systems by means of the gravitational pull they have on their primary stars. Only one of these, Gliese 229B, has been directly imaged and analyzed with enough detail to confirm its similarities to Jupiter. Gliese 229B has roughly 40 times more mass than Jupiter yet is very similar to it in radius.

The interplay of a planet's mass, size, and age can be subtle (*Figure 4*). For example, soon after forming, Saturn would have been much *larger* than Jupiter because of the expansion of gaseous hydrogen in Saturn's weaker gravity. (By extrapolation, hydrogen-rich planets much less massive than Saturn would very likely have been too "fluffy" to persist in the early solar system.) After 4.6 billion years of evolution, cold hydrogen-helium objects about four times the mass of Jupiter have the largest size. For objects of 13 or 14 Jupiter masses, a brief hiatus in contraction occurs when a thermonuclear energy source comes into play: the fusion of deuterium with protons to form helium-3. This vestige of starlike behavior would have occurred in Gliese 229B long ago, and consequently we do not anticipate finding deuterium in its atmosphere. By contrast, Jupiter retains essentially all the deuterium that was present in the nebula from which it formed. Astronomers have begun to use 13 Jupiter masses as a convenient criterion for deciding whether an object should be classified as a giant (Jovian) planet or as a more starlike brown dwarf.

OBLATENESS AND THE INTERIOR

Radius-mass relationships may give important clues to the bulk compositions of the giant planets, but they do not tell us whether dense cores are present, nor do they reveal interior structures in more detail. Such information comes not from direct observations of the planetary interiors, but from the way these worlds accelerate natural and artificial satellites in their vicinity. From a gravitational standpoint, a perfectly spherical planet would act as if all its mass were concentrated in a single, central point. In such cases the motions of a nearby satellite or spacecraft would obey the classical Keplerian laws of motion.

However, to be perfectly spherical (*Figure 5a*), a planet must not rotate and its interior must be in a state of *hydrostatic equilibrium*, in which all forces and pressures are in balance. Once the planet begins to rotate, its shape and the shape of its external gravitational potential, or field, deform and become oblate (*Figure 5b*). Astronomers express the extent of the oblateness as a value, J_2, called a *harmonic coefficient*.

Since all of the giant planets rotate very fast, more rapidly than the Earth does, they are decidedly *not* spherical. Their liquid interiors respond readily to the strong centrifugal forces induced by rotation, producing substantial and rather obvious equatorial bulges (*Table 1*). The size of the bulge, when compared to the magnitude of the rotational perturbation that causes it, gives us insights into the distribution of density in the planet's interior.

In addition to the basic response J_2, rotation induces higher-order perturbations on the planet's gravitational potential, which are designated J_4, J_6, and so on. These progressively more-delicate distortions reflect the distribution of mass and density in the planet's outer layers, but they can only be detected and measured very near the planet's surface. The close flybys made by the Pioneer and Voyager spacecraft have provided such data, as have a number of elegant studies of the motions of the rings of Saturn and Uranus. To date, the values of J_2, J_4, and J_6 have been determined for Saturn, and the planet's true figure is derived from the sum of these distortions.

Analyzing the harmonic components of the giant planets' gravity fields helps us to understand the relationships between density and pressure in their interiors. To interpret these relationships properly, we must also understand the behavior of one of the main constituents, hydrogen, at various combinations of temperature and pressure (*Figure 6*).

In the observable regions of giant-planet atmospheres, at a pressure of about 1 bar, hydrogen forms a molecular gas at temperatures ranging from 165° K (Jupiter) to 134° K (Saturn) to 76° K (Uranus). Deeper down, where pressures are higher, the gas increases in temperature and density according to the thermodynamic law for *adiabatic compression* (compression of a gas without the loss or gain of heat). Still deeper, at pressures exceeding 100,000 bars, the gas begins to resemble a hot liquid. The transition is gradual because temperatures everywhere are well above hydrogen's critical point at 13 bars and 33° K. If the atmospheres of the giant planets were cooler than 33°, the transition would be distinct, with gaseous layers above the 13-bar pressure level and liquid layers below.

About 10,000 km below Jupiter's cloud tops, liquid hydrogen reaches a pressure of 1 million bars (1 megabar) and a temperature of about 6,000° K (roughly equal to that of the Sun's photosphere). At these values, hydrogen's molecular and atomic bonds begin to break. This process yields an entirely new phase: liquid metallic hydrogen. It consists of ionized protons and electrons, as in the Sun's interior. However, the temperature of matter with the density inside Jupiter is about one-thousandth what it would be inside the Sun. Thus, the hydrogen acts not at all like a gas, but more like molten metal. This exotic form of hydrogen, like other metals, is both an electrical conductor and opaque to visible radiation. The strong magnetic fields of Jupiter and Saturn are apparently generated by electrical currents coursing through this material. Modern laboratory experiments, which use intense reverberating shock waves to squeeze hydrogen into a metallic state, have confirmed that most of the interior of Jupiter must be in the form of liquid metallic hydrogen.

Saturn is less massive than Jupiter, so its internal pressures are not as great. Even so, Saturn should also possess liquid metallic hydrogen in its deep interior. Uranus and Neptune, on the other hand, can contain only a small fraction of hydrogen, roughly 15 percent by mass (*Figure 3*), which is mostly limited to their outer layers. Thus, hot liquid hydrogen probably exists in these planets, but it is confined to the molecular phase (with the possible exception of small pockets of liquid metallic hydrogen buried deep within the "ice" layers).

Helium is an important tracer of internal processes in giant planets, because it is unlikely to have been separated from hydrogen as these worlds formed. However, the abundances of helium in the giant planets' atmospheres are not all the same and not all equal to the solar value. The Voyager spacecraft determined that Saturn's atmosphere has about one-fourth the expected abundance of helium, while Jupiter and Uranus's atmospheric helium fractions roughly equal the solar value. Prior to the Voyager flybys it was predicted that, at sufficiently low temperatures, helium atoms become almost insoluble in metallic hydrogen. We think that the critical temperature for the onset of this hydrogen-helium separation may be close to that in the interior of Saturn. However, accurate calculations are difficult, because while hydrogen is ionized in the deep interiors of Jupiter and Saturn, helium is not. In theory, high-pressure interactions between bare (fully ionized) hydrogen and helium nuclei should lead to phase separation. Our uncertainty concerns whether this relatively straightforward model can be applied to the real interactions between ionized hydrogen and neutral helium in Saturn's interior. If separation *has* occurred inside Saturn, the helium may have migrated into the metallic-hydrogen core. The process by which helium might sink deeper into Saturn is thought to resemble terrestrial virga rain showers in desert regions: helium droplets form where the metallic hydrogen is cool enough, "rain" out, and then redissolve at deeper, hotter levels.

The theorized cross-sections of giant planets in our solar system and elsewhere (*Figure 7*) derive from our knowledge of compression curves (*Figures 3,4*), the gravitational harmonics of the four giant planets in our solar system (*Figure 5*), and the phase relationships of hydrogen (*Figure 6*). The cross-section for Jupiter shows the estimated location of the transition from molecular to metallic hydrogen, as well as a possible rock or rock-ice core of up to 10 Earth masses. The Galileo probe, which measured the atmospheric composition to a pressure of 20 bars, showed that Jupiter's envelope contains substantial amounts of other elements in addition to helium — more than would be the case for strictly solar composition. This is also indicated by models based on the gravity harmonics. Saturn,

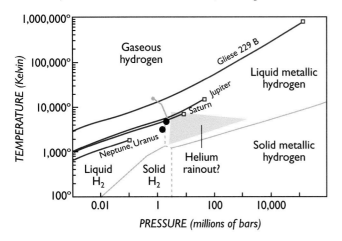

Figure 6. **When subjected to great pressure in laboratory experiments, hydrogen (black dots) shows evidence of metallic behavior. Those results have helped physicists in constructing a phase diagram (tan lines) that delineates the domains of liquid metallic and molecular hydrogen, the main components in Jupiter and Saturn. In the helium "rainout" zone, a solar-composition mixture of metallic hydrogen and helium cannot exist in equilibrium. Also plotted are approximate interior temperature profiles for the Jovian planets and for the brown dwarf Gliese 229B.**

with about one-third the mass of Jupiter, appears to have a larger core. Pressures sufficient to convert hydrogen into its metallic form are not achieved until almost midway to its center.

Interestingly, possible core masses deduced for Jupiter and Saturn are comparable to the *total* masses of Uranus or Neptune. The latter two planets do appear to be mainly "cores," composed of rock and ice in roughly solar proportions with rather indistinct layering. Both are topped by hydrogen-rich atmospheres accounting for about 1 to 2 Earth masses. Hydrogen never reaches pressures greater than about 100,000 bars and thus remains in nonmetallic, molecular form in the outer 5,000 km of both planets. Below this low-mass hydrogen envelope, both Uranus and Neptune appear to consist of a hot, highly compressed liquid with a water-dominated "ice" composition. (Although the term "ice" denotes a combination of water, methane, and ammonia under the high temperatures and pressures deep within a giant planet, this mixture actually will be a hot, liquid soup of various chemical species derived from these molecules.) This interior region makes up the bulk of both planets, up to 10 to 15 Earth masses in each case.

With a mass about half of Jupiter's, the extrasolar giant planet 51 Pegasi B orbits much closer to its central star than Mercury's distance from the Sun. As a result, its outer hydrogen layers are heated to temperatures much greater than 1,000° K. We do not know if 51 Pegasi B is a hydrogen-rich body like Jupiter and Saturn; instead it could conceivably be a giant rocky planet corresponding to the bottom curve in Figure 3. If it is Jupiter-like, however, heat from the star that it orbits, 51 Pegasi A, has probably caused the outer layers of this object to expand well beyond the size of Jupiter (*Figure 7*).

HEAT FROM WITHIN

All planetary bodies generate *some* heat (thermal-infrared energy) deep in their interiors. For the small terrestrial planets, this energy comes mostly from the slow decay of radioactive isotopes incorporated when the planets formed. Such a modest upwelling of heat would be difficult to detect from space, because it is overwhelmed by sunlight's warming effect on inner-planet surfaces.

In the outer solar system, the situation is different. There solar heating contributes much less to a planet's atmospheric

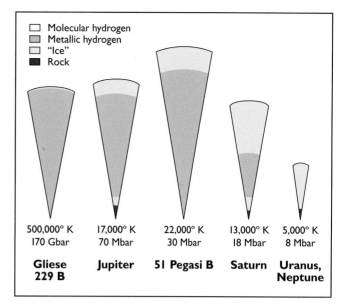

Figure 7. The cross-sections of solar system's largest planets, the extrasolar planet 51 Pegasi B, and the brown dwarf Gliese 229B, with mass decreasing toward the right. As is evident here and in Figure 4, the most massive objects are not the largest ones. The atmosphere of Gliese 229B, like Jupiter's, contains substantial amounts of methane and water, yet liquid metallic hydrogen lies only about 200 km below its surface. Note the wide range of estimates for these objects' core temperatures and pressures.

energy budget (*Figure 8*). Consequently, heat driven to the surface from the deep interior is relatively easy to detect. Jupiter's intrinsic infrared glow was first discovered by Frank Low in 1966, and subsequent observations from aircraft, ground-based telescopes, and spacecraft have yielded progressively better heat-flow data for all four giant planets. After taking careful account of the fraction of solar energy absorbed in the planet's atmosphere and then reradiated into space, observers concluded that Jupiter, Saturn, and Neptune have detectable internal sources of heat. This has proved to be one of the most interesting revelations of modern planetary science.

Some clues to the origin of the giant-planet heat flows can be obtained by computing the specific luminosity of each planet, that is, the average power released per unit of planetary mass (*Figure 9*). The luminosity of the Sun, powered by the ther-

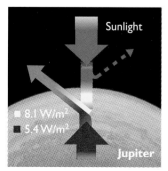
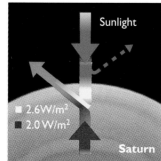
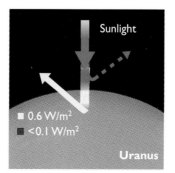
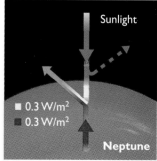

Figure 8. Heat flow in the Jovian planets. The blue rays denote incident solar energy, some of which (dashed) is scattered back into space by the planet's atmosphere. The remaining fraction is absorbed and converted into infrared energy (yellow) deeper in the atmosphere. Together with the infrared energy flowing out from the planet's deep interior (red), this is the heat observed escaping at the surface of each planet. No interior heat has been detected coming from Uranus, even though Voyager 2 flew by at close range in 1986. However, it must surely be present because of radioactive decay in the planet's presumably chondritic (rocky) core.

monuclear fusion of hydrogen, is enormous compared with any other solar-system object. At the other extreme are primitive rocky bodies, such as the carbonaceous chondrite meteorites, whose internal heat comes only from the radioactive decay of what little uranium, potassium, and thorium they contain. The Earth's specific luminosity lies very close to the carbonaceous-chondrite value, suggesting that most of its internal heat derives from radioactive decay.

Among the giant planets, only anomalous Uranus lies close to the very low power-to-mass ratio of carbonaceous chondrites. Jupiter, Saturn, and Neptune release far more energy per kilogram of mass, so they must derive their luminosities from something other than radioactive decay. The thermonuclear fusion of

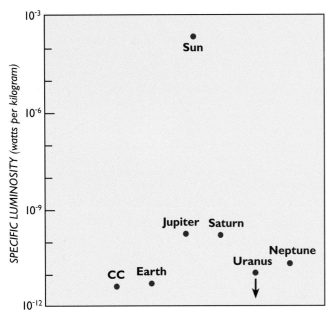

Figure 9. With the exception of Uranus, the Jovian planets emit more energy per unit of mass than other solar-system bodies do. The value labeled *CC* (carbonaceous chondrite) is that expected for a body of approximately chondritic (rocky) composition.

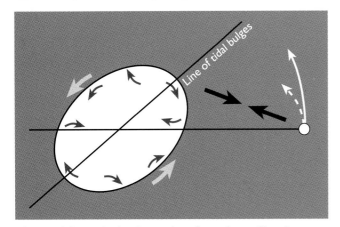

Figure 10. The gravitational attraction of a nearby satellite raises symmetric tidal bulges on a giant planet. Because the planet is rotating with respect to the satellite, fluid currents (red arrows) occur in its interior. This motion results in a net loss of energy, as dissipated heat, that very gradually slows the planet's rotation. A fraction of this energy also serves to speed up the satellite, thus moving its orbit outward. (The bulges' size and angular displacement, or phase lag, from the planet-satellite line are shown much exaggerated.)

hydrogen, which powers the Sun, is not possible inside the giant planets because their calculated central temperatures are nowhere near the 10,000,000° K required for hydrogen fusion to occur, or even the 500,000° K needed for deuterium fusion.

Therefore, by elimination, only one process could be responsible for the luminosities of Jupiter, Saturn, and Neptune. Energy is liberated when mass in a gravitationally bound object sinks closer to the center of attraction, that is, when the object becomes more centrally condensed. In effect, potential energy becomes kinetic energy. This process causes the temperature within a gaseous body to rise. Gaseous stars like the Sun become so hot as they contract that hydrogen fusion begins to occur. At that point they cannot radiate the accumulating heat into space fast enough and the contraction stops. In liquid bodies (giant planets), or even solid bodies (terrestrial planets), pressures depend only slightly on temperature, so they contract as they slowly cool off. Giant planets and brown dwarfs evolve in this manner, slowly radiating heat to space. This slow, self-regulated coupling of contraction and cooling, called the *Kelvin-Helmholtz mechanism*, probably represents the last phase of the giant planets' violent, high-temperature origin. By itself, the Kelvin-Helmholtz mechanism may be insufficient to explain Saturn's luminosity. The extra energy may be coming from the sinking of dense components (helium, or perhaps ice) toward the planet's center.

Although the escape of primordial heat from Jupiter, Saturn, and Neptune is very slow, it appears to be enough to destabilize the hydrogen layers throughout much of their interiors. This causes the hot, liquid hydrogen to convect at the rate of a few centimeters per second. Thus, the convective overturn seen at cloud tops of Jupiter and Saturn probably has deeper "roots" and may extend virtually to these planets' centers!

Deep inside Jupiter and Saturn, convection creates and sustains powerful dynamos in the metallic-hydrogen zones, which in turn generate immense magnetic fields (see Chapter 4). Uranus appears to lack the heat flow required to induce such deep-seated convection, so perhaps it is no coincidence that the planet's enigmatic magnetic field differs markedly in its geometry from those of Jupiter and Saturn. Researchers have attempted to simulate the conditions inside Uranus and Neptune through shock-compression experiments on water and other materials considered likely to exist in the planets' interiors. Their simulations show that high pressure and temperature cause a modest increase in the materials' electrical conductivity. Therefore, even though neither Uranus nor Neptune has a metallic-hydrogen layer, slow convection within their interior fluids may be reason enough for their global magnetic fields observed by Voyager 2.

One clue to dynamic processes at work inside a giant planet comes, oddly enough, from noting how the orbits of its satellites evolve over time. The connection is as follows: A moon's gravitational field creates small, periodic tidal distortions in the shape of its parent body (as the Moon does on Earth). The planet's tidal bulges are not directly "under" the satellite, because all the giant planets rotate faster than their major moons orbit them (*Figure 10*). As the bulges slide backward in longitude to remain with the satellite, they encounter resistance and generate "friction" in the form of fluid currents in the planet's interior. Consequently, a displacement or phase lag develops between where the bulges should be and where they are, and its magnitude is proportional to the amount of energy being dissipated in the internal currents.

Meanwhile, the planet's altered shape (and thus its gravity field) can, over a long time, alter the motions of nearby satellites. If Jupiter were an ideal fluid body, it would have no internal dissipation of energy. The tidal bulges induced by Io would have no phase lag, Jupiter's corresponding effect on the orbit of Io would be zero, and we would be unlikely to see volcanism on that satellite today (see Chapter 17). However, given the heat now seen coming out of Io, we can calculate the rate of tidal evolution of its orbit and, working backward, set some interesting limits on Jupiter's effective internal viscosity. It turns out that the fraction of tidal energy converted into heat in Jupiter's fluid interior is several hundred times less than the corresponding fraction of the tidal energy generated in Earth's interior by the Moon. The situation is much the same for the other Jovian planets.

The interesting question is not why the Jovian planets dissipate tidal energy at such a low rate (after all, they have liquid interiors), but rather why the dissipation is so *large*. We expect the fluid viscosities within the giant planets to be similar to that of water at room temperature. Such values are many orders of magnitude too small to account for the tidal effects we observe among giant planets and their satellites. Apparently, some additional source of dissipation is involved. Perhaps sluggish phase transitions in the planet's deep interior are responsible for the required tidal phase shift. However, as yet we have little clue as to where inside Jupiter this might be taking place.

MAKING GIANT PLANETS

Many aspects of the process of forming giant planets remain unclear, but a coherent picture has started to emerge as we learn more about their interiors. Because Jupiter and Saturn contain so much hydrogen and helium, they must have formed when the entire solar system was still enveloped in these gases. Astronomers who observe very young solar-type stars estimate that their gaseous envelopes start to dissipate after about 10 million years. Because the solar system contracted out of a rotating cloud, angular momentum spread the shrinking system into a disk. In the cooler parts of the disk, away from the primordial Sun, solid particles aggregated, settled into the plane of the disk, and formed asteroid-sized *planetesimals*. Outside a "frost line" at a temperature of about 170° K, not very far from the present orbit of Jupiter, the solid particles became much more massive because they were cold enough to incorporate the abundant water ice in their vicinity (*Figure 11*). Farther out, at still lower temperatures, additional ice compounds may have been incorporated in the growing solids.

According to Hiroshi Mizuno's model for formation of the Jovian planets, the icy planetesimals accreted into "trigger" nuclei for eventual Jovian planets. Once a nucleus had grown to 10 or 20 times the mass of the Earth, the gas in its vicinity could no longer remain dispersed in the primordial nebular disk. Instead, the gas succumbed to the gravitational pull of the massive nucleus and was drawn into the growing protoplanet (*Figure 12*). Such captured gaseous envelopes would be hot by virtue of their sudden cascade into the primordial planets, thus providing a source for the Kelvin-Helmholtz cooling of Jupiter and Saturn that is feebly observable today. The rocky remnants of the

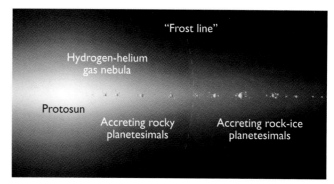

Figure 11. This is how the primordial solar system might have looked (in cross section) at an age of about 10 million years. Rock-ice planetesimals were forming in the outer solar system just before the hydrogen envelope that surrounded the protosun and its planetary disk dissipated.

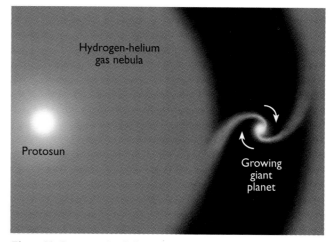

Figure 12. Cosmogonists believe the Jovian planets assumed their immense size when nebular gas collapsed onto rock-ice cores of perhaps 10 or 20 Earth masses. The growth of the cores, and the subsequent sweeping up of gas in their vicinity, must have occurred before the primordial nebula dissipated — an interval of only 1 to 10 million years.

trigger nuclei may form the dense cores at these planets' centers.

Somewhat farther from the infant Sun, at the orbits of Uranus and Neptune, the situation became a race between the accretion of the trigger nuclei and the dissipation of the gaseous nebula. Solid particles were more dispersed out there and orbital periods much longer, so it took longer to aggregate a sizable nucleus. Apparently, the nuclei lost the race, and little gas was left to capture by the time Uranus and Neptune had grown to appreciable sizes.

As appealing as this picture may seem, many details remain unexplained. The formation of a giant planet was probably not such a well-defined two-stage process. Large, icy bodies undoubtedly continued to splash into a protoplanet's gaseous envelope after it drew together, and a certain amount of nebular gas was probably entrained initially during formation of the trigger nucleus. In addition, recent discoveries of extrasolar giant planets such as 51 Pegasi B, which orbit their primaries far inside the frost line, testify to possible incompleteness in our picture of how giant planets form and possibly migrate.

Chemical reactions occurring deep in the interior are poorly understood. These reactions, acting over the age of the solar

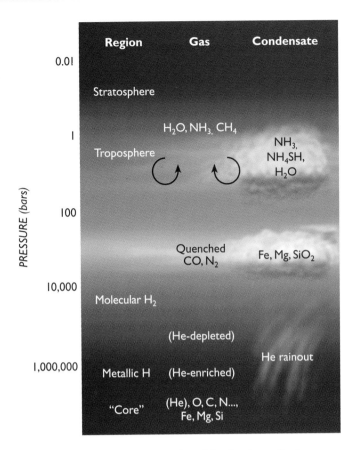

Region	Gas	Condensate
	0.01	

Stratosphere

Troposphere — H_2O, NH_3, CH_4 — NH_3, NH_4SH, H_2O

Quenched CO, N_2 — Fe, Mg, SiO_2

Molecular H_2

(He-depleted)

He rainout

Metallic H — (He-enriched)

"Core" — (He), O, C, N..., Fe, Mg, Si

PRESSURE (bars): 0.01, 1, 100, 10,000, 1,000,000

Figure 13. **A schematic representation of the chemistry of a giant planet, based on calculations by Jonathan Lunine. The dominant species is hydrogen, which exists in molecular form from a pressure level of about 1 million bars up into the stratosphere. At temperatures colder than about 1,000° K, chemical reactions can take longer to reach equilibrium than the time scales for local convective mixing. Thus, "quenched" species that are ordinarily stable only at high temperatures and pressures (such as CO and N$_2$) can be carried upward and become observable in the upper atmosphere.**

system, may have altered the atmospheric composition. For example, we do not yet understand why an excess of methane now exists in all giant-planet atmospheres. Molecules such as carbon monoxide and carbon dioxide have been observed in the upper atmosphere of Jupiter yet are highly unstable there. According to recently developed models, these molecules apparently were created at deeper levels under the influence of higher temperature and pressure, then were carried upward by convection into the observable layers (*Figure 13*). Some have speculated that methane gas, abundant in the gas giants' atmospheres, may decompose at the high temperatures and pressures found deeper down — yielding diamond grains! If these grains sink to the center of the planet, some mechanism (perhaps cometary infall) must replenish the methane over the planet's lifetime. Recent simulations of the high-pressure behavior of methane suggest that other hydrocarbons may be synthesized deep in Uranus and Neptune.

Decades of observation and laboratory simulation have provided important insights into the intricate relations between the molecular chemistry in the giant planets' atmospheres and the composition of their interiors. We now believe that a giant planet is unlikely to maintain a uniform composition throughout its interior. Its atmosphere, envelope, and core are roughly analogous to the crust, mantle, and core of a terrestrial planet, and each may have been similarly modified by chemical processes over geologic time.

Today our insights and assumptions are being severely tested by the discovery of other solar systems. Astronomers can now identify distant relatives of the giant planets either by the faint infrared glows of heat released by their slow contraction, or by their gravitational influence on companion stars. Not only are objects larger than Jupiter being found, but in many cases they lie unexpectedly close to their parent stars (see Chapter 28). The results of these searches, though challenging, continue to give important clues to the physical conditions needed to form the largest planets.

Atmospheres
of the Giant Planets

Andrew P. Ingersoll

Neptune and its Great Dark Spot, as recorded by Voyager 2 in 1989.

THE GIANT PLANETS — Jupiter, Saturn, Uranus, and Neptune — are fluid objects. They have no solid surfaces because the light elements constituting them do not condense at solar-system temperatures. Instead, their deep atmospheres grade downward until the distinction between gas and liquid becomes meaningless. The preceding chapter delved into the hot, dark interiors of the Jovian planets. This one focuses on their atmospheres, especially the observable layers from the base of the clouds to the edge of space. These veneers are only a few hundred kilometers thick, less than one percent of each planet's radius, but they exhibit an incredible variety of dynamic phenomena.

The mixtures of elements in these outer layers resemble a cooled-down piece of the Sun. Clouds precipitate out of this gaseous soup in a variety of colors. The cloud patterns are organized by winds, which are powered by heat derived from sunlight (as on Earth) and by internal heat left over from planetary formation. Thus the atmospheres of the Jovian planets are distinctly different both compositionally and dynamically from those of the terrestrial planets. Such differences make them fascinating objects for study, providing clues about the origin and evolution of the planets and the formation of the solar system.

Naturally, atmospheric scientists are interested to see how well the principles of our field apply beyond the Earth. For example, the Jovian planets are ringed by multiple cloud bands that move quickly, yet somehow remain fixed in latitude. On Earth such bands are disrupted by large transient storms at temperate latitudes and by long pressure waves that are anchored to the continents. Storms on the giant planets can endure for years or centuries, whereas on Earth they last for days or weeks.

Latitudinal banding is rather obvious on Jupiter and Saturn (*Figures 1,2*), which are heated by the Sun most strongly around their equators. On Uranus the heating patterns are dif-

202

Figure 1. Strong east-west winds create a series of latitude-fixed bands on giant Jupiter, seen here as it appeared in February 1979 when imaged by Voyager 1 from a distance of 33 million km. The planet's Great Red Spot, below center, has intrigued astronomers for more than a century.

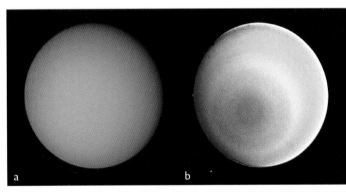

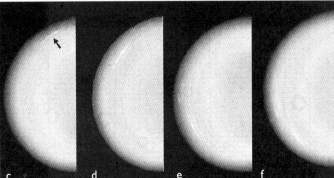

Figure 3. Voyager 2 views of Uranus. The color in *a* has been adjusted to simulate the view the eye would normally see. Features as small as 160 km would be visible — if Uranus had them. In *b*, computer processing has greatly exaggerated both the color and contrast; small doughnut-shaped features are artifacts caused by dust on the camera optics. In *c*, an arrow marks a convective plume at latitude -35° seen rotating around the disk. It is visible only because contrast has been greatly exaggerated.

ferent because the planet effectively spins on its side. During the Uranian year the Sun shines down on one pole then moves over the other, depositing more energy on average in the polar regions than at the equator. Nonetheless, latitudinal banding does exist, albeit weakly, making Uranus look like a tipped-over version of Jupiter and Saturn (*Figure 3*). Neptune also is banded (*Figure 4*). In fact, despite Neptune's great distance from the Sun and lower energy input, its winds are three times stronger than Jupiter's and nine times stronger than Earth's.

Chemistry provides another set of challenges for atmospheric scientists. On Earth, the composition of the air is determined largely by reactions with the solid crust, the oceans, and the biosphere. Even before the advent of life, photodissociation (the splitting of molecules by the ultraviolet component of sunlight) and the escape of hydrogen into space were causing our atmos-

Figure 2. Voyager 2 recorded Saturn, its rings, and four major satellites from a distance of 21 million km in July 1981. Note that cloud bands on Saturn are fewer and lower in contrast than on Jupiter.

Figure 4. Neptune as seen by Voyager 2. The largest discrete feature, termed the Great Dark Spot, is accompanied along its southern edge by bright, high-altitude clouds of methane ice. As on Uranus, the blue color of Neptune is due in large part to the absorption of red light by methane gas.

phere to evolve. The Jovian planets have neither crust nor oceans nor life, and their gravitational fields are so strong that they retain all elements including hydrogen. Yet unstable compounds do form, by a variety of processes that include rapid ascent from the hot interior, photodissociation, condensation (cloud formation), electrical discharge (lightning; see *Figure 5*), and charged-particle bombardment (auroras; see *Figure 6*). As is evident from their multicolored clouds and abundant molecular species, the atmospheric chemistry of the Jovian planets is just as rich as Earth's.

With their internal heat and fluid interiors, the giant planets can be compared with the Sun and stars. On the other hand, we can relate their atmospheric phenomena to those on Earth. Our understanding of terrestrial atmospheric phenomena was inadequate to prepare us for the revelations made in recent years about the giant planets, but by testing our theories against these observations, we can understand the Earth in a broader context.

OBSERVATIONS

Our knowledge of the Jovian planets is derived from Earth-based telescopic observations begun more than 300 years ago, from modern observations from high-altitude aircraft and orbiting satellites, and from the wealth of data returned by interplanetary spacecraft. Pioneers 10 and 11 flew by Jupiter in the early 1970s, followed by Voyagers 1 and 2 in 1979; three of these eventually reached Saturn. Voyager 2 continued past Uranus in 1986 and Neptune in 1989. Comet Shoemaker-Levy 9 became an uninstrumented probe of Jupiter's atmosphere in 1994, preceding the arrival of the Galileo orbiter-probe spacecraft 17 months later.

Basic characteristics like mass, radius, density, and rotational flattening were determined during the first era of telescopic observation. Galileo's early views revealed the four large Jovian satellites that now bear his name. Newton estimated the mass and density of Jupiter from observations of those satellites' orbits. Others, using ever-improving optics, began to perceive atmospheric features on the planet. The most prominent of these, the Great Red Spot, can be traced back more than 150 years, and it may be older still.

Beginning in the late 19th century, astronomers made systematic measurements of Jovian and Saturnian winds by tracking features visually with small telescopes and, later, with photographs made through larger ones. Most recently, tens of thousands of features on Jupiter and Saturn have been tracked with great precision using the Voyager and Galileo imaging systems (*Figure 7*). The constancy of these currents over the centuries spanned by classical and modern observations is one of the truly remarkable aspects of Jovian and Saturnian meteorology. Uranus, on the other hand, is nearly featureless to the eye. Although the planet was discovered in 1781, the first definitive evidence for markings came from Voyager 2's images. Neptune is more photogenic than Uranus and exhibits faint markings that are visible from Earth with modern infrared detectors.

Astronomers began to decipher these planets' atmospheric compositions in the 1930s with the identification of absorptions by methane (CH_4) and ammonia (NH_3) in the spectra of sunlight reflected from their banded clouds. The detection of molecular hydrogen (H_2) followed around 1960. From these data,

observers verified that the proportions of hydrogen, carbon, and nitrogen in Jupiter's atmosphere were roughly consistent with a mixture of solar composition (see *Table 1*). Similar inferences have been made for Saturn, though observations of the compounds that contain these elements (like ammonia) are extremely difficult. The Galileo probe confirmed the spectroscopic estimate for C:H on Jupiter and detected several

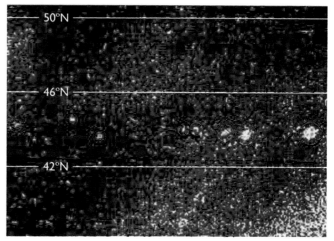

Figure 5. Enormous bursts of lightning (circled patches) were recorded on Jupiter's night side in November 1996 by the Galileo spacecraft. The view is 20,000 km wide. Clouds in the planet's atmosphere are faintly visible as well, illuminated by light from the moon Io. The individual flashes are hundreds of kilometers across and thus dwarf anything comparable on Earth. They lie just below a westward-moving jet at 46° north. Almost all the Jovian lightning seen by Voyagers 1 and 2 also occurred near the latitude of a westward-moving jet.

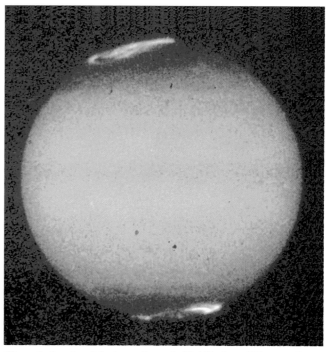

Figure 6. An ultraviolet image from the Hubble Space Telescope shows bright auroral rings encircling the north and south poles of Jupiter. They appear offset from the poles because Jupiter's magnetic field is offset from its spin axis by about 10°. The auroral emissions, which are situated about 500 km above the 1-bar pressure level, are caused by charged particles traveling toward the planet along field lines and striking the upper atmosphere.

Compositions of Outer-Planet Atmospheres

Molecule	Sun	Jupiter	Saturn	Uranus	Neptune
H_2	84	86.4	97	83	79
He	16	13.6	3	15	18
H_2O	0.15	(0.1)	—	—	—
CH_4	0.07	0.21	0.2	2	3
NH_3	0.02	0.07	0.03	—	—
H_2S	0.003	0.008	—	—	—

Table 1. Percentage abundances (by number of molecules) for the giant planets' atmospheres below the clouds. Listed are six of the 11 most abundant elements in the Sun (and the universe). The others are the noble gases neon (Ne) and argon (Ar), and the metals silicon (Si), magnesium (Mg), and iron (Fe), all of which are believed to reside in these planets' cores. Here *Sun* refers to a mixture with the same proportion of elements as the primordial Sun, but cooled to planetary temperatures so that the elements combine to form the compounds listed. Values for Jupiter are from the Galileo probe; others are from remote sensing. All numbers are uncertain in the least significant figure and more so for Jupiter's water content. Dashes indicate unobserved compounds.

elements (sulfur, neon, argon) and isotopes (of carbon, helium, and hydrogen) for the first time.

We study the uppermost atmospheres of these worlds using three main techniques. First, when a spacecraft appears to pass behind a planet (an occultation), electrons in the planet's ionosphere affect the craft's radio signal en route to Earth; the change of electron density with altitude is a measure of temperature. Second, the ultraviolet spectrum of sunlight and starlight that has passed through the upper atmospheres contains information about temperature and composition. And third, by observing excitation processes like auroras (*Figure 6*) and airglows, we learn about the energy sources for high-altitude chemical reactions; these can alter the composition of deeper layers if stable compounds are formed and convected downward.

The temperatures and energy budgets of the giant planets have been studied from Earth for decades, but the best determinations have come from spacecraft. Their infrared and radio-occultation experiments probed from the stratosphere (where

temperature increases with height) down into the troposphere (where the clouds are; *Figure 8*). By comparing the infrared emission with the energy absorbed from sunlight, we can measure the planets' energy budgets and determine the strength of their internal heat sources.

GLOBAL PROPERTIES

The solar composition model does not exactly fit the composition of any planetary atmosphere, but it is a useful standard. The C:H ratio for Jupiter is 2.9 times solar. For Saturn it is more than three times solar, and for Uranus and Neptune the ratios are 30 to 40 times solar. This enrichment of heavy elements relative to hydrogen probably occurred when light gases like hydrogen and helium were blown out of the solar system by the active young Sun. Jupiter and Saturn formed early enough to protect most of their light gases. Uranus and Neptune formed more slowly, presumably because the primordial solar nebula was less dense at its outer edge and took longer there to pull itself together into planet-sized objects (see Chapter 2).

Helium, the second most abundant element in the Sun, has no detectable spectral signature to make its presence known. However, collisions between helium and hydrogen molecules alter the latter's ability to absorb infrared light, an effect that the Pioneer and Voyager instruments could detect and thus give us a kind of "back-door" identification of helium. And we can also combine infrared measurements with the radio-occultation data obtained as the spacecraft passed behind the planets to determine the molecular weight of the atmospheric mixtures. When paired together, these two methods yield helium abundances with an uncertainty of only a few percent. The Galileo probe's result for Jupiter is accurate to better than 1 percent.

The helium abundances tell an interesting story. The values for Uranus and Neptune, 15 and 18 percent by volume, respectively, are consistent with a gas mixture of solar composition, about 16 percent. However, the values derived for Saturn (3 percent) and Jupiter (13.6 percent) suggest that helium has

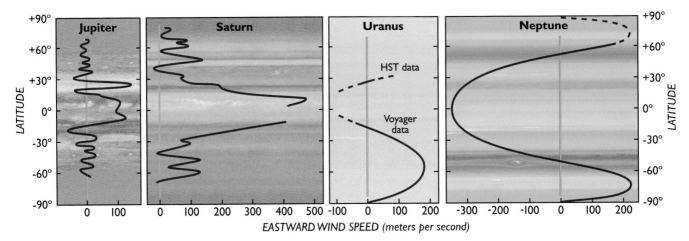

Figure 7. Winds on the giant planets vary in speed and direction with latitude. Positive velocities correspond to winds blowing in the same direction but faster than the planets' internal rotation periods, which are based on observations of their magnetic fields and periodic radio

emissions. Negative velocities therefore, are winds moving more slowly than these reference frames. The equatorial jets are slower than the planets' rotation on Uranus and Neptune, faster on Jupiter and Saturn. In fact, the winds are faster on Saturn than on any other planet.

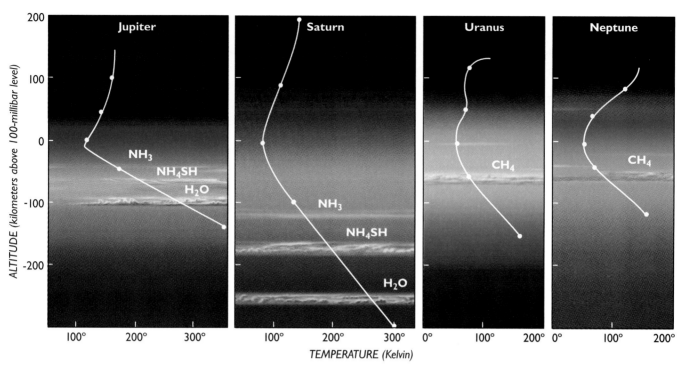

Figure 8. **Pressure-temperature profiles for the upper atmospheres of the giant planets, as determined by Voyager measurements at radio and infrared wavelengths. Altitudes are relative to the 100-mb pressure level, and the dots are spaced to indicate tenfold changes in pressure. For Jupiter and Saturn the altitudes of predicted cloud layers are based on a gaseous mixture of solar composition. For Uranus and Neptune, solar composition grossly underestimates the amounts of condensable gases, and only the methane cloud detected by Voyager 2 is shown. Temperatures are generally lower on planets farther from the Sun, except for Neptune, whose internal heat source makes it as warm as Uranus. The range of cloud altitudes is narrower on Jupiter because gravity is stronger there and compresses the atmosphere more than on the other planets**

been depleted from Saturn's upper atmosphere and probably Jupiter's as well. This depletion is consistent with theories of the planets' internal histories: at sufficiently low temperatures, helium raindrops form in the metallic hydrogen interiors and settle toward the core. On Uranus and Neptune, pressures are not high enough to form metallic hydrogen, so no depletion occurs. And the interior of Jupiter has not cooled down as much as Saturn's, so its depletion has not been as great.

The giant planets radiate to space from a range of altitudes centered in the 0.3- to 0.5-bar pressure range. The temperatures there are 124° K for Jupiter, 95° K for Saturn, and 59° K for both Uranus and Neptune (*Figures 8,9*). Sunlight is absorbed over a broader range, down to pressures of 5 bars or more. The brightness at visible wavelengths determines the planet's *albedo*, which is the fraction of incident sunlight reflected to space. The remaining fraction is absorbed and goes into heating the atmosphere.

From these observations, we infer that Jupiter and Saturn radiate about 1.7 and 1.8 times more heat, respectively, than they absorb from the Sun. Such "infrared excesses" are an indication that each planet is releasing internal heat that remains from planetary formation. Uranus and Neptune emit at the same temperature even though Neptune is 1.6 times farther from the Sun. This implies that Uranus has at most a small internal heat source (its estimated ratio of emitted to absorbed heat is below 1.1), but that Neptune has a relatively large one (its ratio is near 2.6). Such infrared excesses, or lack thereof, provide us with additional insight into internal structures and histories, as explained in Chapter 14.

The ways that temperature and infrared emissions vary with latitude on the giant planets (*Figure 9*) indicate how effective the winds are in redistributing heat away from the subsolar zone. As a reminder, the Sun heats the equators of Jupiter, Saturn, and Neptune more than their poles, while on Uranus the situation is reversed. But in all four cases the temperature dif-

ferences with latitude are small, less than what they would be without winds. Even so, this does not automatically mean that the energy redistribution is taking place within the cloud layers — an interesting question (dealt with more fully later) is whether the redistribution of heat takes place instead in the planets' fluid interiors.

CLOUDS AND VERTICAL STRUCTURE

Temperatures, pressures, gas abundances, cloud compositions — all as functions of altitude and horizontal position — are the principal components of atmospheric structure. Atmospheric dynamics, discussed in the next section, concerns the causes and effects of wind. However, the distinction between structure and dynamics is not always clear. Circulation alters the structure by carrying heat, mass, and chemical species from place to place. The structure, in turn, controls the absorption of sunlight, its re-emission as infrared radiation, and the release of latent heat during condensation — all of which cause the gas to heat up, expand, and set winds in motion.

Atmospheres are self-supporting in that the increase of pressure with depth provides the upward force needed to balance the downward pull of gravity. This balanced state is known as *hydrostatic equilibrium.* The degree of compression of the gas is proportional to the gravitational acceleration present and inversely proportional to its temperature. Thus the outer atmospheres of Jupiter and Neptune are the most compressed and also the thinnest — that is, the altitude range over which pressure increases tenfold is less than on Saturn and Uranus.

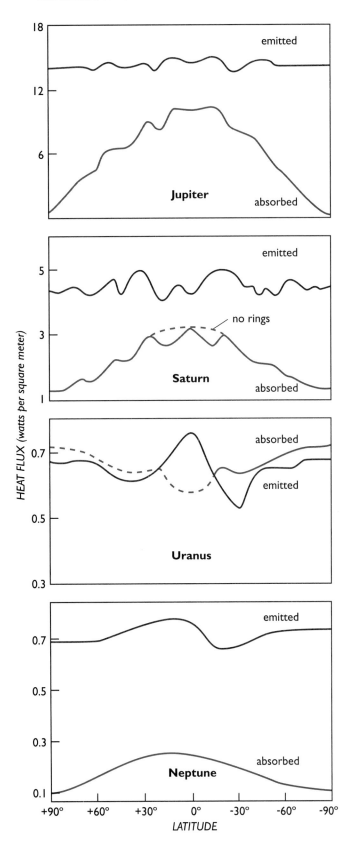

The variation of temperature with depth is controlled by different processes at different altitudes (*Figure 8*). At pressures greater than about 1 bar, rapid convective motions have what is termed an *adiabatic gradient* of temperature. That is, temperature decreases with altitude at the same rate as it would in a rapidly rising parcel of air. The parcel expands and does work at the expense of its own internal energy (and hence temperature). Mixing creates a state where all air parcels resemble one another, and the resulting temperature change with height is then considered adiabatic.

Deep down, heat is effectively trapped in the atmosphere's gases. To move upward, heat must be carried by convection until it reaches a level where the overlying atmosphere is no longer opaque to infrared radiation. For the giant planets, this transition occurs at pressures of 100 to 300 mb, at which point the atmosphere can radiate its heat directly to space. Infrared cooling creates a temperature minimum in this altitude range. Above the 100-mb level, the temperature increases with height because the atmosphere there is absorbing sunlight. Atmospheric gases alone probably cannot absorb energy efficiently enough to account for the rise in temperature. But a haze layer, produced photochemically in the stratosphere, could absorb the extra sunlight required to heat its surroundings.

Higher up, in what is termed the thermosphere, the gas is so thin that even small energy inputs can change its temperature significantly. For instance, the aurora is caused by electrically charged particles striking the upper atmosphere. The particles follow the magnetic field lines that originate far downstream in the wake of the planet in the solar wind flow. The auroral zone is the high-latitude oval where these field lines intersect the atmosphere (*Figure 6*). Spread over the planet, the energy of the incoming charged particles causing the auroras is equivalent to only 0.001 percent of the total incident sunlight, but that is a large input at these altitudes. Solar photons alone would produce temperatures only near 200° K. The cascade of charged particles seems capable of heating the Jovian thermosphere to 1,000° K. The dissipation of electrical currents and upward-propagating waves might also be contributing to the high temperatures, but there are large uncertainties associated with all of these energy sources. The thermosphere sheds its heat principally by conducting it downward to the cooler layers below.

Once the pressure-temperature curve for an atmosphere has been determined, the vertical cloud structure can be inferred.

Figure 9. **A comparison of absorbed solar energy and emitted infrared radiation, averaged with respect to longitude, season, and time of day. Jupiter, Saturn, and Neptune each radiate away more energy than they absorb, implying an internal heat source. This radiation is also distributed more uniformly than the absorbed sunlight, which suggests heat transport across latitude circles at some depth within the planet. The dashed curve for Saturn shows how much sunlight it would absorb without the shadowing caused by its ring system. The dashed curve for Uranus is an extrapolation northward of Voyager's measurements, which assumes that planet's absorption of heat is symmetric with respect to latitude over the course of a Uranian year. Small bumps are due to temperature and brightness differences between adjacent latitude bands.**

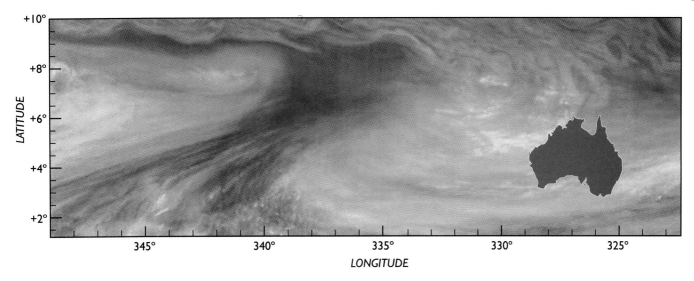

Figure 10 (above). An equatorial hotspot on Jupiter, as seen by the Galileo spacecraft. This mosaic combines three near-infrared images to show variations in altitude: bluish clouds are high and thin, reddish ones are low, and white clouds are high and thick. Dark blue marks a hole in the deep cloud with an overlying thin haze. The silhouette of Australia is shown to scale.

Figure 11 (right). These infrared *(top)* and visible-light *(bottom)* views of Jupiter were obtained, respectively, by the Infrared Telescope Facility on 3 October 1995 and by the Hubble Space Telescope one day later. Holes or thinnings in the upper cloud deck allow radiation to escape to space from warmer layers underneath, creating *hotspots* that appear bright at wavelengths near 5 microns in the infrared. The views match in longitude within 2°, and arrows mark where the Galileo probe entered the atmosphere on 7 December 1995.

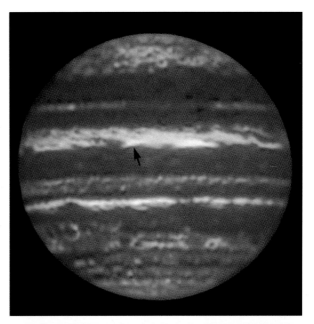

One first computes the altitudes at which the different atmospheric constituents can condense by assuming that the gas mixture has a uniform, known composition, like those in *Table 1.* A particular gas will condense when its absolute abundance (partial pressure) exceeds its abundance at its dew point (the saturation vapor pressure). Since the latter falls rapidly as the gas cools, clouds will form at the coldest layers in the atmosphere. These condensates will collect into drops and rain out in the absence of buoyant updrafts, so they should be found above the cloud-forming levels only as high as upward convection will carry them.

These calculations argue for three distinct cloud layers on Jupiter and Saturn (*Figure 8*). The lowest is composed of water ice or possibly water droplets. Next are crystals of ammonium hydrosulfide (NH_4SH), which is basically a combination of ammonia (NH_3) and hydrogen sulfide (H_2S). At the top we expect an ammonia-ice cloud. These three layers are also present in the colder atmospheres of Uranus and Neptune, but they should lie deeper down — at higher pressures — because the low temperatures necessary for cloud formation exist at greater depths. Even methane will condense if the temperature gets low enough, especially if the gas is abundant. Apparently, such conditions are met on Uranus and Neptune, where a layer of condensed methane overlies the other clouds. Voyager 2's radio signal probed down to the 2-bar level and detected the methane cloud at 1.3 bars but not the ammonia cloud.

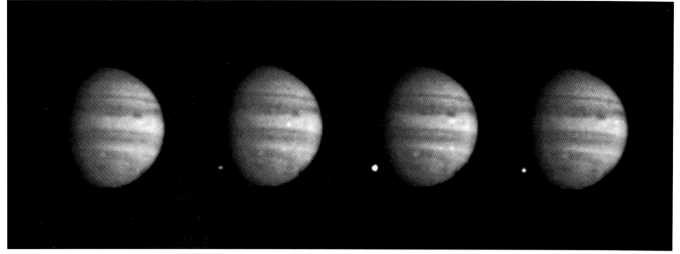

Figure 12. **When fragment W from Comet Shoemaker-Levy 9 struck Jupiter on 22 July 1994, the Galileo spacecraft recorded these green-filtered images every 2.3 seconds, beginning at left. The comet's numerous fragments hit on the dark side of Jupiter near the dawn terminator. At its peak the light from fragment W's annihilation was 1.5 percent of Jupiter's brightness at this wavelength.**

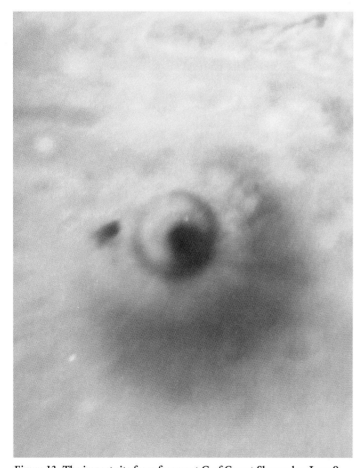

Figure 13. **The impact site from fragment G of Comet Shoemaker-Levy 9, recorded 1.7 hours after G's entry into Jupiter's atmosphere on 18 July 1994. The aftermath was captured by the Hubble Space Telescope after Jupiter's rotation had carried the site into daylight. Dark material, covering an area comparable to Earth, was emplaced within 15 minutes of the impact. The thin ring has been interpreted as either a propagating wave or an outward-spreading flow of debris. The small dark spot to its left marks where fragment D struck some 20 hours (two rotations) earlier.**

The layered-cloud model may be too simple, however. Below the water-cloud base, which for Jupiter is at the 5-bar level, the atmosphere should be well-mixed by convection and its composition constant, matching that of the planet as a whole. Yet the Galileo probe found a rather different situation. While the abundances of NH_3, H_2S, and H_2O it measured were well below solar values at 5 bars, NH_3 and H_2S reached three times solar at the 10-bar level and water's abundance was still increasing when the probe's signal failed near a depth corresponding to 20 bars. Part of this inconsistency is that the probe apparently descended into a "hotspot," a region of intense infrared emission made possible by a gap in the clouds (*Figures 10,11*). Perhaps these hotspots are the deserts of Jupiter, where dry air from high altitudes is forced downward, as occurs over the Earth's subtropical regions. In that case, the weather forecast for a Jovian hotspot is always "sunny and dry."

Jupiter is not dry everywhere — the hotspots are relatively rare (*Figure 10*), and the Galileo orbiter has found atmospheric pockets elsewhere with water abundances 100 times greater than that in the hotspots. Also, since hydrogen and oxygen are the first and third most abundant elements in the Sun (with helium the second), water should be the most abundant compound on the giant planets — assuming everything formed from the same cloud of gas and dust as the Sun. Still, water has been an extremely elusive molecule on the giant planet. Spectroscopic observations from Earth reveal only small amounts of water vapor, because most of it freezes at the cold Jovian cloud tops.

Similar arguments apply to H_2S, which remained undetected from Earth until fragments of Comet Shoemaker-Levy 9 slammed into Jupiter and dredged up material from below the clouds (*Figures 12,13*). These unprecedented data are difficult to interpret because we don't know how much of the material came from the comet and how much from Jupiter's atmosphere. Nevertheless the detected compounds included CO, NH_3, S_2, and various metals, but not SO or SO_2. The implication is that the sulfur (as H_2S) and oxygen (as H_2O) are physically separated in Jupiter's atmosphere.

If the giant planets' atmospheres were in chemical equilibrium, with hydrogen the dominant species, virtually all their carbon would be tied up in methane, all the nitrogen in ammonia, all the oxygen in water, and so on. But in Jupiter's case, we have identified ethane (C_2H_6), acetylene (C_2H_2), carbon monoxide (CO), and other gases that imply disequilibrium. Both C_2H_6 and C_2H_2 are relatively easy to account for, since they form in the upper atmosphere as recombined byproducts of methane photodissociation. These gases are seen on Saturn and Uranus as well.

The existence of carbon monoxide in the upper atmosphere of Jupiter is more interesting, since oxygen is not readily available (water, the principal oxygen-bearing molecule, tends to condense in the lower clouds). Several explanations have been proposed. One is that oxygen ions cascade into the upper atmosphere after being injected into the magnetosphere (as SO_2 molecules) by Io's volcanoes. A second explanation is that CO, which forms and remains stable deep in Jupiter's atmosphere, gets transported upward before it can react with hydrogen and thus be destroyed. If this latter scenario is correct, oxygen should likewise react with phosphorus (in the form of phosphine, PH_3) at depth. But phosphine manages to persist in Jupiter's upper atmosphere without being oxidized, which is something of a mystery. Another puzzle is why methane and germane (GeH_4) have been detected but silane (SiH_4) has not. Perhaps all of Jupiter's silicon is tied up with oxygen and resides in the core. A critical unknown is how fast the high-temperature species are convected upward before they are destroyed by chemical reactions.

One insight into vertical mixing comes from the fact that molecular hydrogen (H_2) exists as two species, one in which the spins of the two nuclei are parallel and one in which they are opposed. The two types react and tend to reach an equilibrium ratio that depends on the temperature — as long as vertical motion is weak. However, in a strong updraft, the ratio of the two species "freezes" at the high temperatures found at depth, reflecting conditions at or above 300° K. By observing how the proportions of the two types vary from place to place, atmospheric scientists can gauge how much vertical motion occurs in the equatorial zones of the giant planets. The rate of mixing is greatest for Jupiter and least for Uranus.

The spectacular colors of Jupiter and more muted colors of Saturn provide further evidence of active chemistry in the atmospheres. These colors correlate with the cloud's altitude (*Figures 10,11*). Bluish regions have the highest apparent temperatures, so they must lie at the deepest levels and are only visible through holes in the upper clouds. Browns are higher up, followed by whites and finally reds. Thus, the Great Red Spot is a very cold feature, judging from its infrared brightness.

The trouble is that all cloud species predicted for equilibrium conditions are white. Therefore, color must arise when chemical equilibrium is disturbed, either by charged particles, energetic photons, lightning, or rapid vertical motion through layers of varying temperature. One possible coloring agent is elemental sulfur, which takes on many hues depending on its molecular structure. Sulfur dominates the appearance of Io, which exhibits many of the same colors as Jupiter. Some scientists believe that phosphine makes the Great Red Spot red, while others have proposed organic (carbon-bearing) compounds to explain almost all of the hues seen in Jupiter's clouds. Since light from solid cloud particles is less diagnostic of their composition than

light from a gas, the coloring agents on Jupiter — and more subtly on Saturn — remain uncertain.

Both the blue-green color of Uranus and the deeper blue of Neptune may have a simple explanation. These planets have so much methane gas in their atmospheres that the gas itself becomes the coloring agent. Methane absorbs red light quite strongly, so the sunlight reflected off the clouds has a blue color.

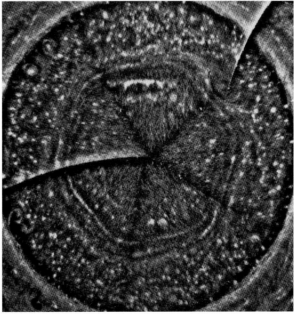

Figure 14. These views of Jupiter *(top)* and Saturn *(bottom)* were constructed by reprojecting Voyager images as if the spacecraft were looking down onto the planets' south and north poles, respectively. In the Jupiter composite, the Great Red Spot is at upper left; in Saturn's, the dark "starfish" and the two radial "cuts" are processing artifacts. The unusual hexagon centered on Saturn's north pole is actually a six-lobed wave in a zonal jet at 76° N latitude. Seen during Voyager 2's 1981 flyby, it rotated with the same period as Saturn's interior (it did not drift eastward or westward in longitude). Although the origin of the hexagon is uncertain, it may be related to the storm just outside the hexagon in the upper-right quadrant.

Also, these planets' sulfur and ammonia clouds are colder and lie much deeper than those on Jupiter and Saturn, so light reflected from the clouds has to pass through more overlying atmospheric gas on its way to our eyes. The color differences between Uranus and Neptune have to do with the depths of the clouds and the color of the cloud particles themselves.

Thus coloration is a subtle process, involving disequilibrium conditions and trace constituents. The correlation with altitude presumably reflects the pressure-temperature conditions need-

ed to drive specific chemical reactions. For example, higher altitudes receive more sunlight and charged particles. Some regions may provide locales for intense lightning and others for intense vertical motion. A different question is why these processes should be organized into large-scale patterns that last for years and sometimes for centuries. This question involves the dynamics of the atmospheres, to which we now turn.

ATMOSPHERIC DYNAMICS

The dominant dynamic features seen in the giant planets' atmospheres are counterflowing eastward and westward winds called *zonal jets* (*Figure 7*). In this respect, Uranus and Neptune resemble the Earth, which has one westward air current at low latitudes (the trade winds) and one meandering eastward current at mid-latitudes (the jet stream) in each hemisphere.

Figure 15. **The names used to designate cloud bands on Jupiter (upper panel) and Saturn have evolved over a century of telescopic observation. These illustrations have south at the top, as they would appear in a backyard (inverting) telescope. Amateur astronomers provide much of the continuity in the historical record of Jovian and Saturnian cloud motions.**

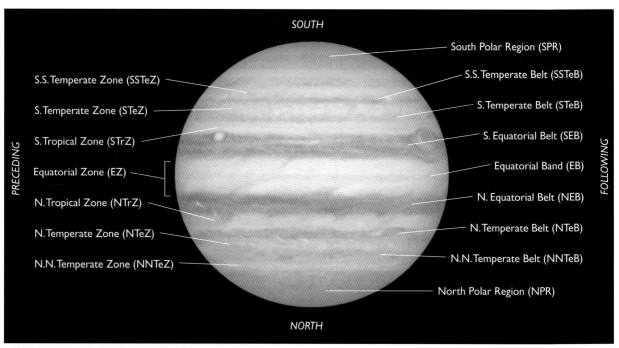

Jupiter has five or six of both kinds in each hemisphere, and they are steadier than on Earth. Saturn seems to have fewer sets of such currents than Jupiter, but they move faster. In fact, the eastward wind speed at Saturn's equator is about 500 m per second — about two-thirds the speed of sound there. As on Earth, these winds are measured with respect to the rapidly rotating planets beneath them. Even supersonic wind speeds would be small compared to the equatorial velocity imparted by rotation. (For the Jovian planets, which lack solid surfaces, the rotation rate of their interiors is deduced from that of the magnetic fields generated in their metallic cores; see Chapter 4.)

The zonal winds of Earth are the weakest of any in the solar system, even though the energy available to drive them is greater than on any other planet (Venus is closer to the Sun, but it reflects such a high fraction of the incident sunlight that its absorbed power per unit area is less than Earth's). The total power per unit area (sunlight and internal heat averaged over the surface) at Earth is 20 times that at Jupiter and 400 times that at Neptune. Yet the winds of Neptune are stronger than those of Jupiter, which in turn are stronger than those of Earth. The reason for this seeming paradox may be that, on Earth, sunlight and internal heat create the small-scale turbulence that acts to dissipate the large-scale winds. The turbulence is weaker in the outer solar system, so the winds are stronger. A balloon-borne observer drifting along at 1,000 mph in Neptune's equatorial jet might have a smooth ride.

Without continents and oceans, a planet's rotation tends to produce a pattern of zonal (east-west) currents and banded clouds that remain confined to specific latitudes. One exception is Saturn's polar "hexagon," but this may be merely a wavelike disturbance caused by a storm at that latitude (*Figure 14*). Rotation has two other effects. On Jupiter, Saturn, and Neptune it smears absorbed solar energy into a band around the equator. And on all planets, not just the Jovian ones, it creates the *Coriolis force*, which acts at right angles to the wind.

Uranus provides perhaps the best example of the importance of rotation. During the Voyager encounter in 1986, the Sun was almost directly overhead at Uranus's south pole. The equator was in constant twilight, and the north pole had been in darkness for 20 years (one-fourth of the Uranian "year"). Without the Coriolis force, circulation would have been dictated solely by the need to redistribute heat: winds in the upper atmosphere would blow away from the sunlit south pole and converge on the night side at the north pole, with a return flow underneath. But the Coriolis force is apparently redirecting the flow at right angles to this hypothetical, pole-to-pole circulation, creating the zonal banding seen weakly in Voyager images. Even so, the problem remains that over a Uranian year the planet's poles receive more energy from sunlight than they can emit as infrared radiation. Therefore, to maintain thermal equilibrium the atmosphere *must* transport heat equatorward, via some unseen mechanism that cuts across latitude circles. Small-scale eddies, below the resolution of the Voyager images, are one possibility; transport within the clear atmosphere above the clouds is another.

The infrared instrument on Voyager measured the temperature of the gas above the clouds on each of the four Jovian planets (*Figure 9*). But instead of a smooth gradient from equator to pole, the curves are bumpy, and the latitude zone receiving the most sunlight is not necessarily the planet's hottest place. A circulation that simply brings heat from warm areas to cold ones would not produce such a distribution. Rather, Voyager investigators believe that heat is being pumped from cold places to hot ones, as in a refrigerator. The power source for this "global refrigerator," they claim, is the kinetic energy of the zonal jets. The steepest gradients of temperature should then be located at the latitudes of the zonal jets (*Figure 7*), and so they are. However, the energy source needed to maintain the jets themselves remains unspecified.

The giant planets' zonal winds are remarkably constant in time. Ninety years of modern telescopic observations reveal no changes in the east-west jets of Jupiter and Saturn. In fact, the basic patterns are so unchanging that astronomers have assigned names to individual latitude bands, calling the dark ones belts and the bright ones zones (*Figure 15*). During the four months

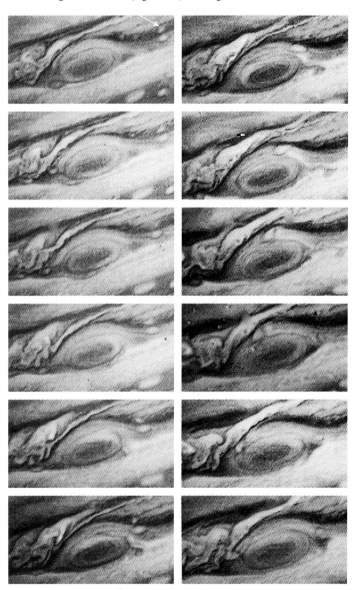

Figure 16. A blue-light image of the Great Red Spot was taken every 20 hours over a period of about two weeks in this Voyager sequence, which begins at upper left, continues down each column, and ends at lower right. Note the small bright clouds that encounter the giant storm from the east, circle counterclockwise around it in 400-km-per-hour winds, and partially merge with it along the southern boundary.

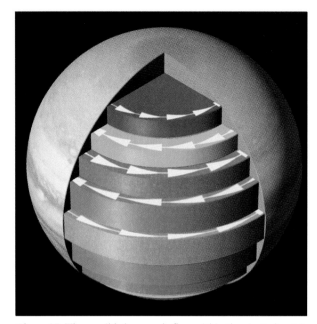

Figure 17. **The possible large-scale flow within the giant planets' fluid interiors. Each cylinder has a unique rotation rate, and zonal winds may be the surface manifestation of these rotations. The tendency of fluids in a rotating body to align with the rotational axis was observed by Geoffrey Taylor during laboratory experiments in the 1920s, and was applied to Jupiter and Saturn by Friedrich H. Busse in the 1970s. Such behavior seems reasonable for Jupiter and Saturn if their interiors follow an adiabatic temperature gradient.**

between the two Voyager encounters of Jupiter, the planet's zonal velocities changed by less than the measurement error (about 1.5 percent). This is remarkable for several reasons. First, although the zonal jets on Jupiter correlate with the latitudes of the colored bands, the bands often change their appearance dramatically in a few years — while the jets maintain their velocities. Second, as revealed in the superbly detailed Voyager and Galileo photographs, an enormous amount of eddy activity accompanies the zonal jets. Small eddies appear suddenly along the boundary between eastward- and westward-moving streams. They last only about 1 to 2 days, as they are quickly sheared apart by the counterflowing winds. Larger eddies, including a group of long-lived white ovals and the Great Red Spot in Jupiter's southern hemisphere, manage to survive by rolling with the currents. Voyager repeatedly observed smaller spots encounter the Red Spot from the east, circulate around it, then partially merge with it (*Figure 16*). How the zonal jets and the large eddies can exist amid such activity is something of a mystery.

We learned an important fact about the eddy motions from a statistical analysis of Jupiter's winds. During the Voyager flybys, the motions of tens of thousands of 100-km-wide cloud features were tracked. We were intrigued by localized eddies, which moved at about one-fourth the mean speed of the surrounding zone. Their motions were not entirely random. Cloud parcels moving *into* a jet from another latitude were typically headed in the same direction, and thus carried the same eastward or westward momentum as the jet itself. Parcels moving *out of* a jet typically had the opposite sign of momentum (opposed the flow). The net result is that such eddies help to

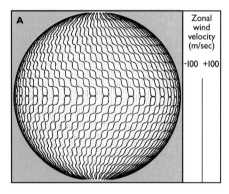
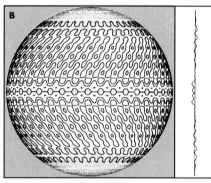
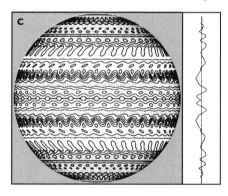
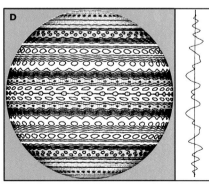
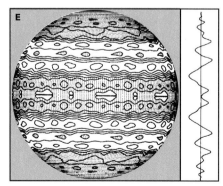
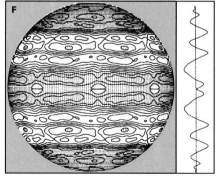

Figure 18. **A computer model of Jovian atmospheric circulation made in 1978 by Gareth P. Williams. He assumes that all the activity is taking place in a thin "weather layer" of atmosphere (as is the case on Earth); the fluid interior plays a passive role. This model first produces small-scale eddies (*a*), which become unstable (*b*) and give up their energy to zonal jets (*c,d*) that eventually dominate the flow (*e,f*).**

Eddies with deep roots demonstrate the same behavior, so remaining questions center on the depths of both the eddies and the zonal flow. Although the energy source for these motions is uncertain, motions in the weather layer are presumably driven by sunlight, and deeper motions are driven by Jupiter's internal heat.

maintain the jets, not weaken them. Yet no matter how many eddies came and went, the jet speeds did not change. So there must be other, unseen eddies — turbulence on a smaller scale, perhaps — that draw enough of their energy from the jets to keep the system in balance.

If one regards the zonal jets as an ordered flow and the eddies as chaotic, these observations suggest that (in Jupiter's case) order arises from chaos. Similar interactions among eddies and mean flows occur in the Earth's atmosphere and oceans. In all cases, the eddies get their energy from buoyancy (hot fluids rising and cold ones sinking). Almost all of this buoyant energy is dissipated as heat. On Earth less than 1 percent of it goes into powering the atmosphere's jets, but on Jupiter the fraction exceeds 10 percent. What causes this fundamental difference is unknown, but eddies do appear to drive zonal flows in a wide range of situations on rotating planets. In fact, years before the Voyagers' arrival, atmospheric scientists proposed this mechanism as a way to explain Jupiter's zonal flow (*Figures 17,18*).

The constancy of Jupiter's zonal jets is remarkable in light of this incessant eddy activity. If all the action were somehow confined to the planet's observable atmosphere, the eddies would be

doubling the kinetic energy of the jets every 75 days. Obviously, that is not the case. Perhaps the zonal jets have remained stable for nearly 90 years because the mass involved in their motions is many times greater than that of the eddies. That is, if the eddies were confined to cloudy layers, at or above the 5-bar pressure level, the jets would have to extend down to much greater pressures — possibly even through the planet and out the other side!

Such behavior is not as unlikely as it first seems. A rapidly rotating sphere with a fluid, adiabatic interior should exhibit two types of small-amplitude motions (relative to the basic rotation). The first are rapidly varying waves and eddies. The second are steady zonal motions defined by a series of cylinders coaxial to the spin axis (*Figure 17*), with each cylinder turning at a

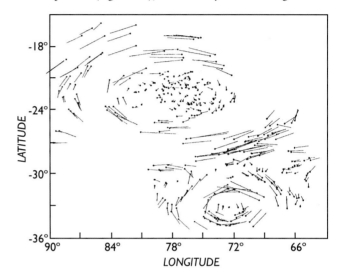

Figure 19. **A wealth of detail appears in a Voyager 1 mosaic** *(below)* **of the Great Red Spot and a white oval (known as BC since its formation more than 60 years ago). The diagram** *(right)* **shows wind vectors in this region, determined from changes observed over a 10-hour period. Each dot marks the position of the cloud feature measured; an attached line points in the direction of flow, and its length is proportional to the wind velocity.**

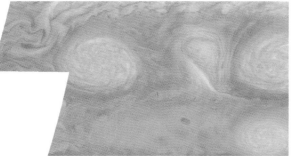

Figure 21. Natural- *(lower panel)* and false-color (infrared) views of Jovian white ovals as seen by the Galileo spacecraft. The larger two ovals, named DE and BC, formed in 1938 and merged into a single oval (BE) in early 1998. Flow around these ovals is counterclockwise, and the pear-shaped region between DE and BC has clockwise flow. Colors in the upper panel indicate that the clouds in the ovals are high while those in the pear-shaped region are low.

Figure 20. The Galileo orbiter acquired this infrared mosaic of the Great Red Spot in June 1996. It is color-coded so that pink and white indicate the highest clouds (in its center) and blue from the deepest ones (around its periphery) — spanning an altitude range of more than 30 km.

unique rate. The observed jets on the giant planets, as far as we know, could be the surface manifestations of an internal clockwork of nested cylinders on a huge scale.

The problem with such speculation is that we lack information about winds far below the visible cloud tops. A totally different approach is to treat Jupiter simply as a larger version of the Earth, whose atmosphere has a fixed lower boundary (our planet's surface). Computer simulations of the Earth's atmosphere have also yielded realistic zonal wind patterns when applied to Jupiter (*Figure 18*). This is somewhat surprising, since the models assume an atmosphere less than 100 km thick and no outward heat flow from the interior.

Clearly, the lack of data on vertical structure within the Jovian planets allows us plenty of "latitude" in trying to explain the zonal jets of both Jupiter and Saturn. Ultimately, we may learn which is correct by comparing the secondary effects predicted by each model with actual observations.

LONG-LIVED STORMS

Theorists have proffered a similarly wide range of theories concerning long-lived oval "storms," which were well known on Jupiter many decades before being discovered on Saturn and

Neptune by Voyager. Jupiter's legendary Great Red Spot (GRS) covers 10° of latitude (*Figures 19,20*), which makes it about as wide as the Earth. The GRS takes about 7 days to spin around, with winds along its outer edge reaching 400 km per hour. The three white ovals slightly to the GRS's south (named BC, FA, and DE) first appeared in 1938, and spots BC and DE merged in 1998.

These long-lived ovals, which drift in longitude but remain fixed in latitude, tend to roll in the boundary region between opposing zonal jets. The circulation around their edges is almost always counterclockwise in the southern hemisphere and clockwise in the northern hemisphere, indicating that they are high-pressure centers. They often have cusped tips at their east and west ends. During the Voyager encounters in 1979 both the GRS and the white ovals had intensely turbulent regions extending to their west, but when imaged by Galileo in the mid-1990s the white ovals were crowded together with the turbulent regions caught in the middle (*Figure 21*).

Although Saturn is normally rather bland, the Voyagers discovered that it too can display long-lived storms (*Figure 22*). Moreover, the planet occasionally erupts with an outbreak of bright clouds obvious even through backyard telescopes. These disturbances seem to occur every 30 years, which is close to the orbital period of Saturn. The most recent one began in September 1990 as a discrete white spot north of the equator. It quickly spread around the planet and filled the region from the equator to 25° north latitude (*Figure 23*), then faded a few months later. The cause of these periodic eruptions is unknown. They are apparently triggered during summer in Saturn's northern hemisphere — but they have not been observed when summer moves to the southern hemisphere 15 years later.

Neptune's large oval structures oscillate more and drift farther than Jupiter's ovals. During the Voyager encounter in 1989, Neptune had two major dark features, dubbed the Great Dark Spot (GDS) and Small Dark Spot (*Figures 24,25*). As Voyager 2 watched, the GDS stretched and contracted while rocking and rolling with a period of 8 days (*Figure 26*). Surprisingly, such behavior is consistent with the simplest type of hydrodynamic flow, that of a frictionless fluid (no shearing forces) moving in two dimensions (motion confined to a plane). The success of this simple model lends further support to the idea that small-scale turbulence is nearly absent in the outermost giant planets, since turbulence is three-dimensional and tends to produce shearing forces. Meanwhile, the GDS was also drifting equatorward at 15° per year. At this rate it should have reached the

Figure 23. **Storms on Saturn, as captured by the Hubble Space Telescope. The disturbed equatorial band in the black-and-white view is the result of an eruption that began on 25 September, 1990. Such major onsets happen about every 30 years, during summer in Saturn's northern hemisphere, but as yet there is no explanation why they appear to follow such a cycle. A smaller storm broke out in September 1994; seen in the color image, it had an east-west extent of some 13,000 km, about one Earth diameter. The bright storm clouds are crystals of ammonia ice that condense in an upwelling of warmer air, similar to a terrestrial thunderhead. Some 200 years ago an equatorial disturbance of this type enabled William Herschel to make the first estimate of Saturn's rotation period.**

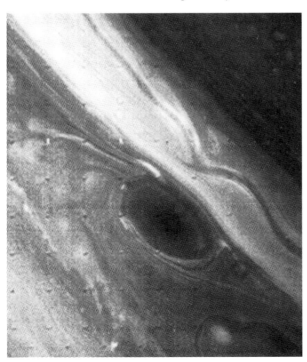

Figure 22. **This oval spot, photographed on two successive Saturn rotations, shows anticyclonic rotation (clockwise in the northern hemisphere) around its periphery. North is to the upper right. The spot's latitude is 42.5° north, and its long axis spans 5,000 km.**

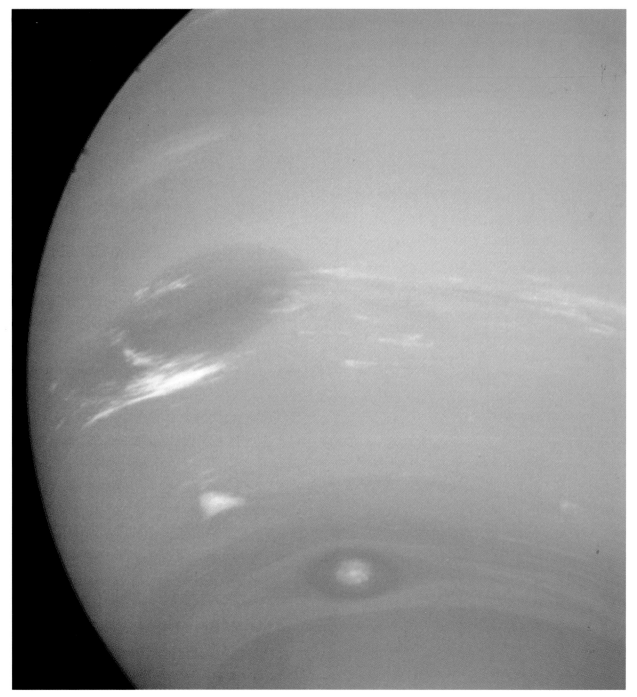

Figure 24. The spots and streamers seen on Neptune by Voyager 2 moved at rates that depended on their latitude. The Great Dark Spot, left of center at latitude 22° south, circled the planet in 18.3 hours. The eyelike Small Dark Spot, near the bottom at 55° south, took only 16.0 hours. The bright cirrus-type patch between them at 42° south latitude was dubbed "Scooter" because its 16.8-hour trips around Neptune were faster than those of other bright clouds. As Voyager watched, Scooter changed shape from round to square to triangular. Neptune's clouds are only barely discernible from Earth.

Figure 25 (left). A Voyager closeup of Neptune's Small Dark Spot, also known as D2, shows knots of bright, methane-ice clouds in its center. The spot may have been rotating clockwise, unlike the much larger Great Dark Spot. If so the material within the dark oval was descending.

equator (where there is a strong westward jet) a year or two later. More recent observations with the Hubble Space Telescope suggest that it did not survive this passage (*Figure 27*).

Any theory of the GRS, GDS, and other ovals must explain their longevity, which is a twofold problem. The first concern is maintaining the hydrodynamic stability of rotating oval flows. An unstable oval would break up into waves and eddies that would disperse the initial energy over large distances. In that case they would last only a few days, which is roughly the circulation time around their edges (or the lifetime of smaller eddies that are pulled apart by the zonal jets). The second concern is their source of energy. A stable eddy without an energy source to power it will eventually run down, though on Jupiter this could take several years and on Neptune conceivably much longer.

How do theorists account for the long lifetimes of the giant planet's oval storms? The "hurricane" model postulates that these structures are giant convective cells extracting energy from below (the latent heat released when gases condense). The

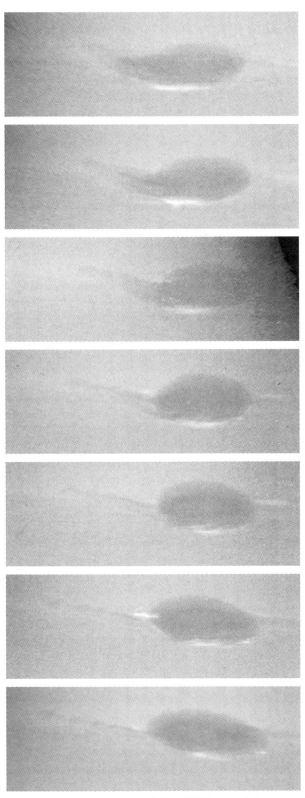

Figure 26. Neptune's Great Dark Spot (GDS), recorded on seven successive rotations of the planet over 4½ days. This sequence shows most of the spot's 8-day oscillation cycle: the GDS is initially elongated toward upper right; then it contracts; and finally it elongates again, this time toward upper left. As noted by Lorenzo Polvani and his colleagues, these features can be reproduced in a simple hydrodynamic model — though modeling the development and collapse of the dark plume at left remains problematic. Note the bright methane-ice clouds along the spot's southern edge, which did not change position over time.

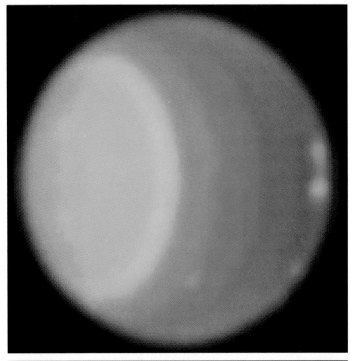

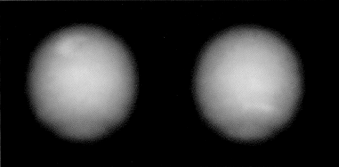

Figure 27. The Hubble Space Telescope turned its gaze to Uranus *(top)* in 1997 and Neptune *(bottom, left and right)* in 1995. The appearance of both planets had changed dramatically since Voyager 2's flybys. Seen in a false-color infrared composite, Uranus displayed a string of bright clouds (along right edge) to the north of its equator — at latitudes completely hidden in darkness when the spacecraft visited in 1986 — and a distinct south-polar "hood." Neptune, meanwhile, showed no trace of its large dark spots.

"shear instability" model holds that they draw energy from the sides (the shear in the zonal currents). Still another model argues that they gain energy much as the zonal jets do: by absorbing smaller, buoyancy-driven eddies. When simulations are run on high-speed computers, the surprising result is that all three hypotheses seem to work. Oval-shaped features form spontaneously in a variety of circumstances that mimic all the likely energy sources, density distributions, and other conditions that have not been completely measured. They roll between the zonal jets just as Jovian vortices do. They grow by merging (*Figure 28*), the largest spot tending to consume all

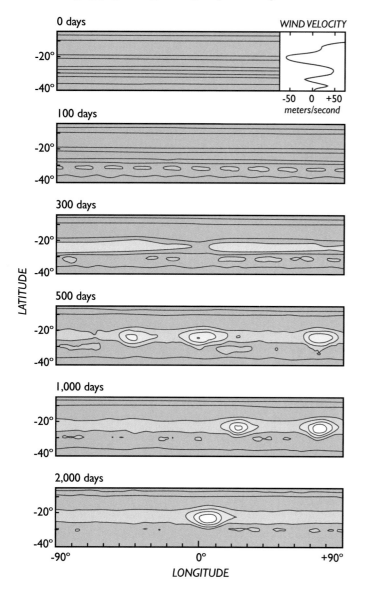

Figure 28. In this computer simulation by Timothy Dowling and the author, Jupiter's atmosphere is assumed to be a single shallow layer of fluid on a rotating spheroid. Supporting the fluid from below is a much deeper layer, in which different latitudes rotate at different rates (see *Figure 17*). This creates a "washboard" topography at the base of the upper layer. Lines denote contours of constant pressure and thus essentially represent streamlines of the flow. East-west velocity is plotted at upper right. Energy is pumped into the zonal flow, which initially breaks into numerous small eddies because it is unstable. But in time the eddies merge, ultimately yielding a single dominant vortex.

the others in its latitude band. Several groups have observed stable isolated ovals in laboratory simulations of flows (*Figure 29*).

Faced with this unavoidable success, we ponder what the atmospheric models have in common that makes them all work. They all share the equations of fluid mechanics — Newton's laws applied to atmospheric motion on a rotating planet. Perhaps the growth of a vortex by merging represents the natural outcome in a rotating fluid. As such, the giant-planet atmospheres are the simple examples; the more-complex Earth may lack a Great Red Spot because its continents and oceans tend to disrupt zonal flows, and its large equator-to-pole temperature differences make the atmospheric flows continuously unstable. By contrast, the giant planets' atmospheres appear to be connected to enormous thermal reservoirs with massive inertia. Temperature differences encountered at a given altitude are therefore smaller, and the flows are steadier — at least on the largest scales. If their interiors really are in differential rotation (*Figure 17*), the inertia of these flows might also contribute to the steadiness of the atmospheres.

But which models best represent a specific planet like Jupiter? Here observations are the key. One sifts through Voyager images to see if vortices come in all size ranges (they do) and whether the small ones behave differently from the large ones (the small ones are rounder). One makes detailed measurements of winds and temperatures, and compares these quantitatively to our computers' output. Eventually some models will fail to fit the data, and at that point we will have learned something new about Jupiter.

As an example of this ongoing process, consider the mapping of hidden topography using the conservation of angular momentum. When a rotating fluid parcel flows over a topographic rise, its vertical thickness decreases and its horizontal area increases. Like a spinning figure skater throwing out her arms, the fluid parcel reacts by spinning more slowly. A topographic depression produces the opposite effect. On giant planets "topography" arises from the flow below the clouds, which produces an uneven interface at cloud base. Spinning parcels are observed at the cloud-top level, and we assume that the gas in between (the cloudy layer) has been stretched or squashed accordingly. The result of this study is that the layer below the clouds does not act like a flat, solid surface. The GRS appears to be relatively shallow to the north, and it increases steeply in depth to the south (*Figure 30*). The inferred zonal jets below the clouds are comparable to those measured in the upper layers.

MOTIONS IN THE INTERIOR

In closing, let us speculate on possible roles that the giant planets' fluid interiors may play in their atmospheric dynamics. Consider, for example, whether an interior could help maintain the poles and equator at the same surface temperature, even though the input of sunlight varies with latitude. An adiabatic atmosphere has constant temperature at each altitude level, regardless of latitude, and vertical convection is normally very effective at producing and maintaining such an adiabatic state — especially when the atmosphere's heat sources are located below its heat sinks. This is the arrangement at least for Jupiter, Saturn, and

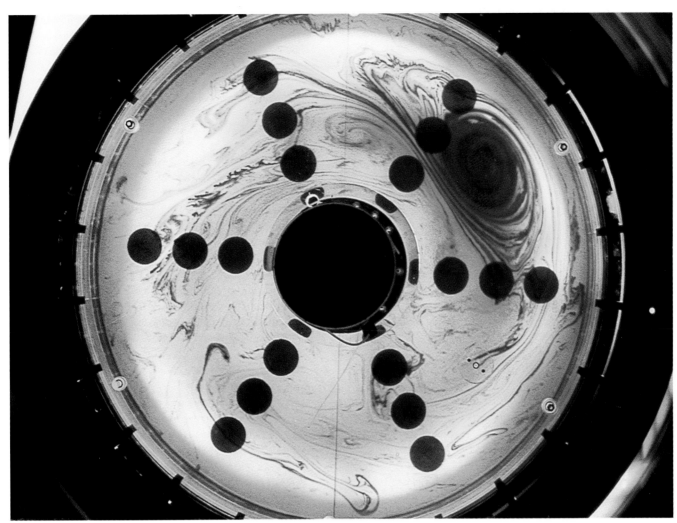

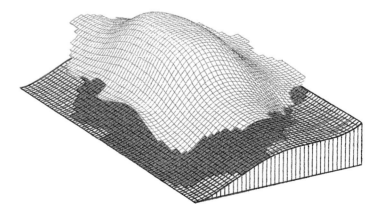

Neptune, whose interiors generate a great deal of heat and whose atmospheres lose it (via infrared radiation) to space.

Actually, we need to know these atmospheres' *net* heat loss: the infrared energy they emit to space minus the sunlight they absorb. For Jupiter, Saturn, and Neptune, the net heat loss is greatest at the poles and least at the equator; to compensate, heat must somehow be transported poleward. (The reverse is true at Uranus; see *Figure 9.*) But this transfer does not appear to be taking place in the cloud layers — that would disrupt the well-ordered banding. So we expect, but cannot yet know for sure, that small departures from a strictly adiabatic state have arisen in the interior to drive the internal heat flow poleward. These departures are too small to manifest themselves, but the result is that the atmosphere is effectively short-circuited by the interior.

The situation is very different on Earth. Here substantial poleward transport of heat *must* take place in the atmosphere. Absorption of sunlight by the oceans and the subsequent release of latent heat from the condensation of water in the atmosphere above them leads to a net heat gain in the tropics. Radiation to space leads to a net heat loss in the polar atmosphere. This combination sets up appreciable temperature gradients with latitude, and obvious mixing takes place across latitude circles. Consequently, Earth's atmosphere is decidedly *not* adiabatic.

Figure 29. This laboratory simulation shows a spinning cylindrical tank, in which a circular zonal jet has been created by injecting or removing fluid using the 18 circular ports. Over time the zonal jet forms, becomes unstable, and breaks up into small eddies that eventually coalesce into a single, long-lasting vortex. Red dye, injected into the vortex after it formed, is unable to escape.

Figure 30. Topography can be inferred for the Great Red Spot (GRS) of Jupiter. This shows the computed shape of two constant-pressure surfaces, the upper one at the cloud tops (0.5 bar) and the lower one at the cloud base (5.0 bars). The lower surface is flat and shallow to the north (right) but slopes down to the south (the vertical scale is greatly exaggerated). The dome in the upper surface indicates that the GRS is a high-pressure center.

Another key speculation is how the Jovian planets' zonal velocities change with depth. According to the thermal wind equation of meteorology, the change of velocity with depth is proportional to the change of temperature with latitude. Since no significant temperature gradients were found on either Jupiter or Saturn by the Pioneer and Voyager instruments at the deepest measurable levels, the velocity change with depth must therefore be extremely small, and so the winds measured at the cloud tops must persist below. A rough estimate is that the eastward wind speeds (relative to the internal rotation rate) persist well below the region that is affected by sunlight, which is mostly absorbed in the clouds. This leaves internal heat as the obvious source for deep motions.

The Galileo probe provided our only direct measurement of winds below the tops of Jupiter's clouds. The winds were inferred from the Doppler shift of the probe's radio signal as it descended. The probe entered at 6.5° north latitude, where the cloud top wind is eastward at 100 m per second (*Figures 7,11*). To almost everyone's surprise, the winds *increased* with depth to 180 m per second at the 5-bar level and then held steady to 20 bars, where the probe died. Since this is below the clouds and below the level that is heated by sunlight, it is tempting to conclude that the winds are driven by internal heat. Yet we cannot be sure. The probe results do imply that the winds run deep: they are qualitatively consistent with inferences based on the figure-skater effect (*Figure 30*) and with the idea of differentially rotating cylinders extending throughout the interior (*Figure 17*). But these models do not identify the energy source that maintains the winds. If they are just coasting without friction, the winds might not need a large energy source to continue in their present state essentially forever.

A complete theory of these deep motions is still a long way off. The problem is similar to modeling the motions in the Sun's interior, but the computational challenges are greater. The giant planets are cold objects that radiate their energy slowly, so their interiors take a long time to reach thermal equilibrium. Yet certain atmospheric phenomena (turbulent waves and eddies, for example) change very rapidly. Even the largest computers cannot simulate both the slow, large-scale processes and the fast, small-scale ones simultaneously. To cope at all, meteorologists must use approximations for the fast processes that convey their net effect on the slow ones. Developing such models, even for the well-observed Earth, is something of a black art.

Consequently, despite our great progress in understanding the atmospheres of the giant planets, many fundamental questions remain unanswered: What energy source maintains the zonal winds? Do the winds and the long-lived ovals really have deep roots, or are they confined to the cloud zone? What accounts for the planet-to-planet differences in the zonal wind patterns, and why do the winds increase as one moves away from the Sun? What accounts for the colors of the clouds? Is Jupiter really dry, or did the Galileo probe just hit a dry spot?

The most interesting questions revolve around the profound differences between the giant planets and Earth. We speculate that the cause of these differences may lie in the fluid interiors beyond our reach. But our speculations may turn out to be wrong — the entire circulation may be taking place in the giant planets' thin outer atmospheres. Further analysis of existing data and more theoretical work should resolve many of the questions. But the best answers will come from more capable telescopes dedicated to the giant planets, more orbiting spacecraft, and multiple probes that penetrate deeper than ever before.

Planetary Rings

Joseph A. Burns

An over-the-pole portrayal of Saturn and its dynamic, complex ring system.

DO NOT know what to say in a case so surprising, so unlooked for, and so novel," confessed Galileo Galilei when Saturn's rings apparently vanished in 1612. At the time, only two years after he had discovered them, the rings were actually just presenting a slim, edge-on view (as they do every 15 years or so when Earth passes through the ring plane). Since their nature as a flat disk encircling Saturn had not yet been realized, Galileo was confounded. The great Italian scientist was actually never sure of their precise form — indeed, at first he believed the rings to be separate bodies, only to think six years later that they were two great arms or handles stretched toward Saturn. Many such interpretations were put forth by skywatchers in the early 17th century (*Figure 1*) as the poorly viewed Saturnian system varied its configuration due to changing Sun-planet-Earth orientations. Not until 1659 did Christiaan Huygens correctly infer the disklike nature of Saturn's "appendages."

Saturn stood as the only planet with a known ring for more than 3½ centuries. That changed in 1977, however, when nine narrow rings were detected around Uranus when the planet occulted a star. Shortly thereafter, in 1979, Voyager 1 discovered a faint band circling Jupiter, and in the early 1980s stellar occultations revealed a set of incomplete "ring arcs" around Neptune. Thus, in but a few years, astronomers had replaced the question of why Saturn alone has rings with another query: why are planetary rings so commonplace, yet so individualistic, in the outer solar system?

Following this rapid growth in our knowledge of ring systems, we have begun to address which properties are fundamental to all planetary rings and which are specific to some in particular. This chapter will first outline some dynamical processes that govern the overall structure of planetary rings, then describe the ring systems encircling Jupiter, Uranus, and Neptune before turning to Saturn's. Along the way will be speculation on the possible origin of planetary rings and their long-term stability, raising the issue of which other planets have had — or will have — rings. Throughout we will find that these planetary ornaments continue to be just as surprising and novel as they were to Galileo.

GENERAL PROPERTIES AND EVOLUTION

Through a modest telescope Saturn's rings appear to be a continuous sheet of matter, broken by at most a single division. Their smooth, nearly opaque appearance motivated Pierre Laplace and Jean Cassini, as well as other early astronomers and mathematicians, to ponder whether the rings were solid, liquid, or particulate. The answer came when James Clerk Maxwell demonstrated theoretically that the rings had to be "comprised

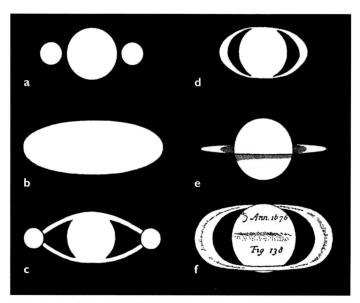

Figure 1. **Our view of Saturn's rings has improved remarkably during the past four centuries, as typified by these early telescopic impressions:** *a*, a drawing by Galileo (1610); *b*, a sketch by Gassendi (1634); *c*, Fontana's view (1646); *d*, a drawing by Riccioli (1648); *e*, Huygens's sketch (1655); *f*, Cassini's drawing (1676), showing his division.

of an indefinite number of unconnected particles," a revelation that won him the Adams Prize in 1857. Four decades later James Keeler confirmed Maxwell's idea by noting that sunlight reflected off the rings was Doppler-shifted such that the ring particles must occupy individual orbits about Saturn, with the innermost ones moving somewhat faster than their outer counterparts.

How might such a system have evolved dynamically? A swarm of particles moving about a central mass will develop in three stages. First, they will swiftly flatten to a thin disk in a specific plane (*Figure 2*). To understand why this flattening happens, consider the fate of two nearby particles orbiting an isolated, spherical planet on separate inclined and elliptical paths. Slight additional forces, caused for example by the planet's oblateness or by an exterior perturbing body like a satellite, will force the orbital planes of the two particles to drift (or precess) gradually relative to a mean plane called the *Laplace plane*. Since the effects of oblateness are strongest near the planet, there the Laplace plane is an extension of the planet's equator; farther out, it lies in the perturbing satellite's orbital plane. As their orbits gradually drift with respect to one another, the two particles can occasionally collide, which reduces the relative velocity between them. In particular, the bodies' motions out of the Laplace plane are lessened until both objects lie essentially in that plane.

For a system of many particles, out-of-plane orbital motion is damped rapidly and continually, since any particle not moving "in step" with the others will pass through the system's mean orbital plane twice every orbit and likely strike another member of the planetary ring. The frequency of these collisions depends on what fraction of the disk's area is filled with particles; for a nearly opaque system like Saturn's main rings, a renegade particle will collide with another during virtually every passage through the plane, or on the average once every few hours. This process is apparently very effective, for the vertical thickness of Saturn's ring system — just tens of meters — is but a tiny fraction of its vast breadth (more than 200,000 km). Saturn's rings are as thin as a sheet of tissue paper spread across a football field.

The mean plane mentioned above depends on the combined influence of planetary oblateness, nearby satellites, and the Sun.

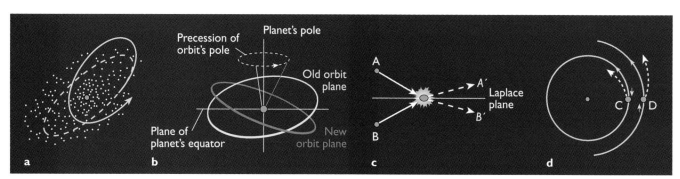

Figure 2. **The gross evolution of a circumplanetary cloud.** In a swarm of particles shown surrounding a central planet (*a*), the dotted line represents the inclined, elliptical orbit of an undisturbed particle. The solid line depicts the orbit of the same particle as its nearly planar path has been gradually modified by perturbations. As seen in a side view (*b*), the pole of an inclined orbit precesses about the pole of an oblate planet's equator; the inclination remains constant, but the orbit's orientation continually shifts (two orbits are shown). Because of relative drifts, the orbits of two nearby particles A and B can intersect (c) to allow particle collisions, which preferentially occur near the Laplace plane. These collisions reduce out-of-plane velocities so that the cloud slowly flattens (dashed lines A', B'). Later in the evolution (d), particles are assumed to have circular orbits. Particle C, nearer the planet, moves faster than (and therefore passes) D. Whenever they collide, equal and opposite forces (colored arrows) push on the particles; C becomes slowed, causing it to drop along the (dashed) path closer to the planet, while D moves farther away.

Since the strength of each gravitational perturbation depends on the distance to the perturber, the mean plane is thus actually a slightly warped surface, looking like a snapped-down hat brim extending out from the planet's equator. The warp amounts to less than 1 km for most planetary rings, since the rings are close to their planets and since the disturbing satellites lie near their planets' equatorial planes. The one exception is Neptune's outermost (Adams) ring: it is pulled more than 200 km off the equatorial plane by massive Triton, which occupies a highly inclined orbit.

Even after the first stage of dynamical evolution ends and the cloud of particles has collapsed to a flattened disk, individual ring members will continue to interact systematically with one another (*Figure 2d*). Under the action of gravity's increased tug, particles close to a planet always move faster than those farther away. Thus, inner particles gradually overtake their outer, slower-moving neighbors, producing a gradient in velocity termed *differential* or *Keplerian shear*. The ensuing collisions tend to retard the inner particles, causing them to drop nearer the planet, while the outer objects are driven away from it. So, after flattening to a thin disk, planetary rings slowly spread out. If unimpeded, this radial diffusion should lead to a featureless disk of material with gradually fading outer edges, unless their peripheries are restrained in some way. The third and final stage in a ring's internal evolution, which is not ever fully achieved, would occur when the ring particles become so widely separated that they no longer collide. After that, the ring does not change dynamically unless external forces act upon it.

Actually, this process of out-of-plane damping never produces a perfectly thin layer, because collisions jostle particles out of the plane. In particular, differential shear motion brings particles of radius r together with a typical relative velocity of ωr, where ω is the orbital angular velocity about the planet. Any collision reorients the relative velocities of the particles involved as well as damping them slightly. Therefore, the vertical component of ωr causes a ring at least several particles thick. The scattering of particles by mutual gravitational interactions can also inflate the ring if the objects are large enough, at least 10 m in Saturn's rings. For a ring containing particles of various sizes, collisions of large and small objects may deflect the latter vertically by many times their radii, while the former may be confined to a relatively thin layer near the central plane.

Low-velocity collisions cause an unbounded ring to diffuse radially inward and outward, as mentioned earlier, at a rate proportional to the square of the particles' radii. The orbits of ring particles also decay gradually under the influence of both *Poynting-Robertson drag* (a decay caused by impacts with photons of sunlight) and *plasma drag* (a drift away from synchronous orbit due to collisions with magnetospheric plasma). However, the rates of orbital evolution caused by these two processes are *inversely* proportional to particle size; that is, bigger objects are affected less. Consequently, any extant narrow ringlets, if formed eons ago, cannot be consist of particles that are too small or too large.

The model of a flattened, nearly opaque disk of particles orbiting a central object pertains to more than just planetary rings. Probably it also is a good representation for one phase in the evolution of the protoplanetary nebula from which the planets supposedly developed (see Chapter 2). However, large objects cannot grow within rings because the latter almost always lie close to planets, within what is known as the *Roche limit*. The Roche limit is $2.456\, R(r'/r)^{1/3}$, where r' is the planet's density, r that of the orbiting object, and R the planet's radius. Anywhere inside this distance a "fluid" satellite (lacking internal friction) can no longer remain intact but instead gets torn apart by planetary tides. At around the same position, particles cannot be held together by self-gravity once brought together in collisions. Nevertheless, solid satellites, because of their material strength, can exist inside the Roche limit; above a certain size, however, they too will be broken apart by tidal forces. In connection with ideas about ring origin, if a large satellite were to rupture, the resulting fragments would occupy similar orbits and continually chip away at one another until the largest members were no larger than some tens of kilometers in diameter.

DYNAMICAL CONCEPTS

The realization that virtually all planetary rings lie within the Roche limits of their planets has naturally prompted the long-held popular speculation that ring systems arise when comets or satellites stray too close to a planet and break apart due to tidal stress. However, these same powerful tides should also have prevented material initially within the Roche limit from accreting in the first place. In other words, if a disk of debris extended to great distances from a just-formed planet, material far away could have accumulated into satellites, whereas that within the Roche limit would not have grown so easily (*Figure 3*). In addition, whatever existed very close to the planet might even have been lost to its surface by Poynting-Robertson and atmospheric drag. Thus any primordial circumplanetary disk should ultimately develop into a system with exterior satellites, rings, and a gap just above the planet's surface — much like what we see today around the giant planets.

As just stated, when far enough from the central planet, a disk of individually orbiting objects will gradually aggregate into a few larger bodies. However, dynamicists Stuart J. Weidenschilling and Donald R. Davis have proposed that, near the Roche limit, the large aggregations neither fully coalesce nor entirely disrupt. Rather, groups of particles cluster together temporarily — only to be torn asunder by tides once they have grown large enough or spin fast enough. In this view, Saturnian ring members are "dynamical ephemeral bodies," changing their individual components every few weeks. In Saturn's F

Figure 3. **The dynamical behavior of a swarm of particles (*a*) orbiting Saturn as numerically simulated by Heikki Salo, including mutual gravity and energy dissipation. Collective gravitational wakes form about 100,000 km from the planet's center (*b,c*); these degrade into stable, self-gravitating clumps beyond about 140,000 km (*d*). As theoretically derived by Roche for fluid bodies, cohesionless clusters are able to retain their integrity beyond this distance but are torn apart by tides when closer to the planet.**

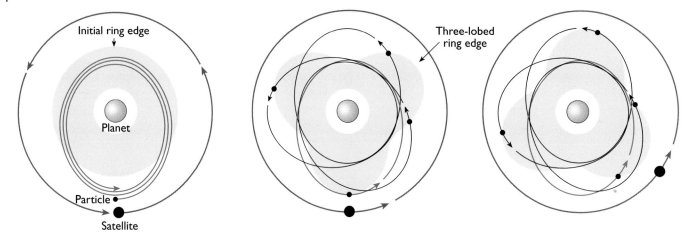

Figure 4. These diagrams show the effect of a hypothetical 3:1 resonance between a satellite and a particle. In any such resonance, a particle at the ring's edge (left panel) completes exactly three orbits in the time the satellite takes to make one (blue path). Thus, as seen from space, when the satellite next returns to the bottom of the left panel, the particle will have gone thrice around the planet (green paths). The satellite's repeated gravitational tugs force the particle into an elliptical orbit, which reaches its greatest distance from the planet as the particle passes beneath the satellite. That is, because of the threefold faster motion of the ring particle, three peaks will occur around a circumference. Therefore, the edge assumes a three-lobed shape, as seen from the satellite's reference frame (center panel). The thin black line shows the particle's path as viewed in this rotating frame. To visualize this motion, note the system's appearance a short time later (right panel), during which the particles have continued to move along their elliptical tracks. The net effect of their orchestrated motion is that the three-lobed edge appears to co-rotate with the satellite.

ring, bright clumps have been observed to form quickly and then dissipate gradually.

Another key orbital distance is the *synchronous radius*, R_S, at which a satellite orbits in the same time as its planet rotates. It can affect the evolutionary paths of ring particles in two ways. First, planetary tides push satellites outward when they orbit beyond R_S but inward when they are closer; thus the orbits of large, nearby objects can gradually collapse toward planets. Second, electromagnetic forces, which may overpower gravity for small particles (say, those much less than 1 micron across), reverse direction in a relative sense at R_S. As such, material may either accumulate there or move away from it.

Satellites undoubtedly play a pair of fundamental roles in sculpting the planetary rings we see today. First, they gravitationally perturb the orbits of ring particles, especially at resonances (discussed below). In this way satellites account for much of the understood gross form of Saturn's rings. They confine the narrow Uranian rings, produce the Neptunian arcs, and likely influence the structure of the Jovian ring. Second, the largest objects embedded within a ring (which could themselves be called satellites, though we will call them "mooms") can serve as either sources or sinks for ring particles. Moreover, collisions with high-speed interplanetary projectiles may drive debris off these mooms, and over time many new small particles are created at the expense of large ones. In combination, these effects influence the particles' total surface area and may account for some of the observed brightness structure of planetary rings.

Resonant orbits are conspicuous features of the solar system's makeup today, in that matter is often either absent from, or preferentially present in, such locations. Resonances between two objects occur at those positions where the orbital period of one of them is an integer fraction of, or "commensurate with," the other's period (*Figure 4*).

Numerous examples exist where resonances seem to determine structure. For instance, the *Kirkwood gaps* (see Figure 2 in Chapter 25) in the asteroid belt contain substantially fewer minor planets than do adjacent locations; these gaps occur where the orbital period of Jupiter (the primary perturber of the asteroid belt) and local orbital periods form a simple ratio like 2:1 or 5:2. Numerical studies have shown that orbits at the 3:1 Jovian resonance become chaotic and develop large eccentricities; over time some asteroids traveling along such elongated ellipses can collide with a planet like Mars or Earth. Resonant positions are not always vacant, however, and indeed sometimes they may be unusually populated: a family of asteroids called Hildas have orbital periods two-thirds that of Jupiter, while literally thousands of Trojan asteroids match Jupiter's period and share its orbit.

The Cassini division, which separates the two major classical rings of Saturn, is like a Kirkwood gap, since a particle at its inner edge (the B ring's outer boundary) would have a period one-half that of the moon Mimas. Objects at the periphery of Saturn's A ring have orbital periods that are in a 7:6 resonance with the co-orbital satellites Janus and Epimetheus. In addition, many low-density regions in Saturn's A ring are located at positions resonant with several moonlets, all of which lie just exterior to the ring system.

Resonance patterns are also seen in satellite systems. For example, the orbital periods of Jupiter's satellites Io, Europa, and Ganymede have a ratio of 1:2:4. In any resonance, given configurations are repeated periodically. For the just-mentioned Laplace resonance of the Galilean satellites, these alignments continually reinforce (through gravity) the eccentricity of Io's orbit and ultimately play a fundamental role in creating its vigorous volcanism (see Chapter 17). Resonances and near-resonances are likewise common among the Saturnian and Uranian moons, where some suspect that they have played a role in heating the interiors of enigmatic Enceladus and Miranda.

Despite these many examples, however, Voyager pictures of planetary rings show that the situation is more complicated: there is *not* a simple one-to-one correspondence of resonant locations with ring structure. Even the Cassini division of Saturn's rings, while generally containing less material than its surroundings, is filled with many ringlets.

In the past the dearth of particles in resonant ring gaps was thought to be caused by a buildup of perturbations at such positions, as the involved particles and satellites periodically repeated their relative configurations. In a sense the mechanism is much like rhythmically pushing a swing until its occupant arcs high above the ground, whereas randomly timed pushes would accomplish little. Since orbits are most disturbed near resonances (so the argument goes), particles there run into one another more frequently and more violently than average, which inhibits the accumulation of material. However, this is probably only a small part of the story. Except in a very narrow band to either side of the exact resonance, adjacent particles are almost equally perturbed and march side-by-side along similarly affected orbits.

A more likely cause was proposed in the late 1970s by Peter Goldreich and Scott Tremaine, who predicted that *spiral density waves*, like those thought to operate in galaxies, would originate at resonant locations in circumplanetary disks. Their mechanism operates much like that just described, but with an additional twist. As before, the gravitational attraction of a satellite in the ring plane alters the orbits of ring particles, making them elliptical. In a populous disk this perturbation causes particles to bunch up, especially near resonances (*Figures 5,6*). However, the combined gravity of these clumped particles pulls on the rest of the disk, producing condensations and rarefactions elsewhere and ultimately giving rise to a spiral-shaped density wave that propagates radially. If the perturbing satellite lies outside the ring, this wave moves outward. An entrained ring particle, if allowed to proceed without interference, would eventually return to its original position. However, the rings are dense enough to make collisions probable, which robs the particles of energy and angular momentum, causing them to "fall" inward toward the planet. Over time, this process can clear a wide gap just outside the resonance position.

The physics deduced by Goldreich and Tremaine also causes a satellite and a nearby planetary ring to repel each other. In fact, "shepherd satellites" were originally proposed to account for the narrow rings surrounding Uranus, constraining their edges from the otherwise inevitable spreading (*Figure 7*). The shepherding mechanism received substantial vindication when Voyager 1 discovered two satellites herding the narrow outer F ring

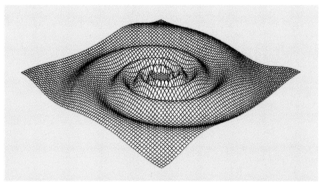

Figure 6. If resonant perturbations are induced in rings by a satellite with an inclined orbit, the out-of-plane tugs induce a vertical bending wave, as shown in this computer schematic of a two-armed spiral. The vertical displacement of these waves, about 1 km (shown greatly exaggerated here), is 10 to 100 times greater than the physical thickness of Saturn's rings.

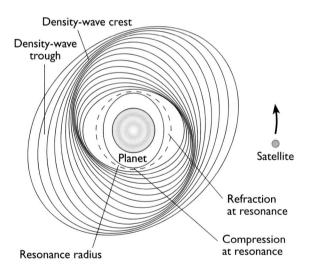

Figure 5. The creation of a spiral-density wave begins when a two-armed spiral wave is excited at the 2:1 resonance (shown dashed) with the exterior moon. The ovals represent particle paths as seen in the frame rotating with the orbiting satellite. Their long axes become less and less well aligned with the satellite's direction at greater distances from the resonances. (This plot is closely related to *Figure 4* except the resonance is now a 2:1, rather than 3:1, and the paths of particles just outside the resonance are shown.) The clustering that occurs in the orbital paths induces coherent oscillations in neighboring particles as they drift past by Keplerian shear. Actual spiral waves are much more tightly wrapped than shown here.

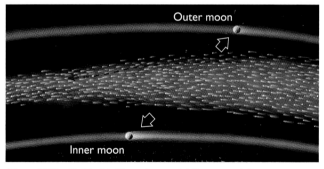

Figure 7. The Goldreich-Tremaine model for constraining narrow rings employs "shepherd satellites" that force a group of particles to travel along narrow paths. Moving slower than the ring interior to it, the outer satellite attracts particles going by. This attraction is slightly greater after particles have passed the satellite, because when slipping by it their paths were pulled somewhat closer (not shown) to the shepherding object. This extra force causes particles to lose energy and "fall" closer to the planet (red dashed line). Conversely, the faster-moving inner satellite adds energy to nearby particles and kicks them onto higher orbits (blue dashed line). Together, these forces herd the particles into a narrow ring. The same process can explain how an embedded satellite could open a gap in a disk, by spreading the system apart.

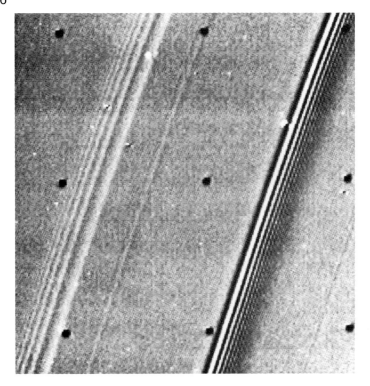

of Saturn, and then again when Voyager 2 pinpointed yet another pair astride Uranus's ε ring. The spacecraft also discovered dozens of spiral-density wave trains, primarily in Saturn's outer A ring. Through very similar processes, inclined satellites can pull ring material up out of the ring plane, producing a local rippling of the surface that propagates away from the resonance location. However, only a few of these bending waves have been identified in Saturn's system (*Figure 8*).

We can gain considerable insight into the nature of the planets' ring systems by matching actual ring characteristics against the dynamical concepts described previously, and also by comparing the known systems with each other (*Figure 9, Table 1*). For example, most rings are very thin relative to their extent. Essentially all of them lie within their planet's respective Roche

Figure 8 (left). **Recorded by Voyager 2 in Saturn's A ring, these two wave features are very tightly wound and resemble watch springs. Both are caused by the satellite Mimas. The outer (left) one, a spiral density wave, is a series of particle-density fluctuations propagating outward (toward left). The inner one (right), a spiral bending wave, propagates inward as a train of vertical corrugations. This image is roughly 1,000 km across. The regularly spaced pattern of black dots is a camera artifact used to correct geometrical distortion.**

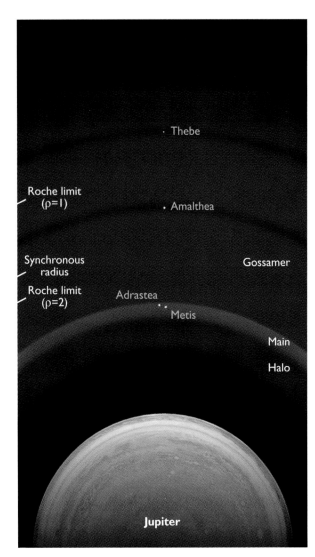

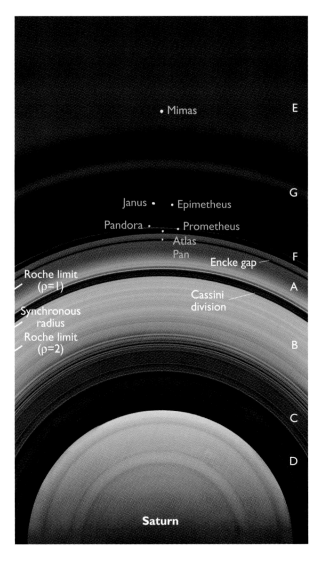

limits and, for the most part, the synchronous-orbit radius. The opacity of rings differs markedly between systems and within a system: nearly opaque in places, yet diaphanous elsewhere. Almost certainly, a range of particle sizes exists in each ring. The material in Saturn's rings is highly reflective, as are most of Saturn's icy moons. Yet particles in the other ring systems are dark, which suggests that silicate- or carbon-rich material — not ice — dominates their surfaces. Small perturbing satellites border (and mingle with) all the ring systems.

As we shall see, all of these characteristics challenge our understanding of the rings' current state, origin, and evolution. Their study has been helped tremendously by the Pioneer and Voyager spacecraft, whose flybys in the 1970s and 1980s provided detailed observations unobtainable from Earth. This wealth of information has been supplemented in the 1990s by telescopic observations and imaging of the Jovian ring by the Galileo spacecraft.

Figure 9 (below, facing pages). The known planetary ring systems are compared by making the planets' radii all the same size. Illustrated here are the distribution of ring material, nearby satellite locations, the synchronous-orbit radius R_s, and the Roche distances for satellites having either the density of 1 g/cm³ (that of water) or 2 g/cm³.

Properties of Planetary Ring Systems

	Jupiter	Saturn	Uranus	Neptune
Width of main and narrowest structure (km)	7,000 < 100	20,000 < 0.01	100 < 0.01	< 50 < 15
Thickness (km)	< 30 (halo: 10^4)	0.01–0.1	0.01–0.1	30
Optical depth	1–6×10^{-6}	0.1–2	0.1–2.3	0.1–0.4
Albedo	< 0.05	0.2–0.8	0.015	0.03
Particle sizes	10^{-3} mm	cm–5 m	10 cm–10 m	cm–m (?)
Surface mass density (g/cm²)	10^{-10}–10^{-3}	10–100	10–100	?
Total mass (g)	10^{11}–10^{16}	10^{21}–10^{22}	10^{18}–10^{19}	?

Table 1. Each outer planet has a ring system whose characteristics are distinctly different from the others'. *Optical depth* refers to the fraction of light able to penetrate an obstructing layer; for small values, it is essentially the fraction of photons that cannot pass through. The typical *particle sizes* listed are based on numerous assumptions and should not be taken too literally. *Surface mass density* is the optical depth multiplied by the average particle's size and density, and *total mass* is the ring's surface mass density times its area. The extensive ring system around Saturn, if compressed to a single object, would form a small satellite no more than 100 km across.

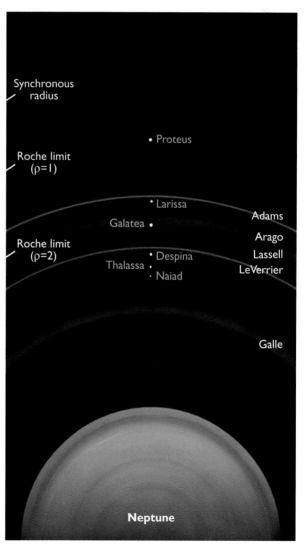

THE RING OF JUPITER: FAINT YET DISTINCT

As Pioneer 11 traversed Jupiter's magnetosphere in 1974, its counts of high-energy particles decreased when the spacecraft came closest to the planet's center, a distance of 1.6 Jovian radii (R_J) or about 114,000 km. This unexpected falloff prompted speculation that the spacecraft might have passed under the path of a hitherto undetected ring or a satellite, which would absorb charged particles that it encountered. Most scientists dismissed this possibility, since ground-based observations had not revealed any such object and other explanations of the reduced counts were available. Five years later Voyager 1 was programmed to look along Jupiter's equatorial plane while traversing it. In doing so, the spacecraft detected a faint belt of material encircling the planet.

In analyzing observations taken by Voyager 2, investigators distinguished three components for the Jovian ring, a morphology confirmed by Galileo's observations in 1996–97 (*Figure 10*). The main band starts abruptly at 1.81 R_J and ends more gradually at 1.72 R_J, 7,000 km closer to the planet. Its thickness is no more than 30 km. Even in the brightest part of the ring, matter covers a very small fraction of space (its optical depth is a few times 10^{-6}). The Jovian ring's most novel feature, a toroidal halo with a comparable optical depth, arises at the main ring's inner edge and rapidly expands inward to a full thickness of roughly 20,000 km; it is faintest where closest to the planet and farthest from the equatorial plane. Exterior to the main band is the so-called gossamer ring, which is even more tenuous. Its brightness decreases outward from the main ring until it fades into the background beyond Thebe's orbit at 3.1 R_J. In Galileo observations the gossamer ring is seen to have at least two components, one inward of — and presumably derived from — Amalthea, and the other connected with Thebe (*Figure 10b*). Thanks to improved infrared detectors, all components of Jupiter's ring are now observable from Earth (*Figure 11*).

The main ring exhibits three characteristics. First, it is always brighter inward of a satellite's orbit, possibly due to material drifting into such zones from the satellite. Second, it has brightness fluctuations that might be produced by vertical or horizontal waves. Finally, it thickens vertically with decreasing orbital radius, which presumably indicates the growing importance of electromagnetic forces on the small grains that make up most of the visible ring.

Voyager and Galileo found that the Jovian ring is many times brighter in forward-scattered light (which is redirected slightly by diffraction off its path from the Sun) than in back-scattered (reflected) light. This implies that a significant fraction of the ring particles are only 1 or 2 microns across. Such small objects have correspondingly short lifetimes because Poynting-Robertson drag, plasma drag, and perhaps other electromagnetic effects

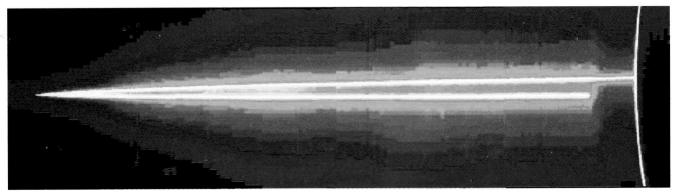

Figure 10a. As the Galileo spacecraft lay hidden in Jupiter's shadow on 8 November 1996, it captured a nearly edge-on view of the planet's sunlit ring. Rendered here with exaggerated color-coding to indicate relative brightness, that image shows the relatively bright main ring and, surrounding it, the tenuous halo. The unusual halo probably results from the "levitation" of micron-size particles out of the ring plane by electromagnetic forces. Because of shadowing, the ring appears truncated close to the edge of Jupiter's disk at far right.

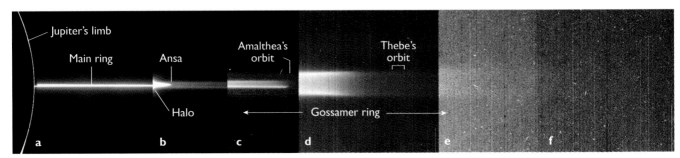

Figure 10b. Galileo was in Jupiter's shadow and only 0.15° from the planet's equatorial plane when it recorded this six-image mosaic of the Jovian ring system in September 1997. The tiny ring particles forward-scatter sunlight quite strongly, and phase angles here range from 177° to 179°. Exposure times vary between frames. The overexposed ansa of the main ring (in frame *b*) is surrounded by the halo, visible as a faint smudge. Extending faintly from the ansa is the Amalthea gossamer ring, which disappears abruptly very near this small satellite's orbit (*c*). Frame *d*, a much longer exposure, displays the broader Thebe gossamer ring, which likewise appears to end near Thebe's orbit. The bright top and bottom edges of both gossamer rings agree roughly with the maximum vertical excursions that their respective moons make away from the midplane, additional evidence that Amalthea and Thebe are sources of ring particles.

cause them to leave the system within 1,000 years or even less. Meanwhile, in a comparable time they are shattered by micrometeoroid impacts or eroded more gradually by the energetic particles that are abundant in Jupiter's magnetosphere.

Given its short lifetime, the dust around Jupiter must be regenerated continually if the ring is a permanent feature. Most likely, fast-moving projectiles from outside the Jovian system are bombarding boulder-size objects within the ring, chipping off flecks of dust in the process. The impacting projectiles are probably interplanetary micrometeoroids but could even be volcanic dust drawn out of Io's rarified atmosphere and transported inward.

The magnetospheric plasma that surrounds Jupiter continually sweeps past the ring particles and creates a negative electric potential on their surfaces. Because of this charging, micron-sized ring particles feel a force due to the planet's magnetic field about one-thousandth that from the planet's gravity. At most places in the ring, such small perturbations will induce comparably small oscillations about a circular path. However, plasma drag causes particles from the main ring to evolve inward, where the Jovian magnetic field and its consequent effects on charged particles intensify. When grains reach the inner regions of the main ring they experience electromagnetic forces that vary in time with the same periods as their orbits. At such *Lorentz resonances*, an initially circular orbit becomes substantially eccentric and inclined. Grains that drift farther inward encounter another Lorentz resonance near the halo's interior boundary, where their inclinations and eccentricities are increased even more.

Throughout this evolution the grains are continually being ground down and thus subjected to ever-larger electromagnetic perturbations. Eventually they become minuscule enough (about 0.03 micron across) that electromagnetic forces acting on them overpower the effects of gravity. Once this point is reached, the particles are swiftly yanked totally out of the ring plane to travel along field lines into the planet's atmosphere. Because the Jovian ring is relatively uncrowded, this microscopic dust gets pumped into the halo before interparticle collisions can dampen its motion. Thus, the Jovian halo does not flatten as other rings do.

Jupiter's main ring contains a variety of particle sizes. Very tiny grains are by far the most plentiful, and these are almost surely impact debris. Their unseen parent bodies are responsible for the absorption of energetic particles observed by Pioneer 11. These larger objects are dark and red (much like the nearby small satellites), probably contaminated by the material from Io that pervades Jupiter's inner magnetosphere. The largest fragments may be the outcome of the catastrophic breakup of a small satellite, or they may just represent uncompleted accretion of a satellite located within the Roche limit. Whatever their origin, these dominant blocks have gradually spread apart from differential shear and drag, filling the main region populated by dust. The satellites Adrastea (a 15-km-wide object that skirts the ring's outer boundary) and Metis (more than 40 km across and located a little further into the ring) may be merely the largest of these mooms.

Despite its insubstantial nature, the tenuous band around Jupiter has illuminated our understanding of planetary ring systems because its sheerness and the smallness of its constituent particles accentuate certain phenomena. In particular, it is

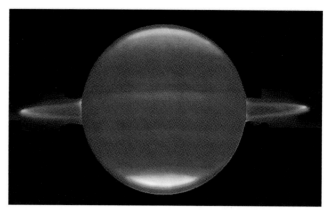

Figure 11. Jupiter's tenuous ring can be seen by groundbased telescopes, but only barely. This is a false-color composite of two nights of observations with the 10-m Keck I telescope in August 1997, when Jupiter's ring plane was nearly edge on to Earth. It was recorded at 2.27 microns, an infrared wavelength at which methane in Jupiter's atmosphere absorbs so strongly that the planet's disk appears no brighter than the faint ring. Shining due to reflected sunlight, the ring remains visible but is still quite dim. To be apparent here its image still had to be enhanced considerably.

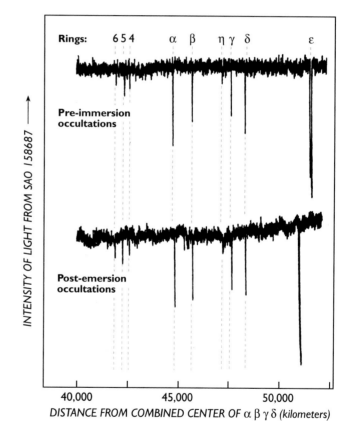

Figure 12. In this set of discovery data, light from the star SAO 158687 was recorded before (upper trace) and after (lower trace) its occultation by Uranus's disk on 10 March 1977. Dips in the tracing are due to each of the nine Uranian rings; apparent fluctuations in the star's brightness are caused by system "noise." Note that the starlight is diminished at nearly the same distances on either side of the planet, implying the rings are nearly circular. Ring ε is the most eccentric and shows different widths on opposite sides of the planet. Most rings obscure the star very well and rather abruptly, so they must be quite opaque and have well-defined edges. In order to have these properties, small satellites must exist nearby to prevent the rings from spreading.

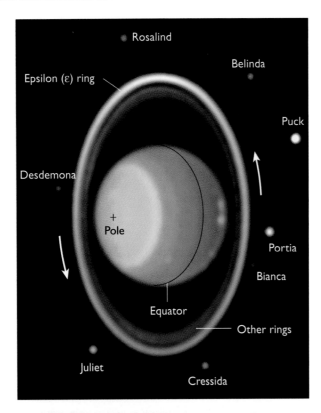

apparent that small satellites supply much ring material, with the remainder coming from relatively large moons embedded in Jupiter's ring. Accordingly, we wonder whether the other ring systems are also fed from as yet unseen reservoirs of material.

URANUS'S RINGS: NARROW AND NEAT

The way in which a star's light is extinguished as a planet passes in front of it tells much about the planet's atmospheric properties and also permits, by the length of the chord traced behind the disk, a measurement of the planet's shape. With this in mind, several teams of astronomers trekked to observatories surrounding the Indian Ocean to witness the occultation of a star by Uranus on 10 March 1977. They received an unexpected bonus when a series of slender rings revealed their presence around the planet by interrupting the starlight. The best observations (*Figure 12*) came from a group watching the occultation from high above the Indian Ocean aboard the Kuiper Airborne Observatory.

Figure 13 (left). **A view of the inner Uranian system, taken by the Hubble Space Telescope on 28 July 1997, shows the intimate relation between the planet's rings and small satellites. The image is a false-color composite of three infrared frames, wavelengths at which Uranus appears relatively dim but the rings and moons do not.**

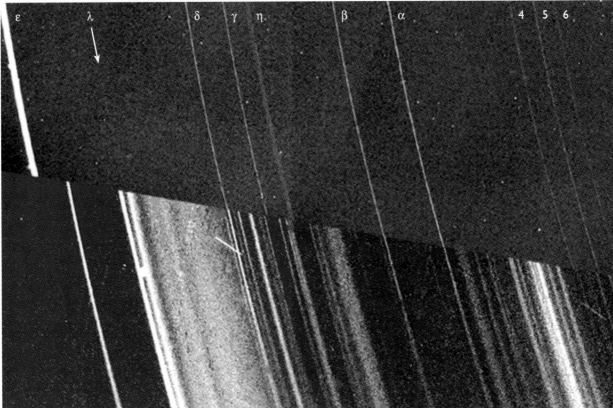

Figure 14. Compare these images of the Uranian ring system. Voyager 2 acquired the upper one the day before it passed Uranus in January 1986; at the time sunlight striking the ring particles was reflected back toward the camera. Computer enhancement makes the rings appear much brighter than they really are. (To see the faint λ ring, put your eye close to the page bottom and look toward the arrow.) In the lower image, taken somewhat later, Voyager 2 was looking almost directly back toward the Sun's

direction (the phase angle is 172.5°). This backlighting of the rings dramatically enhanced the visibility of any micron-size dust particles they contain. Voyager 2's motion during the exposure caused some smearing of the detail, especially near the bottom edge. The nine rings discovered from Earth can be discerned with relative ease (note the poor segment match for the markedly eccentric ε ring). However, many other ringlets visible here do not correspond to known features in the system.

During the 1977 occultation the rings attenuated starlight substantially at approximately symmetric points about the planet; thus observers concluded that the Uranian rings are densely packed and are almost circular. The observations revealed a total of nine rings, which were named in order of increasing distance from Uranus: 6, 5, 4, α, β, η, γ, δ, and ε. Over the years numerous subsequent stellar occultations have refined the shapes and relative locations of these rings — which are 3 billion km from Earth — to a precision of a few hundred meters! Compared to their circumferences (some 250,000 km), the rings are remarkably narrow; most do not exceed 10 km in width, and only the outermost one, ε, spans as much as 100 km.

At least six Uranian rings are inclined with respect to the planet's equatorial plane, but typically by no more than a few hundredths of a degree. They are also slightly out-of-round, with eccentricities of 0.001 to 0.01, and have variable widths. The shapes of several rings seem to pulsate slightly; one "breathes" in and out, while another changes its eccentricity somewhat. The ε ring, among the few not inclined, is by far the most massive and eccentric one in the system. Its distance from Uranus varies by about 800 km and its width changes proportionately: from 20 km (where closest to Uranus) to 100 km (where farthest from it).

A planet's oblateness causes elliptical and inclined rings to precess slowly about the planet. In the Uranian case, the rings precess as rigid structures, which is surprising since oblateness should drive a ring's inner edge more rapidly than its outer boundary, smearing each ring into a circular band in no more than a few hundred years. Apparently, therefore, the rings' eccentricities must be continually induced, either by satellites or by the ring material itself. However, while models that employ this concept seem generally correct, they still fall short of matching our observations completely.

Uranian ring boundaries, like many features in the Saturn system, are remarkably crisp, suggesting the ineffectiveness of the differential shear that should force particles to collide and thus diffuse smoothly. According to current ideas, sharply defined rings can be explained only if satellites bound the edges in some way, either by shepherding or by trapping ring particles along so-called horseshoe orbits that almost precisely duplicate the satellite's path. If nearby, such moons need be only a few kilometers across and could thus have escaped detection by Voyager 2's instruments and ground-based telescopes. At Uranus we find a close tie between small ring-moons and the rings themselves (*Figure 13*), a characteristic also shared by the Jovian and Neptunian systems. However, the spacecraft did confirm, in remarkable detail (*Figure 14*), our impressions of the rings inferred from ground-based observations. In addition, Voyager 2 discovered two faint rings of its own: λ, a narrow strand between the δ and ε rings, seems to contain predominantly small particles and may be longitudinally variable; 1986U2R begins about 1,500 km interior to ring 6 and has a radial extent of about 3,000 km. The spacecraft also found a pair of small satellites that shepherd the ε ring (*Figure 15*) and probably also constrain edges of rings δ, γ, and perhaps λ.

According to visual and radio observations, the nine "classic" rings are composed mainly of meter-size boulders and contain scarcely any dust at all. These strands are among the darkest objects in the solar system, reflecting just a few percent of the feeble sunlight that strikes them, and they have neutral colors. Such dark color, which the rings share with the numerous small moons that Voyager found nearby, may result from the magnetospheric bombardment of surfaces that contain some organic molecules.

While in Uranus's shadow, Voyager 2 briefly peeked sunward toward the Uranian rings, which presented quite a startling image in forward-scattered light (*Figure 14*). The entire region inward of λ is filled with faint dust having optical depths of 0.0001 or less. It is organized into many slender ringlets, among which the nine rings are unexceptional members. Most of the ringlets are remarkably narrow, which is surprising since Poynting-Robertson and plasma drag should be driving small grains into Uranus on relatively short time scales — in no more than a million years for a micron-sized grain. However, the actual lifetimes are probably much shorter, because the outermost vestiges of Uranus's atmosphere should extend into the ring region and thus hasten the orbital collapse of particles near the planet. Mechanisms like those at work in the Jovian ring system probably account for the continued presence of Uranian dust. A major puzzle is what causes the ringlet structure, with theorists now concentrating on the roles that embedded moonlets play as both sources and sinks for dust particles.

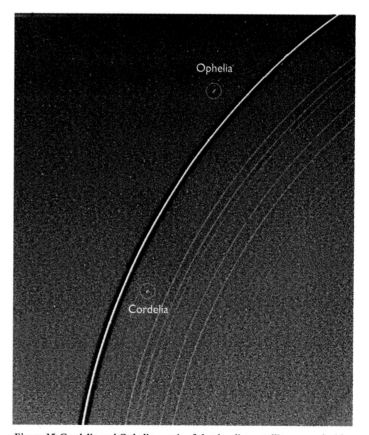

Figure 15. Cordelia and Ophelia, a pair of shepherding satellites on each side of Uranus's ε ring, keep the ring particles in place through resonant gravitational forces. No other shepherds for the other rings were spied by Voyager 2 during its 1986 flyby. The λ ring, located just outside the orbit of Cordelia, is too faint to be visible in this image.

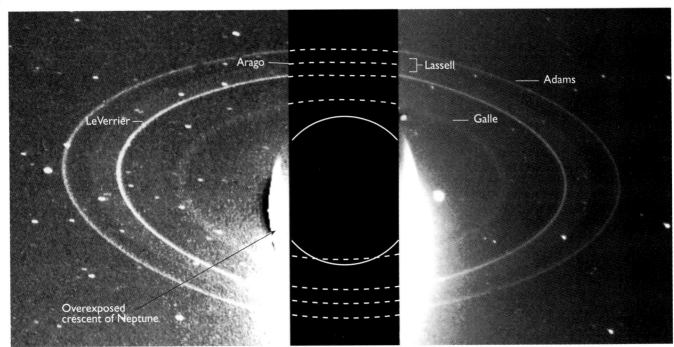

Figure 16. A pair of Voyager 2 portraits of Neptune's ring system, which appears backlit by feeble sunlight. The ring arcs seen in *Figure 17* were not captured in either of these images; by unfortunate coincidence they were on the opposite side of the planet when these frames were taken 1.5 hours apart.

Figure 17. A long-exposure, forward-scattered image of Neptune's outer two rings shows that the ring arcs are in fact clumps of material within the outermost ring, Adams, which is about 50 km wide. The direction of motion is clockwise; the longest arc (Fraternité) is trailing. Neptune's highly overexposed crescent is at lower right.

NEPTUNE'S ENIGMATIC RINGS AND ARCS

Soon after the late-1970s' discovery of rings around Jupiter and Uranus, planetary astronomers began to search in earnest for material around Neptune using the stellar-occultation technique that had been successful at Uranus. Scientists were surprised that the first half dozen occultations found nothing. They were even more startled when, on 22 July 1984, an event was seen on only one side of the planet. The detection was clearly real because two teams of observers separated by about 100 km saw similar events: it appeared as though the starlight was being obscured by a moderate optical-depth, 10-km-wide strip of material with abrupt edges at about 50,000 km above Neptune's cloudtops. Over the next few years, similar detections occurred but only about 10 percent of the time. The hit-or-miss nature of the occultation data clearly indicated that Neptune's rings, at least as probed from the Earth, were in fact just partial arcs.

Modelers were delighted when Voyager observations in 1989 revealed the odd ring morphology (*Figures 16,17*). Neptune's outermost ring, now named Adams, was narrow and continuous, as a well-behaved ring should be. However, it also contained three and perhaps five longitudinally confined arcs (the main ones have been named Fraternité, Egalité, and Liberté). These arcs varied in angular extent from 1° to 10°, and all lay within a 40° strip of longitude. Neptune has four other continuous rings: Galle and Lassell are broad, while LeVerrier and Arago are narrow. One more, which is discontinuous, shares the orbit of the ring-moon Galatea just interior to the Adams ring. The Neptunian bands, especially Adams and its arcs, seem quite dusty, much like Saturn's disturbed F ring. Dust is also present throughout Neptune's inner magnetosphere, even in its polar regions, perhaps because of the planet's highly tilted magnetic field (see Chapter 4) or the odd orbit of Triton. The ring material is very dark, like that of the Uranian rings, and it may be red.

This intimate mix of small satellites and dusty rings, all situated within the Roche zone, is also present in the systems surrounding Jupiter and Uranus. Detailed numerical simulations by Joshua Colwell and Larry Esposito, based on an initial model by Jeffrey Cuzzi and the author, show how disrupted satellites can provide the source of the Neptune rings. If true, the ring systems probably undergo continual change in appearance.

The orbital geometry and phasing of the ring arcs may be due to gravitational effects from the 150-km-wide moon Galatea, which circles Neptune only about 1,000 km inside the Adams ring. The moon and ring have orbital periods in an exact 42:43 ratio. According to ring specialist Carolyn Porco, the moonlet perturbs the ring's dusty debris in two ways. A *Lindblad resonance* constrains particles from moving inward or outward, forcing them to occupy a thin band. Because its orbit is inclined 0.03° with respect to the ring plane, Galatea also induces a *corotation resonance* that keeps the particles confined to short segments spaced about 4° apart. (Theorists were fooled initially by the arcs' apparent separations of 12° to 13°, not realizing that these were simple multiples of shorter segments.) It is not so clear, however, why so few resonant sites are populated.

SATURN'S SYSTEM: RINGS GALORE

Ground-based observations at infrared, visible, and radio wavelengths have provided considerable information on the properties and overall form of Saturn's celebrated rings. These results were all but eclipsed by the findings of the Voyager spacecraft, which resolved previously unimaginable architecture and taught us much about the size, shape, and distribution of particles in the system.

Saturn's rings are much more elaborate and complex than any of the other systems (*Figures 18–20*). The bright, classically recognized components A and B are separated by the Cassini division, which was first noticed in 1675. Optical depths average 0.5 to 0.7 for the A ring but rise to between 1 and 2 for the B ring. Interior to these is the crepe or C ring, which was recognized in 1850 and has an average optical depth of roughly 0.1. Voyager 1 detected some material, including a few dusty, narrow, widely spaced bands, inside the C ring that extends at least halfway to the planet; however, this D ring is essentially undetectable from Earth.

Outside the traditional system lies the E ring, which is so tenuous (optical depth: about 0.00001) that it is scarcely more than a slight concentration of debris in the satellites' orbital plane. It becomes visible only when the ring system is viewed approximately edge-on, as occurred in 1995–96; then one can see that material extends to at least Rhea's orbit and thickens to perhaps 30,000 km at its outer edge (*Figure 21*). This ring is composed almost entirely of 1-micron grains, which have correspondingly short lifetimes and thus were only recently injected into the system. The E ring's density peaks near the orbit of Enceladus. Consequently, this uniquely smooth and bright satellite might well be the source of the ring's material, which perhaps is debris thrown off the satellite by meteoritic collisions.

In 1979, Pioneer 11's crude imaging system located the slender F ring just 4,000 km beyond the A ring. The spacecraft also found that charged particles were intermittently absent from this region (see Chapter 4), which hinted that undiscovered moonlets or rings might also lurk in the vicinity as well. When the Voyagers arrived in 1980 and 1981, they indeed found several new satellites skirting the edge of the rings and confirmed the existence of two more that had just been properly identified by terrestrial astronomers. Seen in detail (*Figure 22*), the F ring turned out to be a contorted tangle of narrow strands. The Voyagers also glimpsed a ring of their own, named G, which is located about 2.8 Saturnian radii out, is surprisingly narrow (8,000 km wide), and is even more tenuous than the E ring.

When viewed edge on, Saturn's main rings remain faintly visible with a brightness corresponding to a ring that is about 1.4 km thick (*Figure 23*). However, this turns out to be something of an illusion. Several lines of evidence suggest the actual thickness of most segments of the A and B rings is only several tens of meters. Moreover, the apparent thickness of the ring overall is due principally to the fact that the F ring's inclination makes it extend above the plane of the main system. Other possible contributors include a localized thickening due to vertical ripples (bending waves), an "atmosphere" enveloping the rings, and an overall warping of the ring's Laplace plane due to Titan and the Sun. Surprisingly, the times at which the west and east ring

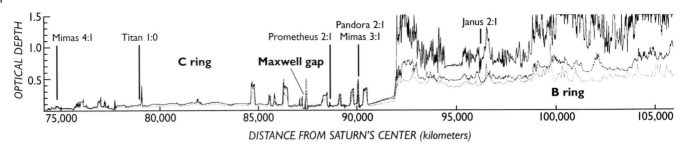

Figure 18. Saturn's bright, overexposed limb is visible through its rings as seen by Voyager 1 from 1.6 million km above and beyond the planet. The C ring is scarcely visible as it crosses the planet's limb. The material in parts of the B ring is so closely packed that it entirely screens the planet in places; radial spokes near the top center of the B ring are bright in this forward-scattered light. The outermost and threadlike F ring displays its characteristic brightness variations.

Figure 19. The rings of Saturn, in a computer-enhanced image from Voyager 1. On the far left, the thin F ring is accompanied by one of the two shepherding satellites that skirt its edges. Interior to this lies the bright A ring (the darkest gap within it is called the Encke gap; black spots are camera artifacts). As seen here, the Cassini division contains four ringlets, each about 500 km across. The broad B ring does not exhibit as much structure because it is more uniformly opaque — though at higher resolution (as in *Figure 22*) it is very complex. The inner C ring shows many narrow ringlets up to the point where it ends abruptly.

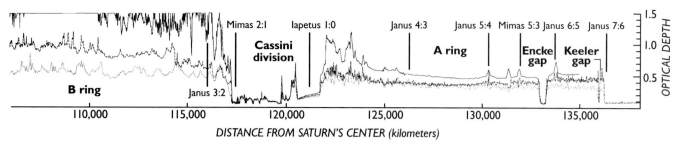

DISTANCE FROM SATURN'S CENTER (kilometers)

Figure 20 (above, facing pages). The optical depth of Saturn's rings, as determined by occultations at three wavelengths. Various named features and resonance locations are labeled. In visible light (solid line), which is affected by particles larger than 1 micron, the rings' opacity can vary greatly even over short distances. The dashed and dotted curves were obtained when the rings interrupted the Voyager spacecraft's radio signals at wavelengths sensitive to particles larger than about 1 and 4 cm, respectively. Separations in the three signals' strengths indicate the area covered by particles in the intervening size range. Differences in the structural character of the A, B, and C rings are even more apparent in these occultation profiles than in the images.

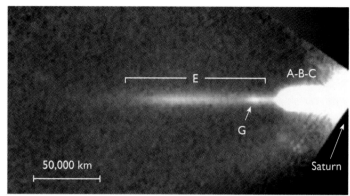

Figure 21. Whenever Saturn's rings appear nearly edge on as seen from Earth, observers take advantage of the much-reduced glare from the main bands to search for faint rings and satellites. Saturn's E and G rings extend beyond the classical ring system in this infrared (2.27-micron) image obtained on 10 August 1995 by the Keck I telescope. The G ring is so faint it was never before been seen from a ground-based telescope. The E ring's brightness peaks very near the orbit of Enceladus, the moon believed to be responsible for its microscopic particles. Although not obvious here, the E ring gets wider vertically as the distance from Saturn increases.

ansae appeared edge on differed by nearly 50 minutes in August 1995, indicating that if the ring's brightness is due to seeing the edge of a disk, that disk is warped.

More information on the system's thickness comes from what is termed an *opposition effect:* the rings surge in brightness whenever the phase (illumination) angle approaches 0°, that is, when we observe the system from precisely the same direction as incoming sunlight. The effect may be a consequence of the particles having rough surfaces, but more likely at such times we simply can no longer see them shadowing one another. In the latter case, calculations show that the particles must be separated by five to 10 times their size and be at least several layers thick. However, it is probable that the largest objects (which represent only a small fraction of the total surface area) do in fact reside in a more confined, central layer. As described earlier, collisions or gravitational scattering by such large ring members can thicken a ring.

Scattering might also be responsible for an interesting brightness variation that has been noticed primarily in the A ring. When the rings are fully open (that is, when one of Saturn's poles is tipped 26° toward us), the A ring appears about 10 percent brighter in the two quadrants approaching the Earth-Saturn line. This enhancement reaches 20 percent when the rings have intermediate tilts (about 10°) but then decreases as the rings approach their edge-on configuration. Perhaps the particles are not spherical and have become rotationally locked by tides, such that their long axes all point toward Saturn.

The particles are almost certainly too small for tides to exert such control, and in any event interparticle jostling would disrupt any well-ordered orientation. Instead, the largest ring members probably act to cluster particle orbits. When viewed from above, the pattern would contain a series of short arcs like the blades of a turbine. Similar structures are produced by the natural tendency of ring particles to form elongate clusters driven by Kepler shear (*Figure 3*). Such structures would appear "broadside" to our view in the dawn and evening quadrants,

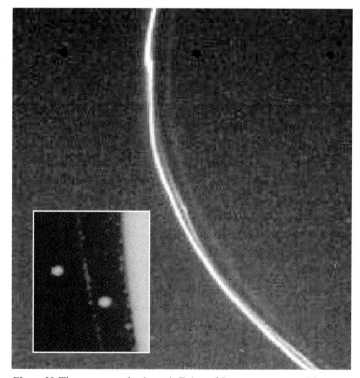

Figure 22. The narrow and enigmatic F ring of Saturn can appear either as a smooth and featureless ribbon or, as here, a set of knotted and braided strands. This clumpy appearance is caused in part by the shepherding satellites Pandora and Prometheus (inset; left and right, respectively), but other processes appear to also be at work.

Figure 23. Saturn's resplendent ring system was reduced to a razor-thin line when Earth passed through the system's plane on 10 August 1995. This view from the Hubble Space Telescope, taken four days earlier, shows Saturn's disk at left and a cluster of satellites — Mimas, Tethys, Janus, and Enceladus — near the west ansa at right.

and thus reflect more light. This clustering mechanism would be most effective when the rings are only moderately filled, as the A ring is.

THE COMPOSITION OF SATURN'S RINGS

During the 1970s, telescopic observers used infrared spectroscopy to determine that water ice dominates the ring particles' outer layers. However, a slight reddening of the rings in visible light indicates some surface contamination, perhaps from trace impurities, micrometeoroid debris, or radiation-induced modification of the ice's crystalline lattice. Clathrate compounds (in which methane or ammonia molecules are packed into ice's crystal lattice) may be present as well, because these are presently indistinguishable from water ice in near-infrared and visible spectra. The optical properties of ice also match our observations of how the rings vary in brightness with changing wavelength and phase angle.

Infrared and microwave measurements imply that the ring particles' interiors (just like their surfaces) are probably water ice as well. Notably, ice-coated *metallic* particles would also satisfy these measurements. However, cosmochemically speaking, metal just doesn't make sense (see Chapter 2). Not only should water ice be the dominant solid condensate in the outer solar system, but also the densities of nearby Saturnian satellites and of the particles themselves are not far from 1 g/cm³. Therefore, Saturn's ring particles are thought to be dirty snowballs throughout. Curiously, somehow the icy exteriors of Saturn's rings and satellites remain relatively pristine, especially compared to the black cometary material that must be bombarding them and the darkened rings of other planets.

Particles in the C ring and within the Cassini division have noticeably different properties from those elsewhere in the system (*Figure 24*). They are generally less red (more neutral in color), larger, darker, and do not scatter light forward as well as the average material in the A and B rings. The eccentric C ringlets have distinct optical properties as well. Such segregation of particle compositions — and its maintenance over geo-

Figure 24. An exaggerated-color picture of Saturn's classical rings (upper panel). Notice that the "blue" of the C ring and the Cassini division are similar, but that broad-scale, fairly smooth color variations exist across the B ring as well as between the A and B rings. Also note the striking green band in the outer portion of the Cassini division. These color differences are partly explained by multiple-scattering effects and partly by intrinsic color differences. The bottom panel shows a true-color representation of the same part of the rings, as they would appear to the unaided eye if illuminated by a daylit sky.

logically long times — is a further indication that diffusive processes are less effective than once believed. Abrupt changes in properties across ring boundaries suggest that differences among the system's components may have survived from primordial time.

We can estimate particle sizes using several observational techniques. The cooling rate of Saturn's ring particles as they enter the planet's shadow implies that, if solid, they are at least 1 or 2 cm across and likely even larger. Moreover, much of the ring population must exceed a few centimeters in size because the system as a whole is remarkably reflective to ground-based radar transmitting at the two frequencies most sensitive to particles larger than about 1 and 4 cm, respectively. This observation, combined with the rings' darkness at radio wavelengths, led to the conclusion that a many-particle-thick layer with a "power law" size distribution (containing many centimeter-size particles for every meter-size one) would satisfy the observations. Given how radio signals transmitted through the rings by the Voyager spacecraft, we find that particles in the A and C rings fit a power-law size distribution ranging from about 1 cm up to 5 or 10 m. This size range seems to vary somewhat from region to region (*Figure 20*).

However, this model requires several caveats. First, while little dust is evident, undoubtedly some bodies in the A and B

Figure 26. Voyager 2 was able to examine Saturn's rings from their unilluminated side. This eerie false-color view is a composite of images taken through green, clear, and violet filters. In most places some sunlight peeks through the ring system, but the densest part of the B ring is visible primarily because of Saturn-shine.

Figure 25. A 6,000-km-wide section of the outer B ring. Most of the structure visible in this region consists of features several hundred kilometers wide. Finer structures down to 15 km in width can be seen throughout the image, and these vary with time and longitude. The relatively vacant Cassini division is at upper left.

rings are quite small, particularly those continually being generated by micrometeoroid impacts. Second, to account for all the intricate ring structure, a few massive objects hundreds of meters to several kilometers across, like the moonlet Pan, are probably present. Furthermore, because the unilluminated sides of the rings are extremely cold (about 55° K), some scientists argue that a single layer of material, which would shadow portions of particles more effectively, is required. Finally, however unlikely, the existence of metallic objects larger than a few centimeters cannot be ruled out.

SATURN RING DYNAMICS

The Voyagers' high-resolution images disclosed Saturn's rings to be much more finely divided than anticipated. Rings lay within rings: upward of 1,000 have been counted in images (*Figure 25*), but the Voyager occultation results (*Figure 20*) revealed that many more "ringlets" are present at scales of less than 10 m. Even the so-called gaps are crammed with material — the Cassini division, for example, contains perhaps 100 ringlets. Therefore, such classically identified divisions merely denote relative absences of matter when compared to adjacent portions of the A and B rings.

Brightness variations in the A and B rings are due principally to the spacing of ringlets. Saturn's disk can be seen through most of the "opaque" B ring (*Figures 18,26*), and only a few places are truly impenetrable to light. Surprisingly, the C ring

exhibits a more organized structure than either A or B. This arises in part from its lower optical depth, so there is less multiple scattering of light among close-packed particles to mask details from our view. This region is also structurally different, as seen in *Figure 20*.

The highly organized structure seems to conflict with the notion of diffusional spreading, which should produce a smoother overall distribution. Perhaps the multitude of ringlets in the B ring is caused by an inability of the disk material to move as a true fluid, what dynamicists term a *viscous instability*. Alternatively, the intricate structure of the rings may be due to large, unseen objects (mooms) that influence the motions of material near them within the disk. Eccentric rings may also be telling us that large objects reside nearby in the disk. Such ringlets occur in the C ring and Cassini division, and the F ring itself is out of round.

As mentioned earlier, some prominent features of the rings appear to be associated with satellite resonances and gravitational interactions. Spiral density waves have been identified across the outer A ring (*Figure 8*), and a few are present in the outer Cassini division and in parts of the B ring. These in-plane clumpings are due primarily to Janus and other small satellites near the main rings. A few vertical bending waves, out-of-plane corrugations caused by Mimas, are seen in the A ring with amplitudes of a few hundred meters; another due to Titan is in the C ring.

Ironically, these elaborate wave features are among the few things in ring morphology that we understand at all. Nevertheless, much of the radial structure found in rings B, C, and elsewhere does not seem to correlate simply with known satellites. Some strongly perturbed locations contain material, while others do not; some gaps occur at resonant positions, but most do not.

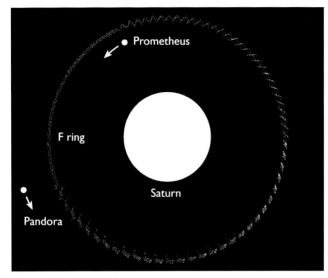

Figure 27. A computer simulation by Mark Showalter and the author of the ring perturbations produced by shepherd satellites. The inner, faster-moving satellite has a circular orbit and creates a trail of smooth, sinusoidal waves. The sluggish outer moon, on an eccentric orbit, creates waves that contain discrete clumps and travel ahead of it. The spacings of some structures in Saturn's F ring compare favorably with those that should develop due to the shepherd satellites assumed in this model.

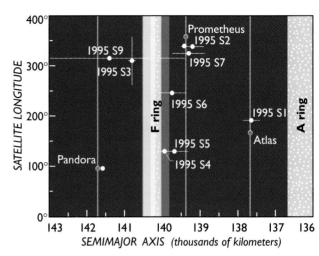

Figure 28. No fewer than three satellites and the F ring crowd the space within 6,000 km of the A ring's outer edge. During the 1995–1996 series of ring-plane crossings, moons not seen since their discovery by the Voyager spacecraft were recovered (white dots) — but not necessarily where expected (gray dots). The wandering of Prometheus is particularly puzzling because its orbit had been well determined using Voyager images. Observers suspect that some of the "satellites" spotted in their images, labeled here with provisional designations, are merely temporary clumps of material within the F ring.

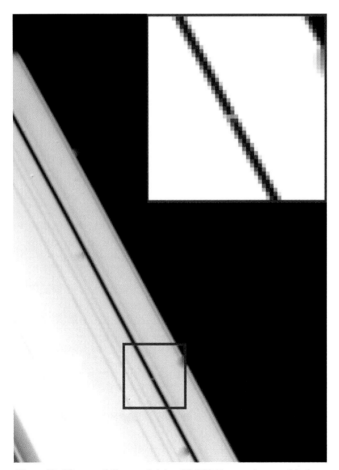

Figure 29. After carefully scrutinizing 30,000 Voyager images, Mark Showalter turned up the ring-moon Pan lurking in the Encke gap of Saturn's rings. The existence of this 20-km-wide moonlet had been predicted from the gravitationally perturbed scalloping seen along the gap's edges.

Dynamicists fare somewhat better in understanding why certain rings begin and end where they do. The boundary between the B ring and Cassini division is located near the 2:1 resonance of Mimas, the largest inner satellite. In fact, the B ring's perimeter forms an ellipse centered on Saturn, with a variation in radial position of about 140 km, just as it should if forced by interactions with Mimas. Numerous transient, noncircular features, roughly 20 km wide, occur in the same region and are presumably caused by waves and instabilities. The outer edge of the A ring seems to have a six-lobed petal shape that, along with its location, shows it to result from a 7:6 resonance with Janus. It is not surprising that the major outer boundaries of rings A and B coincide with the locations of the two strongest satellite resonances in the system. However, no explanation can yet account for the inner edges of the A, B, and C rings.

Satellites Pandora and Prometheus herd the particles in the narrow F ring (*Figure 22*) and may play a role in its clumpy appearance. Because their orbits are slightly eccentric, these two satellites periodically perturb ring particles as they mutually drift past one another (*Figure 27*). Whether perturbations from the same satellites account for the "braiding" of the F ring is still an open question. (The strands making up the F ring may appear interwoven, but the paths of individual particles in the separate braids do not actually pass through one another; rather the entire structure orbits as a whole and only gradually distorts by Keplerian shear motion.)

The F ring forward-scatters visible light very efficiently, just as Jovian ring particles do, indicating that it too consists principally of micron-size grains. These are perhaps generated in the collisional turmoil produced within the strands by the shepherds, as well as by unseen (but suspected) embedded moonlets. Brightness clumps have been seen to form in just a few days and then to fade into the background over much longer times in Voyager images; several probably similar structures were visible in images taken during the 1995–96 ring-plane crossings, where they were originally misidentified as unknown satellites (*Figure 28*). Perhaps related is the fact that Prometheus was found in 1996 to be 20° away from its well-predicted position.

While moonlets in the F ring and Cassini division are only suspected, one has been found in the Encke gap by a clever piece of detective work. Portions of the gap's inner and outer edges appear scalloped in Voyager images. The wavelength of the scalloping, together with the precise locations of density undulations called gravitational wakes in the nearby ring, implicated a satellite orbiting near the gap's center. Such an object, if at least 5 km across, would be able to pry apart a disk of material through the Goldreich-Tremaine mechanism. From the characteristics of the disturbances, Mark Showalter was able to predict which images of the decade-old Voyager archive might contain the moonlet (*Figure 29*). His successful search was important because it established that all ring systems, including the most magnificent, contain at least one embedded moonlet 10 km or more in diameter. This seems to be *prima facie* evidence against a ring origin by incomplete accretion (one of the possibilities discussed in the next section).

Narrow rings and "empty" gaps are not the only places that unexpected behavior occurs. Curious, sporadic features dubbed *spokes* occur near the densest part of the B ring, about 104,000 km from Saturn's center, and extend outward almost to the edge of the Cassini division (*Figures 18,30*). Each spoke appears as a pair of opposing triangles, like an hourglass, with the narrowest part located near the synchronous-orbit radius. Spokes form swiftly as radial strips. As they develop, one boundary always remains roughly radial, while the other assumes the differentially sheared profile of the initial radial strip. From their reflective properties, we know that these enigmatic features consist of microscopic grains. This fact and the long-lasting radial edges (which rotate at the same rate as the planet's magnetic field does) suggest that electromagnetic effects are at work. For example, small particles may become charged and thus levitate off larger ring bodies. This notion is reinforced by the spokes' location near the synchronous-orbit radius and by their approximate symmetry about that position.

It is fitting that Saturn's rings, as the first discovered set, remain the most ornate and interesting system. Fortunately they are also the best observed and, as the Voyager findings become more fully assimilated, are sure to provide many fundamental insights into the nature of all planetary rings. The Cassini orbiter, part of a mission that is due to arrive at Saturn in 2004, will scrutinize the intriguing and elegant structures of this complex system over a four-year period and should teach us much about the important processes going on there.

THE ORIGIN OF PLANETARY RINGS

Until then, theorists will continue to ponder the reasons why planets have rings at all, and why some worlds lack them completely. Two possible modes for the origin of planetary rings — the tidal breakup of a satellite and incomplete accretion — were mentioned briefly at the outset of this chapter. The first of these

was espoused by Edouard Roche in the mid-19th century. If applied to a satellite drawn in and torn apart by planetary tides, Roche's mechanism could explain only ring systems located within the synchronous-orbit radius but would not apply to the outer reaches of Saturn's rings. However, as hinted at by the easy disintegration of Comet Shoemaker-Levy 9, the rings could also be the disrupted remains of a comet or other interplanetary body that strayed inside the Roche limit. Either way, some fragments would still be tens of kilometers across — even after eons of collisional grinding. Such large remnants could be creating the elaborate structure in Saturn's rings or fashioning the narrow Uranian ribbons.

New planetary ring systems may soon develop elsewhere in the solar system. The Martian moon Phobos, residing well inside the planet's synchronous-orbit radius, is being drawn inexorably inward due to tides. If Phobos remains intact, it should strike Mars in about 50 million years. More likely, it will fracture or, at the very least, be denuded of its loose surface covering and thereby produce another faint ring system. An even more rarified dust belt probably circles Mars today, though it has not been detected: this would be debris blasted off Phobos and Deimos that remains in orbit about Mars. The orbit of Neptune's giant satellite Triton is also collapsing, though dynamicists do not expect Triton to come dangerously near Neptune for billions of years. Given enough time, therefore, planetary rings will be commonplace and not at all the rarity they were once thought to be. Indeed, following the discovery of rings about Uranus, Jupiter, and Neptune, some researchers have maintained that rings may have — or may still — encircle Venus, Earth, and even the Sun itself!

The breakup of a close-in satellite may not be caused by tides but instead by the impact of a large object. Since a planet's gravity acts to draw in interplanetary material in its vicinity, the frequency of collisions (and their impact speed) will be augmented near the planet. By determining the packing density of craters on distant Saturnian and Uranian satellites and extrapolating this number into the ring locale, the late Eugene Shoemaker argued that any Mimas-sized satellites within 2 planetary radii were pummeled into pieces many times over in the solar system's early history. At Mimas's distance, the debris ring created by these cataclysmic collisions swiftly reaccumulated back into a single object, but the debris from a shattered satellite closer in — within the Roche limit — cannot reaccumulate and would form a permanent ring.

The second general hypothesis for the origin of planetary rings was first proposed by Laplace and the metaphysicist Immanuel Kant at the end of the 18th century. In their view, Saturn's rings formed from the same circumplanetary nebula as the planet's satellites — much as the planets themselves accumulated in the primordial solar nebula. In fact, the essential properties of their nebular model of planetary origin were derived by observing Saturn's resplendent entourage. In the modern version of nebular theory (see Chapter 2), giant planets form in the outer parts of a circumstellar cloud wherever local instabilities become sufficiently large and dense. The end result is a large, gaseous protoplanet at the center of its own flattened disk of gas and dust. It is interesting to note in this regard that only the gaseous outer planets — but not the rocky terrestrial ones — have ring systems.

In regions beyond the Roche limit, the composition of accumulating objects may have depended upon their distance from the protoplanet; the latter's collapse generated considerable heat, which governed where specific volatile compounds could have condensed out of the nebula. High temperatures in the neighborhood of today's planetary rings would have prevented the condensation of water for several million years after the systems first formed. Apparently, Jupiter and Uranus lost their gaseous disks before cooling completely, leaving only less-volatile materials like silicates from which to assemble their rings. In contrast, Saturn cooled earlier, which allowed water vapor to condense into the magnificent rings we see today.

Material that accretes inside the Roche limit cannot grow without bound. Particle sizes will represent a balance between disruptive tidal stresses and the attraction of gravity and inter-

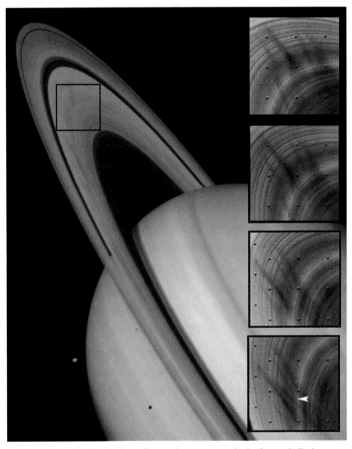

Figure 30. The dark, shadowy fingers known as *spokes* in Saturn's B ring occur sporadically in time, with a preference toward the morning ansa. Depending on illumination and viewing geometry, their brightness changes relative to their surroundings; for example, seen here in back-scattered light they appear dark (see also *Figure 18*). This indicates that spoke regions are relatively dusty. Insets at right show the swift formation of a new, radially aligned spoke (arrow in last panel) among a number of already-existing ones in the center of the B ring over a span of 35 minutes. (The regularly spaced black dots are reference marks used for geometric calibration of the imaging system.)

particle "stickiness." In this regard, the apparent absence of many large particles in Saturn's rings could be meaningful. Indeed, the identification of many objects larger than 1 km or so in Saturn's rings would be very unsettling to the nebula model, making the discovery of Pan quite provocative.

Finally, the very notion that planetary ring systems formed billions of years ago must be viewed with some skepticism, because several lines of evidence argue that the present-day rings — or at least parts of them — are quite young. (Clearly, the *dust* in all these systems is youthful.) To make this case for Saturn's rings, we can use the angular momentum transferred in spiral density waves to compute the times needed for nearby satellites to evolve away from the ring edge and for the A ring to be dragged down into the B ring. In each case ages are much less than that of the solar system. The apparently pristine surfaces of Saturn's rings suggest relatively short exposure ages. The lifetimes of ring particles undergoing erosion are invariably brief (though quite model dependent), and calculations demonstrate that the Uranian ε ring cannot be restrained by shepherds for the age of the solar system.

If all this is accepted at face value, then planetary rings must indeed be young: typical estimates are a few hundreds of millions of years. Perhaps they are the consequence of recent catastrophic events. However, the number of interplanetary intruders is continually decreasing, as more and more of them either strike planets or are completely ejected from the solar system. If catastrophic events have become less likely over time, how can the rings have had a recent origin? Perhaps our models are wrong or incomplete — explanations equally unpalatable to those of us who contrive them!

However they came to exist, the austerely beautiful rings encircling Jupiter, Saturn, Uranus, and Neptune remain what Galileo called a "most extraordinary marvel." As our knowledge of them has improved, we view them as perhaps more individualistic than the planets they surround. Uranus's narrow bands, made of dark boulders, reside within an extensive dusty disk and display an intriguing dynamical structure. The icy snowballs of Saturn's rings are baroque in their organization and variety. Jupiter's ring is a mere wisp that must be continually generated from adjacent moonlets and also perhaps from unseen parent bodies. Neptune's ring arcs illustrate the odd outcomes that satellites can produce on rings. We have learned that the catchall "rings" actually encompasses a wide variety of ephemeral clumps, arcs, braids, and other features that are only just becoming understood.

The dynamical processes now occurring in planetary ring systems may provide an appropriate analog for events in the early solar system or in distant spiral galaxies. The rings' detailed structure could be a fossil record of an intermediate stage in the accretion of orbiting bodies. Thus, planetary rings are more than just striking, exquisite phenomena — in a very fundamental way, they may represent the solar system's ancient beginnings.

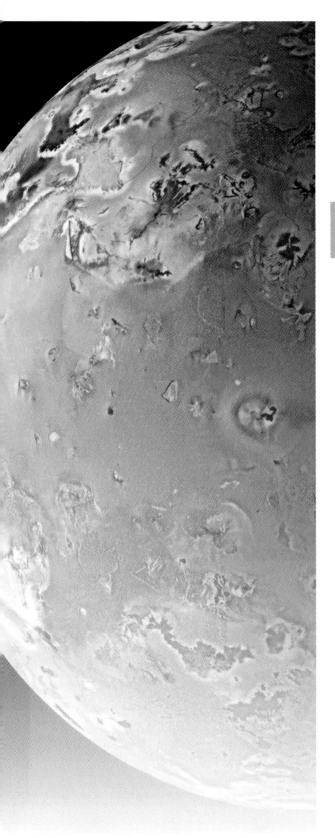

Io

Torrence V. Johnson

WITHOUT QUESTION, THE Jovian moon Io is one of the most remarkable bodies in the solar system. It orbits closer to the giant planet's cloud tops than our own Moon does to the Earth. Intense radiation belts bathe the satellite with an incessant shower of energetic electrons, protons, and heavier ions. Io is the most volcanic body known, with lava flows and volcanic calderas dominating its sulfurous landscape and geyserlike plumes towering hundreds of kilometers above the surface. Gases emitted by this volcanic activity form a thin, patchy atmosphere and a global frost of sulfur dioxide. These volcanic gases are in turn the ultimate source for a neutral "cloud" of atoms orbiting with Io around Jupiter and a huge, doughnut shaped torus of ions emitting ultraviolet light.

We now know that the energy to power all the volcanism comes from Io's gravitational interactions with Jupiter and neighboring satellites through the complex phenomenon of tidal heating. As explained later, the reality and magnitude of this power source were realized only shortly before the Voyager 1 spacecraft began its historic Jupiter encounter in 1979. But before exploring Io's many dynamic phenomena and their causes, let us first establish some basics about the satellite and its surroundings.

The Galilean satellites — Io, Europa, Ganymede, and Callisto — are a remarkably diverse set of worlds (*Figure 1*). Together with tiny Amalthea, a trio of moonlets discovered by Voyager (Metis, Adrastea, and Thebe), and Jupiter's tenuous ring, they form one of the four known regular satellite systems, the others being those of Saturn, Uranus, and Neptune. These systems are characterized by satellites and rings with circular orbits confined to the planet's equatorial plane. They are, in effect, solar systems in miniature.

241

A false-color Galileo image of Io, the variegated and wildly volcanic moon of Jupiter.

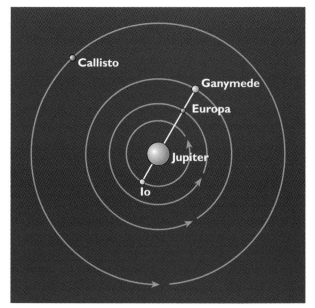

Figure 2. The Galilean satellites have orbits (shown here to scale) that are relatively close to Jupiter — a key factor in their formation and evolution. The innermost three share a resonant orbital relationship: when Europa and Ganymede are closest to one another, Io is always on the opposite side of Jupiter.

Figure 3 (below). On the basis of gravitional and magnetic-field data, researchers believe the four Galilean satellites have very different interiors. The varying densities of metal, rock, and ice have caused them to separate into distinct layers within Io, Europa, and Ganymede. But Callisto remains only partly differentiated, with rock and ice mixed in varying proportions throughout most of its interior.

Figure 1 (above). This "family portrait" includes the edge of Jupiter, with its Great Red Spot, and a same-scale montage of the planet's four largest moons (from the top): Io, Europa, Ganymede, and Callisto.

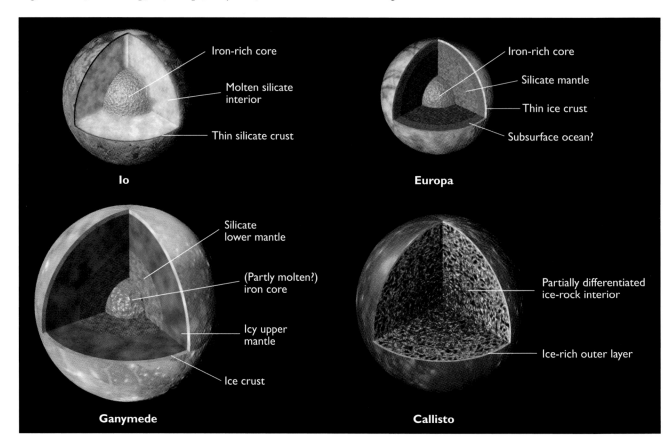

For more than 3½ centuries following their discovery in 1610, the Galilean satellites remained tantalizing points of light in astronomers' telescopes, tiny disks barely discernible even under the best atmospheric seeing conditions. Still, their relatively rapid motions around Jupiter fascinated astronomers from the outset. The quartet's discovery provided strong support for the still-heretical Copernican solar system, and in 1676 the Danish astronomer Ole Rømer used timings of their eclipses by Jupiter's shadow to make the first accurate measurement of the speed of light.

All four Galilean satellites are relatively large, roughly comparable to our Moon in size and mass, and they thus perturb each other's orbits significantly. In the late 1700s, Pierre-Simon de Laplace deduced that the orbital periods of Io, Europa, and Ganymede maintain a nearly perfect 1:2:4 ratio (*Figure 2*). This knowledge permitted Willem de Sitter and R. A. Sampson to make reasonably good estimates of the satellites' masses in the 1920s. Much more accurate values are now available from the tracking of spacecraft as they pass nearby, but the old values were in general accurate to 20 percent or better.

By contrast, the moons' diameters were very uncertain prior to 1970 due to their small angular size as seen from Earth, but stellar occultations and spacecraft imagery have improved this situation considerably. Mass and diameter estimates have yielded their mean densities, which in turn provide the most direct evidence of basic differences in their bulk compositions. The densities of the inner two moons mark them as essentially rocky, silicate-rich bodies, Io (3.5 g/cm^3) being slightly more dense than Earth's Moon and Europa (3.0 g/cm^3) slightly less so. These two may be lunarlike in bulk properties, but they differ from the Moon greatly in surface appearance and evolution.

Meanwhile, the low densities of Ganymede and Callisto (1.9 and 1.8 g/cm^3, respectively) strongly suggest that much of their mass is something other than rock, most likely water in some form (*Figure 3*). These are precisely the densities that should have resulted from condensation of solar-composition gas at temperatures where water ice is stable. Such bodies would have approximately equal proportions of silicates and water.

More compositional evidence has come from the remarkably diagnostic spectra acquired at near-infrared wavelengths with ground-based and airborne telescopes (*Figure 4*). Deep absorptions near 1.5 and 2.0 microns in the light reflected from Europa and Ganymede are due to water ice. In fact, the strength of these absorptions and the high albedos suggest large amounts of relatively "clean" ice. On the other hand, Callisto's weaker absorptions and low albedo argue for a "dirty" surface, which has proved to be dark, hydrated (clay-rich) material that is almost free of water ice in some areas but mixed with more ice or frost in others.

Why are there such significant variations in density and surface compositions among the Galilean satellites? In the early 1950s, Gerard P. Kuiper suggested that Jupiter was very hot after it formed, perhaps enough so to prevent lighter elements from condensing or cause them to boil off the inner satellites. Recent investigations suggest that, indeed, Jupiter's starlike infancy could have produced enough heat to make the inner satellites significantly warmer than their outer siblings. James Pollack, Fraser Fanale, and their coworkers modeled early Jovian history and

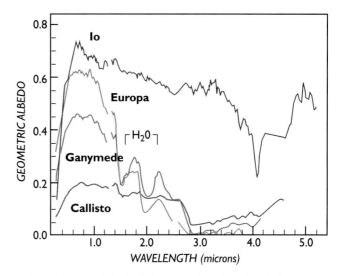

Figure 4. Studies of Jupiter's largest satellites from Earth show that they have distinctive spectral signatures. Absorptions in the near-infrared reveal the presence of water ice on the surfaces of Europa, Ganymede, and Callisto — but not Io.

identified, for the most plausible cases, a period of some 100 million years when conditions would have permitted the formation and retention of water ice where Ganymede and Callisto are now, while allowing only higher-density (though probably water-enriched) silicates to exist at Io and Europa. Theorists do not yet agree on all aspects of satellite formation, and other factors certainly must have played a role. But the striking variation in the amount of condensed volatile material (mostly water ice) with increasing distance from Jupiter is strong circumstantial evidence that the infant planet's immense gravitational field and central heat source played a dominant role in determining the nature of its satellites — just as the Sun's must have on a vaster scale.

OF TIME AND TIDES

Once formed, these moons evolved geologically and geophysically in ways dictated largely by energy — how much heating they experienced, when, and how fast they lost that heat. The major thermal source in the early solar system, at least initially, was the energy released by colliding planetesimals that served as building blocks for the final object. Another early source was the rapid radioactive decay of short-lived radionuclides, primarily aluminum-26. Longer-lived radioisotopes of uranium, potassium, and thorium have provided significant heating throughout solar-system history.

A third heat source was, and still is, the tidal energy created by the mutual gravitational interactions of bodies that orbit one another closely. This goes beyond the liquid-ocean tides we are familiar with on Earth — it can involve tidal distortion of the whole body. Any object placed in the gravitational field of another will experience forces that tend to distort it, because the attraction is stronger on one side than the other. If the body is liquid, it will assume a somewhat elongated shape with the long axis pointed toward the gravity's source. If rigid, the body will gradually assume a similarly stable elongated shape over time. However, if the gravitational field changes the body will flex in response. This flexing produces friction and heat inside the body — just as a rubber ball will get hot if constantly squeezed and released.

Jupiter, of course, has the strongest gravitational field of any planet. So tidal friction would have been an obvious source for heating the Galilean satellites, except for one thing: from Earth their orbits all appeared to be perfectly circular. To dynamicists that geometric constancy meant no change in the attraction, no work, and no heating! Thus it came as a surprise when Stanton Peale, Patrick Cassen, and Raymond Reynolds recognized the consequences of this interesting fact: while Io's *average* orbit is circular, at any given time it must be a little eccentric. The resonant orbital relationship discovered by Laplace means that Europa and Io give each other a little "tug" each time the two are near, which prevents them from having purely circular paths.

Even though the magnitude of Io's forced orbital eccentricity is relatively small, never more than 0.004, merely being so close to massive Jupiter means that the moon must experience large tides throughout its 42.5-hour "day" (*Figure 5*). In fact, Peale

and his colleagues calculated that tidal heating might be sufficient to produce active volcanism on Io. Just one week after their findings were published in March 1979, Voyager 1's cameras captured images of geyserlike eruptions rising hundreds of kilometers above the flow-covered surface of Io. Rarely has a scientific prediction been so rapidly and spectacularly confirmed!

How much tidal heating is actually going on? Soon after Io's volcanoes were discovered, infrared observations revealed that thermal energy was being released from Io's hottest areas. The total infrared power is nearly 100 *trillion* watts, about 2 watts for every square meter if averaged over all of Io and more than 100 times the Moon's outward heat flow. Subsequent observations and studies have refined this result, but the best current value for Io's heat flow (2.5 W/m^2; *Table 1*) is still very close to the original estimate.

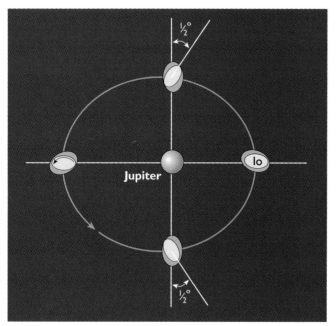

Figure 5. Io's dynamic activity stems from an orbital resonance with nearby Europa that forces Io into a slightly eccentric orbit. Ordinarily, Jupiter's strong gravity would keep one hemisphere of the satellite facing the planet at all times. But the forced eccentricity ($e = 0.004$) makes Io travel at different velocities along its orbit, and the side facing Jupiter nods back and forth slightly (by an angle of about $\frac{1}{2}°$) as seen from the planet. Although slight, this movement causes tidal forces inside the satellite that generate heat through friction, and much of the interior remains partially molten as a result.

Heat Flows Compared

	Watts/m²	Remarks
Io	>2.5	Global average
Earth		
average	0.06	Global heat flow through crust
geothermal area	1.7	Wairakei, New Zealand
Moon	0.02	Average of Apollo 15 and 17 sites

Table 1. Tides raised within Io by Jupiter generate (through friction) prodigious amounts of heat — enough to power its vigorous volcanoes.

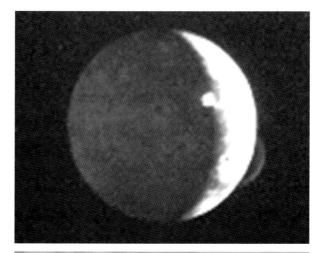

Figure 6. The dome-shaped eruptive plume of Pele (*upper panel*, *lower right*) rises more than 300 km above the limb of Io in the 1979 "discovery picture" from Voyager 1 that first alerted navigation engineer Linda Morabito to the satellite's dynamic nature. A second plume, from Loki, is the sunlit spot along the day-night terminator. Pele was still a major eruptive center 17 years later (*lower panel*), as its 400-km-high plume appears silhouetted against the clouds of Jupiter on 24 July 1996. This false-color view from the Hubble Space Telescope combines an ultraviolet-filtered image (in which the eruption is obvious) and a violet one (in which it is absent). On the basis of this color dependency, astronomers believe the plume probably consists of some combination of fine dust particles and sulfur dioxide gas.

This much power creates a problem for simple theories of the satellites' resonant orbits. Dynamicists initially believed that the current situation represents a balance between heat dissipated within Io (which causes its orbit to shrink) and the tides raised on Jupiter *by* Io (causing the orbit to enlarge). However, the amount of tidal dissipation within Jupiter required by this equilibrium model has proved to be at odds with other theoretical estimates and with the current position of Io's orbit (it should be farther out). This has led to more complex, "disequilibrium" theories. One of these holds that Io heats up sporadically or cyclically over time, its interior alternately melting then freezing throughout.

Whether constant or occasional, all that escaping heat argues for a very interesting interior. When the Galileo spacecraft swept by Io in December 1995 at an altitude of just 900 km, careful tracking of the spacecraft's altered velocity and trajectory provided a precise measurement of the moon's gravitational field. At the same time instruments were collecting information about the interaction of Io with Jupiter's powerful magnetic field and the charged ions and electrons in its magnetospheric environment. The gravity-field data provided strong evidence that Io has a dense core of iron and iron sulfide (FeS) extending about halfway to its surface and overlain by a mantle of partially molten rock and a relatively thin rock crust. This arrangement came as largely no surprise: theoretical models incorporating volcanic activity and tidal heating suggested that much of the satellite must have been molten at some point, and perhaps many times. Such a scenario provides ample opportunity for the separation and segregation of denser phases into a core.

The possibility that a magnetic field might arise from such a core structure has been suggested by a number of geophysicists. The results so far are ambiguous, however, ironically because of the strength and complexity of the ways that Io interacts with its environment. While approaching Io, Galileo's magnetometer recorded large oscillations in the ambient magnetic field, indicating the presence of electric currents from ions flowing around Io and being picked up from its thin atmosphere. Other measurements demonstrated that Io has a rather dense ionosphere, at least over some regions. "Beams" of electrons were found flowing up and down the field lines linking Io to Jupiter.

As explained in Chapter 4, space physicists are divided on whether all these data can be explained without requiring a magnetic field intrinsic to Io.

IO'S VOLCANOES

Of course, there is no longer such uncertainty about how Io's hot interior is affecting its surface. Primed by the theoretical prediction of tidal heating, geologists scanned Io's disk for evidence of volcanic activity as Voyager 1 approached the Jovian system in early 1979. Once the moon was resolved with enough detail to show small features, they realized that the impact craters and basins so ubiquitous on most planetary surfaces were just not there. Instead the landscape was covered with irregularly shaped pits and calderas of all sizes, many accompanied by obvious lava flows, and large diffuse deposits surrounding what appeared to be volcanic vents. Voyager's infrared spectrometer determined that the largest of these features had temperatures of 300° to 600° K, far above the frigid noontime highs of 110° to 120° K found elsewhere on Io's surface. This clearly indicated the presence of volcanic heating. The flyby's crowning discovery came when a ghostly umbrella-shaped cloud of dust was discovered rising over 300 km into space (*Figure 6*). Seven more plumes were found almost immediately.

Nowhere else in the solar system do volcanic processes so dominate everything we see as on Io (*Figure 7*). When the International Astronomical Union had to designate a "theme" for place names on Io, the decision was easy. Fire goddesses and gods and mythological allusions to fire and volcanoes from many diverse traditions and cultures dot our maps of Io. For example, the first-discovered plume is now known as Pele, named for the legendary Hawaiian volcano goddess.

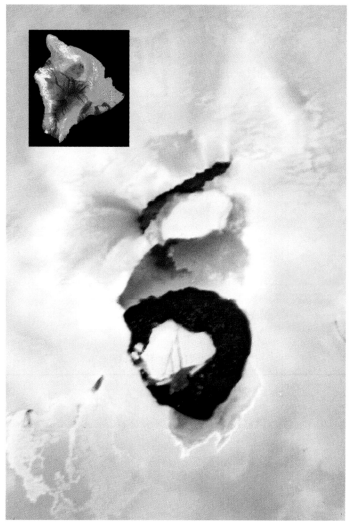

Figure 8. A comparison of the Loki region on Io with the island of Hawaii (inset). Both are the result of volcanic activity, but Loki is considerably larger. The large black area may be a lake of liquid rock and sulfur 200 km across. Northeast of the lake is Loki's large, elongated fissure, spouting a gray plume from its left end. Computer enhancement makes the fissure and lake look black, even though they are actually more reflective than the Moon.

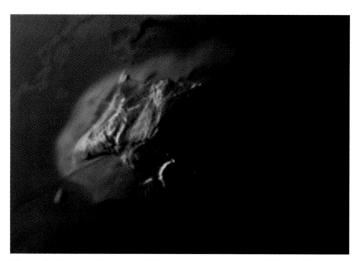

Figure 9. Haemus Mons, near Io's south pole, measures 200 by 100 km and is roughly 10 km tall. It is an example of the topography that can be glimpsed on Io under favorable lighting conditions (near the terminator, as here, or against the limb).

The most common volcanic features on Io are craters or calderas. From Voyager data alone more than 200 calderas larger than 20 km in diameter were counted. These look very similar to their terrestrial counterparts but are generally much bigger. (Earth has 3½ times more surface area than Io but only 15 calderas of this size.) At least one lava lake on Io, Loki Patera, is larger than the entire above-water portion of the island of Hawaii (*Figure 8*). It is important to draw distinctions between Io's volcanoes and its active plumes. Volcanoes occur where molten material breaches the surface, producing an eruption and subsequent lava flows. Although the plumes superficially resemble the ejecta from large volcanic explosions (such as Mount St. Helens), they are probably more closely related to geysers, as discussed later.

Following the discovery of volcanism on Io, geologists began a lively debate concerning the nature of its volcanism, particularly whether it was produced by molten rocks (silicate volcanism) as on Earth, or whether the features seen were due to molten sulfur. Although the density of Io clearly shows that most of it consists of rock and iron, the idea of sulfur volcanism was appealing for several reasons.

First, although it may seem an exotic material from our terrestrial perspective, sulfur is in fact very abundant in the solar system and, for that matter, the universe. Earth is well endowed with sulfur, though most of it probably lies hidden in the core as iron sulfide. This makes the element a minor player in terrestrial eruptions, relegated to sulfurous gases and deposits around volcanic vents. In contrast, abundant sulfur dioxide (SO_2) frost had been discovered spectroscopically on Io's surface, and sulfur dioxide gas was identified in its thin atmosphere by Voyager's infrared spectrometer. Ultraviolet sunlight readily breaks this molecule apart into sulfur and oxygen atoms, thus providing a ready source of elemental sulfur.

Second, most of the surface of Io is much more reflective than normal volcanic rocks and has a yellow-brown color suggestive of sulfur, which can absorb light strongly in the blue portion of the spectrum. In fact, spectroscopists had suspected the element's presence on Io even before the Voyagers arrived, particularly after the 1976 discovery of sulfur ions concentrated around Io's orbit. Color variations in the floors of calderas and along lava flows on the surface were attributed to different forms of sulfur, called *allotropes*, that crystallize at varying temperatures. (The stability of these allotropes as flows cool to Io's ambient cold temperatures has been challenged, however.) In addition, the temperatures inferred from Voyager data for hot areas on Io were too low molten silicates, which form at 1,000° K or more, but a good match to molten sulfur.

On the other hand, the best argument in favor of silicate volcanism is simply that the upper crust of Io must be strong enough to support mountains 5 to 10 km high (*Figure 9*), and deep steep-walled calderas many kilometers deep. Solid sulfur, particularly when hot, is not strong enough. Furthermore, some vulcanologists argued that the temperatures seen on Io could be explained by silicate lava topped by a somewhat cooler, solidified crust. In their view, the surface coloration is due to relatively thin veneers of sulfur-bearing compounds and sulfur dioxide frost.

Ironically, telescopic observers here on Earth have helped resolve this conundrum. Since most of Io's surface is actually quite cold, it radiates very little energy at mid-infrared wavelengths (from 5 to 10 microns) in accordance with the Planck function, which dictates how much radiation comes from an object of a given temperature. But volcanic hotspots emit most of their heat at precisely these wavelengths (*Figures 10,11*). Thus, although they cover only a small fraction of Io's surface, the hotspots actually produce most of its radiation at mid-infrared wavelengths. This allows astronomers on Earth to monitor and even pinpoint volcanic eruptions on Io from 5 AU away!

In 1986 a team studying Io from an observatory atop Mauna Kea noted a striking increase in the moon's infrared brightness compared with its "normal" volcanic output. They concluded that the outburst was probably due to a brief eruption of lava with a temperature of at least 900° K. This is much too hot to be molten sulfur, and it thus provided the first direct evidence that silicate lava was responsible for at least some of Io's volcanic activity. Since then similar events at even higher temperatures have been observed, meaning that in places Io has the hottest surface in the solar system (except the Sun's).

After Galileo reached Jupiter in late 1995, its instruments likewise began to monitor the volcanic regions. Spectacular

images of Io in Jupiter's shadow show many bright spots due to glowing hot lava, indicating that very high temperatures (from 700° to 1,800° K) are common in most volcanic areas (*Figure 12*). Galileo's maps of thermal emission at various infrared wavelengths and its spectra of individual hotspots also indicate that many areas are hotter than is reasonable for sulfur. The spacecraft discovered at least a dozen new volcanic centers during its first few orbits around Jupiter. Thus it appears that very hot lava — almost certainly some form of silicate magma — is involved in most, if not all, of the eruptions on Io (*Figures 13,14*).

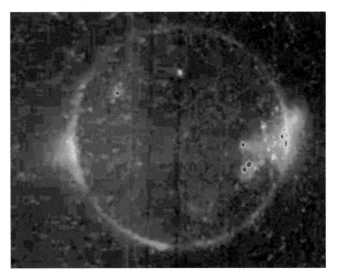

Figure 12. This image was acquired by Galileo while Io was in eclipse (in Jupiter's shadow) in May 1997. The small red or yellow spots mark the sites of high-temperature magma erupting onto the surface in lava flows or lava lakes. The glow on the left edge extends some 800 km above the active volcano Prometheus, whose eruptive plume appears only 75 km high in sunlight. The diffuse glow on the right side reaches a height of 400 km, even though no known eruption corresponds to that location.

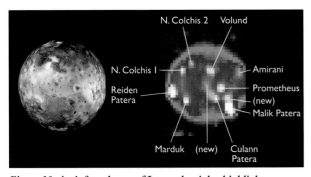

Figure 10. An infrared map of Io, on the right, highlights hotspots present in September 1996. This is a dayside view, with the sunlit clouds of Jupiter forming the dark-blue background. At least 11 spots are apparent; two were newly discovered, and the others corresponded to known volcanic features. On the left is a Voyager mosaic of Io with the same viewing geometry for comparison purposes.

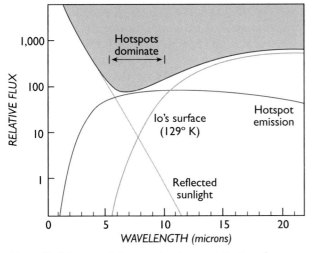

Figure 11. Io's spectral signature involves contributions from three major components, but hotspot emissions dominate from 6 to 10 microns. Observations in this range can be used to monitor the hotspots from Earth as they rotate into and out of view.

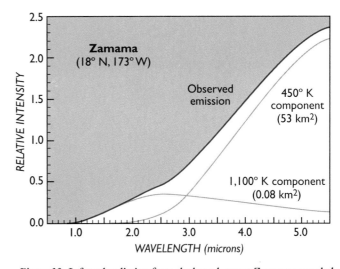

Figure 13. Infrared radiation from the large hotspot Zamama recorded in June 1996 while Io was in Jupiter's shadow, can be fit best by assuming two sources of heat. Most of the eruption, an area about the size of Manhattan Island, was a cooling "crust" at a temperature of 450° K. But red-hot silicate lava, radiating at 1,100° K from a city block's worth of cracks, "breakouts," and the vent itself, dominated the output at shorter wavelengths.

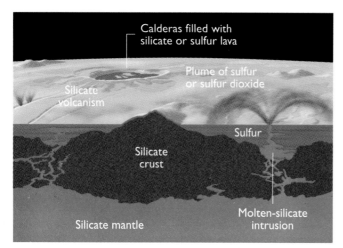

Figure 14. A schematic depiction (not to scale) of most of the major phenomena found on Io. As described in the text, at least three distinct types of active volcanism appear to be reworking the satellite's surface and outer layers of crust.

Still, in one sense, both sides of the "sulfur/silicate wars" were right. Galileo's observations also reveal large changes in the appearance of some areas without any corresponding spike in temperature. These may well represent a cooler, sulfur-driven volcanism of some type. Also, sulfur allotropes may indeed be responsible for some of Io's garish colors. The Galileo camera is sensitive in the red part of the spectrum where the Voyager cameras were "blind," and its color images of Io show strikingly red deposits around currently active volcanic areas (*Figure 15*). The evidence suggests that these fade back to the prevailing yellow-brown hue with time. Such traits are what we would expect if high-temperature sulfur allotropes were erupting onto the surface, then reverting to normal sulfur as they cool.

There are several ways to estimate how rapidly lava flows are covering Io's surface, and the level of volcanism is staggering by terrestrial standards. The total lack of impact craters in spacecraft

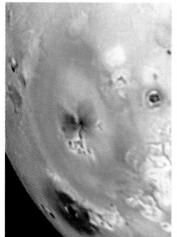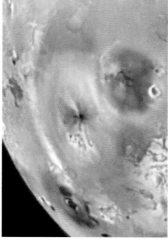

Figure 15. A dramatic eruption occurred on Io in the months between April (left) and September (right) 1997, creating a new dark spot 400 km across surrounding the volcanic feature Pillan Patera. In these Galileo images the fresh outflow contrasts markedly with the reddish ring that has been deposited by the giant plume of Pele. However, it shows similarities to the darkish spot that marks Pele's center and to Babbar Patera at lower left.

images provides one important constraint. If craters are being created on Io at the same rate as on the Moon, Laurence Soderblom and I have calculated that a global surface layer at least 1 millimeter thick must be produced every year, on average, to account for the crater-free Voyager images. If the impact rate is faster, as the late Eugene Shoemaker believed, then Io must be renewing its surface by as much as 1 cm per year. Another approach is to look at the total heat radiated from the volcanic areas. If most of this comes from silicate lava, as we now believe, the total volume of lava needed to bring that much heat to the surface also equates to a global resurfacing rate of a little over 1 cm per year. This corresponds to at least 500 km³ of erupting lava annually — more than 100 times the rate of the entire Earth!

IO'S PLUMES

The fountainlike plumes seen arching over many of Io's eruptive centers are among the most impressive and beautiful sights in the solar system (*Figure 16*). The satellite's low gravity (about one-sixth of Earth's) and its lack of an appreciable atmosphere allow volcanic dust and gas to rise unimpeded to great heights before falling slowly back to the surface. These discharges frequently form symmetrical mushroom-shaped plumes that leave circular or oval deposits on the surface. When the plumes were first discovered, Voyager's planetary geologists believed that these must be explosive volcanic eruptions. However, significant problems with this idea quickly became apparent. First, explosive eruptions on Earth tend to be brief, violent affairs, while the plumes on Io are remarkably stable and long lived. Seven of the eight volcanic plumes seen by Voyager 1 were reobserved by Voyager 2 four months later, and six were still erupting. They thus seemed more like planetary-scale fountains than scaled-up versions of Mount St. Helens. Second, calculations show that even violent eruptive blasts on Io could not throw material so high. The exit velocities of ejecta in terrestrial volcanoes rarely exceed about 100 m per second, whereas the plume heights on Io imply velocities at least five to 10 times faster!

The solution came from appealing to another terrestrial phenomenon: geysers. Although geysers are relatively small-scale events on Earth, the lack of atmospheric pressure and low gravity on Io changes the picture considerably. Geophysicist Susan Kieffer has calculated that Old Faithful in Yellowstone National Park would rise to an impressive 35 km if it were erupting on Io. What would power such a spectacle? On Earth, geysers are driven by rapid phase changes in water (from liquid to steam, for instance). Io is very dry, by contrast; apparently most of its water was baked or boiled away long ago.

Sulfur and sulfur dioxide, both seemingly ubiquitous on Io, can potentially serve as a working fluid for a geyser system if they are brought into contact with hot material below the surface. Kieffer and her colleagues concluded that SO_2-powered geysers could generate the vent velocities required for a typical plume (*Figure 17*). In their model, liquid sulfur dioxide comes into contact with heated rock or magma at relatively shallow depths. If there are fractures in Io's upper crust, the superheated sulfur dioxide will rise rapidly, boiling and adding energy to the process. By the time it reaches the surface, the mixture will begin to "snow" out, producing a high-velocity column of cold gas and

Figure 16. A Galileo false-color composite image shows two volcanic plumes on Io as they appeared on 28 June 1997. The one seen along the bright limb (and at left) is erupting 140 km high over Pillan Patera, named after a South American god of thunder, fire, and volcanoes. The second plume, seen near the day-night terminator (and at lower right) is Prometheus, the Greek fire god. Conceivably, Prometheus has been continuously active since it was first glimpsed in 1979.

frost grains as it emerges into the near vacuum above Io's surface. This SO₂ snow, shot out into the vacuum of space, is what our cameras see; the gas is too tenuous to be observed in sunlight.

To keep the geyser running, a large and steady supply of sulfur dioxide is required. There are a number of variations to this basic idea, depending on the initial conditions and materials involved. For instance, the largest plumes require the highest exit velocities. Heating sulfur dioxide to an even higher temperature in the subsurface reservoir is one way to achieve this extra speed; using sulfur as the fluid is another.

When Galileo began observing Io in 1995, some 17 years after the Voyager flybys, one surprise was how few plumes were evident. Prometheus, observed by both Voyagers, was still erupting, and an impressive new plume had appeared over a volcanic fea-

ture known as Ra Patera. However, Pele's large plume, seen by Voyager 1 but not Voyager 2, was apparently absent. One possibility was that some of the volcanic fountains might still be there but just hard to see — "stealth plumes," in effect. Kieffer's original analysis suggested that such a case might arise if the SO₂ reservoir started at a very high temperature. Then the erupting gas would expand so rapidly that it never produced snow, just a high-velocity jet of cold gas that would be hard to detect.

Pele, at least, apparently behaves this way; in 1996 the Hubble Space Telescope captured its plume in silhouette against Jupiter's clouds even though Galileo's camera did not see anything in scattered sunlight (*Figure 6*). Additionally, images taken while Io was eclipsed by Jupiter show "auroras" of glowing gas (energized by charged particles from Jupiter's magnetosphere) over areas that displayed no obvious eruptions. Since

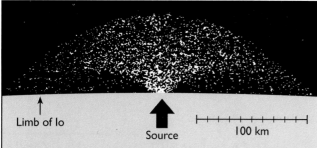

Figure 17. Compare the Voyager 1 image of Prometheus' plume along the limb of Io with the computer simulation below it. The latter assumes that material is ejected along ballistic trajectories at a speed of 0.5 km per second and at angles of at least 55° from horizontal. The plume's actual base in both cases is 7 km below the "surface" because the source vent is turned 5° toward us from the limb.

Figure 18 (below). Observations from spacecraft and Earth-based telescopes have recognized at least 80 hotspots on the surface of Io. *Table 2* lists the characteristics of the most prominent ones.

Major Volcanoes on Io

Plume or source	Location long.	lat.	Height (km)	Width (km)
Kanehekili	37°	–18°	—	—
Amirani	116°	+25°	95	220
Maui Patera	122°	+20°	90	230
Malik Patera	128°	–35°	—	—
Prometheus	154°	–2°	75	270
Culann Patera	160°	–19°	—	—
Zamama	173°	+18°	—	—
Volund	176°	+23°	100	125
Marduk	209°	–27°	70	195
Pillan Patera	244°	–12°	—	400
Pele	256°	–18°	400	1,200
Loki (plume)	303°	+18°	200?	400
Loki Patera	309°	+13°	—	—
Aten	312°	–48°	300?	1,200?
Ra Patera	325°	–8°	—	400
Surt	338°	+45°	300?	1,200

Table 2. The characteristics of major volcanic centers on Io (also identified in *Figure 18*). By early 1998, a total of more than 80 active volcanoes had been identified. The heights and widths refer to eruptive plumes observed in Voyager images. Pele, the first plume so discovered, has subsequently been observed by Galileo, Hubble Space Telescope, and ground-based infrared telescopes.

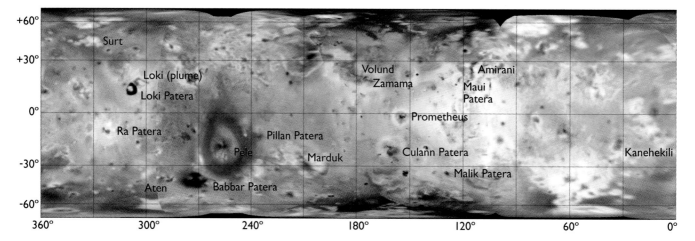

Figure 19. What vistas await would-be travelers to the Galilean satellites?
Clockwise from upper left are the surfaces of Callisto, Ganymede, Io,
and Europa. Each scene shows Jupiter at its correct relative size.

Io's plume activity increased later in Galileo's mission, several factors might be affecting both the number of plume sources and their visibility at any given time.

IO'S ATMOSPHERE

With volcanoes constantly belching molten rock and gas (*Figure 18*, *Table 2*), a significant atmosphere would seem to be a logical byproduct. Yet the first direct measurements of Io's atmosphere showed that there wasn't one — at least not much of one, anyway. In 1971 a very bright star slid behind the satellite as viewed from Earth. By timing how long the event lasted at different locations along the occultation's path, astronomers determined a very accurate diameter for the moon. In addition, their recordings showed that the star's light was cut off essentially instantaneously. If Io had had an appreciable atmosphere, the star should have dimmed gradually before disappearing. Since that did not happen, Io must not have an atmosphere with a surface pressure any greater than about 1 microbar — one-millionth the sea-level pressure on Earth.

Two years later, data from the Pioneer 10 spacecraft suggested that Io might indeed have a thin atmosphere. Changes in the radio signal as the spacecraft passed behind Io (another type of occultation experiment) showed an unexpectedly high concentration of electrons close to the moon's surface. Io seemed to have an ionosphere, which in turn implied the presence of neutral atmospheric gases to provide a source of electrons and ions. Under the right conditions even a very tenuous atmosphere (one consistent with the star-occultation results) can produce the ionospheric densities inferred from the Pioneer data. At the time, the composition of the atmospheric gases involved was nothing more than guesswork.

The next major step in understanding Io's atmosphere was the detection of neutral sodium and potassium "clouds" associated with Io's orbit and the discovery of ionized sulfur in the Jovian magnetosphere. These phenomena likely result from material being knocked off Io by magnetospheric ions, a process known as *sputtering*. Does the material come directly from the surface, in which case Io's atmosphere must be very thin? Or, does it escape from the upper layers of a "thicker" atmosphere (still less than one one-millionth as dense as Earth's)?

The Voyager 1 encounter provided important clues to resolve the debate. Its infrared spectrometer detected SO_2 gas above at least one volcanic source, Loki Patera. Meanwhile, ground-based spectra had shown that a frost of sulfur dioxide coated much of the surface. Both were obvious sources for atmospheric gases, though Voyager's results implied that the atmosphere might be extremely localized, with much higher densities at midday and over active vents. The Galileo image of nighttime auroral glows over some known plumes (*Figure 12*) adds support for a "patchy" distribution of gas. It appears that no more than about 10 percent of Io's surface has dense gas over it at any given time.

Galileo's eclipse images of Io show three major phenomena. One is a thin glowing layer of gas close to the surface, outlining Io's disk. This is probably a gas that does not freeze easily, possibly oxygen. Next, there is intense emission from small areas centered on known volcanic features; this is red-hot lava glowing in the dark, and the pictures help to pinpoint their locations and determine their temperature. Finally, diffuse patches of light are evident, some associated directly with plumes and some not. These emissions apparently come from volcanic gases, probably sulfur dioxide, lingering near volcanic vents.

As this atmospheric detective work demonstrates, astronomers continue to use many tools in their probing of Io and its kaleidoscopic activity. Thanks to new observations and techniques, we have a renewed awareness of the importance of all four Galilean satellites to solar-system studies (*Figure 19*). Earth-based spectroscopy has helped us decipher their surface compositions and revealed vast orbiting clouds related to Io. Pioneer and Voyager spacecraft transformed our view of these objects from dots of light into places. The latest stage in Jovian exploration began in December 1995, when the Galileo spacecraft went into orbit around Jupiter. How fitting that an artificial satellite named after the discoverer of these worlds has become our latest window on their amazing diversity.

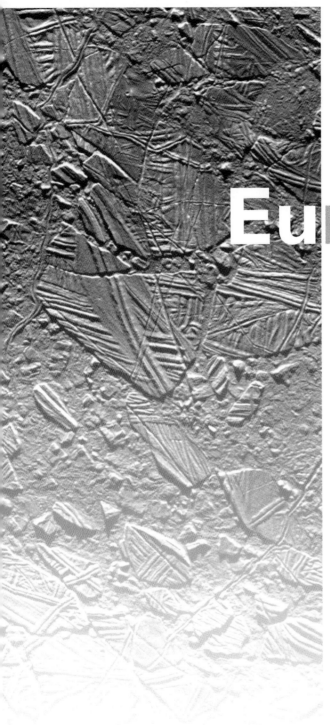

18

Europa

Ronald Greeley

Ice "rafts" collide in the icy crust of Jupiter's moon, as captured by the Galileo orbiter.

THE **J**OVIAN SATELLITE Europa is, like Mars, a world for which we have a special fascination. It has about the same size and density as Earth's Moon, but the similarities end there. Europa's surface is among the brightest in the solar system, a consequence of sunlight reflecting from an icy surface. As we'll soon see, the exotic coloring, texture, and detail visible on that surface continue to pose significant challenges for planetary geologists. There is much we don't understand, and there is much yet to learn.

What we *do* know about Europa stems from centuries of telescopic observation — beginning with Galileo's discovery of four Jovian moons in 1610 — combined with spacecraft data and sophisticated computer models. Modern telescopic observations of the solar system were ushered in some 40 years ago by Gerard Kuiper and others, who showed that Europa's crust has a predominantly water-ice composition. Later measurements revealed a compositional dichotomy. Most of the leading hemisphere (facing forward along its orbit) appears to be covered with water frost, while the trailing hemisphere is markedly darker and redder. Subsequently, it was found that ions in the Jovian magnetosphere overtake Europa as they orbit around Jupiter and cause preferential bombardment on the moon's trailing hemisphere. Sulfur derived from Io is a key component of this bombardment, and the implantation of sulfur ions may well account for the trailing side's lower albedo and reddish color.

Spacecraft exploration of the Jovian system began with the Pioneer and Voyager flybys in the 1970s. Pioneers 10 and 11 confirmed that Europa is immersed in magnetospheric radiation and also obtained close-up images of this small world — but not with sufficient resolution to see any useful detail. The Voyager spacecraft returned the first high-resolution images from the outer solar system. Because of their flyby trajectories,

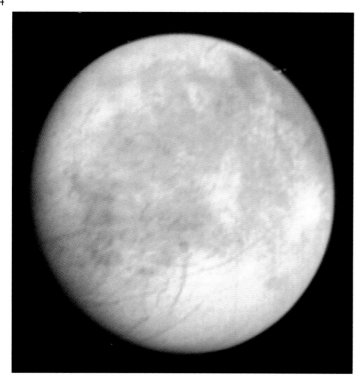

Figure 1. Voyager 1 provided the first detailed views of Europa in March 1979. This image, a false-color composite, reveals the moon's three basic surface types: mottled terrain, bright plains, and linear features.

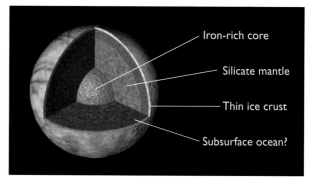

Figure 2. Europa's interior may consist of a metallic core, rocky mantle, and outer water crust, part of which could be liquid. This model is one of several possibilities based on Europa's size, density, external gravitational field, and characteristics of its magnetic field.

however, they recorded only snapshot-quality views of Europa, mostly of the side facing away from Jupiter. Despite these limitations, the Voyager pictures showed a bizarre, oddly smooth-looking globe whose surface resembles a string-wrapped baseball without its cover (*Figure 1*). Fewer than a half dozen impact craters were seen, suggesting a geologically young surface. Some researchers even suggested that Europa was experiencing active resurfacing.

In late 1995 the Galileo spacecraft began orbiting Jupiter and completed three close flybys of Europa during its two-year-long primary mission before undertaking eight more in its extended mission. Galileo brought unprecedented instrumental sophistication to its scientific studies: images with resolutions measured in meters, and the means to assess surface composition, temperature, and physical state from ultraviolet through infrared wavelengths.

REVEALING EUROPA'S INTERIOR

Scientists use such remote-sensing data to characterize the outermost veneer of Europa. But to really understand what we see on the moon's exterior, we must consider this world from the inside-out. Prior to the Galileo mission, our ideas about the interior characteristics of Europa were rather unconstrained by measurements. Dynamicists had shown that Europa is subjected to the same internal heating generated by tidal flexing that occurs inside Io, but with less intensity (see Chapter 17). However, they could not determine with certainty whether this energy was sufficient to melt the interior or to trigger surface changes.

Galileo has provided our first indirect "look" under Europa's icy exterior. Tracking the position of the spacecraft during each close flyby of Europa yielded information on the moon's gravity field and thus on the large-scale geophysical characteristics of its interior. During its close passes the spacecraft also detected hints of a magnetic field. Researchers do not agree whether this field is intrinsic to the moon or induced inside it by the extended magnetic field of Jupiter (see Chapter 4). But its existence does help constrain the nature of Europa's interior.

John D. Anderson and his colleagues have used Galileo gravity data to develop a suite of models to explain Europa's inner structure. In one case, an outer shell of water and ice, possibly 100 to 200 km thick, surrounds a homogeneous core of metal and rock. However, Galileo's detection of a magnetic field would seem to indicate a metallic core, presuming that the field is generated by Europa itself. Although many uncertainties and assumptions remain, these researchers find that the best fit involves a three-layer model (*Figure 2*). Its components are a metallic core about 1,250 km in diameter, a rocky mantle, and an ice/water "crust" some 150 km thick.

This layered model has interesting implications for Europa's past. At some stage of its evolution, this moon must have been molten throughout and become differentiated. During that process, dense, metallic elements sank inward to form the core. Lighter, volatile-rich compounds floated to the surface, leaving intermediate-density rocky materials in the middle. The lighter compounds were mostly water, forming a global ocean whose exterior froze to form a crust of ice. It is not known if the entire ocean froze, or if the surface has since frozen to some depth over a body of liquid water. A number of theorists believe the magnetic field found at Europa is not generated in a core but is instead induced by the passage of Jupiter's field through a salty (and thus conductive) liquid layer below the surface.

Another possibility is that the ocean froze solid and then remelted from the bottom up, either planetwide or locally. Present data do not allow us to determine if any of the water remains liquid today, and the question probably will not be answered until we send spacecraft to Europa carrying instruments (such as radar) capable of detecting such liquids.

Whatever its true interior structure, Europa holds a unique place in the family of solar-system objects. This moon displays a mix of characteristics that make it something of a transitional world between the rocky planets of the inner solar system and the predominantly ice-rich satellites of the outer solar system. Like its neighbor, Io, Europa also experiences interior heating from tidal forces and could be geologically active today.

A COMPLEX GEOLOGIC RECORD

We assume that the surface of Europa should manifest some of the moon's internal geologic activity, past or present. Voyager images allowed geologists to outline the basic physiography of Europa. The surface represents a mix of bright plains, mottled terrains, and disrupted regions. All of these exhibit an overprint of crisscrossing ridges and linear markings, some extending for thousands of kilometers. These regions are subdivided further according to their albedo and color properties, as well as geometric patterns of features assumed to be of tectonic origin.

The globe's color and spectral properties differ from place to place, in ways that cannot be explained as simple mixtures of water ice (or frost) and sulfur. Silicate minerals or other non-ice components must also be present, and these most likely are of internal origin. Galileo's imaging and multispectral mapping show that the plains in Europa's northern hemisphere can be

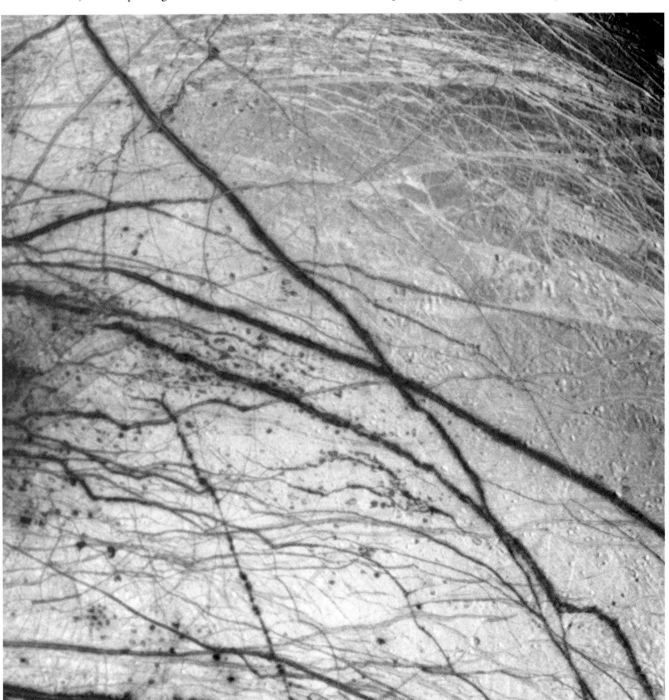

Figure 3. A false-color composite of visible and infrared images from the Galileo spacecraft covers a part of Europa centered at 45° north, 221° west. Triple bands, other linear features, dark spots, and mottled terrain appear reddish brown, while at infrared wavelengths the plains subdivide into bright and dark (bluish) units. The area shown is 1,260 km across.

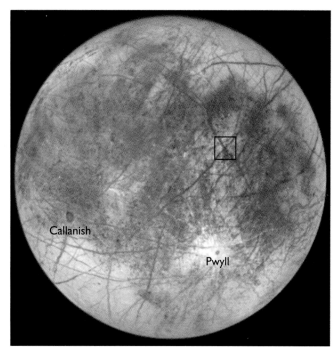

Figure 4. The trailing hemisphere of Europa, which faces the direction opposite its orbital motion, is generally darker than the other half and dominated by mottled terrain. Notable features are the impact craters Pwyll, surrounded by bright ejecta, and Callanish. Prominent triple bands form an "X" (box) and are shown in higher resolution in *Figure 10.*

Figure 5 (below). A 70-km-wide region of the thin, disrupted, ice crust in the Conamara region of Europa shows the interplay of surface color with ice structures. Crustal plates up to 13 km across have been broken apart and "rafted" into new positions, mimicking the disruption of pack ice on polar seas during spring thaws on Earth. The size and geometry of these slabs suggest their motion was enabled by water or soft ice not far below them. Color has been exaggerated to show areas that have been blanketed by a veneer of ice particles ejected during the Pwyll impact some 1,000 km to the south. The small craters probably were formed at the same time as the blanketing occurred, by large intact, blocks of ice thrown up in the impact explosion. The unblanketed surface has a reddish brown color due to mineral contaminants carried and spread by water vapor released from below the crust when it was disrupted.

divided into infrared-bright and infrared-dark units (*Figure 3*). To help calibrate its instruments, the spacecraft also mapped ice deposits in Antarctica through the same filters during one of its flybys of Earth prior to reaching Jupiter. However, the terrestrial and Europan ice fields have different appearances, which have been attributed to variations in ice grain size, composition, or both. The Galileo color data also indicate that various dark features, such as the dark spots and mottled zones, represent a single, dark, reddish component mixed with ice. Given its near-infrared spectrum, this material could include deposits of salts, such as the magnesium-sulfate compound hexahydrite ($MgSO_4 \cdot 6H_2O$).

The *bright plains* appear to be the basic surface unit on Europa from which many of the other terrains are derived. High-resolution Galileo images show that in some places these plains consist of multiple sets of ridges and grooves as small as a few tens of meters wide. In low-angle sunlight, the bright plains exhibit plateaus up to 10 km wide and tens of meters high. Numerous pits and depressions a few kilometers across pockmark the surface here and there. The ridges, grooves, plateaus, and pits suggest that the bright plains formed when the ice crust was repeatedly deformed by tectonic processes, perhaps accompanied by local collapse. The plateaus could be low-density materials, such as water ice, supported from below by a mixture of water ice and higher-density materials such as salts. Alternatively, the plateaus could be tectonic features supported in relatively rigid ice of the same density.

Mottled terrain, as the name suggests, includes an irregular patchwork of darker zones, with irregular margins grading into the bright plains. The patches range in size from 50 km to more than 500 km across (*Figure 4*). We now suspect that mottled terrain forms in response to tectonic forces. For example, a region on the trailing hemisphere of Europa named Conamara Chaos (*Figure 5*) comprises bright plains that have been broken into plates with intervening areas of chaotic terrain. The plates are as small as a few km across and stand 25 to 200 m above the chaotic surface. Many of the plates have been moved into new positions, and some appear to be tilted — as if being submerged into the chaotic material. The low-lying chaotic terrain is darker

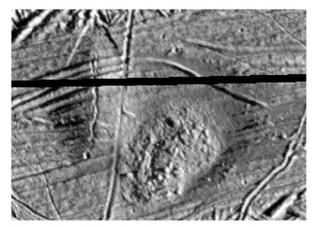

Figure 6 (above). These dome-shaped structures near Conamara Chaos may be manifestations of subsurface upwelling, perhaps the consequence of diapirs or mantle convention. The feature at left has a medial fracture; the one at right has been disrupted by extrusion from below or some other process. Illumination is from the right (east), and the area shown is about 14 by 20 km. The black line is an image artifact.

Figure 7 (right). Geologists don't yet understand what causes the deformation of Europa's crust, but one or more of these processes may be involved.

Figure 8 (lower right). The anti-Jovian hemisphere of Europa bears a dark, wedge-shaped zone (left panel, arrowed), which is in some ways analogous to sea-floor spreading on Earth. Extensional forces caused breaks within the crust, which later were filled in with low-albedo material. Running through the dark wedge are parallel ridges that display bilateral symmetry, just as our mid-ocean ridges do. When the dark material is removed by computer, the fractured plates (color-coded for clarity) can be reassembled to show how they might have appeared prior to the extensional episode (right panel).

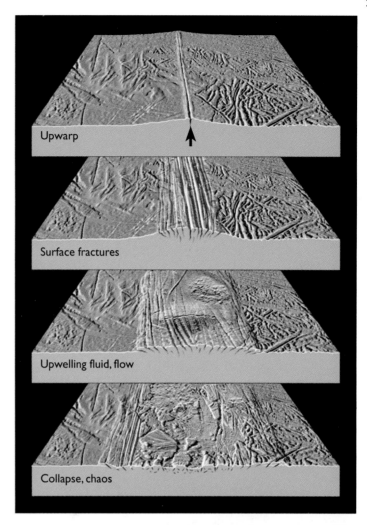

Upwarp

Surface fractures

Upwelling fluid, flow

Collapse, chaos

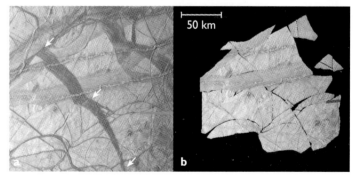

50 km

than the disrupted plates and probably contains a greater proportion of non-ice material. This suggests that the crust of Europa consists of an ice-dominated exterior layer overlying a zone containing a significant fraction of something else. If the relationships seen in Conamara Chaos are representative of mottled terrain elsewhere, then mottled terrain forms at the expense of the bright plains.

What causes this disruption of the icy crust? Other landforms near Conamara Chaos may provide some clues (*Figure 6*). The region displays gentle, up-bowed surfaces with central fractures, dome-shaped masses with chaotic surfaces, and distinctive domes surrounded by moatlike depressions. These all suggest upward warping of the crust by some force applied from below. Many of these features are darker than the surrounding terrain. They could represent different stages of deformation resulting from buoyant, low-density masses (called diapirs) rising toward the surface, convective upwelling, or intrusions of "molten" material (*Figure 7*). Any low-viscosity material that becomes fluidized below the surface, even water-ice slush, will suffice.

The process implied by some of these possible mechanisms is called *cryovolcanism*. Many of the icy outer-planet satellites show evidence for resurfacing, similar to what happens on

Earth when a volcano erupts and covers surrounding terrain. However, the resurfacing on those worlds is accomplished by an eruption of water, slush, and other volatile-rich materials. On Europa, cryovolcanism requires enough localized heating either to melt or mobilize the ice-rich materials presumed to exist below the surface. If this is occurring, then the crustal disruption that forms the plates we see could also result from localized heating, such as convection within ductile ice or a liquid-water zone.

Taking these observations and interpretations into account, we think that mottled terrain represents areas of the crust that have been modified by local disruption. This breakup has allowed the exposure or upward percolation of a subsurface zone containing salts, clays, or some other dark colorant.

Figure 9. This high-resolution Galileo image, measuring 15 by 18 km, reveals detail in Europa's complex ridged terrain. The prominent ridge has a large fracture running along its center and is flanked by narrow fractures. Crosscutting sets of ridges with different orientations suggest that the forces responsible for surface deformation have changed direction over time.

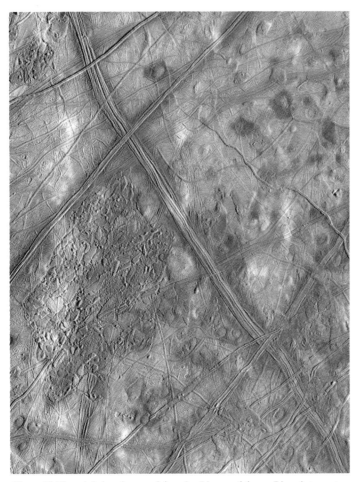

Figure 10. The triple bands named Asterius Linea and Agave Linea intersect just above a zone of disrupted terrain named Conamara Chaos. Dark components of the triple bands have the same spectral properties as the dome-shaped structures and dark zones in the chaotic terrain. Whitish patches are ejecta from Pwyll crater. This false-color composite of violet, green, and near-infrared Galileo images covers an area of about 295 by 320 km.

Whether this subsurface zone is global or merely local in extent is not known.

Interestingly, the region of Europa in greatest disarray is its anti-Jovian hemisphere. As first seen on Voyager images, this area was characterized by wedge-shaped dark bands more than 100 km long. The bands appear to represent places where the crust experienced extensional stress — effectively, where it has been pulled apart. One such wedge (*Figure 8*) has ridges and grooves in a bilateral symmetry similar to sea-floor spreading centers on Earth. We suspect that the Europan wedges are areas where the crust has been separated repeatedly, with the intervening area filled with darker material. As with mottled terrain, this suggests that some non-ice material exists abundantly below the brighter surface crust. Careful mapping shows that the individual wedges in this "pull-apart" terrain fit back together like pieces of a jigsaw puzzle, with about 8 percent of the "original" surface missing. This suggests that parts of the crust have been consumed, that they have been deformed beyond recognition, or that Europa has undergone a net global expansion.

LINEAR FEATURES

Ever since the Voyager flybys, geologists have been fascinated with the wide variety of linear features observed on Europa. Some ridges, for example, have a cycloidal (scalloped) shape and appear to be concentrated toward the south polar region. Elsewhere, dark stripes cross the surface in complex patterns. However, we now realize that much of the classification of linear features seen in Voyager images was affected by lighting conditions, viewing geometry, coverage, and resolution. For example, Galileo pictures show that some cycloidal ridges grade into ridges lacking the characteristic scallop pattern. Moreover, when resolved at high resolution, many seemingly individual features appear as multiple ridges and grooves cutting across each other (*Figure 9*). Perhaps, in response to periodic tidal flexing of the crust, some of Europa's ridges were formed along fracture lines by repeated extrusion of ice or liquid water. This mechanism could explain the multiple parallel structures seen on many of the ridge sets.

Most intriguing of all are Europa's *triple bands* (*Figure 10*), which in Voyager images comprise two parallel dark zones separated by a bright central stripe. They have a wide variety of morphologies (*Figure 11*): a typical three-stripe form, those that grade into single bright bands or single dark bands, and varieties in which the dark components are discontinuous. Triple bands range in width from a few kilometers to nearly 20 km, and they can extend 1,000 km or more. When imaged in very oblique sunlight, the bright central bands of most triple bands are found to be ridges, many of which have a medial depression. Some of the central bright bands consist of complex multiple ridges and grooves.

The color and albedo variations across triple bands show several interesting relationships. For example, the outer margins of the dark stripes are not sharp but instead fade into the surrounding plains. Using crosscutting relationships to determine the bands' relative ages (*Figure 12*), we believe that the dark stripes lighten over time. Spectrally, the low-albedo compo-

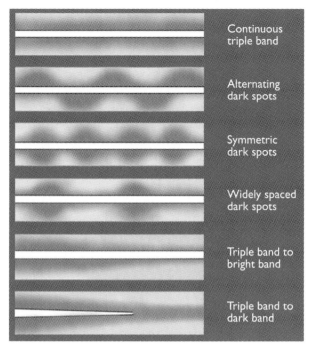

Figure 11. On the basis of Voyager data, geologists initially believed that the triple bands that crisscross Europa's surface could be explained solely by tectonic forces. However, higher-quality Galileo observations show that bands exhibit a wide variety of configurations and morphologies. Consequently, the underlying causes are probably more complex.

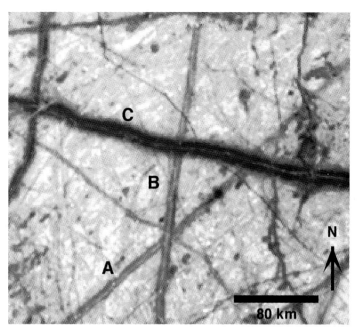

Figure 12. Crosscutting relations seen for these triple bands show that the sequence of formation was *A*, *B*, then *C*. Over time the darker (non-ice) component in these and other triple tends to brighten and take on an appearance more like that of the plains in which they occur.

nents appear similar to the dark material found in the mottled terrain. There is also a suggestion that the central bright ridge of a triple band may flatten as it ages.

Many ideas have been proposed to explain how triple bands form and evolve. Prior to the Galileo mission, most speculation involved various tectonic processes. In what is termed the block-faulting model, the surface experienced extensional forces, which caused it to slump down between parallel sets of fault lines. The resulting grabens, or sunken valleys, were then flooded by icy slush containing non-ice components (which would explain the dark stripes). Other scenarios involve fracturing in the crust, followed by the intrusion of hydrated silicates such as clays. The water they contained separated out to form relatively pure ice in the central bright band. However, the higher-resolution views afforded by Galileo apparently rule out both of these possibilities because the outer boundaries for the dark segments are not sharp, as would be characteristic of these processes.

The triple bands' diffuse outer boundaries suggest variations of those themes, as well as some other possibilities. If upwelling currents brought heat to the surface in a focused way, the overlying ice crust would tend to sublimate away and leave behind whatever contaminants it carried in the form of a lag deposit. The deposits would become most concentrated where the heating was greatest and diminish with distance outward from the heat source. In this model's final stage, relatively pure, ductile ice extrudes from below to form the central ridge.

Another idea that might explain the triple bands involves an extreme form of cryovolcanism: explosive venting by geysers. Some planetary scientists have suggested that carbon dioxide

or other volatiles might rise through the ice crust and erupt forcefully. If this occurred along a linear fracture, it could lead to a continuous line of geysers, somewhat analogous to the "curtain of fire" observed along fissures in some Hawaiian eruptions. In the case of Europa, material jets upward through a newly opened fracture, then the non-ice components fall back to the surface and form deposits around the fissure — with thicker concentrations along the fracture and grading to diffuse, feathered edges farther out. As the volatile propellant becomes depleted, the explosive phase ceases. Then cleaner ice squeezes out through the fracture to form the central ridge system, a process perhaps aided by repeated tidal flexing of the crust.

Regardless of the triple bands' origin, we want to learn what causes them to degrade over time. Conceivably the central ridge complex could gradually settle and slump as the ice deforms under its own weight. But such *viscous relaxation* is unlikely to occur in features only a few kilometers wide — at least in the absence of heat. Brightening of the dark stripes is also problematic, but at least three mechanisms are possible: some erosional process removes the non-ice components; the darker material sinks into the crust because it absorbs more solar energy than the surrounding ice; or the material becomes masked by a deposit of frosts. These last two seem most plausible.

Prior to the Galileo mission, various investigators mapped the ridges and other linear features on Europa. Then they attempted to relate the patterns to models of deformation that might result from synchronous rotation, nonsynchronous rotation, global volume changes, or tidal stressing. However, Voyager coverage was insufficient to map all the features, so these ideas didn't advance very far. With the better coverage made available by Galileo, the overall pattern seems most consistent with that expected from nonsynchronous rotation. That is, if Europa's rotation and revolution periods did not match at some time in the past, Jupiter's powerful gravity would have induced enough stress in the icy crust to trigger widespread fracturing.

THE CRATERING RECORD

Scattered among these various terrains are numerous impact craters. These excavations provide a great deal of insight into the nature and history of the moon's icy surface. Details of their shape and their ejecta deposits give clues to the target material hidden beneath the surface. The size and frequency of craters allow us to infer the surface's relative age. Galileo images reveal more craters than previously seen, and they confirm that many features seen poorly in Voyager images are indeed the consequence of impacts.

The 30-km-wide crater Manann'an (*Figure 13*) lies near Europa's trailing hemisphere. Here the impact penetrated through the icy crust, scattering subsurface material across the terrain for hundreds of kilometers beyond the rim. The crater Pwyll was first suspected to be an impact structure from Voyager images. Its bright rays of ejecta can be traced more than 1,000 km across the surrounding terrain. In some areas the ejecta splashed on and banked against the medial ridges of triple bands. In moderate-resolution images (*Figure 14*), Pwyll displays a dis-

continuous crater rim some 26 km across, a complex central peak, and a dark aureole extending about 10 km beyond the rim.

As with Manann'an, the Pwyll impact is thought to have penetrated through the crust to excavate darker material, which formed the dark skirt around the crater. More importantly, Pwyll appears to have formed in a location where the crust was particularly thin. Galileo images show that its floor lies at the same level as the surrounding terrain, suggesting that it filled immediately with slushy material. Here and there large chunks of material — Europan icebergs? — appear to protrude from the crater's floor. If Pwyll formed in the recent past, these remarkable data provide the strongest evidence yet that liquid water or an ice-water slush underlies portions of the Europan crust.

The appearances of Manann'an and Pwyll, when combined with observations of the mottled terrain and the most likely origins for the low-albedo material in the triple bands, imply that a zone containing a higher proportion of non-ice material — possibly the clays and salts discussed earlier — exists at a depth of a few kilometers.

Several circular dark spots were identified during the initial exploration of Europa by Voyager. Termed *maculae* (Latin for "dark spots"), these features could be either impact scars or some type of volcanic or tectonic formation. One of them, named Callanish, is about 100 km in diameter and includes a series of concentric fractures, some of which are sunken valleys (grabens). The central zone has rugged, hummocky relief, while scattered around the fracture zone are numerous small craters resembling secondary impacts. Tyre (*Figure 15*), which is similar in appearance to Callanish, consists of concentric fracture rings outlining a structure nearly 200 km across. Many of these rings are superposed on sets of older, linear features. In turn, Tyre is cut by triple bands and several narrow bright bands less than a few hundred meters wide. These relationships sug-

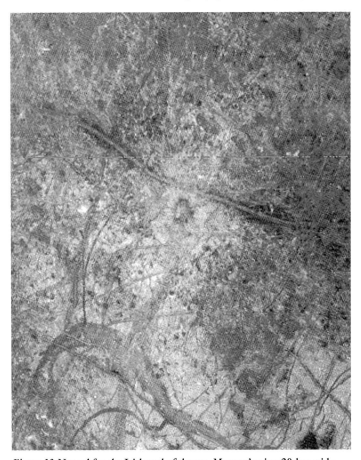

Figure 13. Named for the Irish god of the sea, Manann'an is a 30-km-wide crater centered at 2° north, 239° west. Note the dark band within the crater bowl. Since Manann'an lies in mottled terrain, this band is perhaps a remnant of a darker subsurface zone excavated during the impact event.

Figure 14. The 26-km diameter of Pwyll crater is defined by the ring of knobs outlining the crater rim; note also the central peak complex. Dark ejecta can be traced one crater radius from the rim, but farther out the ejecta appear quite bright (see *Figure 5*).

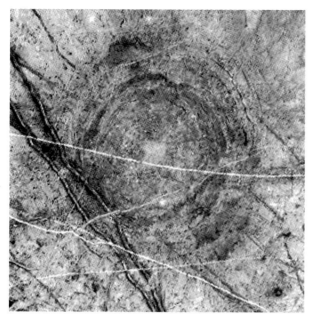

Figure 15. Tyre Macula is a multiringed structure 140 km across. Geologists believe this is an impact crater emplaced among bright, ridged plains and subsequently crosscut by several triple bands and bright bands.

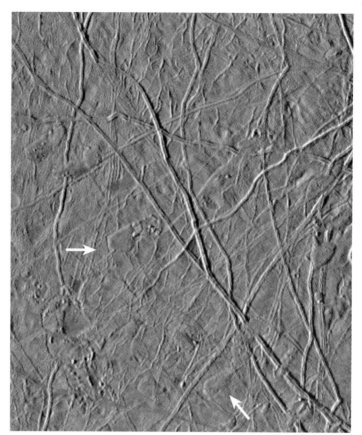

Figure 16. A portion of the leading hemisphere of Europa, seen here with sub-kilometer resolution, displays ridged plains and various flowlike features (arrows). Some of the latter appear to have breached the ridges locally.

gest that this part of the Europan crust was tectonically modified both before and after Tyre's formation.

The morphological characteristics of Callanish and Tyre resemble the topographically subdued impact features on Ganymede called *palimpsests*. At least two possibilities have been proposed to account for the low relief of palimpsests. First, while a typical bowl-shaped crater could form following impact into an ice-rich target, the ice would relax (flow) with time, reaching equilibrium with the surrounding terrain. Relaxation can be very effective in degrading large, ice-rich features. Alternatively, if the impact occurred in a thin, brittle crust overlying more ductile ice or liquid, then the excavated cavity might have collapsed immediately without preserving a crater form.

In addition to large impact features, small ones (10 km across or less) are found all over Europa. Some of these are probably clusters of secondaries from Pwyll and other large craters. Others are isolated and appear to be primary impacts. For example, a crater about 3½ km in diameter is seen near the dark wedge in *Figure 8*. Its pronounced low-albedo halo and radial ejecta suggest that the impactor penetrated a dark substrate.

Considerable controversy has developed regarding the ages of Europan surfaces. In the absence of crust samples to provide radiometric dates, crater size-frequency distributions ("crater counts") provide our only means to estimate the ages of planetary surfaces. This approach works well on the Moon, where crater counts are calibrated with surface ages obtained from Apollo and Luna samples. Difficulties arise, however, when extrapolations are made to other planetary systems because of uncertainties in the populations of potential impactors (especially in the outer solar system), the effects of scaling for gravity and target properties (ice versus rock), and other uncertainties and assumptions.

We have two ways to assess Europan surface ages using crater counts. The first is to estimate the current rate of colli-

sions taking place within the Jovian system by observing objects such as Comet Shoemaker-Levy 9, then extrapolating backward in time. This approach suggests relatively frequent impacts and implies that Europan surfaces could be as young as a few million years.

The second approach employs the cratering record preserved on heavily scarred Ganymede and Callisto. It assumes that the largest impact basins preserved on Ganymede represent the final stages of heavy bombardment within the solar system, when the Moon's youngest basins were formed. The crater counts for Ganymede's younger surfaces are then matched to the lunar record to establish the collision rate through time. This long-term rate must be adjusted because objects typically strike Europa faster than they do Ganymede, a consequence of Europa's proximity to Jupiter. This approach yields surface ages of about 3 billion years — three orders of magnitude older than estimates deduced from the first approach! Unfortunately, this is the current state of affairs, and the means for obtaining more accurate ages remain elusive.

CRYOVOLCANISM

If the geyser model is valid for the formation of triple bands, then cryovolcanism has undeniably taken place on Europa. Domes and similarly localized features (*Figure 6*) could likewise be extrusions of material from the interior. But do larger flows and outpourings exist? One region imaged by Galileo (*Figure*

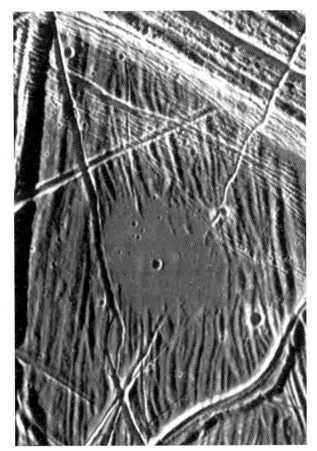

Figure 17. A smooth, dark "puddle" lies atop older ridged and grooved terrain. It could be a frozen pond of material erupted onto the surface from below, or the location of a local hotspot that melted ice-rich surface materials. This view reveals details only a few tens of meters across, and the Sun is illuminating the scene very obliquely from the left.

16) exhibits flowlike lobes, some 20 km long, that appear to cut across or through ridges. Unfortunately, we have no high-resolution views of these particular features. If these are flows, then they appear to have been very viscous and sluggish at the time of their emplacement because the flow fronts and margins are an estimated 100 m high. No source vents are apparent; presumably they have been covered over, as are most fissure vents for basalt flows on Earth. On the other hand, these features could mark where the surface was modified by upward warping. In either case, the flowlike features would probably reflect localized heating.

We see evidence of possible local heating elsewhere on Europa. In one such instance a flat surface 3 to 5 km across is superposed on the surrounding ridged and grooved terrain (*Figure 17*). The lack of flow margins suggests that the material was very fluid at the time of its emplacement. This "puddle" may represent the eruption of a fluid onto the surface from below, or it may be the manifestation of a hotspot that melted or mobilized surface ice without involving an eruption from below.

While the formation of domes, flowlike features, and other landforms may have been driven by near-surface heat sources, such conditions do not necessarily exist today. This is one reason why resolving the crater-count issue is so important. Abundant craters superposed on smooth Europan surfaces imply that those plains formed billions of years ago. Yet the paucity of craters seen on some flowlike features suggests ages so young, geologically speaking, that eruptions could be ongoing today. Unless we turn up something remarkable in Galileo's data, the issue is likely to remain unresolved until future spacecraft are flown to Europa.

LIFE ON EUROPA?

Evidence of hotspots on Europa signals not only the possibility of liquid water but also of environments capable of supporting organic evolution. The presence of tidally generated thermal energy inside Europa, whether now or in the distant past, satisfies one of three conditions considered fundamental for the development of biology. The other two are the presence of water (preferably liquid) and the availability of compounds to enable organic evolution. An inventory of the solar system shows that there are relatively few places where these conditions appear to coexist (see Chapter 26). Europa is one of them.

We know that the water ice seen on the surface of Europa probably extends to a depth of some 150 km. Interior models suggest that the heat generated by tidal flexing, together with radioactive decay from rocky matter deeper down, might be sufficient to melt the ice and provide a source of liquid water. This leaves the issue of the presence of organic compounds. Although we have no direct observations of them on Europa, almost surely organic materials have been delivered there by comets and meteorites. Thus, all three conditions for biologic development appear to be met. However, meeting these simple requirements does not mean that the "magic spark" ever occurred to spawn organic evolution and perhaps life itself. Still, the prospect is tantalizing enough to justify the high priority for its further exploration.

The Galileo Europa Mission, or GEM, continues the operation of that spacecraft through the end of this century. The first phase involved eight close flybys of Europa, which amassed at least 20 times more remote-sensing data than Galileo gathered during its nominal mission. We now possess single-wavelength and color images of nearly the entire globe at a resolution of 2 km or better. Some areas have been seen at resolutions of a few hundred meters and selected locations down to 6 m. Part of the strategy includes searching for active geysers and changes on the surface that occurred since the Voyager flybys or during Galileo's four years of operation. Although the probability of such discoveries is low, GEM has completed the reconnaissance of Europa and set the stage for its further exploration.

Future spacecraft will almost certainly go into orbit around Europa. The payloads might include a camera, instruments to map surface compositions and topography, and a radar system capable of penetrating the ice crust to search for liquid water and, if present, to determine its depth. If liquid water exists there, then following missions might include landers and vehicles capable of melting through the ice crust to explore the subsurface liquid zone. These missions are technologically ambitious, but they appear feasible. More importantly, they could yield very exciting results about a moon that has fascinated researchers for centuries. Europa's rich evolutionary history, bizarre environment, and exobiologic potential combine to create a fascinating target for our future explorations.

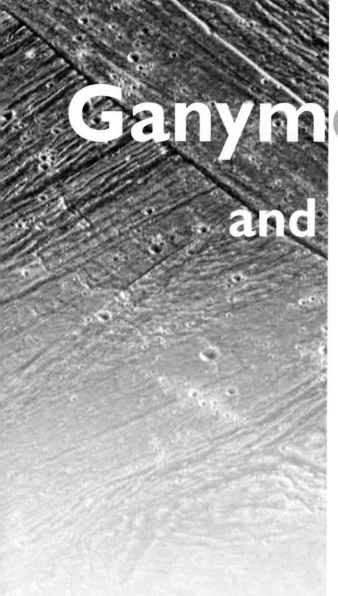

19

Ganymede and Callisto

Robert T. Pappalardo

Engimatic "grooved" terrain covers much of Ganymede, the largest moon in the solar system.

OUR VIEW OF the heavens was changed forever in 1610, when the giant planet Jupiter was discovered to have moons of its own. Callisto, Ganymede, Europa, and Io were first seen through the telescopes of Simon Marius, who proposed their names, and Galileo Galilei, who came to recognize their true nature as moons orbiting Jupiter. Discovery of the Galilean satellites, as they would become known, foreshadowed an end to the Earth-centered universe of Aristotle and Ptolemy, a doctrine that could not withstand scientific rigor. The Copernican Revolution was at hand.

Now, nearly four centuries later, spacecraft have visited Galileo's worlds. No longer just telescopic points of light, these moons have been revealed as incredible and unique places. Io's volcanoes of molten rock and sulfur compounds erupt so regularly that we can track the changes across its face. Europa's icy countenance may hide a subsurface ocean of liquid water. The outer two Galilean satellites, Ganymede and Callisto, are like twin siblings separated at birth that have gone on to lead very different lives.

Ganymede and Callisto shared similar infancies, having formed in close proximity within the great eddy of gas and dust that spawned Jupiter. A rain of rogue planetesimals continually showered the proto-satellites, aiding their steady growth. In this cold planetary neighborhood, far from the hot Sun and far enough from the embryonic warmth of colossal Jupiter, ice could accumulate along with rock. Their similar sizes and densities (*Figure 1, Table 1*) imply that each formed as a mixture of about 60 percent rock and 40 percent ice.

Astronomers can analyze the fingerprint of sunlight reflected from these moons with telescopes on the ground and in space. Missing are those wavelengths of light absorbed by materials on the satellites' surfaces. In these absorption spectra we can readi-

Figure 1 (above). Even from afar, the distinction between Ganymede and Callisto is clear. Ganymede — the solar system's largest satellite — flaunts bright and dark terrains, distinctive ray craters, and hazy white polar caps. In contrast, Callisto's surface is more uniformly dark and cratered.

Table 1 (left). The bulk properties of Ganymede and Callisto, the two largest satellites of Jupiter.

Ganymede and Callisto Compared

	Diameter (km)	Mass (10^{26} g)	Density (g/cm³)	Distance from Jupiter (km)
Ganymede	5,261	1.482	1.94	1,070,000
Callisto	4,820	1.077	1.84	1,883,000

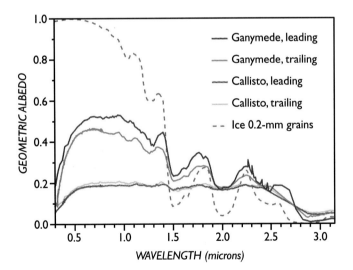

Figure 2. Earth-based reflectance spectra of Ganymede and Callisto divulge the distinctive infrared fingerprints of water-ice: twin dips at wavelengths of 1.6 and 2.0 microns. Callisto's lower albedo (reflectivity) and shallower ice absorption bands attest to its having less surface ice than Ganymede. Both satellites show a positive slope across the visible part of the spectrum, from 0.3 to 0.8 micron, so they would appear slightly reddish to our eyes. Differences between the satellites' leading and trailing hemispheres may be due to redistribution of ice by micrometeorites, which preferentially strike the leading hemisphere.

ly identify the telltale infrared signature of frozen water (*Figure 2*). Average temperatures of about 110° K have kept water ice on their surfaces over billions of years of solar-system history. Earth-based spectra of Ganymede and Callisto indicate a rocky component mixed into the ice. Spacecraft-based instruments are needed to determine what that contaminant is and to search for other types of ice.

Despite the family resemblance, Ganymede and Callisto form a distinct dichotomy, one hinted at by telescopic observations and demonstrated dramatically by spacecraft data. Crude images from the Pioneer 10 and 11 spacecraft, which passed by the Galilean satellites in 1973 and 1974, suggested that Ganymede's surface is a patchwork of bright and dark areas. True reconnaissance of the Jovian system awaited the arrival of Voyagers 1 and 2 in 1979. Their cameras revealed features smaller than a kilometer across. Ganymede exhibited large expanses of dark and bright terrains, the latter criss-crossed by ridges and grooves torn by internal geologic forces. In stark contrast, Callisto showed enormous tracts of dark, heavily cratered terrain and few signs of internal activity.

In December 1995 the Galileo spacecraft swung into orbit about Jupiter, and during its multiyear mission a suite of instruments has relayed remarkable information on Ganymede and Callisto with each close pass. These "siblings," it turns out, are even more dissimilar than we supposed. What might cause them to be so disparate? Are the differences only skin deep? Examining the surface characteristics of each satellite in turn, then probing their interiors, will reveal the true nature of the Ganymede-Callisto dichotomy.

GANYMEDE'S SURFACE: THE DARK TERRAIN

Even from afar, the division of Ganymede's surface into bright and dark terrain is apparent. Planetary scientists generally consider that dark terrain contains a greater fraction of rocky material, while bright terrain is probably more ice-rich. A closer look at Ganymede shows that these regions are also different in their degree of deformation and the number of craters they bear (*Figure 3*). This moon is sprinkled with impact features of diverse forms and ages, their morphologies recounting the history of the surface. Frosts coat the polar latitudes with a thin icy shroud, indirectly telling of interactions with the harsh space environment. Ganymede's unique geologic characteristics help to unravel the tumultuous history of the solar system's largest satellite.

Ganymede's dark terrain is densely pockmarked with impact craters. Generally, a more heavily cratered planetary surface is an older one, peppered over time by the collision of asteroidal and cometary debris (see Chapter 6). The prevalence of craters in the dark terrain suggests a surface nearly as old as the solar system itself, in excess of 4 billion years. However, local variation in crater density means that portions of dark terrain have been reshaped by later events.

Arcing across the dark terrain are long depressions called *furrows* — troughs about 5 to 10 km in width. Furrow sets commonly trace the outlines of roughly concentric circles, resembling planetary-scale bull's-eyes. A model by William McKinnon and H. Jay Melosh suggests that most furrows resulted from huge impacts very early in Ganymede's history. During the waning stages of accretion, warm, mobile ice lay just 10 km or so below the cold, outermost layer of Ganymede, its brittle lithosphere. Island-size chunks of space debris would occasionally wander toward Ganymede at typical speeds of 16 km per second, 27 times faster than a Concorde jet. The resulting explosion blasted through the relatively thin brittle lithosphere. Slushy warm subsurface ice then flowed inward to fill the central crater (*Figure 4*). Giant slabs of the cold surface material were dragged toward the crater as well, breaking

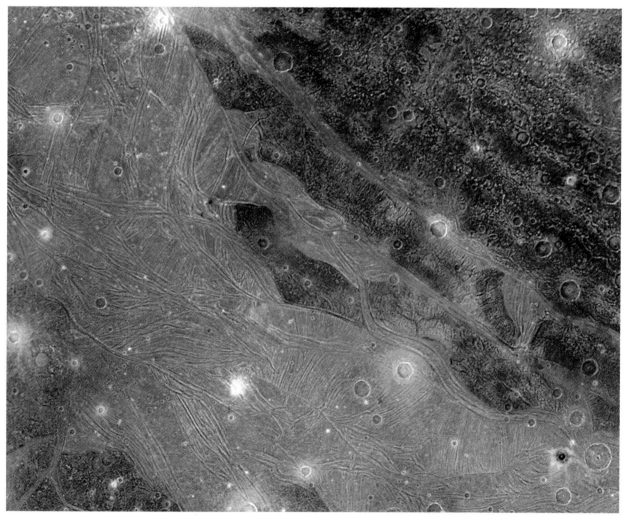

Figure 3. Voyager images of Ganymede tell of a complex history along the boundary between Uruk Sulcus (lower left) and Galileo Regio (upper right). Bright swaths of parallel grooves, spaced 5 to 8 km apart, and patches of smooth terrain dominate Uruk Sulcus. Furrows snake across the more heavily cratered dark terrain of Galileo Regio. Some major grooves lie along the older furrow trends, suggesting inheritance of structures when grooved terrain formed. Crater types scattered across the region include bright-rayed, dark-floored, dark-halo, pedestal, pit, and dome craters.

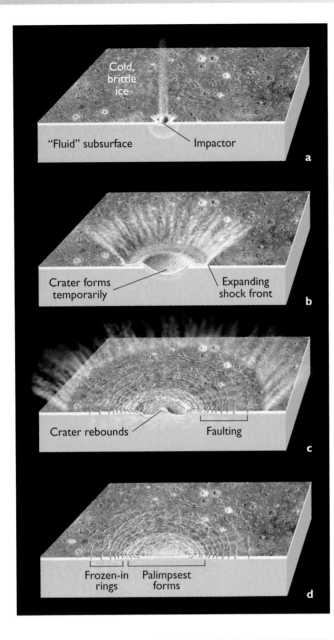

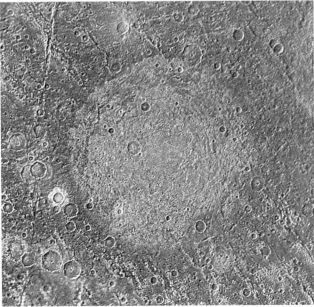

along faults to create the arcuate troughs. Other furrows on Ganymede radiate like the giant spokes of an ill-defined wheel, and these are probably related in origin to the ancient collisional calamities.

Dark terrain on Ganymede is punctuated by bright, low-relief patches termed *palimpsests*, named after reused parchment from which earlier writing is incompletely erased. Like older text still partially visible, traces of Ganymede's furrows are sometimes discernible beneath the outer portions of a bright palimpsest disk (*Figure 5*). Palimpsests probably formed early in Ganymede's history, when large impacts completely penetrated the brittle lithosphere and tossed out the slushy ejecta that covered the older furrows.

As the Galileo spacecraft sped past Ganymede during its first encounter in June 1996, its camera focused on an appropriately named region of dark terrain, Galileo Regio. The resulting high-resolution images are a study in black and white that might have been painted by an abstract artist (*Figure 6, top*). These pictures are confusing to the eye — and even harder to interpret — because the intricate patchwork makes the relationship between brightness and topography enigmatic. Additional data were needed to clarify this confusing scene, so scientists and engineers reprogrammed the spacecraft to photograph Galileo Regio again from a slightly different vantage point during its second flyby of Ganymede.

The resulting stereo imaging allows us to perceive the three-dimensional shape of the landscape (*Figure 6, bottom*). A relationship between brightness and elevation is evident: high-standing furrow rims, crater rims, and isolated hills are bright; low-lying regions such as crater and furrow floors are dark. Streaks of dark material have spilled down the furrow slopes. It seems that darker material has moved downslope over time, into the valleys and hollows.

In dark terrain, some slopes tilted toward the Sun are noticeably darker than slopes that face away. Even 5 AU from the Sun, ice is only marginally stable against sublimation to space. Being a bit warmer, Sun-facing slopes can rapidly lose their ice, with a thin lag of dark rocky material left behind. Moreover, because a dark surface absorbs more heat than a bright one, sublimation-darkened regions become even warmer, speeding the process along. In contrast, brighter and colder slopes should retain ice and frost. This may explain the segregation of "dark" terrain into a checkerboard of very dark and very bright patches.

Figure 4 (top). A large, fast-moving projectile *(a)* can pierce a thin, cold brittle ice lithosphere into warmer, "fluid" icy material beneath. Ejecta splashed onto the surface can create a palimpsest *(b)*. The slushy subsurface ice, unable to support crater topography, flows inward toward the rebounding crater. Dragged with it is the brittle surface layer, which breaks along concentric faults *(c)*. The surface retains this record of a warm past long after the satellite's brittle lithosphere has cooled and thickened *(d)*.

Figure 5 (bottom). Memphis Facula, 350 km across, is the most prominent palimpsest on Ganymede. Preexisting furrows seem to show through its outer portion, and dark-floored craters have apparently punched through into underlying dark terrain. These characteristics hint that much of the bright palimpsest is a thin disk, though bright material may extend well beneath its central scalloped depression.

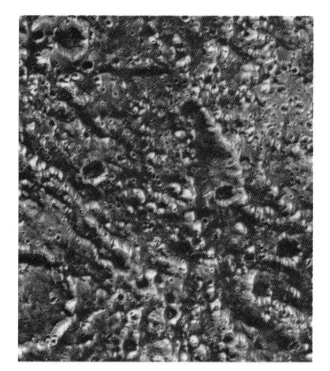

Figure 6. The "dark" expanse called Galileo Regio (upper panel) is actually a complex checkerboard of bright and dark patches. Relationships become clearer in a computer-generated perspective view of the area (lower panel), in which red represents the highest topography and blue the lowest. The darkest regions, such as furrow floors, are generally low-lying, while bright features correspond to ridges and islandlike knobs. The 20-km-wide crater Ea (lower left) has an elevated center, for reasons that remain uncertain.

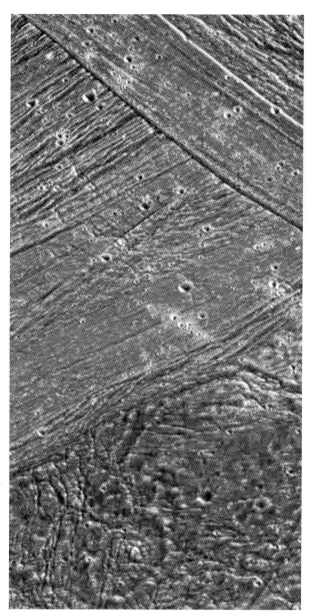

Figure 7. The difference between Ganymede's major terrain types is highlighted along the boundary between dark Marius Regio (bottom) and bright, grooved Philus Sulcus (center). A 17-km-wide swath of southern Nippur Sulcus (upper right), which slices across Philus Sulcus, may have been overrun by icy volcanic material.

Spectra from the Near-Infrared Mapping Spectrometer (NIMS) aboard Galileo confirm that brighter regions of Ganymede are rich in water ice. In addition, an absorption at a wavelength of 4.25 microns indicates scattered patches of even more volatile carbon dioxide (CO_2) ice. Ganymede's dark material contains abundant clays and might crunch under some future astronaut's boot like frozen mud. An absorption at 4.57 microns also suggests organic molecules rich in carbon, hydrogen, oxygen, and nitrogen, a mixture known as *tholin*. Tholins were probably delivered by the impact of organic-bearing comets.

The dark terrain was probably stained first by the primordial debris from which Ganymede accreted, then by the rain of cometary and asteroidal impactors that followed. Overall,

Galileo observations suggest that a clay-rich veneer — perhaps only meters deep — drapes a brighter, icier substrate. Down-slope movement and sublimation probably concentrated the dark rocky layer, while impact craters stirred it into the underlying crust. Enormous ancient collisions spewed cleaner icy material to the surface from depths of many kilometers to create the palimpsests. Partially fluid ejecta of palimpsests and other immense impacts paved over some patches of dark terrain, wiping out preexisting features.

The boundary between dark and bright grooved terrain can be quite abrupt and remarkably straight (*Figure 7*). The dark terrain alongside can be rutted with fractures up to 2 km wide. These cut across the earlier, wider furrows and thus postdate them. Moreover, they commonly align with ridges and grooves

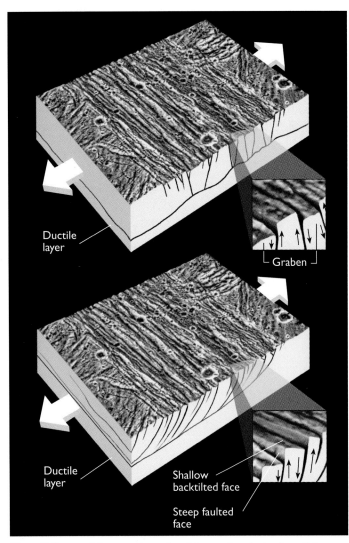

Figure 8. Ganymede's grooved terrain might consist of sets of parallel graben (upper panel), which form when a brittle layer is pulled apart atop a ductile substrate such as warm ice. Another possible faulting style is tilt blocks (lower panel). These result when parallel normal faults face the same direction, producing mountains inclined like toppled dominoes.

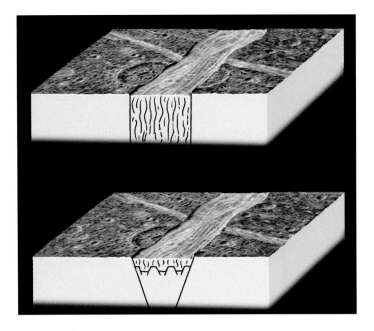

of neighboring bright terrain. This is more than coincidence: both fracture systems were probably torn apart by similar tectonic forces.

BRIGHT GROOVED TERRAIN

Ganymede's bright regions are crisscrossed by sets of parallel ridges and valleys termed *grooved terrain* (*Figure 3*). Tens of kilometers wide and hundreds long, these corrugated swaths divide the older dark terrain into polygonal tracts. Groove lanes of distinct orientations crosscut one another, so they must have formed over an extended period. Compared to the dark terrain, the sparser crater frequency on grooved terrain is consistent with relative youth. Its absolute age, however, is very uncertain, with estimates ranging from 4 billion years to only hundreds of millions of years. Profound events have transpired here, offering essential clues to Ganymede's tumultuous history.

Different from mountain ranges like the Appalachians (which were compressed into accordionlike folds), grooved terrain apparently has been shaped by tectonic stretching of the moon's surface. The process of extension can break the brittle lithosphere along inward-sloping faults, creating long down-dropped valleys called *graben* (*Figure 8*). Extensional faulting also might cause mountain-forming slabs to rotate and tilt, like a stack of books that falls over as its bookends are removed. Viewed from above, Ganymede's grooved terrain is reminiscent of a series of tilted blocks or parallel graben.

These tectonic styles are different from those on neighboring Europa, where long bands have been yanked apart completely, with new material welling up in between (see Chapter 18). On Ganymede, we see little evidence that preexisting craters or furrows were sliced and completely separated across lanes of grooved terrain. Instead, it appears that bands of dark terrain have been stretched and somehow converted into bright terrain (*Figure 9*).

How might this happen? To deduce the answer, planetary scientists must account for certain fundamental characteristics of bright terrain: ridge-and-groove topography, relative brightness, and occasional flat, smooth areas. The brightness of the grooved and smooth terrains can be explained if icy volcanism has covered up the ancient dark surface. Such *cryovolcanism* could bury and erase the older landscape. Some craters in bright terrain have dark halos around them. Geologists envision that these impacts punched through the bright icy surface to excavate dark material from beneath. This supports the idea that older dark terrain has been concealed by icy volcanic flows.

The extension and faulting that creates ridges and grooves might even trigger icy volcanism, in the way that faulting and volcanism occur together within terrestrial rift zones, as in east-

Figure 9 (left). If Ganymede's grooved terrain formed by complete separation and spreading of the lithosphere, it should be possible to reconstruct older craters and furrows across opposing sides of a groove lane (upper panel). This is commonly true of bands on Europa, but not on Ganymede. Instead, Ganymede's grooved terrain probably formed by a combination of tectonic stretching and icy volcanism (lower panel). Some large impacts have punched through the bright layer, excavating dark material from below that rings the craters with diffuse, dark halos.

ern Africa. It is not obvious, however, how cryovolcanic material would get to the surface. A water "magma" would be more dense than the surrounding ice and so should sink rather than rise. But just as a shaken bottle of soda froths when opened, dissolved gases can drive a water melt upward, allowing it to flow out onto the surface of an icy moon. It is also possible that liquid water is not involved at all. Warm, mobile ice — buoyant compared to its cold surroundings — could simply rise to the surface and "erupt."

If California were on Ganymede, the Voyager cameras would have resolved San Francisco, while Galileo's camera would discern the Golden Gate Bridge. This is dramatically illustrated in *Figure 10*, which compares Voyager's view of grooved terrain within Uruk Sulcus to the same location as seen by Galileo. The Galileo images reveal pervasive faulting on very fine scales. This intense deformation signals a fierce and violent past, during which severe icequakes shuddered through the satellite. Galileo has confirmed that grabenlike structures make up some grooved terrain, and it discovered that rotated tilt blocks (*Figure 11*) also rake the surface. Tilt blocks imply that some groove lanes were stretched 10 times farther apart than previously believed. Some areas also show signs of strike-slip faulting — the motion that occurs when lithospheric blocks lurch past one another, as happens along the San Andreas fault on Earth.

The Galileo images make us wonder: are the larger ridges and grooves that were seen by Voyager real or an artifact of lower resolution? Galileo stereo imaging of Uruk Sulcus shows that both scales of deformation are genuine, with the smaller-scale topography discovered by Galileo superimposed on the longer wavelength undulations imaged by Voyager. Geologists can account for this configuration if the ice a few kilometers down was relatively warm when the grooved terrain was stretched apart by tectonic forces, developing long-wavelength undulations like pulled taffy. Meanwhile, cold, near-surface ice broke along faults to create the small-scale ridges.

Although we anticipated icy volcanism on the basis of the Voyager observations, it is surprising how little direct evidence for it can be found in Galileo's higher-resolution images. For example, few smooth areas exist in Uruk Sulcus, and no flow lobes or eruptive centers are seen. Perhaps the Uruk region was initially brightened by icy volcanism, then torn beyond recognition by faults. Some planklike swaths in other regions do appear smoothed by icy volcanism, such as along the border of Nippur Sulcus (*Figure 7*). Elsewhere we observe unusual scalloped depressions that might be local sources for eruption of icy material (*Figure 12*). Still, the overall paucity of identifiable cryovolcanic features points to the dominance of faulting in shaping grooved terrain.

The ubiquity of these grooves has profound implications for understanding Ganymede's adolescent years. On Earth, subduction zones and fold belts are large-scale compressional

Figure 11 (right). Like dirty panes wiped clean, Galileo images are windows into the geologic history of Uruk Sulcus. Grabenlike grooves are visible at upper right, and dominolike tilt-blocks dominate the grooved terrain at lower left. A zone cutting diagonally across the mosaic center probably marks a strike-slip fault. En-zu is a 7-km-wide, dark-ray crater; the remains of its dirty impactor lie strewn across the surface.

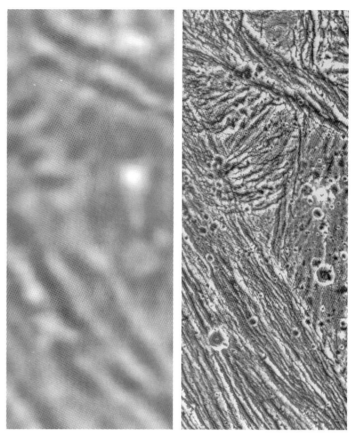

Figure 10 (above). A portion of Uruk Sulcus as imaged by Voyager 2 in 1979 *(left)* and 12 times closer by Galileo in 1996. Note the fine-scale structure revealed in grooved terrain. In these views the Sun is high overhead, so brightness differences are emphasized. The largest craters here — only 2 km across — have noticeably bright icy rims and dark floors.

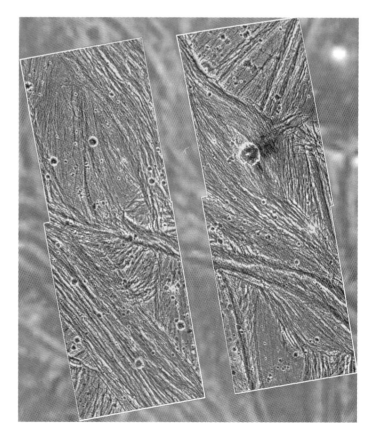

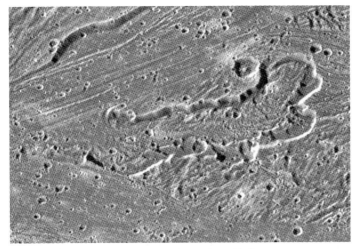

Figure 12. Scalloped depressions like this 20-km-wide feature are uncommon on Ganymede. They may be places where icy lavas have poured out onto the surface to form bright, smooth swaths.

regions that accommodate extension elsewhere on the globe as part of the plate-tectonic process (see Chapter 12). However, Ganymede's widespread grooved terrain is principally extensional in origin, pointing to expansion of the entire moon. Volcanism during this time recounts an interior warmed to near the melting point of ice. The concentration of extension, strike-slip faulting, and volcanism in narrow lanes intimates that buoyant slabs of warm material ascended through the satellite's icy interior toward the surface. Because grooved terrain postdates dark terrain and most palimpsests, this fierce internal strife must have occurred after the satellite accreted and its brittle lithosphere thickened. We will return later to the global ramifications of grooved-terrain formation in the context of the Ganymede-Callisto dichotomy.

OTHER SURFACE CHARACTERISTICS

Impact craters serve as probes of a planet's lithosphere. By examining crater morphologies, scientists gain insight into the characteristics of both the projectile and its planetary target.

On Ganymede such scars range from small bowl-shaped craters, through those with more complex morphologies, to large basinlike features and palimpsests. Some are surrounded by a low pedestal, probably a sheet of icy debris that surged outward from the impact crater as it formed (*Figure 13*). Dark ray craters, such as En-zu in Uruk Sulcus, may portray the remains of a dark projectile that was strewn across the surface upon impact. Other craters have dark floors. Rocky particles and meteoritic dust might have collected there over time; alternatively, the impact explosion may have volatilized ice and concentrated the dirty remains of the projectile and target. Large, relatively fresh craters commonly show bright icy rays, some extending many hundreds of kilometers from the impact site. Such bright-rayed craters are conspicuous in global images of the satellite. Rays ultimately fade with age, through mixing into the surface by the bombardment of micrometeorites and the high-energy particles that bathe the satellite.

Many midsize and large craters, those roughly 20 to 100 km across, have an intriguing central pit. One model suggests that these pits are the frozen-in-place end stage of an impact splashed into an icy target. In another hypothesis, they were formed by especially slow-moving projectiles. A central dome is situated within the pit in craters wider than about 60 km. Paul Schenk has proposed that deeply penetrating impacts allowed warm subsurface ice to move up into the crater center and onto the surface, where it chilled rapidly. Pit and dome craters are shallow in depth relative to their diameters, and the floors of some are bowed upward. When these craters formed, the lithosphere may have been warm enough for the crater topography to rebound slowly, like a baking cake that bounces back from the finger poke of an anxious cook.

As noted earlier, palimpsests probably formed long ago when large impactors pierced the thin, brittle lithosphere of Ganymede. Galileo images indicate that a bright palimpsest boundary probably corresponds to the edge of the continuous splash of ejecta tossed onto the surface during such a collision. Later, as the icy lithosphere cooled and thickened, large impacts could not penetrate through it but instead created huge basins more analogous to those on the rocky worlds. Gil-

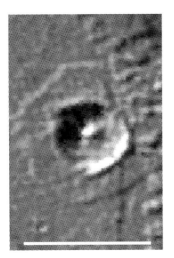
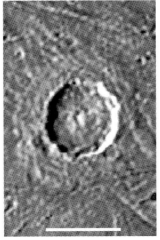
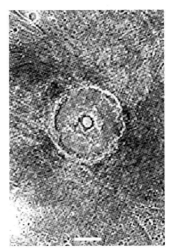
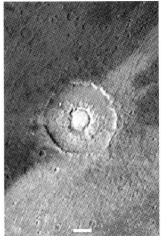

Figure 13. Ganymede shows a diversity of craters. The four seen here have been scaled so their rims match in apparent size; the scale bars represent 30 km. Small craters (5 to 20 km across, *far left*) have central peaks, and some show low "pedestals" around them. Larger craters have pits, and domes appear in still larger craters (*right*). Similar crater forms are found on Callisto.

Figure 14. Measuring 800 km across its concentric escarpments, Gilgamesh is the largest preserved impact basin on Ganymede. Its creation followed the formation of grooved terrain. By that time, Ganymede's brittle lithosphere had thickened so that a depressed scar was left, rather than a low-relief palimpsest like Memphis Facula (*Figure 5*). Beyond its smooth outer ejecta blanket, secondary craters mark sites where chunks of debris were tossed from the impact; many are strewn in radial chains.

Figure 15. At near-polar latitudes on Ganymede, bright frosts paint patches of craters and kilometer-scale ridges. Here in the northern hemisphere, frosts are located on slopes that face north (toward upper left) and east, which tend to be the coldest.

gamesh is the largest and most recent of Ganymede's basins (*Figure 14*). The giant impact scar might have formed as recently as hundreds of millions of years ago — or it could be billions of years old. The age of Ganymede's surface is an important issue that we will return to later.

Whether ancient or young, the entire surface of Ganymede exists in a harsh environment beyond the realm of our ordinary experience. Although ice does sublimate, the process occurs too slowly to create a substantial shielding atmosphere; consequently, ultraviolet sunlight and high-energy charged particles constantly bombard the icy surface. The imparted energy can tear water molecules apart. Once separated, the water's hydrogen atoms can escape to space more easily than the heavier oxygen atoms can, and the latter get left behind.

Using Earth-based telescopes, John Spencer and Wendy Calvin have found the fingerprint of molecular oxygen (O_2) in the infrared spectrum of Ganymede. Hubble Space Telescope observations have found signs of ozone (O_3) as well. However, the surface is probably too warm for oxygen frosts to exist. Instead, the oxygen may be trapped inside ice crystals, a kind of cryogenic alchemy that occurs when high-energy charged particles or photons rip through the upper microns of the surface. Galileo observations confirm that a tenuous atmosphere of atomic hydrogen surrounds the satellite, and that ionized

hydrogen (protons) are streaming away at supersonic speed. Oxygen, however, was not observed — a confirmation that most of it is left behind to accumulate within the icy surface.

Ganymede does have a thin dusting of frost near its poles. The underlying terrain is discernible as if through a bright haze. Close up, icy patches are seen concentrated on the colder north-facing slopes of craters and ridges (*Figure 15*). In places, frosts may blanket hillocks to depths of several meters. (Future explorers may find the makings of a good ski run!) We suspect that the polar frosts are predominantly water ice, perhaps tinged with frozen carbon dioxide. Some of it may have migrated to the poles after sublimation from the warmer equatorial regions.

A more important factor in polar-cap formation may be bombardment of the icy surface by energetic particles. Galileo made the remarkable and unanticipated discovery of a magnetosphere at Ganymede. One implication is that charged particles coursing through Jupiter's immense magnetosphere are funneled onto Ganymede's poles. Crashing down onto the satellite, these particles "sputter" and redistribute water-ice molecules, brightening the polar surface overall. As we will see, this is another of the many dissimilarities between Ganymede and its sibling Callisto.

THE SURFACE OF CALLISTO

Callisto has been facetiously called the "boring" Galilean satellite because images show it to be a heavily cratered world with scant signs of the faulting or cryovolcanic resurfacing that would signal past geologic activity. Thanks to Galileo, we now realize that Callisto is far from boring — it is, in fact, quite puzzling and mysterious in many ways.

Voyager's global-scale images revealed a surface densely packed with craters tens of kilometers in size. Callisto is also scarred by tremendous multiringed structures. The largest of

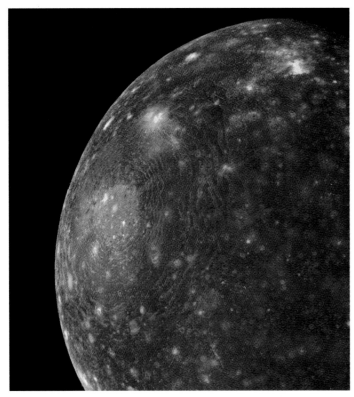

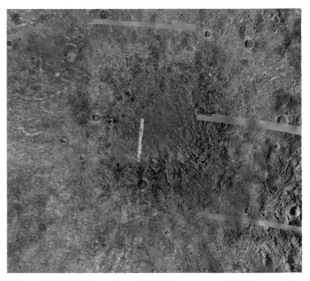

Figure 17. The oblong shape of the feature named Lofn suggests that the impactor which formed it struck Callisto at a shallow angle. The color overlay shows infrared data related to surface composition: Lofn's bright, 300-km-wide palimpsest is ice-rich (red) compared to the surrounding dirty plains (blue).

Figure 18. Galileo images resolve the rings of Valhalla into sinuous troughs spaced about 50 km apart. Their bright rims and dark floors are reminiscent of furrows on neighboring Ganymede. The crater chain Svol Catena, which slices across the large crater Skuld (far left), is likely the remains of a comet or asteroid that split into pieces before striking Callisto.

Figure 16. The impact which formed the Valhalla multiringed structure blasted deep into Callisto. Ice from the cleaner interior is now exposed in its central palimpsest, 600 km in diameter.

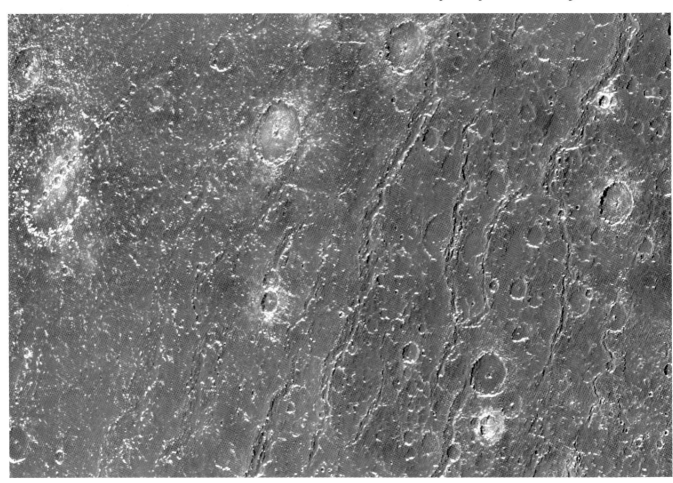

Figure 19. At a small scale, Callisto's surface is surprisingly smooth and sparsely cratered. The small craters that do exist, such as those beside this 10-km-wide ring of Valhalla, have knobby eroded rims. These craters have probably disintegrated over time.

these is Valhalla (*Figure 16*). Its bright central plains cover an area the size of Colorado, while its bull's-eye of rings dominates an entire hemisphere of the satellite and could span the contiguous United States. Like Ganymede's palimpsests, the bright centers of Valhalla and other impact structures on Callisto may be regions where cleaner ice has been excavated from beneath the dirtier surface layers (*Figure 17*).

The character of Valhalla's rings changes with radial distance and to some degree along their circumference. Many of them resemble graben, while others are outward-facing fault scarps (*Figure 18*). Analogous to Ganymede's furrows, the rings are probably impact-related. The formation of Valhalla probably triggered faulting of a cold, brittle lithosphere as it was dragged inward atop a layer of warmer ductile ice. Smooth material seen in Voyager views along the bases of Valhalla's scarps hinted that limited episodes of cryovolcanism might have taken place on Callisto. However, Galileo found no sign of icy volcanism there — only blankets of smooth, dark material. The smooth, dark blanket is seen nearly everywhere, interrupted only where bright crater rims poke up through it (*Figure 19*).

Most planetary scientists were confident that Galileo's high-resolution pictures of Callisto would reveal a myriad of small impact craters, as befitting an ancient, static surface. This prediction proved quite wrong. At small scales, the surface is remarkably crater-poor. The smaller craters that do exist range in appearance from fresh to quite subdued. This suggests that they are somehow filled in and degraded over time. Several possible explanations exist for this subdued appearance. The ejecta thrown out by impacts can soften or bury topographic relief, an effect observed on our Moon. Callisto's surface appears much

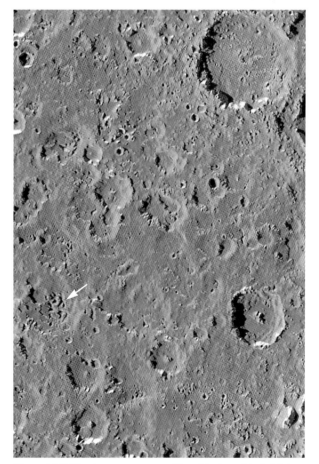

Figure 20. Callisto's craters are muted by a blanket of dark debris. Irregular crater rims (center) and kilometer-sized pits (arrowed) attest to the slow removal of ice through sublimation and subsequent collapse of the surface. This area is in Callisto's southern hemisphere, so it is a mystery why bright frosts occur on the craters' sunnier, north-facing slopes.

Figure 21. This 20-km-wide *catena* (crater chain) on Callisto formed by the near-simultaneous impact of a line of comet fragments. Unlike the strings of secondary craters tossed out by large impacts, this chain is not aligned radially to a large basin. Still visible are remnants of ridges that once separated the individual craters. Featureless smooth debris along the base of the bright crater rims attest to billions of years of erosion.

more degraded than the lunar landscape, but its composition is also very different. Perhaps resurfacing by impact ejecta is more efficient on ice-rich Callisto.

The bright rims and central peaks of the large craters provide another clue, as these craters probably penetrated to bright ice that lies below the dark surface. Some rims appear fragmented into isolated hills, as if they eroded in place. One possibility is that sublimation has slowly robbed exposed rims of their ice, leaving behind a dark lag deposit. Numerous kilometer-sized pits on Callisto may mark where the surface collapsed after being severely undermined by gradual ice loss (*Figure 20*). Some of the ice extracted through sublimation would escape to space, but some of it should be redeposited in bright, cold patches, especially at high latitudes. Isolated patches of very bright material in many of Callisto's craters may be these frost deposits.

Downslope movement can redistribute dark material on Callisto, at least to some extent. During its third orbit, Galileo imaged an unusual chain of craters (*Figure 21*). The smooth, dark talus piled along their bases derived from erosion of the crater walls. Elsewhere on Callisto, tongues of dark material have streamed downhill into the crater floors. Some of these landslides were evidently jostled loose by nearby impacts, as small craters are found near them.

What *is* all this mysterious dark stuff coating Callisto? Spectra from the NIMS instrument suggest that hydrated minerals (clays) are widespread and that tholins are also present, as on Ganymede. Another confirmed compound is sulfur dioxide (SO_2). The SO_2 is concentrated on the hemisphere that always faces forward as Callisto orbits Jupiter. Therefore, it could be related to the impact of micrometeorites, which preferentially strike a satellite's leading side.

In stark contrast to Ganymede, Callisto shows few signs of internal geologic activity. It has no grooved terrain, no smooth lanes of bright material, and only rare signs of fractures cutting through its dark, cratered surface. Some features — specifically the multiringed structures, palimpsests, and central-dome craters — hint that Callisto's ice was warm and soft enough to flow in the ancient past, but that this mobile ice resided at substantial depth. Nonetheless, Callisto has a very "young" surface as seen at high resolution. Its small craters evidently are being erased much more efficiently than on neighboring Ganymede, and the processes that have muted Callisto's topography in the past might still be ongoing at a slow rate today.

GANYMEDE AND CALLISTO: NATURE OR NURTURE?

Why should two sibling satellites, so similar in origin, size, density, and composition, have such contrasting surface geology? Did some small difference in original characteristics cause their life histories to diverge? Or instead did some momentous event affect one satellite but not the other? This question is essentially one of "nature versus nurture." The answer may come from our inferences about the internal states of Ganymede and Callisto, based on Galileo's measurements of their gravitational and magnetic fields.

As Galileo soared by each moon, engineers on Earth tracked its velocity very precisely through slight Doppler shifts in the frequency of its radio signal. Changes in Galileo's velocity are a measure of the gravitational field surrounding each satellite, which is in turn related to the distribution of mass within the satellites.

Analysis of the gravity data by John Anderson and Gerald Schubert shows that Ganymede is highly differentiated. Evidently, this moon was once warm enough for its icy component to separate completely from its rock and metal. Intense heating must have melted the ice in an internal flood that may have lasted hundreds of millions of years. Where warm water met frigid space, an ice skin formed that thickened over time. Deeper down, rocky material was released from its icy matrix and sank to form a mantle. In fact, the gravity data suggest that Ganymede may have an iron-rich core (*Figure 22*). This would mean internal temperatures were once so high that metals melted and sank out of the rocky mantle.

In contrast, Callisto could not have melted throughout and certainly was never hot enough to form a metallic core. Galileo's gravity data indicate that this moon is only weakly differentiated. While some shallow melting and slow sinking of rocky chunks through warm ice may have occurred, its rock and ice are not completely separated. Instead, the interior may grade from an icier near-surface to a rockier center. Compared to Ganymede, Callisto has had a cool and uneventful internal history.

Another probe of the interior comes from magnetic field and charged-particle measurements. Galileo discovered that Ganymede has an internally generated magnetic field, with a strength on its equator of about 750 nanoteslas (about 1 percent that of Earth). This field is strong enough to carve out its own magnetospheric bubble within Jupiter's much larger magnetosphere. Hubble Space Telescope observations confirm that Ganymede's poles glow at far-ultraviolet wavelengths — the satellite has aurorae of its own. At Callisto, Galileo detected no sign of an *intrinsic* magnetic field. However, magnetic disturbances were noted in the moon's vicinity, the type that might arise from a field induced in a global, subsurface ocean by Jupiter's magnetosphere. Taken at face value, the data imply a global, salt-laced layer of water at least 10 km thick and no more than about 100 km deep. How such an ocean might have come to exist and persist within Callisto, in light of the incompletely differentiated interior implied by gravity data and the moon's static-looking surface, remains a mystery.

Ganymede's intrinsic magnetism confirms the existence of an iron core. Assuming the field is generated by an internal dynamo driven by convection, the core must be at least partially molten. This is not what geophysicists expected. Before the Galileo mission, they believed that the bulk of Ganymede's internal activity took place nearly 4 billion years ago, and that the satellite should have frozen solid long ago. But apparently Ganymede has not yet cooled from its period of intense activity. An intriguing possibility is that Ganymede's heating and differentiation occurred relatively late in solar-system history, perhaps less than a billion years ago. Moreover, the induced magnetic field detected inside Callisto raises the speculation that Ganymede might also be hiding a subterranean ocean. The magnetic field induced in its briny waters could easily be masked by the moon's intrinsic magnetism.

Heating and differentiation have left significant imprints on the surface of Ganymede. If this moon began as a uniform mixture of ice and rock, ice in its deep interior would be compressed by the great pressure of overlying matter. Pressurized ices have denser molecular structure than the ice in our freezers. During differentiation, dense ices were melted and displaced upward, ultimately refreezing closer to the surface in lower-density forms. This process, along with warming of the interior, would create a net expansion of the satellite and global-scale extension of its surface. The grooved terrain we see today is thus the frozen-in manifestation of a tumultuous past.

Galileo's observations of Callisto tell a very different story. Its lack of an intrinsic magnetic field is consistent with a cold, inactive interior; likewise, gravity data imply only partial segregation of its ice and rock. A relatively cold, weakly differentiated Callisto is commensurate with the stark picture painted by its debris-laden landscape. It seems that the collective process of heating, melting, and differentiation is the fundamental difference between these satellites. It led to widespread deformation on Ganymede but essentially none on Callisto.

Planetary scientists have pondered and debated the cause of the Ganymede-Callisto dichotomy ever since Voyager images of these seemingly paired worlds first reached Earth. Ganymede, slightly rockier than Callisto, was endowed with a slightly greater percentage of the radiogenic elements that generate heat as they decay into more stable isotopes. Moreover, its larger size ensured that late-arriving impactors would strike its surface slightly faster and deliver more kinetic energy, compared to similar events onto neighboring Callisto. Ganymede had the evolutionary edge with slightly more heat. However, the extreme differences we see in these satellites' surface and internal structure attest to something more profound in their histories.

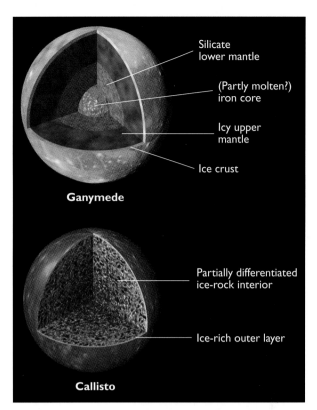

Figure 22. **Beneath its complexly deformed surface and thick icy shell, Ganymede is inferred to be differentiated into a rocky mantle and metallic core. In contrast, Callisto has remained largely unmodified since its formation roughly 4.5 billion years ago, its interior a mix of rock and ice in roughly equal proportions.**

A likely suspect is tidal heating. The three inner Galilean satellites — Io, Europa, and Ganymede — orbit Jupiter in a silent rhythm known as the Laplace resonance. All three are locked in a state of perpetual tug-of-war owing to their mutual gravitational interactions. They are forced into orbits that are somewhat eccentric rather than circular, a departure from symmetry that causes tidal flexing within them as they revolve about Jupiter. This orbital kneading is most pronounced for Io and has made it the most volcanically active body in the solar system (see Chapter 17). On Europa, tidal heating may be sufficient to maintain a water ocean within the satellite today (see Chapter 18).

Ganymede's eccentricity is presently too small to cause much tidal heating. Nor could it ever have been large enough, if the satellite trio have spent their entire lives in the Laplace resonance. However, such may not always have been the case. Renu Malhotra calculates that other resonances could have preceded the current configuration, temporarily pumping Ganymede's eccentricity to values much higher than at present. The result might have been enough tidal heating to trigger global differentiation. Ganymede's warm core hints that this heating may have happened in the last billion years of solar-system history — an exciting possibility that warrants further study. Meanwhile, lonely Callisto, left out of the orbital dance, cannot have been tidally heated during its lifetime.

It seems that the orbital push and pull of their siblings, Io and Europa, drove Ganymede onto an evolutionary road distinct from that of Callisto. Nurture has won over nature in resolving the fate of these two satellites.

The Huygens probe will descend to Titan's murky surface in December 2004.

Titan

Tobias Owen

FIRST GLIMPSED AS a tiny "star" accompanying Saturn (*Figure 1*), Titan was discovered by the Dutch astronomer Christiaan Huygens in March of 1655. Huygens was using an extremely long-focus telescope of his own design that represented the cutting edge of optical technology. However, once the satellite's orbit was worked out, its brightness measured, and its size roughly determined, little more of substance was learned about Titan for nearly 300 years. In the 1940s, Gerard P. Kuiper (another Dutch astronomer) began a systematic spectral survey of the planets and their satellites. He discovered that Titan's spectrum exhibits the same absorption bands of the gas methane (CH_4) that were by then well known in spectra of Jupiter and Saturn. Kuiper concluded that Titan must have an atmosphere (*Figure 2*).

The possibility that the largest satellites in the outer solar system might have atmospheres had been explicitly suggested some 20 years earlier by Sir James Jeans, who had studied how ther-

Figure 1. The Hubble Space Telescope captured Titan near the disk of Saturn on 6 August 1995, when the planet's rings were nearly edge-on as seen from Earth. In this four-color composite image, note the satellite's orange hue and the shadow it casts on Saturn's clouds.

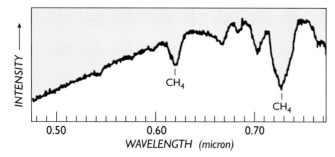

Figure 2. **The light reflected from Saturn's moon Titan bears the characteristic spectral signature of methane: deep absorption bands near the wavelengths of 0.62 and 0.73 micron. This spectrum was obtained in 1978, using the 2.1-m telescope at the McDonald Observatory in Texas, the same telescope used by Gerard P. Kuiper to discover Titan's atmosphere in 1944.**

mally energized gases can escape from the planets. Jeans realized that to maintain an envelope of gas over the lifetime of the solar system, an object simply needed sufficient condensed mass to provide a strong gravitational field and a temperature low enough to keep colliding gas molecules from reaching escape velocity. Thermal velocities are higher for gases of low molecular weight like hydrogen and helium, so these are lost to space with greater ease. However, methane was quite a reasonable candidate for the atmosphere of distant Titan. This gas consists of cosmically abundant carbon and hydrogen, has a molecular weight of 16, and condenses only at extremely low temperatures.

But how much methane did Titan have, what other gases were present, and how did they get there? Unlike Jupiter and Saturn, Titan has a relatively weak gravitational field and was certainly not expected to retain large amounts of hydrogen and helium. Were there clouds on Titan like those on the giant planets? And what was the surface like? Earth-based observers struggled with these and other questions, finding various, often contradictory answers. Titan is more than 1 billion km away and appears as a barely resolvable disk even in our best telescopes – a serious handicap in trying to unravel its mysteries. By 1980, two quite different models for Titan's atmosphere were in vogue, suggesting surface pressures of 20 millibars and 20 bars!

Fortunately, help was on the way. The Voyager 1 spacecraft flew past Titan in November 1980 and at its closest was only 4,000 km away, about one-tenth the altitude of a geosynchronous satellite above the Earth. All the instruments on board were brought into play, and we learned more about Titan in those few days than in all the preceding 325 years. Voyager's flood of information is still being analyzed, augmented by images from the Hubble Space Telescope and by ground-based radar and infrared observations.

ORIGIN AND INTERIOR

With a diameter of 5,150 km, Titan is bigger than Mercury and ranks second in size among satellites. Nevertheless, it is but one of several large, icy objects in the outer solar system that include Ganymede and Callisto at Jupiter, Triton at Neptune, and the Pluto-Charon system (see Chapters 17–19, 21). Their similarities and differences (and their relationship to the smaller icy satellites, the Centaur objects such as Chiron, and to classical

comet nuclei) are far from clear at present. What we do know is that all of these objects contain a mixture of water ice and silicates, with the ice apparently garnering ever-greater amounts of frozen, trapped, and adsorbed gases when it formed at lower temperatures, farther from the Sun.

But it is not just heliocentric distance that determines the final composition of one of these bodies. Where the object formed with respect to its parent planet is also extremely important. Each giant planet had a major influence on the gas and dust in the solar nebula surrounding it. We can thus envision planetary "sub-nebulae," in which conditions varied as a function of distance from the primary planet in a manner analogous to variations in the solar nebula at different distances from the embryonic Sun. In both cases, temperature was a key parameter. The formation of the giant planets produced large amounts of heat; gravitational potential energy was turned into kinetic energy when portions of the solar nebula "condensed" into the much smaller volume of a planet. Calculations show that Jupiter radiated so much thermal energy early in its history that it must have glowed. Some of this planet's primordial heat is still escaping today (see Chapter 14).

We see the consequences of this early, hot phase of Jupiter in the variation of density now found among its Galilean satellites: Io and Europa, situated closest to the planet, are much denser than the more distant Ganymede and Callisto, which have bulk compositions much like that of Titan. Thus, the Jovian system provides an analogy to the inner and outer planets. The bodies richest in highly volatile compounds (hydrogen and helium for the planets, water ice for the satellites) formed farther from their primary and thus in cooler regions. In the case of Jupiter, this trend produced dramatic results. For the Saturnian system the effect is more subtle, since Saturn has less than one-third the mass of Jupiter and therefore the heat generated during its formation was correspondingly less. Instead of a distinction between rocky inner satellites and icy outer ones, we find icy bodies throughout the Saturn system, from the ring particles to distant Phoebe. But other volatiles, borne by the ices in these bodies, should reflect the fact that the planet at the system's center was hot when it formed. Thus, substances such as methane, ammonia (NH_3), nitrogen, carbon monoxide, and argon must have been more prevalent in the materials that formed the more distant satellites.

Similarly, the generally lower temperatures in the Saturnian system compared with Jupiter explain why Titan has an atmosphere, while Ganymede and Callisto do not – despite the fact that the three satellites' similar densities suggest similar compositions. Laboratory studies have shown that crystallizing water ice rapidly loses its ability to trap and retain gases at higher temperatures. In other words, if the ice within Ganymede and Callisto condensed at temperatures above 100° K, as seems likely its gas content would have been markedly depleted.

In contrast, Titan probably formed from gas-rich ice that had condensed at lower temperatures. Heat generated during the accretion process would have been more than sufficient to liberate gases from the infalling ices. It could also have driven reactions among them, forming new species. Some volatiles must also have been contributed by impacting comets originating outside the Saturn sub-nebula. The resulting mixture of gases then became Titan's earliest atmosphere. Cometary delivery would not have worked on Ganymede and Callisto; the greater impact velocities there, caused by Jupiter's powerful gravitational field, would have driven the liberated gas into space.

Additional internal heat was generated later from the decay of short-lived radioactive isotopes in the rocky component of Titan. Just as Earth and the other terrestrial planets melted and differentiated, so too must this distant satellite have formed a core of dense, rocky material that became surrounded by a mantle of ice. Thus, Titan obtained a secondary, outgassed atmosphere from the same general mechanism that produced the early atmospheres on Mars, Venus, and Earth (see Chapter 13).

Since ice assumes different crystalline structures corresponding to the ambient temperature and pressure, we expect Titan's icy mantle to consist of layers of these different forms of ice. Depending on the amount of heat generated in the satellite's interior and the rate of its release, a layer of liquid water might exist even now between the mantle and the icy crust.

This discussion of Titan's origin has not considered where the material that formed the satellite came from or how it got organized in such a way that one large satellite arose instead of several smaller ones. These are problems whose solutions are still unknown. As stated above, the current consensus is that the giant planets and their satellites formed from sub-nebulae much as the Sun and planets formed from the primordial solar nebula. But there is at least one significant difference. It is commonly assumed that no material came from outside the solar nebula to affect the forming Sun and planets. In contrast, material from outside the planetary sub-nebulae must have been incorporated into individual planets and satellites. The extent to which this occurred is one of the many unresolved problems that require a new mission to the Saturn system for their solution.

THE ATMOSPHERE: COMPOSITION AND EVOLUTION

Before Voyager's arrival, ground-based astronomers had amassed considerable evidence that Titan's atmosphere contained particles in addition to gas. Polarization measurements, Titan's brightness as a function of wavelength, and large variations in the strength of its methane absorption bands all indicated that the atmosphere was not clear. But what were these particles, and how were they distributed? As Voyager 1 approached in the fall of 1980, it quickly became apparent that opaque haze completely enshrouded the satellite (*Figure 3*). No hint of the surface of Titan could be seen, even in the many high-resolution pictures obtained during the spacecraft's closest approach. Yet the atmosphere's aspect showed a sharp demarcation, with the northern hemisphere noticeably darker than the southern. Just how those airborne particles know where the satellite's equator is remains an unsolved problem of Titanian meteorology. The aerosols also appeared to be layered in altitude, as indicated by images of the satellite's limb, and there was a dark northern cap.

More surprises came from the experiments designed to study atmospheric structure and composition. The dominant gas surrounding Titan is molecular nitrogen (N_2), just as on Earth. In fact, the satellite is enveloped by about 10 times more nitrogen

Figure 3. Opaque layers of particles in Titan's atmosphere (left) prevented Voyager 1 from seeing the satellite's surface during its 1980 flyby. Note the lighter color of clouds over the southern hemisphere and the dark hood at the north pole. Voyager 2 looked back at a crescent Titan the following year (right); the extension of blue light around the moon's night side is due to scattering by smog particles in the sunlit portion.

than we are, yielding a surface pressure 1.5 times greater than at sea level here. Thus, neither of the two extreme pre-Voyager models for the atmosphere proved correct. Yet Donald Hunten's high-pressure, nitrogen-rich model fits the run of pressure and temperature in Titan's atmosphere remarkably well. It was simply necessary to set the satellite's surface at the 1.5-bar level rather than lower down at 20 bars.

Methane, the one gas identified with certainty before Voyager arrived, turns out to be a minor constituent with an abundance of a few percent. High above Titan's surface, nitrogen and methane molecules are being broken apart continuously by the Sun's energetic ultraviolet photons and by the bombardment of electrons from Saturn's magnetosphere. The fragments of these molecules then recombine to form an impressive variety of trace constituents, some of which were detected by the Voyagers' infrared spectrometers (*Figure 4, Table 1*). This list includes familiar gases such as acetylene (C_2H_2), propane (C_3H_8), and hydrogen cyanide (HCN). You may be surprised to find molecular hydrogen (H_2) among them since, as mentioned, this gas can easily escape from Titan. But hydrogen is being continuously generated by the break up of CH_4, and the abundance of H_2 found in Table 1 represents the balance achieved between production and escape.

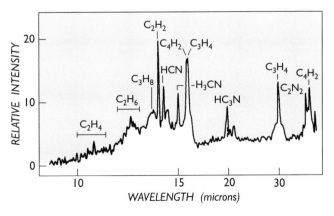

Figure 4. A section of the infrared spectrum recorded by the spectrometer on Voyager 1 shows numerous sharp peaks due to gases in the satellite's atmosphere.

Composition of Titan's Atmosphere

Major constituent		Percent
Nitrogen	(N_2)	82 – 99
Methane	(CH_4)	1 – 6
(Argon?)	(Ar)	<1 – 6

Minor constituent		Parts per million
Hydrogen	(H_2)	2,000
Hydrocarbons		
Ethane	(C_2H_6)	20
Acetylene	(C_2H_2)	4
Ethylene	(C_2H_4)	1
Propane	(C_3H_8)	1
Methylacetylene	(C_3H_4)	0.03
Diacetylene	(C_4H_2)	0.02
Nitrogen compounds		
Hydrogen cyanide	(HCN)	1
Cyanogen	(C_2N_2)	0.02
Cyanoacetylene	(HC_3N)	0.03
Acetonitrile	(CH_3CN)	0.003
Dicyanoacetylene	(C_4N_2)	condensed
Oxygen compounds		
Carbon monoxide	(CO)	50
Carbon dioxide	(CO_2)	0.01

Table 1. In addition to nitrogen, Titan's atmosphere contains an appreciable amount of methane, whose abundance varies with altitude and is still poorly determined. The presence of argon has been deduced only indirectly – there may be only a tiny trace of it, in which case the nitrogen fraction would be larger.

Usually Titan resides within Saturn's magnetosphere, but it sometimes lies in the boundary (magnetosheath) region or even outside in the undisturbed solar wind. Voyager 1 found that the satellite has no appreciable magnetic field of its own (see Chapter 4), so ions and magnetic fields can interact directly with its upper atmosphere. Neutral gases become ionized through impacts with high-velocity ions and electrons. Once that happens, the magnetic field picks up the newly ionized gases and sweeps them away. Thus, Titan must also be losing gases other than hydrogen through such interactions.

Here we have another puzzle: Titan's methane and hydrogen are constantly being broken apart, with some fragments escaping into space while others form new constituents that condense in the cold atmosphere and precipitate to the surface. At the present rate of destruction, all of the methane now in the atmosphere will be gone in just a few million years. This is a tiny period of time compared to the 4.5-billion-year lifetime of the solar system, so there must be a source of methane that replenishes the atmosphere. Could it be comets? Volcanoes? Underground springs? We simply don't know.

At Titan we have found an atmosphere that is still evolving from a primitive hydrogen-rich state. This is not too surprising, given the extremely low temperature at the satellite's surface. Voyager measured a value of 94° K, only a few degrees warmer than the surface temperature expected for a body that far from the Sun with Titan's reflectivity and no atmosphere at all. At this low temperature, the vapor pressure of water is vanishingly

small. What this means for Titan is that water ice on the surface cannot participate in the atmospheric photochemistry, because as a solid it does not undergo the dissociation by sunlight that would provide oxygen atoms to interact with the methane, hydrocarbons, and nitriles. If we could magically move Titan closer to the Sun, say, to the orbit of Mars, the character of its atmosphere would immediately change. Warmer surface temperatures would drive plenty of water vapor into the atmosphere, and the resulting supply of oxygen would rapidly convert methane and its byproducts to carbon dioxide (CO_2), exactly the dominant carbon-carrier now found on Mars.

Meanwhile, back on the real, present-day Titan, some carbon dioxide is in fact present, as is carbon monoxide (*Table 1*). But given the dearth of oxygen, how is this possible? There are at least two solutions to this apparent paradox. The carbon monoxide could be primordial — it may have been trapped in the ices that formed the satellite, then released into the atmosphere as Titan formed. Electrons from Saturn's magnetosphere, which continually bombard Titan's atmosphere, can break CO apart and leave its oxygen in an excited state. If one of these oxygens encounters a methane molecule, it reacts to yield a hydroxyl (OH) radical; this can in turn combine with another CO molecule to make CO_2, while the hydrogen escapes. In other words, if CO was present on Titan from the outset, the formation of a small amount of CO_2 is predicted even in the absence of water vapor. Alternatively, the CO could arrive episodically, borne to Titan by impacting comets.

To distinguish between these choices, it will be necessary to study the relative abundances of noble gases such as argon and neon and the ratios of major-element isotopes. To reach the level of precision we need, these studies require a mass spectrometer that is taken to Titan by a spacecraft. However, we can gain an idea of the power of this approach by using our telescopes here on Earth to measure the relative abundances of the stable isotopes of hydrogen.

The most common hydrogen atom consists of a single proton and a single electron in orbit about it. The next heavier isotope, deuterium, has a neutron in addition to the proton, which doubles the nuclear mass. Chemically speaking, the two isotopes are nearly identical, since reactions usually involve only the electron. But the physical behavior of hydrogen and deuterium can be very different. Therefore, a study of the relative abundances of these two isotopes can help us to unravel various physical and chemical processes that have taken place in a planetary atmosphere.

In Titan's atmosphere most of the hydrogen is in the methane, and spectroscopists can now discriminate between the deuterated form of this molecule (CH_3D) and CH_4. It turns out that deuterium occurs in Titan's methane molecules some four to eight times more frequently than it does in methane in the atmospheres of Jupiter and Saturn. Less than half of this can be attributed to processes acting on the satellite's atmosphere since it formed. The remainder must be telling us something about the origin of the methane we see today. Comets from the Oort cloud (see Chapters 5,24) contain water ice with a deuterium-to-hydrogen value twice that in ocean water on Earth and about four times that found on Titan. This high enrichment was the result of ion-molecule reactions in the cloud of gas and dust that ultimately condensed to form cometary ices. How similar is the ice that formed Titan to the ice in comets? We don't know, but we presume that Titan's methane must have derived, at least in part, from deuterium-enriched ice. The methane falls short of the deuterium excess in comets, because hydrogen in the ice that formed Titan had an opportunity to exchange deuterium with hydrogen gas in the Saturn sub-nebula. So we speculate that deuterium is much less enriched in Titan's icy crust and mantle than it is in comets. If the methane in Titan's atmosphere was formed during accretion and later cometary impacts by carbon combining with the hydrogen in Titan's ice, it would exhibit just the intermediate deuterium abundance we observe. But many other scenarios are also possible at this point. To deduce the true origin of this fascinating atmosphere we must have data on isotope abundances in other elements on Titan, as well as better information about comets and the atmospheres of other planets.

THE ATMOSPHERE: VERTICAL STRUCTURE, HAZES, AND CLOUDS

Titan currently possesses the largest unexplored surface in the solar system. The ubiquitous haze of aerosols that gives the satellite its fuzzy tennis-ball appearance in *Figure 3* is completely opaque to visible light. To understand the unusual properties of this haze layer and to find out if there is any hope to glimpse that hidden surface, we first need to know how the temperature in the atmosphere varies with height – a trend that was determined by both the radio occultation experiment and by the infrared spectrometer on the Voyager spacecraft. The combined results (*Figure 5*) show that temperature decreases with altitude from the surface until reaching the tropopause, about 42 km up. At this point, a reversal occurs, and temperature begins to increase with height, eventually peaking at 175° K, some 80° *higher* than at the surface. This temperature turnaround results primarily from the ultraviolet sunlight entering Titan's upper atmosphere. Its absorption not only contributes to the photochemical processes described in the previous section, but also excites the resulting molecules and warms the dark material making up the aerosol.

On Earth, urban smogs usually form within 1 km of the planet's surface. But the particulate haze around Titan extends to an altitude of roughly 200 km, with a detached and much more tenuous second layer about 100 km higher. Over the dark north-polar cap the main cloud is more extended, and in Voyager images the two layers appear to blend together. Actually, Titan's atmosphere contains relatively few particles per unit volume of gas, but the haze is so deep that it completely hides the surface from our view. We now realize that some of this aerosol material is simply condensed forms of the gases shown in *Table 1*. With the exception of hydrogen and CO, all of the minor gases will condense at the tropopause temperature (about 71° K). Indeed, all of these species (except CO) were discovered in Titan's stratosphere – not in the lower atmosphere, where they would not exist as gases.

But the aerosol layers seen in Voyager images are more than just simple condensation products, which would be white or

gray. The dirty-orange color suggests that some additional chemistry is occurring, transforming simple molecules into more complex substances. Some of these end-products are probably polymers, chainlike structures in which the same molecular configuration repeats again and again. Both hydrogen cyanide and acetylene form dark polymers that could certainly contribute to the observed effects. Laboratory experiments that combine the principal ingredients of Titan's atmosphere using a variety of energy sources to drive the reactions have little difficulty in producing a long list of suitably dark organic compounds. But it has not yet proved possible to achieve a close match between the laboratory simulations and what is found in the satellite's atmosphere.

A SATELLITE WITH FLAMMABLE SEAS?

Thus, what we see around Titan is a thick photochemical smog whose exact composition remains unknown, though it is probably a rich mixture of many compounds. One thing is clear, however: this aerosol cannot stay suspended indefinitely. Voyager observations indicate that, the airborne particles average about 0.2 to 1.0 micron in diameter. As the particles grow larger, they precipitate out of the atmosphere and onto the satellite's surface. What happens next depends on the local temperature and topography. It could be quite remarkable! For example, methane could in principle condense as a liquid on Titan's frigid surface. A careful analysis of how temperature changes with

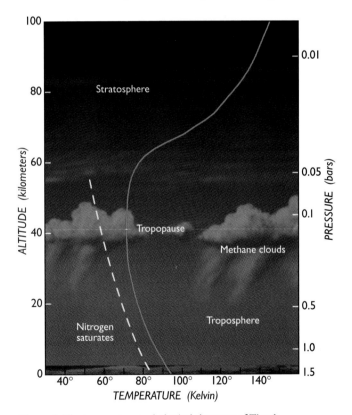

Figure 5. The temperature and physical character of Titan's atmosphere vary markedly with altitude. Note in particular that the temperature reverses near the tropopause, about 42 km above the surface. The lower atmosphere is just barely too warm to allow the condensation of liquid nitrogen, but clouds of methane ice and a drizzle of liquid ethane are distinct possibilities.

altitude just above the surface indicates that this is unlikely, but thin clouds of methane crystals may form in the lower atmosphere (*Figure 5*).

The story for ethane is more interesting, even though this gas is just a trace constituent in the atmosphere. As condensed ethane falls through Titan's lower atmosphere it may form clouds or hazes, but at the surface it should be liquid. One can calculate how much ethane has been produced on Titan over the entire history of the solar system (it is the most abundant

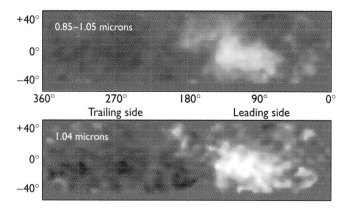

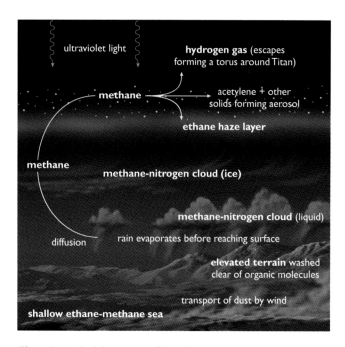

Figure 6. These maps of bright and dark features on the surface of Titan were created from near-infrared images obtained with the Hubble Space Telescope in 1994. (No data were collected for the polar regions.) While the composition of this surface remains unknown, we can clearly see that it is not homogeneous. The bright "continent," roughly the size of Australia, shows up better at the longer infrared wavelength in the lower panel, and other studies have detected it at wavelengths extending from 1 micrometer to 4 centimeters. A cross in each panel marks the entry site for the proposed Huygens probe.

Figure 7. Patchy lakes or seas of liquid hydrocarbons could be driving an ethane-based weather cycle on Titan, with towering clouds and frequent rain. This cycle could also be supplying the atmosphere with a steady supply of methane, part of which is converted into more complex compounds. Titan does not appear to be covered by a global sea, but the extent and depth of liquid on its surface remains uncertain.

byproduct in the photochemical destruction of methane). The result is that this remarkable moon could be covered by a global ocean of ethane with an average depth of up to several kilometers! This ocean would contain nitrogen and methane dissolved within it, along with all of the other products shown in *Table 1*. The bottom could be lined with a mixture of insoluble aerosols (only the very "fluffy" ones will float) and CO_2 ice. Any surface topography protruding above the ethane sea would also be coated by this mixture.

However, we know that Titan is not covered by a global ocean, because we have learned how to see through the haze layer with Earth-based instruments. Radar can penetrate the haze of Titan as easily as it sees through the cloud cover of Venus (see Chapter 8). Duane Muhleman and three colleagues obtained the first indication that Titan's surface is not homogeneous by means of radar observations in 1990 that revealed a bright spot on the satellite's leading hemisphere. At about the same time, other observers realized that it was possible to sense the surface at infrared wavelengths in between the strong absorptions caused by Titan's methane (*Figure 2*). The Voyager cameras recorded only visible light, but it is well known that infrared radiation can pass through hazes made of small particles. In 1994 this fact was used with the Hubble Space Telescope at wavelengths near 1 micron to produce the first map of the satellite's surface (*Figure 6*). The bright, continent-size feature centered at latitude 10° south and longitude 70° west must be the same one that produces the high radar reflectivity. It certainly proves that there is not a global ocean on Titan, but what can it be? The radar brightness suggests that this is rough terrain. The roughness could also produce the contrast at infrared wavelengths, if the expected covering of precipitated aerosols is patchy.

So what has happened to all that ethane? It appears that our theories for the chemistry on Titan are incomplete, or else there must be some unknown process that periodically recycles the ethane, converting it back to methane. We know ethane is being produced, and it must condense on the surface, as will propane. Thus we expect seas, lakes, and ponds of liquid hydrocarbons, even rivers flowing over a surface whose "bedrock" most probably consists of water ice. Except for this ice and a possible frosting of solid CO_2 in some places, everything in this exotic landscape is highly flammable! This includes possible wind-blown drifts of aerosols that have precipitated from the atmosphere, accumulating in depressions and converting ponds into organic-rich swamps. There may even be the equivalent of a hydrologic cycle on Titan, in which ethane takes the place of water (*Figure 7*). Evaporation from the lakes and seas, followed by condensation in the atmosphere could lead to rainfall. The resulting rivers could sculpt the icy landscape just as water erodes the rocky terrain on Earth. All this will be happening on a surface pockmarked by impact craters. This ensemble of processes and materials may lead to bizarre land forms that are unique in the solar system!

What would it be like to sit in a boat, rocking in the hydrocarbon seas of Titan? Dusky and cold, for sure. The atmosphere is so dense that the surface and near-surface environments experience little change in temperature with varying latitude, time of day, or even at different seasons. But the horizontal visibility

Figure 8. **If all goes well, in December 2004 a probe called Huygens will descend through Titan's murky atmosphere and relay its findings to Earth via the Cassini orbiter. The probe is equipped with a variety of scientific sensors to measure the physical properties of the moon's atmosphere; it also carries an imaging device to return pictures of Titan's surface.**

might be very good if you weren't trapped in an ethane fog bank. There is some suggestion that the lower atmosphere will be quite clear because the aerosol layers are concentrated where they are produced, at high altitudes (*Figure 5*). Titan absorbs about 80 percent of the sunlight incident upon it; the smog particles absorb some of the visible and ultraviolet light, while methane absorbs in the infrared. Nevertheless, at noon the distant Sun's light, filtered through the smog, casts an orange glow over the local scene that would be about a thousand times brighter than the illumination of the full Moon on the Earth, and a thousand times fainter than our daylight.

Will there be waves? Winds in the upper atmosphere appear to move in the same direction as Titan's rotation at roughly 100 m per second, but we have no information yet about winds near Titan's surface. Indeed, we are not completely certain of the existence of the seas! Ethane may undergo further reactions, producing more complex substances so efficiently that there is not enough of it left to form large bodies of liquid. Yet lakes (or at least ponds) seem highly likely. Despite the tantalizing glimpses we have managed to obtain from Earth, we still lack the definitive observations that will tell us how much liquid is present, what its composition is, and how it is distributed over Titan's surface.

THE NEXT STEP

Titan is a member of a class of volatile-rich icy planetesimals that include everything from comet nuclei to the cores of the giant planets. Comets are capable of delivering these volatiles to the atmospheres of the inner planets, so the relevance of studying Titan may be greater than we might think. In Titan, we can examine a secondary atmosphere produced by the degassing of the volatiles trapped in such a planetesimal, with an additional contribution from comets. The similarities and differences between these gases and those found in the class's other members — especially comets — will tell us something about the conditions and processes involved in their formation.

As we have seen, there appear to be opportunities to reach back in time to study an evolving, oxygen-poor atmosphere in which complex chemical reactions are occurring today that may resemble some of the first steps along the path from chemistry to biology on the early Earth. Since Titan lacks liquid water, we cannot expect to find examples of the famous "primordial soup" from which life is thought to have arisen on Earth. But Titan can at least supply "primordial ice cream," and that may be enough to give us some useful insights. (It is perhaps worth stressing that there is a long and as yet uncharted passage from chemical evolution to biology. *Life* on Titan seems ruled out by the exceedingly low surface temperature, which must slow chemical reactions to unproductive rates.)

This is a rich harvest indeed, and to reap it we must return to Titan. To this end, NASA and the European Space Agency (ESA) are now pursuing a joint mission called Cassini-Huygens. Its orbiter, NASA's contribution, has been designed to carry a radar experiment for mapping the surface of Titan right through the smog layers, much as American and Soviet radar systems have mapped the surface of Venus. The orbiter also carries cam-eras and infrared and microwave spectrometers much more capable than their counterparts aboard Voyager. But the most anticipated results will surely come from Huygens, ESA's instrumented probe. Descending slowly through the atmosphere for almost three hours, the probe will determine directly the composition and isotopic ratios of the gases and aerosols around it. A sensitive camera is to record the scene below at many different wavelengths (*Figure 8*). The descent will be slow enough that the probe might even survive its landing – be it a splash or a thud – and make additional measurements at the surface.

If everything goes according to plan, all of this information plus volumes more about the rest of the Saturnian system will be sent back to Earth beginning in late 2004. At that point, we will know much better just what it is that this mysterious moon is trying to tell us about our solar system's early history.

Triton, Pluto, and Charon

Dale P. Cruikshank

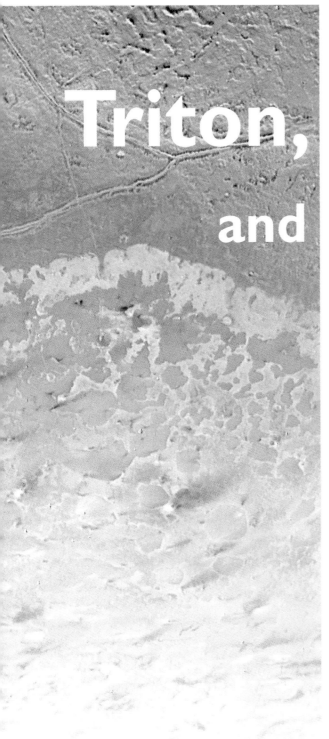

Complex ridges and enigmatic "cantaloupe" terrain on the icy surface of Triton.

BEYOND ABOUT **30** AU, some 4½ trillion km from the Sun, there are no more major planets. Neptune is the last of the gas giants, the final outpost in a progression of large worlds. But the solar system does not end with Neptune. Outside its realm we encounter another class of planetary body. Pluto, its satellite Charon, and Triton, the largest satellite of Neptune, mark the gateway to the region of the outer solar system defined and populated by the small icy bodies of the Kuiper belt (or disk) and the Oort cloud. Moreover, as this chapter will describe, Pluto and Triton share a number of unique physical properties that set them apart from the icy satellites of the other giant planets.

TRITON: DISCOVERY AND EXPLORATION

Within a few days of the discovery of Neptune on 23 September 1846 by Johann Galle and Heinrich d'Arrest, astronomers were training their telescopes on the newfound planet. William Lassell, a prominent English brewer and amateur astronomer with an unusually large (24-inch) telescope, discovered Neptune's first satellite, shining feebly at magnitude 13.6, on 10 October. He further noted that it was orbiting Neptune backward, or retrograde, the first satellite ever found with this unusual property (see Chapter 23). Curiously, Lassell's find had no name for more than 60 years; even as late as 1908, a popular book about the solar system referred to Neptune's nameless moon. Shortly thereafter it acquired the name Triton, one of the nymphs attending Neptune in Greek mythology. Lassell would go on to find a moon of Saturn (Hyperion) and two moons of Uranus (Ariel and Umbriel), but Triton may have been his most important discovery.

The fact that Triton was visible at all gave astronomers an early indication that this is a body of significant size. Under any

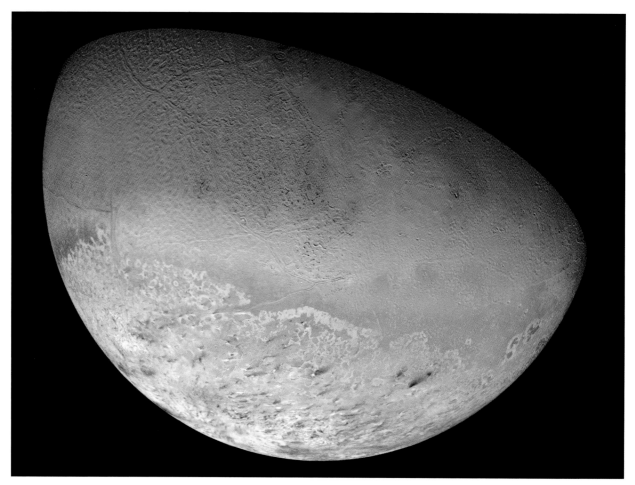

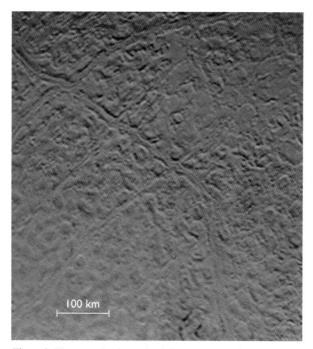

Figure 1. The portion of Triton viewed by Voyager 2 at high resolution is geologically unique in several ways. This photomosaic is centered near 20° north, 0° west; the equator runs approximately through the center of the bright swath across the middle of the image.

Figure 2. Voyager scientists coined the name "cantaloupe terrain" to describe the large tracts of roughly circular depressions (cavi) that cover much of Triton's surface. On the basis of its relationship to other geologic features (such as the bright, fresh frost near the bottom), cantaloupe terrain is considered the oldest landform on Triton. The region shown is the eastern portion of a province named Bubembe Regio.

plausible assumptions, Triton would be a planet-class object, but attempts to measure its diameter and mass up through the 1980s were fraught with difficulty and great uncertainty. Crude infrared spectra eventually gave hints of an ice-covered surface. Even so, as the Voyager 2 spacecraft approached Neptune's vicinity in 1989, we knew little more about Triton than we had a century before.

Voyager arrived traveling at a relative velocity of 27 km per second. So even though it passed within 39,800 km of Triton's center, the flyby offered only a fleeting opportunity to scrutinize this mysterious body. With a diameter of 2,710 km, Triton proved to be smaller than the Moon, Titan, or Ganymede — but larger than Pluto. Its relatively high reflectivity (geometric albedo), 72 percent at visible wavelengths, is consistent with a surface covered with various ices, as will be discussed later.

The slight gravitational tug exerted on Voyager by Triton yielded the first estimate of the satellite's mass; this value, together with the newly determined diameter, yielded a mean density of 2.05 grams per cubic centimeter, indicating an interior that is a mix of rocky material and ice.

But the most stunning information sent by Voyager 2 from Triton were the views of this icy world's extraordinary surface, images that reveal a recent history of very unusual geologic

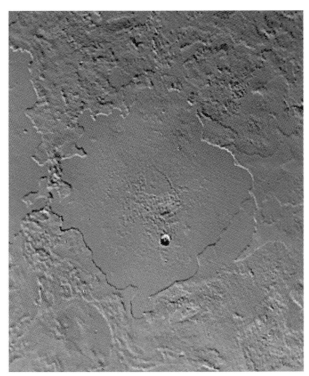

Figure 3. Roughly 180 km across, Ruach Planitia is one of several plains on Triton that apparently underwent episodes of flooding by cryovolcanic fluids, freezing, remelting, and collapse. A large pit and an associated cluster of smaller depressions is clearly seen. Note the well-defined edges of the smooth terraces around the basin's perimeter.

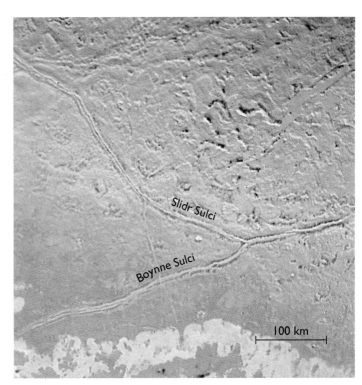

Figure 4. The complex linear ridges Boynne Sulci and Slidr Sulci merge in a Y near Triton's equator. These ridge systems, typically 12 to 15 km wide, are reminiscent of cracks in the surface of Jupiter's moon Europa (see Chapter 18). Their double structure may be due to the pressure of ices welling up from below through a crack in the crust. Enigmatic "cantaloupe" terrain dominates the surrounding icescape.

activity (*Figure 1*). Although some of the features on Triton looked familiar from earlier views of the icy satellites of Jupiter, Saturn, or Uranus, other utterly new landforms were seen for the first time.

Only a few impact craters, the largest of which is 27-km-wide Mazomba, are recognized on Triton, implying that the surface is relatively young. Nearly all other landforms appear to have been created by the eruption of material from the interior, in a variety of processes collectively termed *cryovolcanism* to denote the curious emplacement of "lava" made not of molten rock, but of liquid or slushy ices.

Lacking the ancient, impact-scarred landscape commonly found on other planetary bodies, Triton instead exhibits a vast, rugged surface, with a texture that bears an uncanny resemblance to a cantaloupe's skin (*Figure 2*). Most likely shaped by the upwelling movement of fluid materials from the interior, this "cantaloupe" terrain is characterized by large, roughly circular dimples (called *cavi*) some 25 to 30 km across, with the whole region crisscrossed by long, interconnected ridges. Pits and chains of pits also contribute to the rough texture.

Triton exhibits large smooth regions as well; for example, the floors of walled plains called *planitia* are rather flat and surrounded by stair-step terraces some 150 to 200 meters high (*Figure 3*). In other tracts of smooth terrain, bulbous and lobate deposits (cryovolcanic outflows?) fill valleys.

The geologic structures that we now see probably do not date from the earliest times of Triton's accretion, but rather from the time after its capture by Neptune. The tidal energy

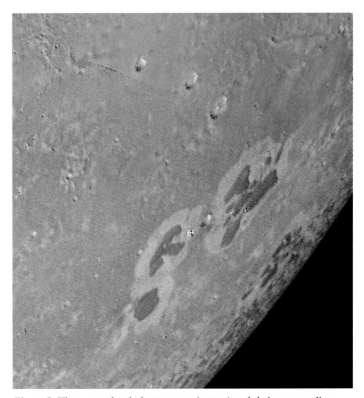

Figure 5. These complex dark structures (guttae) and their surrounding brighter aureoles make up features called maculae. Measuring 100 to 200 km across, maculae remain enigmatic and appear to be unique to Triton's surface. Nearby are three impact craters; the largest, Mazomba, is 27 km in diameter.

dissipated within Triton during that dynamic event, and the subsequent forcing of its orbit into a circle, caused significant — and possibly complete — internal melting. As the crust froze over, it formed a new surface over the entire body. We will return to Triton's origin later, but for now think of this melting epoch as the second period of Triton's geologic history. The satellite's surface at that time was probably a featureless cryovolcanic plain or perhaps a global expanse of cantaloupe terrain.

A third postulated interval took place when various features of the present surface, such as the planitia, were emplaced as material explosively erupted from the interior. Other landforms took shape when material of unknown composition and consistency oozed from fractures in the surface. The long cracks and ridges also appeared at this time (*Figure 4*). In a fourth period, thick deposits of icy material were laid down, probably when a dense but transient atmosphere condensed in the extreme cold.

Voyager has shown us that a number of unexplained processes are still shaping the surface of Triton (*Figure 5*). But impacts from large comets are playing only a minor role. Only about 15 impact craters are found on the third of Triton imaged with moderate or high resolution by Voyager. Their shapes resemble the impact scars seen scattered across icy surfaces elsewhere in the outer solar system; some are simple bowl-shaped craters, while those larger than about 11 km diameter tend to have central peaks. Some of the smaller, poorly seen ones might not be impacts at all but rather collapse pits or blowouts caused by other processes. Why so much territory should have so few impacts is not well understood; there is at least the suggestion that Triton has undergone extensive, geologically recent, and perhaps ongoing resurfacing.

AN ICY VENEER

Overall, Triton's surface consists of slightly pinkish bright regions and dark regions that are pale blue to neutral in color. Voyager did not carry any instruments to tell us the composi-

tion of the surface (see Chapter 1), so we depend upon spectroscopic observations from Earth to reveal just what lies on the frigid landscape. Coincidentally, at about the time Voyager was launched (1977), it finally became possible to make the first determinations of Triton's composition using infrared spectroscopy from ground-based telescopes. Those early observations showed the presence of frozen methane (CH_4). By the time Voyager reached the Neptune system 12 years later, the techniques had improved enough to reveal a band attributed to molecular nitrogen (N_2), which is ubiquitous on Earth but had not previously been seen on any other planetary body.

Unfortunately, infrared spectroscopy tells us directly only about the composition of the thin surface veneer that reflects the incident sunlight. However, such exposures often belie the chemistry of the basic materials from which a body is made and the processes that have taken place there. At present we know of five molecules on Triton's surface: methane and nitrogen, as already noted, plus carbon monoxide (CO), carbon dioxide (CO_2), and water (H_2O) (*Figure 6*). All of these molecules are volatile to varying degrees, depending strongly upon the temperature. On Earth, water exists simultaneously as a gas, liquid, and a solid (ice), but the other four are gases here. However, at Triton's surface temperature (38° K), all five are frozen. The spectroscopic evidence shows that about half of Triton's surface is covered by a combination of N_2, CO, and CH_4 mixed at the molecular level and frozen together as a "solid solution." Water ice serves as the hard bedrock of Triton's crust and mantle, and

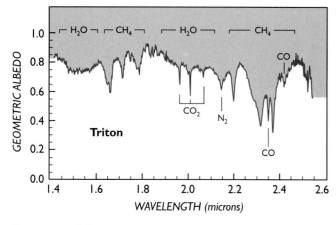

Figure 6. Five different kinds of ice are revealed in this near-infrared spectrum of Triton's surface, each having its own characteristic and diagnostic pattern of absorption bands. Analysis of this spectrum reveals that much of Triton's surface is covered with a mixture of solid nitrogen and trace amounts of methane and carbon monoxide, all frozen together. These molecules are also the main constituents of Triton's atmosphere.

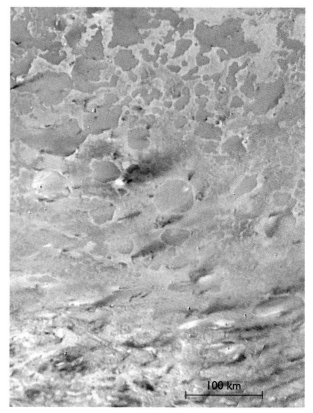

Figure 7. Dark streaks seen on Triton during the Voyager 2 flyby were tens to hundreds of km long and preferentially pointed northeast, away from the south polar cap. Their exact origin is not clear, though they could be smaller, lower-velocity counterparts of the eruptive plumes seen elsewhere.

the spectral data show that patches of frozen water and carbon dioxide cover a little less than half of Triton's surface.

Voyager images of Triton show a pattern of more than 100 elongated streaks that look like relatively thin deposits of dark material on the lighter background ices (*Figure 7*). Remarkably, most of them appear to point away from the geographic south pole, which currently is in full sunlight, giving the impression that they were deposited by wind coming from the pole's direction. They emanate abruptly from specific points, often some topographic feature, then thin out toward their ends. Some are only a few kilometers in length, while others extend more than 100 km. The composition of these streaks cannot be determined yet, but their dark color suggests a powder of rock particles or perhaps some carbon-rich material. (Almost any solid matter superimposed on a bright, icy background will appear dark.)

WISPS OF FRIGID WIND

When Voyager flew past Triton, two experiments on board revealed the presence of a thin atmosphere composed primarily of nitrogen, confirming the earlier spectroscopic observations. An atmosphere with a surface pressure of about 15 microbars (0.000015 times the sea-level pressure on Earth) is consistent with the presence of N_2 ice on the surface, because nitrogen is the most volatile of the five detected constituents. Tiny amounts of CH_4 and CO may also occur in the atmosphere, but they have not yet been identified.

Voyager images of the limb of Triton also showed discrete clouds and a haze layer hovering roughly 13 km above the surface. This is most likely a layer of photochemical smog, perhaps similar to the opaque orange haze that enshrouds Saturn's satellite Titan (see Chapter 20). The presence of the haze layer indicates that molecules other than nitrogen occur in Triton's atmosphere. The haze probably consists of a complex of organic molecules produced as ultraviolet sunlight and charged particles from Neptune's magnetosphere break the N_2 and CH_4 molecules into fragments, which then recombine into more complex organic and nitrile (CN-bearing) compounds. These byproducts probably include hydrogen cyanide (HCN) and ethane (C_2H_6), among others, and they should precipitate onto the surface at a rather high rate. Such species have not been detected so far, but they may add a touch of color to the portions of Triton's icy surface where they accumulate, particularly if they have joined together in larger molecules and polymers. Direct bombardment of the icy surface with ultraviolet light and energetic particles will drive additional reactions, so the surface should be covered with a rich organic chemistry.

The fact that we have not detected these more complex molecules adds strength to the contention that Triton's surface is both geologically very young (from the absence of large numbers of impact craters) and chemically "fresh." Something is happening on Triton to bury or destroy the complex material, perhaps recycling it back to the simpler molecules from which it formed.

Because of the orbital geometry of both Triton and Neptune, Triton experiences seasonal extremes in a cycle some 688 years long (*Figures 8,9*). At the present time the southern hemisphere is experiencing an "extreme summer," with the Sun directly overhead near latitude 50° south. The last time the Sun reached

this far south was in the year 1350, and in about 2090 the Sun will be over latitude 46° north. A significant change in the temperature distribution on Triton must surely occur during these extreme seasonal flips. Nitrogen ice on the surface tends to stabilize the temperature because it sublimates (evaporates) in sunlight, absorbing heat in doing so. It then precipitates in colder regions and releases its latent heat of condensation. Nitrogen, carbon monoxide, and methane all appear to migrate in this way throughout the seasonal cycle. The water-ice bedrock and local patches of carbon dioxide are left behind because they do not

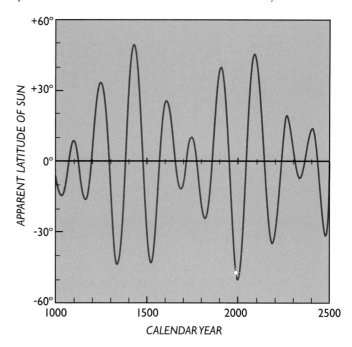

Figure 8. Dynamicists believe Triton undergoes wide seasonal swings and dramatic climate changes, because the latitude on Triton where the Sun is directly overhead can change from far north to far south (and vice versa) over several decades. A symbol indicates the present subsolar latitude.

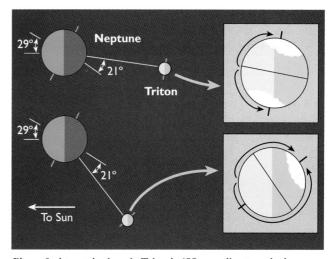

Figure 9. At certain times in Triton's 688-year climate cycle the Sun is almost directly over the equator (upper half), which causes atmospheric gases to migrate toward a pair of polar caps. But at the seasonal extremes (lower half) one cap basks in constant sunlight and disappears, while the other, in shadow, accumulates large deposits of ice condensing from the atmosphere.

Figure 10. Although unrecognized as such at the time of Voyager 2's flyby, an erupting plume appears in this image of Triton. The plume's dark column, seen here obliquely, rises almost vertically to a height of 8 km before drifting downwind (right) for more than 100 km. This plume has since been named Mahilani.

sublimate readily, even in full sunlight, at Triton's distance from the Sun. Heavier molecules created in the atmosphere or in the surface ices are also left behind, only to be covered over during the next winter. Thus, as the volatile ices move from one region of the satellite to another, then back again, the surface is renewed and remains bright.

The movement of these atmospheric gases means that there must be winds on Triton, and winds can carry material across the surface. The direction of movement depends upon where the icy source is sublimating into gas and where it is being redeposited. The explanation for the 100 or so dark streaks then becomes clearer: dark particulate material from Triton's surface is somehow picked up by the winds and carried along, in a more or less uniform direction. But when were these dark streaks deposited?

It came as a great surprise when it was recognized that Voyager 2 had imaged evidence of this airborne transport in action! At least four active plumes were seen projecting from Triton's surface into the atmosphere (*Figure 10*). Stereoscopic analysis of the images show dark columns rising nearly vertically to a height of 8 km, then trailing off nearly due west as a long, diffuse, opaque cloud. The two plumes seen most clearly were later named Hili and Mahilani; the latter seemed to change over a 90-minute interval, suggesting that the wind in that locale was blowing at about 15 meters per second. The dark streaks elsewhere on the surface are most probably the result of similar plumes ejecting particulate material from some source under Triton's surface.

Voyager's extreme good fortune has given us a new perspective on a world that was formerly thought to be cold and utterly dead, geologically speaking. What powers the plumes, and what is the material lofted into the atmosphere? How small must the particles be to rise so high and be spread downwind for 100 km or more? There are no clear answers yet. One good possibility is that a greenhouse effect of sorts may be at work within the frozen surface. Nitrogen ice is highly transparent, enabling sunlight to penetrate to depths of several meters. If the buried nitrogen ice is warmed by just 2°, sublimation will generate enough gas pressure to burst through the surface, overcome Triton's weak gravity, and propel a column of ice and dark particles several kilometers high into the tenuous atmosphere.

Sometime between June 1980 and May 1981, astronomers noticed a change in the character of Triton's methane-absorption bands, indicating that something had masked or diminished the surface exposures of CH_4 ice previously manifested in its spectrum. Perhaps some ice or frost other than methane snowed out of the atmosphere, covering regions where methane was previously visible. Whatever the cause, something had changed on Triton to affect its reflectivity in a major way.

Thus Triton should be viewed as a dynamic world, with ample evidence of recent geologic activity, a chemically active surface and atmosphere with a strong seasonal pattern of change, plumes of material jetting into the high atmosphere, and perhaps the wholesale redistribution of its surface ices as recently as in the early 1980s. Had it remained in an independent orbit around the Sun, we might well have dubbed it a planet. After all, that's what followed the discovery of the like-size body we now call Pluto.

PLUTO AND CHARON

The idea that a distant, ninth planet might exist emerged shortly after the discovery of Neptune. The person most identified with a search for "Planet X" was Percival Lowell. He had established a private observatory in Flagstaff, Arizona, primarily for studies of Mars, but the idea of a trans-Neptunian planet also fascinated him. Lowell's astronomers conducted a photographic search for such a planet for years without success. After a hiatus, the search resumed in earnest in 1929, some 13 years after Lowell's death. Clyde W. Tombaugh, a young amateur astronomer from Kansas, was hired to take an exhaustive series of photographs and to examine them for objects that were not stars. Tombaugh exposed 14-by-17-inch plates by night, and by day he examined the star images in an instrument called a blink comparator. By rapidly alternating a magnified view of two photographs of the same area of sky taken a few hours apart, any object that was not a distant star appeared to move against the background star field.

In searching for "Planet X," Tombaugh chanced upon a comet and a great number of asteroids. But the real reward for his diligence and patience with the tedious work came on 18 February 1930, when a starlike, 15th-magnitude image danced left and right in the blink comparator in a section of sky near the star Delta Geminorum (*Figure 11*). The newly discovered object soon proved to lie beyond Neptune's orbit, and it was named Pluto, after the gloomy god of the underworld of the dead. Initially, astronomers believed Pluto's discovery had validated the rationale for the search, namely, their perception that unexplained motion of Neptune was due to the gravitational influence of a massive unseen planet. Nevertheless, for more than 45 years after the discovery of Pluto, repeated attempts to measure its size and mass failed even when the world's largest telescopes were used. Careful measurements of the planet's brightness from night to night indicated that it rotates once every 6.4 days (*Figure 12*). But by the mid-1970s little more was known about Pluto than some details about its distant elliptical orbit.

Then came two discoveries that began to change our understanding of the most distant planet in a profound way. The first,

Figure 11. When Clyde Tombaugh compared this photograph taken on 29 January 1930 with another taken six days earlier, he discovered that a 15th-magnitude speck had moved. His discovery of Pluto was announced six weeks later on 13 March. The bright star at right is Delta Geminorum.

Figure 12 (right). As Pluto rotates on its axis, its brightness as seen in telescopes from Earth varies by about 30 percent. This graph shows the planet's brightness (its magnitude at the blue-light wavelength of 4400 angstroms) as measured in a four-year interval. The undulating curve indicates that Pluto's surface has a number of large bright and dark regions; the curve repeats exactly every 6.3872 days — the time Pluto takes to rotate once.

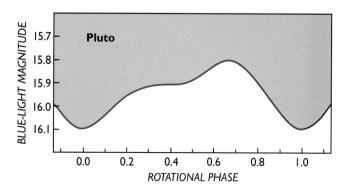

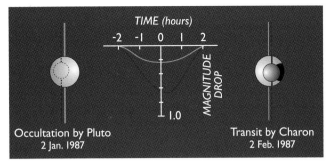

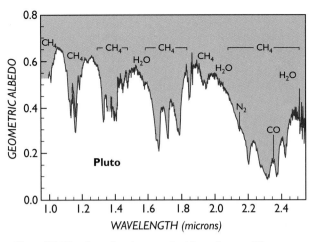

Figure 13. When Charon transits Pluto (right), the drop in total light is greater than when it slips behind the planet (left). This difference can exceed 0.4 magnitude (as plotted at center) and is due to the combined effects of Charon's shadow and darker surface. Although the two objects were not seen separately during such events, astronomers could monitor subtle changes in their combined brightness to derive the dimensions, separations, and albedos of this remarkable planet-satellite pair. In particular, both objects have bright, ice-covered surfaces, but Charon is significantly darker than Pluto.

Figure 15. Pluto's surface is covered with a mixture of four different kinds of ice — the same ones Triton has, with the exception that carbon dioxide has not been found. Comparison of this spectrum with that of Triton (*Figure 6*) suggests that Pluto has about twice as much methane on its surface, and possibly less nitrogen ice, as Triton has.

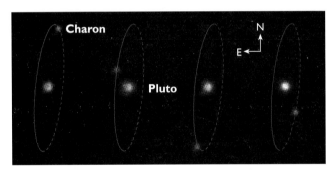

Figure 14. Four blue-light images taken with the Hubble Space Telescope in 1994 show Charon in locations corresponding to about half an orbit around Pluto. The planet's north pole is toward the right, and the solid portion of the orbit is the half closest to Earth.

in 1976, was that frozen methane exists on Pluto's surface. (The detection of methane was itself a triumph of spectroscopy; at that time only water ice and frozen carbon dioxide had been detected elsewhere in the solar system.) The presence of icy material on Pluto meant that its surface must be highly reflective. So, given its measured overall brightness and its known distance from Earth and Sun, astronomers at last realized that Pluto must be smaller than our own Moon.

The second discovery would prove even more important. On 22 June 1978, as he methodically measured the positions of Pluto on astrometric photographs, James Christy noticed that several images of the planet were all uniformly distorted in one direction, as though a very close star were interfering with Pluto's image. After inspecting other astrometric plates, Christy quickly concluded that Pluto must have a satellite. The newly discovered satellite was named Charon, after the repulsive old boatman in Greek mythology who ferried souls across the River Styx to Pluto's realm in the underworld.

Even though the planet and satellite were not clearly resolved from one another, astronomers could still estimate the radius and period of Charon's orbit. Armed with this information, and Kepler's laws of planetary motion, Robert S. Harrington made the first direct calculation of Pluto's mass: only about one-fifth that of our Moon.

The detection of a satellite around Pluto changed forever the way we think of this small member of the solar system. It soon became clear that Charon orbits Pluto in the same 6.387 days that Pluto takes to rotate, indicating that the planet and its satellite are locked in what is called synchronous rotation and revolution. Like our Moon, Charon keeps the same face directed toward Pluto throughout its revolution around the planet. But in contrast to the Earth, which spins 29½ times during each lunar cycle, one side of Pluto always faces Charon and vice versa. If hypothetical Plutonians happened to live on the hemisphere directed away from Charon, they wouldn't even know that Pluto had a moon. Conversely, for inhabitants of the Charon-facing side, the moon would always be in the sky at the same place, unchanging its location over the decades.

Another surprise was the discovery that Charon moves north-south in the sky, indicating that the pole of its orbit lies essentially in the ecliptic plane. This implied that Pluto's rotation axis is tipped very close to the ecliptic plane as well, much like the situation with Uranus.

Planetary astronomers soon realized that sometime in the mid-1980s Charon's orbital plane would be seen edge-on from Earth. That is, the satellite would appear to pass in front of Pluto (a transit) and, a half revolution later, pass behind it (an occultation). We immediately appreciated our good fortune: this alignment occurs only twice during Pluto's 248-year trek around the Sun, and we had chanced upon the satellite just before these "mutual events" began. Had Charon been discovered in, say 1993, we would have had to wait until the 22nd century to witness the next series of overlappings. Thus, with hope and anticipation, a small cadre of photometrists and spectroscopists began to monitor the feeble, combined light from Pluto and Charon. (Recall that during the 1980s, ground-based telescopes rarely showed Charon at all — and never with sufficient resolution to permit the study of these two bodies individually.)

A transit or occultation in progress in the Pluto-Charon system would be evident as a drop in the pair's combined light, as either part of Pluto's surface or Charon's became obscured (*Figure 13*). The exact onset of the event "season" would depend critically on the dimensions of the two bodies, which had only been crudely estimated, and on the uncertain orbital geometry of Charon. Richard Binzel made the first certain observation of a grazing occultation on 17 February 1985; the combined light from Pluto and Charon diminished only 3 to 4 percent as the edge of Charon slowly passed in front of Pluto. Within a few days two other teams of astronomers confirmed that the eagerly awaited mutual events had begun.

Throughout 1985 and 1986 the occultations and transits were only partial, but in 1987 the first total events occurred as Charon passed fully behind and then in front of Pluto. Observations of complete occultations and transits, which contain the most telling information about the distant pair, continued during 1988. From a growing body of precise photometric observations, combined with refined estimates of Charon's orbital radius, David Tholen and Marc Buie deduced that Pluto's diameter is just two-thirds that of Earth's Moon. Subsequent refinements have yielded a value of 2,302 ± 14 km. Charon proved to have almost exactly half the diameter of Pluto itself: 1,186 ± 20 km. (There is some evidence that Charon may have different surface scattering properties and could in fact be 10 to 20 km larger.) The combined density of Pluto and Charon is 2.0 g/cm³, indicating a composition that combines 60 to 70 percent rocky material with 40 to 30 percent ice, probably mostly frozen water.

Dynamicists suspected from the beginning that the locking of Pluto's rotation and Charon's revolution period around the planet must require that the mass of the satellite be rather large relative to Pluto's. Mutual-event observations confirmed this suspicion: Charon has about 0.12 the mass of Pluto, a much higher ratio than for any other planet-satellite pairing. Previously, the Earth-Moon system had always been considered most like a "double planet." But that distinction has now passed to Pluto and Charon.

Further understanding of Charon's orbit has come from the exquisite angular resolution of the Hubble Space Telescope (HST) images, which clearly separate Pluto from its satellite (*Figure 14*). One series of HST observations showed Pluto wobbling around the system's *barycenter*, the center of gravity common to both Pluto and Charon. Tholen and Buie determined that the distance between the two bodies averages 19,636 ± 8 km. Surprisingly, however, Charon's orbit is not perfectly circular but rather very slightly eccentric (0.0076) — a small but significant deviation in a system so completely tidally locked. A relatively recent collision of a sizable body with Charon may have jogged its orbit.

The discovery of this slight ellipticity complicates Charon's gravitational interaction with the distant Sun. When Douglas Lin calculated the dynamical effects of solar tides on Charon, the result was surprising: the ongoing dissipation of tidal energy within Charon may be generating enough heat to soften or perhaps even liquefy some of the icy material inside. Over time some of the most volatile compounds, such as N_2 and CO, might have been driven off Charon entirely — if they were there in the first place.

COMPOSITIONAL CLUES

From their overall densities, we strongly suspect that Pluto and Charon are composed of a mixture of rocky and icy materials. Even though it only probes the surface to a very shallow depth (say a few millimeters), spectroscopy can still tell us much about the exact surface composition and thus the chemical processes that have taken place. The discovery of frozen methane (CH_4) on Pluto in 1976 was followed by the detections of frozen nitrogen (N_2), carbon monoxide (CO), and most recently water ice (*Figure 15*). In terms of the molecular species discovered so far, Pluto seems very similar to Triton, except that Triton has the carbon dioxide (CO_2) that Pluto apparently lacks. Much of Pluto's surface, like Triton's, must be covered with large crystals of frozen nitrogen, within which is much of Pluto's methane and probably its carbon monoxide. But Pluto's spectrum indicates that there is another reservoir of pure CH_4 elsewhere on its surface, and that much more CH_4 exists on Pluto than on Triton.

A significant characteristic of Pluto's surface is the patchwork of light and dark regions. That much could be divined from the large change in overall brightness as the planet rotates (*Figure 12*) and from mutual-event data (*Figure 16*). But it became clearly evident in images obtained with the Hubble telescope

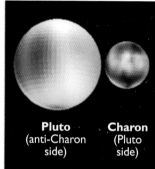

| **Pluto** (Charon side) | **Charon** (anti-Pluto side) | **Pluto** (anti-Charon side) | **Charon** (Pluto side) |

Figure 16. Mutual-event data and values of Pluto-Charon's combined light curve have been combined to derive these synthesized maps of Pluto and Charon. Pluto has a generally dark equatorial band with light-colored polar regions, the southern cap being brighter. Charon shows some detail but overall is darker than Pluto. These general characteristics compare well with spatially resolved photographs from the Hubble Space Telescope (seen below).

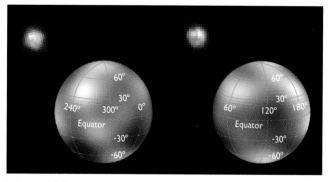

Figure 17. The Hubble Space Telescope has resolved several crude details on Pluto's surface. The raw blue-light images (small disks) are part of a series of 12 obtained in 1994 to generate global maps (large disks) of the planet's surface. The numerous bright and dark features show surprisingly high contrast.

(*Figure 17*). Bright areas are rather obviously covered with N_2 and other ices, but the nature of the dark regions is less clear. They could be exposures of rocky material. More likely, the Sun's ultraviolet light has driven reactions within the ice mixture to yield dark, reddish compounds. As in the case of Triton, cosmochemists predict that a host of complex hydrocarbons and nitriles can be created from the known molecules on Pluto's surface. Polymers of these compounds often have distinctive colors, typically red, orange, and black. Complex organic matter of this kind might be added to Pluto and other outer-solar-system bodies by colliding comets, which carry organic molecules that formed in the interstellar grains of ice and dust before there was a solar system.

The Pluto-Charon mutual events afforded a few opportunities to derive a spectrum for Charon itself; the observer's trick was to subtract the spectrum of Pluto (obtained while Charon was eclipsed) from the spectrum recorded when both bodies were in view. Surprisingly, Charon's surface is covered in large part by ordinary water ice, making it quite different from that of Pluto. But other molecules could easily have escaped detection, and Charon's frozen surface might still comprise a few percent methane and sizable portions of nitrogen and carbon monoxide. Furthermore, Charon's surface is too dark to be purely water ice; something else must be dirtying the surface. Also, unlike Pluto, Charon appears to be a neutral gray. Perhaps the darkening agent is a carbon-rich "dirt," or it might be an assortment of complex organic molecules that we have not yet identified.

The presence of various ices on Pluto leads to the expectation that the planet has a tenuous atmosphere composed of some of those same molecules. Just how much atmosphere might exist depends very strongly on the surface temperature, which is very difficult to measure directly. Spectral data suggest a value of 40° K for the bright regions, though the dark areas are probably a few degrees warmer. Such theoretical arguments and circumstantial observational evidence were borne out on 9 June 1988, when teams of astronomers scattered across the South Pacific watched as Pluto briefly hid an obscure 12th-magnitude star from view. The star disappeared and reappeared gradually, not abruptly, signaling that Pluto is surrounded a tenuous atmosphere. Since nitrogen is the most volatile ice identified on Pluto's surface, it must dominate the atmosphere, again mimicking Triton's case. Small amounts of methane and carbon monoxide are also likely atmospheric constituents, as is a photochemical haze of hydrogen cyanide (HCN), acetylene (C_2H_2), ethane (C_2H_6), and other complex molecules that should gradually fall to the surface. So far, these additional molecules have not been identified in Pluto's spectrum, but the search continues. In any event, the atmosphere of Pluto is very tenuous, with a surface pressure of perhaps 50 microbars.

The characteristics of Pluto's surface and atmosphere observed over the last two decades may not be constant throughout the planet's centuries-long traverse around the Sun. Its orbit is considerably eccentric (0.25); Pluto was 29.68 AU from the Sun at its perihelion in 1990, but when it reaches aphelion in 2114 that distance will have increased by two-thirds to 49.54 AU. The slowly varying heliocentric distance changes the temperature of the surface and atmosphere. In fact, there is some suspicion that the atmospheric pressure may drop abruptly in the next decade or two, possibly depositing a new veneer of frost on the already icy surface. A fresh layer might in turn raise the overall reflectivity significantly, causing even less sunlight to be absorbed and lowering the temperature by several degrees. Maybe the entire planet will turn uniformly white as the entire, already pitifully thin, atmosphere collapses in a global freeze-out!

ORIGINS

The similarities between Triton and Pluto are too close to ignore: they have comparable sizes, bulk densities, surface compositions, temperatures, and heliocentric distances (at least when Pluto is near perihelion). Yet their differences are also striking: Triton is the satellite of a major planet, circling in a retrograde orbit that suggests it began as an independent body before being captured by Neptune's gravitational field. Pluto is one component of a "planetary binary" (with Charon), both sharing an inclined and eccentric orbit in a 3:2 resonance with Neptune.

Numerous researchers have suggested various scenarios in which Triton and Pluto-Charon interacted with one another in the distant past. In one past scheme Pluto was a satellite of Neptune ejected during an encounter that put Triton in its present retrograde orbit. The discovery of Charon seriously complicates the ejection hypothesis, however, as does the 3:2 resonance between the orbits of Neptune and Pluto-Charon. Computer simulations of the motions of these worlds back to the origin of the solar system show that they have never been close, nor can they ever be (*Figure 18*).

Other hypotheses for the formation of the Pluto-Charon binary include one scenario in which a rapidly rotating Pluto

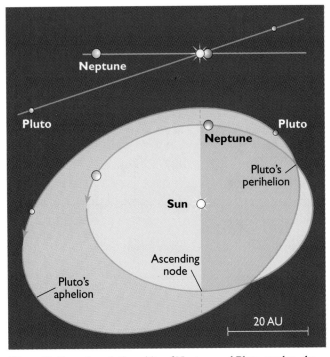

Figure 18. Even though the orbits of Neptune and Pluto overlap, the bodies can never come near one another. Their orbital periods maintain a 3:2 ratio, such that their conjunctions (tan symbols) always occur with Pluto near aphelion. Blue symbols show the planets' locations in mid-1998.

Figure 19. Triton could well have become a member of Neptune's satellite system by crashing into another moon upon its arrival from deep interplanetary space. The force of such an impact could have slowed Triton enough to become captured by the planet's gravity. Thereafter, the marauding moon might have remained hot or even molten for a billion years — even as it swept up other moons in its vicinity.

shed material that coalesced to become Charon. Another is that Pluto and Charon accreted from the solar nebula together as a binary pair. However, both concepts have drawbacks in terms of the pair's mechanical properties, as well as their apparent compositional differences. Deeper insight may bring these theories back to the serious attention of planetary dynamicists in the future, but for now another idea has emerged as the most likely one.

Alan Stern, William McKinnon, and Jonathan Lunine have proposed that Pluto formed in a near-circular, low-inclination heliocentric orbit, probably beyond Neptune's position. A great many other icy planetesimals also accreted in the solar nebula beyond Neptune, becoming the original population of

the Kuiper belt. The gravity of Neptune perturbed these bodies as they accumulated, resulting in frequent collisions among them. Eventually Pluto managed to garner considerable mass. Later, the powerful impact of a fairly large planetesimal with Pluto resulted in the formation of Charon. This hypothetical impact may also explain why Pluto's rotational axis is tipped so extremely.

It is likely that Triton also formed by accretion in the Kuiper belt, only to be captured as it ventured too close to the early Neptune. Calculations by Peter Goldreich and three colleagues offer a remarkable scenario in which Triton goes from a "free flyer" in its own heliocentric orbit to captured satellite. They speculate that when Triton encountered Neptune, the planet already had its own family of moons. Triton collided with one of them (having perhaps a few percent of Triton's mass, similar to Saturn's Tethys or Rhea) and smashed it to bits. The collision did not destroy Triton, but it was slowed enough to be captured into an eccentric, retrograde orbit around Neptune (*Figure 19*).

The planet's strong gravity distorted the shape of Triton greatly, a tidal tug of war that gradually robbed the new arrival of orbital energy. In a process that may have taken a billion years, Triton's orbit slowly shrank from a looping ellipse to near-perfect circularity; its inclination also decreased, though it remained in its original retrograde path. Meanwhile, the remaining moons of Neptune were enormously disturbed; most of them were "acquired" by Triton through collisions. Other satellites could have been ejected entirely or redirected into eccentric, inclined orbits (such as Nereid occupies at present). We now know that Nereid has water ice on its surface, a trait similar to the satellites of Saturn and Uranus, and one consistent with Nereid's having been an original satellite of Neptune.

An important component of the Goldreich scenario is that Triton was heated to its melting point by the tidal interaction with Neptune, and its interior may have remained soft and warm for almost a billion years. But such a thorough cooking is at odds with the presence of highly volatile material on its surface and in its atmosphere. Perhaps the nitrogen, carbon monoxide, and methane arrived after Triton cooled, delivered by infalling objects from the Kuiper belt. Or, despite the traumatic interactions with Neptune and its satellites, Triton may have never become warm enough to lose all of the volatiles it originally possessed.

However they came to be, we suspect that Triton and Pluto have something of a shared past. On the basis of their physical similarities and their proximity to the Kuiper belt, Triton and the Pluto-Charon binary are ever-more frequently being regarded as very large members of the Kuiper belt (see Chapter 5). Pluto's status as a planet is thus challenged. However, because it was found as a result of a search for a new planet, and because it has been called one for nearly 70 years, Pluto will probably retain that traditional designation for the indefinite future — at least in the minds of most of us.

Midsize Icy Satellites

William B. McKinnon

THE **SPACE AGE** has brought into focus some remarkable similarities among the giant planets Jupiter, Saturn, Uranus, and Neptune. For example, all four are now known to have rings. The four also possess systems of *regular* satellites, those that move in circular, equatorial orbits and that travel in the same direction as the planet rotates. Furthermore, with the 1997 discovery of two Uranian satellites in remote, highly eccentric, and inclined orbits, we now realize that the giant planets all possess *irregular* satellites as well.

Each giant planet, with its various rings and moons, is a miniature planetary system. The richness and complexity of these systems is what attracts our interest, for in their formation, structure, and evolution we glimpse some of the possibilities for planetary systems elsewhere in the galaxy. At the same time, the satellites themselves are natural laboratories for studying planetary evolution and processes.

In our solar system, the giant planets all lie sufficiently far from the Sun that the moons that circle them are, except those closest to Jupiter, composed at least in part of water ice. Other chapters deal individually with the largest of these moons; this one focuses on midsize icy satellites, those sufficiently large to have formed under gravity's grip into spheres and complex enough to display individual geologic personalities. Generally, this class involves icy bodies roughly between 400 km and 1,600 km in diameter (*Figure 1*).

ORIGINS OF REGULAR SATELLITES

Regular satellites share a common plane and sense of orbital angular momentum, which indicates a common origin. These same characteristics led the famed French mathematician Pierre-Simon Laplace in 1796 to propose that the planets formed, or coalesced, from a single rotating disk of gas and small dust parti-

Much of the outer solar system's history is written in the icy surfaces of satellites like Tethys.

Figure 1. Each of the four giant planets possesses a system of satellites, which are depicted here in order from their primaries and to the same scale. These moons span a wide range of sizes, albedos (reflectivities), and surface characteristics.

cles orbiting the young Sun. Today we call this primordial birthplace the *solar nebula*. Few cosmogonic ideas from the 18th century have survived intact, but Laplace's brilliant deduction is considered to be essentially confirmed, since we can now observe similar nebulae rotating around young stars in the star-forming regions of nearby molecular clouds (see Chapter 2). Similarly, the regular satellite and ring systems of the giant planets are taken as direct evidence that at some point (probably very early) in the history of these planets each was surrounded by a rotating disk of gas and dust, a *protosatellite nebula*.

Such nebulae fit in well with modern ideas of giant planet formation, in that a rotating gas and dust disk is a natural, if not inevitable, consequence of the growth (accretion) of a giant planet. Consider Jupiter and Saturn. Although mostly hydrogen and helium, deep inside each is an "inner planet" composed mainly of heavier elements such as oxygen, carbon, nitrogen, iron, silicon, sulfur, and magnesium (see Chapter 14). Under normal temperature and pressure conditions, these elements would form rocks, metal, and ices. They represent what are essentially the cores of Jupiter and Saturn, which formed by the accretion of rock, metal, and ice planetesimals. When these cores became large enough, on the order of 10 Earth masses or more, their gravity began to draw in gas from the surrounding solar nebula at ever increasing rates. This rapid, runaway inflow of gas supplied the atmospheric envelopes and the bulk of the masses of Jupiter and Saturn. However, the gas could not flow directly onto the cores of Jupiter and Saturn, because it contained the

orbital angular momentum it had when it was in orbit about the Sun. As Laplace realized, though, a rotating spherical cloud can instead contract under gravity while conserving its angular momentum by collapsing in a direction parallel to its rotation axis. The natural result is a flattened nebular disk (*Figure 2*).

Planetary scientists are undecided as to whether protosatellite disks form directly during the inflow phase of solar nebular gas, or whether the young giant planet must first shrink from a more distended state, leaving behind a protosatellite disk (what the late James Pollack termed a "spinout" disk). In either case, the protoplanet becomes much more luminous than Jupiter or Saturn is today, and the inner parts of their protosatellite disks are heated well above the temperature of the local solar nebula. These higher temperatures, in conjunction with the higher pressures caused by the compacting disk, allow the gases and much of the solid matter derived from the solar nebula to vaporize and interact chemically. The stage is then set for the classic condensation and accretion scenario.

First, as the protosatellite nebula cools, particles condense out from the vapor state and stick together. They become larger by colliding and sticking, forming an ensemble of small bodies that begin to attract each other by mutual gravity. In relatively short order (due to the rapid circulation of bodies about Jupiter or Saturn), the large ensemble of small bodies accretes into a small ensemble of larger bodies — icy satellites. The final stage in this drama is the dissipation of the protosatellite nebular gas, which probably can occur only after the solar nebula itself vanishes during the young Sun's T Tauri phase (see Chapter 2).

The characteristics of each satellite system — the number, sizes, and compositions of its member bodies — therefore depend on details specific to each protosatellite disk. Indeed, the search for a general theory of satellite formation is one reason Pollack's spinout-disk model is presently favored by some researchers. Consider Uranus and Neptune. Although classed as gas giants, they are in fact 20 times less massive than Jupiter, and they are not even mostly hydrogen and helium by mass. Their atmospheres contain an abundance of these gases, but they are in bulk composed of heavier elements, especially the ice formers oxygen, carbon, and nitrogen. Thus, Uranus and Neptune are not unlike the cores of Jupiter and Saturn postulated above, but ones whose growth ended before copious gas could flow in from the solar nebula. This is the key point. The regular satellites of Uranus and Neptune could not have been formed in a disk formed from the direct inflow of solar nebula gas and solids, because that inflow did not occur. Nevertheless, when Uranus and Neptune accreted they would have been hotter and larger than they are today, and as they lost energy and shrank, they too could have left spinout disks behind.

The large tilt of Uranus's spin axis and to a lesser extent that of Neptune imply that the final assembly of these worlds involved the off-axis collisions of some rather massive bodies, possibly as large as Earth in Uranus's case. This has led to the alternate and intriguing hypothesis that a massive collision blasted out the disk that, in one fell swoop, gave birth to the Uranian satellites. The shock waves resulting from such an impact would have heated the atmosphere of proto-Uranus and the material sent into the orbiting disk to rather high temperatures, perhaps above 10,000° K. This would definitely have changed the chemistry of the material that condensed to form satellites. Such a special origin for one satellite system may seem a bit *ad hoc*, but in our solar system such chance effects play a major role in determining outcomes.

In considering the regular satellites of the giant planets, one is struck by the range in sizes (*Figure 1*). All of Jupiter's Galilean satellites are relatively massive; Ganymede and Callisto are comparable to the planet Mercury. Saturn's system, in contrast, is dominated by one similarly large moon, Titan. The closeness in sizes is even more remarkable when enough water ice is added to rocky Io and mostly rocky Europa so that all five great moons have "solar" proportions of ice and rock (meaning a mixture based on the same composition as the Sun). Does this imply that an icy satellite can grow to some special or limiting size and not beyond? We suspect that this may be the case, but we are not sure.

It is intriguing that Jupiter has four major satellites while Saturn has only one. Jupiter is nearly 3½ times more massive than Saturn, and the ratio of the masses of their atmospheric

envelopes is even closer to four. The number of major satellites may simply be an expression of the mass of condensable material available to accrete into satellites in these systems. Regardless, Saturn's Titan — despite its large size — would not have had the gravitational reach to accrete *all* the solid material in an extended protosatellite disk. Hence, the remainder apparently accumulated into a number of more modest sized bodies, ranging between approximately 400 and 1,500 km in diameter, the middle-size icy satellites of Saturn.

Amalthea, the 270-by-150-km moon whose orbit lies inside that of Io, may have formed in a similar way from leftover rock and metal around Jupiter. Amalthea is relatively dark and red as satellites go, but this may be merely a coating of sulfur-rich material derived from Io. The little moon's proximity to Jupiter

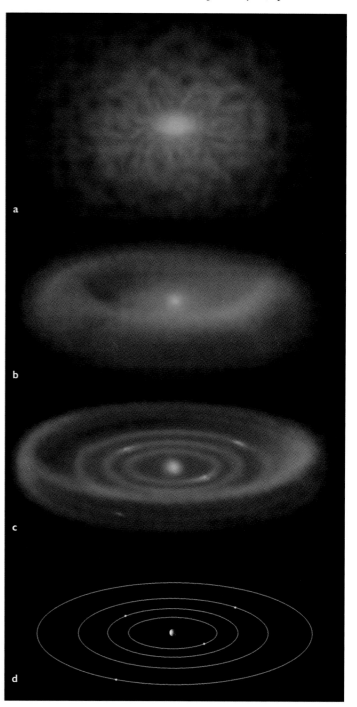

Figure 2. Regular satellites formed in much the same way that the planets did, but on a smaller scale. The process began (*a*) with a giant protoplanetary core enveloped in gas and dust flowing in from the surrounding solar nebula (though the flow may have been small for Uranus and Neptune). This envelope flattened into a disk-shaped nebula (*b*) that was hot and dense near the protoplanet, cooler and more diffuse (flared) farther way. Particles of rock, ice, and organic matter condensed (*c*), settling toward a midplane at the planet's equator. The particles accreted into a hierarchy of small bodies that ultimately combined to yield a sparse set of orbiting satellites (*d*) before the nebular gas dispersed.

Figure 3. Voyager 1 images of crater-pocked Mimas reveal two different hemispheres of this inner, 400-km-diameter satellite of Saturn. Craters are spaced closely and overlap each other, indicating that the surfaces seen are ancient. The relatively large, 130-km-diameter crater Herschel is named after Mimas's discoverer. Several fissures or grooves (best seen in the right image) may be a consequence of Herschel's formation, heat-driven expansion of the crust, or tidal forces from Saturn.

Properties of Icy, Midsize Satellites

Satellite	Diameter (km)	Density (g/cm³)	Geometric albedo
Saturn system			
Mimas	398	1.1	0.8
Enceladus	498	1.0	1.0
Tethys	1,060	1.0	0.8
Dione	1,120	1.4	0.6
Rhea	1,525	1.2	0.7
Iapetus	1,440	1.0	0.04–0.5
Uranus system			
Miranda	472	1.2	0.3
Ariel	1,158	1.7	0.4
Umbriel	1,170	1.4	0.2
Titania	1,580	1.7	0.3
Oberon	1,525	1.6	0.2
Neptune system			
Proteus	420	—	0.07
Nereid	340	—	0.2

Table 1. Midsize satellites in the outer solar system fall crudely into groupings having diameters of 500, 1,000, and 1,500 km. However, size alone is not a good indicator of geologic activity. *Geometric albedo* is the ratio of an object's brightness at 0° phase angle to the brightness of a perfectly diffusing disk of the same apparent size.

continually subjects it to a harsh impact and radiation environment. Amalthea's ragged shape suggests that this could well be the collisional remnant of a somewhat larger moon. In any event, Jupiter has no other major (or icy) moons and will not figure further in this chapter.

The Uranian protosatellite disk was surely less massive than Saturn's, so we should probably not be surprised that no Titans orbit Uranus. Rather, Uranus's regular satellite system in its essentials resembles the intermediate satellites of Saturn — right down to number and sizes of moons. The Neptunian system, which one might think ought to be similar to that of neighboring Uranus, is actually lorded over by singular Triton. Despite being circular and relatively near Neptune, the orbit of Triton is retrograde, a dynamical state of affairs that has long pointed to its being a captured body (see Chapter 21). Neptune's regular satellite system in fact only exists much closer to the planet, for reasons probably related to Triton that will be discussed later.

That we can meaningfully discuss any of these satellites is due first and foremost to NASA's highly successful Voyager project. Voyager 2, in particular, completed a reconnaissance of all four giant planets with its encounter with Neptune in August 1989. The images and other data obtained during these flybys is backed up by the painstaking observations of Earth-based astronomers working to interpret the feeble light reflected by these distant bodies.

SATURN'S SYSTEM, SANS TITAN

The "classical" satellites of Saturn, those known prior to the Space Age, are, moving outward from the planet: Mimas, Enceladus, Tethys, Dione, Rhea, Titan, Hyperion, Iapetus, and Phoebe. (Their ordering can be easily remembered using the mnemonic "MET DR. THIP".) Of these, the regular, midsize ones occur in pairs that increase in dimension farther from Saturn: Mimas and Enceladus (roughly 500 km in diameter), Tethys and Dione (roughly 1,000 km), and Rhea and Iapetus (roughly 1,500 km).

Voyager images of little Mimas (*Figure 3*) reveal an icy sphere covered with impact craters. Most are less than 30 km across, but one striking central-peak crater, Herschel, exceeds 100 km. Mimas is a rather small body, and, however menacing Herschel may appear, its morphology is actually typical of many craters of its size (such as Copernicus or Tycho on Earth's Moon).

What lessons can we take away from Mimas? One is that Mimas does indeed seem ice rich. In fact, Voyager views of the Saturn's icy satellites confirmed that they are all, with the anticipated exception of Iapetus, highly reflective (*Table 1*). These satellites' infrared spectra had revealed strong absorption bands of water ice in the years prior to the Voyagers' arrival. The beautiful definition of these spectra indicated that the ice on these satellites, again with the exception of Iapetus, was nearly pure. The densities of middle-size satellites are remarkably low, close to that of pure ice in several cases. These Saturnian moons, along with Saturn's ring particles, are the iciest group of bodies in the solar system.

Surface temperatures on the Saturnian satellites are under 100° K. We can see from Mimas that cold water ice is a reasonably strong material. Although never as strong as igneous rock, water ice is able to absorb the punishment of a hypervelocity impact, form a crater, and retain that crater shape for geologic lengths of time. Indeed, when we see a body as cratered as Mimas, we surmise that it has been collecting craters for billions of years.

That a relatively small satellite can be so dominated by impact craters is not surprising to planetary geologists. Small bodies have higher area-to-volume ratios, so they lose heat faster than larger ones do. Ice-rich worlds contain less of naturally occurring radioactive elements like uranium, thorium, and potassium-40, so any heat derived from their radioactive decay is modest. Much larger bodies, such as the Moon or Mercury, appear geologically dead (though their deep interiors are probably still hot), so what hope does little Mimas have? Theoretical models of Mimas's thermal evolution show that it is in fact presently cold and inert throughout. Thus, Voyager scientists were quite startled by the first high-resolution images of Enceladus, the next moon out from Saturn and hardly larger than Mimas.

Enceladus (*Figure 4*) is a very different world from Mimas. Although some parts of its surface are cratered, none are saturated with craters as on Mimas. Instead, large expanes are relatively smooth, nearly free of craters, and crossed by numerous ridges, folds, and fissures. In one remarkable location, a set of craters has been cut by a fault, after which part of the terrain dropped and was flooded with fresh ice. The resurfaced unit is, at the image's resolution of about 2 km, essentially uncratered.

The implication is clear: Enceladus has been tectonically and volcanically active in recent geologic times. It is hard to believe that this activity continued over the eons until the geologically recent past, then simply stopped. Rather, it is more logical to assume that Enceladus is active today. Evidence to support this truly exciting possibility comes from Enceladus's very high reflectivity (*Table 1*); its surface is the brightest of any satellite in the solar system. Such an albedo indicates a very fresh (frosty) icy surface, possibly one caused by a historically recent deposit of frost particles from an ice volcano. The eruption from a single vent could easily coat the entire satellite because of the very low surface gravity there. That a very small icy body can be so active is one of the major enigmas of solar-system science, one to which we will return after discussing the next regular satellite pair.

As we move outward to Tethys and Dione, the diameters involved double. Compared with inner Mimas and Enceladus, these satellites have four times the surface area and nearly an order of magnitude more mass. Thus we might expect a greater diversity of geologic expression, and this is borne out in Voyager imaging. Both Tethys and Dione contain a variety of geologic terrains, ranging from ancient and crater saturated to more modestly cratered, and therefore younger, resurfaced terrains. One side of Tethys is heavily cratered, including the 400-km-wide basin named Odysseus (*Figure 5*). Odysseus is slightly larger than Mimas altogether, and it attests to the fact that the only factor limiting the crater size on a body is probably the size of the body itself. Unlike Herschel on Mimas, Odysseus is a highly flattened or relaxed structure that follows the curvature of Tethys's surface. Its central region appears complex and contains a distinct mountain ring.

One region of Tethys includes a broad, roughly circular plain with distinctly fewer craters (*Figure 6*), most of which are smaller than elsewhere on the satellite. Similar but more extensive plains are seen on Dione (*Figure 7*), as are a set of troughs and fissures near the day-night terminator. Some of the troughs are reminiscent of sinuous rilles on the Moon. Lunar sinuous rilles are the traces of the extraordinarily long, but now collapsed, lava tubes that helped to transport basaltic magma the great distances necessary to create the lunar maria by volcanic flooding (see Chapter 10). Whether or not the troughs on Dione are collapsed lava tubes, there is little doubt that the plains seen on

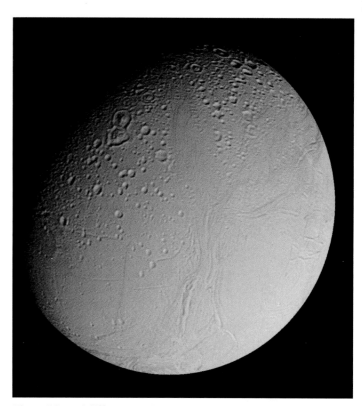

Figure 4. Several different terrain types have been identified on Enceladus, seen in a false-color mosaic of Voyager 2 images. Counts of craters per unit area on these terrains vary by at least a factor of 500, so they must have widely varying ages. Planetary geologists infer that Enceladus has undergone several episodes of resurfacing due to internal activity. The floors of the largest craters appear to bow upward. This suggests that considerable heat is rising from the interior, making surface ice warm enough to flow or creep.

Figure 5. Heavily cratered Tethys is dominated by the highly flattened, 400-km wide basin Odysseus. The vertical right edge is an artifact.

Figure 6. In Voyager 2's best global view of Tethys, features as small as 4 km across are visible. The boundary between heavily cratered terrain (upper half) and a more lightly cratered plain (lower right) is easily discerned. An enormous trench or multiply faulted valley, Ithaca Chasma, stretches diagonally across the cratered terrain. Over 100 km wide in places and up to several km deep, Ithaca Chasma roughly follows a great circle centered on Odysseus for at least three-fourths of Tethys's circumference. Its formation is probably intimately connected with that of Odysseus.

Figure 7. Icy plains stretch from pole to pole on Dione, but the abundance of small impact craters they bear shows the moon's extensive resurfacing occurred very long ago. Numerous bright streaks and a few fractures can be seen near the day-night terminator (at left). An extensive pattern of streaks ("wisps") appears in other, lower-resolution views, a pattern best explained by fault systems.

Dione, Tethys, and Enceladus are due to some form of icy volcanism. Perhaps surprisingly, planetary geologists do not think the "lava" in this case was liquid water but rather supercold solutions of ammonia and water.

Why ammonia? To understand this we need to return to Saturn's protosatellite nebula. Most of the nitrogen in this nebula would have existed as ammonia vapor, which does not condense easily. As the nebula cooled down from its initially high-temperature state, particles of metal, rock, and water ice solidified. At Saturn's distance from the Sun, nebular temperatures probably continued to drop, and somewhere around 150° K ammonia became incorporated in the water-ice crystal lattice and condensed (snowed) out as a *hydrate*. This much is straightforward physical chemistry. Once incorporated into the body of an icy satellite, ammonia hydrate can play a key role in that body's geologic history by forming a mobile fluid phase that can erupt onto the satellite's surface. Essentially, ammonia is a super-antifreeze that suppresses the complete freezing of an aqueous solution by nearly 100° K!

As an icy satellite warms up, say, by the decay of radioactive isotopes, heat is at first carried to the surface by conduction. However, once warmed to nearly its melting point, water ice is able to creep while still in its solid state (like a glacier) and transport heat by convection. This modulates the temperature rise inside the satellite and, in all but extraordinary circumstances,

prevents the ice from reaching its melting point. Without melting, there is no lava and no volcanism. Without volcanism, there is no obvious way for a homogeneous ice-and-rock satellite to differentiate, that is, to segregate its ice and rock components, and so form the beautiful ice-dominated surfaces we see on Saturn's midsize satellites.

Ammonia changes all this. A satellite's internal temperature does not have to be driven to 273° K for melting to occur. Temperatures need only reach a moderate 176° K (much too cool for solid-state convection), whereupon an ammonia-rich melt forms, coexisting with unmelted water ice. This mixture with the lowest melting point, termed a eutectic fluid, is slightly buoyant compared to the remaining water ice, so it is free to percolate upward. Eventually it accumulates in amounts large enough to erupt onto a satellite's surface.

Our experiences with household cleaning fluids, which are often dilute solutions of ammonia, might lead us to think that ammonia-water lava is very fluid. However, this eutectic melt is actually highly concentrated and rather viscous (sluggish); it behaves a great deal like much-hotter basaltic magma (*Figure 8*). So perhaps it would not be too surprising if eruptions of supercold ammonia-water melt lead to volcanic structures similar to those found on the Moon and other terrestrial planets.

Years ago, planetary chemist John Lewis predicted that if we were to discover evidence of volcanic activity on an icy satellite,

Figure 8. Ammonia-water eutectic melt, a probable outer solar system lava, is mobile even at 176° K and flows like cold honey (or basaltic magma).

Figure 9. Mosaic of the north polar region of Rhea. Features as small as 1 km can be made out in this heavily cratered landscape. The surface is well peppered with small craters under 20 km in diameter. Many poorly defined but parallel linear elements can also be made out, and a so-called megascarp, possibly indicating an episode of global contraction, runs vertically through the center of the mosaic.

we should immediately suspect the presence of ammonia. The irony is that evidence of volcanism has been found, and ammonia is suspected — but a positive identification has yet to be made! The Voyager spacecraft did not carry instruments to determine surface composition, for example. Ground-based spectroscopy, which in most cases measures the light from an entire hemisphere at once, has not yet detected ammonia absorption bands in the light from any satellite in the solar system. Ammonia is a rather volatile compound, however, and the hard vacuum and harsh irradiation environment typical for these satellite surfaces drives off ammonia molecules in a geologically brief time. NASA's Cassini mission, due to orbit Saturn in 2004, should be able to identify and map *local* areas where ammonia ice is still preserved, such as around fresh impact craters or in the youngest plains on Enceladus.

The geologic mobility that ammonia affords a middle-size icy satellite plausibly explains the ancient volcanic plains of Tethys and Dione. Continuing volcanic activity on diminutive Enceladus is another matter altogether, though. Clearly needed is a long-lived and sufficiently strong internal heat source, for otherwise Enceladus should have ended up like Mimas by now. The answer is almost certainly *tidal heating*. As it travels along an eccentric orbit a satellite's body is slightly distorted by the planet's gravity, and this repetitive tidal flexing heats the interior. Enceladus's orbit is only slightly eccentric (about 0.005), but this deviation from a pure circle is maintained by a dynamical resonance with Dione. Every time Dione goes around Saturn, Enceladus goes around twice. This 2:1 orbital matchup is the same type of resonance that couples the Galilean satellites Io, Europa, and Ganymede. The consequences for Io (and Europa)

are spectacular, of course (see Chapters 17,18). Although the tidal heat produced in Enceladus is more modest, calculations show it should be able to at least episodically warm the satellite's interior above the ammonia-water eutectic temperature.

Does Dione get anything out of its relationship with Enceladus? Not much, unfortunately. Dione is much more massive than Enceladus, and so it acts as a sort of gravitational anchor in the resonance. Enceladus's orbit is the more distorted one, and in any case Dione's greater distance from Saturn ensures that its tidal flexing is much less. Nevertheless, the resonance and its effects on Enceladus exemplify one of the principal findings of studies of icy satellites. Regular satellite systems are quite compact compared to the planets, and thus tides and tidal heating can be an important — if not the most important — driver in a satellite's geophysical and geological evolution. Such tidal heating can and has subverted our notions of what size body should be geologically active and for how long.

The correlation of body size with geologic activity (as exemplified by the terrestrial planets) is weakened further by the final and largest pair of middle-size satellites, Rhea and Iapetus. Rhea is the most massive object in this group and, according to conventional wisdom, ought to be geologically diverse. Voyager 1's encounter with Rhea was the closest pass to any of Saturn's midsize satellites. The spacecraft's spectacularly detailed images reveal a hemisphere not bristling with complexity but instead completely saturated with impact craters (*Figure 9*). Some images with much lower resolution do hint at resurfaced areas elsewhere on the satellite as well as wisps (or fractures) similar to those seen on Dione. Overall, however, Rhea appears geologically inert.

We are not sure what accounts for this conundrum. One possibility is that Rhea's size and mass, perversely, work against its being geologically active. Lacking a gravitational resonance with any other moon, Rhea relies on radiogenic heating to power any internal activity. Most of Rhea's geologic history can be characterized as cooling from an earlier, warmer epoch. As its deep ice cools under pressure, however, it can convert to a denser polymorph, ice II, with a corresponding loss in volume of over 20 percent. The pressures necessary to form ice II early on generally only occur in objects of Rhea's size and mass or greater. The internal volume loss would throw Rhea's surface layers into strong compression, closing any volcanic conduits and shutting off the possibility of resurfacing. Careful examina-

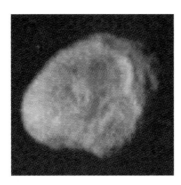

Figure 10 (right). **Irregularly shaped Hyperion, seen face-on in this Voyager 2 image, measures 330 by 260 by 215 km. One or more major collisions have, in all likelihood, blown away much of its original volume. This missing material did not reaccumulate on Hyperion, because it was scattered by Titan's gravity and swept up by other Saturnian satellites (mainly Titan but with some going to Rhea).**

Figure 11. **The two-faced nature of Iapetus was apparent even to early telescopic astronomers. The moon's bright terrain consists of dirty ice, and the dark terrain is surfaced by carbon-rich molecules. Although Iapetus was not seen well by the Voyager spacecraft, it will be explored more thoroughly when the Cassini orbiter reaches Saturn in 2004. The north pole lies about halfway between the bright crater at the center of the image and the large crater on the terminator.**

tion of the Rhea images indicates numerous ridges and "megascarps," which are plausibly related to this global compression.

Rhea's surface is peppered with many craters less than 20 km in diameter. So numerous are these smaller scars that Voyager scientists felt they resulted from a distinct type of impact, which they dubbed population II. If comet crashes created the preponderance of larger craters, population I, what made population II? A favored hypothesis is that the latter's impactors were indigenous to the Saturnian system.

One specific source may be the moon Hyperion (*Figure 10*), which orbits just outside Titan. Hyperion is irregularly shaped but hardly small; its long axis is about 330 km — only slightly less than the diameter of Mimas. A moon of this size would not have accreted naturally into such a shape. Rather, Hyperion has probably suffered one or more catastrophic collisions, and what we see now is probably only a remnant of the original object. Notably, Hyperion seems to be tumbling chaotically as it orbits the ringed planet. Its unpredictable rotation is due to its irregular shape and highly elliptical orbit.

The late Eugene Shoemaker estimated that each of Saturn's innermost moons was probably struck by enough large bodies early in its history to have shattered completely a few times, reaccreting in orbit after each disruption. Hyperion cannot reaccrete after disruptions or even normal cratering events, however, because it is locked in a gravitational resonance with much more massive Titan. Any debris thrown very far off Hyperion is scattered by Titan's gravity. While most of this debris is swept up by Titan, a small fraction comes crashing down on Rhea, the next moon in. The fundamental lesson learned from all of this is that the impact environment among the intermediate and smaller icy worlds is dynamic and often catastrophically violent.

Iapetus, the outermost of the middle-size satellites of Saturn, stands as one of the enduring puzzles of the solar system. This moon was discovered in 1671 by Jean-Dominique Cassini, who noted that it became markedly brighter and dimmer (to the point of invisibility) as it circled Saturn. Cassini realized that if the rotation of Iapetus were tidally locked to its planet, as the Moon's is to Earth, then the "leading" hemisphere (which faces in the direction of orbital motion) must be much darker than the "trailing" hemisphere.

Voyager did not approach Iapetus closely, but even the relatively low-resolution images sent back reveal a truly bizarre world divided into rather bright and very dark halves (*Figure 11*). The bright terrain is apparently heavily cratered, much like Rhea, and occupies much of the trailing hemisphere and the northern polar region (the only pole imaged). The dark region has an albedo of only 4 percent, one-tenth that of the bright terrain. It is so dark that no details can be made out in the Voyager images. Moreover, the darkened zone is centered precisely on the leading hemisphere. A dichotomy of this sort is unlikely to result from an internal process like volcanism. Rather, it is precisely the one expected from the bombardment of its leading hemisphere by dark material. As further evidence, white spots appear in the dark unit near its border with the bright (as if the coating were thin and could be penetrated by impacts), and dark spots occur in the bright terrain near the same transition (as if craters could preferentially collect dark material).

A favored version of the meteoritic hypothesis posits that the particles bombarding the leading side of Iapetus have been ejected from craters on Phoebe, the next moon out. Phoebe is very dark and, because of its retrograde orbit, thought to be a captured object (see Chapter 23). Dark ejecta from Phoebe should be spiraling in toward Saturn and striking Iapetus's leading side at more than 6 km/sec (*Figure 12*). This particular dynamical arrangement serves to rationalize why Iapetus's appearance is unique in the solar system. Unfortunately, ground-based spectrophotometry has shown that Iapetus's leading hemisphere is dark and spectrally reddish, whereas Phoebe is dark and spectrally neutral. Thus if "Phoebe dust" is to blame for contaminating Iapetus, either something must be modifying the dark contaminant at the time of its impact or something else in Iapetus's ice must account for the redness.

Dark, reddish material is a recognized but poorly understood chemical component of many bodies in the outer solar system, particularly comets and certain classes of dark asteroids (unfortunately, those from which we have no meteoritic samples). It is thought to be organic material, specifically, macromolecular combinations of carbon, hydrogen, oxygen, and nitrogen. Organic material from the solar nebula may have been directly incorporated into Iapetus's icy mix, but then we would expect more rock than is indicated by the satellite's low density. A more plausible hypothesis is that Saturn's protosatellite nebula was cool enough at Iapetus's distance for methane (CH_4) to condense and be incorporated into the moon. When shock-heated by impacting Phoebe dust, the mixture of volatile ices on Iapetus could be converted into reddish tarry compounds and thus account for the satellite's amazing dichotomy. All of this is rather conjectural, to be sure, and no consensus exits on what is going on at Iapetus — except that the dark material is organic rich and that the bright-dark pattern implies some impact control. Iapetus is naturally a prime target for the upcoming Cassini mission.

THE DOMINION OF URANUS

The planet Uranus was known to possess five regular satellites prior to the Voyager encounter. The smallest of these, inner Miranda, was discovered in 1948 by Gerard P. Kuiper. Given their distance from the Earth, Miranda and its siblings are extraordinarily difficult to study. Not until the 1980s were astronomers able to determine accurate diameters for four of them. As a group, they turned out to be remarkably similar in size to the intermediate satellites of Saturn, but considerably less reflective (*Table 1*). Infrared spectra showed that water ice exists on all five objects, but the subdued spectral signatures mean a dark, spectrally neutral material, such as carbon-rich dirt, is mixed with the surface ice. Accurate masses and densities for the five would have to await Voyager 2's 1986 encounter, during which 10 smaller satellites were also found circling inside Miranda's orbit.

With the exception of Miranda, the middle-sized satellites of Uranus turn out to be a rather dense group, at least when compared with their Saturnian counterparts. We even find some tendency toward increasing density with distance from the planet, a trend opposite that exhibited by the Galilean satellites or the planets as a whole. Consequently, the rocky fraction of the

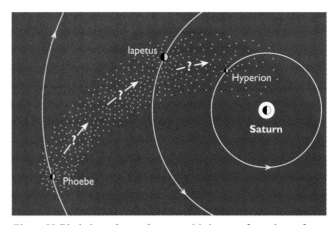

Figure 12. Black dust, thrown by meteoritic impacts from the surface of retrograde-moving Phoebe, slowly spirals toward Saturn over time and may be the source of dark material on the leading hemisphere of Iapetus. One remaining puzzle, however, is reconciling why Phoebe's surface is neutral black, while the dark face of Iapetus is reddish.

Uranian moons must be at least 50 percent, again except for Miranda. This indicates that the protosatellite disk about Uranus was relatively ice poor, which in turn suggests two possible scenarios. One is that the cooler Uranian disk inherited largely unprocessed material from the solar nebula. In such a primitive mixture, carbon and nitrogen would have been locked up as volatile and largely uncondensable CO and N_2, rather than as methane and ammonia. The other assumes that Uranus's protosatellite disk was thrown out by a giant impact, in which case shock energy would have converted methane and ammonia from the Uranian atmosphere into CO and N_2.

The latter scenario is appealing in that the shock processing of methane may also yield amorphous carbon. This could be the ubiquitous, spectrally neutral, darkening agent on the planet's satellites and ring particles. Carbon-bearing material from the solar nebula, by comparison, should have been dark and red, which is not seen in the Uranian system. In either scenario, the inner nebula at Uranus was probably warm and dense enough for CO and N_2 to convert into methane and ammonia (which are chemically more stable) during the nebula's lifetime. This could account for the increasing iciness on average of the Uranian satellites closer to the planet.

The Uranian system is tipped close to 90° with respect to the plane of the solar system. Voyager 2's encounter took place near the peak of summer in Uranus's southern hemisphere, with the planet's south pole pointed very nearly at the Sun. This meant that there would be no satellite-hopping tour as at Jupiter and Saturn. Instead, Voyager 2 had to dive through the bull's-eye and settle for a good view of only one or two satellites. Furthermore, a pass close to Uranus was necessary to redirect the spacecraft to Neptune, which in turn dictated that only tiny Miranda would be seen at close range. Would Miranda turn out to be an austere, cratered ball like Mimas, or a wild world like Enceladus? Saving the best for last, we will consider the Uranian suite of moons from the outside in.

Titania and Oberon are both in the 1,500-km-diameter class. Oberon was poorly imaged (*Figure 13*), but we can discern a cratered landscape with bright ejecta around some of the younger craters and intriguing dark patches on the floors of

Figure 13. As Voyager 2 found, both Oberon (*upper left*) and Umbriel (*above*) are heavily cratered and thus must have ancient crusts. About half the size of Earth's Moon, Oberon displays numerous impact craters surrounded by bright rays of ejected material. Dark patches within some craters may have formed when carbon-bearing icy lava erupted from within Oberon's interior. A broad, 11-km-high mountain — probably the central peak of a large crater or basin, pokes out on the lower-left limb. Umbriel is only two-thirds as large as Oberon. It appears rather uniformly dark, except for a single, bright-floored crater at top (near the equator) 160 km in diameter and a bright central peak in a nearby crater, both of which plausibly represent frost deposits.

Figure 14. Titania, Queen of the Fairies in Shakespeare's *A Midsummer Night's Dream*, is also the largest Uranian moon. Much of the surface visible here appears to be a densely cratered plain, spectacularly cut by several fault-bounded canyons. The longest is at least 1,500 km long and up to 100 km wide. It descends in multiple steps to depths of several km. A large impact crater or basin is situated along the equator at top.

others. The latter are reminiscent of the inky coating on Iapetus, though Oberon's dark stuff is neither red nor as dark (albedo: about 20 percent). Oberon's leading hemisphere is slightly redder overall, but this may be due to an accumulation of reddish particles from the pair of outlying, retrograde satellites.

Titania, in comparison, is crossed by a spectacular set of fault-bounded valleys or canyons that are up to several kilometers deep (*Figure 14*). These canyons indicate the work of extensional forces. Either Titania has expanded, perhaps due to internal heating, or its outer layers have cooled over a more static interior, which would have caused the surface shell to contract and fracture. Although distinct and seen to crosscut many craters, the faults are still probably quite ancient. The crater population itself seems to be deficient in larger examples compared with Oberon, which may indicate an episode of resurfacing long ago.

The unequivocal evidence for widespread surface extension and perhaps resurfacing on Titania, and the muted evidence for some similar faulting on Oberon, stand in stark contrast to what we find on similarly sized Rhea. Titania and Oberon are much rockier than Rhea and so must have experienced greater and longer-lasting radiogenic heating. Furthermore, if Titania or Oberon became differentiated, the resulting rocky core would be rather large. The ice mantle above would not experience pressures great enough to convert ice I to ice II, and there would not have been a compressional clamp-down on surface activity. Titania's geologic history is thus more in keeping with, or represents a step up from, that of relatively rock-rich Dione.

Moving in toward Uranus, we come to the roughly 1,000-km-wide bodies Umbriel and Ariel. Umbriel's rather dark, low-contrast surface, together with Voyager 2's long-range images (*Figure 13*), severely hampers geologic analysis. Nonetheless, Umbriel appears to be an ancient, very heavily cratered body with a barely discernible pattern of surface faults.

Apparently inactive for a very long time, Umbriel is the best candidate among the midsize satellites of Uranus to be a primitive, undifferentiated body. If so, what allowed Dione (Umbriel's near twin in size and mass) to remain geologically active for so long?

Ariel (*Figure 15*) displays all the evidence for geologic activity that Umbriel lacks. Moderately cratered plains are broken up by an extensive pattern of fault-bounded canyons and valleys. Much more extensive than those observed on Titania, Ariel's canyons isolate the older terrain into polygonal blocks or mesas. Moreover, the canyon floors and portions of the

Figure 15. A four-image mosaic of Ariel shows networks of deep valleys and smoother areas that have been volcanically resurfaced. Ariel's exterior is younger, brighter, and geologically more complex (with the possible exception of Miranda) than any other of Uranus's satellites. Near the bottom are spectacular canyons, which themselves are filled with smoother, volcanic flows. Up to several kilometers thick, these flows spill over the brinks of the cratered plateaus toward the top.

plains have been resurfaced by a much smoother material. These flows have a rounded cross-section, and sinuous troughs run down their middles, which indicate that the vast eruptions involved an icy volcanic material more viscous than that seen on the Saturnian satellites.

The range of crater densities found on its various terrains implies that Ariel must have remained tectonically and volcanically active for an extended period. Today it has the brightest (iciest) surface of any of the Uranian satellites. More importantly, the activity was too extensive to be ascribed solely to radiogenic heating. The consensus view is that tidal heating must have been involved, yet there are no orbital resonances among the Uranian satellites at present. Resonant couplings with Umbriel and possibly Titania may have existed in the past, however. If so, Ariel's orbital eccentricity would have been driven

up. This would in turn have resulted in tidal heating and, plausibly, the geologic activity whose record we see today.

Miranda is the sole Uranian moon in the 500-km-diameter class, and by chance it became the best-resolved satellite of the entire Voyager mission. Nothing, however, could have prepared scientists for what the spacecraft revealed (*Figure 16*). Miranda's surface is a bizarre amalgam of terrains unique in the solar system. Amid its bright cratered terrain are three large and mostly darker regions, termed coronae, with oval or trapezoidal shapes. The cratered terrain is familiar, resembling the rolling, heavily cratered landscape of the lunar highlands. However, the coronae are different — very different.

The first reaction among Voyager experimenters to this seeming geologic jumble was that Miranda offered the first direct evidence for the catastrophic breakup and reassembly of

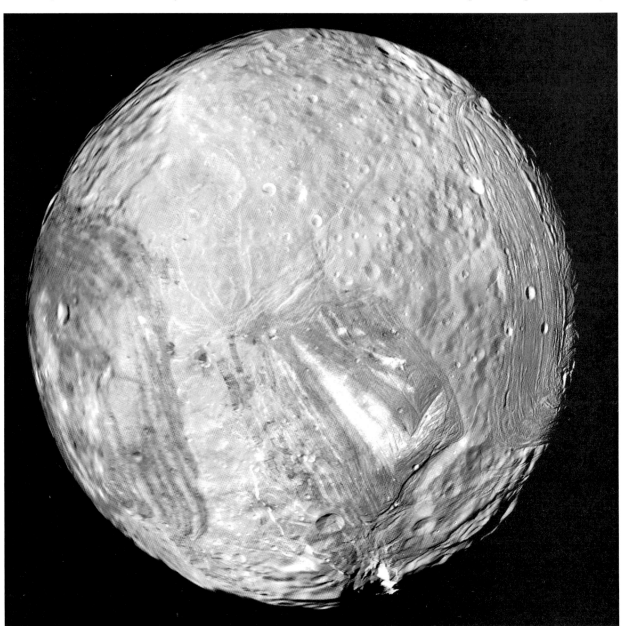

Figure 16. Miranda is a world frozen in mid-upheaval. Its ancient cratered terrain is rent by deep fault valleys, and three distinct new terrains can be seen. Termed coronae, they are polygonal in outline and formed from various combinations of tectonic disruption and icy volcanic extrusion. Some of the volcanic materials are intrinsically darker; in other places faults expose darker material situated deeper down.

an icy satellite. The coronae thus would represent rockier, carbon-rich chunks that once lay in the deeper interior of proto-Miranda. But the coronae do not look like exhumed fragments, nor are they sinking back into the interior. Instead, careful geologic evaluations have shown that extensional stresses led to uplift and faulting, which fashioned great cliffs and troughs (*Figure 16, bottom*). Fissure-fed icy eruptions then created parallel volcanic ridges that coalesced to form the coronae's interiors (*Figure 17*).

The icy volcanic materials that erupted on Miranda were clearly quite viscous, much like those on Ariel. Some linear ridges on Ariel resemble Miranda's ridges. Thus, the compositions of the volcanic materials on these two bodies are probably similar, as to some extent were their volcanic styles. Miranda's great faulted terrains, so initially striking, today seem quite comparable to Ariel's extensive canyon system. Indeed, we now realize that extensional faulting is the norm on the midsize satellites of Uranus and, with the possible exceptions of Rhea and Iapetus, on those of Saturn.

Miranda is thus a world caught in mid-metamorphosis, attempting to transform an ancient, presumably primordial moon into something differentiated and resurfaced. The only plausible heat source to drive such activity is the dissipation of tidal energy. If so, it must have been a transient event, as Miranda is not now active nor does it share any orbital resonance. According to dynamicists William Tittemore and Jack Wisdom, Miranda most likely had a resonant relationship with more massive Umbriel (and possibly Ariel) in the geologic past. This coupling would have left Miranda's orbit with the otherwise unexplained 4° inclination it retains today.

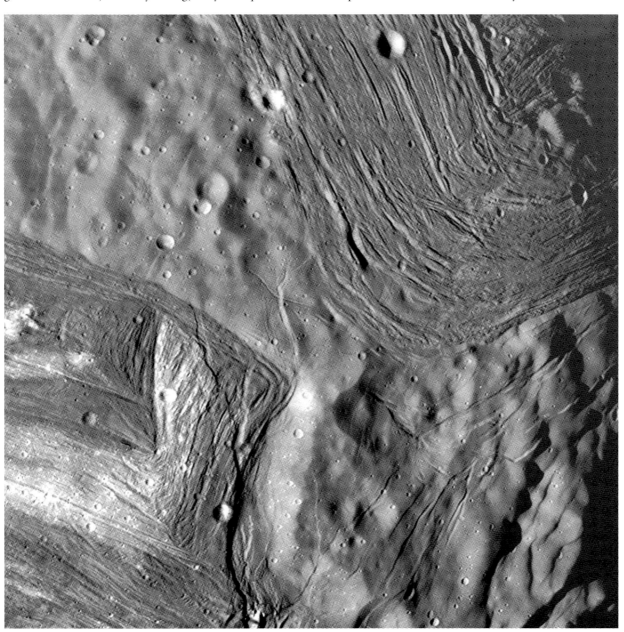

Figure 17. From 30,000 km away, Voyager 2's telescopic camera captured details on Miranda as small as 500 m across — the best resolution for any object studied by the spacecraft. A close-up of the "chevron," formally named Inverness Corona, shows well the ridges and grooves within its sharp, trapezoidal boundaries. To its upper right is a more uniformly toned corona that appears to be a concentric pattern of volcanic ridges and grooves. The area shown measures about 150 km on a side.

Figure 18. Voyager found six previously unknown moons circling Neptune. The largest, Proteus, is 420 km across and thus in the size class of Mimas, Enceladus, and Miranda. Unlike these, however, Proteus is a primitive body — dark, heavily cratered, and irregularly shaped.

Overall, the most exciting aspect of our initial exploration of the Uranian system is that, notwithstanding the poor images of some, less than 50 percent of each satellite has been seen at all. What remarkable features and processes might be observed by a future mission, if properly timed? Such a mission could take advantage of the lighting during the spring or fall equinox, where a satellite's entire surface is illuminated during an orbital cycle. Equinoxes at Uranus occur every 42 years, the next in 2006.

MAYHEM AT NEPTUNE

Neptune lacks a system of regular, intermediate-size satellites. However, six small satellites orbiting close to the planet, plus big, retrograde-orbiting Triton and distant, eccentric Nereid, speak to us of a catastrophe in Neptune's past.

The sextet of inner satellites discovered by Voyager 2 appear to be dark, primitive bodies, rather like the innermost satellites of Uranus. The largest of these, Proteus (*Figure 18*), is actually the size of Mimas and Miranda. It orbits at a distance of 4.75 Neptune radii, which again is similar to the orbital positions of Mimas and Miranda (when measured in planetary radii). Although dark, Proteus would seem to mark the beginning of a family of middle-sized icy satellites. So where did they all go?

The answer probably lies with comparatively massive Triton, which orbits at a distance of 15 Neptune radii — squarely in the domain occupied by regular satellites at Jupiter, Saturn, and Uranus. As discussed in the previous chapter, Triton's inclined, retrograde orbit indicates that it is not a regular moon created in a circumplanetary nebula. It is instead an interloper, a large rock-ice body born in solar orbit and captured by Neptune long ago.

Triton's orbit after capture would have been very eccentric. Calculations by dynamicist Peter Goldreich and others have shown that solar tidal perturbations to this original eccentric orbit would have periodically brought Triton in as close as 5 Neptune radii. If a satellite system comparable to Uranus's originally existed around Neptune, these original moons would have been cannibalized, scattered, or ejected altogether by Triton. Even objects very close to Neptune would have had their orbits destabilized. The original inner moons would have smashed into one another, with the resulting debris able to reaccrete only after tidal torquing from Neptune circularized Triton's orbit. Some dynamicists think Proteus is one of this second generation of moons.

Nereid, Neptune's outermost moon, has the most eccentric orbit of any planetary satellite. While its motion is not retrograde, Nereid is usually considered a captured body. Voyager 2 did not get close enough to photograph any surface details on Nereid. However, it is comparable in size to Mimas. Its albedo of about 20 percent is like those of the darker middle-sized satellites of Uranus — but unlike those of the dark asteroids and comet nuclei that are usually considered candidates for capture. Nereid may in fact be one of Neptune's original satellites, placed into its distant orbit by gravitational interactions with Triton. This idea could be tested with new spacecraft observations. How wonderful it would be to find a survivor from Neptune's early days of chaos.

23

Small Worlds: Patterns and Relationships

William K. Hartmann

The Martian moon Phobos, one of many small solar-system worlds with unknown origins.

UNTIL ABOUT A century ago, the solar system seemed like an orderly place, with just a few well-defined classes of inhabitants. The planets were the eight large objects going around the Sun. Of 13 satellites that had been found by 1840, all were intermediate in size and moving around their respective planets in direct, or prograde, orbits (in the same direction that the planets themselves circle the Sun). The solar system of the 1800s was also populated by asteroids, or "minor planets," much smaller than planets or moons. These all grouped neatly between Mars and Jupiter, thus fitting the prediction by Titius and Bode that this niche would be filled by some sort of planetary body. A fourth class of objects, comets, seemed different than the other classes; they had fuzzy comas and enormous diffuse tails.

However, some problematic objects soon appeared that messed up this otherwise neat picture. In 1846, Neptune's large satellite, Triton, was discovered in a retrograde orbit. Triton seemed perhaps only a bizarre exception, since the next six satellites found all moved in direct orbits. This group, however, included moons the size of asteroids, like Phobos and Deimos (discovered circling Mars in 1877) and Amalthea, Jupiter's fifth satellite (discovered in 1892). Were these asteroid-sized moons really more related to asteroids than to the big moons? The situation became more confused in 1898 with the discovery of Phoebe, the small, retrograde, outermost satellite of Saturn. Astronomers pondered how, or why, a satellite would travel "backward." By 1960 the list of known satellites had grown to 30, four of which moved in the retrograde direction and six of which were less than 100 km across. Today we know of satellites that grade down in size to glorified ring particles. At the distribution's other end, two moons are larger than the planet Mercury, and seven are larger than Pluto! Thus, satellites have

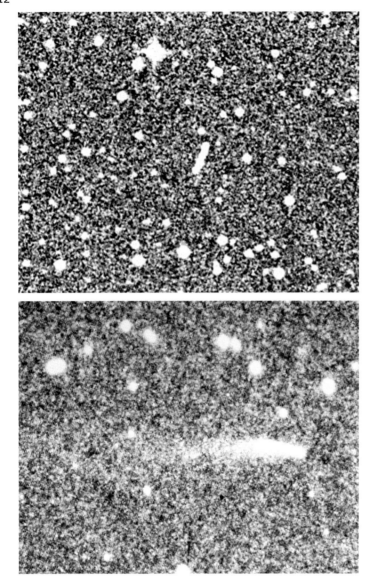

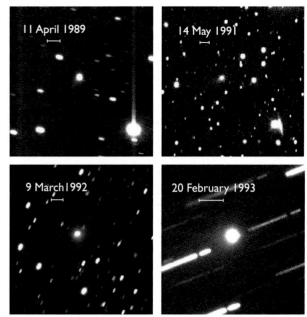

Figure 2. Long-exposure images of "asteroid" 2060 Chiron taken in red light show its faint, extended coma of gas and dust. The scale bar in each frame represents 100,000 km at Chiron's distance.

Figure 1. **A remarkable example of how distinctions between comets and asteroids have become blurred. The photograph at top shows the asteroid number 4015 as it appeared when "discovered" in 1979. The photograph above, taken 30 years earlier, shows the same object with a diffuse tail, which at the time led to its designation as Comet Wilson-Harrington.**

become an unholy mixture of planet-size, asteroid-size, prograde, and retrograde bodies.

Meanwhile, asteroids began popping up in unexpected parts of the solar system. In 1898, two astronomers independently discovered 433 Eros, whose perihelion lies inside the orbit of Mars. Asteroid 624 Hektor, spotted in 1907, turned up in the same orbit as Jupiter but 60° ahead of the planet. Numerous other asteroids were later found sharing Jupiter's orbit, averaging either 60° ahead of or behind the planet. They became known collectively as *Trojan asteroids* and were named after characters in Homer's Trojan War epics. The two orbital positions they occupy are named Lagrangian points, after Joseph Lagrange, the French dynamicist who discovered the gravitational effects that define them. In 1932, asteroid 1862 Apollo was found passing inside the Earth's orbit. By then it had become clear that asteroidal bodies resided not just in their "belt" but elsewhere in interplanetary space as well.

The neat physical distinctions between asteroids and comets also began to break down as astronomers argued about emerging differences between the two groups. A comet, by definition, must contain ices. By the 1970s, spectroscopic observations showed that comets emit gases derived from icy material in their interiors. Most observers expected that bare cometary nuclei would look like the "dirty snowballs" envisioned by Fred Whipple: white with a smattering of dust. But telescopic studies during the 1980s, and the Giotto probe's close-up images of Halley's nucleus, revealed cometary nuclei as black or dark brown, as dark as carbonaceous asteroids! Moreover, as comets recede from the Sun, they lose their comae and tails, becoming telescopically indistinguishable from asteroids. Thus, a comet found while inactive might be mistakenly cataloged as a minor planet. One celebrated example involves asteroid 4015, a pointlike object "discovered" in 1979 (*Figure 1*). But a check of old photographs showed that it was really the periodic comet 107P/Wilson-Harrington, which had lost all traces of its coma and tail in the years since being spotted in 1949.

It had been assumed that all "true" asteroids had surfaces whose spectra are dominated by stony, metallic, or sooty carbonaceous materials. But some researchers raised the issue that many asteroids with stony-looking surfaces might contain ices trapped inside. As evidence for this, they noted that the spectra of many large, black asteroids in the outer asteroid belt indicated minerals with chemically bound water. Furthermore, many carbonaceous meteorites contain carbonates, apparently created when liquid water percolated through the interiors of their parent asteroids. Such bodies probably formed with ice in them, which later melted and seeped through internal fractures.

The blurred distinctions among small bodies came to the fore with Charles Kowal's 1977 discovery of an object orbiting between Saturn and Uranus with a period of 51 years. Because it had no cometary characteristics, it was cataloged as an asteroid, 2060 Chiron. Chiron's colors were soon found to be similar to those of other blackish asteroids and comet nuclei of the outer solar system. Yet in 1988, as Chiron drew closer to the Sun, it unexpectedly doubled in brightness. A few months later, observers noticed a fuzzy coma surrounding it, and by 1991 they had detected emission from gaseous CN molecules (*Figure 2*). "Asteroid" Chiron had turned into a comet!

Yet Chiron is unlike any traditional comet. With a diameter of about 200 km, it is 12 times larger than the nucleus of Halley's Comet. Its unstable orbit means that Chiron is not long for this system. It comes close to Saturn every 10,000 years or so, and such interactions will cause its orbit to evolve rapidly. If one of these passes throws it into the inner solar system, it could become the biggest and brightest comet in history!

Chiron turned out to be just the first of a group of bodies found in elliptical orbits that cross the paths of giant planets, in the general region from Saturn to Neptune, roughly 20 to 50 AU from the Sun (*Figure 3*). These came to be called *Centaurs*, named after the man-horse beasts of Greek mythology. *Table 1* gives details for the first seven discoveries.

Besides the Centaurs, observers have discovered another group of still more distant objects, generally located beyond Neptune. Sixty-eight of these were found between 1992, when sensitive instruments first made their detection possible, and mid-1998. Most are Centaur-sized, an estimated 100 to 300 km across. From the statistics of the sky area searched and the discovery rate, David Jewitt and Jane Luu estimate that 35,000 objects in this size range might orbit 30 to 50 AU from the Sun, in what is called the *Kuiper belt*. As explained in Chapter 5, on this fringe of the primeval solar nebula there was not enough material to form planets. *Table 2* gives details for some examples of Kuiper-belt objects.

Dynamicists have pointed out that some Kuiper-belt objects have the same 2:3 orbital resonance with Neptune that Pluto does. That is, although they travel in different orbits, each completes two circuits around the Sun in the time taken by Neptune for three.

Instead of getting hung up in the Neptune resonance, Centaurs apparently have been scattered inward from the Kuiper belt by gravitational interactions. In dynamical terms, Centaurs thus relate to the Kuiper belt as Earth-crossing asteroids relate to the main asteroid belt. By coming inward as far as Uranus or Saturn, they give us a more readily observable sampling of the Kuiper belt. But their surface properties are soon altered by cometary activity if they venture too far inward.

Comets are often billed as the most pristine objects in the solar system, having been long in the distant deep freeze of interplanetary space. However, comets derived from the Oort cloud actually originated much closer to the Sun (see Chapter 5). Furthermore, cometary activity itself alters and blows off the pristine surface layers, and even a gradual reduction in orbital radius exposes a body to stronger sunlight that can cause the loss of the lowest temperature ices. Thus, neither Centaurs nor any active comets are as pristine as members of the Kuiper belt, which have been in their distant original locations ever since they formed.

Interestingly, even though the Centaurs and the Kuiper-belt objects are almost certainly comets, in terms of ice content, they are named and numbered according to asteroid traditions. Like the first Centaur, Chiron, most of them would develop cometary characteristics if they came close enough to the Sun.

Centaur Asteroids

Name	Diameter (km)	Distance from Sun			Color
		Closest	Mean	Farthest	
2060 Chiron	200	8.4	13.6	18.9	Neutral, dark
5145 Pholus	200?	8.7	20.2	31.8	Very red
7066 Nessus	80?	11.8	24.6	37.5	Red
1994 TA	30?	11.7	16.8	21.9	?
1995 DW$_2$	70?	18.9	24.9	31.0	Red
1995 GO	90?	6.8	18.1	29.4	Very red
1997 CU$_{26}$	300?	13.0	15.7	18.5	Red, dark

Table 1. The first seven known Centaurs, a class of objects that occupy eccentric orbits in the outer solar system.

Kuiper-belt Objects Compared

Name	Diameter (km)	Distance from Sun			Color
		Closest	Mean	Farthest	
1992 QB$_1$	280?	40.9	44.3	47.6	Red
1993 FW	290?	41.5	43.5	45.6	Slightly red
1993 SC	250?	32.2	39.9	47.5	Red
Pluto	2,300	29.7	39.5	49.3	Slightly red

Table 2. Many objects are now being found every year in orbits beyond Neptune. Here three early discoveries are compared with Pluto, which itself lies within the Kuiper Belt.

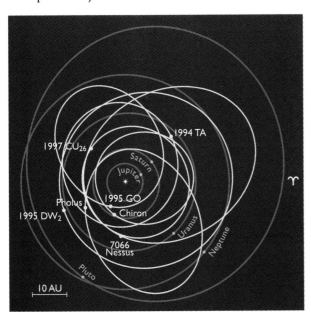

Figure 3. The orbits of the first seven Centaurs, small bodies that occupy the realms of Saturn, Uranus, and Neptune. Their orbits are unstable, which means they must have recently arrived from elsewhere in the outer solar system — almost certainly from the Kuiper belt beyond Neptune.

AN OVERVIEW AT THE TURN OF THE 21ST CENTURY

How can we make more sense of this welter of icy and rocky, prograde and retrograde bodies? We can rephrase that question in a more productive way. What patterns of relationships can be found among the thousands of interplanetary bodies and small moonlets that would help group them according to origin? Do smooth progressions exist from one class to another class? What is the underlying organization? The reason for asking such questions is that these bodies hold clues to the earliest conditions and processes at work both during our solar system's formation and thereafter inside the larger planets and moons.

As a first step, we must consciously drop the old semantic traditions based strictly on visual criteria, to free our minds from archaic stereotypes. In the past, astronomers looked at the solar system in a way that I call "the nine-planets gestalt." In this view, the planets were supremely important; all other bodies were looked upon as less important and less interesting. This hierarchical ordering — and particularly the disinterest accorded small objects — paved the way toward misleading distinctions among comets, asteroids, and certain moons.

Now we are experiencing the breakdown of this gestalt. The Sun's family is seen instead as a complex system of worlds, 25 of which exceed 1,000 km in diameter. Pluto is looking more and more like merely the largest member of the Kuiper belt. One might accurately describe our planetary system as eight major planets and a host of smaller worlds that were trapped in specific locations during the planet-forming process. While satellites might seem less "important" than planets, we realize today that some satellites are larger or more geologically active than Mercury or Pluto. Rather than distinct categories, we are finding more of a continuum of geologically interesting worlds and worldlets among the smaller bodies.

Take the specific example of comets versus asteroids. In the Victorian tradition asteroids were starlike points, while comets were big fuzzy blobs. Telescopically, asteroids were rocky bodies studied by geologist-astronomers, while comets' gaseous emissions were studied by gas spectroscopists. At most meetings of professional astronomers during the 20th century, asteroids were discussed by one group and comets by a generally different group. This actually proved quite detrimental; it has taken decades to realize that these objects might share similar materials, or to allow the possibility of a smooth transition from one group to the other, based on ice content.

Today we have a clearer sense of the continuum of compositional types among planetesimals, ranging from rock-and-metal "asteroids" to ice-dominated "comet nuclei" — and that the specific type probably depends mostly on the temperature at which a given object formed, which depends in turn on its initial distance from the Sun.

SMALL-BODY ZONES

Looking at things in this way, we can roughly divide the solar system into a sequence of what might be called "primordial zones," depending on the compositions of bodies smaller than the giant planets (*Figure 4*). The giant planets must be discounted because their initial, solid-body compositions were modified by the accretion of massive, hydrogen-based atmospheres and mantles from the surrounding nebular gas. The inner solar system, from the Sun out to the mid-asteroid belt, about 2.5 AU away, is a zone of rocky and metallic bodies. Farther away, a remarkable change occurs: the outer half of the belt is dominated by black objects composed of carbon, various carbon-rich compounds, plus ordinary rock-forming minerals. I refer to this transitional region, located at 2.5 to 2.7 AU, as the "soot line;" it marks the point beyond which the primordial solar nebula was cold enough for black carbonaceous compounds to condense, and they now dominate the colors of the aggregate solid material even if it is mostly silicates or metal.

Somewhere nearby, perhaps around 3 to 4 AU, is what could be called the "frost line," beyond which it was cold enough for water to freeze. This ice was added to the silicate-metallic-carbonaceous mix. At large enough distances it dominated the total composition. (Note that, at the distance of Saturn and beyond, other volatile compounds such as carbon dioxide, methane, and ammonia could freeze into additional ices.) Interestingly, bodies beyond the frost line are not white. They remain black or dark brown because they are colored by the carbonaceous material. Laboratory studies have shown that adding just a few percent of sooty matter to ice will make the entire mixture look very dark.

Another trend involves the role of the frozen water. Among black carbonaceous asteroids in the outer belt, ice has apparently melted in the past, percolating through the rock and creating hydrated minerals that have been detected spectroscopically. Carbonaceous meteorites, probably from this region, show deposits of carbonates left behind in their fractures, much as a carbonate "bathtub ring" forms in a bathtub. Hydrated minerals are not evident in asteroids very far from the Sun; they might contain more ice, but it never melted to yield mobile, reactive liquid water.

This is not the end of the story of the color trends and zones in the solar system. As early as 1980, Jonathan Gradie and

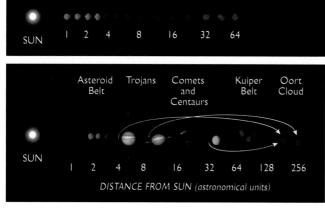

Figure 4. The top panel shows the initial zonal structure of the solar system. Rocky and metallic bodies formed close to the Sun. Beyond 2.5 AU, dark carbonaceous-icy objects dominate, but there is some trend toward more brownish and reddish hues farther outward. The bottom panel shows how subsequent migrations altered the initial zonal structure. The Oort cloud was created when the gravity of the giant planets ejected a mixture of ice-rich bodies nearly out of the solar system. Planetary perturbations also winnowed the populations of the main asteroid belt and the inner Kuiper belt.

Joseph Veverka predicted that organic materials (those having C-H bonds) form readily at large solar distances and cause objects to appear redder. (To an astronomer, "red" means not just that something looks red, but more specifically that it reflects more light at longer, redder wavelengths than at shorter, bluer ones.) Supporting observations do indeed reveal a rough color of this type. The Trojan asteroids, in the orbit of Jupiter at 5 AU, have a range of colors, but they average more brownish or reddish brown than the black asteroids just beyond the soot line around 3 AU. They may contain water ice in their interior, but any ice near their surfaces has apparently long been vaporized by impacts and exposure to the Sun.

An important discovery of the 1990s is that the rough color trend toward redder objects continues outward into the zone of the Centaurs and Kuiper belt. Astronomers were excited to discover that some of these objects are in fact the reddest interplanetary bodies yet seen. Centaur 5145 Pholus, for example, is roughly as red as Mars. The ruddy color is not due to oxidized iron minerals, as on Mars, but rather seems to be associated with abundant organic compounds — possibly by reactions among carbonaceous materials, frozen methane (CH_4), and other ices bearing the C-H bond. These reactions may be driven by high-energy cosmic rays, as has been simulated in laboratory experiments that yield molecules such as C_2H_6, HCN, CH_3NH_2, and other species. Methane-related ices would be stable only at very cold temperatures, so they would come into play only in distant objects that had rarely or never looped into the inner solar system. Telescopic observations have begun to validate the laboratory tests: in 1997 the spectrum of the reddish Kuiper-belt object 1993 SC revealed the presence of frozen methane and ethane (C_2H_6).

If this scheme is correct, such objects might lose their reddish colors if they are perturbed into orbits closer to the Sun. Their exotic ices would be easily sublimated by solar heating, and the red-colored carbonaceous dust might be blown off their surfaces by cometary activity or impacts, or it might be cloaked by a fresh veneer of frost. Differences in color among Centaurs and Trojans might thus depend on factors such as their orbital and collisional history. This hypothesis would thus explain why Kuiper-belt objects and Centaurs are the reddest objects, with comets less so. It is also consistent with fact that Chiron, the only Centaur known to show cometary activity, is not red or brown, but neutral black.

The color trends do not necessarily continue out into the unseen Oort cloud, because that swarm was populated with a grab-bag mixture of different types of comets from the regions of the four giant planets. Perhaps some of them lacked enough methane-related ices to produce red colors. Thus, we cannot assume there is a single type of "pristine, Oort-cloud comet."

To summarize, there appear to be three simplifying principles at work. First, the farther from the Sun a body originated, the colder it was and the more ice it acquired (water ice beyond 3 or 4 AU, and frozen methane, ammonia, and carbon dioxide beyond that). Second, some process (organic synthesis?) seems to produce more frequent reddish colors among the most distant objects than among nearer ones. Third — and this one remains speculative — these objects eventually lose their red colors when they are perturbed inward toward the Sun.

OBSERVATIONAL QUESTIONS

These principles set up a framework to discuss additional questions about the history of the solar system. For example, did the Trojan asteroids form around 5 AU from the Sun, where they have remained trapped for 4½ billion years? Or are they a grab-bag of comets that wandered in more recently from the Kuiper belt or Oort cloud, only to be captured into the Lagrangian points? A first step in answering this would be to compare the Trojans' colors and spectral classes with those of known Kuiper-belt objects. So far, the Trojans seem less red, on average, than the Centaurs and their Kuiper-belt kin (*Figure 5*); no one has seen a Trojan as red as Pholus or 1995 GO. This suggests that Trojans have accompanied Jupiter from the outset. However, if red coloration fades after exposure to stronger sunlight, perhaps Trojans are captured Centaurs after all. Clearly, more data are needed to reveal the Trojans' source region.

A second question is whether there are some simple measurements that help us clarify the role of the ices and colored sooty material on bodies too faint to be studied spectroscopically. The answer seems to be yes. It appears that different bodies in the outer solar system can be systematically segregated in terms of whether their surfaces show pure ice, dirty ice, or dark soil. *Figure 6*'s vertical axis plots difference in reflectivity, or albedo, at roughly yellow light in the visible spectrum (*V*) and at the near-infrared wavelength of 1.25 microns (called *J* for historical reasons). Similarly, the horizontal axis represents the difference between *J* and *K* (2.2-micron) reflectivity. Defined this way, extremely red objects plot at upper right and bluish-white ones at lower left. Only bodies in the outer solar system, dominated as they are by carbonaceous soils and ices, are included, which permits us to see relationships among satellites, the most distant "asteroids," and comets — all on one graph.

Think of the outer solar system as made of two basic materials, bright ice (white) and dark dirt (gray or red-brown). Probably the dominant physical quality that determines position on this diagram is the ratio of these two materials. Other characteristics

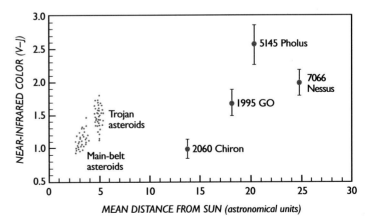

Figure 5. The color of interplanetary bodies tends to redden with increasing distance from the Sun, a characteristic perhaps due to the presence of organic matter in the outer solar system. Here "color" means the difference between an object's apparent brightness at visual (*V*) wavelengths and the near-infrared (*J*) band at 1.25 microns; objects considered red plot toward the top and neutral gray ones near the bottom. The ranges for main-belt and Trojan asteroids are shown schematically.

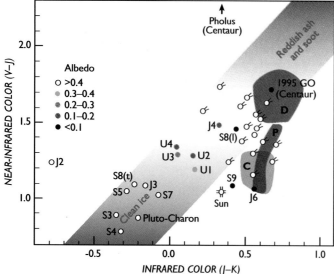

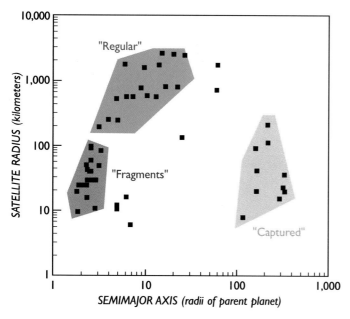

Figure 6. Small bodies in the outer solar system can be grouped according to three parameters: color in the visible and near infrared (*V* minus *J*); color further in the infrared (*J* minus *K*, at 2.2 microns); and visual albedo (surface reflectivity). In this diagram bright ices cluster at lower left and dark reddish brown carbon-rich soils at upper right. Therefore, an object's plotted position may indicate the ice-to-soil ratio in its surface materials. Grayish intermediate objects include Jupiter's satellite Callisto (J4) and the Uranian satellites Ariel, Umbriel, Titania, and Oberon (U1 to U4). Active and inactive comets cluster at upper right among the dark-brown asteroids, evidence that led to the successful prediction that the nucleus of Halley's Comet would be very dark. The most distant asteroids fall into one of three spectral classes. C objects are neutral black and dominate the outer edge of the asteroid belt, while D types are reddish brown and are common farther out among the Trojans. P denotes what appears to be a transitional class.

Figure 7. Satellites of the outer solar system form relatively compact clusters when grouped according to size and distance from their parent planet. "Regular" moons tend to be large and occupy prograde orbits. "Fragments" are very small inner satellites, often found near or even within a planet's ring system, that may be the collisional remnants of larger predecessors. "Captured" objects have distant, unusual orbits; probably they were once interplanetary wanderers that became captured after venturing too close to the planet they now orbit.

are no doubt involved, but they exhibit second-order effects. The more bright ice on a surface, the more blue-white its color (tending toward the lower-left corner); the more dark soot and organic matter, the redder its color (upper-right corner).

Thus, simply by observing gross color characteristics — in the absence of harder-to-get spectra — we can get an idea of surface composition among the outer solar system bodies. Observers Dale Cruikshank, David Tholen, and I first began to fill in this diagram in the 1980s. We used it to successfully predict the low albedo of Halley's comet (at a time when comets were erroneously believed to be bright due to their ice content), and to predict the moderate reflectivities of the Uranian satellites.

More recently, it has been used to explore the qualities of the faint Centaurs and Trojans. One result, which initially seemed surprising, is that none of the Trojan asteroids, Centaurs, Kuiper-belt objects, or comets studied to date have the blue-white colors suggesting clean ice; they are instead dominated by dark, sooty dust. In a few cases we also know that these bodies are dark. Only certain larger satellites show evidence of clean ice, probably because they have been heated by tidal or other effects, so that the interior ice melted, the dust sank, and fluids erupted onto the surface and froze as fresh ice layers. Clear evidence of this resurfacing process has been found on Europa and Enceladus, two of the brightest, whitest satellites in the solar system (labeled J2 and S2, respectively, in *Figure 6*).

On the other hand, small bodies like comet nuclei have no such heating mechanism to melt the ice and sequester the dark soil. In fact, just the opposite might be happening: micrometeoritic sandblasting probably vaporizes the ice and leaves the "dirt" behind, creating a dark surface. Add the occasional production of reddish organics, and we have the whole gamut of colors shown in *Figure 6*. It is perhaps a crude theory, but an instructive one that unites diverse phenomena in a simple way.

In 1998 astronomers Stephen Tegler and William Romanishin reported observations of Centaur and Kuiper-belt objects that raise new questions about color trends and relationships. The 16 objects they surveyed do not follow a smooth reddening trend with greater distance but instead fall into two distinct color groups, one fairly neutral reddish-grey and the other quite red. Most members of the grayish group are Centaurs, while seven of the eight red objects reside in the Kuiper belt. These results are consistent with the idea that Centaurs are objects that, after being thrown from the Kuiper belt into orbits closer to the Sun, have "burned off" a thin crust of reddened material, exposing the greyer inner material.

RELATING SMALL MOONS TO ASTEROIDS AND COMETS

This organizational framework can also be applied to the smaller moonlets of the solar system — bodies that are the size of modest asteroids. Do they show similar colors (and compositions, by inference) to one or more groups of asteroids and comets, or are they distinct? It is an especially illuminating question for the outermost moons of Jupiter, Saturn, Uranus, and Neptune, and the two moonlets of Mars (*Figure 7*). The orbits

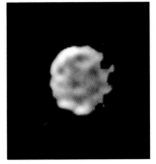

Figure 8 (above). Voyager 2 took this image of Saturn's dark, outermost satellite Phoebe in 1981 from a distance of 2.2 million km. It reveals the moon's spherical form and cratered surface, but other details are indistinct. Phoebe's surface reflects no more than 6 percent of the sunlight that strikes it.

Figure 9 (right). Himalia (arrowed) is only 180 km across, yet it is by far the largest of Jupiter's eight outermost satellites. The 14th-magnitude speck was recorded in a long-exposure image that left Jupiter itself greatly overexposed.

of these moons suggest that they were captured as they wandered in from interplanetary space. For example, about half of them move in retrograde directions around their planets — an even split that would be expected only if they were approaching from random directions when snared.

If these really are captured asteroids or comets, measurements of their color or spectrum might clarify where they originated. Do they match the colors expected for errant asteroids thrown out of the main belt? Are they more like Trojans? Could they have been drawn from the interloping Centaurs? Or, as a group, are they different from any of these populations? Interestingly enough, they all have very low albedos, thus matching the sample of objects that originated outside 2.5 AU. Yet their color data are controversial. Of Jupiter's eight captured moons, only the brightest few have been studied in this way, and the results of different observers disagree. Some have reported them to be neutral black, others see them as reddish-brown, similar to Trojans. A minor embarrassment to planetary science is that even Phobos and Deimos, the closest moons beyond our own, have large color uncertainties (it is hard to separate their light from the red glare of Mars). Only a few years ago they had been confidently reported as neutral black. But new observations, including Hubble Space Telescope data, suggest a closer match to the redder Trojans. Phoebe, the retrograde moon of Saturn, is a neutral black object (*Figure 8*), similar in color to outer main-belt asteroids or to Chiron. So far, after analyzing about half of these 12 probably-captured moons, observers do not find color correspondences to typical objects from either the inner or outer

asteroid belt. But they could match a sample drawn from the Trojans, or perhaps the Centaurs, that have partially lost their red coloration.

Perhaps the scenario went something like this: as Jupiter and the other outer planets formed, they possessed extended atmospheres or even circumplanetary nebular disks that slowed and captured moderately reddish-brown asteroids passing too close. But the gas drag was great, and the entire first wave of captured objects spiraled too far inward and crashed into their host planets. At the very end of the planets' growth, after the extended atmospheres dissipated, there was an "opportune moment" of perhaps a million years or so when a satellite could be captured without spiraling to oblivion.

Something else important was happening during, or just before, this brief window. As Jupiter and the other giants gained their massive, hydrogen-rich outer layers by feasting on the surrounding nebula, their gravitational influence grew very rapidly. Jupiter, for example, cleared a zone that initially was 2 AU wide and later extended all the way from the outer edge of the asteroid belt nearly to Saturn. Meanwhile, Saturn, Uranus, and Neptune were clearing their own zones and scattering the dark-colored planetesimals in all directions.

These two circumstances combined to create a unique, brief epoch when the outer planets were able to scatter untold thousands of small, dark planetesimals that had originated everywhere from the outer asteroid belt (3 AU from the Sun) to Pluto (some 40 AU). It is easy to imagine, therefore, that a few could have been captured by fortuitous approaches to planets. More importantly, these would have been a mix of black to reddish-

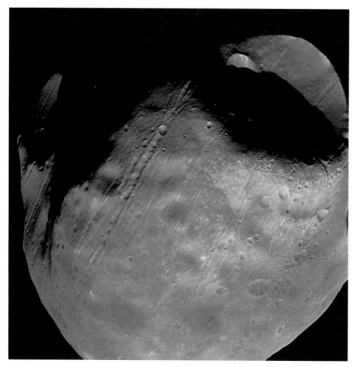

Figure 10 (above). The two Martian moons Deimos (lower left) and Phobos (lower right) are reproduced at the same scale as the asteroid 951 Gaspra. All three have irregular outlines shaped by past collisions. Their surfaces look remarkably different, perhaps due to gross compositional differences but more likely the result of their varied impact histories. Phobos and Deimos reflect only a few percent of the sunlight striking them and are much darker than many other asteroids in the inner asteroid belt beyond Mars. Their spectra resemble those of certain C-type asteroids common to the outer asteroid belt and to Jupiter's Trojan asteroids; these objects are believed to have a composition similar to carbonaceous chondrite meteorites. This suggests that perhaps Phobos and Deimos were at one time farther from the Sun, but Jupiter perturbed them into orbits that allowed their capture by Mars — a theory that is inviting but remains unproven.

Figure 11 (left). Phobos must have nearly shattered from the force of the impact that excavated the dramatic 10-km crater Stickney (upper right). The blow created a complex of fractures, some of which appear on the surface as pit-lined grooves in this composite of Viking 1 views from 1978. Tidal interaction with Mars is forcing the orbit of Phobos to decay gradually, and this moon is doomed to strike the planet in about 40 million years. This unstable situation strengthens the argument that Phobos is a captured asteroid.

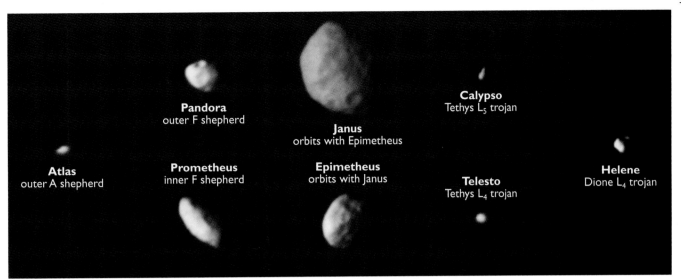

Pandora
outer F shepherd

Janus
orbits with Epimetheus

Calypso
Tethys L$_5$ trojan

Atlas
outer A shepherd

Prometheus
inner F shepherd

Epimetheus
orbits with Janus

Telesto
Tethys L$_4$ trojan

Helene
Dione L$_4$ trojan

Figure 12. **Shown to scale in this composite of Voyager images are eight tiny satellites known to orbit Saturn. All of them either "share" orbits or "guard" rings. Epimetheus and its companion moon Janus may be two halves of what was once a single object.**

brown planetesimals — exactly what we see today among the likes of Phoebe, Phobos, Deimos, and the outer Jovian moons. This scenario is admittedly very sketchy; more dynamical work is needed to understand how the asteroid belt acquired large clearings known as the Kirkwood gaps (see Chapter 25), how Trojan asteroids were captured in Lagrangian clouds, and how the orbits of early interplanetary bodies evolved.

MANY MOONS, SPARSE DATA

It is easy to speculate about the origins of small moons because in most cases we know so little about them. The outermost Jovian moons, which orbit at the limit of the planet's gravitational sphere of influence, make a strange constellation of two distinct groups. Leda, Himalia, Lysithea, and Elara all have prograde orbits about 11.5 million km from Jupiter, with inclinations between 26° and 29°. On the other hand, Ananke, Carme, Pasiphae, and Sinope move in retrograde orbits about 22 million km from Jupiter, with inclinations between 16° and 33°. Conceivably only two asteroids were captured initially; then each underwent a collision that broke it into fragments that assumed similar orbits. Alternatively, the planet's capture mechanism may have been highly selective, allowing objects to end up in only one characteristic prograde orbit and another retrograde orbit. The exact history of Jupiter's outer moons remains a mystery. Despite visits to the giant planet by several spacecraft, they are simply too small and too far from the inner moons to be studied in detail (*Figure 9*).

Dark Phoebe lies on the outskirts of Saturn's gravitational grip and lends strong support for the capture hypothesis. It travels in a retrograde orbit some 26 million km across and inclined 30° to Saturn's equator. (Other major Saturnian moons are larger, more icy, and have prograde orbits with inclinations not exceeding 15°.) Curiously, Phoebe and Chiron have very similar sizes and surface colors. Chiron does come close to Saturn but is unlikely to be captured. Even so, it may offer a present-day example of the kind of interplanetary wanderers among which Phoebe once traveled.

Mars's two small satellites, Phobos and Deimos (*Figures 10, 11*), with diameters of roughly 22 and 12 km, respectively,

look much like asteroids from the pool of Trojan-like carbonaceous (and icy?) objects that formed somewhere beyond the main belt. Phobos and Deimos are extremely interesting and accessible targets for exploration. That was the specific intent of the Soviet Union's Phobos mission. Its two well-instrumented spacecraft were dispatched from Earth in 1988, but one was lost part way to Mars. The other lost attitude control after its first few approaches to Phobos. Provocatively, data from Phobos 2 suggest that water molecules may be leaking from the moon's interior — even though its surface is dry and ice free. If so, Phobos may some day be classified as a defunct comet nucleus!

Two new finds came in September 1997, when the 16th and 17th moons of Uranus were discovered by a quartet of astronomers using the famed 5-m Hale telescope on Palomar Mountain in California. At 20th and 22nd magnitude, these are the dimmest moons yet seen using a ground-based telescope. The brighter object, provisionally named Sycorax, has an estimated diamter of 120 km; Caliban may be half that size. More importantly, the satellites have highly eccentric, retrograde orbits that put them an average of 12,200,000 and 7,200,000 km, respectively, from the planet. This is more than 10 times farther out than Uranus's other moons. As noted earlier, such irregular orbits suggest that these bodies were captured by the planet.

An entire class of small moonlets was completely unknown prior to the era of interplanetary spacecraft. They are the ones closest to the giant planets, in and near the ring systems (*Figure 12*). Unlike their outerlying, captured siblings, these are believed to have formed from the disk-shaped clouds that spawned the giant planets' large satellite systems. However, they do have one interesting connection with wandering interplanetary bodies: they are the most likely to be hit by them. As passing bodies are pulled into collision paths with Jupiter and the other giants, they become most concentrated — and attain the highest velocities — in the zones closest to the planets. This means that small, inner moonlets are the most susceptible to being "sandblasted" by micrometeorites or smashed apart in a larger collision. Any number of them might be the reaccumulated remains of larger moons destroyed in the distant past.

In fact, shattered moonlets probably created the ring systems themselves! Dynamicists believe that the rings are maintained by small flecks of debris that spiral toward the planet after being blasted off nearby moonlets, sometimes gradually and sometimes catastrophically (see Chapter 16). According to these ideas, the prominence or even existence of ring systems may depend on the most recent destruction of an ill-fated moonlet.

SUMMARY

This discussion has looked at the small bodies of the solar system in a new way. We put aside the old observational categories of asteroids, comets, and moons, considering these bodies instead as a collection of planetesimals with a range of initial compositions determined mainly by their original location in the solar system.

The planetesimals that formed in the inner asteroid belt were primarily rocky objects, like ordinary meteorites. Some odd surface types seen among them today may be associated with major collisions that split them and revealed interior materials such as metallic cores. Those that formed beyond about 2.5 or 4 AU incorporated two important additional components: very dark, carbon-rich "soot" and bright ices. As long as the dark material remains evenly mixed in the icy material, it dominates the colors of all such bodies. Thus, all outer-solar-system planetesimals were probably initially blackish to dark brown. On some larger bodies the ice melted, perhaps allowing the black sooty dust to settle inward and leaving the surface ices clean and bright white. The outermost bodies of the original solar system, in the Kuiper belt, incorporated low-temperature ices, such as methane. These compounds perhaps fueled the production of reddish-colored organic compounds, thus explaining a rough trend toward redder colors at distances farther from the Sun.

Those icy bodies that chance to pass through the warmth of the inner solar system, where their ices sublime and produce tails of gas and dust, are still called comets. But the classical distinctions between comets and asteroids have all but disappeared. We can point to distant "asteroids" that would become comets if brought nearer the Sun, and we suspect many "extinct" comets — their ices long ago depleted — now masquerade as asteroids.

In short, the processes of planetary evolution have conspired to present us with a rich variety of miniworlds, in which we are only beginning to understand patterns and relationships — the clues to their origin and history.

Comets

John C. Brandt

The spectacular Comet West, as it appeared in predawn twilight in March 1976.

EVEN THOUGH ASTRONOMERS have observed and studied comets for thousands of years, we remain deeply intrigued by these enigmatic interplanetary wanderers. During the past decade, our long-standing fascination with comets was rekindled by the appearance of three incredibly diverse examples of such objects. In July 1994, a score of fragments from Comet Shoemaker-Levy 9 (officially designated D/1993 F2) crashed into Jupiter with spectacular results. Comet Hyakutake (C/1996 B2) came very close to Earth in the spring of 1996, its bluish plasma tail stretching for more than 55° across the northern sky. Comet Hale-Bopp (C/1995 O1; *Figure 1*) then capped this parade with an impressive visual display that lasted for months despite the fact that it never came closer than 1.3 AU to Earth.

Understandably, these appearances triggered great public interest, and they were watched closely by scientists as well. Most comets remain dormant on the distant fringes of the solar system, becoming visible only when perturbed inward toward the Sun. Consequently, these objects have remained virtually unchanged since planetary formation, and they thus represent key pieces in our efforts to understand the origin and early evolution of the solar system. Moreover, comets do strike Earth from time to time, and in eons past they may have been a major source of some of our planet's volatiles. For these and other reasons, the scientific study of comets is increasing, after decades of relative obscurity.

HALLEY'S LEGACY

Despite these recent spectacles, it is Halley's Comet that holds the key to much of what we know about comets. No other object of this type has been studied in such detail. Multiple spacecraft were sent to Halley in the mid-1980s, producing

Figure 1 (above). Comet Hale-Bopp displayed impressive plasma (blue) and dust (white) tails during its widely observed appearance. Visible to the unaided eye for many months, here it drifts among the Milky Way's stars as seen from near Sedona, Arizona, in March 1997.

Figure 2 (below). Halley's Comet, as recorded on 13 May 1910, 24 days after perihelion. This photograph captured the comet's 45°-long tail, as well as the planet Venus (at lower left). The structured feature below the main tail is a disconnected plasma tail.

Figure 3. An apparition of Halley's Comet is depicted in the celebrated Bayeux Tapestry, which commemorates the Norman Conquest in 1066. The comet made a close and impressive approach to Earth that year, about the time that William the Conqueror invaded England from Normandy (France). Its appearance was considered an evil omen for King Harold of England; in fact, Harold was killed later that year during the Battle of Hastings. The Latin inscription *Isti Mirant Stella* translates "They marvel at the star."

extensive *in situ* measurements and our first resolved images of a comet's solid central body, its *nucleus*. An extensive, world-wide network of ground-based observers amassed large volumes of data. These extraordinary studies are a blessing and a curse. For now we must rely heavily on the Halley results, but we also recognize that all comets are probably quite unlike this object in detail. Rather, nature has taught us that every comet is an education.

Observations of Halley's Comet go at least as far back as Chinese records of its appearance in 240 BC, and it has been observed at each perihelion passage for more than two millennia. Until the time of the 18th-century astronomer Edmond Halley, however, comets were tacitly assumed to pass through the inner solar system only sporadically, and no serious attention was given to the possibility of their periodic return. Halley used Isaac Newton's then-new theories of gravitation and planetary motion to compute the orbits of several comets. He noted that the orbits of those observed in 1531, 1607, and 1682 were quite similar, and he assumed that the sightings probably referred to the same object at successive apparitions. On this basis, Halley boldly announced that it would return again in 1758–1759. Its return occurred as predicted, and later the comet was named in his honor and officially designated 1P, indicating its rank as the first-known periodic comet.

Figure 4. Halley's Comet on 8 March 1986. This beautiful photograph shows the distinctive colors of its dust and plasma (ion) tails. This photograph was obtained from Easter Island as part of the International Halley Watch. During that week, three spacecraft flew through the comet's inner coma and passed very close to its nucleus.

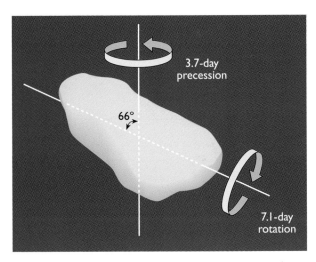

Figure 6. The nucleus of Halley's Comet rotates around its long axis with one period and precesses around another axis tipped 66° away with a different period. This combination of motions means a "day" on the nucleus varies from 2.4 to 5 Earth days.

Figure 5. The elongated nucleus of Halley's Comet, as seen in a composite of 60 images from the Giotto spacecraft. Resolution varies from 800 m at lower right to 80 m at the base of the jet at upper left. The Sun is toward the left, and material in the bright jets streams sunward. Measuring 16 by 8 km, Halley's nucleus is much larger than that of a typical comet.

The physical study of Halley's Comet was begun by Friedrich W. Bessel at the 1835 apparition, and astronomers worldwide made organized observations on its next return in 1910 (*Figure 2*). We owe its repeated naked-eye visibility over many apparitions to fortunate circumstances. First, its passage through perihelion is often favorably placed from Earth's perspective. The orbit of Halley's Comet has an average period of 76 years, with a perihelion 0.59 AU from the Sun (inside Venus's orbit) and an aphelion of 35 AU (beyond Neptune's). Second, the comet is large, and the resultant activity is unusually spectacular. These factors have combined through the centuries to make Halley's Comet readily visible and memorable. In the popular view, it has been associated with numerous historical events. Among its many historical representations are the depiction in the Nuremberg Chronicles (the apparition of 684 AD), the Bayeux tapestry (1066 AD; *Figure 3*), and the naturalistic fresco by Giotto di Bondone in the Arena Chapel of Padova, Italy (1301). Ironically, the comet was seen poorly during its most recent apparition (*Figure 4*) because it passed through perihelion on 9 February 1986, almost exactly on the side of the Sun opposite Earth.

Thus, Halley's Comet is unique. It possesses an unquestioned place in human history and is, at the same time, a large, dependable object with exciting scientific possibilities. This is the only comet that exhibits the entire range of classic, observable phenomena and occupies a predictable orbit. Not surprisingly, therefore, efforts to launch the first spacecraft toward a comet inevitably focused on Halley. In March 1986, aided by a worldwide network of astronomers, a small armada of spacecraft sped by the comet — some at very close range. The dozens of experiments they carried undertook a vast array of investigations, from detecting plasma waves on the fringe of the cometary activity to obtaining images of the icy core itself.

The appearance of Halley's nucleus was not what most researchers expected (*Figure 5*). It is larger, darker, and more irregularly shaped than our preconception of it, which had been of something roughly spherical with a diameter of about 5 km and an albedo upward of 60 percent. In spacecraft images the irregular shape is clearly visible and bears some resemblance to a peanut or a potato roughly 16 km long and 8 km across at its widest. The surface is not smooth but shows features that can be likened to hills and craters. It is also very black — darker than coal or black velvet, with an albedo close to 4 percent. This makes it one of the darkest bodies in the solar system.

Infrared radiation emanating from the surface indicates a temperature of approximately 330° K. This is close to what we would expect for a slowly rotating body with a dark, dusty crust situated 0.8 AU from the Sun, as the comet was during the spacecraft encounters. However, a surface of sublimating ice should be roughly 100° colder. Thus, we may reasonably infer that the sublimation of ices takes place below the surface.

The bright jets visible in the images originated from a limited number of locations on the comet's surface (perhaps one-tenth of the total area) and were confined to the sunward side. The jets appeared bright, presumably from sunlight reflecting off particles of dust dragged off the nucleus by the expanding, freshly sublimated gas. Once free of the surface, the gas quickly expands laterally, but the dust grains retain the jets' original configuration as they coast outward. The rapid turn-on of activity at sunrise and turn-off at sunset imply that the crust is thin where it overlies the jets but is thicker elsewhere.

Despite the close inspection by spacecraft, the rotation state of Halley's nucleus remains uncertain. We can rule out a simple rotation around a single axis, because the light curve produced as the spinning nucleus reflects sunlight is complex. While images obtained by the Vega and Giotto spacecraft were very helpful, they represent only three snapshots in time. The best interpretation is a messy combination of rotation and precession (*Figure 6*).

Dust measurements contained surprises too. The compositions of the dust particles encountered by the spacecraft fall into three general categories. The first consist almost entirely of the elements carbon (C), hydrogen (H), oxygen (O), and nitrogen (N), which came to be called "CHON" particles. The second type has a silicate mineralogy similar to rocks in the crusts of Earth, Moon, Mars, and most meteorites. The third and most common type consists of mixtures of the other two groups. Compositionally, these particles are similar to primitive meteorites called carbonaceous chondrites, except they are enriched in the CHON elements. They bear chemical and perhaps physical similarities to tiny particles of cosmic dust collected by research aircraft flying in Earth's upper atmosphere.

The measurements registered many more very small grains (each only about 10^{-17} g) than expected. The proportion of these tiniest particles rose with increasing distance from the nucleus. This trend is consistent with dust particles that are relatively large and fragile when they leave the nucleus but which fragment as they move outward into space. The small grains may be the remnants of the original icy-organic-dusty particles (discussed later) from the nebula that formed the solar system.

The spacecraft encountering Comet Halley also deduced the concentration and composition of cometary gases encountered along their paths. During its 1986 perihelion passage, the molecules of gas surrounding Halley's nucleus had the following proportions: about 80 percent water (H_2O); about 10 percent carbon monoxide (CO), as determined from a sounding-rocket experiment; about $3\frac{1}{2}$ percent carbon dioxide (CO_2); a few percent polymerized formaldehyde (($H_2CO)_n$); and trace amounts of other substances. The discovery of polymerized (chained) formaldehyde was especially exciting. These organic molecules, which have the general form

$$\begin{array}{ccccccccc}
& H & & H & & H & & H & \\
& | & & | & & | & & | & \\
H - & C & - O - & C & - O - & C & - O - & C & - O - \\
& | & & | & & | & & | & \\
& H & & H & & H & & H &
\end{array}$$

when terminated with suitable "end caps" like hydrogen, accurately reproduce the series of mass peaks recorded by mass spectrometers while flying through the comet. Other polymers or combinations may also provide a good match to the spacecraft data. Thus, for now it is best to think of polymerized formaldehyde as the most likely surrogate for a class of complex organic compounds present in comets.

GENERAL CHARACTERISTICS

Although our intensive observations of Halley's Comet turned up numerous surprises, ultimately those data dovetailed well with the large body of cometary studies amassed over the previous decades. Admittedly, the range of features and processes we have been able to observe are heavily biased toward bright comets, which invariably get most of the attention. Still, all evidence available at present indicates that the ultimate source of all cometary phenomena is its *nucleus*, a lump of snow and dust with a typical dimension of roughly 1 to 10 km. The ability of this "dirty iceball" to interact with solar radiation and the solar wind to produce such spectacular features is remarkable indeed.

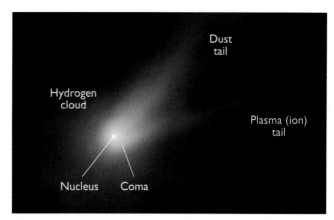

Figure 7. **A schematic representation of the basic parts of a comet. The components are not shown to scale; in reality, a nucleus is typically no more than 10 km across, whereas the tails and hydrogen cloud often extend outward for 10,000,000 km or more.**

Let's examine the comet's main features, as seen schematically in *Figure 7*.

Tails. To the eye, comets' most distinguishing features are their tails, which routinely stretch across space for tens of millions of kilometers — and occasionally for 1 AU (about 150 million km) or more. Photographs of comets usually show two distinct kinds of tail (*Figures 1,4*): one containing dust and the other plasma (ions and electrons). These can be found alone or together in a given comet. The *dust tail* appears pearly yellow because the light reaching us from it is reflected sunlight. The *plasma tail* looks blue because ions of carbon monoxide (CO^+) within it fluoresce in the presence of sunlight. This emission peaks at a wavelength of about 4200 angstroms.

Usually observed as sweeping arcs, dust tails typically have a homogeneous appearance and lengths ranging from 1 to 10 million km. The vast majority of the dust grains are no bigger than 1 micron across, the size of smoke particles. Once released from the nucleus, this dust moves away from the nucleus in ways controlled by its ejection velocity, the Sun's gravity, the outward push of solar radiation pressure, and the masses of individual particles (*Figures 8,9*). Sunward-pointing features, the so-called *antitails* seen in comets Arend-Roland (C/1956 R1), Kohoutek (C/1973 E1), Halley, and Hale-Bopp (*Figure 10*), among others, are not directed at the Sun at all. They merely result from our seeing a dust tail projected ahead of the Earth-comet line.

Plasma tails are usually straight, contain a great deal of fine structure, and attain lengths roughly 10 times that of their dusty siblings — up to 100 million km. The plasma races outward almost directly away from the Sun, lagging the true antisolar direction by a few degrees in the sense opposite that of the comet's motion. Locally the plasma becomes concentrated into thin bundles called rays or streamers. Such ubiquitous details provide convincing evidence that a magnetic field threads the tail's entire length. Consisting of a dense, cold mixture of electrons and molecular ions, the plasma-tail streamers seem to be rooted in a limited zone on the Sun-facing side of the nucleus. Their turning and lengthening provide a good hint as to the characteristics of the magnetic field that entrains them. Plasma tails routinely become detached from the comet's head during

Figure 8. The spectacular dust tail of Comet West (C/1975 V1), as it appeared in predawn twilight in March 1976. Note the tail's delicate, lacy structure, created by countless dust particles shed from the nucleus over many days. About a week later the nucleus abruptly split into four pieces.

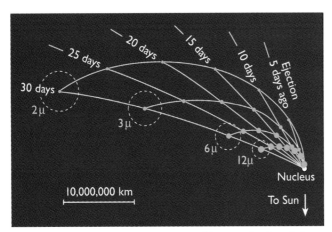

Figure 9. The dust particles in a comet's tail become distributed along lines of equal time since their ejection from the nucleus *(synchrones)* and equal particle size *(syndynes)*. However, this simple ordering becomes smeared because dust particles are ejected with different velocities in many directions. Consequently, small grains have a greater range of possible trajectories (dashed circles) than large ones do.

what is termed a *disconnection event*, or DE, to be explained more fully later. During a DE, part or all of the old plasma tail drifts away and a new one forms.

A third tail, dominated by the element sodium, was discovered unexpectedly in Comet Hale-Bopp. It was a long (almost 7°) and quite narrow (10' wide) feature that ran very close to the line from the Sun through the comet (*Figure 11*). Apparently atoms of sodium were accelerated rapidly by the radiation pressure of sunlight, a process involving the absorption and rerelease of photons, which produced the element's characteristic yellow-line emission. Sodium had been detected in past comets but never in such a striking extension. Astronomers suspect that the exceptional size and intrinsic brightness of Comet Hale-Bopp undoubtedly contributed to the detection.

Hydrogen cloud. As impressive as these tails might be, they are not a comet's largest feature. In 1970, observations made above Earth's atmosphere at the Lyman-alpha wavelength of 1216

Figure 10. Comet Hale-Bopp displays a pronounced *antitail,* the spike of dust apparently pointed sunward (down) from the nucleus, in this image taken from Chile on 5 January 1998. The antitail is a projection effect, as all of the comet's dust was indeed streaming away from the Sun.

Figure 11. The left-hand image, which records the fluorescence (D-line) emission from sodium atoms, clearly shows the thin, straight sodium tail that accompanied Comet Hale-Bopp in April 1997. Compare its shape and orientation with the traditional plasma and dust tails seen in the right-hand image, which isolates light emitted by water vapor.

angstroms (deep in the ultraviolet) indicated that comets Tago-Sato-Kosaka and Bennett were surrounded by huge tenuous clouds of hydrogen atoms. Similar clouds have accompanied several other comets and span many million kilometers, making them substantially larger than the Sun (*Figure 12*). Given the strength of the ultraviolet emission, astronomers estimate that bright comets in the vicinity of Earth's orbit can produce more than 10^{29} hydrogen atoms per second! This gas cannot originate directly from the icy nucleus because its observed outflow speed is roughly 8 km per second, about 10 times faster than predicted for material simply sublimating (evaporating) from the nucleus' surface. Instead, most of this hydrogen probably comes from the dissociation by sunlight of hydroxyl radicals, OH, which themselves are derived from molecules of water.

Coma. Nearer to the nucleus is a denser spherical envelope of gas and dust called the coma. This gas flows away at an average speed of 0.5 to 1.0 km per second and extends 100,000 to 1,000,000 km out from the nucleus. It is the outflow of coma gas that drags dust particles off into space. Comas usually do not appear until comets come to within about 3 AU of the Sun, at which point a comet's water ice begins to sublimate vigorously.

How can such seemingly prodigious outflow rates be put in perspective? A bright comet must be losing a great deal of water to generate more than 10^{29} hydrogen atoms per second. Comet Hale-Bopp, a very large comet, was losing 1,000 metric tons of dust and 130 metric tons of water per second in late March. By late April, these values had dropped to 900 and 80 metric tons, respectively. While these numbers seem huge, the

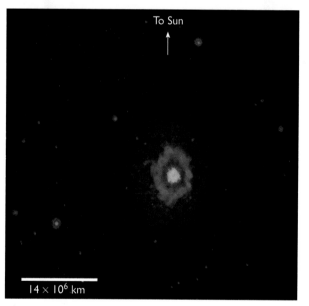

Figure 12. A color-coded, far-ultraviolet image of Comet Hale-Bopp taken on 31 March 1997 by the Midcourse Space Experiment spacecraft. It reveals clouds of neutral hydrogen (purple), oxygen (light blue), and carbon (pink). Such huge "coronas" of hydrogen and oxygen have been observed around comets before, but the envelope of carbon is rarely seen.

material is spread over a very large volume of space. For Hale-Bopp, the density works out to roughly one very small dust particle per cubic centimeter in the coma. For most comets, stars are easily seen through the tails and their light is dimmed in the coma only if seen very close to the nucleus. Remarkably, a comet usually sheds only 0.1 to 1 percent of its mass per passage through the inner solar system.

Gases within the coma, as determined by spectroscopy from Earth and by mass spectrographs aboard spacecraft, are principally neutral atoms and molecules (*Figures 13,14*). Dozens of individual species and their ions have been detected, many of which are chemically related. However, astronomers suspect that a more limited number of discrete compounds lie frozen or trapped within the nucleus. Once warmed by sunlight and released into the coma, these "parent molecules" either break down or react to form a wider assortment of species (*Table 1*). Relatively complex compounds like HCN, CH_3CN, and

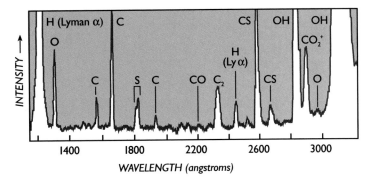

Figure 13. An ultraviolet spectrum of Comet Bradfield (C/1979 Y1), obtained by the International Ultraviolet Explorer satellite. It reveals emissions from a wide variety of simple compounds that were present (even in small amounts) in the comet's coma.

Figure 14. These spectra of Comet Hale-Bopp, acquired on 13 April 1997, show scans (vertical dimension) through the comet's coma and tail. The emission strength is denoted by color, with red strongest. The upper panel shows that emissions from cyanogen (CN) and diatomic carbon (C_2) dominate the coma. By comparison, abundant ionized carbon monoxide (CO^+) in the plasma tail is responsible for the tail's distinctive blue color. Neutral sodium (Na) is present in both spectra. The symbol [O] denotes "forbidden" oxygen lines, emissions that can only be detected when the ambient gas density is low.

$(H_2CO)_n$ can give rise to a host of byproducts. Emissions from heavier metals like iron and aluminum, present in dust, begin to appear in the coma as a comet nears the Sun.

Nucleus. Until the advent of space missions, we had no certain way to study a comet's heart directly. A coma's gas and dust are quite effective at hiding the embedded nucleus from our view. Very few comets have been studied far enough from the Sun for observers to be confident that no coma is present. Because the nucleus of Halley's Comet is hardly typical, much hard work and many assumptions are needed to infer the sizes of other comets. Such was the case with Comet Hale-Bopp. By assuming that the comet's coma was not too complex, astronomers could model its brightness and electronically subtract the coma from images taken with the Hubble Space Telescope (*Figure 15*). The residual signal thus represented the brightness of the nucleus alone. One further assumption, an albedo of 4 percent, yielded an effective nucleus diameter of 27 to 42 km — reaffirming our belief that Hale-Bopp is a very large comet.

Our searches have begun to include comets with sizes outside the traditional range of 1 to 10 km. On the large end, sensitive ground-based observations are picking up icy bodies with diameters of roughly 100 to 400 km with heliocentric distances of 30 to 50 AU. These objects almost surely mark the inner edge of the much more extensive Kuiper belt, a vast reservoir of comets beyond the orbit of Neptune (see Chapter 5). The small end of the size scale, those with diameters less than 2 km, are faint and hard to detect. A few have been discovered simply because they came very close to Earth. Several were discovered very near the Sun by coronagraphs on orbiting spacecraft. Finally, observations with the Hubble Space Telescope are starting to identify small nuclei, using the same technique employed to determine the radius of Comet Hale-Bopp. If a major population of small

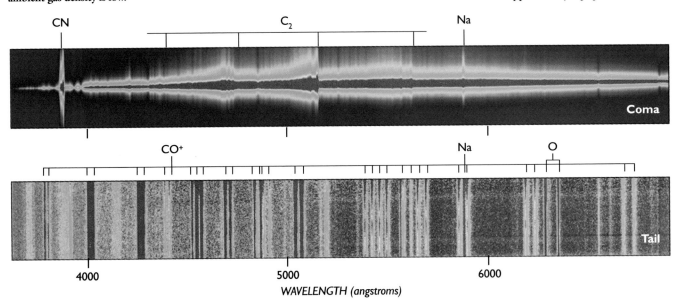

Table 1. A surprising variety of atoms and molecules (left) and ions (right) have been found in the comas and tails of comets. Researchers suspect that many of them are the photodissociated byproducts of a limited set of volatile and organic "parent" molecules. Other elements, like heavier metals, derive from silicate grains carried off the nucleus and into space by expanding gas.

Known Atoms, Molecules, and Ions in Comets

H, OH, O, S, S_2, OCS, H_2S, H_2O, HDO, H_2CO, $(H_2CO)_n$, H_2O^+, OH^+, H_3O^+
C, C_2, C_3, CH, CN, CO, CO_2, CS, CH_4, C_2H_2, C_2H_6 CO^+, CO_2^+
NH, NH_2, HCN, CH_3CN, N_2, NH_3 CH^+, CN^+, N_2^+
Na, Fe, K, Ca, V, Cr, Mn, Co, Ni, Cu, Si, Mg, Al, Ti C^+, Ca^+

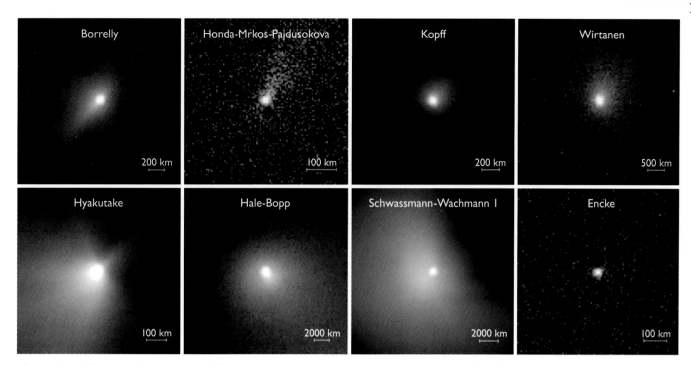

Borrelly	Honda-Mrkos-Pajdusokova	Kopff	Wirtanen
200 km	100 km	200 km	500 km
Hyakutake	Hale-Bopp	Schwassmann-Wachmann I	Encke
100 km	2000 km	2000 km	100 km

Figure 15. Hubble Space Telescope views of faint comets passing in Earth's general vicinity. By subtracting out the brightness in each image assumed to be coming from the coma, astronomers can now estimate the size of cometary nuclei with much greater confidence. Of those shown here, 45P/Comet Honda-Mrkos-Pajdusokova has the smallest apparent nucleus (440 to 680 meters across), while Hale-Bopp and Schwassmann-Wachmann 1 have the largest (30 to 40 km).

comets exists, they should continue to be discovered over the next few years.

Much more controversial is the contention by researchers Louis A. Frank and John B. Sigwarth, based on imagery from orbiting spacecraft, that tens of thousands of house-sized "comets" strike Earth's upper atmosphere daily. First put forward in the mid-1980s and debated anew since 1997, their proposal holds that these objects have masses of 20 to 40 tons and a low density, are composed of very pure water, and are encased in such a way that prevents them from sublimating as they cross the inner solar system. Apart from questions about the soundness of Frank and Sigwarth's data, the putative objects must have extraordinary properties. They have not been detected by sensitive seismometers placed on the Moon by Apollo astronauts. They must be very pure water ice, as even a very small fraction of dust particles would produce sky-filling meteor showers. Yet to avoid telescopic detection, the objects' proposed carbon-based coatings must be so black as to seem physically impossible. Most astronomers find the combination of so many extraordinary properties highly unlikely. If the spacecraft data are sound, an alternative explanation is required.

PHYSICAL MODELS

Any comprehensive and acceptable theory of comets must explain not only their major features (*Figure 7*) but also the ways in which they change with heliocentric distance. Such a theory should explain existing observations in a simple way and,

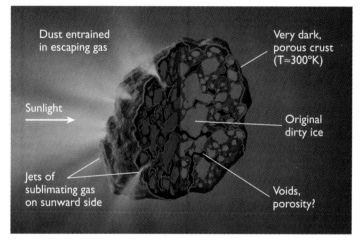

Figure 16. Fred Whipple's "icy conglomerate" model, as extended by Armand Delsemme, and spacecraft observations of Halley's nucleus are the basis for this idealized rendering of a cometary nucleus.

if necessary, point out critical measurements needed for its validation. While we believe that our general understanding of comets is in good shape, experience shows that additional surprises can easily occur.

The basis of current theory is the "icy conglomerate" model of the nucleus, proposed by Fred L. Whipple in 1950 and later extended by Armand Delsemme (*Figure 16*). As a comet approaches the inner solar system, all the sunlight it absorbs goes into heating the nucleus. Closer in, the surface layers become warm enough to trigger the sublimation of ices. Then almost all solar radiation goes into maintaining that conversion process. As the ices vaporize, a dusty crust forms that insulates the deeper layers and regulates the sublimation process (now occurring a few centimeters below the surface). Irregularities in the materials cause sublimation to occur faster in some areas, a situation that can produce jets and ultimately the irregular shape and surface features of the nucleus.

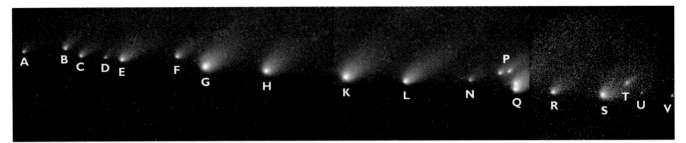

Figure 17. Comet Shoemaker-Levy 9, as imaged by the Hubble Space Telescope in January 1994, about six months before the comet's fragments all struck Jupiter.

Historically, the masses of comets could be estimated only crudely by assuming a size and a density (usually 1 g/cm³). However, some indirect determinations of the mass of Halley's nucleus, when taken with its larger-than-expected dimensions, imply an overall density of about 0.25 g/cm³. If such data can be believed — and for now they should be accepted only with some reservation — Halley's nucleus must be a porous and perhaps fragile structure despite its size.

The fact that some comets do fragment, as did West in 1976 and Shoemaker-Levy 9 in 1992, lends support for the view that nuclei are fragile collections of smaller, loosely bound building blocks. Some theorists maintain that these sub-nuclei have sizes comparable to those of the observed fragments, hundreds of meters or a kilometer across. Others have argued that the nuclei could basically be rubble piles, agglomerations of units with a variety of sizes. However, we have little understanding of why such splittings occur. Moreover, most comets do *not* break apart, at least in obvious ways, implying that they may be coherent solids. All in all, cometary nuclei may come in a variety of structures.

Comet Shoemaker-Levy 9 provided a spectacular example of a ruptured nucleus (*Figure 17*). It was an unprecedented phenomenon that would have generated a flurry of activity even if the pieces had not collided with Jupiter. Most analyses indicate a diameter of under 10 km for the parent nucleus and 0.5 to 2 km for the sub-nuclei. The ease of disruption of the parent body during its close approach to Jupiter in July 1992 implies a material of very low strength, or possibly even "rubble-pile" aggregates. The fragments' collisions with Jupiter did not reveal much about the comet itself, even though spectra of the events showed emissions of silicon, magnesium, and iron that probably came from the cometary bodies.

Physical structure aside, the fact that comas appear when comets are near 3 AU from the Sun is most consistent with water ice being the principal constituent of the nucleus, at least in its outer layers. Spectroscopic observations and direct measurements have confirmed the predominance of "water," meaning both H_2O and its derivatives OH, OH^+, H_2O^+, and H_3O^+. Telltale emissions from molecules such as cyanogen (CN) and diatomic carbon (C_2) are often the first emissions observed as a comet approaches the Sun. This makes sense if the ice occurs as a *clathrate hydrate*, in which minor constituents are trapped in cavities within the water-ice crystal lattice. Thus, the sublimation of water ice may control the release and escape of all substances from the nucleus.

Other minor constituents provide important observational tracers. The CO^+ found in plasma tails may derive from CO in the clathrate lattice or from the breakdown of a more complex molecule. Polymerized formaldehyde may explain why the surface of Halley's Comet is so extremely dark. Formaldehyde (H_2CO) is a likely parent molecule, and a coating of polymerized formaldehyde on the surface (either produced long ago or more recently by exposure to ultraviolet light) should be very dark. It could also be that the nucleus' surface, while intrinsically dark, is made to appear more so by the multiple reflections (and absorption) of light in a porous crust.

If this overall view is correct, then the outer crust of a comet near the orbit of Earth should have a temperature of about 300° K, as was observed for Halley, and its sublimating ice about 215° K. (Curiously, some places on Earth are this cold, one example being Plateau Station in Antarctica. Such frigid locales may provide the opportunity for testing instruments designed to probe inside the nucleus on some future mission to a comet.)

The dominance of water ice may not apply to comets approaching the inner solar system for the first time or throughout a nucleus' interior. First, consider a dynamically "new" comet, which often appears abnormally bright at heliocentric distances greater than 3 AU. Its water-ice lattice can store no more than 17 percent of the number of molecules forming the lattice itself. If other substances (like carbon dioxide) exceed this value, they must exist outside the clathrate hydrate and thus control their own thermodynamic destinies. Since CO_2 and most other plausible minor constituents sublimate at lower temperatures than water does, their presence in a dynamically fresh comet could cause it to "turn on" quite far from the Sun.

What about the overall composition of comets? Water is certainly important, but is it always near the 80 percent value measured for Halley? Probably not. Comets that pass close to the Sun show much lower $H_2O:CO_2$ ratios than Halley did. Qualitatively, this is easy to explain. When the outer dust crust of a comet is heated, a wave of thermal energy travels inward, raising the temperature of successively deeper layers. For comets very near the Sun, the strong thermal wave may cause some deeplying pristine ices to sublimate. Consequently, the composition derived for Halley, a comet that has passed through the inner solar system many times, may not be representative of the pristine ices in comets.

Again, such compositions are not determined by measuring the nucleus directly (though this may happen in the future). Rather, we must extrapolate based on what we discover in the coma, and complexities arise from those assumptions. For example, we cannot be sure that the wide range of molecules detected in the coma indeed exists in the nucleus. Moreover,

The composition of the coma varies with distance from the surface of the nucleus. We see that CN is created from some parent molecule relatively near the nucleus and destroyed by exposure to sunlight farther out. The chemistry is sufficiently complex that we cannot determine the CN's true precursor from among multiple parent-molecule candidates.

The expanding gaseous coma carries with it dust from the nucleus liberated by the sublimation of the ices. Giotto's measurements at Comet Halley revealed many small particles not previously suspected, a revelation that has complicated our notion of what cometary "dust" really is (*Figures 18,19*). Most likely, they began as grains of refractory matter, which solidifies at high temperatures, in interstellar clouds. Once inside the protosolar nebula, they acquired a coating of ice. These, in turn, stuck together to form the fluffy grains observed in the coma. The comet dust collected in Earth's upper atmosphere comprises only the larger aggregates because the smallest particles are blown out of the solar system by the Sun's radiation pressure.

INTERACTIONS WITH THE SOLAR WIND

The neutral molecules continuously produced by the sublimation of ices flow away from the nucleus and outward into the coma. This gas, once ionized, ultimately interacts with the solar

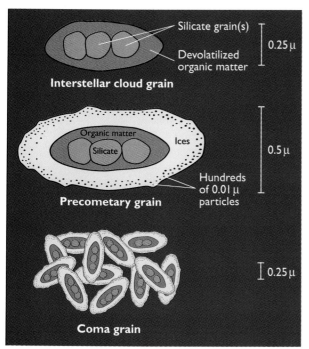

Figure 18. Cometary dust may begin as a silicate grain mantled with organic compounds (top), the type of particle thought to be typical of interstellar clouds. Such grains became coated with various ices in the proto-solar nebula (middle), which then collected into a loose agglomerate (bottom). Compare this porous cluster with the interplanetary dust grain seen below.

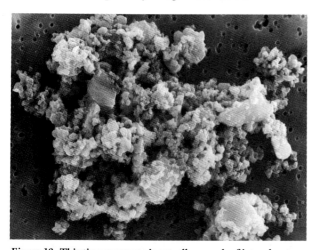

Figure 19. This tiny aggregate is actually a speck of interplanetary debris swept up by the Earth. It was collected by a special research airplane flying through the stratosphere at an altitude of about 18 km. The particle's tiny spheres may have once had ice between them, and the entire assembly — only a few microns long — was probably ejected from a comet's nucleus.

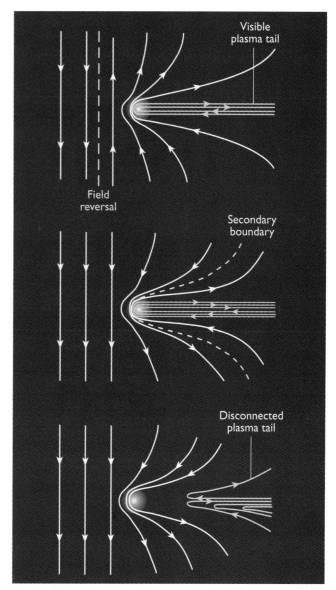

Figure 20. In 1978 Malcolm Niedner and the author proposed that disconnection events result when a comet's tail crosses a sector boundary (polarity reversal) in the solar wind's magnetic field. The field lines compress and wrap around the comet *(top)* as they encounter the ionized coma. At the point of reversal (arrows denote field direction), the field lines "pinch off" ahead of the nucleus and slide around it *(middle)*, carrying the tail away *(bottom)*.

wind, the importance of which was demonstrated by Ludwig Biermann in the early 1950s. Actually, the existence of a solar wind was then unknown; it was inferred by Biermann from observations of comets' plasma tails. Then, in 1957, Hannes Alfvén discovered that the magnetic field carried along by the solar wind plays a vital role in cometary interactions. Some of the ions in the coma become trapped on the magnetic field lines. This causes the field, now burdened with more mass, to decelerate in the vicinity of the comet and wrap around the nucleus like a folding umbrella, forming the plasma tail. In this scenario, the plasma tail is normally connected to the region near the nucleus by this "captured" magnetic field.

These phenomena can be photographed because trapped molecular ions serve as visual tracers of the field lines. The most obvious ion is CO^+, which becomes evident in the plasma tail when carbon monoxide becomes abundant in the coma, about 1½ AU from the Sun. Direct measurements by spacecraft sent

to three comets have also verified this physical picture. For example, when the International Cometary Explorer passed through the tail of Comet 21P/Giacobini-Zinner in September 1985, it confirmed that a reversal of magnetic polarity occurs as predicted in the central, denser tail. Shock fronts were detected around Giacobini-Zinner, Halley, and Grigg-Skjellerup because comets are ionized obstacles in the solar wind. A shock front lowers the wind's speed and allows it to flow smoothly around the comet.

Disconnection events (DEs) occur when the plasma tail's attachment to the region near the nucleus is disrupted (*Figure 20*). Such events are relatively common: 19 fairly obvious episodes were recorded during Comet Halley's most recent appearance, and a dramatic DE occurred in the tail of Comet Hyakutake (*Figure 21*). We now realize that DEs occur when the polarity of the solar-wind magnetic field changes at crossings of the heliospheric current sheet (also called a sector boundary). In this situation, adjacent field lines within the tail cross and reconnect, severing the connection to the near-nuclear region on the sunward side.

Direct measurements by the Ulysses spacecraft have established that comets are exposed to different solar-wind environments depending on their heliocentric latitude. In the

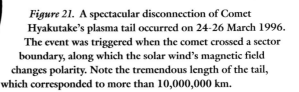

Figure 21. **A spectacular disconnection of Comet Hyakutake's plasma tail occurred on 24-26 March 1996. The event was triggered when the comet crossed a sector boundary, along which the solar wind's magnetic field changes polarity. Note the tremendous length of the tail, which corresponded to more than 10,000,000 km.**

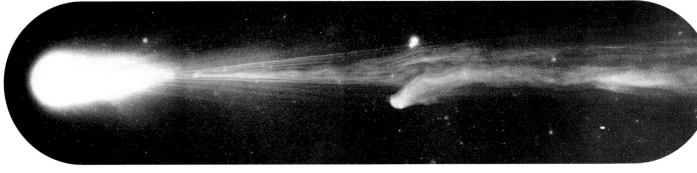

equatorial region, the average solar-wind speed is about 450 km/sec and its density relatively high, though both speed and density can be quite variable. However, the situation changes at latitudes too high to be crossed by the current sheet. In the polar regions, the solar-wind speed is approximately 750 km/sec, its density is relatively low, and variations in speed and density are small (see Chapter 3).

Comet Hale-Bopp's passage almost directly over the north pole of the Sun near perihelion provided an ideal probe of these different solar-wind regimes. First, the comet experienced DEs only when exposed to current-sheet crossings in the equatorial solar latitudes, as expected. Second, the deviation of its plasma tail from the true Sun-comet line reflected a

higher speed in the polar region and a lower speed in the equatorial region. Third, the plasma tail appeared less disturbed in the polar region's steadier solar wind and more so in the equatorial region (*Figure 22*).

The solar wind's interaction with comets is the likely explanation for the X-rays unexpectedly observed in Comet Hyakutake by the Röntgen X-ray Satellite in March 1996 (*Figure 23*). A search of archival data has turned up similar X-ray emission in other comets, and the phenomenon was repeated during Comet Hale-Bopp's perihelion passage in 1997. Thus, X-ray emission may be a common feature of comets. Some theorists suspect that the X-rays arise due to charge exchanges among ions in the coma and the solar wind. However, no proposed

Figure 22. Comet Hale-Bopp provided a remarkable probe of the solar wind over a wide range of heliocentric latitudes. When recorded on 2 March 1997 *(left)*, the comet was almost directly over the Sun's north pole. The plasma tail looks relatively undisturbed and the dust tail below it appears smooth.

By 2 May *(right)*, the comet had moved closer to the Sun's equatorial plane. The dust tail is still smooth, but the fainter plasma tail to its lower left appears much more disturbed. These images were obtained as part of a worldwide observing effort called the Ulysses Comet Watch.

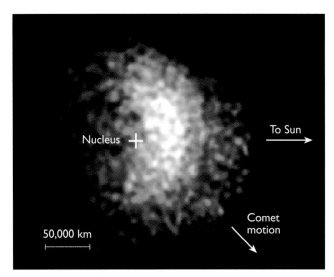

Figure 23. When the orbiting Rosat observatory examined Comet Hyakutake on 27 March 1996, it recorded a surprisingly strong glow of X-rays emanating from the coma's sunward side. Note that the peak emission is not centered along the Sun-comet line. This indicates that the X-rays are probably not a response to direct sunlight but perhaps instead to an interaction with the solar wind.

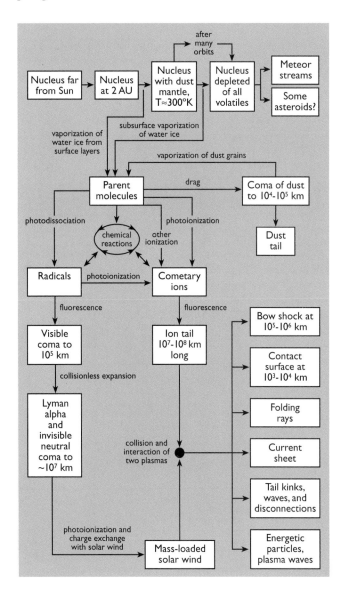

mechanism has yet accounted for the combination of steady emission and impulsive events observed.

A summary of the physical processes believed to be important in comets is given in *Figure 24.* The basic model of the nucleus originated by Whipple and our understanding of the solar-wind interaction developed by Biermann and Alfvén have been severely tested by the space missions. As our knowledge has expanded, these concepts have remained sound.

THE ORIGIN OF COMETS

Here theory attempts to answer the question, "Where do these icy bodies originate?" Much of the information on this subject comes from the study of their orbits (see Chapter 5). Even without our knowing their orbital histories, however, comets would seem to be a natural byproduct of the physical processes responsible for the creation of the solar system. Measurements made directly by spacecraft and remotely from the ground have yielded key isotopic ratios, particularly $^{12}C:^{13}C$, $^{14}N:^{15}N$, and $^{32}S:^{34}S$, that are all consistent with solar-system values. Therefore, we can safely conclude that comets were created with the planets and not in a nearby region of the interstellar medium.

The real difficulty lies in deducing the initial formation zone. The very distant Oort cloud is the current source of long-period comets for the inner solar system. But comets are unlikely to have formed there, thousands of AU from the Sun, because the concentration of matter in the outermost regions of the solar nebula was undoubtedly much too low to permit objects of such size to accumulate. Instead, theorists now view the Oort cloud as a steady-state reservoir that loses comets by gravitational disturbances and gains them from a denser inner cloud with up to 10 times more comets and mass.

To the best of our knowledge, the original "comet factory" was in the realm of the solar nebula occupied by Jupiter, Saturn, Uranus, and Neptune, and in the Kuiper belt just outside the orbit of Neptune. There the icy mantles of primordial dust grains were not sublimated away, as occurred nearer the Sun. Over time these ice-coated grains collided and accumulated, producing countless objects that became the building blocks of comets and the major planets. However, the comets' residence among the outer planets was short lived, as gravitational perturbations soon dispersed them into the Oort cloud or out of the solar system entirely. The Kuiper belt was left more or less intact, and it remains the source of most short-period comets (those taking less than 200 years to travel around the Sun).

One of the most vexing problems in cometary origins is the question of these objects' dust-to-gas ratio. The spectra of comets are rather alike, indicating generally similar composi-

Figure 24. Features and processes involved in a comet's interaction with sunlight and the solar wind are shown schematically. A comet making a single pass through the inner solar system should evolve to a "nucleus with dust mantle" and exhibit the phenomena in the boxes at middle and lower right. If captured into a short-period orbit, the comet ultimately evolves to a dormant or extinct end state. However, many comets never evolve this far because they chance to pass near a massive planet, the gravitational perturbation from which may eject it from the solar system altogether.

tions. However, while Comet Hyakutake showed very little dust, Comet Hale-Bopp effused so much dust that its plasma tail near the nucleus became difficult to observe. How could this be? At present, we have no convincing answer. The traditional view is that objects in the Kuiper and Oort reservoirs have existed in a deep freeze and undergone little or no change for billions of years. However, collisions among the cloud's members, irradiation by cosmic rays, and heating by passing stars may have caused significant alterations. Thus, comets may be ancient relics, but they are not entirely pristine. Nor are they likely to have been drawn from a single compositional mix.

Wherever they came from, comets probably were an important source of the atmospheres of the terrestrial planets (see Chapter 13). In addition, they may well have supplied the original organic molecules necessary for the development of life (see Chapter 27). Just how much of Earth's volatile inventory was supplied by comets remains an often-debated issue among planetary scientists. The ratio of deuterium (D) to hydrogen (H) in Halley's Comet, as measured by the Giotto mass spectrometer, was initially reported to be the same as terrestrial ocean water (within the errors of measurement). However, later revisions of the Giotto data, together with D:H values derived spectroscopically for other comets, indicate ratios roughly twice that here on Earth. If comets did indeed supply a significant fraction of Earth's air and water, then logically they could also have been the source of complex organic species like polymerized formaldehyde. These ideas, though unproved, serve to illustrate the breadth of the scientific interest in comets — and some of the excitement too.

THE FATES OF COMETS

Comets are transient by nature — they come and go. Most of those that venture into the inner solar system make but a single pass before returning to the Oort cloud. Those that linger in short-period orbits have varying fates. Some are tossed out of the solar system altogether after passing near a massive planet. Others crash *into* a planet or satellite, or pass so close to the Sun that they vanish in a cloud of dust.

Those that do not go violently suffer a slow decline into oblivion. For a typical comet with a diameter of 2 km, a passage through perihelion causes the loss of its outer layers to a depth of about 3 m. Sooner or later, all the ices sublimate and are lost to space, leaving behind a mass of fluffy fragments. If the comet initially had a rocky core, the end result becomes a member of an extinct-comet class of asteroids. In fact, dynamicists estimate that roughly one-third to one-half of all asteroids that cross or approach Earth's orbit are extinct comets (see Chapter 6).

Even when the comets themselves are no longer observable, we see reminders of their existence. The smallest particles jetted free of the nucleus are blown in the antisolar direction by the Sun's radiation pressure and are eventually driven out of the inner solar system. However, somewhat larger particles, not as

Some Comets Associated with Meteor Showers

Comet	Period (years)	Associated meteor shower	Date of maximum	ZHR
1861 I (Thatcher)	410	Lyrid	21 Apr	15
1P/Halley	75.7	Eta Aquarid	5 May	35
1P/Halley	75.7	Orionid	21 Oct	30
109P/Swift-Tuttle	134	Perseid	12 Aug	80
21P/Giacobini-Zinner	6.6	Draconid (Giacobinid)	9 Oct	20
2P/Encke	3.3	Taurid	3 Nov	10
55P/Tempel-Tuttle	33.2	Leonid	17 Nov	15
3200 Phaethon	1.4	Geminid	13 Dec	90
8P/Tuttle	13.6	Ursid	23 Dec	10

Table 2. **Debris shed by a periodic comet can create a shower of meteors in our atmosphere each time Earth crosses the comet's orbit. The asteroidal object 3200 Phaethon does not display a coma or tail, though it occupies a elongated, cometlike orbit. ZHR stands for zenithal hourly rate, the number of meteors an observer would see per hour in a dark sky under ideal conditions.**

strongly affected by radiation pressure, continue to orbit the Sun. The sunlight they reflect is seen in the night sky as a faint glow called the zodiacal light. The largest solid particles are probably produced when a comet has spent a long time near the Sun. Gravitational perturbation disperses the fragments along the comet's orbit. If the path of this debris stream happens to intersect Earth's orbit, some of these remnants will strike our upper atmosphere, producing meteor showers (*Table 2*). Most of the debris responsible for meteors is believed to be light and fluffy (as in *Figure 19*). Fragments liberated from a comet's interior would perhaps be denser, but no meteorite in our possession has characteristics that appear "cometlike."

Because we still have so much to learn about comets, we intend to visit them again with spacecraft. Three such missions are currently under development. The first, NASA's Stardust mission, should pass within 100 km of the nucleus of Comet 81P/Wild 2 in 2004, taking detailed photographs of surface features and collecting a sample of the coma to be returned to Earth two years later in a sealed capsule. NASA's second mission, Contour (short for Comet Nucleus Tour), is an ambitious project to visit comets Encke in 2003, Schwassmann-Wachmann 3 in 2006, and d'Arrest in 2008. Contour will fly close to the nuclei of these well-known periodic comets, recording spectra and analyzing dust coming from them. Finally, the European Space Agency's Rosetta mission will rendezvous with Comet 46P/Wirtanen in 2011. Cameras will be used to select a suitable target site for the lander, called Champollion, to be launched separately for arrival in 2012.

The last few years have provided dramatic advances in the study of comets. However, the progress needs to be sustained by the continued collection and archiving of ground-based observations and by sending spacecraft like Stardust, Contour, Rosetta, and Champollion to additional comets. We can only hope that such efforts will continue the halcyon days that began with Halley's apparition and continued into the 1990s. Perhaps historians will someday look back on this period as the era when cometary studies came of age.

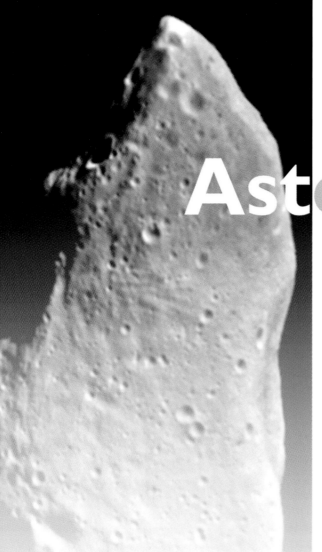

25

Asteroids

Clark R. Chapman

The asteroid 951 Gaspra, like a number of others, has now been studied by spacecraft at close range.

EARLIER THIS CENTURY, asteroids were dismissed as uninteresting planetary dregs or as "the vermin of the skies." However, we now realize they hold important clues concerning the nature of the planetary system and its earliest history. Moreover, public curiosity about asteroids, sparked by a hypothesis linking them to the dinosaurs' demise and the potential catastrophic end of our own civilization, has paralleled a developing appreciation among scientists of their relevance to the evolution of life on Earth and to the dawning age of human exploration and utilization of the planets.

The asteroids are a multitude of "minor planets" orbiting the Sun at distances ranging from inside the Earth's orbit to beyond Saturn's. They predominate in a large torus (the *main belt*), which is located well beyond the orbit of Mars and has a volume exceeding the sphere of interplanetary space inside Mars' orbit (*Figure 1*). Almost as many asteroids, called *Trojans*, orbit at Jupiter's distance from the Sun, roughly 60° ahead of or behind the giant planet. While individually small, asteroids are very numerous. By the late 1990s over 8,000 had been discovered, numbered, and named, and vastly more of similar sizes remain undiscovered. Additional objects, orbiting beyond Saturn (like 2060 Chiron) and in the Kuiper belt beyond Neptune, share the nomenclature of asteroids — even though they are more closely related to comets (see Chapter 5).

ORIGIN OF ASTEROIDS

Apparently, the asteroids and their precursors were planetesimals just like those growing elsewhere in the solar nebula during planetary accretion. Before they could form a planet, however, these bodies became gravitationally perturbed into tilted, elongated orbits. So instead of slowly accumulating into a single whole, asteroids began to strike each other at speeds of

338

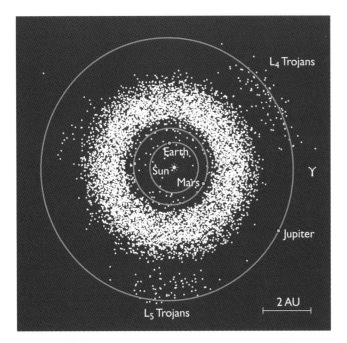

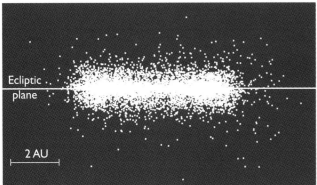

Figure 1. A computer snapshot of more than 8,000 numbered asteroids, positioned as they would have appeared as seen from above (upper panel) and along the ecliptic plane on 1 January 1998. The loose clusters situated ahead of and behind Jupiter along its orbit are the Trojan asteroids. Note that the inner margin of the main belt is not circular but instead concentric with the rather eccentric orbit of Mars.

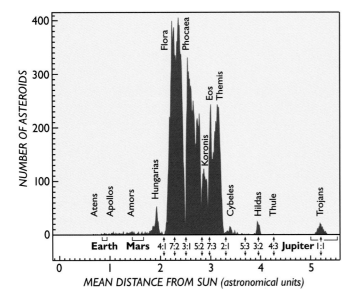

kilometers per second, often resulting in catastrophic fragmentation and disruption rather than coalescence. This process of hypervelocity collisional destruction continues today, though at a much diminished rate.

What kinds of gravitational perturbations affected planetesimals in the asteroidal zone more than those in other planetary zones? We cannot be sure, but two plausible scenarios have been advanced. One idea is that some large planetesimals in Jupiter's zone were gravitationally scattered by close approaches to that partly formed giant into eccentric orbits that penetrated the asteroidal region. They might have passed near most of the asteroids before once again encountering Jupiter and being ejected from the solar system. Close passes to asteroids may have stirred up their velocities and altered their orbits. Most early asteroids might have been destroyed either by collisions with such Jupiter-scattered planetesimals or collisions with each other.

An alternate idea holds that distant gravitational forces from Jupiter itself stirred up and depleted the asteroids, during the period when the Sun was divesting the primordial solar system of its nebular gases. We see today numerous spaces (the *Kirkwood gaps*) and other lacunae in distributions of asteroidal orbital elements that are due to orbital commensurabilities and resonances with Jupiter (*Figure 2*). Resonant effects occur at fixed locations today, but might have swept through the asteroidal region while the solar system was losing mass early in its history. Such resonant interactions with Jupiter might have pumped up asteroidal velocities. Either way, it seems likely that massive Jupiter was responsible for the absence of an asteroidal planet.

The reservoir of asteroids persists today, having endured these destructive and dissipative forces for billions of years. Small "leakages" do occur, however, creating stray bodies throughout the inner solar system. Some of these have orbits that come near or cross the Earth's, thus supplying the meteorites that fall from the skies. Presumably, a great many small bodies were left wandering throughout the solar system after the planets formed. Those in orbits that made close approaches to planets risked hitting them (the final dribble of accretion), falling into the Sun, or ejection from the solar system. Only in orbits far from any planet or strong gravitational resonances could remnant bodies have remained until now. The asteroid belt is one such place. Another may exist within the orbit of Mercury — if material ever condensed and accreted there in the first place and managed to survive collisional destruction. Other bodies, such as the Trojans, are protected in special resonant orbits.

Figure 2 (left). The population of the more than 22,000 known asteroids is shown for various distances from the Sun. Our census of asteroids is very strongly biased by observational effects: small, bright bodies orbiting near the inner edge of the asteroid belt are favored, while very dark asteroids farther out are underrepresented. Also indicated are locations where an asteroid's orbital period forms a simple ratio with Jupiter's and is subject to a gravitational resonance with the giant planet. Inside roughly 3.5 AU, these resonances are the locations of gaps in the asteroids' distribution, called *Kirkwood gaps*. One very apparent gap is located at about 2.5 AU, where an asteroid orbits the Sun three times for every revolution Jupiter makes. Beyond 3.5 AU, resonances correspond not to gaps but to isolated *groups* of asteroids, at ratio values of 3:2, 4:3, and 1:1.

CHAPTER TWENTY-FIVE

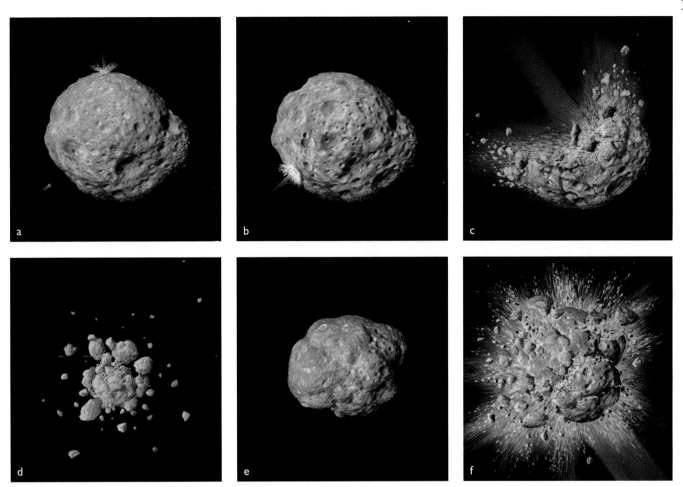

A final category of remnant planetesimals consists of comets, the subject of the preceding chapter. Some comets formed in locations like the Kuiper belt beyond Neptune. Those objects in transition from the Kuiper belt to eventual membership among the Jupiter-family comets are called Centaurs and are given asteroidal designations. Other planetesimals in the outer solar system were ejected from their place of origin by close planetary encounters. Those that did not escape the solar system have been preserved for eons in the Oort cloud, the deep-freeze of distant interplanetary space. Chance perturbations by passing stars and molecular clouds, followed by encounters with the outer planets, bring a few comets into the inner solar system each year. Here their ices sublimate, become ionized, and produce the flashy comas and tails that are their hallmarks. After a cosmically short time, a comet's volatiles are depleted, it becomes dormant, and eventually it dies. The remnant body is then termed an "asteroid."

As discussed more fully in Chapter 23, we can only speculate about whether asteroids and comets might originally have been the same type of planetesimal. Most asteroids remain in nearly their initial orbits, where any exposed ices would long since have sublimated away. Prior to entering the inner solar system, "new" comets are better preserved than asteroids, but the site of their origin has been lost due to their orbital wanderings. Comets probably formed with more ices and volatile compounds than did asteroids and certainly had a better chance of preserving them. While a few comets may have been implanted into the asteroid belt and some asteroids may have been ejected by Jupiter into

Figure 3. Stages in the fragmentation history of a moderately large asteroid. Originally composed of strong rock, the asteroid is cratered *(a,b)*, then catastrophically fragmented by a more energetic impact *(c)*. Most of the ejecta fail to reach escape velocity, and the body reassembles *(d)*. Later impacts further fragment the body, converting it to a gravitationally bound pile of boulders *(e)*. Finally *(f)*, a sufficiently gigantic collision completely disrupts and destroys the asteroid; its remnants then become scattered through space, forming an asteroid family.

comet-like orbits, the two classes of remnant planetesimals are fairly distinct. It is rather ironic that they are roughly equally represented among those bodies that venture near the Earth.

COLLISIONAL EVOLUTION

Since their formation, the dominant factor in the asteroids' evolution — by far — is what happens during those rare times when they encounter one another while hurtling through space. Asteroids are minuscule compared with the immensity of the torus of space through which they travel. Yet there are enough of them moving sufficiently fast that major collisions are inevitable during the lifetime of the solar system for all but a lucky few asteroids. The typical collision velocity is about 5 km per second, involving a projectile that is most likely negligibly small compared to the target. Laboratory experiments and simulations suggest that for the largest impacts a typical asteroid might experience, the total energy is much more than sufficient to fracture and fragment an object having the material strength of solid rock. The only bodies that might be expected to survive such collisions more or less unscathed are (1) those with a cohesive strength exceeding that of iron and (2) the very largest asteroids, whose interiors are strengthened by gravitational compression.

In rare "super-catastrophic" collisions, most fragments achieve the escape velocity of the target body and are lost to space. Unless an asteroid is quite small, lesser collisions probably cause fragmentation but not complete destruction. After such marginal collisions, the target object's gravity may keep most fragments from escaping, and they soon coalesce back into a single body (*Figure 3*). If a rapidly spinning asteroid were hit off center, the resulting angular momentum might be too great for a single body to form again, resulting instead in a binary or multiple system.

A super-catastrophic collision provides sufficient energy and momentum to disperse the fragments into independent, but still similar, heliocentric orbits. These pieces rarely or never meet again, and consequently the asteroid population gains a family of smaller members at the expense of the larger target body. Several such groupings of asteroids in similar orbits were discovered by the Japanese astronomer Kiyotsugu Hirayama about 80 years ago and are now called *Hirayama families*. Among the thousands of smaller asteroids discovered in the decades since, astronomers have recognized and tabulated several dozen families in all. The largest families are accompanied by distinctive bands of impact-generated dust detectable by spacecraft (*Figure 4*).

The spectra of members of a single Hirayama family offer clues to the interior composition of its now-shattered precursor asteroid. For example, objects belonging to the populous

Figure 4. Tenuous traces of asteroidal dust, here color-coded blue, were discovered in 1983 by the Infrared Astronomical Satellite. The broad bands straddle the ecliptic plane and correspond to doughnut-shaped clouds of dust particles associated with the Koronis and Themis families in the outer asteroid belt. According to theorists, the dust bands result from the gradual erosion by collisions among a family's members. The red (cold) emission at upper right is from distant interstellar clouds, and the thin line (arrowed) below the dust band is debris lying along the orbit of the periodic comet 10P/Tempel 2. North is up and east to the left.

Themis, Eos, and Koronis families have similar spectra, which implies that their respective precursors were homogeneous throughout. Other groupings exhibit a variety of spectral types, which are rarely easy to put back together into a single precursor body that makes physical and geochemical sense. This raises questions about (1) whether smaller families are truly collisional by-products, (2) the mineralogical interpretations for some spectral types, or (3) the geophysical and cosmochemical models assumed for asteroidal formation.

Our theoretical models now allow us to estimate confidently the frequencies and energies with which asteroids of different sizes collide with each other. However, we know less well how the energy and momentum in such huge collisions (beyond anything in our experience) are partitioned into fracturing the material, heating it up, ejecting fragments into space, and so on. The history of collisional evolution of asteroids must lie somewhere between the following two scenarios.

First, it may be that asteroids have been broken up into generations of successively smaller fragments, gradually grinding themselves down to meteoroids. Eventually they are ground to dust that is swept out of the asteroid belt, out of the solar system, or into the Sun. In this case, which assumes efficient conversion of impact energy into ejecta velocities, the present-day asteroids might represent a small remnant of a much larger earlier population. The larger asteroids still existing would be considered the lucky few that have by chance escaped destruction.

The second alternative, which increasingly seems to be the correct one, assumes that asteroids are comparatively difficult to fragment and disrupt into families. In this case the number of asteroids larger than 40 km in diameter has not changed much over the eons. However, individual asteroids have been damaged repeatedly, as depicted schematically in *Figure 3*. When such asteroids are smashed apart, the ejecta velocities are low and their fragments often reaccumulate into a gravitationally bound collection of rubble. Some of these asteroids will be spun up by repeated collisions or yield multiple-body systems following an energetic off-center impact. Small fragments may wind up as satellites orbiting the main, reaccreted body. In the most battered asteroids, their components have likely been strongly shocked and rearranged, making it more difficult to read the evidence about primordial events.

Small asteroids, as fragments of larger ones, may be little more than orbiting rubble piles. Indeed, radar observations of some small Earth-approaching bodies suggest that many of them, including 4769 Castalia (*Figure 5*) and 1620 Geographos (*Figure 6*), are compound or rubblized objects barely held together by gravity. In such cases, however, impacts may not be the cause. Instead, they may have been torn apart by internal tidal forces during a near-miss of Earth or Venus, after which they drew back together. An analogy would be how an individual fragment of Comet Shoemaker-Levy 9 reaccreted following its tidal disruption by Jupiter in 1992.

Our spectral insights are sometimes confounding. It remains a puzzle how one very large asteroidal precursor — the *parent body* of the large, apparently metallic asteroid 16 Psyche — could have been stripped to its core, while the similar-size object 4 Vesta has apparently preserved its thin basaltic crust nearly intact. Although the collisional environment over 4½ bil-

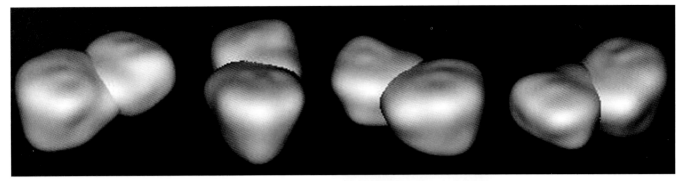

Figure 5 (above). These are computer-reconstructed models of what apparently is a small, double asteroid, designated 4769 Castalia. They are derived from the echo delays and frequency shifts introduced when radar pulses from Arecibo Observatory were bounced off the asteroid during its close approach to Earth in August 1989. The two blobs, which rotate around each other every 4 hours, appear distinct in the radar echoes, though Castalia may simply have a strongly dumbbell shape.

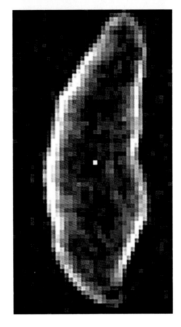

Figure 6 (left). The cigar-shaped silhouette of 1620 Geographos, as determined from radar observations made in August 1994. The asteroid measures 5.1 by 1.8 km, making it the most elongated object known in the solar system. This appearance may have resulted when Geographos came very close to a planet. Computer models show that tidal forces during such passages would stretch and distort a rubblized body into this shape.

lion years should have been the same for both, Psyche's exposed core implies torrential bombardment rates, whereas Vesta's crust has endured only moderate damage. Possibly, there was an early but short-lived period of bombardment by a population of large (cometary?) projectiles, which by chance struck only some asteroids. Those lucky enough to escape intact became reservoirs for most of the asteroid materials that remain today.

This "battered-to-bits" scenario, dominated by large projectiles that are no longer common, may explain some other puzzles. For example, iron meteorites appear to represent pieces of the metallic cores of more than 60 separate differentiated parent bodies, all of which must have been shattered at some point. What became of the mantles of those bodies, which would have consisted mostly of the mineral olivine? Few olivine-dominated asteroids remain today. Were all the others destroyed by later collisions? This speculative scenario remains the subject of continuing research.

The asteroids continue to collide with each other today. Smaller impacts crater and crack their surfaces, which are gradually covered over again by regional or even global blankets of debris from the larger cratering events. Infrequent large impacts destroy smaller asteroids or reassemble the configurations of larger ones. Even rarer super-catastrophic collisions yield new Hirayama families; if some of these fragments are sprayed into resonant orbits, they can be quickly perturbed into elliptical paths that cross the orbits of other planets. Such small asteroids and meteoroids lead a transitory existence lasting only a few million years before they strike the Sun or a planet or become gravitationally yanked into a radically different trajectory.

PHYSICAL CHARACTERISTICS

This collisional and dynamical evolution is just the tail end of the accretionary processes that gave rise to the terrestrial planets earlier in the solar system's history. In all probability, asteroids are cold, dead, airless bodies whose destinies have been shaped solely by external forces since the first few tens of millions of years of solar-system history. They have thus been spared the

processes that have virtually destroyed records of primordial history on the Earth, Moon, and other large bodies. A large planet experiences high temperatures, pressures, chemical alteration, and crustal motions in response to vast reservoirs of primordial and ongoing radioactive heat. Yet the large surface-area-to-volume ratios of the comparatively tiny asteroids dictate that they lose internal heat to space rapidly.

Nevertheless, clues from meteorites suggest that many asteroids were heated at such prodigious rates during their first few million years that they were warm enough for liquid water to permeate their interiors. Some were heated to metamorphic temperatures and, indeed, some even underwent partial melting, causing internal segregation of materials — the heavy metals sinking and the more buoyant lavas erupting onto their surfaces. In order to understand these primordial processes, and to sort out the truly primitive planetesimals from those that underwent early modification, we must investigate what asteroids are like.

We turn first to the astronomical evidence, gleaned from study of the time-variable behavior and spectral characteristics of reflected and emitted radiation coming from their distant, starlike images. Asteroid sizes are now well established (*Figure 7*). The largest, 1 Ceres (discovered in 1801), is about 930 km across and constitutes more than a quarter of the mass of all main-belt asteroids combined. The next largest, 2 Pallas and 4 Vesta, are each a bit over 500 km in diameter. Still smaller ones

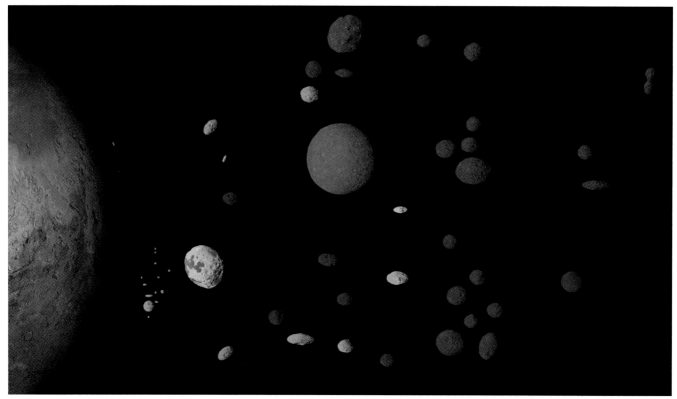

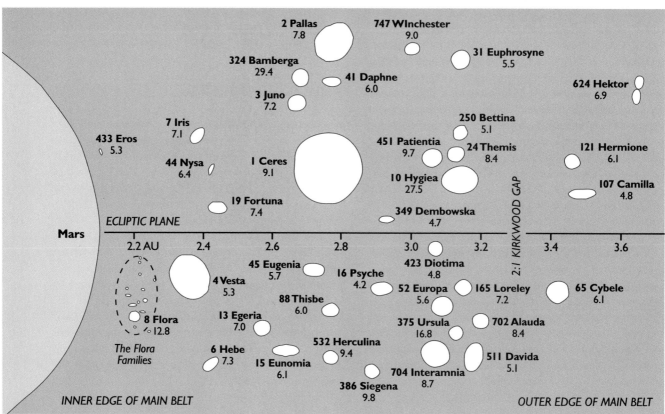

Figure 7. This representation of the physical properties of interesting asteroids includes most of the asteroids larger than about 200 km in diameter. They are portrayed with their correct colors, albedos, and relative sizes and shapes (the limb of Mars is shown to scale for size comparison). The bodies are positioned at their correct relative distances from the Sun. Asteroids located near the top or bottom of the diagram occupy relatively eccentric or inclined orbits (or both), while those shown near the ecliptic plane move in relatively circular, non-inclined orbits. Rotation periods, in hours, are given in the lower panel. Among the special smaller asteroids indicated are members of the Flora families larger than 15 km in diameter, but this illustration would be hopelessly cluttered if all asteroids of comparable size were shown — an estimated 1,150 asteroids in the main belt alone have diameters larger than 30 km, yet only five Flora family asteroids attain that size. Note the possible contact-binary Trojan asteroid 624 Hektor at upper right.

are increasingly numerous, grading down to countless kilometer-size asteroids and smaller boulders too small to detect (unless they happen to pass very close to the Earth).

Asteroid shapes and configurations are a matter of great interest. They generally range from roughly spherical to elongated and irregular, bespeaking a fragmental origin. This is inferred from variations in asteroids' brightness as they spin. Typical rotational periods are about 9 hours, with extremes from under 3 hours to several weeks. As described earlier, the shapes of many asteroids may reflect a "rubble-pile" internal structure, resulting from their collisional history. Some asteroids may even have a binary or compound structure.

Although they are difficult to detect with ground-based telescopes, small satellites may commonly orbit asteroids. The one unambiguous case is Dactyl, a 1.5-km-wide object revealed by the Galileo spacecraft to be orbiting 243 Ida. Suggestive evidence of other such situations comes from asteroidal light curves that mimic eclipsing binaries and from unexpected blinkouts of stars when asteroids pass by. A few asteroids, like 288 Glauke (which takes 48 days to rotate once), have extraordinarily long rotation periods. It is very improbable that they could have been almost "stopped in their tracks" by a chance collision. One idea is that spins might somehow be slowed by tidal drag of a satellite. But this theory fails for 253 Mathilde, which has a 17-day period but was found to lack satellites when seen at close range. Direct optical searches, using coronagraphic (masking) techniques and adaptive optics, have hunted without success for asteroidal satellites. Hence, we cannot yet estimate what fraction of asteroids are binary or multiple systems.

COMPOSITIONS

Asteroids differ significantly in color. Nicholas Bobrovnikoff made this important discovery some 70 years ago at Lick Observatory. Yet the full import of his revelation was not appreciated until the 1970s, when asteroid spectrophotometry blossomed. Spurred by new detector technology, astronomers have recorded spectra of sunlight reflected from asteroids ranging from ultraviolet wavelengths out to the mid-infrared, where heat radiated from the warm surface begins to overwhelm the corresponding reflected component of sunlight. Not only do the visible colors of asteroids differ, but many of their spectra exhibit absorption bands that are due to different minerals and hydrated compounds. Some examples appear in *Figure 8*. The diversity of colors and spectra implies different surface compositions. If we could relate these derived mineralogies to the different types of meteorites being studied by cosmochemists (see Chapter 26), then we could tie the meteorites' implications about primordial environments and events to bodies in particular parts of the solar system.

There are two chief types of meteorites. Those whose non-volatile chemical elements occur in roughly "cosmic" abundances are inferred to be relatively unaltered condensates from the primordial solar nebula. Those highly enriched or depleted in certain elements appear to have been created from material greatly modified by processes of "planetary" evolution. One such process, called differentiation, was the melting and physical segregation that occurred early in Earth's history; metals

sank to form a core, and lighter materials floated and cooled to form the crust (*Figure 9*). Most primordial types of meteorites are called chondrites, while most of the geochemically altered ones occur in the iron, stony-iron, and achondrite classes.

An important conclusion about asteroids is that both the primordial and altered mineralogies are represented in the main-belt population. At least one large asteroid, 4 Vesta, has a surface mineralogy similar to basaltic achondrites — meteorites physically and chemically similar to the basaltic lava flows common on the surfaces of the Earth and Moon. Many other asteroids are extremely black and show infrared absorption bands indicative of the carbon-rich, hydrated mineralogy of the most primitive meteorites — the carbonaceous chondrites. Therefore, some asteroids apparently melted and became geochemically differentiated, just like the larger terrestrial planets, while in others the initial chemistry has been preserved more or less intact with only modest warming. It remains a profound mystery how some asteroids could have been so dramatically altered while others of similar size in nearby orbits escaped major modification.

The simple fact that some small asteroids melted after they formed establishes important constraints on thermal conditions and processes operating in the early solar system. It had been supposed that the high surface-area-to-volume ratio of small bodies would have allowed any internally or externally generated heat to radiate away quickly, keeping them relatively cool. But evidently some bodies became much warmer than can be ascribed to this traditional view. More exotic thermal sources must be considered, including intense pulses of heat due to the decay of short-lived radionuclides. For example, a nearby supernova explosion probably infused the solar nebula with copious amounts of aluminum-26 (and may have even triggered the onset of the solar system's formation). However, in order for the heat from its decay to melt an asteroid, aluminum-26 had to be incorporated into an accreting body within just one or two million years of its synthesis, before it had decayed away. Thus the telescopic detection of basalt absorption bands in Vesta's spectrum, an indication of igneous activity, supports the notion that our planetary system formed with unexpected rapidity.

Such ground-based determinations of asteroids' colors and albedos, augmented by data from the Infrared Astronomical Satellite, which measured thousands of their reflectivities and sizes, have been used to group the asteroids into taxonomic classes. Members of each class probably have roughly similar, though not identical, surface mineralogy (*Table 1*).

More than three-fourths of the asteroids are extremely dark (with typical geometric albedos of 3.5 percent) and, so far as we can tell, have mineralogies analogous to carbonaceous chondrites. There are significant spectral variations among these so-called C-type asteroids and their rarer siblings, the B-, F-, and G-types. About two-thirds of them appear hydrated, probably indicating that they were heated to the point that liquid water soaked their minerals, as is commonly noted in carbonaceous meteorites. The dry C-types may never have had water percolate through them, may have been heated so much that any hydrated minerals reverted or metamorphosed, or may never have contained much H_2O to start with. Two classes of very dark asteroids show reddish colors. Such P- and D-type spectra are common for bodies near and beyond the outer edge of the

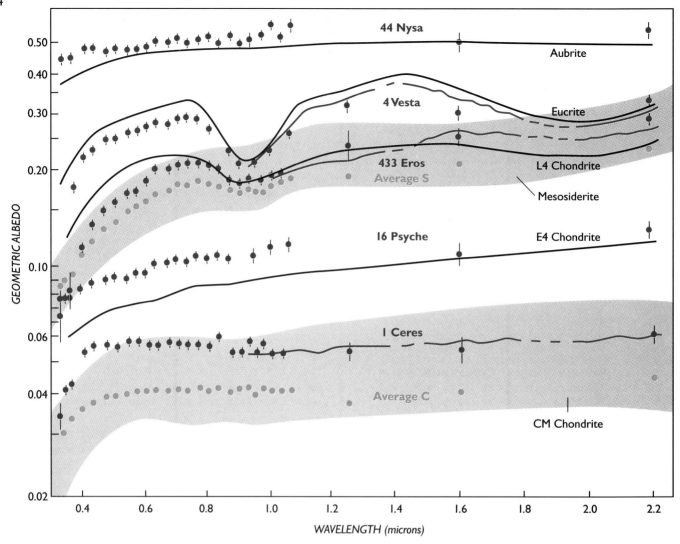

Figure 8 (above). The principal means for deducing the compositions of distant asteroids is to compare their reflectance spectra with those of meteorites, seen here for visible through infrared wavelengths. Asteroid data, in shades of reddish brown, consist of points with error bars (from filter spectrophotometry), lines (from Fourier spectroscopy), and shaded circles (average values for the S- and C-type asteroids). Laboratory measurements of meteorite powders are reproduced in shades of gray; two classes occupy the ranges of values indicated by wide bands. Investigators deduce surface mineralogy primarily from the shapes of these curves rather than from the objects' precise albedos. Evidently, the diverse mineral assemblages found in our meteorite collections are also represented in the asteroid belt.

Figure 9 (below). A schematic representation of successive stages in the evolution of an asteroid that is heated early in its history. The original body of primitive composition *(left panel)* is heated to the point that constituent iron separates and sinks to its center, forming a core *(middle).* Partially melted rock from the mantle floats upward through cracks in the crust, erupting onto the surface as basaltic lava flows. As heat radiates away, the body cools, the iron solidifies, olivine accumulates in the deep interior, and crustal magmas solidify. Repeated collisions fragment the mantle and crustal rocks into a "megaregolith" and ultimately eject the rocks, exposing the iron core and any embedded rocks *(right).* Most asteroids were not heated much beyond the first stage; 4 Vesta reached stage *2*, but was not fragmented thereafter. Some M- or S-type asteroids *(right)* may be the parent bodies for iron and stony-iron meteorites.

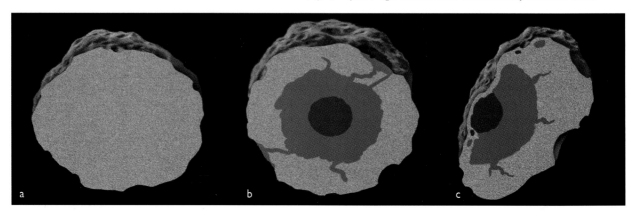

Asteroid Characteristics by Type

Type	Albedo	Reflectance spectrum	Meteorite analog(s)
C	0.03–0.07	Fairly flat longward of 0.4μ; UV and sometimes 3μ absorption bands	Carbonaceous chondrites (CM)
B	0.04–0.08	C-like, but slightly brighter and more neutral in color	Carbonaceous chondrites (?)
F	0.03–0.06	Flat (neutral color) with no ultraviolet absorption	?
G	0.05–0.09	C-like, but brighter and with very strong ultraviolet absorption	Carbonaceous chondrites (?)
P	0.02–0.06	Linear and slightly reddish (like M, but with very low albedo)	None
D	0.02–0.05	Redder than P's, especially longward of 0.6μ; very low albedo	None (kerogens?)
T	0.04–0.11	Reddish, esp. at shorter wavelengths; intermediate between D and S	?
S	0.10–0.22	Reddish shortward of 0.7μ; weak to moderate absorptions near 1μ and 2μ	Stony-irons and ord. chondrites
M	0.10–0.18	Linear and slightly reddish (like P, but with moderate albedo)	Irons; enstatite chondrites
E	0.25–0.60	Linear, flat or slightly reddish	Aubrites
A	0.13–0.40	Strong absorptions in UV and near 1.1μ due to olivine	Brachina
Q	moderate	Like S, but with stronger absorptions	Ordinary chondrites (unweathered)
R	mod. high	Like S, but with stronger absorptions (particularly due to olivine)	?
V	mod. high	Like S, but with stronger absorptions (particularly due to pyroxene)	Basaltic achondrites

Table 1. Astronomers find that asteroids exhibit a number of characteristics that can be used to subdivide them into taxonomic classes. The most important of these are the shape and slope of their reflectance spectra (the listed albedos are typical but do not define the classes). Several other types have been defined, but these are the most notable ones. The symbol μ stands for microns.

main belt, including many Trojans and some of Jupiter's small outer satellites (see Chapter 23). We can speculate that these colors may be due to "ultraprimitive" organic compounds, perhaps like the non-icy components of comets.

Roughly one-sixth of the asteroids have moderate albedos (typically 15 to 20 percent) and reddish colors. The spectra of such *S-type* asteroids imply assemblages of iron- and magnesium-bearing silicates (pyroxene and olivine) mixed with pure metallic nickel-iron. Unfortunately, these spectral characteristics are shared by two radically different types of metal-bearing stony meteorites: (1) the ordinary chondrites, which are metamorphosed but otherwise relatively unaltered primitive objects believed to have formed closer to the Sun than the carbonaceous chondrites; and (2) the stony-iron meteorites, which are enriched in metal and other compounds due to extensive melting and geochemical fractionation within their parent bodies. The most straightforward interpretation of S-type reflectance spectra, especially for a handful of well-observed objects like 8 Flora, suggests that their metal content is higher than that in ordinary chondrites, that their olivine content is also a bit higher, and that other mineralogical inconsistencies exist.

However, if all S-types were interpreted as stony-irons, we would be left with no main-belt parent bodies for the abundant ordinary chondrite (OC) meteorites. It had been hypothesized that these might be derived from small, faint (and thus unsampled) asteroids. However, recent spectral surveys of faint main-belt asteroids by Richard Binzel and his collaborators have failed to find OC-like bodies. Notably, Binzel's spectra of even smaller Earth-approaching asteroids reveal a continuum between S-types and those with ordinary-chondrite spectra, termed *Q-types*. Perhaps some property of the fragmented surface debris (regolith) on larger, older asteroids is OC-like material that has been modified by some unknown process to yield S-type spectra. (As described later, evidence does suggest that some kind of "space weathering" is occurring on S-type asteroids over time.) The smallest asteroids are relatively young fragments with little gravity to retain regoliths. Thus, the Q-types may simply be large OC meteorites with unmodified spectral properties.

A third spectral grouping is also compositionally ambiguous. So-called *M-type* asteroids have moderate albedos. Their spectra exhibit the characteristics of metallic nickel-iron, with no hint of absorption bands from silicate minerals. Many of these objects may well be wholly metallic, perhaps the remnant cores of differentiated precursors stripped of their rocky mantles and crusts by collisions. However, this same spectral signature is shared by enstatite chondrites, stony meteorites that consist of flecks of nickel-iron embedded in a clear matrix of the magnesium-rich silicate mineral enstatite. These primitive meteorites formed in a highly reducing (oxygen-poor) environment, perhaps close to the Sun or within a protoplanet. Since enstatite is colorless and lacks absorption bands, an enstatite-chondritic asteroid would have a spectrum just like that of a nickel-iron asteroid. Radar soundings appear to resolve this ambiguity: the M-type objects 16 Psyche in the main belt and two small near-Earth bodies are unusually metal-rich. However, a few M-types exhibit spectral hints of hydrated minerals, a characteristic not typical of either iron-nickel meteorites or enstatite chondrites. So the M class may contain objects with a variety of compositions.

In addition to the common types described above, observers have identified a number of rarer low-albedo asteroid classes (*Table 1*). In addition, several percent of asteroids are oddballs of one sort or another, like olivine-rich 349 Dembowska. As smaller, fainter asteroids are surveyed, more have been found with this same pure-olivine (*A-type*) spectrum. Several dozen small asteroids have reflectance spectra similar to that of Vesta. Designated *V-types*, these might be fragments of Vesta's crust ejected during a major crater-forming event (*Figure 10*).

Only within the last quarter century have researchers realized that the compositional distinctions among asteroids show a remarkably distinct distribution with distance from the Sun

(*Figure 11*). While examples of most types span a large range of solar distances, there is a clear progression from high-albedo type E to S in the inner belt, then to M and C in the middle and outer belt. Members of the P class occur mostly near and beyond the belt's outer edge, and D-types dominate among the Trojans. This gradation with heliocentric distance likely reflects properties of the primordial solar nebula, with high-temperature minerals preferentially condensing nearer the Sun and cometlike mixtures of ices and organics farther out. However, the variation could conceivably reflect subsequent evolutionary processes. For example, collisions among asteroids are more energetic closer to the Sun. Electrical induction from an early, intense solar wind (a hypothesized alternative to aluminum-26 heating) would have been more effective closer to the Sun as well.

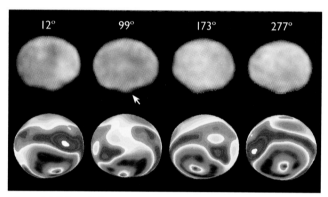

Figure 10. Yellow-light images of 4 Vesta obtained in May 1996 with the Hubble Space Telescope reveal a very large, circular cavity (arrowed) near the asteroid's south pole. Labels denote the central longitude in each panel. The accompanying colored maps show that the 460-km-wide basin has a pronounced central peak and that its floor (deep blue) lies 12 km below ground level. The impact may have excavated enough crustal materials to form a widely scattered family of "Vestoids" — though it remains to be understood how objects 5 to 10 km in diameter could be ejected at such high velocities.

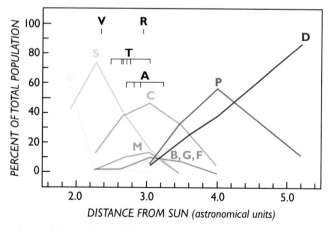

Figure 11. Asteroids of different compositions are systematically distributed with respect to their distance from the Sun. A telescopic survey made in eight colors and data from other observational programs have been corrected to eliminate biases against dark, fainter asteroids. This makes it possible to derive the fraction of asteroids above a certain size (determined by observational limitations and varying with distance) within each class. The letter designations, which refer to different spectral types, are summarized in *Table 1*.

GEOLOGY FROM SPACECRAFT

The preceding sections describe nearly everything known about asteroids based on telescopic observations alone. However, during the 1990s our knowledge of these bodies grew tremendously — indeed, asteroid science finally came of age — when the first spacecraft encounters took place. Ironically, our first close-up data came from Galileo, a mission dedicated to studying the Jupiter system. Since it was well equipped to evaluate the geology and mineralogy of planetary surfaces, Galileo was directed to make serendipitous visits to the S-type, main-belt asteroids 951 Gaspra and 243 Ida.

A third encounter involved the Near Earth Asteroid Rendezvous (NEAR) spacecraft. The trajectory to its primary target, 433 Eros, carried NEAR past 253 Mathilde, a C-type object and thus a representative of the most common type of asteroid. Mathilde's numerous huge, gaping craters (*Figure 12*) exceed the dimensions, relative to the target body, of craters observed on any other object in the solar system. For its size Mathilde must contain relatively little mass, because NEAR's trajectory was barely altered despite the close flyby. All indications point to a body whose overall density is about 1.3 g/cm^3. Mathilde would thus seem an excellent candidate for rubble-pile status. Unfortunately, neither the huge craters nor the low density nor anything else in the Mathilde data explain the anomaly of its exceedingly slow, 17-day rotation period.

Galileo's flyby of the S-type asteroid 951 Gaspra occurred on 29 October 1991. Gaspra is a member of the populous Flora family, which occupies the inner portion of the main belt. Measuring 18 by 11 by 9 km, Gaspra is stunningly angular and nonspherical (*Figure 13*). Although some views of it make it look as though it is composed of two distinct objects crammed together, other views emphasize planar surfaces, suggesting that portions of the original body were sheared off by large impacts. At the limit of Galileo's image resolution, Gaspra shows evidence of crustal cracks. Similar features appear prominently on the Martian satellite Phobos, where they are thought to result from tidal and impact forces.

The most striking feature of Gaspra is its population of craters. Superficially, its surface resembles those of other planets and satellites. However, closer inspection shows that it is peppered only with small craters; those of medium and large size are essentially absent. The ratio of small-to-large craters is high — but no higher than that for the relatively young maria of the Moon. The surprising thing is that we expected Gaspra to be saturated with craters of all sizes, since the cratering rate in the asteroid belt is so much higher than on the Moon. Either Gaspra's surface is very strong (made of metal), so that even large projectiles produce only small craters, or else Gaspra is too young for larger craters to have accumulated.

Galileo researchers were intrigued to find that ridges on Gaspra were slightly bluer than the flat "lowlands" between them. A few small craters also look slightly bluish. One tentative explanation is that loose material migrates downslope, in which case the freshly exposed surfaces on the ridges, as well as young craters, reveal Gaspra's true colors, whereas some kind of "space weathering" has gradually reddened the surface elsewhere over time. These tentative conclusions, based on features near the resolution limit of the Gaspra data, were strongly reinforced when Galileo flew by the larger S-type asteroid, 243 Ida, on 28 August 1993.

Even more than Gaspra, Ida's croissant-like shape, 56 by 15 km, and narrow central waist suggest that it is two objects combined (*Figures 14,15*). Indeed, one half is more heavily cratered than the other. Unlike Gaspra, Ida's surface is essentially saturated with a mix of large and small craters, making it look rather like Phobos or the surface of the Moon. Galileo's sharp pictures reveal not only linear grooves, as on Gaspra, but also a scattering of large boulders or blocks several tens of meters across. These and other characteristics have led Galileo geologists to conclude that Ida, again unlike Gaspra, is apparently covered with a deep layer of regolith. Indeed, Ida could well be essentially a "megaregolith" throughout — a real-world example of one of the rubble piles imagined by theorists.

An unexpected discovery was a spherical, 1.5-km-wide moonlet, later named Dactyl, orbiting Ida. Although asteroidal satellites had been suspected from some unconfirmed Earth-based observations during the 1980s, their reality was widely debated. The probability of satellites forming, either originally or following collisions, was deemed low. Conceivably Dactyl could be a rare anomaly. But its existence, together with recent computer simulations of asteroid collisions, now suggests that small satellites might be fairly common.

Dactyl provided another bonus: the opportunity to use its orbital period, together with Kepler's Third Law of motion, to determine Ida's mass and, in theory, its bulk density. Thus, the potential existed for resolving the "S-type conundrum," at least for one such asteroid. A density near 5 g/cm^3 would confirm the longstanding interpretation of S-types as stony-iron objects; a value near 3.5 g/cm^3 would imply that S-types are space-weathered ordinary chondrites. If Ida contained internal voids, as would be expected for a rubble pile, then the densities might be less — as low as 3.5 for a stony-iron and perhaps 2.5 for an ordinary chondrite.

Ida's volume proved to be 16,000 ± 2,000 km^3, a fairly accurate determination. Unfortunately, the orbit of Dactyl was, coincidentally, almost edge-on to Galileo's view, which prevented a unique determination of its orbital period. Nevertheless, the little moon could not be too tightly bound to Ida, because it would soon collide or be deflected by Ida's irregular gravity field into a much different orbit. The orbit could not be distant from Ida, because escape would then be likely. These constraints limit the bulk density of Ida to the range of 2.0 to 3.1 g/cm^3, strongly favoring the ordinary-chondrite hypothesis. Yet we have no way to know how much of Ida's interior is empty space, so a stony-iron composition cannot be completely ruled out.

Fortunately, color images appear to have settled the matter. Spectrally, most of Ida's surface resembles a typical S-type asteroid, as anticipated from ground-based observations. However, small regions are less red and show much more prominent absorption bands due to pyroxene and olivine — spectral characteristics approaching those of ordinary chondrites. These anomalous regions are located where small, deep craters have recently penetrated Ida's surface. Others appear to be patches of material ejected from a large, fresh crater named Azzurra. In other

Figure 12. The NEAR spacecraft obtained close-up views of the C-type, main-belt asteroid 253 Mathilde during a flyby in June 1997. The asteroid is 60 km long, and its shape is dominated by the rims of huge craters. On the hemisphere seen in sunlight during the brief encounter are at least four craters with diameters exceeding the average radius of Mathilde itself — a dense profusion of large craters unmatched among the planets, satellites, and asteroids imaged to date by spacecraft.

Figure 13. The angular, 18-km-long asteroid 951 Gaspra was the first ever to be encountered by a spacecraft. Note its numerous small craters and the dearth of moderate- or large-sized ones, which are more abundant on the asteroids Mathilde and Ida.

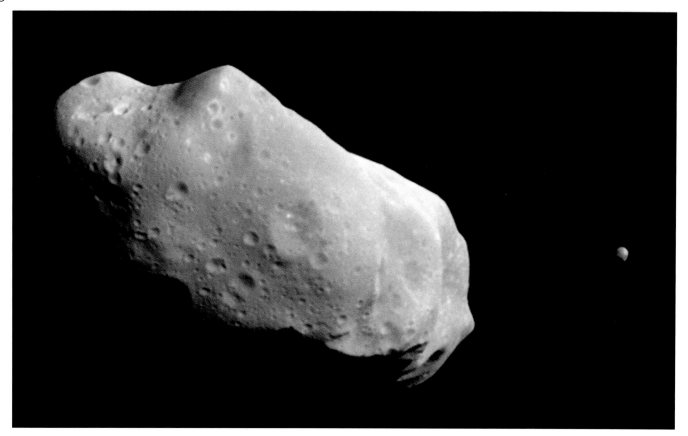

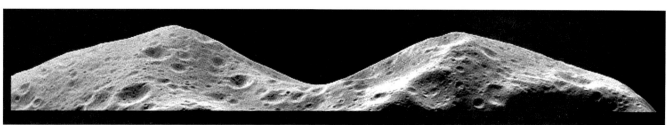

Figure 14 (upper panel). A false-color representation of 243 Ida and its moonlet, Dactyl, as recorded during Galileo's 1993 flyby. Note that the bluer patches often correspond to fresh, small craters. The blue patch on the limb is probably ejecta from a large, recent crater — named Azzurra — on the other side. Ida is 56 km long.

Figure 15 (lower panel). The Galileo spacecraft's highest resolution image of Ida shows a portion of its irregular limb.

words, recently exposed material bears a spectral resemblance to ordinary chondrites, but older terrains have grown redder and adopted the spectral characteristics of typical S-type asteroids. There can be little doubt that some form of "space weathering" is gradually modifying how mineral grains reflect sunlight. Most likely, hypervelocity micrometeorites are striking Ida and altering the optical properties of the outermost regolith grains. Similar color changes have been observed on meteorite samples subjected to simulated micrometeorite impacts (laser pulses).

This convergence of data about Gaspra and Ida, though tentative, yields an inherently satisfying conclusion: S-type asteroids, which dominate the inner main belt, could be composed of the minerals in ordinary chondrites, Earth's most abundant meteorites. Of course, spectroscopists have identified a wide variety of S-type asteroids, so S-types may actually include both ordinary-chondrite *and* stony-iron objects, as well as other non-carbonaceous mineral assemblages.

The space-weathering hypothesis will be tested, once again and in more depth, when NEAR goes into orbit around 433 Eros, yet another S-type body. Once again (but at much higher resolution), we will see if fresh bedrock looks like OC material and ages into the S-type spectrum typical for Eros. However, NEAR also carries gamma-ray and X-ray instruments that can directly measure the chemistry and mineralogy of surface materials. Furthermore, NEAR — as an orbiter — can obtain a bulk density for Eros that should rule out one of the two principal hypotheses for S-type materials. Of course, the "S-type conundrum" will not be fully resolved until our spacecraft have visited other S-type asteroids besides Eros.

EARTH-APPROACHING ASTEROIDS

NEAR's primary objective, 433 Eros, is more than just a random, accessible asteroid. It is an Earth-approaching asteroid or, more specifically, an *Amor:* an object whose orbit comes within 1.3 AU of the Sun. Asteroids with orbits that actually overlap Earth's are called *Apollos.* Over 400 Apollos and Amors have been discovered so far, as well as two dozen *Atens*, which have semimajor axes of less than 1 AU. There may be nearly 2,000 Apollo-Amors, 1 km or greater in diameter, capable of crossing the Earth's orbit.

Until the 1980s, it was a mystery how to get much material from the main belt into Amor- or Apollo-type orbits. Dynamicists had explored the complexities of resonances, collisions, and perturbations by Mars or Jupiter, but these all fell short of providing the observed influx of small asteroids and meteorites. Then theorist Jack Wisdom discovered that the mathematics of chaos, applied to asteroids orbiting near the 3:1 Kirkwood gap (*Figure 2*), readily explains the extrication process. He showed that such orbits undergo dramatic changes at unpredictable times, inducing sudden increases in eccentricity that permit them to cross the orbit first of Mars, then of Earth. Most specialists now believe that the most common meteorites — the ordinary and carbonaceous chondrites, as well as basaltic achondrites — are derived from S-, C-, and V-type asteroids, respectively, lying near the 3:1 Kirkwood gap.

Asteroids perturbed into planet-crossing orbits are transient objects, typically lasting for only a few million years before crashing into the Sun, striking a terrestrial planet, or being ejected from the solar system after getting too close to Jupiter near their aphelia. While most Earth-crossers are believed to be fragments of main-belt asteroids, derived (like smaller meteorites) via resonance-driven escape hatches, some must be dead or dormant short-period comets. Observational surveys show that, apart from differences plausibly ascribed to their smaller sizes, the Earth-crossers have colors indistinguishable from those of main-belt asteroids near the 3:1 resonance. If comets generally resemble C- or P-type asteroids in color, as suspected, then up to a few tens of percent of the asteroids in Earth's vicinity *could* be of cometary origin.

Recent discoveries of very small Earth-approaching asteroids (chiefly by the Spacewatch program at Kitt Peak, Arizona) suggest that there may be an unexplained "excess" of objects with diameters near 10 m. In other words, while we expect to find scores of many small objects for every large one, the ratio of 10-m to 100-m objects seems surprisingly high; it is much larger for example than the ratio of 10-km to 100-km asteroids. Some astronomers have hypothesized that a unique population of very small asteroids orbits preferentially close to the Earth. However, others counter that the observed size distribution parallels the relative numbers of lunar craters that would be made by objects between 100 m and 1 km across. The same crater distribution has been found on Mars and on the asteroid Gaspra, so the supposed excess of 10-m objects is likely a more general characteristic of interplanetary debris, at least in the inner solar system. Why so many of them exist at all, however, is not well understood.

There are further questions about whether the Earth-crossers' distribution of sizes and types is the same now as it was in early post-primordial times, or whether the mix varies over time. Undoubtedly, Earth occasionally experiences "showers" of asteroids. These bombardment episodes would follow a catastrophic collision of two large objects near a resonance escape hatch in the main belt, or the collisional fragmentation of an asteroid already in an Earth-approaching orbit. The compositions of the bodies involved in such random collisions would presumably affect the relative proportions of meteorite types falling to Earth thereafter. However, asteroid (and comet) showers probably heighten the infall rate only modestly, and changes in the mix likely cannot happen much more rapidly than the million-year dynamical lifetimes of Earth-crossing objects.

The earliest populations of remnant planetesimals delivered a final veneer to the nearly completely formed planets and may have contributed much to their crusts and hydrospheres. There have been conjectures about whether volatile-rich comets and asteroids contributed water and organic compounds, necessary for life to get started on Earth. Certainly, they contributed many *craters*, which remain the dominant topographic features on all but the most geologically active terrestrial bodies (see Chapter 6). Crater-forming impacts have probably affected and even interrupted the evolution of life numerous times over the last half billion years. One well-documented episode occurred 65 million years ago, an impact powerful enough to excavate a 180-km-wide crater in southern Mexico and to trigger a worldwide conflagration and mass extinction. It is unknown whether an asteroid or a comet was responsible for this cataclysm. However, the resulting demise of the dinosaurs opened an opportunity for mammals, and ultimately human beings, to flourish on Earth.

It is also uncertain what role asteroids may play in our future. Impacts on Earth large enough to threaten the future of civilization may occur every several hundred thousand years, equating to one chance in several thousand sometime during the 21st century. We cannot yet know whether a dangerous asteroid will strike Earth during the next 100 years — because the next collision will most likely involve an object that we have not yet discovered. For that reason, many astronomers support the full implementation, and completion, of an observational survey to identify nearly all Earth-approaching asteroids larger than 1 km across.

If the survey is funded and the pace of discovery quickens, there may be many cases of a period of uncertainty, during which a future impact with Earth seems possible, before improved observations and calculations can prove that a newly discovered asteroid will not impact our planet. Such an interval may last just one day, as happened with 1997 XF_{11} in March 1998, or even a year or two. Once 90 percent of the civilization-threatening objects are discovered and put in the "safe" category, the odds of being suddenly struck by an unknown asteroid during the next year will have been reduced from one chance in several hundred thousand to only one chance in several million.

Eros cannot strike Earth in the immediate future because it doesn't come close enough to the Sun to actually intersect our orbit. However, recent computer integrations show that the perihelion of Eros's orbit wanders in and out, as do the perihelia of many other Earth-approachers. There is roughly one chance in 10 that, some millions of years from now, Eros will end its existence by crashing into the Earth. Having a diameter of 35 km, Eros is much larger and more massive than whatever struck Earth 65 million years ago. The consequence of that unfortunate outcome seems fairly obvious: life would cease to exist here.

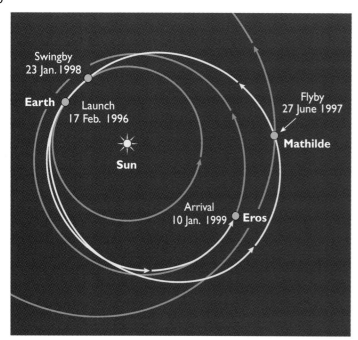

Figure 16. The Near Earth Asteroid Rendezvous (NEAR) spacecraft makes a 3-year cruise through interplanetary space before reaching its primary objective, 433 Eros. In 1997 NEAR encountered and made brief studies of 253 Mathilde, a C-type asteroid.

Table 2. Well over 20,000 asteroids have been found in the past two centuries. These listed here rank among the most interesting, for the reasons noted briefly at right.

Hopefully, human beings will have colonized other planets or learned how to deflect such large, asteroidal projectiles long before Eros threatens us. Nevertheless, the preceding scenario presents us with a philosophical touchstone. Had the evolution of mammals and the human species since Earth's last cosmic catastrophe 65 million years ago taken just a few million years longer, Eros might have wiped us all away before we had evolved to the point of being able to develop the technological capacity to survive. These are the larger issues in the backdrop of NEAR's in-depth explorations of 433 Eros (*Figure 16*).

Asteroids are much more likely to be helpful, rather than destructive, during the 21st century. They can serve as humankind's steppingstones to the planets. Some of the small, Earth-approaching asteroids found to date are easily accessible with existing rockets. From them, future astronauts may mine materials necessary for living and working in space, including water, organic compounds, and metals, and it will be far cheaper to obtain these directly from nearby asteroids than to hoist them up from Earth.

For now, we should think of asteroids collectively not as friend or foe, but rather as one of the remaining frontiers in our ongoing exploration of the solar system. What we have discovered thus far is at some times enlightening and at others perplexing — but always interesting (*Table 2*). No longer considered the "vermin of the skies," asteroids can teach us much about where our solar system has been, and where it is likely to go.

Characteristics of Selected Asteroids

Number	Name	Year found	Discoverer	Diameter (km)	Spectral type	Mean dist. from Sun (AU)	Orbital period (years)	Orbital eccen.	Orbit incl. (°)	Remarks
1	Ceres	1801	G. Piazzi	933	G	2.768	4.60	0.08	10.6	First asteroid found
253	Mathilde	1885	J. Palisa	57 × 50	C	2.646	4.30	0.27	6.7	NEAR flyby in 1997
433	Eros	1893	G. Witt, A. Charlois	35 × 13	S	1.458	1.76	0.22	10.8	NEAR flyby in 1999
243	Ida	1905	J. Palisa	56 × 15	S	2.862	4.84	0.04	1.1	1993 flyby; has satellite
624	Hektor	1907	A. Kopff	225	D	5.203	11.86	0.02	18.2	First Trojan found
951	Gaspra	1916	G. Neujmin	18 × 9	S	2.209	3.28	0.17	4.1	Galileo flyby in 1991
1221	Amor	1932	E. Delporte	1	?	1.919	2.66	0.44	11.9	Archetypal Earth-approacher
1862	Apollo	1932	K. Reinmuth	1.5	Q	1.471	1.78	0.56	6.4	Archetypal Earth-crosser
2101	Adonis	1936	E. Delporte	(1)	?	1.874	2.57	0.76	1.4	Second Apollo found
1937 UB	Hermes	1937	K. Reinmuth	(1)	?	(1.644)	(2.11)	(0.62)	6.1	Found near Earth; now lost
2062	Aten	1976	E. Helin	0.9	S	0.967	0.95	0.18	18.9	Archetypal Earth-crosser

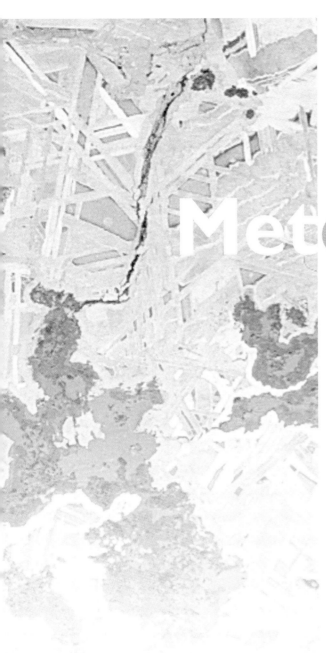

Meteorites

Harry Y. McSween Jr.

ACH YEAR AS it swings around the Sun, the Earth sweeps up perhaps 78,000 tons of extraterrestrial matter, consisting mostly of microscopic dust particles but including a significant proportion of larger objects called *meteorites*. Scientific research on meteorites has been going on for more than two centuries, so it may come as a surprise that there are so many recent revelations about them. One leap in understanding has been made possible by the thousands of meteorites collected in Antarctica during the last few decades (*Figure 1*), among which are a number of types never before seen. Another is the suite of powerful new observational and analytical techniques brought to bear on meteorites already well studied. Tiny particles of interplanetary dust have also joined

351

Patterns of iron-nickel alloys in the acid-etched surface of an iron meteorite.

Figure 1. This iron meteorite was found sitting atop Antarctic ice in 1978. The counter at left provides documentation for field photographs of meteorites. Antarctic field parties sometimes recover hundreds of meteorites in a single season.

Figure 2. This carbonaceous chondrite specimen, Allan Hills 77307, was found in Antarctica. The meteorite's rounded shape was sculpted during its flight through the Earth's atmosphere. The cracked surface is due to weathering in the Antarctic environment. The cube is one centimeter on a side.

Figure 3. A cut surface of the Allan Hills 77278, Antarctica, ordinary chondrite shows numerous round, frozen spherules of silicate melt, called chondrules. The origin of chondrules remains a perplexing aspect of these meteorites. The specimen measures 8 centimeters across.

the ranks of meteorites in providing "ground truth" about solar-system bodies. The recognition that a few types of meteorites appear to be samples of identifiable asteroids or planets, and that some micrometeorites may even derive from comets, has transformed these extraterrestrial materials from museum oddities into keys that can unlock the vaults of cosmic memory.

TYPES OF METEORITES

The classification of meteorites is complicated and seems rather arcane. For the purposes of this chapter, these objects can be viewed as comprising four broad groups. The first, the *chondrites*, are relatively unaltered materials that were probably stored within small planetesimals (asteroids) for a long time after their formation in the early solar system. Chondrites take their name from a Greek word meaning "seeds," an allusion to the appearance of the tiny quenched droplets of once-molten silicate, called chondrules, they commonly contain (*Figures 3,4*). Chondritic meteorites come in many varieties, including the *ordinary chondrites* (the most abundant type of meteorite), *enstatite chondrites* (meteorites that formed under very reducing conditions), and *carbonaceous chondrites* (meteorites especially rich in carbon-bearing organic matter). These various chondrite types constitute 86 percent of the meteorites now falling to Earth (*Table 1*).

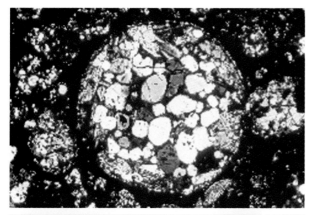

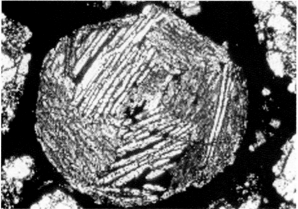

Figure 4. These thin sections of carbonaceous chondrites contain tiny spheres called chondrules. The one at the top, from a 1970 fall in Isna, Egypt, is 0.5 mm across and displays individual crystals of the minerals olivine and pyroxene. The 1-mm-wide chondrule below it, composed of thin olivine crystals, is from a stone found in Smarkona, India, in 1940. The crystals appear artificially colored due to the transmission of polarized light.

Figure 5. Chondrites can be grouped into chemical types (presumably representing different parent asteroids), and petrologic types (referring to the intensity of their thermal metamorphism or aqueous alteration). Tinted boxes indicate the type combinations that are known to exist.

Known Meteorites (1997)

Class	Falls	Finds	Total	Antarctic finds
All meteorites	933	2487	3420	8982
Chondrites*	803	1700	2503	8497
H chondrites	305	860	1165	4622
L chondrites	340	641	981	3190
LL chondrites	73	93	166	401
E chondrites	15	12	27	52
C chondrites	33	28	61	160
Achondrites*	73	49	122	391
Eucrites	28	12	40	147
Howardites	15	9	24	54
Diogenites	10	0	10	85
Aubrites	9	2	11	4
Ureilites	5	16	21	38
SNC meteorites	4	2	6	6
Lunar meteorites	0	2	2	11
Stony-iron meteorites*	12	57	69	29
Pallasites	5	36	41	10
Mesosiderites	7	21	28	13
Iron meteorites*	45	681	726	65
Octahedrites	27	481	508	56
Hexahedrites	6	52	58	2
Ataxites	0	43	43	7

Table 1. Japanese glaciologists found the first handful of Antarctic meteorites in December 1969. Since then, Antarctica's windswept ice fields have provided an unmatched source of extraterrestrial material — including most of the lunar and Martian achondrites. Among the meteorites found elsewhere, note that fewer than one-third of them were actually seen to fall from the sky. This tabulation excludes known pairings — that is, two or more meteorites from the same fall event. The total counts for each category (*) include ungrouped members not otherwise listed.

Chemical type		Chondrule texture	1 Absent	2 Sparse	3	4	5	6
					Abundant / distinct		Increasingly indistinct	
Ordinary chondrites	H							
	L							
	LL							
Carbonaceous chondrites	CI							
	CM							
	CR							
	CO							
	CV							
	CK							
R chondrites	R							
Enstatite chondrites	EH							
	EL					(not yet found)		

< 150°C   < 200°C   400°C   600°C   700°C   750°C   950°C

← Increasing aqueous alteration   Increasing thermal metamorphism →

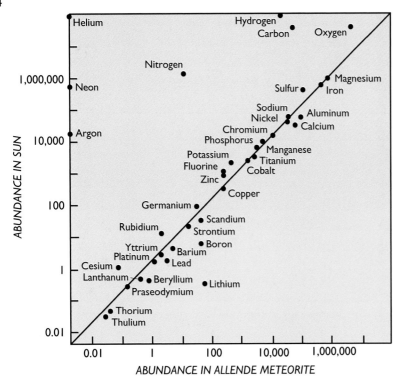

Figure 6. Concentrations of 39 chemical elements in the Allende, Mexico, carbonaceous chondrite, plotted against their concentrations in the solar atmosphere (using logarithmic axes). In both cases, the abundances are relative to 1 million silicon atoms. Clustering of points along the diagonal line indicates that element abundances in chondrites are similar to those in the Sun, except for volatile elements that plot well above the line.

Figure 8. A photomicrograph of a thin section of the Allan Hills 77257, Antarctica, achondrite, several millimeters across. This meteorite is a ureilite, consisting mostly of crystals of the silicate minerals olivine and pyroxene.

Figure 7. The Johnstown, Colorado, achondrite is of a type called diogenite, thought to be a sample of asteroid 4 Vesta. This sawed surface, 10 centimeters wide, shows clasts of igneous rock that have been broken by impact and then recemented together on the asteroid parent body.

Many chondrites have been heated and recrystallized, a process known to geologists as metamorphism. Other chondrites have experienced alteration by water, another process common to terrestrial rocks. The classification for chondrites depends on their chemical composition (which identifies them as ordinary, enstatite, or carbonaceous) and their metamorphic or alteration history (*Figure 5*). The chemical groups are commonly subdivided; for example, the ordinary chondrite group consists of high-iron (H), low-iron (L), and low-iron, low-metal (LL) types.

Distinct chondrite groups differ somewhat in appearance and composition, but all share a common chemical characteristic: they contain nonvolatile elements — those that exist as solids at high temperatures — in approximately the same relative proportions as in the Sun (*Figure 6*). Think of chondrites as a sort of solar sludge — a cooled, crystallized sample of the Sun, depleted only in volatile elements like hydrogen, helium, oxygen, and nitrogen that are difficult to condense into solid form. These primitive meteorites contain an unparalleled record of the solar system's formation and earliest evolution.

Achondrites, literally "meteorites without chondrules," make up the second group. They result from partial melting within asteroids that presumably were themselves once chondritic. Various achondrite groups (*Figures 7,8*) formed either as the crystallized products of once-molten silicates or as the solid residues left behind after such magmas were extracted. The partial melting of chondrites produces magmas with compositions that differ from the original rocks', so achondrites no longer retain the distinctive solar chemistry. Like chondrites, however, they are composed mostly of silicates — minerals that combine silicon, oxygen, and other metals. These meteorites are more akin to lavas found on Earth, but they formed under different conditions and from different starting materials. Taken together, achondrites comprise about 8 percent of the meteorites falling to Earth.

Another 7 percent of meteorite falls are *iron meteorites*, composed mostly of metallic alloys of iron and nickel, which also formed by melting within planetesimals. In this case, dense metallic liquids segregated from lighter silicate magmas or solids and sank to form cores within these bodies, a process called differentiation. Once sequestered into cores, the irons cooled slowly enough to form exquisite intergrowths of metal phases called Widmanstätten patterns (*Figure 9*). *Stony-irons* are complex mixtures of metal and silicates. Crystallization at the boundaries between cores and the overlying silicate mantles created pallasites (*Figure 10*), intimate mixtures of metal and gem-quality olivine (magnesium silicate). Alternatively, combinations of metal and fragments of achondrite, called mesosiderites, formed when already differentiated asteroids collided. Stony-irons are relatively uncommon, comprising only about 1 percent of fallen meteorites. Altogether, these igneous meteorites — the achondrites, irons, and stony-irons — reveal the complex

Figure 9. Acid etching of iron meteorites reveals a Widmanstätten pattern, an intergrowth of several alloys of iron and nickel. This slice of the Maltahöhe, Namibia, meteorite shows large, intertwined bands of the low-nickel alloy kamacite and high-nickel taenite. The Widmanstätten pattern, named for the individual who first observed it in 1808, forms by diffusion of nickel atoms into solid iron during slow cooling within an asteroidal core.

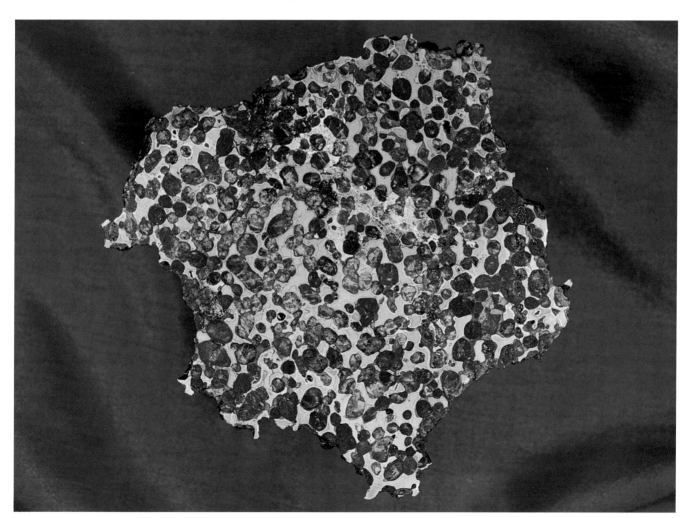

Figure 10. This polished section of the Springwater, Canada, pallasite shows rounded olivine crystals embedded in iron-nickel metal. The mixture of silicates and metal, which normally should separate because of pronounced differences in density, is thought to have occurred as the meteorite solidified at a core-mantle boundary. Pallasites are among the most beautiful of all meteorites.

Figure 11. The Lafayette, Indiana, nakhlite shows exquisitely preserved fusion crust. During the stone's rapid transit through the atmosphere, air friction melted its exterior. The lines trace beads of melted rock streaming away from the leading edge. Lafayette is a planetary achondrite, thought to be from Mars.

Table 2 (right). A small but growing number of meteorites are now recognized to have reached Earth from the Moon and Mars. Unless otherwise noted, all locations are in Antarctica. Asterisks (*) denote individual stones from the same fall, and bold type indicates falls that were witnessed.

Lunar Meteorites

	Designation	Discovery location	Discovered	Mass (g)
1	Y 791197	Yamato Mountains	20 Nov. 1979	52.4
2	Y 793169	Yamato Mountains	8 Dec. 1979	6.1
3	Y 793274	Yamato Mountains	3 Jan. 1981	8.7
4	ALHA 81005	Allan Hills	18 Jan. 1982	31.4
5	Y 82192	Yamato Mountains	13 Jan. 1983	36.7
	Y 82193*	Yamato Mountains	13 Jan. 1983	27.0
	Y 86032*	Yamato Mountains	9 Dec. 1986	648.4
6	EET 87521	Elephant Moraine	20 Dec. 1987	30.7
7	Asuka 881757	Nansen Ice Field	20 Dec. 1988	442.1
8	MAC 88104	MacAlpine Hills	13 Jan. 1989	61.2
	MAC 88105*	MacAlpine Hills	13 Jan. 1989	662.5
9	Calcalong Creek	Calcalong Creek, Australia	1960	19
10	QUE 93069	Queen Alexandra Range	11 Dec. 1993	21.4
	QUE 94269*	Queen Alexandra Range	10 Dec. 1994	3.2
11	QUE 94281	Queen Alexandra Range	12 Dec. 1994	23.4
12	EET 96008	Elephant Moraine	14 Dec. 1996	53
13	Dar al Gani 262	Al Jufrah, Libya	23 Mar. 1997	513
14	Dar al Gani 400	Al Jufrah, Libya	10 Mar. 1998	1,425

Martian Meteorites

	Designation	Discovery location	Discovered	Mass (g)
1	Chassigny	Haute Marne, France	**3 Oct. 1815**	4,000
2	Shergotty	Gaya, Bihar, India	**25 Aug. 1865**	5,000
3	Nakhla	Alexandria, Egypt	**28 June 1911**	40,000
4	Lafayette	Tippecanoe County, Indiana	before 1931	800
5	Governador Valadares	Minas Gerais, Brazil	1958	158
6	Zagami	Katsina province, Nigeria	**3 Oct. 1962**	18,000
7	ALHA 77005	Allan Hills	29 Dec. 1977	480
8	Y 793605	Yamato Mountains	14 Nov. 1979	16
9	EETA 79001	Elephant Moraine	13 Jan. 1980	7,900
10	ALH 84001	Allan Hills	27 Dec. 1984	1,930.9
11	LEW 88516	Lewis Cliff	22 Dec. 1988	13.2
12	QUE 94201	Queen Alexandra Range	16 Dec. 1994	12.0
13	Dar al Gani 476	Al Jufrah, Libya	1 May 1998	2,015

Figure 12. This interplanetary dust particle, only about 0.01 mm (10 microns) across, was collected in the stratosphere on an aircraft wing. It consists of tiny grains of silicates and other minerals similar to those in chondrites. Some particles are thought to have been derived from comets.

thermal histories of small, differentiated planetesimals like those that assembled into the Earth and its neighboring planets.

The third group of meteorites is made up of *planetary achondrites,* rocks launched by impacts from the surfaces of large bodies like the Moon and Mars. Lunar achondrites are mostly samples of the ancient highlands, though a few are basaltic lavas from the maria. In contrast, the Martian meteorites are mostly igneous rocks from terrains younger than 1.3 billion years old (*Figure 11*). These are often referred to collectively as SNCs ("snicks"), an achronym for shergottites, nakhlites, and chassignites. The SNCs reveal details of the inner workings and geologic history of a planet that is distinct in composition from the Earth, and they provide data on the complex responses of water and atmospheric gases during Mars's protracted evolution. The known lunar and Martian achondrites together number only two dozen meteorites (*Table 2*), comprising only a small fraction of a percent of meteorite falls.

Micrometeorites, more often called *interplanetary dust particles,* are miniature specks of material. Many of these are lofted off asteroids during impacts or ejected from comets as their ices sublimate. The Ulysses and Galileo spacecraft also discovered that a stream of interstellar dust is coursing through the solar system at present. Micrometeorites can be collected in the upper atmosphere by trapping them on the leading edges of airfoils on high-flying aircraft. Some also are recovered from polar ice packs. Most of these dust motes are so small that powerful microscopes are required to study them (*Figure 12*).

Dust particles that pass through our atmosphere are usually melted in transit, making it difficult to discern their sources. Some micrometeorites contain the same minerals and have similar chemical compositions to known meteorite groups, and probably represent powder scraped off the same parent bodies that provide the major meteorite classes. However, a significant fraction of these tiny objects exhibit no such lineage and appear to have been derived from distant objects whose locations are unknown to us.

Comets passing through the inner solar system generally have higher orbital velocities than asteroids do. Consequently, cometary dust gets hotter than asteroidal dust when it enters our atmosphere and thus releases any trapped gases more efficiently. We can therefore use a dust particle's helium content as one way to identify those grains that were in cometary orbits. Some chondritic porous particles, so named to distinguish their composition and texture, apparently enter Earth's atmosphere at very high speeds and thus likely originate from comets.

A very useful way to classify most meteorites utilizes their oxygen isotopic compositions. This system, largely the work of Robert Clayton and his coworkers, interrelates the proportions of oxygen's three isotopes: ^{16}O (the most common), ^{17}O, and ^{18}O. When values for the ^{17}O:^{16}O ratio are plotted against ^{18}O:^{16}O, compared to the ratios in a terrestrial standard, most chondrite classes form distinct clumps (*Figure 13*). Achondrites and planetary achondrites tend to be smeared along straight lines with the same slope, which reflects the partitioning of oxygen isotopes into silicate minerals during melting and crystallization. Meteorites from the same parent object define distinct lines displaced from those of other sources. For example, all Martian meteorites plot on one line, even though they may have varying mineral compositions. Stony-irons and even some irons can be classified using this system, by analyzing the oxygen in their inclusions of silicate minerals.

GETTING TO EARTH

As already noted, the great majority of meteorites are asteroidal fragments. Yet in only four cases have the atmospheric trajectories of found meteorites been determined well enough to calculate their pre-arrival orbits in space (*Figure 14*). At their farthest reaches, these highly elliptical orbits intersect the asteroid belt. Careful study shows that the belt contains nearly empty gaps, locations where asteroids were apparently scattered away after repeated gravitational interactions with massive Jupiter (see Chapter 25). The gaps were actually cleaned out long ago, but impacts between asteroids near the gaps periodically throw fragments into these "escape hatches," where their orbits are quickly modified by chaotic interactions so that they cross Earth's orbit. We believe these gaps supply most meteorites now falling to Earth. Conceivably, other sources may have predominated in the distant past.

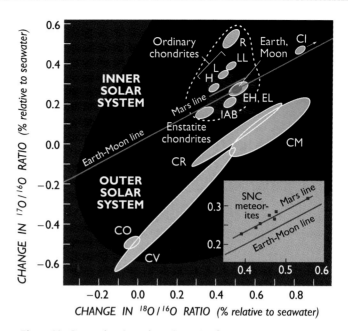

Figure 13. Cosmochemists rely on isotopic "fingerprints" to determine the origin and age of solar-system objects. Ratios of oxygen's three isotopes are particularly diagnostic. Terrestrial and lunar rocks have the same oxygen ratios, arguing that they formed at the same distance from the Sun. But the ratios in various classes of meteorites (shaded ovals) suggest different origin locations, presumably in the asteroid belt. The ratio values for the Martian SNC meteorites (inset) trend along a line that parallels the Earth-Moon line. Achondrites (not shown) also fall on straight lines that parallel that of terrestrial samples.

The linkage between asteroids and meteorites is strengthened by similarities between the spectra of light reflected from their surfaces. For example, the howardites, eucrites, and diogenites — all achondrites with the same oxygen isotopic fingerprint and thus probably from the same source — are spectrally similar to asteroid 4 Vesta. Eucrites and diogenites have different compositions (howardite is a physical mixture of the two), but each group can be linked to distinct regions on Vesta's surface. Moreover, a host of small bodies with similar spectra occupy the orbital space between Vesta and the nearest resonant gap — additional evidence that pieces of Vesta have made their way to Earth. Some other spectral correlations between specific asteroids and certain meteorite classes have been found, though the spectral variation seen among asteroids is greater. Apparently there are many minor planets not currently represented in our meteorite collections.

Once a piece of asteroidal material has been broken into a small enough fragment in space, its interior can be penetrated by cosmic rays. These high-energy particles dislodge protons and neutrons in the target atoms, creating new isotopes. For example, some iron atoms are transmuted into manganese-53, an unstable isotope that decays over time. By measuring the amounts of various cosmic ray-produced isotopes with different decay rates, geochemists can calculate how long a meteorite existed as a small object in space. Such cosmic-ray exposure ages are typically tens of millions of years for stony meteorites and hundreds of millions of years for irons, reflecting the latter's greater ability to survive impacts in space. In either case, however, these times are short compared to the age of the solar sys-

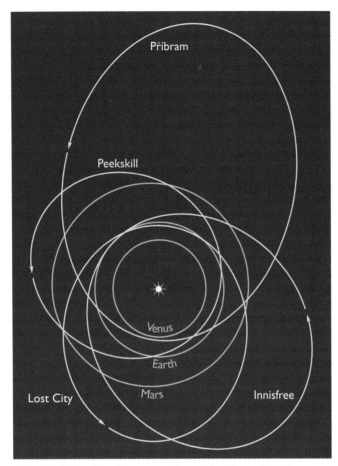

Figure 14. The orbits of four meteorites — all ordinary chondrites — before they struck the Earth. The Peekskill orbit was derived from videotapes of its spectacular nighttime arrival on 9 October 1992. Those for the Příbram (1959), Lost City (1970), and Innisfree (1977) falls were calculated from photographs taken during systematic sky patrols.

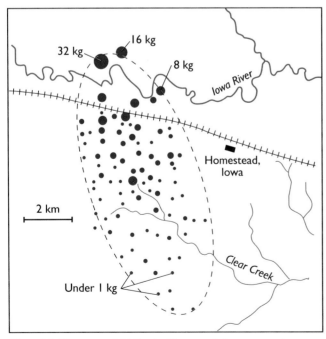

Figure 15. The strewn field for the Homestead, Iowa, meteorite shower of 12 February 1875 illustrates how larger objects carry to the far end of the ellipse.

tem. Meteorites were dislodged from their parent bodies long after they formed, and the race to Earth was relatively quick.

Removal of rocks from the Moon or Mars required much larger and more energetic impacts than those that ejected fragments from asteroids, because the greater gravity of these larger bodies increases the required escape velocity. Once in space, however, these planetary castoffs reach Earth faster than their asteroidal counterparts do. Orbital simulations and cosmic-ray exposure ages indicate that the transit times for lunar and Martian meteorites are typically just a few million years.

Several tons of meteorites that weigh at least 10 grams apiece reach Earth's surface each year. But few of these arrivals are ever found. Most are lost to the oceans, while many others land in remote, uninhabited terrain. Even those that drop onto dry land will weather to an unrecognizable state within a few thousand years. A recovered meteorite whose fall to Earth was witnessed is called a *fall*, whereas one that was discovered at some later time is called a *find*. Irons and stony irons are relatively easy to distinguish from terrestrial rocks, so their find-to-fall ratio is much greater than those of chondrites and achondrites *(Table 1)*.

Falls sometimes occur as showers of meteorites, created when large objects break up during atmospheric passage. If this occurs while they are traversing the atmosphere at a shallow angle, air friction will slow the smaller fragments more than the larger ones. Consequently, the large objects carry farther, spreading the points of impact into an ellipse called a *strewn field* (*Figure 15*). Large meteors that do not fragment during atmospheric transit excavate craters when they violently strike the Earth's surface at high speed (see Chapter 6).

Sometimes meteorites reach Earth in groups or streams (distinct from annual meteor showers, which are associated with comets). For example, meteorites preserved in limestone sediments in Sweden make a strong case that falls were up to 100 times more frequent for a short interval about 500 million years ago. Notably, a great many L chondrites were heavily shocked and degassed 500 million years ago; conceivably a major collision at that time shattered their parent asteroid and sent a cascade of debris toward the inner solar system and ultimately to Earth.

Meteorites sometimes develop distinctive shapes during atmospheric passage. The leading edge usually melts, and ablation of this molten material produces a smooth face (*Figure 11*). Melt droplets streaming along the sides may collect at the opposing face and solidify into a roughly textured material, resulting in a conical shape known as an oriented meteorite. (If the meteorite is not aerodynamically stable, rotation prevents the oriented shape from forming.) Quenching of the molten coating produces a dark, glassy *fusion crust*, which serves as one way to recognize the object as extraterrestrial. Even though its exterior may melt, a meteorite passes through the atmosphere so rapidly that its interior is not heated.

Another common characteristic of meteorites is the iron-nickel metal they usually contain. Even chondrites that might otherwise be mistaken for terrestrial rocks exhibit shiny flecks of metal underneath their fusion crusts. A true meteorite will usually attract a small magnet suspended by a string.

The most important source of meteorite finds is Antarctica. For millions of years meteorites have fallen onto the ice-covered surface, where they become frozen into glaciers that eventually carry them toward the edges of the continent. Wherever ice piles up behind an obstruction, like a mountain range, it is uplifted. Strong winds sweep across the ice and gradually ablate the ice, exposing and concentrating the embedded meteorites on the surface. Meteorite-hunting scientists mount expeditions to these select regions during Antarctic summers, commonly recovering hundreds and sometimes thousands of meteorites in a single season. The arid, windswept Nullabor plains of Western Australia and the Sahara Desert are other rich sources of meteorites; several hundred have been found in each location.

PRESOLAR MATERIALS

All of the various kinds of meteorites allow geochemists to study cosmic processes and chronologies in detail that can otherwise only be guessed at. These extraterrestrial samples provide a level of information not revealed by the remote-sensing measurements from telescopes and spacecraft that define the rest of our knowledge of the Sun and its family.

As noted in Chapter 2, chondrites are the only known materials that can be traced back to the very birth of the solar system. They consist mostly of matter that formed within the flattened cloud of gas and dust that surrounded the infant Sun (the solar nebula). However, they also contain minute amounts of interstellar grains that actually predate the solar system. These exotic grains have been separated by dissolving in acids most of a chondrite sample, leaving behind a fine powdery residue whose interstellar heritage can be recognized by the unusual combinations of noble-gas isotopes it contains.

The first interstellar grains extracted from meteorites were found to be miniature diamond crystals (*Figure 16*). Chemist Edward Anders worked for nearly 20 years to accomplish this separation, a feat he likened to "burning down a haystack to find the needle." The diamonds were tagged with distinctive isotopes of xenon that could only have formed during a supernova. The xenon was created during fusion reactions deep in a dying star's interior, whereas the diamonds must have grown within carbon-rich layers at the surface. Clouds of diamond stardust drifting away from the star during its bloated red-giant phase were presumably overtaken by the energetic, high-speed wind produced by the supernova, and that is when the xenon was implanted into the diamonds.

A few other kinds of stardust have also been recognized in chondrites. The list so far includes graphite, silicon carbide (SiC, better known as the industrial abrasive carborundum), corundum (Al_2O_3), spinel ($MgAl_2O_4$), and silicon nitride (Si_3N_4). The ability to analyze presolar grains like these in the laboratory provides a critical test for astrophysical models that describe how elements are synthesized in stars. Stars shine as a consequence of the fusion of lighter elements into heavier ones. This process creates specific combinations of isotopes, which tell a great deal about the particular fusion processes involved. Although astronomers can use spectroscopy to measure the isotopic compositions of a handful of elements in stars, many more can be analyzed in a meteorite's stardust — and with much greater precision.

Another kind of presolar material recognized in chondrites consists of organic matter. Within frigid interstellar clouds,

Figure 16. **Stardust on Earth. This tiny vial contains more than a trillion grains of microscopic diamond crystals (the small white lump), which were isolated from a chondritic meteorite by acid dissolution. These interstellar grains, older than the solar system itself, formed in the vicinity of another star.**

organic molecules are produced when charged ions collide with gas atoms and interstellar dust grains. Under such conditions, deuterium (the heavy isotope of hydrogen) is preferentially incorporated into the hydrocarbons, as revealed by the high ratios of deuterium to hydrogen measured spectroscopically. The organic compounds in chondrites consist of chains of carbon atoms with complicated branchings, molecules with greater complexity than those in interstellar space. For example, the Murchison carbonaceous chondrites, which rained onto southeast Australia in 1969, contain dozens of amino acids. Such compounds still bear the characteristic deuterium fingerprint of interstellar clouds, indicating that they formed from simpler organic molecules that were already present when the solar system came together.

The tiny motes of stardust and molecules of organic muck found in chondrites thus allow us to test theories about the element factories we know as stars and about refrigerated concentrations of matter drifting between them. They also allow an otherwise unobtainable view of some solid materials that long ago coagulated together to form our solar system.

CHONDRITES AS TIME CAPSULES

Interstellar gas and dust collapsed into a solar nebula bound by its own gravitational forces, the first step toward forming the glimmering stellar centerpiece and its retinue of planets we observe today. Chondrites contain a variety of unusual isotopes that reveal how this collapse may have been triggered. These particular isotopes are the products of radioactive disintegration of highly unstable and very short-lived isotopes that once populated the nebula but are now extinct. One key example is aluminum-26, which decays to magnesium-26 within only a

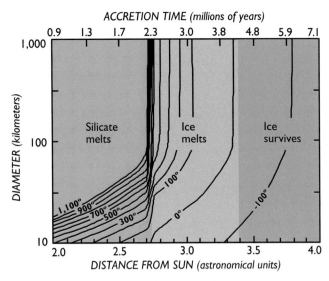

ACCRETION TIME (millions of years)

DISTANCE FROM SUN (astronomical units)

Figure 17. This thermal model for the asteroid belt assumes heating by decay of a short-lived radioactive isotope, aluminum-26. Comparison of the scales at the top and bottom of the figure shows that asteroids farther from the Sun accreted later, and thus incorporated less "live" aluminum-26. Those closer to the Sun were heated by ^{26}Al to higher temperatures, as shown by the contours. For asteroids with diameters of 100 km within 2.7 AU of the Sun, their silicate minerals melted, producing achondrites. Ice melted in bodies located between 2.7 and 3.4 AU, allowing metamorphism and aqueous alteration of chondrites. At great solar distances, asteroids never warmed above the melting point of ice.

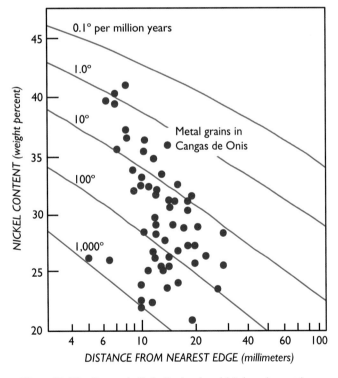

DISTANCE FROM NEAREST EDGE (millimeters)

Figure 18. The Cangas de Onis, Spain, chondritic breccia contains metal grains that cooled at rates varying from 1° to 1000° per million years (shown by contours), as determined by measuring the migration of nickel in them. The slowest cooling rates correspond to burial at depths of perhaps 100 km. Since the asteroid was unlikely to have been much larger than this, it must have been disrupted by impact, and rock fragments from throughout the body reaccreted to form a new, rubble-pile asteroid.

few tens of millions of years. Excess amounts of magnesium-26 have been found in chondrites, located within sites in feldspar crystals that could only have accommodated aluminum when they formed. This means that aluminum-26 had not yet decayed when the crystals formed. To have been incorporated "live," this short-lived isotope must have been synthesized in a supernova just before the nebula's formation. Moreover, the supernova had to be just next door; otherwise the aluminum-26 would have decayed as it drifted about in the expanses between stars. Some theorists have gone so far as to suggest that the expanding shock wave from a nearby supernova compressed the gas and dust ahead of it, triggering its collapse into what became the solar nebula and, ultimately, our solar system.

Once the nebula formed, parts of it were apparently heated to quite high temperatures. Localized jumps in temperature melted balls of dust to form chondrules. Chondrites also contain white lumps, called *refractory inclusions*, that must have formed at high temperatures. These are perhaps the byproducts of dust that was vaporized, then resolidified. Computer simulations of the sequence of minerals condensing from a cooling gas of solar composition predict the very same minerals found in the refractory inclusions. These unusual objects were apparently the earliest solid matter in the nebula, and their ancient age (4.56 billion years, as determined from isotopic dating) pinpoints when the nebula formed.

After that, the condensation sequence became more complex. Most ordinary chondrites endured high temperatures and recrystallized, rendering their chondrules barely recognizable. Many carbonaceous chondrites remained cool, only to have aqueous fluids produce a complex assortment of water-bearing clays, veins of sulfate and carbonate, and other secondary minerals. The difference in the thermal histories of chondritic asteroids simply reflects whether rocky materials accreted dry (as did the ordinary chondrites) or mixed with ices (as did the carbonaceous chondrites). We now suspect that ice, while melting to make liquid water, absorbed the heat and thus prevented most carbonaceous chondrite planetesimals from reaching high temperatures.

Of course, some asteroids got hot enough to melt and differentiate, as revealed by achondrites and irons. How could different bodies be heated to varying temperatures? Spectroscopic studies of the asteroid belt indicate that bodies nearer the Sun were commonly melted, those at intermediate distances experienced metamorphism or aqueous alteration, and objects even further away were never heated significantly. One thermal model (*Figure 17*) explains this observation as a result of incorporating different amounts of aluminum-26, which gives off heat as a byproduct of its rapid decay. Because matter was concentrated closer to the Sun, planetesimals in the inner asteroid belt accreted faster and earlier, incorporating much of the aluminum-26 before it decayed. These bodies thus experienced higher temperatures than asteroids that accreted later, when less aluminum-26 was available. Another possible explanation for planetesimal heating is electromagnetic induction, whereby powerful electric currents were induced within asteroids by a massive solar wind.

Another important process in the early solar system is collision, and evidence of this pummeling can be seen in meteorites. Many chondrites and achondrites are *breccias*, consisting of angular clasts of broken rock now cemented together. Sometimes chondritic breccias contain clasts with metal grains that cooled at greatly varying rates: some metal grains cooled quickly near the parent asteroid's surface, whereas others cooled slowly near its center. The only way to intermix metal grains from such disparate depths was for the asteroid to have been shattered by a large impact. The debris then rapidly reaccreted into a new object that resembled a rubble pile (*Figure 18*).

Smaller impacts were not so violent— merely pulverizing the surface rocks to create a blanket of rock fragments and dust, called a *regolith*. Meteorites formed from compacted asteroidal regoliths are also breccias (*Figure 19*), and their former residence on the surface is revealed by implanted atoms from the solar wind and by tracks in crystals produced by energetic cosmic rays.

PLANETARY INSIGHTS

Terrestrial rocks contain chondritic abundances of most elements. However, relative to chondrites, Earth is significantly depleted in volatile elements, those that vaporize easily. All achondrites share this peculiar characteristic of pronounced volatile-element depletion. Apparently this loss did not occur during differentiation, but instead reflects an inherent property of bodies that formed nearer the Sun. It thus seems logical to conclude that the Earth accreted largely from achondritic planetesimals.

The accretion of the inner planets from already differentiated planetesimals, rather than from chondrites, is a recent revelation with interesting implications. For example, the formation of Earth's core would have been greatly facilitated by accreting bodies having large, perhaps still-molten cores. Also, the core's composition should reflect whether the metals and silicates beneath us separated at low pressures inside planetesimals or deep inside Earth itself, because elemental affinities depend in part on pressure. Seismic studies indicate that the liquid outer core contains a light element, currently unidentified. Oxygen has been suggested as a possibility, because high-pressure experiments show that it is soluble in iron under those conditions. However, if the core was assembled from smaller cores formed at low pressures, sulfur is a more likely candidate since it has greater solubility than oxygen in iron.

The end of planet formation, about 3.9 billion years ago, was marked by especially large impacts (see Chapter 6). The surface of the Moon is pockmarked with huge multiringed basins that testify to this terrible period. On Earth, geologic processes have erased all the impact scars from so long ago. However, this time of cataclysm is recorded by the isotopic ratios found in meteorites. Radioactive potassium-40 decays slowly to argon-40, but the argon (a gas) escapes during impact, thereby resetting the radiometric clock to that time. A large number of meteorites have ^{40}Ar-derived ages of about 3.9 to 4.0 billion years. So we at least know that the meteorites' parent bodies and the Moon suffered a similar fate. The oldest Martian meteorite, designated ALH 84001, has a similar argon-40 age, supporting the idea that this bombardment was a solar-system-wide event.

Lunar achondrites (*Figure 20*) are similar enough to rocks returned from the Moon by Apollo astronauts that their identification is unambiguous. However, there are some important chemical differences between lunar meteorites and Apollo sam-

Figure 19. Found in Antarctica, Allan Hills 78113 is a breccia composed of angular chips of rock fragmented by impact. It once resided within the regolith on its parent asteroid. The cube is 1 centimeter on a side.

Figure 20. This golfball-size breccia, ALH 81005, was found in 1982 during a meteorite-hunting expedition in the Allan Hills region of Antarctica. The white fragments are feldspar-rich rock called anorthosite, which (among other characteristics) proved that the small stone formed on the Moon — the first such meteorite to be recognized.

ples. The global distribution of iron on the Moon's surface, as determined by the Clementine spacecraft, clearly shows that the Apollo missions did not obtain a representative sampling of the crust. By contrast, the mean composition of achondrites from the lunar highlands provides a better match for the average iron content in the global map (see Chapter 10). Thus the abundances of other elements in these meteorites may be more informative than those of the Apollo samples.

The young crystallization ages of all but one of the Martian meteorites, less than 1.3 billion years, is strong evidence that they did not form on asteroids. Only large bodies can retain enough internally generated heat to melt rocks for billions of years, and Mars is really the only likely source on that basis. More convincing evidence for a Martian origin is the trapped gas found in pockets of impact melt in several SNC meteorites (*Figures 21, 22*). These gases match precisely the elemental and isotopic composition of the Martian atmosphere as determined by the Viking landers.

Figure 21. This slab of the Martian meteorite ALH 77005 shows numerous crystals of brown olivine and white pyroxene. It also contains gray patches of impact melt, probably produced when it was ejected from the surface of Mars. The cube is 1 cm on a side.

Figure 22. A microscopic view of Allan Hills 77005 shows its complex igneous texture and several irregular patches of partly crystallized impact melt.

The 13 known Martian achondrites provide considerable information on the red planet's geologic history. The compositions of these mostly volcanic rocks have been used to infer the composition of the mantle source regions that partially melted to produce them. Depletions in certain elements also provide information on the probable composition of the Martian core.

It is astonishing just how much planetary history can be gleaned from a few billionths of a gram of a key isotope. One recent study of eight Martian meteorites suggests that the red planet formed and differentiated quickly, then remained relatively quiet on a global scale in the billions of years thereafter. This rather sweeping conclusion comes from measuring the abundance of tungsten-182, the decay product of the short-lived radioisotope hafnium-182. Geochemists know that tungsten has an affinity for iron, and thus would likely have been drawn into the Martian core, whereas hafnium tends to end up in crustal rocks. The hafnium is long gone, but its one-time presence in the meteorites is betrayed by the residual tungsten-182. Had Mars been more geologically active after forming, with large-scale mantle overturn or Earthlike plate tectonism, the Martian meteorites would now bear more tungsten-184 (its dominant isotope) and less tungsten-182.

Mars is the only planet besides Earth with clear evidence that water once ran on its surface. Here too we have gained valuable

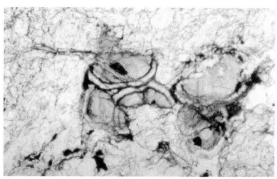

Figure 23. The small chip from the Martian meteorite Allan Hills 84001 (*left*) contains tiny orangish grains of carbonate in some of its fractures. In a greatly magnified thin section 0.5 mm across (*above*), the carbonate globules have orange centers surrounded by black rims containing magnetite and sulfides. This varying composition and other characteristics have led to the suggestion that the globules were formed by ancient Martian microbes. This idea remains controversial, and other explanations for these features have been offered.

insights from Martian meteorites. They contain small amounts of carbonate, sulfate, and clay minerals, which indicate that some water must have been present in the subsurface. Over time, the thin Martian air has become enriched in deuterium several thousandfold (see Chapter 13). Geochemists have found this same deuterium enrichment in water-bearing minerals in Martian meteorites, implying a global cycling of water between the atmosphere and solid planet.

In 1996, David S. McKay and his colleagues at the NASA/Johnson Space Center and Stanford University offered the intriguing proposal that ALH 84001, the oldest Martian achondrite, contains evidence for ancient life. This meteorite is cut by fractures that contain small globules of orange-tinted carbonate (*Figure 23*). McKay and his coworkers argued that tiny iron oxide and sulfide grains in the carbonate had similar sizes and shapes to those formed by bacteria on Earth. These researchers also found traces of organic matter, which they suggested might be the decayed remains of organisms, as well as very small "ovoid and elongated forms" resembling terrestrial microfossils.

All of this evidence occurs within the carbonate globules. However, the formation temperature of the carbonate has generated considerable argument, with some scientists favoring low temperatures and some advocating temperatures much too high to support life. Other researchers have suggested that the organic matter was due to terrestrial contamination by melted Antarctic ice. Still others believe the purported "microfossils" are nothing more than long, narrow crystals of inorganic magnetite, similar in morphology and mode of growth to those known to condense from hot vapors.

Whether ALH 84001 is ultimately shown to contain evidence for life, this meteorite still demonstrates that soluble minerals, microscopic grains, and possibly some organic matter can be preserved in Martian rocks for billions of years. All this offers great hope for a Mars exploration program focused on the search for life.

Lunar and Martian achondrites also demonstrate that large bodies, even planets, can swap rocks during impact events, and calculations indicate that some ejected rocks can travel between

Mars and Earth in just a few years. These findings reopen the possibility of importing life from elsewhere, a concept known as *panspermia*. Even if one rejects the notion that organisms have piggybacked on meteorites, the accretion of a veneer of carbonaceous chondrites or comets on the early Earth may have introduced at least the raw materials for life, in the form of complex organic compounds that include amino acids and other biomolecules (see Chapter 27). Thus, life on Earth did not necessarily have to start from scratch by first synthesizing complex organic molecules.

Meteorites are sometimes called "the poor man's space probe," because they provide so much information about the solar system yet cost very little to obtain (especially when compared to the cost of spacecraft exploration or even astronomical observations). Chondritic meteorites contain tiny, intact stardust grains that predate the solar system, as well as organic molecules that retain the isotopic stamp of interstellar space. The chondrules, refractory inclusions, and other grains that accreted to make chondritic meteorites were among the first solid materials to form in the solar nebula. Achondrites, irons, and stony-irons reveal information about the differentiation of planetesimals heated to high temperatures, and their parent bodies were probably the building blocks of the Earth and its neighboring planets. Planetary achondrites tell us about geologic (and, some have argued, even biologic) processes on other worlds.

If we had no meteorites, our view of the new solar system would be much less thorough, lacking quantitative information on the timing of major events, the chemistry and mineralogy of the matter that formed the nebula, or the nature of internal processes within planetesimals, planets, and stars that are not visible through observations of their surfaces. There is great synergy between the laboratory study of interstellar and interplanetary grains, asteroidal samples, and planetary rocks, on the one hand, and the remote-sensing observations and analyses of extraterrestrial bodies on the other. Understanding our solar system requires much more than imaging and remote sensing, and to that end meteorites continue to provide an incredible window to our distant past.

Life in the Solar System

Gerald A. Soffen

THE STORY OF life in the solar system is an exciting blend of scientific thought and human curiosity about "what's out there?" It is a chronicle woven from 20th-century advances in organic chemistry, astrophysics, and geology, together with the spectacular discoveries made during the solar-system explorations begun by NASA in the 1960s. At its heart is our urge to explore the planets and find out if life on Earth is the end product of a seemingly random set of actions that occurred only on this planet — or if it happened elsewhere.

Our story really begins in the 19th century, when Charles Darwin proposed his theory of the origin of species. Thanks to his work, the word "evolution" came to symbolize a whole new way of thinking about the creation of life. Yet in Darwin's time studies of the emergence of life were in such a state of affairs that he confided in an 1863 letter to J. D. Hooker, "It is mere rubbish, thinking of the origin of life, one might as well think of the origin of matter."

With the turn of the century came new ideas and new directions needed to understand the underpinnings of life. British geneticist John B. S. Haldane and Russian chemist Aleksandr I. Oparin set the biological sciences on a crucial path of inquiry when they speculated on the conditions of the primeval Earth and the nature of the life forms that arose. In 1924 Oparin wrote, "…at the present time it is regarded as highly improbable that free oxygen was contained in the original Earth's atmosphere." Haldane took that thought a step further, arguing that since the primitive atmosphere contained little or no oxygen, "the first precursors of life must have obtained the energy which they needed for growth by some other process than oxidation." He ventured even further afield, speculating on the basis of early life: "The first living…things were probably large molecules synthesized under the influence of the Sun's radiation, and only capable of reproduction in the medium in which they originated.

365

Meteoritic bombardment and geologic upheavals wreak havoc with primitive life on the young Earth.

Figure 1. **Meteoritic bombardment and geologic upheavals more than 3½ billion years ago may have released so much carbon dioxide into Earth's early atmosphere that surface temperatures approached the boiling point of water. Such extreme conditions might have stalled the evolution of life, perhaps constraining its development to the relative security of volcanic vents near the ocean floor.**

Each presumably required a variety of highly specialized molecules before it could reproduce itself, and it depended on chance for its supply of them."

It wasn't long before chemists and biologists took steps to recreate the conditions of the early Earth. In 1953 Stanley Miller and his mentor Harold Urey performed a historic experiment. They produced amino acids and sugars by exposing methane, ammonia, and water to electrical excitation, simulating the prebiotic conditions that might have existed when life formed on this planet.

These and other experiments stimulated scientific interest in extraterrestrial life in the early 1960s and ultimately led to NASA's Viking mission in 1975–76 to search for signs of indigenous organisms on Mars. That mission did not find definitive evidence of life, but it did spur scientists in a wide variety of disciplines to investigate further the question of life in the solar system, beginning with more specific studies of Earth's incredible biological diversity.

LIFE ON EARTH

Within the solar system, Earth is the only planet known to harbor life. The origin and evolution of the terrestrial biosphere comprise a complex interplay of astrophysics, geology, chemistry, and biology. We are familiar with the astrophysics involved, from the discussion in Chapter 2 about the formation of the solar system. Geologic and chemical forces shaped the emerging Earth and its atmosphere. Biology and biochemistry allow us to identify and classify the life that evolved here. The interplay stems from using multidisciplinary research to answer specific questions about the conditions of the early Earth, and about how organisms survived and flourished in the unique environments they encountered.

There is no universal agreement on a single definition of life. Terrestrial life is largely recognized by two characteristics: its ability to reproduce and its ability to evolve through mutation and natural selection. We have also come to realize that life requires three things: water, energy, and access to organic materials. Clearly the Earth provides this template, with its oceans, sunlight, chemical and volcanic energy, and a rich collection of organic compounds. When we look at those organics, one important fact stands out: terrestrial life is made up of hydrogen, oxygen, carbon, and nitrogen — the most abundant elements of the cosmos. Using these elemental "building blocks" and other trace elements, researchers have investigated many possible life-initiating chemistries in the attempt to characterize life.

Most biologists who study the origin of life and its chemical processes think that all terrestrial beings resulted from a single sequence of events. The general sequence began with the formation of the solar system, continued through the development of conditions favorable for life, prebiological organic synthesis, self-replicating molecules, and ultimately the formation of the first organisms.

The early Earth was not a hospitable place for life (*Figure 1*). The oceans as we now know them did not exist. The surface was constantly subjected to a rain of comets and asteroids. This violent bombardment very likely resulted in a planetary surface quite hostile to biogenesis. Some scientists have suggested that primitive life may have been initiated more than once in the chaos of those early millennia, only to be wiped out by changing conditions.

Yet the bombardment was not contrary to the ultimate formation of life. As incoming objects were cratering the surface of our planet, they also delivered water and a rich assortment of organic materials. Christopher Chyba and the late Carl Sagan make a very good case for comets having been major contributors to the formation of Earth's oceans and the delivery mechanism for organics. As we shall discuss later, if the comets and asteroids could deliver these substances to Earth, it's more than likely that they delivered them to other places in the solar system as well. The implications of this scenario are intriguing.

As Earth cooled and the heavy bombardment subsided, solar and chemical energy led to the synthesis of more of the organic compounds being delivered from comets, and of those formed by interactions between Earth's early atmosphere and its surface. These components, ultimately concentrated in a watery solution with an inorganic chemical milieu, combined to form the very complex macromolecules necessary for life. A very popular theory today is that one of those molecules, ribonucleic acid (RNA), exhibited a kind of molecular evolution. The first stage was a primitive RNA with catalytic properties that made a "nucleotide soup." These RNA molecules self-replicated using recombination. Mutation allowed new functions and niches. By using certain cofactors, the molecules developed a spectrum of enzymatic activity and eventually began to synthesize proteins. Another possible synthetic pathway might have been the early formation of amino acids and their polymerization into "active protoproteins" that could catalyze reactions and form cell-like vesicles. Some scientists think that lipids were important in membrane formation.

Beyond these ideas the picture is very incomplete. Somehow, at least 3.85 billion years ago, Earth's first self-replicating "biological entity" emerged. It began to reproduce and evolve through mutation and natural selection; certain advantageous mistakes during reproduction produced inexact copies of the ancient organism that proved to be better survivors in that particular world. Initially, in making more of itself, the first organism had only to use the chemical materials and sources of energy surrounding it. However, the evolving system was driven by the second law of thermodynamics. As entropy increased, complexity emerged to cope with less available free energy. The surrounding milieu was consumed and changed, and the general state of energy decreased. The new creature and its progeny adapted to the unsteady environment and survived whatever

crises it encountered. These earliest organisms developed an efficient internal chemistry that led to a path of continuous evolution. They became the dominant form, and their success inhibited the emergence of other variants.

This sequence explains how life might have formed but does not answer *where* on Earth all this took place. Oceans have long been widely regarded as the birthplace of life. For many years, researchers (beginning with Darwin) surmised that primitive organisms somehow emerged in a languid tidal pool at the edge of an infant ocean warmed by the Sun. However, that ancient pool was probably not the only "garden of Eden" where life arose. It is just as likely that life began in a deep ocean environment or in a geothermal region, where conditions would be conducive to the processes needed. Moreover, there were significant threats to surface life on the early Earth. One was bombardment by a continual rain of debris, which created violent conditions of heating, cooling, and atmospheric change. Another was the inadequacy of sunlight itself. Astronomers believe that the young Sun was much dimmer than it is now, providing roughly two-thirds of the radiant energy that we bask in today.

Given these problems, is it possible — or more likely — that life formed not on the surface, but in the deeps away from the constantly changing surface conditions? Did organisms form in hydrothermal systems near underwater volcanoes (as in *Figure 2*)? All the necessary conditions would have been met in such locales: water, warmth, and organic matter. Or did life arise deep underground, where the heat sources were more steady and the organics plentiful? As we discuss later, we now know of huge colonies of microorganisms that thrive near deep-sea volcanic vents without benefit of sunlight; others are perfectly capable of living in deep rock formations.

Wherever it emerged, terrestrial life persisted and succeeded. At least one form survived all of the disasters of time, eventually

Figure 2. Did life on Earth begin deep in its oceans? This "black smoker," a mineral-laden hot spring lying 2.5 km below the surface of the Pacific Ocean, may be similar to the primordial oases that provided the energy and nutrients necessary for primitive life forms. This historic first-ever photograph of a black smoker was taken in 1979, after the submersible craft *Alvin* literally bumped into it. The geyser's toppled chimney rests behind the smoke column, and some of *Alvin*'s equipment is visible in the foreground.

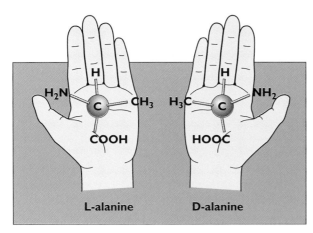

Figure 3. A schematic illustration demonstrates the "left-" and "right-handed" stereoisomers of the amino acid alanine. The Murchison meteorite has an excess of left-handed alanine molecules, the kind of disparity that may have influenced Earth's early stereochemistry.

giving rise to the many millions of species observed today. The evidence is strong that all known contemporary life stemmed from a single occurrence of successful formation. This is because all terrestrial life has the same biochemistry, uses the same genetic code, and has the same set of organic *stereoisomers.* Stereoisomers are versions of a single molecule that are mirror images of one another (*Figure 3*). Organic compounds *not* created by biologic activity contain equal numbers of these isomers, a mixture termed *racemic.* Terrestrial organisms, on the other hand, use only "left-handed" amino acids to make their

proteins, completely rejecting the "right-handed" versions of these same molecules. This handedness, or chirality, is universally accepted as a fingerprint of biological activity.

Interestingly enough, there may be an interstellar link to the left-handedness of life on Earth. The evidence comes from the Murchison meteorite, which landed in Australia in 1969. Like many meteorites of its type, known as carbonaceous chondrites, Murchison contains a wealth of organic material. In its interior geochemists found 16 amino acids, the basic building blocks of proteins. These chemicals did not come from living organisms but result instead from natural reactions that are probably widespread in the universe. However, the compounds show a significant excess of left-handed amino acids (called L-enantiomers) over right-handed ones (D-enantiomers). One problem in performing this delicate research has been the very real possibility of the meteorite becoming contaminated by Earth's life, which is based on these same L-enantiomers. However, the relative abundances of nitrogen isotopes in Murchison's amino acids suggest that they indeed have an extraterrestrial source.

This has profound implications for the origins of life on Earth. While life most certainly formed here, it now appears very possible that it is based on the chemistries of organic compounds formed well before Earth accreted. Clearly, more studies must be done to confirm this idea, but it does give us one more reason to look for life elsewhere in the solar system. If the flux of meteorites and comets that bombarded the planet delivered a set of blueprints for life on Earth that were clearly left-handed, those same blueprints could be present on other solar-system bodies. Therefore, if life (or its aftermath) is discovered somewhere else in the solar system, it may well have the same chirality as life here. The possibilities are intriguing, indeed.

LIFE'S EXTREMISTS: TEMPLATES FOR LIFE ELSEWHERE?

Because we know that the early Earth exhibited extreme conditions, and because life originated, survived, and even flourished under those conditions, it seems logical to assume that it could gain a foothold in similar environments on other planets. To understand how it might do so on Mars, for example, we look at life found in similar environmental niches on Earth and use it as a way of characterizing those habitats. Remarkably, whole classes of organisms called *extremophiles* exist in some of our planet's most inhospitable locales. These entities make up a domain of life called Archaea (cyanobacteria), which include some of the earliest forms of life on Earth.

Organisms that live in very cold environments are called psychrophiles. They exist quite well at frigid temperatures and may thaw for only a few weeks in the summer. Large numbers of them form colonies in the Antarctic sea ice. Photosynthetic eukarya, algae, diatoms, and bacteria make up a large fraction of these colonies and account for the observed colors of the ice. Methane vented at low temperatures at the ocean bottom forms a gas hydrate, a kind of ice on the seafloor. Scientific expeditions in the Gulf of Mexico using deep-sea submersibles discovered flat, eyeless, pink worms swarming in one of the methane outcrops (*Figure 4*). It appears that the worms live off resident microbes, which gain *their* energy from metabolizing methane.

Figure 4. In 1997 a dense colony of centipede-like worms was discovered living on an outcrop of methane hydrate at the bottom of the Gulf of Mexico. The worms apparently feed on bacteria, which in turn use the methane hydrate as their source of energy for metabolism.

Acidophiles and alkaliphiles live in environments that are extemely acidic (pH below 5) or alkaline (above 9). Found among coal deposits, acidophiles play biological tricks to keep the acid outside their cell membranes. Alkaliphiles, found in soda springs, have special enzymes that function near the cell surface to maintain neutrality inside the cell. Similarly, organisms called halophiles live where seawater has evaporated; they have developed special osmotic mechanisms that allow them to survive among the dry salt crystals.

Many organisms not only survive under extraordinary conditions but sometimes require special conditions for their survival. *Sulfollbus acidocaldarius* is an Archaean organism that grows well at temperature around 85° C. Some bacteria grow at even higher temperatures, exceeding the boiling point of water. *Pyrolobus fumatii,* found in the walls of the "smokers" near deep-sea vents, grows at temperatures of 113° C but does not do well below 90° (which apparently is too cold for it). Conversely, the optimum temperature for *Polaromonas vavuolata* is 4° C, and it cannot reproduce at temperatures above 12° C. *Methanopyrus,* an ancient, seafloor-dwelling thermophile, produces its own methane to survive.

Recently, there has been attention to a group of microorganisms known as lithoautotrophic microbes. These "bugs," as microbiologists like to call bacteria, live at the bottom of the ocean near igneous rocks and obtain their energy from hydrogen, methane, and hydrogen sulfide produced by geochemical reactions of undersea volcanic activity. Their nutrient carbon and nitrogen sources come from dissolved substances in the ocean. Geologists and microbiologists joined forces to explore subsurface environments, examining core samples from a kilometer below the surface. There they have found a whole population of lithoautotrophs, which apparently had remained completely isolated from the surface for millions of years. Does this open the door for subsurface ecologies on other worlds? One thing to remember is that many of these vent-dwellers do not require sunlight. Their energy is derived from chemical reactions — a process called chemosynthesis. This has profound implications for life on places such as the Jovian moon Europa, where the surface is covered with ice but the interior may be warm enough to allow life-sustaining liquid water.

Crytoendoliths are organisms living on or in rocks. The interiors of some porous rocks such as sandstones support whole colonies of lichens, fungi, green algae, and a wide variety of bacteria. Yeasts, amoebas, and cyanobacteria make their home in tiny crevices beneath the rock surface. Consider such an existence from a microbe's scale. A 1-micron-long bacterium living 1 cm below the surface of a rock is buried at a depth 10,000 times its own dimension. It may well have all of the necessities for life and be part of a colony of millions that depend on chemosynthesis or radiation from sunlight for their energy needs. (Sunlight easily penetrates 1 cm into many transparent crystals.)

Until very recently, we knew little of these exotic environments. They and the ecologies they support have been out of our convenient reach. As our studies probe further, we realize that historically we have had a very limited view of what forms life might take. Extremophiles could be the template for life found on other planets. At the very least, we must be prepared to expand our notion of where to look for life –- and that it may

not exist in the shallow oceans, pools of water, and geologic surface units where we've come to expect it.

There exists a great deal of circumstantial evidence that our search for life elsewhere will not go unrewarded. Dozens of simple organic compounds have been discovered in interstellar space by sensitive spectroscopic techniques, with more compounds added each year. Carbonaceous chrondrites like the Murchison meteorite contain several percent organic material, including numerous amino acids. Laboratory simulations of primitive planetary environments can synthesize the precursor organic chemicals of living things with relative ease. Finally, with the discovery and further study of life in Earth's most extreme environments, we continue to arm ourselves with the tools needed for our search for life elsewhere.

THE LURE OF MARS

Of all the solar system's planets and moons, the place where we have always anticipated finding evidence of life — past or present — is Mars. By the late 1960s, based on our existing knowledge of life on Earth, the cosmic abundance of the elements, and the fundamentals of organic chemistry, numerous biochemists fully anticipated finding the molecular precursors of life or, with long odds, perhaps some primitive organisms on Mars.

So NASA took up the challenge, and in the summer of 1975 two identical Viking spacecraft were launched from the Kennedy Space Center to explore Mars. Each spacecraft consisted of an orbiter and an attached lander. The Viking orbiters mapped the Martian surface and identified landing sites that were safe (not too rough) yet still held promise for the detection of terrestrial-type life. The landers took the first pictures from the surface (*Figure 5*), performed tests for living microor-

Figure 5. **A partial view of the landscape surrounding the Viking 1 lander includes a series of trenches excavated by the surface sampler. The tall white mast is topped with a suite of miniaturized weather instruments.**

Figure 6. At one point, Viking 2's sampling arm pushed aside a rock to obtain soil not exposed to large doses of ultraviolet radiation from the Sun. It was thought that the protected soil would provide a safe habitat for Martian organisms, but none were detected by the spacecraft's experiments.

ganisms, and characterized the chemical and physical nature of the surface and atmosphere. The first Viking lander reached Mars on 20 July 1976, and the second arrived several weeks later on 3 September. Both Viking missions were very successful, lasted for more than five years, and performed every experiment perfectly (though Viking 2's seismometer failed to deploy). Little did we know at the time that the Viking landers would provide our only data from the surface of the red planet for two decades.

The Vikings found a world with no seismic activity, a weathered (oxidized) surface covered with volcanic debris, and seasonally changing polar caps. The landers determined the planet's seasonal weather patterns and revealed that the Martian surface material has an unusual oxidizing quality. Many surface samples were tested from around the spacecraft, from under nearby rocks and from digging with the scoop down to depths of 30 cm (*Figure 6*).

A trio of biology instruments tested samples for evidence of metabolism, growth, and photosynthesis – all hallmarks of terrestrial life. The pyrolytic-release experiment sought life forms able to assimilate radioactively "tagged" carbon dioxide or carbon monoxide. A second test, the labeled-release experiment, put the carbon tracer in a nutrient-laden broth that was "fed" to the surface samples. In the gas-exchange experiment, an organically rich "chicken soup" was added to samples in the expectation that metabolic gases would be produced. When subjected to these three tests, the surface material did exhibit an unusual oxidizing quality that mimicked positive responses in some situations. But in the end the Viking biology team concluded that nothing in their instrumental results "could only be explained by the presence of living organisms."

Apart from their tests for life, the Viking landers also carried a critical experiment that searched for organic material of any sort. Meteorites and comets should be delivering organic material to the Martian surface, even if there were no indigenous life on the planet. Thus the organic-analysis team assumed that its task was to distinguish between the organic compounds indigenous to Mars from those coming in on the shattered meteorites from the asteroid belt.

Since very little organic material should exist on the surface, due to the destructive nature of ultraviolet sunlight, the experiment team designed the most sensitive test known to laboratory science at that time. The sensitivity of the instrument reached down to the range of 0.0000001 percent (one part in a billion). In every place on Earth where life exists, this instrument would have detected either the organisms themselves or their organic residue.

Once on Mars, the twin instruments worked perfectly. They analyzed various atmospheric volatiles absorbed in the surface dust: CO_2, CO, N_2, and NO. They measured trace gases like argon, neon, krypton, xenon, and their isotopes — which would later be critical in establishing the existence of meteorites from Mars (see Chapter 26). The instruments even detected the freon used to clean their components during assembly. But they found no organic compounds, no methane, no hydrocarbons, no amino acids, no polycyclic compounds — nothing! The complete lack of organic compounds at either landing site was one of the most important discoveries of the Viking mission, and certainly not one we anticipated.

After combining their results, the biology team made the statement that the Viking results do not permit any final conclusion about the presence of life on Mars. However, a National Academy of Sciences' panel published a report concluding that the Viking results lowered the possibility of life on Mars. Further exploration of the question must await samples of Mars returned to Earth laboratories.

MARS AFTER VIKING

Viking's biology experiments stimulated a brilliant line of research in chemical evolution to understand the origin of life on Earth. It now appears that the organic world of carbon chemistry was very tightly coupled to the inorganic world of aqueous and gaseous chemistry. Reactions involving salts, ions, metals, and especially the early formations of clays were part of the events that led to the first large polymers of organic molecules. Information-bearing molecules followed, which finally led to the processes of self-replication that are considered the hallmark of living organisms.

Another important factor for continued study is the role of water in the Martian environment. Liquid water may be the key to searching for life on Mars. We know that water ice exists at the poles beneath the surface, and it forms a surface frost in the Martian winter. This in itself maintains Mars as a strong candidate for either extant or fossil life. However, we have not determined how deep the polar-ice layer is, or if there is any liquid water at some depth. Drilling will not be easy! The permanent frozen water at the poles is extremely interesting because it could act as a "cold trap" where organic molecules would condense. Moreover, due to the presence of water, the oxidizing agents revealed by the Vikings might be absent.

Figure 7. **The surface of Mars as seen by the Mars Pathfinder lander (foreground) in 1997. By design, Pathfinder landed amid a variety of terrain and rock types that were carved and deposited by a cataclysmic flood billions of years ago. The two hills on the horizon, nicknamed Twin Peaks by mission scientists, lie about 1 km from the lander.**

We continue our explorations of the red planet with an eye toward understanding its geology, and from that clues to its origin and evolution. The successful Mars Pathfinder and the Sojourner roving craft in 1997 showed us another part of Mars. The highly detailed views of their landing site in Ares Vallis, an ancient floodplain, revealed a landscape endowed with rocks of various textures, colors, and compositions (*Figure 7*). Pathfinder's camera saw ice clouds in the atmosphere, and other experiments recorded wind speeds and temperatures. Mars Global Surveyor was placed in orbit around Mars in late 1997 to map the planet's geology, topography, and mineralogy at high resolution. Future missions will include landers as well as orbiters. The exploration strategy for Mars involves polar exploration, drilling to reach the permafrost, and sample-return missions.

I have devoted most of my life to thinking about the question of Martian biogenesis. In my role as Viking project scientist, I began with an optimistic view of the chances for life on the surface of Mars — if it were similar to Earth life. I now believe that our expectations of possibly finding an Earthlike desert, while plausible at the time of the Viking missions must be corrected. Clearly if we continue to search for life on Mars we will have to develop a more complex strategy about in situ searching and sample return. Viking, Pathfinder, and Mars Global Surveyor, it must be stressed, were surface experiments only. If life formed in some subterranean hydrothermal vent system and remained there, as occurs here on Earth, the first round of probes had no chance to discover it. If life exists in rocks, as on Earth, those rocks have not been found. If it existed in the oceans that may have covered parts of Mars' surface in the past, then its remains have not *yet* been found. Thus, the search for life on Mars has — in a very real sense — barely scratched the surface.

THE ALH 84001 METEORITE

There is one experiment for life on Mars that has generated much recent interest: the ongoing tests of rocks found on Earth that came to us from Mars. The most famous of these meteorites is called ALH 84001 (*Figure 8*). It was found in the Allan Hills region of Antarctica in 1984, some 16 million years after being excavated from the surface of Mars during an impact event, and after spending another 13,000 years in the Antarctic ice before being discovered. In late 1993, isotopic assays confirmed that the 4½-billion-year-old rock came from Mars.

In many ways, the analysis of ALH 84001 typifies the interplay of chemistry, geology, and biology that characterizes the search for the origins of life on Earth — experience that will be important in future studies of life on other worlds. Discovered within ALH 84001 were tiny bits of carbonate minerals scattered along fractures within the meteorite (*Figure 9*). In 1996 a group of scientists led by David S. McKay announced that the rock bears chemical and mineral characteristics that they claim are most log-

Figure 8. In 1996 a team of researchers argued that this squash-size, 1.9-kg meteorite contains vestiges of primitive life that thrived on Mars billions of years ago. ALH 84001, its official designation, was discovered in 1984 in Antarctica and is one of only 13 known Martian meteorites.

Figure 9. The carbonate "globules" (colored brown) in the Martian meteorite ALH 84001, in which a team of scientists claim have found evidence for life, are no more than 250 microns (0.25 mm) across. Their round forms are rimmed with minerals rich in magnesium (black) and iron (white).

ically explained as the fossils of long-dead Martian microbes. For example, ultramagnified images show clumps of rounded and elongated "things" no more than 100 nanometers (0.1 micron) long (*Figures 10,11*). The carbonates also contain clusters of tiny magnetite crystals. Since some terrestrial bacteria produce similar-looking strings of magnetite beads, biologists accept such perfect, well-ordered grains as evidence of biologic activity.

Not surprisingly, other research teams dispute that conclusion and have offered a variety of nonbiological interpretations. Microbiologists argue that the minuscule size of the putative "nannofossils" creates interior volumes too small to contain the genetic essentials of a living cell. (This is, in itself, a topic of debate, as some researches have reported that microbes this small do exist on Earth.) Carbon- and sulfur-isotope studies have yet to prove the existence of life or of biologic activity within the rock. However, regardless of the final outcome, ALH 84001 has spurred a burst of study in the very fine points of chemistry and microbiology. The methods used in its analysis will most certainly prove invaluable when the time comes to search for Martian life forms either in rocks on the planet's surface or in samples returned to Earth by spacecraft.

OTHER PLANETARY NICHES

An amazing discovery in the past few decades is that extremely complex organic materials are formed when mixtures of simple gases such as methane, ammonia, water, carbon monoxide, and carbon dioxide are exposed to various sources of energy. These sources can be as varied as a spark discharge (mimicking lightning), ultraviolet radiation (sunlight), or conditions of high temperature and pressure (volcanism). The organic compounds created number in the many hundreds — and most are *prebiotic,* the precursors of living organisms. If organic substances form so easily, it is easy to imagine that they are commonplace on other planets and satellites in our solar system.

Even if planets did not evolve an organic chemistry of their own, an abundance of basic organic building blocks may have been delivered by comets and asteroids. Comets are dominated by the ices of condensed gases (like H_2O, CO, and CO_2) and simple organic compounds like hydrogen cyanide (HCN). But we suspect that the molecular mix is actually much richer. For example, the Giotto spacecraft detected dozens of organic compounds when it flew through the coma of Comet Halley in 1986 (see Chapter 24). Comets are the most unaltered bodies of the solar system, and thus they offer an opportunity to examine some of the pristine solid materials from which the planets accreted. As discussed earlier, organic materials from comets may have contributed to the "primordial soup" that gave rise to life here on the Earth. So biochemists are understandably enthusiastic about mounting a mission to return samples of one or more comets to their laboratories for analysis. They could then answer many questions about the materials and conditions that existed in the period of planetary history that preceded biology.

No matter how rich the "primordial soup" might have been on the worlds around us, it is important to remember that life is a system of highly organized chemical processes. The existence of organic compounds, even complex ones, in itself falls far

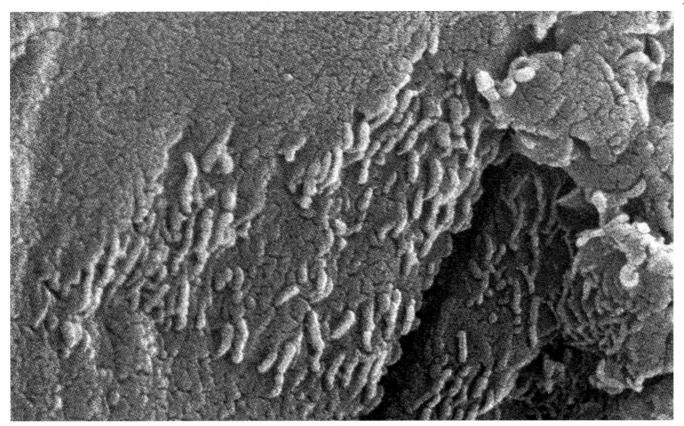

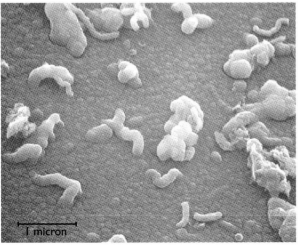

Figure 10 (above). Could the elongated forms in this image, enlarged some 100,000 times, be microscopic fossils of past life on Mars? They bear a strong resemblance to simple organisms that were widespread on Earth nearly 3½ billion years ago, though their terrestrial analogs were 100 times larger.

Figure 11 (left). Biologists have questioned whether the elongated features seen in ALH 84001 (upper panel) are large enough to be the fossilized remains of Martian microbes. Shown in the bottom panel at the same scale are similar features, thought to be biogenic, found on a basalt surface near the Columbia River in the Pacific Northwest. These scanning electron microscope images are reproduced at 15,000 times their actual size.

short of having biology. An analogy might be having a few musical instruments and wishing for a symphony.

Some of our neighboring worlds appear to be extremely inhospitable to life. We have visited the Moon and concluded that there is no life there. Mercury has essentially no atmosphere, and its surface alternates between extreme hot (on the sunlit side) and extreme cold (on its night side). Some have speculated that the twilight zone along the day-night terminator could be an interesting region to search for organic debris, but no one expects life to exist on Mercury.

Venus has a massive atmosphere, 90 times denser than Earth's, that consists mostly of carbon dioxide gas and clouds of sulfuric-acid droplets. Because heat cannot easily escape, the mean surface temperature is well above 700° K. Given the strongly acidic environment, the existence of organic compounds on the surface seems impossible, since in that environment they would all break down. No traces of organics were detected by the Pioneer Venus and Venera atmospheric probes. This absence, combined with the dearth of water in Venus's atmosphere, strongly suggests that the planet harbors no living organisms (at least as we know them).

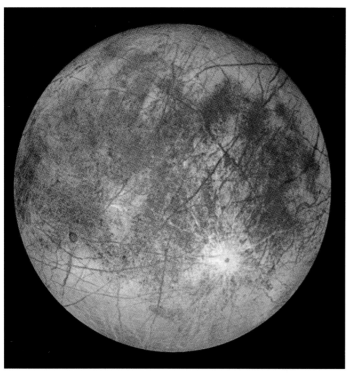

In striking contrast with Earth, the giant planets Jupiter, Saturn, Uranus, and Neptune are composed mostly of gases and ice. These planets are so massive that the envelopes of hydrogen and helium they acquired eons ago remain bound to them by gravity, and thus their elemental composition is rather similar to the Sun's. The giant planets themselves are not believed to be inhabited by living organisms. Nevertheless, scientists who study primitive organic chemistry are anxious

Figure 12 (left). The trailing hemisphere of Jupiter's ice-covered satellite Europa exhibits darker areas of rocky material derived from the interior, implanted by impact, or both. Long, dark lines are fractures in the crust, some of which are more than 3,000 km long. The bright feature containing a central dark spot in the lower third of the image is Pwyll, a young impact crater some 50 km in diameter.

Figure 13 (below). A bloated Jupiter, glowing from the residual heat of gravitational collapse, hangs in the sky of primordial Europa. The giant planet's warmth may have been sufficient to keep Europa's surface liquid, perhaps long enough for life to develop. Eventually temperatures dropped and the surface froze. Did life retreat into a global ocean under the ice crust, where it was sustained not by sunlight but by internal heat?

to learn about the possible reactions taking place in these planets' atmospheres.

It is not the gas giants themselves but rather some of their moons that have attracted the attention of biologists and biochemists. Topping the list is Europa, a moon of Jupiter that is covered by water ice. Voyager flybys and recent Galileo spacecraft data have ignited a great deal of interest in the possibility of life somewhere on this frozen world. The frigid surface is saturated with cracks and crevices that indicate melting and refreezing (*Figure 12*). Either internal volcanic activity or the gravitational pull of Jupiter could cause heating, which in turn would drive the surface changes we see. These processes could also be driving life, because activity also suggests the possibility of a subsurface of liquid water (*Figure 13*).

There is much circumstantial evidence from the Galileo mission that points to a liquid ocean beneath Europa's ice crust (see Chapter 18). If an ocean exists, it is an extremely interesting site for organic chemistry or perhaps primitive life forms made possible by chemical interactions like those that took place on early Earth. We already know from our inventory of terrestrial extremophiles that parts of Europa could be hospitable to such entities as the cold-loving psychrophiles. If a warmer environment exists nearer the core of this moon, it allows the possibility for deep ocean dwellers and vent creatures similar to those we have found on Earth today. Plans are being made to probe the surface and subsurface of Europa by small spacecraft early in the next century. For now, however, the prospects for life on Europa remain in the realm of theory and speculation.

Even more speculative, but hardly less interesting, is the possibility of life on Titan, Saturn's largest satellite. Titan is enveloped by a dense, opaque atmosphere of nitrogen, methane, and traces of ethane, acetylene, ethylene, and cyanide. The surface of Titan is expected to have oceans of liquid hydrocarbons, and its atmosphere contains what appear to be aerosol hazes made up of organic polymers (see Chapter 20). Perhaps Titan's cold, murky landscape holds clues to the pathways that cause organic chemical processes to trigger the actual formation of some kind of exotic living organisms. In any case, we hope to learn much more about what lies beneath Titan's thick haze in 2004. If all goes as planned, the Cassini spacecraft will arrive at Saturn and send its Huygens probe through Titan's atmosphere and onto its surface.

SEARCHING ELSEWHERE

Interest in life in the solar system has expanded to life in the cosmos. This broadened interest has spawned a new field: *astrobiology*. Its practitioners study life in the universe, its origins, distribution, chemical and evolutionary processes, planetary environments, ecosystems, and the use of humans and robots for this exploration. This work is accelerating now that we have confirmed the existence of planets around other stars (see Chapter 28). None of these planets seems Earthlike, though almost everything we know about them is an inference based on their orbit or presumed mass. Nonetheless, their discovery has reawakened the interest in searching for extraterrestrial life.

Assuming that our Sun is an average star and that planetary systems are numerous, we can imagine that many planets of the cosmos have evolved and nurtured living organisms. Some, like Earth, may have developed technological civilizations (*Figure 14*). If so, they might well communicate with each other using high-powered radio transmitters. We now have the technology to listen for such signals, and a handful of programs worldwide are engaged in this Search for ExtraTerrestrial Intelligence (SETI). This electronic eavesdropping is fraught with uncertainty. However, SETI's proponents maintain that the payoff is so large that a modest effort is warranted — and most scientists agree.

Of course, not all life may be able to actively signal its existence, so we must search for other signs. What assumptions and tests would we make to find life on extrasolar planets? Studies of our own early solar system focus on such aspects as the temperature regimes in the solar nebula and their effect on the habitable zones where life could form. So for now we can simply extrapolate what we know of our own system to the formation of other systems and habitable zones.

Another issue is that our own Moon may have played a role in stabilizing the obliquity (tilt) of Earth's spin axis. Mars has apparently no such stabilizing influence, and geologic evidence suggests that the obliquity of Mars has undergone wide swings in the past. A wildly varying obliquity would make it more difficult to maintain the stable climate needed to sustain life once it formed. Also, a planet's land-to-sea ratio, and its placement in its own system's habitable zone may play larger roles in sustaining life than the obliquity.

Finally, giant planets may play a role in creating and sustaining life. Jupiter, with its mass and gravity, has swept away much of the debris that remained after planetary formation. Because of that cleansing, we are now much less susceptible to total annihilation because of a major impact. On the other hand, Jupiter and the other giant planets continue to redirect some of that debris in our direction. Thus, the existence or lack of a giant planet around another star could prove crucial to the development and endurance of life in its solar system.

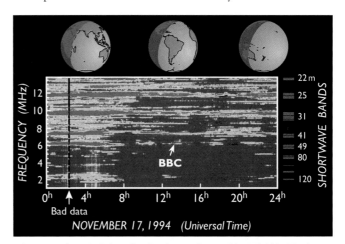

Figure 14. A typical day of radio chatter detected by NASA's Wind spacecraft, which at the time was 204,000 km from Earth. Note how emissions fall into discrete bands, matching those allocated to commercial shortwave broadcasts (at right, with wavelengths in meters); red indicates a signal at least three times stronger than the background of natural cosmic sources. The received signals are most intense when Asia, Africa, and Australia are in view. One isolated burst (arrowed) corresponds to BBC transmissions from Delano, California.

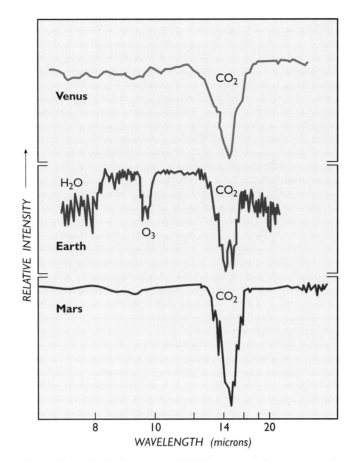

Figure 15. Is this the fingerprint of life? These near-infrared spectra of Venus, Earth, and Mars share prominent absorption features due to carbon dioxide, but only Earth's bears the signatures of water and ozone. An Earthlike spectrum from another world may someday provide us with our first evidence of extraterrestrial biology.

These and other assumptions will be used by the coming generation of astrobiologists to devise the spectroscopic, observational, and inferential tests necessary to determine the likelihood of life on another planet (*Figure 15*). However, there will be many challenges, most of them a consequence of the great interstellar distances involved. Perhaps the greatest challenge will be to understand why we exist at all. In cosmic terms, we have barely begun to understand ourselves and our origins. Truly, the story of life is as complex as the events that influenced its formation. It is a long story, and many more chapters must be written before we understand the processes that have combined to form life here or elsewhere.

28

Other Planetary Systems

R. Paul Butler

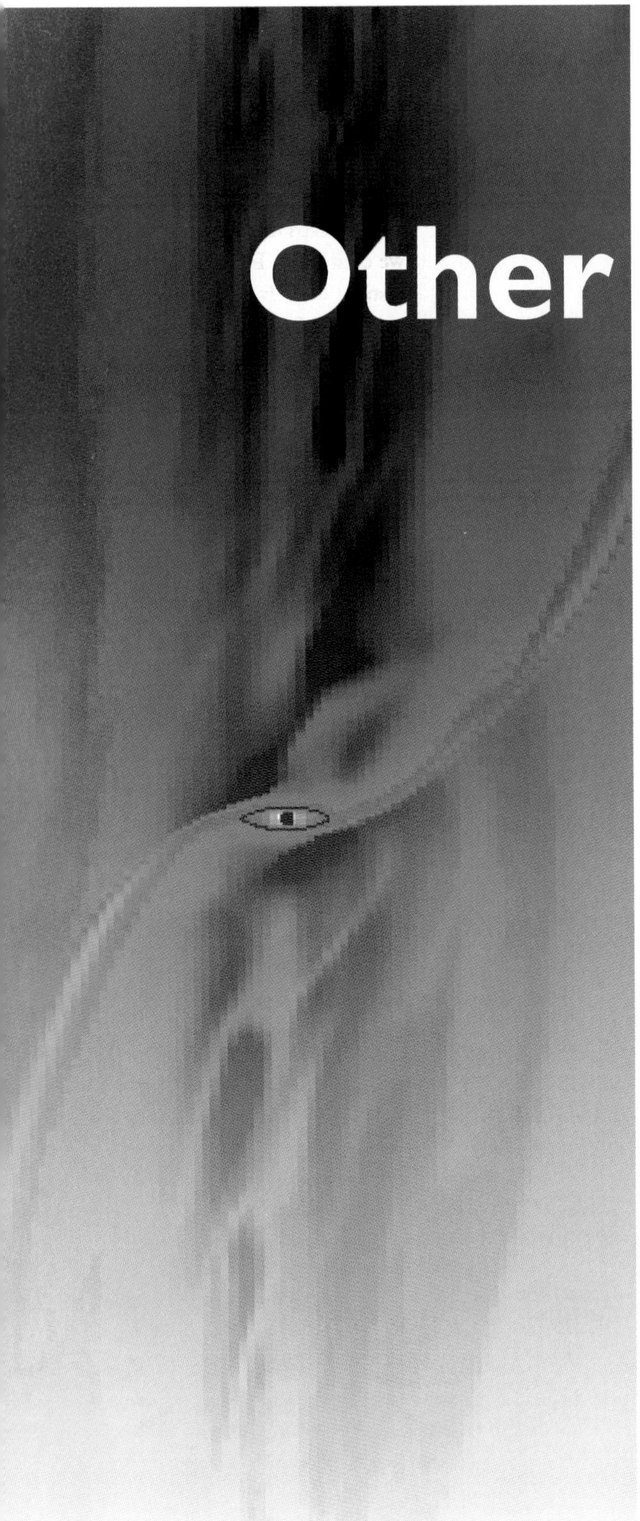

A computer simulates the birth of a Jupiter-size planet around another star.

MODERN ASTRONOMY REVEALS to us, for the first time in history, scenes from one end of the cosmos to the other. We have picturesque views of planetary surfaces in our own solar system — as this book amply demonstrates — and panoramas of adolescent deep-field galaxies swarming near the limit of the observable universe. Beyond providing pretty pictures, astronomy places our world and our brief human lives in their true contexts: as vanishingly tiny subplots in a truly enormous cosmic play. The curtain opens with a Big Bang synthesis of the chemical elements that eventually lead to self-replicating, competitive structures of molecules we call "life." While we humans play out our brief bit parts, we yearn to grasp the overall plot. Naturally, we wonder whether there are worlds beyond those of our solar system. Are they numerous or rare? How many of them have conditions ripe for biology?

These are not new questions. In the fourth century BC, the Greek philosopher Epicurus spoke boldly of the infinite worlds that logically followed from the infinite number of "atoms" that he postulated. His contemporary, Aristotle, differed, seeing Earth as the unique center of a perfect crystalline sky. Aristotle's Earth-centered cosmos dominated Western thought for more than 1,500 years. The notion of other worlds took hold again only after Copernicus yanked Earth from its central position and placed it in orbit around the Sun with other planets. Soon various thinkers realized that the stars might be distant suns and therefore might have planets of their own.

For centuries thereafter, detecting these *extrasolar planets* seemed beyond all possibility. Shining by reflected light, such objects should be roughly a billion times (perhaps 22 to 25 magnitudes) fainter than their host stars. Moreover, they would appear separated by less than a few arcseconds, at best, from even the nearest stars in our stellar neighborhood. If extrasolar planets exist, they are lost in the glare surrounding a

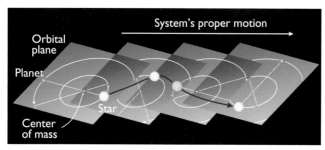

Figure 1. A visible star and its unseen companion orbit their shared center of mass. The system moves through space as well, causing the star to trace a wobbling path across the plane of the sky. Observers seek to detect extrasolar planets astrometrically, by tracking the star's tiny motions with precise positional measurements.

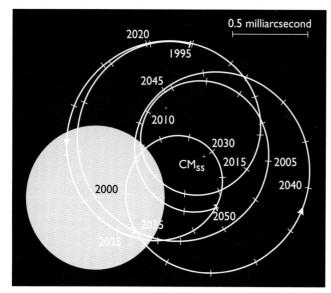

Figure 2. Alien astronomers tracking the Sun's motion from a distance of 33 light-years toward the North Ecliptic Pole in Draco would see this complex wobble after removing the effects of proper motion and parallax. A cross marks the solar system's center of mass.

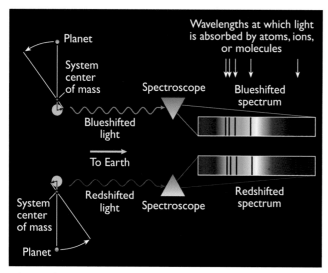

Figure 3. A star approaches and recedes from Earth while orbiting the center of mass it shares with its unseen companions. This induces periodic Doppler shifts in the spectral lines that emerge from the star's atmosphere. Tracking these minute wavelength shifts with state-of-the-art spectrographs enables astronomers to determine the star's reflex motion and in turn to estimate the masses of unseen planets.

star's image. For these reasons, detecting a planet's visible or infrared light directly has long been a pipe dream. (However, as described later, astronomers are on the verge of a revolution in planet-spotting technology.)

In the meantime, a number of indirect observing strategies have been devised over the last century, and especially over the last few years, to overcome the extreme difficulties in detecting planets orbiting even the nearest of the stars. Indirect techniques focus on the small perturbations that a planet imposes on its host star.

For example, Jupiter and the Sun are locked into a gravitational dance around their common center of mass. Since the Sun is 1,050 times more massive than Jupiter, their center of mass is 1,050 times closer to the Sun — a point about 50,000 km outside the solar sphere. Simply by moving along its orbit, Jupiter causes the Sun to circle around that point. Alien astronomers could infer the presence of Jupiter by noting that the Sun is periodically wobbling relative to the background stars (*Figures 1,2*). Viewed from 33 light-years away, the Sun would appear to travel around a tiny circle with a diameter of only 0.001 arcsecond (1 milliarcsecond) every 12 years. That's the apparent size of a hula hoop on the surface of the Moon as seen from Earth.

If the aliens were fortunate enough to be viewing our solar system edge on, they could deduce Jupiter's existence by observing that the Sun becomes about 1 percent fainter for several hours every 12 years (Jupiter's orbital period) as the giant planet crosses in front of the Sun. This technique is known as *transit photometry*. Planets in small orbits, especially those with orbital periods of a few days to a few weeks, have a much-enhanced chance of being detected this way. Several research groups are looking for the transit signature of a planet by monitoring clusters of thousands of stars or selected eclipsing binary stars. European and American groups are also designing space-based systems to monitor thousands of stars in rich clusters simultaneously. Such satellites could be built with enough sensitivity to detect the much smaller signal from Earth-sized planets crossing in front of host stars.

WOBBLING PLANET DETECTIONS

Over the last 20 years observers have most often searched for extrasolar planets using *Doppler spectroscopy*. Sometimes called radial-velocity spectroscopy, this indirect detection method has paid off spectacularly. Just as a leashed dog can jerk its heavier owner around in circles, a gravitationally bound planet will swing its star around their shared center of mass in a small mirror image of its own orbit. Such a stellar wobble, *or reflex motion*, betrays the existence of an unseen orbiting body. The size of the wobble yields the planet's mass. The time the star takes to complete one wobble is the planet's orbital period.

The challenge rests in detecting the tiny stellar movements, which are best detected by the Doppler effect they impose on a star's light (*Figure 3*). As a star approaches the observer, its light waves become very slightly compressed, shortening the wavelengths toward bluer colors. Conversely, as the star recedes from Earth, the wavelengths are slightly lengthened, or redshifted. These Doppler shifts are excruciatingly tiny. The Sun wobbles by about 12.5 m per second, so a clear, reliable detection of a

Jupiter-induced wobble requires measurement precision of about 3 m per second, or easy jogging speed. This causes the wavelengths of starlight to change by a mere one part in 100 million. Until recently such precision was far beyond reach, but it is now possible.

The instrumental heart of the Doppler-spectroscopy technique is a high-resolution spectrometer. The spectrum of a Sun-like star is filled with rich detail in the form of absorption lines (small dark gaps in the otherwise continuous rainbow spectrum). These lines convey information about the chemical composition, temperature, pressure, magnetic activity, and spin rate of the star, and they also provide the wavelength markers for Doppler measurements. A change in the position of these lines reflects a change in the line-of-sight velocity of the star in space.

For several decades the precision of astronomical Doppler measurements had been stalled at about 500 m per second. The problem was that very, very small changes in a spectrometer's condition (such as temperature, pressure, and illumination of the optics) produced spurious "instrumental" Doppler shifts that are much larger than the signal of a planet. Two different solutions have overcome these difficulties. One was the introduction of "super-stabilized" spectrometers. The other was to combine a control spectrum (typically that of vaporized iodine or hydrogen fluoride) with the starlight prior to entry into the spectrometer. A computer model then uses the known spectrum to monitor and allow for any variations in the spectrometer.

In 1991, even as astronomers struggled to perfect their spectrometers, the first system of extrasolar planets was discovered quite unexpectedly. While using the giant 300-m-wide

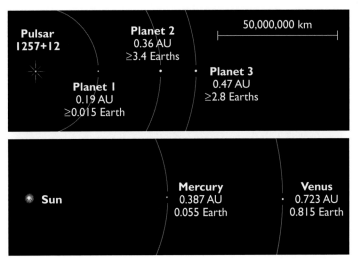

Figure 4 (above). Radio astronomers have found strong evidence that at least three planets circle PSR 1257+12, a fast-spinning pulsar in the constellation Virgo (upper panel). Its outer two worlds have masses not much greater than Earth's. However, the pulsar's wind of energetic particles and radiation strips the planets of any atmospheric gases, leaving their surfaces parched and inhospitable. The inner planet takes just 23 days to orbit the pulsar, the middle one 67 days, and the outer one 95 days. For comparison, Mercury (lower panel) takes 88 days to circle the Sun.

Figure 5 (below). Eleven solar systems at a glance. The stars are scaled to their relative sizes, though these are known only approximately. L_{Sun} is luminosity compared to the Sun's output; M_{Jup} is the companion's mass compared to Jupiter's. At this scale all the planets would be much smaller and farther from their stars than shown. Planets with eccentric orbits have horizontal bars showing how near and far they venture from their suns.

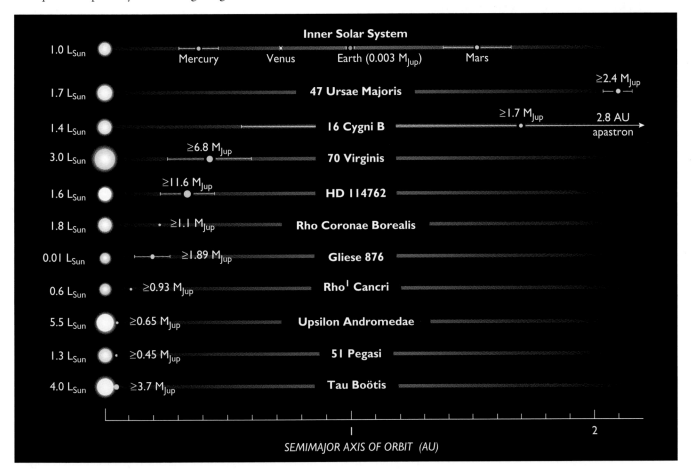

Known Planets of Sunlike Stars

Star name	Distance (light-years)	Spectral type	Star mass (Suns)	Orbital semimajor axis (AU)	Orbital period (days)	Orbital eccentricity	Doppler semiamplitude (m/sec)	Minimum mass (Jupiters)
14 Herculis	59	K 0V	0.80	2.5	1,619	0.35	80	3.3
47 Ursae Majoris	46	G 0V	1.1	2.1	1,098	0.10	46	2.4
16 Cygni B	72	G 2.5V	1.0	1.7	802	0.67	44	1.7
70 Virginis	59	G 4V	0.95	0.47	116.6	0.40	308	6.8
HD 114762	90	F 9V	1.15	0.36	83.9	0.34	613	11.6
Gliese 876	15	M 9V	0.32	0.20	61.1	0.37	220	1.89
Rho Coronae Borealis	57	G 0V	1.0	0.23	39.6	0.04	67	1.1
Rho¹ Cancri	44	G 8V	0.85	0.11	14.64	0.03	77	0.93
Upsilon Andromedae	57	F 7V	1.25	0.056	4.61	0.10	74	0.65
51 Pegasi	50	G 2.5V	1.0	0.051	4.23	0.01	55	0.45
Tau Boötis	49	F 7V	1.25	0.045	3.31	0.006	469	3.7

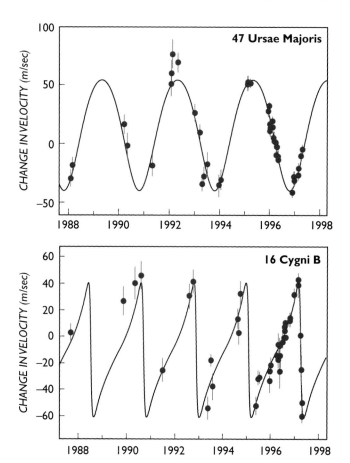

Table 1. The planets found around nearby, solar-type stars through mid-1998 exhibit a wide range of orbital characteristics and likely masses. Not included here are the three planets surrounding the pulsar designated PSR 1957+12.

radio telescope near Arecibo, Puerto Rico, Alex Wolszczan found that a trio of small planets orbits the pulsar PSR 1257+12 (*Figure 4*). Pulsars are dead stars, the rapidly spinning, neutron-star remnants of supernova explosions. They beam intense radio waves along their magnetic field lines, and from our perspective these radio waves appear to blink on and off, like a lighthouse, as the pulsar spins. Ultraprecise timing of PSR 1257+12's signals against atomic clocks indicated extremely small wobbles in its motion, the telltale signature of orbiting companions. Pulsar timing is the only technique currently able to detect Earth-mass planets.

The Doppler technique succeeded soon thereafter, and since 1995 extrasolar planets have been unambiguously detected around 11 Sunlike stars (*Figure 5, Table 1*). Michel Mayor and Didier Queloz discovered the first planet orbiting a normal star, 51 Pegasi. The search conducted by Geoffrey Marcy and the author revealed the next six new planets, one of which (around 16 Cygni B) was found independently by William Cochran and Artie Hatzes. The planet circling Rho Coronae Borealis was found by a team of nine astronomers led by Robert Noyes. One

Figure 6 (above). Velocity curves for the two stars with planets in relatively long-period orbits. Vertical bars show the measurement errors. The curve that fits the data for 47 Ursae Majoris best is nearly sinusoidal, revealing a nearly circular orbit. The curve for 16 Cygni B shows the distinctive signature of a much more elliptical orbit.

Figure 7 (right). An unavoidable ambiguity occurs in spectroscopic estimates of planetary masses because the same radial-velocity signal can be produced either by a lower-mass planet orbiting in a plane near the line of sight, or by a a more-massive planet in an orbit seen nearly face on. Because of this, some recently discovered stellar companions may be brown dwarfs, not planets. Arrow length represents velocity.

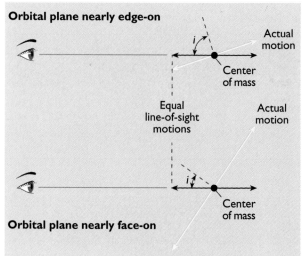

additional object, discovered by David Latham and others, orbits the star HD 114762 in Coma Berenices. It has a large mass, more than 11 Jupiters, which suggests that it is perhaps not a planet but instead a substellar brown dwarf (see Chapter 14).

The orbital period of each planet is the wobble cycle time. From that period and Kepler's Third Law, the orbit's semimajor axis (the planet's average distance from its star) can easily be determined. If the velocity varies like a perfect sine wave, we know the orbit is circular; if not, the skewness of the velocity curve can be analyzed to find the eccentricity of the elliptical orbit (*Figure 6*).

The amplitude of the Doppler variations tell us the planet's minimum mass, not its actual mass (*Figure 7*). If the orbit were seen nearly edge on, the Doppler shifts would yield the star's full orbital velocity. However, the viewing angle is unknown, and orbits that are strongly tilted with respect to our line of sight will produce a subdued Doppler shift. Thus, we can only determine the mass of the planet multiplied by the sine of the orbital inclination ($M \sin i$), where i remains unknown. Our inferred mass will most likely be smaller than the real one. Luckily this problem is not too great. For randomly oriented orbit planes, the true mass of a planet will be, on average, $\frac{1}{2}\pi$ times the minimum mass we determine. So the true mass very probably lies within twice the value we measure.

A PLANETARY ZOO

The extrasolar planets discovered around Sunlike stars have surprised more than a few astronomers. They all have roughly Jupiterlike masses, with values of $M \sin i$ between 0.4 and 12 Jupiters. Of course these stars could also have smaller planets that are currently beyond our detection limits, but larger ones are unlikely. Any companion having a mass of 10 to 80 Jupiters would have been very easily detected. Apparently, nature rarely makes planets much larger than Jupiter. Why that is so remains unclear.

What's more astonishing is that six of these Jupiter-mass planets orbit remarkably close to their host stars, within 0.25 AU. This is much closer than Mercury's mean distance from the Sun (0.39 AU). The Doppler search technique preferentially identifies such tightly bound planets, because their orbital velocities are very fast. No one expected to find any giant planet so close to a star, and no one knows for sure how they could have gotten there.

By comparison, the planet orbiting 47 Ursae Majoris seems more similar to our Jupiter. It has a minimum of 2.4 Jupiter masses, its orbit is almost circular, and its orbital radius of 2.1 AU corresponds to the inner edge of our asteroid belt. Such a planet, if placed in the solar system, would look like Jupiter's big brother.

Also puzzling are the six known planetary companions that have dramatically elliptical orbits (*Figure 8*). The companions to 16 Cygni B, 70 Virginis, Gliese 876, and 14 Herculis have eccentricities of 0.67, 0.40, 0.37, and 0.35, respectively. (The largest eccentricities in our solar system, for Mercury and Pluto, are about 0.2.) Why do some stars have planets in such strongly elliptical orbits? Indeed, how common are *circular* planetary orbits? The arrangement of our own solar system has always seduced theorists into including just enough physical processes

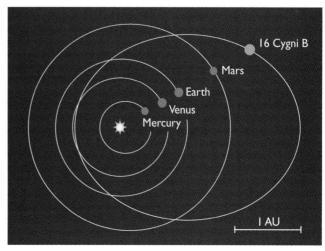

Figure 8. The planets in our solar system have nearly circular orbits, but nearly half of the known extrasolar planets do not. The most extreme case is the planet of 16 Cygni B (eccentricity 0.57). Its orbit is shown superposed on the inner solar system.

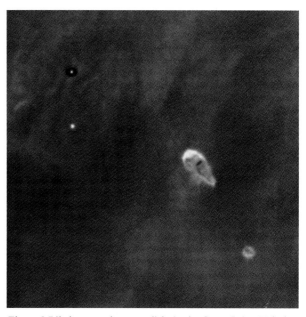

Figure 9. Likely protoplanetary disks in the Great Orion Nebula, photographed by the Hubble Space Telescope. The largest blob contains a dark, nearly edge-on disk about the size of Pluto's orbit. The bright part of the blob is gas that has been evaporated from the disk's surface by radiation from hot stars outside the frame. At least two other circumstellar disks dot this small field.

in their calculations to yield circular orbits as "predictions" for solar systems everywhere.

When the new discoveries showed that ellipses happen, theorists scurried back to their blackboards and soon developed some beautifully sensible elaborations on standard planet-formation theory. Pawel Artymowicz and Patrick Cassen have reanalyzed the birth of planets in the kind of protoplanetary disks that are observed around newborn, Sunlike stars (*Figure 9*). Their calculations show that protoplanets exert gravitational forces on the disk that trigger spiral density waves, not unlike tightly wound arms in spiral galaxies. These spiral waves in turn exert subtle gravitational forces back on the planet, pulling it away from pure circular motion. In the course of a million years the departures from circular motion can build up to be quite significant, leading to eccentric orbits. It is not yet clear whether spiral waves can induce eccentricities as high as 0.6. Nonetheless, this theory echoes the wild historical past of planet hunting. The first photographs of "spiral nebulae" (now known to be galaxies), taken by James Keeler at Lick Observatory in the early 1900s, were originally interpreted as planetary systems caught in the act of formation.

A second theory for the large orbital eccentricities seems equally likely. Suppose, for a moment, that Saturn had grown to a larger mass as it condensed around the infant Sun. Indeed, all four giant planets might have attained larger masses if the protoplanetary disk had contained more material or had lasted longer. In that case our solar system would have started out with four true heavyweights, each exerting greater gravitational force on the others than it does now. They would perturb each other's orbits until they overlapped. The planets would then gravitationally scatter one another and assume chaotic, unpredictable final orbits — almost like the break of balls in billiards (*Figure 10*).

The real planetary pool table, however, is not flat but rather has a central depression produced by the gravity of the host star. The scattered Jupiters careen around the depression for a while. Eventually the situation stabilizes when planets collide and merge, fall into the star, or are flung out of the system entirely. What's left will be just a few planets in eccentric orbits that are far enough apart not to affect each other much.

Nature may have provided us with an excellent example of a chaotic planetary system in the making. The southern-sky star Beta Pictoris is surrounded by a large disk of dust and gas that appears nearly edge on to us. First seen in 1984, the disk has a clearing in its center roughly 8 billion km (50 AU) across. Astronomers believe one or more planets may reside in this clearing, and they may be sizable. The inner part of the disk is "warped" — it occupies a different plane than the part farther out (*Figure 11*). There is strong circumstantial evidence that the warping is due to gravitational force of a planet with a mass anywhere from $\frac{1}{20}$ to 20 times the mass of Jupiter, depending on its exact location within the disk (or the central clearing). This body

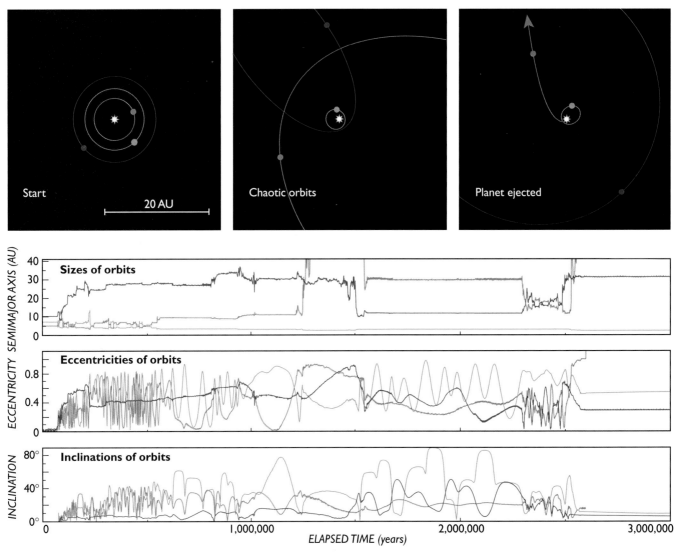

Figure 10. Survival of the fittest. In this computer simulation (*upper panels*), three Jupiter-mass planets start in nearly coplanar orbits 5.0, 7.3, and 10.2 AU from the Sun. As they perturb each other, their orbits begin to cross and evolve chaotically after only about 100,000 years. Eventually one planet is usually thrown out of the system (right), leaving the other two in eccentric orbits. The *lower panels* show how the semimajor axes, eccentricities, and inclinations of the three orbits change drastically and unpredictably before one planet is lost.

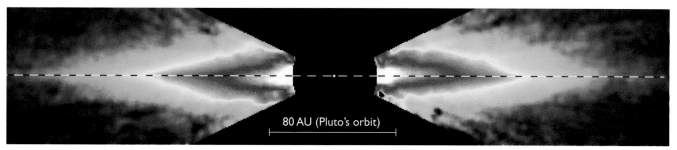

80 AU (Pluto's orbit)

is apparently not very close to the star itself, because we do not detect Doppler variations in the spectrum of Beta Pictoris. However, the warp's tilt may be telling us that the planet occupies an inclined orbit. If that proves true, this system may be in the throes of a "planetary shakeout" like that depicted in *Figure 10*.

The gravitational-scattering model can explain even the most extreme eccentricities seen among the extrasolar planets. If this is what really happened at 16 Cygni B and the other eccentric companions we should eventually detect other remaining planetary billiard balls in these systems, presumably orbiting much farther from each star. A variation on this theory is that a binary-star system would perturb planets orbiting tightly around either star. Indeed, 16 Cygni B is accompanied at a great distance (at least 880 AU) by the more massive star 16 Cygni A, which may be responsible for the large eccentricity of the planet around star B.

MYSTERIOUS SHORT-PERIOD PLANETS

The most mysterious and controversial new planets in the menagerie are the six "hot Jupiters" in circular orbits closer than 0.25 AU from their stars. These are the companions of Tau Boötis, 51 Pegasi, Upsilon Andromedae, Rho¹ Cancri, Rho Coronae Borealis and Gliese 876. Their orbital periods are 3.3, 4.2, 4.6, 14.6, 39.6, and 61.1 days, respectively, compared to Mercury's 88-day orbit around the Sun. The planets' minimum masses range from 0.45 Jupiter (for 51 Pegasi) to 3.7 Jupiters (for Tau Boötis). Five have essentially circular orbits.

How did these giant planets end up so close to their stars? They starkly contradict standard solar-system formation theory (see Chapter 2), which dictates that giant planets form at least several AU from the host star. There in the cool, outer regions of a protoplanetary disk, small ice grains of frozen water and methane can survive, collide, stick together, and grow to become planetary embryos that gather into the cores of giant

Figure 11. **The Hubble Space Telescope obtained this 22-arcsecond-wide image of the dusty disk around Beta Pictoris, a type-*A* star 63 light-years away. A subtle warp (from lower left to upper right) in the disk's inner zone may betray the gravitational effects of a Jupiter-mass planet closer in.**

planets. If this scenario is correct, how did these six extrasolar giant planets arrive so absurdly close to their stars, where ices are not stable? These planets either formed right where they now reside, with abundant rocky grains obviating any need for ices, or they formed farther out and migrated inward.

The leading theory comes from Douglas Lin, Peter Bodenheimer, and Derek Richardson, who borrowed heavily from the work of William Ward. In their view, a protoplanet in its natal disk *must* move inward, for two reasons. First, the disk material itself is spiraling into the young star due to viscosity (fluid friction) within the disk. This general flow will drag protoplanets inward. Young solar-type stars do show clear spectral signs of accretion in their excess ultraviolet and infrared emissions. So, inward migration can hardly be avoided. Second, protoplanets will have resonant orbital periods with those of the spiral waves they set up in the disk. This causes the protoplanets to lose angular momentum even if the disk was not itself draining inward. Inevitably, they will be dragged toward the stellar furnace.

This inward planet migration raises enormous conflicts, as yet unresolved. The six short-period planets apparently halted their inward migration and "parked" in orbits just short of destruction (*Figure 12*). How did they stop? Moreover, why aren't all giant planets found close to their stars? Why didn't Jupiter and

Figure 12. **Possible events near a newborn star. The star is still surrounded by a massive, dark disk of gas and dust, but the star's magnetic field has swept the disk's innermost region clear. A giant planet has become gravitationally stabilized just inside the disk's inner edge. Farther out, a second giant planet has cleared an empty ring in the disk; the ring is bounded by spiral density waves that the planet has set up, which in turn will shepherd it inward. This portrayal is not to scale.**

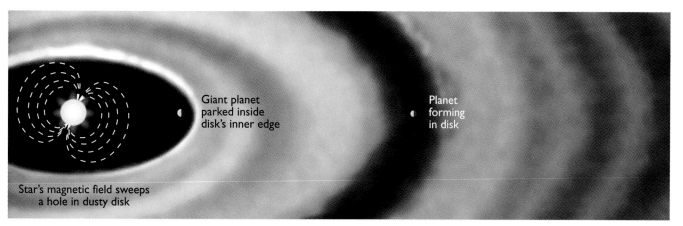

Giant planet parked inside disk's inner edge

Planet forming in disk

Star's magnetic field sweeps a hole in dusty disk

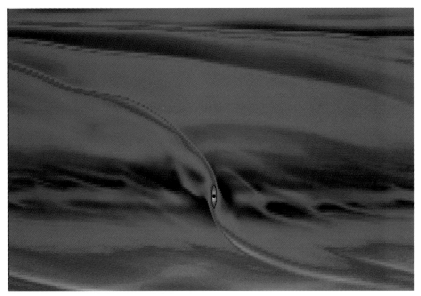

Figure 13. **The birth of a Jupiter. In this computer simulation, gas continues to pour onto a newly condensed giant planet even after the planet has swept a ragged gap in a star's protoplanetary disk.**

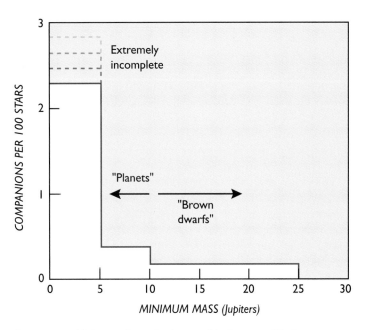

Figure 14. **In addition to planets having roughly the mass of Jupiter, some astronomers are discovering much higher-mass companions orbiting stars. Objects in the range of 10 to 80 Jupiter masses are often classed as brown dwarfs. Even though they are much easier to detect, they are proving to be rarer than lower-mass stellar companions — suggesting that we are seeing different classes of objects that form by different methods.**

its siblings move inward? Maybe they did! One theory suggests that a number of giant protoplanets formed at least 5 AU out, migrated inward, and met their fate in the Sun. Jupiter and the rest of the Sun's familiar planets were only the last in line, left stranded when the rest of the protoplanetary disk cleared out.

Some preliminary resolutions to this migration dilemma are emerging. The protoplanetary disks around young Sunlike stars probably develop a hole or a gap in their centers with a diameter about 10 times larger than the star itself. This central hole is cleared by the spinning star's magnetic field, which entrains the hot gas

(especially the ions and electrons), either flinging it outward or forcing it to flow along magnetic field lines down onto the star. The hole provides a safe parking spot for the inwardly migrating planet. Once it emerges from the inner edge of the disk, the planet is no longer dragged farther inward. A protoplanet may be fortunate enough to remain in the clearing until the disk disperses, thus allowing it to orbit there happily ever after.

The magnetic-gap parking mechanism cannot explain the planets around Rho[1] Cancri or Rho Coronae Borealis, whose orbital radii are 0.11 and 0.23 AU, respectively. These distances are well outside the postulated clearings. So we are left with a mystery: If orbital migration brought these objects inward, why didn't they continue? Maybe they were left stranded when the protoplanetary disk dissipated away. Maybe they formed right where they are, without any orbital migration at all. Perhaps especially massive protoplanetary disks carry enough ice-free material (rock dust and iron dust) to form giant planetary cores.

Why doesn't our solar system have a giant planet in close to the Sun? Perhaps Jupiter formed near the end of the lifetime of our protoplanetary disk, as the viscosity waned and died out. Perhaps our disk never had enough gas and dust to exhibit much viscosity in the first place, so it stayed put until the heat of the newborn Sun blew it away. Indeed, protoplanetary disks cover a wide range of masses — from a few Jupiters up to hundreds, as deduced from the radio emission by their dust. So the diversity of observed planets may represent a sequence of protoplanetary disk masses or lifetimes, and perhaps chemical compositions.

A challenge to the very existence of the six shortest-period extrasolar planets came in early 1997. High-resolution spectra of 51 Pegasi suggested that the *shapes* of the star's spectral lines, not their apparent wavelengths, are what change with a 4.2-day period. This would imply that the photosphere of the star is undergoing a complex oscillation of a sort never seen in the Sun nor anticipated theoretically. More importantly, it suggested that the Doppler periodicity of 51 Pegasi could be explained without invoking a planet.

However, more careful study revealed no such oscillatory "overtones" in the Doppler measurements of 51 Pegasi. Instead, the shapes of the spectral lines remain constant, and the putative planet neatly explains the single, clear Doppler period of 4.2 days. The spectrum of Tau Boötis underwent similar scrutiny. This star appears to host the most massive and closest-orbiting "hot Jupiter." Its Doppler-shift cycle is the strongest yet detected — so strong, in fact, that it could have been discovered many years ago had anyone been looking! Tau Boötis showed no variations in the shapes of its spectral lines at all. Thus, oscillations cannot explain a star's Doppler shifts, and the existence of a planet remains the only viable conclusion.

PLANETS OR BROWN DWARFS?

How can we be sure that the eight companions discovered by the Doppler approach are really planets as such, rather than extremely low-mass brown dwarfs? Sometimes considered "failed stars," brown dwarfs are objects with too little mass to shine by their own nuclear power, a threshold achieved at 0.08 solar mass (about 80 Jupiters). In today's parlance, the distinction between a "brown dwarf" and a "planet" with the same mass in the same orbit depends on how the body formed. A planet grows from dust and gas accreting in a circumstellar disk (*Figure 13*). A brown dwarf forms the way a star does, by a fragment of a gas cloud collapsing in on itself before the gas has the chance to develop a protoplanetary disk around another star.

Some good evidence bearing on this question comes from the distribution of the newly-discovered objects' masses (*Figure 14*). The most numerous companions have the lowest masses, below 5 Jupiters. This is absolutely remarkable, because such objects are the most difficult to detect with the Doppler technique (or any other). In fact, we are probably overlooking many companions with masses below 5 Jupiters because they induce Doppler shifts in their stars too small to be detected. In contrast, we are hardly missing any companions between 20 and 70 Jupiter masses, because their Doppler amplitudes would be so strong.

Thus, two populations of low-mass objects are found around Sunlike stars. There is a spotty population having masses between 10 and 70 Jupiters; some astronomers would call these brown dwarfs. The striking concentration in mass below 5 Jupiters suggests a separate class that may have formed by a different method. The term "planetary" seems quite appropriate for these. Three of them (the companions of 47 Ursae Majoris, Rho Coronae Borealis, and Rho¹ Cancri) have circular orbits in situations where the orbits could not have been circularized by tides from the system's star.

THOUGHTS ON PLANETARY DIVERSITY

With only 14 extrasolar planets confirmed, we lack a good statistical sample against which to compare our own solar system. For example, all the extrasolar planets found so far by the Doppler technique have orbital periods of less than 5 years. This is not a reflection of planetary systems in general but rather due to the limited duration (about 10 years) of the hunts to date. With time and improved Doppler precision, more planets may be found in slower, longer orbits farther from their stars.

However, the meager sample is already sending us a warning that our solar system may not be the norm. Suppose that gravitational scattering of planets is common in newborn solar systems. After all, many signs give testimony to heavy bombardment by bodies in crisscrossing orbits when our solar system was young. The cratered faces of the Moon and other bodies, the Moon's very formation, and the extreme axial tilt of Uranus all tell of a violent, impact-racked infancy. The neat, racetrack orbits of today's middle-aged solar system are the crash survivors from its reckless youth. Our system may be unusual in that small planets like Earth were not flung away by giants falling into eccentric orbits, as may have happened around 16 Cygni B.

An outcome of tidy, circular orbits may require special starting conditions. The near-circular orbit of Jupiter actually promotes the stability of circular orbits among the other eight

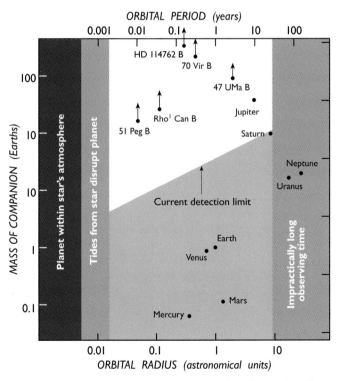
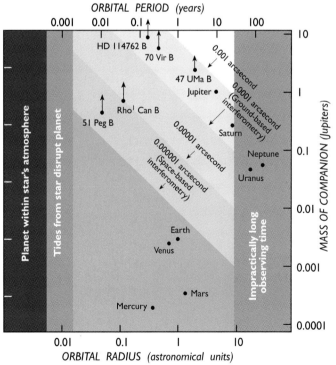

Figure 15. **Complementary means for detecting the reflex motion of a Sunlike star in a planetary system. For current Doppler-based searches (left panel), the limiting sensitivity is shown as a diagonal line, along with the five best finds to date. This technique favors planets in short-period, high-velocity orbits. Note that, in principle, a planet with Jupiter's mass and orbit could be discerned. The right-hand panel shows the sensitivity of astrometric searches for stars 33 light-years**

(10 parsecs) away. This technique favors planets in wide, and therefore long-period, orbits. While out of reach today, wobbles seen in the positions of HD 114762, 70 Virginis, and 47 Ursae Majoris may soon be detectable by ground-based interferometers. In all likelihood, however, neither Doppler spectroscopy nor astrometry will be able to reveal analogues of Earth.

planets. Simulations show that, were Jupiter in an eccentric orbit, Earth and Mars would likely have been flung out of the solar system long ago. There would be no terrestrial planet in the habitable zone, and no readers to reflect on this fact. The existence of intelligent life may depend on both Jupiter and Earth being in mutually stable orbits. Thus we will find ourselves in such a circularized system, no matter how unusual it may be.

The handful of planets discovered so far around Sunlike stars show orbital properties profoundly more diverse than those of the worlds around us. Even the relatively normal-seeming planets of 47 Ursae Majoris and 16 Cygni B have masses of at least 1.6 Jupiters, showing that our own Jupiter is not the largest planet that nature can form. The Doppler shifts of 70 Virginis and 16 Cygni B vary in nonsinusoidal cycles that are clearly the result of

Figure 16 (above). NASA's proposed Space Interferometry Mission involves several telescopes mounted on a 10-m-long truss. The telescopes' locations would be adjustable, to keep the light path stable to within 10 angstroms. This orbiting interferometer should resolve objects separated by only 0.01 arcsecond. SIM may reach orbit as early as 2005.

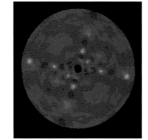

Figure 17 (right). A four-planet system 33 light-years from Earth stands out in this simulated image from a futuristic planet finder. Current designs produce two images of each planet.

bodies orbiting in ellipses according to Kepler's laws. Isaac Newton, were he resurrected, could examine these Doppler data and immediately recognize pure orbital motion stemming from his laws of gravity. Planets are surely orbiting those stars, and they have eccentricities greater than those of all major planets in our solar system.

It took humanity more than 2,200 years to develop the technology proving that Epicurus was right about the existence of many worlds. The next round of questions should be answered much more quickly. What fraction of stars have planets? What fraction of planetary systems resemble ours? Finally, do small, Earth-sized planets commonly occur in a star's habitable zone, where temperatures are right to allow life?

Today more than 1,000 stars are being surveyed by the precision Doppler technique using the world's most powerful optical telescopes. Meanwhile, several groups are surmounting the technical hurdles needed to combine the beams from two or more telescopes, known as an *interferometer.* These devices are capable of extraordinary resolution. Within a few years interferometric systems should be working with giant telescopes in Hawaii, Arizona, and Chile. These systems will be used to make precise measurements of stellar positions, revealing the presence of planets through astrometry (*Figure 15*).

The following generation of interferometers will be space based. Getting above the blurring effects of Earth's atmosphere will allow us to measure stellar positions with a much higher level of precision. Among the first of these will be the Space Interferometry Mission, or SIM (*Figure 16*). While ground-based interferometers should eventually detect planets comparable in size to Neptune (about 1/20 the mass of Jupiter), SIM should be able to detect something the size of Earth (about 1/300 of a Jupiter mass).

Ground-based interferometers are now serving as technological testbeds for SIM. In turn, SIM will help define the subsequent generation of space-based interferometers. Two such concepts are NASA's Terrestrial Planet Finder (TPF) and the European Space Agency's Darwin. Unlike their predecessors, TPF and Darwin are being designed to actually *see* planets orbiting nearby stars. They will combine the beams from several telescopes in a so-called "nulling interferometer." Such an instrument cancels out the central star's light, allowing its faint nearby planets to be seen directly (*Figure 17*). This critical achievement will allow us to take spectra of planets and identify the chemical constituents in their atmospheres. The detection of oxygen, carbon dioxide, water vapor, and methane in the atmosphere of an Earth-mass planet orbiting another star would be a discovery of historic proportions.

The technological hurdles for these projects are formidable, and it is difficult to predict when any of them might see fruition. But the questions they may answer are no less profound. What is our place in the universe? Are there other Earths? Is our heartwarming solar system a freakish twist in the cosmic script or merely some common plot device used over and over throughout the solar neighborhood? We do not know — but we will.

Planet, Satellite, and Small-body Characteristics

Characteristics of the inner planets

	MERCURY	VENUS	EARTH	MOON	MARS
Mass[1] (g)	3.302×10^{26}	4.865×10^{27}	5.974×10^{27}	7.349×10^{25}	6.419×10^{26}
Mass[1] (Earth = 1)	0.0553	0.8150	1.000	0.0123	0.1074
Reciprocal mass[2]	6,023,600	408,500	332,900	27,065,000	3,098,700
Equatorial radius (km)	2,440	6,052	6,378	1,738	3,396
Equatorial radius (Earth = 1)	0.382	0.949	1.000	0.272	0.532
Ellipticity[3]	0.0	0.0	0.0034	0.002	0.0069
Mean density (g/cm[3])	5.43	5.20	5.52	3.34	3.91
Equatorial surface gravity (m/s[2])	3.70	8.87	9.78	1.62	3.69
Equatorial escape velocity (km/s)	4.4	10.4	11.2	2.4	5.0
Sidereal rotation period	58.6462 d	243.02 d	23.9345 h	27.3217 d	24.6230 h
Obliquity (tilt of equator to orbit)	(0.1°)[4]	177.4°[5]	23.45°	6.67°	25.19°

Characteristics of the outer planets

	JUPITER	SATURN	URANUS	NEPTUNE	PLUTO
Mass (g)	1.898×10^{30}	5.685×10^{29}	8.683×10^{28}	1.024×10^{29}	1.32×10^{25}
Mass (Earth = 1)	317.710	95.162	14.535	17.141	0.002
Reciprocal mass	1,048	3,499	22,910	19,314	150,000,000
Equatorial radius[6] (km)	71,492	60,268	25,559	24,766	1,150
Equatorial radius[6] (Earth = 1)	11.209	9.449	4.007	3.883	0.180
Ellipticity	0.0649	0.0980	0.0229	0.017	(0.0)
Mean density (g/cm[3])	1.33	0.69	1.318	1.638	(2.0)
Equatorial surface gravity (m/s[2])	23.12	8.96	8.69	11.00	0.66
Equatorial escape velocity (km/s)	59.5	35.5	21.3	23.5	1.1
Sidereal rotation (internal)	9.9250 h[7]	10.6562 h[8]	17.240 h[9]	16.110 h[10]	6.3872 d
Obliquity (tilt of equator to orbit)	3.12°	26.73°	97.86°[5]	29.56°	119.6°[5]

Characteristics of planetary orbits

	Mean distance from Sun (AU)	(10⁶ km)	Sidereal period (years)	(days)	Synodic period (days)	Mean orbital velocity (km/s)	Orbital eccentricity	Orbital inclination to the ecliptic
MERCURY	0.3871	57.91	0.24084	87.968	115.88	47.87	0.206	7.00°
VENUS	0.7233	108.21	0.61518	224.695	583.92	35.02	0.007	3.39°
EARTH	1.0000	149.60	0.99998	365.242	—	29.79	0.017	0.00°
MARS	1.5237	227.94	1.88071	686.930	779.94	24.13	0.093	1.85°
JUPITER	5.2026	778.30	11.85652	4,330.60	398.88	13.07	0.048	1.30°
SATURN	9.5549	1,429.39	29.42352	10,746.94	378.09	9.67	0.056	2.49°
URANUS	19.2184	2,875.04	83.7474	30,588.7	369.66	6.83	0.046	0.77°
NEPTUNE	30.1100	4,504.50	163.7232	59,799.9	367.49	5.48	0.009	1.77°
PLUTO	39.5447	5,915.80	248.021	90,589	366.72	4.75	0.249	17.14°

Characteristics of planetary magnetospheres

	MERCURY	VENUS	EARTH	MOON	MARS
Solar-wind density (amu/cm³)	35–80	16	8	8	3.5
Magnetic moment (gauss cm³)	5.53×10^{22}	$<3 \times 10^{22}$	7.91×10^{25}	0	$<2 \times 10^{22}$
(Earth = 1)	0.0007	<0.0004	1.00	0	<0.0002
Surface magnetic field (gauss)	0.003	$<2 \times 10^{-5}$	0.31	<0.003	<0.0003
Dipole tilt and sense	+14°	—	+10.8°	—	—
Observed size of magnetosphere (R_p)	1.4	—	8–12	—	—
(km)	3,600	—	70,000	—	—
Maximum ion density (cm⁻³)	(1)	—	1–4,000	—	—
Composition of magnetospheric ions	H⁺	—	O⁺, H⁺	—	O⁺, H⁺
Source of magnetospheric ions	solar wind	—	ionosphere	—	—
Source strength (ions/second)	?	—	2×10^{26}	—	—

	JUPITER	SATURN	URANUS	NEPTUNE	PLUTO
Solar-wind density (amu/cm³)	0.3	0.1	0.02	0.008	0.008–0.003
Magnetic moment (gauss cm³)	1.58×10^{30}	2.37×10^{28}	3.95×10^{27}	1.98×10^{27}	?
(Earth = 1)	20,000	600	50	25	?
Surface magnetic field (gauss)	4.28	0.22	0.23	0.14	?
Dipole tilt and sense	−9.6°	0.0°	−59°	−47°	?
Observed size of magnetosphere (R_p)	50–100	16–22	18	23–26	?
(km)	7×10^6	1×10^6	5×10^5	6×10^5	?
Maximum ion density (cm⁻³)	>3,000	(100)	3	2	?
Composition of magnetospheric ions	Oⁿ⁺, Sⁿ⁺	O⁺, OH⁺, H⁺	H⁺	N⁺, H⁺	?
Source of magnetospheric ions	Io	rings, moons	atmosphere	Triton	?
Source strength (ions/second)	$>10^{28}$	2×10^{26}	10^{25}	10^{25}	?

Notes and explanations

[1] Planetary masses do not include satellites.

[2] The mass of the Sun divided by the mass of the planet (including its atmosphere and satellites).

[3] The ellipticity is $(R_e - R_p)/R_p$, where R_e and R_p are the planet's equatorial and polar radii, respectively.

[4] Values in parentheses are uncertain by more than 10 percent.

[5] By IAU convention, each planet's north pole is the one lying north of the ecliptic plane; however, Venus, Uranus, and Pluto are considered to have retrograde rotation.

[6] Since the outer planets have no solid surfaces, these are the radii at the 1-bar pressure level in their atmospheres.

[7] Jupiter's equatorial (System I) rotation period is 9.841 hours.

[8] Saturn's equatorial rotation period is 10.233 hours.

[9] Uranus's equatorial rotation period is about 18.0 hours.

[10] Neptune's equatorial rotation period is about 18.8 hours.

[11] V_O is an object's magnitude in visible light at opposition.

[12] Caliban and Sycorax are the provisional names for S/1997 U1 and S/1997 U2, respectively.

[13] Although the discovery of this gap is often ascribed to J. Encke, J. Keeler was likely the first to observe it.

[14] Chiron displays a coma and is considered a periodic comet (95P).

[15] Table compiled from data supplied in January 1998 by the IAU Minor Planet Center.

[16] Table compiled from data supplied by Frank Wlotzka (Max Planck Inst. für Chemie).

[17] Table compiled from data supplied by R. Grieve (Natural Resources Canada).

Satellite of Earth

Name	Discoverer	Year of dis-covery	Magni-tude (V_o)	Mean distance from Earth's center (km)	Sidereal period (days)	Orbital inclina-tion (°)	Orbital eccen-tricity	Radius (km)	Mass (g)	Mean density (g/cm^3)
Moon	?	?	−12.7	384,400	27.322	18.3−28.6	0.05	1,738	7.35×10^{25}	3.34

Satellites of Mars

Name	Discoverer	Year of dis-covery	Magni-tude (V_o)	Mean distance from Mars's center (km)	Sidereal period (days)	Orbital inclina-tion (°)	Orbital eccen-tricity	Radius (km)	Mass (g)	Mean density (g/cm^3)
Phobos	A. Hall	1877	11.3	9,377	0.319	1.082	0.0151	13.1 × 9.3	1.08×10^{19}	1.9
Deimos	A. Hall	1877	12.4	23,463	1.262	1.791	0.0003	7.8 × 5.1	1.80×10^{18}	1.8

Satellites of Jupiter

Name	Discoverer(s)	Year of dis-covery	Magni-tude (V_o)	Mean distance from Jupiter's center (km)	Sidereal period (days)	Orbital inclina-tion (°)	Orbital eccen-tricity	Radius (km)	Mass (g)	Mean density (g/cm^3)
Metis	S. Synnott	1979	17.5	127,960	0.295	(0)	<0.004	30 × 17	?	?
Adrastea	D. Jewitt, E. Danielson	1979	19.1	128,980	0.298	(0)	(0)	102 × 7	?	?
Amalthea	E. Barnard	1892	14.1	181,300	0.498	0.40	0.00	125 × 64	?	?
Thebe	S. Synnott	1979	15.7	221,900	0.675	(0.8)	0.02	58 × 42	?	?
Io	Galileo, S. Marius	1610	5.0	421,600	1.769	0.04	0.04	1,821	8.93×10^{25}	3.53
Europa	Galileo, S. Marius	1610	5.3	670,900	3.552	0.47	0.01	1,565	4.80×10^{25}	2.97
Ganymede	Galileo, S. Marius	1610	4.6	1,070,000	7.155	0.21	0.00	2,634	1.48×10^{26}	1.94
Callisto	Galileo, S. Marius	1610	5.6	1,883,000	16.689	0.28	0.01	2,403	1.08×10^{26}	1.85
Leda	C. Kowal	1974	20.2	11,094,000	238.72	27	0.15	(5)	?	?
Himalia	C. Perrine	1904	14.8	11,480,000	250.57	28	0.16	(85)	?	?
Lysithea	S. Nicholson	1938	18.4	11,720,000	259.22	29	0.11	(12)	?	?
Elara	C. Perrine	1905	16.8	11,737,000	259.65	28	0.21	(40)	?	?
Ananke	S. Nicholson	1951	18.9	21,200,000	631	147	0.17	(10)	?	?
Carme	S. Nicholson	1938	18.0	22,600,000	692	163	0.21	(15)	?	?
Pasiphae	P. Melotte	1908	17.0	23,500,000	735	148	0.38	(18)	?	?
Sinope	S. Nicholson	1914	18.3	23,700,000	758	153	0.28	(14)	?	?

Satellites of Saturn

Name	Discoverer(s)	Year of dis-covery	Magni-tude (V_o)	Mean distance from Saturn's center (km)	Sidereal period (days)	Orbital inclina-tion (°)	Orbital eccen-tricity	Radius (km)	Mass (g)	Mean density (g/cm^3)
Pan	M. Showalter	1990	(19)	133,583	0.575	(0)	(0.0)	(10)	?	?
Atlas	R. Terrile	1980	18.0	137,640	0.602	(0)	0.003	18 × 14	?	?
Prometheus	S. Collins and others	1980	15.8	139,350	0.613	(0)	0.002	74 × 34	1.4×10^{20}	0.27
Pandora	S. Collins and others	1980	16.5	141,700	0.629	(0)	0.004	55 × 31	1.3×10^{20}	0.42
Epimetheus	R. Walker and others	1966	15.7	151,422	0.695	0.34	0.009	69 × 53	5.5×10^{20}	0.63
Janus	A. Dollfus	1966	14.5	151,472	0.695	0.14	0.007	99 × 76	2.0×10^{21}	0.65
Mimas	W. Herschel	1789	12.9	185,520	0.942	1.53	0.020	199	3.7×10^{22}	1.12
Enceladus	W. Herschel	1789	11.7	238,020	1.370	0.02	0.004	249	6.5×10^{22}	1.00
Tethys	G. Cassini	1684	10.2	294,660	1.888	1.09	0.000	529	6.1×10^{23}	0.98
Telesto	B. Smith and others	1980	18.5	294,660	1.888	(0)	(0)	15 × 8	?	?
Calypso	D. Pascu and others	1980	18.7	294,660	1.888	(0)	(0)	15 × 8	?	?
Dione	G. Cassini	1684	10.4	377,400	2.737	0.02	0.00	560	1.1×10^{24}	1.49
Helene	P. Laques, J. Lecacheux	1980	18.4	377,400	2.737	0.2	0.01	16	?	?
Rhea	G. Cassini	1672	9.7	527,040	4.518	0.35	0.00	764	2.3×10^{24}	1.24
Titan	C. Huygens	1655	8.3	1,221,850	15.945	0.33	0.03	2,575	1.34×10^{26}	1.88
Hyperion	W. Bond	1848	14.2	1,481,100	21.277	0.43	0.10	185 × 113	?	?
Iapetus	G. Cassini	1671	10.2−11.9	3,561,300	79.330	7.52	0.03	720	1.6×10^{24}	1.0
Phoebe	W. Pickering	1898	16.4	12,952,000	550.48	175.3	0.16	115 × 105	?	?

Satellites of Uranus

Name	Discoverer(s)	Year of discovery	Magnitude (V₀)	Mean distance from Uranus's center (km)	Sidereal period (days)	Orbital inclination (°)	Orbital eccentricity	Radius (km)	Mass (g)	Mean density (g/cm³)
Cordelia	Voyager 2	1986	24.1	49,752	0.335	0.08	0.000	(13)	?	?
Ophelia	Voyager 2	1986	23.8	53,763	0.376	0.10	0.010	(16)	?	?
Bianca	Voyager 2	1986	23.0	59,166	0.435	0.19	0.001	(22)	?	?
Cressida	Voyager 2	1986	22.2	61,767	0.464	0.01	0.000	(33)	?	?
Desdemona	Voyager 2	1986	22.5	62,658	0.474	0.11	0.000	(29)	?	?
Juliet	Voyager 2	1986	21.5	64,358	0.493	0.07	0.001	(42)	?	?
Portia	Voyager 2	1986	21.0	66,097	0.513	0.06	0.000	(55)	?	?
Rosalind	Voyager 2	1986	22.5	69,927	0.558	0.28	0.000	(29)	?	?
Belinda	Voyager 2	1986	22.1	75,256	0.624	0.03	0.000	(34)	?	?
Puck	Voyager 2	1985	20.2	86,004	0.762	0.32	0.000	77	?	?
Miranda	G. Kuiper	1948	16.3	129,800	1.413	4.22	0.003	240 × 233	6.59×10^{22}	1.20
Ariel	W. Lassell	1851	14.2	191,240	2.520	0.31	0.003	581 × 578	1.35×10^{24}	1.67
Umbriel	W. Lassell	1851	14.8	266,000	4.144	0.36	0.005	585	1.17×10^{24}	1.40
Titania	W. Herschel	1787	13.7	435,840	8.706	0.10	0.002	790	3.53×10^{24}	1.71
Oberon	W. Herschel	1787	13.9	582,600	13.463	0.10	0.001	760	3.01×10^{24}	1.63
(Caliban)[12]	B. Gladman and others	1997	22	7,168,900	579	139.68	0.082	(30)	?	?
(Sycorax)[12]	P. Nicholson and others	1997	20	12,213,600	1,289	152.67	0.509	(60)	?	?

Satellites of Neptune

Name	Discoverer(s)	Year of discovery	Magnitude (V₀)	Mean distance from Neptune's center (km)	Sidereal period (days)	Orbital inclination (°)	Orbital eccentricity	Radius (km)	Mass (g)	Mean density (g/cm³)
Naiad	Voyager 2	1989	24.7	48,227	0.294	4.7	0.0003	(29)	?	?
Thalassa	Voyager 2	1989	23.8	50,075	0.312	0.2	0.0002	(40)	?	?
Despina	Voyager 2	1989	22.6	52,526	0.335	0.1	0.0001	(75)	?	?
Galatea	Voyager 2	1989	22.3	61,953	0.429	0.1	0.0001	(80)	?	?
Larissa	H. Reitsema, Voyager 2	1989	22.0	73,548	0.555	0.2	0.0014	104 × 89	?	?
Proteus	Voyager 2	1989	20.3	117,647	1.122	0.6	0.0004	218 × 201	?	?
Triton	W. Lassell	1846	13.5	354,760	5.877	156.8	0.000	1,353	2.15×10^{25}	2.05
Nereid	G. Kuiper	1949	18.7	5,513,400	360.14	27.6	0.753	170	?	?

Satellite of Pluto

Name	Discoverer	Year of discovery	Magnitude (V₀)	Mean distance from Pluto's center (km)	Sidereal period (days)	Orbital inclination (°)	Orbital eccentricity	Radius (km)	Mass (g)	Mean density (g/cm³)
Charon	J. Christy	1978	16.8	19,636	6.387	0.00	0.01	625	1.6×10^{24}	(1.7)

Rings of Saturn

Name	Distance from Saturn's center (Rₛ)	(km)	Radial width (km)	Thickness (km)	Optical depth	Total mass (g)	Albedo
D	1.11–1.24	66,000–73,150	7,150	?	(0.01)	?	?
C	1.24–1.52	74,500–92,000	17,500	?	0.05–0.35	1.1×10^{24}	0.12–0.30
Maxwell gap	1.45	87,500	270				
B	1.52–1.95	92,000–117,500	25,500	0.1–1	0.8–2.5	2.8×10^{25}	0.5–0.6
Cassini division	1.95–2.02	117,500–122,200	4,700	?	0.05–0.15	5.7×10^{23}	0.2–0.4
A	2.02–2.27	122,200–136,800	14,600	0.1–1	0.4–0.5	6.2×10^{24}	0.4–0.6
Encke gap[13]	2.214	133,570	325				
Keeler gap	2.263	136,530	35				
F	2.324	140,210	30–500	?	0.01–1	?	0.6
G	2.72–2.85	164,000–172,000	8,000	100–1,000	10^{-6}	10^{20}	?
E	(3–8)	(180,000–480,000)	(300,000)	15,000	10^{-5}	?	?

Rings of Jupiter

Name	Distance from Jupiter's center (R_J)	(km)	Radial width (km)	Thickness (km)	Optical depth	Total mass (g)	Albedo
"Halo"	1.40–1.72	100,000–122,800	22,800	(20,000)	(10^{-6})	?	(0.05)
"Main"	1.72–1.81	122,800–129,200	6,400	<30	3×10^{-6}	(10^{16})	(0.05)
"Gossamer"	1.81–3.2	129,200–228,000	99,000	3,000	$< 10^{-7}$	?	(0.05)

Rings of Uranus

Name		Distance from Uranus's center (R_U)	(km)	Radial width (km)	Thickness (km)	Optical depth	Total mass (g)	Albedo
1986 U2R		(1.49)	(38,000)	(2,500)	(0.1)	< 0.001	?	(0.03)
6		1.597	41,840	1–3	(0.1)	0.2–0.3	?	(0.03)
5		1.612	42,230	2–3	(0.1)	0.5–0.6	?	(0.03)
4		1.625	42,570	2–3	(0.1)	0.3	?	(0.03)
Alpha	(α)	1.707	44,720	7–12	(0.1)	0.3–0.4	?	(0.02)
Beta	(β)	1.743	45,660	8–12	(0.1)	0.2–0.3	?	(0.02)
Eta	(η)	1.801	47,180	0–2	(0.1)	0.1–0.4	?	(0.03)
Gamma	(γ)	1.818	47,630	1–4	(0.1)	1.3–2.3	?	(0.03)
Delta	(δ)	1.843	48,300	3–9	(0.1)	0.6–0.8	?	(0.03)
Lambda	(λ)	1.909	50,020	1–2	(0.1)	0.1	?	(0.03)
Epsilon	(ε)	1.952	51,150	20–100	<0.15	0.5–2.1	?	(0.02)

Rings of Neptune

Name	Distance from Neptune's center (R_N)	(km)	Radial width (km)	Thickness (km)	Optical depth	Total mass (g)	Albedo
Galle	1.68	41,900	(15)	?	10^{-4}	?	(low)
LeVerrier	2.14	53,200	110	<30	0.01–0.02	?	0.03
Lassell	2.15–2.4	53,200–59,100	4,000	?	10^{-4}	?	(low)
Arago	2.31	57,200	<100	?	?	?	?
Adams	2.53	62,930	<50	<30	0.01–0.15	?	>0.07

Selected comets

Designation	Comet name	Year of discovery	Perihelion distance (AU)	Orbital eccen- tricity	Orbital inclination (°)	Orbital period (years)	Last perihelion	Associated meteor shower(s)
3D	Biela	1772	0.861	0.756	12.55	6.62	(lost)	Andromedids
23P	Brorsen-Metcalf	1847	0.479	0.972	19.28	69.5	1989	
27P	Crommelin	1818	0.745	0.918	29.11	27.5	1984	
6P	d'Arrest	1851	1.347	0.614	19.52	6.52	1995	
2P	Encke	1786	0.333	0.849	11.91	3.29	1997	Taurids
21P	Giacobini-Zinner	1900	1.034	0.706	31.86	6.61	1998	Draconids
1P	Halley	(−239)	0.596	0.967	162.22	75.8	1986	Eta Aquarids, Orionids
96P	Machholz 1	1986	0.125	0.959	59.96	5.24	1996	
29P	Schwassmann-Wachmann 1	1908	5.729	0.046	9.38	14.9	1989	
109P	Swift-Tuttle	1862	0.968	0.963	113.47	135	1992	Perseids
55P	Tempel-Tuttle	1866	0.977	0.906	162.49	33.3	1998	Leonids
8P	Tuttle	1858	1.034	0.819	54.94	13.6	1994	Ursids
81P	Wild 2	1978	1.583	0.540	3.24	6.39	1997	
46P	Wirtanen	1954	1.063	0.657	11.72	5.45	1997	
C/1956 R1	Arend-Roland	1956	0.316	1.000	119.95			
C/1969 Y1	Bennett	1970	0.538	0.996	90.04			
C/1858 L1	Donati	1858	0.578	0.996	116.96			
C/1995 O1	Hale-Bopp	1995	0.914	0.995	89.43			
C/1996 B2	Hyakutake	1996	0.231	0.999	124.95			
C/1965 S1	Ikeya-Seki	1965	0.008	1.000	141.86			
C/1983 H1	IRAS-Araki-Alcock	1983	0.991	0.990	73.25			

The largest asteroids

Number	Name	Year of discovery	Discoverer(s)	Radius (km)	Rotation period (hours)	Albedo	Spectral type	Mean distance from Sun (AU)	(10⁶ km)	Orbital period (years)	Orbital eccentricity	Orbital inclination (°)
(1)	Ceres	1801	G. Piazzi	467	9.08	0.11	G	2.768	413.9	4.60	0.077	10.58
(2)	Pallas	1802	H. Olbers	263	7.81	0.16	B	2.774	414.5	4.61	0.232	34.82
(4)	Vesta	1807	H. Olbers	255	5.34	0.42	V	2.361	353.4	3.63	0.090	7.13
(10)	Hygiea	1849	A. De Gasparis	204	(27.6)	0.07	C	3.136	470.3	5.55	0.120	3.84
(511)	Davida	1903	R. Dugan	163	5.13	0.05	C	3.172	475.4	5.65	0.182	15.94
(704)	Interamnia	1910	V. Cerulli	158	8.73	0.07	F	3.064	458.1	5.36	0.146	17.32
(52)	Europa	1858	H. Goldschmidt	151	5.63	0.06	CF	3.099	463.3	5.46	0.101	7.47
(3)	Juno	1804	K. Harding	134	7.21	0.24	S	2.669	399.4	4.36	0.258	13.00
(87)	Sylvia	1866	N. Pogson	130	5.18	0.04	P	3.499	521.5	6.55	0.082	10.84
(15)	Eunomia	1851	A. De Gasparis	128	6.08	0.21	S	2.644	395.5	4.30	0.187	11.75
(31)	Euphrosyne	1854	J. Ferguson	128	5.53	0.05	C	3.144	472.1	5.58	0.228	26.34
(16)	Psyche	1852	A. De Gasparis	127	4.20	0.12	M	2.922	437.1	4.99	0.138	3.09
(65)	Cybele	1861	E. Tempel	120	4.04	0.07	P	3.432	513.0	6.36	0.104	3.55
(324)	Bamberga	1892	J. Palisa	115	29.43	0.06	CP	2.686	401.4	4.40	0.337	11.10
(451)	Patientia	1899	A. Charlois	113	9.73	0.08	CU	3.062	458.2	5.36	0.075	15.23
(48)	Doris	1857	H. Goldschmidt	111	11.89	0.06	CG	3.111	465.5	5.49	0.073	6.55
(107)	Camilla	1868	N. Pogson	111	4.84	0.05	C	3.495	521.8	6.54	0.080	10.02
(532)	Herculina	1904	M. Wolf	111	9.41	0.17	S	2.777	414.7	4.63	0.175	16.33
(45)	Eugenia	1857	H. Goldschmidt	107	5.70	0.04	FC	2.721	407.1	4.49	0.083	6.61
(29)	Amphitrite	1854	A. Marth	106	5.39	0.18	S	2.553	382.1	4.08	0.073	6.11
(121)	Hermione	1872	J. Watson	105	9.24	0.05	C	3.440	516.3	6.38	0.146	7.58
(423)	Diotima	1896	A. Charlois	104	4.78	0.05	C	3.071	459.0	5.38	0.040	11.23
(13)	Egeria	1850	A. De Gasparis	104	7.05	0.08	G	2.575	385.4	4.13	0.087	16.53
(94)	Aurora	1867	J. Watson	102	7.22	0.04	CP	3.161	472.4	5.62	0.083	7.98
(7)	Iris	1847	J. Hind	100	7.14	0.28	S	2.385	356.9	3.68	0.231	5.52
(702)	Alauda	1910	J. Helffrich	97	8.36	0.06	C	3.191	477.8	5.70	0.026	20.60
(128)	Nemesis	1872	J. Watson	94	39	0.05	C	2.750	411.4	4.56	0.127	6.25
(372)	Palma	1893	A. Charlois	94	17.3	0.07	BFC	3.142	470.1	5.57	0.264	23.87
(6)	Hebe	1847	K. Hencke	93	7.27	0.27	S	2.425	362.8	3.78	0.202	14.77
(76)	Freia	1862	H. d'Arrest	92	9.97	0.04	P	3.412	507.1	6.30	0.164	2.12
(154)	Bertha	1875	P. Henry	92	?	0.05	?	3.182	476.3	5.68	0.091	21.10
(22)	Kalliope	1852	J. Hind	91	4.15	0.14	M	2.910	435.3	4.97	0.101	13.72
(130)	Elektra	1873	C. Peters	91	5.23	0.08	G	3.117	466.6	5.50	0.214	22.89
(2060)	Chiron[14]	1977	C. Kowal	(90)	5.92	0.05	B	13.648	2,051.9	50.42	0.381	6.94
(259)	Aletheia	1886	C. Peters	89	?	0.04	CP	3.146	469.6	5.60	0.113	10.77
(41)	Daphne	1856	H. Goldschmidt	87	5.99	0.08	C	2.762	413.6	4.59	0.275	15.77
(120)	Lachesis	1872	A. Borrelly	87	?	0.05	C	3.116	466.4	5.50	0.063	6.96
(93)	Minerva	1867	J. Watson	86	5.97	0.09	CU	2.755	412.2	4.57	0.141	8.56
(747)	Winchester	1913	J. Metcalf	86	9.40	0.05	PC	2.991	448.5	5.17	0.345	18.18
(96)	Aegle	1868	J. Coggia	85	?	0.05	T	3.053	456.4	5.33	0.137	16.02
(153)	Hilda	1875	J. Palisa	85	5.11	0.06	P	3.979	593.8	7.94	0.144	7.83
(241)	Germania	1884	R. Luther	85	9.00	0.06	CP	3.051	456.3	5.33	0.098	5.52
(790)	Pretoria	1912	H. Wood	85	10.37	0.04	P	3.407	509.5	6.29	0.151	20.55
(566)	Stereoskopia	1905	P. Gotz	84	?	0.04	C	3.389	506.7	6.24	0.101	4.91
(194)	Prokne	1879	C. Peters	84	15.67	0.05	C	2.618	391.3	4.24	0.236	18.50
(54)	Alexandra	1858	H. Goldschmidt	83	7.02	0.06	C	2.708	405.4	4.46	0.200	11.83
(386)	Siegena	1894	M. Wolf	83	9.76	0.07	C	2.897	433.2	4.93	0.171	20.24
(911)	Agamemnon	1919	K. Reinmuth	83	?	0.04	D	5.233	778.1	11.97	0.067	21.81
(59)	Elpis	1860	J. Chacornac	82	13.69	0.04	CP	2.712	405.9	4.47	0.119	8.65
(1437)	Diomedes	1937	K. Reinmuth	82	?	0.03	DP	5.132	760.4	11.63	0.045	20.55
(409)	Aspasia	1895	A. Charlois	81	9.02	0.06	CX	2.578	385.4	4.14	0.070	11.23
(444)	Gyptis	1899	J. Coggia	80	6.21	0.05	C	2.770	414.5	4.61	0.177	10.27
(209)	Dido	1879	C. Peters	80	8.0	0.04	C	3.144	470.4	5.57	0.066	7.18
(185)	Eunike	1878	C. Peters	79	10.83	0.06	C	2.738	409.7	4.53	0.130	23.26

Number	Name	Year of discovery	Discoverer(s)	Radius (km)	Rotation period (hours)	Albedo	Spectral type	Mean distance from Sun (AU)	(10⁶ km)	Orbital period (years)	Orbital eccentricity	Orbital inclination (°)
(804)	Hispania	1915	J. Comas Sola	79	7.42	0.05	PC	2.837	424.7	4.78	0.141	15.39
(139)	Juewa	1874	J.Watson	78	20.9	0.06	CP	2.779	416.6	4.63	0.177	10.94
(165)	Loreley	1876	C. Peters	78	7.23	0.08	CD	3.130	469.2	5.54	0.077	11.25
(334)	Chicago	1892	M.Wolf	78	9.20	0.06	C	3.871	582.2	7.62	0.043	4.66
(354)	Eleonora	1893	A. Charlois	78	4.28	0.20	S	2.797	418.3	4.68	0.115	18.43
(11)	Parthenope	1850	A. De Gasparis	77	7.83	0.18	S	2.452	366.8	3.84	0.100	4.62
(173)	Ino	1877	A. Borrelly	77	6.16	0.06	C	2.745	410.3	4.55	0.207	14.19
(89)	Julia	1866	E. Stephan	76	11.39	0.18	S	2.550	381.8	4.07	0.182	16.14
(39)	Laetitia	1856	J. Chacornac	75	5.14	0.29	S	2.769	414.2	4.61	0.112	10.37
(85)	Io	1865	C. Peters	77	6.88	0.07	FC	2.655	397.0	4.33	0.191	11.97
(536)	Merapi	1904	G. Peters	76	?	0.05	X	3.505	523.6	6.56	0.084	19.37
(488)	Kreusa	1902	M.Wolf,L.Carnera	75	?	0.06	C	3.148	472.0	5.58	0.176	11.50
(150)	Nuwa	1875	J.Watson	76	8.14	0.04	CX	2.979	446.1	5.14	0.131	2.19
(238)	Hypatia	1884	V. Knorre	74	8.88	0.04	C	2.905	434.9	4.95	0.092	12.41
(145)	Adeona	1875	C. Peters	76	8.1	0.04	C	2.671	399.9	4.37	0.146	12.62
(49)	Pales	1857	H. Goldschmidt	75	10.42	0.06	CG	3.081	462.3	5.41	0.236	3.19

Potentially hazardous asteroids[15]

Designation	Name	Abs. mag. (H)	Likely diameter (km)	Minimum distance (AU)	Distance from Sun (minimum) (AU)	(mean) (AU)	Orbital inclination (°)	Orbital eccentricity	Year found	Discoverer(s)
(4953)		14.1	4–9	0.040	0.555	1.621	24.4	0.658	1990	R. McNaught
(4183)	Cuno	14.4		0.038	0.718	1.981	6.8	0.638	1959	C. Hoffmeister
(3200)	Phaethon	14.6	3–7	0.026	0.140	1.271	22.1	0.890	1983	IRAS
(2201)	Oljato	15.25		0.000	0.624	2.173	2.5	0.713	1947	H. Giclas
(4179)	Toutatis	15.30		0.006	0.920	2.512	0.5	0.634	1989	C. Pollas
(1981)	Midas	15.5	2–5	0.000	0.622	1.776	39.8	0.650	1973	C. Kowal
(4486)	Mithra	15.6		0.045	0.743	2.200	3.0	0.663	1987	E. Elst, V. Shkodrov
(1620)	Geographos	15.60		0.046	0.828	1.246	13.3	0.335	1951	A. Wilson, R. Minkowski
(4015)	Wilson-Harrington	15.99		0.049	1.000	2.644	2.8	0.622	1979	E. Helin
1990 HA		16.0	2–4	0.012	0.782	2.571	3.9	0.696	1990	A. Mrkos
(2102)	Tantalus	16.2		0.029	0.905	1.290	64.0	0.299	1975	C. Kowal
(3671)	Dionysus	16.3		0.033	1.003	2.195	13.6	0.543	1984	C. and E. Shoemaker
(5604)		16.4		0.037	0.551	0.927	4.8	0.405	1992	R. McNaught
(8566)		16.5	1–3	0.017	0.857	1.507	38.0	0.431	1996	JPL/GEODSS NEAT
1990 SM		16.5		0.019	0.486	2.125	11.5	0.771	1990	R. McNaught, P. McKenzie
1991 VH		16.5		0.013	0.973	1.137	13.9	0.144	1991	R. McNaught
1993 DQ₁		16.5		0.048	1.036	2.038	10.0	0.492	1993	Spacewatch
1994 CN₂		16.5		0.014	0.952	1.573	1.4	0.395	1994	Spacewatch
(7482)		16.8		0.017	0.904	1.346	33.5	0.328	1994	R. McNaught
(1566)	Icarus	16.9		0.040	0.187	1.078	22.9	0.827	1949	W. Baade
(4769)	Castalia	16.9		0.023	0.550	1.063	8.9	0.483	1989	E. Helin
(1862)		16.25		0.028	0.647	1.471	6.4	0.560	1932	K. Reinmuth
(5693)		17.0	1–2	0.008	0.527	1.272	5.1	0.585	1993	Spacewatch
(7335)		17.0		0.042	0.913	1.771	15.2	0.484	1989	E. Helin
1991 AQ		17.0		0.021	0.495	2.222	3.2	0.777	1991	E. Helin
1996 SK		17.0		0.003	0.495	2.428	2.0	0.796	1996	JPL/NEAT
1997 XF₁₁		17.0		0.000	0.744	1.442	4.1	0.484	1997	Spacewatch
(5011)	Ptah	17.1		0.026	0.818	1.635	7.4	0.500	1960	C. van Houten, others
(4450)	Pan	17.2		0.027	0.596	1.442	5.5	0.586	1987	C. and E. Shoemaker
(5189)		17.3		0.044	0.810	1.551	3.6	0.478	1990	R. McNaught
(7822)		17.4		0.033	0.938	1.123	37.1	0.165	1991	R. McNaught
1991 EE		17.5		0.033	0.844	2.246	9.8	0.624	1991	Spacewatch
1992 UY₄		17.5	1.0–1.5	0.031	1.010	2.656	2.8	0.620	1992	C. Shoemaker
1994 AW₁		17.5		0.036	1.021	1.105	24.1	0.075	1994	K. Lawrence, E. Helin

Potentially hazardous asteroids (continued)

Desig-nation	Name	Abs. mag. (H)	Likely diameter (km)	Minimum distance (AU)	Distance from Sun (minimum) (AU)	(mean) (AU)	Orbital inclina-tion (°)	Orbital eccen-tricity	Year found	Discoverer(s)
1994 PM		17.5		0.027	0.366	1.479	18.0	0.752	1994	C. Shoemaker
1995 SA		17.5		0.008	0.862	2.442	20.4	0.647	1995	Spacewatch
1997 AE$_{12}$		17.5		0.048	0.979	2.218	5.6	0.559	1997	Spacewatch
1997 BR		17.5		0.000	0.927	1.335	17.2	0.306	1997	BAO Schmidt
1998 CS$_1$		17.5		0.015	0.629	1.495	7.8	0.579	1998	BAO Schmidt
(6239)	Minos	17.9		0.028	0.676	1.151	3.9	0.413	1989	C. and E. Shoemaker
(2135)	Aristaeus	17.94		0.014	0.795	1.600	23.0	0.503	1977	S. J. Bus, E. Helin
1978 CA		18.0	0.7 – 1.5	0.027	0.883	1.125	26.1	0.215	1978	H.-E. Schuster
1989 DA		18.0		0.043	0.982	2.160	6.5	0.545	1989	J. Phinney
1994 CC		18.0		0.011	0.955	1.637	4.6	0.417	1994	Spacewatch
1997 BQ		18.0		0.037	0.909	1.746	11.0	0.479	1997	T. Hasegawa
1998 DV$_9$		18.0		0.012	0.988	1.745	8.7	0.434	1998	D. Tholen, R. Whiteley
1937 UB	(Hermes)	18.0		0.003	0.616	1.639	6.1	0.624	1937	K. Reinmuth
(4034)		18.1		0.023	0.589	1.060	11.2	0.444	1986	E. Helin
(4660)	Nereus	18.2		0.005	0.953	1.490	1.4	0.360	1982	E. Helin
(3362)	Khufu	18.3		0.018	0.526	0.990	9.9	0.469	1984	S. Dunbar, A. Barucci
1989 UR		18.5	0.5 – 1.2	0.031	0.696	1.080	10.3	0.356	1989	J. Mueller, D. Mendenhall
1996 FG$_3$		18.5		0.028	0.686	1.054	2.0	0.350	1996	R. McNaught
1996 GT		18.5		0.041	1.013	1.642	3.4	0.383	1996	Spacewatch
1996 RG$_3$		18.5		0.004	0.790	2.000	3.6	0.605	1996	Spacewatch
1997 NC1		18.5		0.035	0.685	0.866	16.7	0.209	1997	JPL/NEAT
1998 HJ$_3$		18.5		0.012	0.509	1.986	6.5	0.744	1998	LINEAR
1998 KN$_3$		18.5		0.019	0.179	1.605	2.4	0.888	1998	LINEAR
(7753)		18.6		0.005	0.761	1.468	3.1	0.482	1988	Y. Oshima
(2101)	Adonis	18.7		0.012	0.441	1.874	1.4	0.765	1936	E. Delporte
(6037)		18.7		0.024	0.636	1.270	3.5	0.499	1988	J. Alu
(8014)		18.7		0.018	0.951	1.747	1.9	0.456	1990	E. Helin
(3757)		18.95		0.026	1.017	1.835	3.9	0.446	1982	E. Helin
1954 XA		19.0	0.4 – 0.9	0.038	0.509	0.777	3.9	0.345	1954	G. Abell
1979 XB		19.0		0.040	0.649	2.262	24.9	0.713	1979	K. S. Russell
1983 LC		19.0		0.023	0.766	2.632	1.5	0.709	1983	E. Helin, S. Dunbar
1986 JK		19.0		0.007	0.894	2.798	2.1	0.680	1986	C. Shoemaker
1989 UQ		19.0		0.015	0.673	0.915	1.3	0.265	1989	C. Pollas
1991 DG		19.0		0.048	0.909	1.427	11.2	0.363	1991	R. McNaught
1991 GO		19.0		0.021	0.663	1.956	9.7	0.661	1991	K. Endate, K. Watanabe
1991 RB		19.0		0.047	0.749	1.450	19.5	0.484	1991	R. McNaught
1993 KH		19.0		0.008	0.850	1.234	12.8	0.311	1993	R. McNaught
1994 RC		19.0		0.045	0.903	2.266	4.7	0.602	1994	E. Helin
1994 XD		19.0		0.018	0.639	2.359	4.3	0.729	1994	Spacewatch
1996 EO		19.0		0.039	0.804	1.341	21.6	0.401	1996	JPL/NEAT
1998 FH$_{12}$		19.0		0.011	0.503	1.092	3.5	0.540	1998	LINEAR
1998 FW$_4$		19.0		0.001	0.678	2.498	3.6	0.728	1998	LINEAR
1998 HL$_1$		19.0		0.037	1.013	1.246	20.0	0.187	1998	LINEAR
1998 KM$_3$		19.0		0.001	0.636	1.730	4.8	0.632	1998	LINEAR
(3361)	Orpheus	19.03		0.013	0.819	1.209	2.7	0.323	1982	C. Torres
(2340)	Hathor	19.2		0.006	0.464	0.844	5.8	0.450	1976	C. Kowal
(6489)	Golevka	19.2		0.038	1.012	2.518	2.3	0.598	1991	E. Helin
1990 UA		19.5	0.3 – 0.7	0.012	0.770	1.721	1.0	0.553	1990	E. Helin
1991 JW		19.5		0.036	0.915	1.038	8.7	0.118	1991	K. Lawrence, E. Helin
1993 VB		19.5		0.005	0.918	1.911	5.1	0.519	1993	R. McNaught
1996 AW$_1$		19.5		0.039	0.737	1.527	4.7	0.517	1996	Spacewatch
1996 JG		19.5		0.013	0.611	1.815	5.3	0.663	1996	R. McNaught
1997 US$_2$		19.5		0.004	0.566	1.674	3.2	0.662	1997	Spacewatch
1998 KJ$_9$		19.5		0.009	0.523	1.441	10.9	0.637	1998	LINEAR

The largest meteorites[16]

Stony meteorites	Date of fall or find	Class	Total mass (kg)	Largest known fragment(s) (kg)	(location)
Jilin, China	1976 Mar 8	H chondrite	4,000	1,770	Jilin, China
Allende, Mexico	1969 Feb 8	CV chondrite	2,000	380	Washington, D.C.
Tsarev, Russia	1968	L chondrite	1,225	284	Moscow, Russia
Norton County, Kansas	1948 Feb 18	Aubrite	1,100	1,000	Albuquerque, N. Mexico
Plainview, Texas	1917	H chondrite	(700)	93	Tempe, Arizona
Long Island, Kansas	1891	L chondrite	564	549	Chicago, Illinois
Knyahinya, Ukraine	1866 June 9	LL chondrite	500	286	Vienna, Austria
Ochansk, Russia	1887 Aug 30	H chondrite	500	115	Kazan, Russia
Wildara, Australia	1968	H chondrite	500	500	Perth, Australia
Etter, Texas	1965	L chondrite	450	68	Mainz, Germany
Paragould, Arkansas	1930 Feb 17	LL chondrite	408	370	Chicago, Illinois
ALHA76009, Antarctica	1977	L chondrite	(407)	196	Tokyo, Japan
Hugoton, Kansas	1927	H chondrite	350	325	Tempe, Arizona
Bjurbole, Finland	1899 Mar 12	L chondrite	330	330	Helsinki, Finland

Stony-iron meteorites					
Brenham, Kansas	1882	Pallasite	4,300	490	Chicago, Illinois
Vaca Muerta, Chile	1861	Mesosiderite	3,800	850	
Huckitta, Australia	1924	Pallasite	2,300	1,400	Adelaide, Australia
Imilac, Chile	1822	Pallasite	920	200	
Bondoc, Philippines	1956	Mesosiderite	888	888	Tempe, Arizona
Brahin, Byelarus	1810	Pallasite	823	80	Moscow, Russia
Esquel, Argentina	1951	Pallasite	755	755	Tucson, Arizona
Krasnojarsk, Russia	1749	Pallasite	700	700	Moscow, Russia

Iron meteorites					
Hoba, Namibia	1920	Ataxite	60,000		find site
Cape York, Greenland		Octahedrite	58,000		
Ahnighito (The Tent)	1894			30,900	New York, New York
Agpalilik	1963			20,000	Copenhagen, Denmark
Savik I	1913			3,400	Copenhagen, Denmark
The Woman	1894			3,000	New York, New York
Campo del Cielo, Argentina		Octahedrite	50,000		
(no name)	1969			18,000	find site
El Mesame)	1969			18,000	find site (location lost)
El Toba	1923			4,200	Buenos Aires, Argentina
El Taco	1962			2,000	Washington, D.C.
Canyon Diablo, Arizona	1891	Octahedrite	30,000	639	Meteor Crater, Arizona
Armanty, China	1898	Octahedrite	28,000	28,000	Urumchi, China
Gibeon, Namibia	1836	Octahedrite	26,000	650	Cape Town, South Africa
Chupaderos, Mexico		Octahedrite	24,300		
Chupaderos I	1854			14,000	Mexico City, Mexico
Chupaderos II	1854			6,800	Mexico City, Mexico
Adargas	(before 1600)			3,400	Mexico City, Mexico
Mundrabilla, Australia	1911	Octahedrite	24,000	12,000	Perth, Australia
Sikhote-Alin, Russia	1947 Feb 12	Octahedrite	23,000	1,700	Moscow, Russia
Bacubirito, Mexico	1863	Octahedrite	22,000	22,000	Culiacan, Mexico
Mbosi, Tanzania	1930	Octahedrite	16,000	16,000	find site
Willamette, Oregon	1902	Octahedrite	15,500	15,500	New York, New York
Morito, Mexico	(before 1600)	Octahedrite	10,100	10,100	Mexico City, Mexico

Terrestrial impact craters[17]

Crater name	Location	Latitude			Longitude			Age (10⁶ years)	Diameter (km)
Acraman	S.A., Australia	32°	1′	S	135°	27′	E	(590)	90
Ames	Oklahoma, U.S.A.	36°	15′	N	98°	12′	W	470	16
Amguid	Algeria	26°	5′	N	4°	23′	E	<0.1	0.45
Aorounga	Chad, Africa	19°	6′	N	19°	15′	E	<345	12.6
Aouelloul	Mauritania	20°	15′	N	12°	41′	W	3	0.39
Araguainha Dome	Brazil	16°	47′	S	52°	59′	W	244	40
Avak	Alaska, U.S.A.	71°	15′	N	156°	38′	W	>95	12
Azuara	Spain	41°	10′	N	0°	55′	W	(40)	30
B.P. Structure	Libya	25°	19′	N	24°	20′	E	<120	3.2
Barringer	Arizona, U.S.A.	35°	2′	N	111°	1′	W	0.049	1.19
Beaverhead	Montana, U.S.A.	44°	36′	N	113°	0′	W	(600)	60
Beyenchime-Salaatin	Russia	71°	50′	N	123°	30′	E	<65	8
Bigach	Kazakhstan	48°	30′	N	82°	0′	E	(6)	7
Boltysh	Ukraine	48°	45′	N	32°	10′	E	88	24
Bosumtwi	Ghana	6°	30′	N	1°	25′	W	1.03	10.5
Boxhole	N.T., Australia	22°	37′	S	135°	12′	E	54	0.17
Brent	Ontario, Canada	46°	5′	N	78°	29′	W	450	3.8
Calvin	Michigan, USA	41°	50′	N	85°	57′	W	450	8.5
Campo Del Cielo	Argentina	27°	38′	S	61°	42′	W	<0.004	0.05
Carswell	Saskatchewan, Canada	58°	27′	N	109°	30′	W	115	39
Charlevoix	Québec, Canada	47°	32′	N	70°	18′	W	357	54
Chesapeake Bay	Virginia, U.S.A.	37°	17′	N	76°	1′	W	35.5	90
Chicxulub	Yucatán, Mexico	21°	20′	N	89°	30′	W	64.98	(170)
Chiyli	Kazakhstan	49°	10′	N	57°	51′	E	(46)	5.5
Chukcha	Russia	75°	42′	N	97°	48′	E	<70	6
Clearwater East	Québec, Canada	56°	5′	N	74°	7′	W	290	26
Clearwater West	Québec, Canada	56°	13′	N	74°	30′	W	290	36
Connolly Basin	W.A., Australia	23°	32′	S	124°	45′	E	<60	9
Couture	Québec, Canada	60°	8′	N	75°	20′	W	430	8
Crooked Creek	Missouri, U.S.A.	37°	50′	N	91°	23′	W	(320)	7
Dalgaranga	W.A., Australia	27°	38′	S	117°	17′	E	0.27	0.02
Decaturville	Missouri, U.S.A.	37°	54′	N	92°	43′	W	<300	6
Deep Bay	Saskatchewan, Canada	56°	24′	N	102°	59′	W	(100)	13
Dellen	Sweden	61°	48′	N	16°	48′	E	89	19
Des	Latvia	56°	35′	N	23°	15′	E	(300)	4.5
Eagle Butte	Alberta, Canada	49°	42′	N	110°	30′	W	<65	10
El' gygytgyn	Russia	67°	30′	N	172°	0′	E	(3.5)	18
Flynn Creek	Tennessee, U.S.A.	36°	17′	N	85°	40′	W	360	3.8
Gardnos	Norway	60°	39′	N	9°	0′	E	500	5
Glasford	Illinois, U.S.A.	40°	36′	N	89°	47′	W	<430	4
Glover Bluff	Wisconsin, U.S.A.	43°	58′	N	89°	32′	W	<500	8
Goat Paddock	W.A., Australia	18°	20′	S	126°	40′	E	<50	5.1
Gosses Bluff	N.T., Australia	23°	49′	S	132°	19′	E	142.5	22
Gow	Saskatchewan, Canada	56°	27′	N	104°	29′	W	<250	5
Goyder	N.T., Australia	13°	9′	S	135°	2′	E	<1,400	3
Granby	Sweden	58°	25′	N	14°	56′	E	470	3
Gusev	Russia	48°	22′	N	40°	13′	E	49	3.5
Gweni-Fada	Chad, Africa	17°	25′	N	21°	45′	E	<345	14
Haughton	N.W.T., Canada	75°	22′	N	89°	41′	W	23	24
Haviland	Kansas, U.S.A.	37°	35′	N	99°	10′	W	<1	0.01
Henbury	N.T., Australia	24°	34′	S	133°	8′	E	(4)	0.16
Holleford	Ontario, Canada	44°	28′	N	76°	38′	W	(550)	2.35
Ile Rouleau	Québec, Canada	50°	41′	N	73°	53′	W	<300	4
Ilumetsa	Estonia	57°	58′	N	27°	25′	E	>2	0.08
Ilyinets	Ukraine	49°	7′	N	29°	6′	E	395	4.5

Crater name	Location	Latitude			Longitude			Age (10⁶ years)	Diameter (km)
Iso-Naakkima	Finland	62°	11'	N	27°	9'	E	>1	3
Janisjarvi	Russia	61°	58'	N	30°	55'	E	698	14
Kaalijarvi	Estonia	58°	24'	N	22°	40'	E	(0.004)	0.11
Kalkkop	South Africa	32°	43'	S	24°	34'	E	<1.8	0.64
Kaluga	Russia	54°	30'	N	36°	15'	E	380	15
Kamensk	Russia	48°	20'	N	40°	15'	E	49.0	25
Kara	Russia	69°	12'	N	65°	0'	E	73	65
Kara-Kul	Tajikistan	39°	1'	N	73°	27'	E	<5	52
Kardla	Estonia	58°	59'	N	22°	40'	E	455	4
Karikkoselkä	Finland	63°	13'	N	25°	15'	E	<1.88	1
Karla	Russia	54°	54'	N	48°	0'	E	<10	12
Kelly West	N.T., Australia	19°	56'	S	133°	57'	E	>550	10
Kentland	Indiana, U.S.A.	40°	45'	N	87°	24'	W	<97	13
Kursk	Russia	51°	40'	N	36°	0'	E	(250)	5.5
La Moinerie	Québec, Canada	57°	26'	N	66°	37'	W	(400)	8
Lappajarvi	Finland	63°	12'	N	23°	42'	E	77.3	23
Lawn Hill	Queensland, Australia	18°	40'	S	138°	39'	E	>515	18
Liverpool	Sweden	63°	0'	N	14°	49'	E	>455	7.5
Logancha	Russia	65°	30'	N	95°	50'	E	(25)	20
Logoisk	Belarus	54°	12'	N	27°	48'	E	(40)	17
Lonar	India	19°	58'	N	76°	31'	E	0.052	1.83
Lumparn	Finland	60°	9'	N	20°	6'	E	(1)	9
Macha	Russia	60°	6'	N	117°	35'	E	<0.007	0.3
Manicouagan	Québec, Canada	51°	23'	N	68°	42'	W	214	100
Manson	Iowa, U.S.A.	42°	35'	N	94°	33'	W	73.8	35
Marquez	Texas, U.S.A.	31°	17'	N	96°	18'	W	58	12.7
Middlesboro	Kentucky, U.S.A.	36°	37'	N	83°	44'	W	<300	6
Mien	Sweden	56°	25'	N	14°	52'	E	121	9
Mishina Gora	Russia	58°	40'	N	28°	0'	E	<360	4
Mistastin	Labrador, Canada	55°	53'	N	63°	18'	W	(38)	28
Mizarai	Lithuania	54°	1'	N	24°	0'	E	570	5
Mjolnir	Norway	73°	48'	N	29°	40'	E	144	40
Montagnais	Nova Scotia, Canada	42°	53'	N	64°	13'	W	50.5	45
Monturaqui	Chile	23°	56'	S	68°	17'	W	<1	0.46
Morasko	Poland	52°	29'	N	16°	54'	E	0.01	0.1
Morokweng	South Africa	26°	28'	S	23°	32'	E	145.0	70
Mount Toondina	S.A., Australia	27°	57'	S	135°	22'	E	<110	4
Neugrund	Estonia	59°	20'	N	23°	40'	E	0	(7)
New Quebec	Québec, Canada	61°	17'	N	73°	40'	W	1.4	3.44
Newporte	North Dakota, U.S.A.	48°	58'	N	101°	58'	W	<500	3.2
Nicholson	N.W.T., Canada	62°	40'	N	102°	41'	W	<400	12.5
Oasis	Libya	24°	35'	N	24°	24'	E	<120	18
Obolon'	Ukraine	49°	30'	N	32°	55'	E	160	17
Odessa	Texas, U.S.A.	31°	45'	N	102°	29'	W	<0.05	0.17
Ouarkziz	Algeria	29°	0'	N	7°	33'	W	<70	3.5
Piccaninny	W.A., Australia	17°	32'	S	128°	25'	E	<360	7
Pilot	N.W.T., Canada	60°	17'	N	111°	1'	W	445	6
Popigai	Russia	71°	40'	N	111°	40'	E	(35)	100
Presqu' ile	Québec, Canada	49°	43'	N	74°	48'	W	<500	24
Pretoria Saltpan	South Africa	25°	24'	S	28°	5'	E	(0.22)	1.13
Puchezh-Katunki	Russia	57°	6'	N	43°	35'	E	175	80
Ragozinka	Russia	58°	18'	N	62°	0'	E	55	9
Red Wing	North Dakota, U.S.A.	47°	36'	N	103°	33'	W	(200)	9
Riachao Ring	Brazil	7°	43'	S	46°	39'	W	<200	4.5
Ries	Germany	48°	53'	N	10°	37'	E	15.1	24
Rio Cuarto	Argentina	32°	52'	S	64°	14'	W	<0.1	4.5

Terrestrial impact craters (continued)

Crater name	Location	Latitude			Longitude			Age (10⁶ years)	Diameter (km)
Rochechouart	France	45°	50'	N	0°	56'	E	186	23
Roter Kamm	Namibia	27°	46'	S	16°	18'	E	3.7	2.5
Rotmistrovka	Ukraine	49°	0'	N	32°	0'	E	(140)	2.7
Saaksjarvi	Finland	61°	24'	N	22°	24'	E	(560)	6
Saint Martin	Manitoba, Canada	51°	47'	N	98°	32'	W	(220)	40
Serpent Mound	Ohio, U.S.A.	39°	2'	N	83°	24'	W	<320	8
Serra da Cangalha	Brazil	8°	5'	S	46°	52'	W	<300	12
Shoemaker (Teague)	W.A., Australia	25°	52'	S	120°	53'	E	1,630	30
Shunak	Kazakhstan	47°	12'	N	72°	42'	E	(12)	3.1
Sierra Madera	Texas, U.S.A.	30°	36'	N	102°	55'	W	<100	13
Sikhote Alin	Russia	46°	7'	N	134°	40'	E	0	0.03
Siljan	Sweden	61°	2'	N	14°	52'	E	368	52
Slate Islands	Ontario, Canada	48°	40'	N	87°	0'	W	(450)	30
Sobolev	Russia	46°	12'	N	138°	54'	E	<0.001	0.05
Soderfjarden	Finland	62°	54'	N	21°	42'	E	(600)	5.5
Spider	W.A., Australia	16°	44'	S	126°	5'	E	>570	13
Steen River	Alberta, Canada	59°	30'	N	117°	38'	W	95	25
Steinheim	Germany	48°	41'	N	10°	4'	E	15	3.8
Strangways	N.T., Australia	15°	12'	S	133°	35'	E	<470	25
Suavjarvi	Russia	63°	7'	N	33°	23'	E	2,400	16
Sudbury	Ontario, Canada	46°	36'	N	81°	11'	W	1,850	250
Suvasvesi N	Finland	62°	42'	N	28°	0'	E	<1	4
Tabun-Khara-Obo	Mongolia	44°	6'	N	109°	36'	E	>1.8	1.3
Talemzane	Algeria	33°	19'	N	4°	2'	E	<3	1.75
Tenoumer	Mauritania	22°	55'	N	10°	24'	W	(2.5)	1.9
Ternovka	Ukraine	48°	1'	N	33°	5'	E	350	15
Tin Bider	Algeria	27°	36'	N	5°	7'	E	<70	6
Tookoonooka	Queensland, Australia	27°	7'	S	142°	50'	E	128	55
Tvaren	Sweden	58°	46'	N	17°	25'	E	>455	2
Upheaval Dome	Utah, U.S.A.	38°	26'	N	109°	54'	W	<65	10
Ust-Kara	Russia	69°	18'	N	65°	18'	E	73	25
Vargeao Dome	Brazil	26°	50'	S	52°	7'	W	<70	12
Veevers	W.A., Australia	22°	58'	S	125°	22'	E	<1	0.08
Vepriai	Lithuania	55°	10'	N	24°	34'	E	>160	8
Vredefort	South Africa	27°	0'	S	27°	30'	E	2,023	(300)
Wabar	Saudi Arabia	21°	30'	N	50°	28'	E	(0.006)	0.12
Wanapitei	Ontario, Canada	46°	45'	N	80°	45'	W	37	7.5
Wells Creek	Tennessee, U.S.A.	36°	23'	N	87°	40'	W	(200)	12

Glossary

absolute magnitude the brightness a solar-system object would have if placed 1 AU from both the observer and the Sun, and seen with a phase angle of 0° (fully illuminated)

accretion the collection of gas and dust to form larger bodies like stars, planets, and moons

achondrite a stony meteorite that lacks chondrules; most achondrites appear to be the products of igneous differentiation

active region a disturbed area on the Sun that can exhibit prominences, sunspots, solar flares, and other magnetic phenomena

adiabatic occurring without the gain or loss of heat

advection horizontal transport of an atmosphere solely by mass motion (like fog moving through a valley)

airglow an emission of light caused by excitation of neutral molecules in the upper atmosphere

albedo the reflectivity of a body; *geometric albedo* is the ratio of an object's brightness at 0° phase angle to the brightness of a perfectly diffusing disk with the same position and apparent size

Amor a class of asteroids having perihelion distances between 1.017 and 1.3 AU

anorthosite a granular igneous rock composed almost wholly of anorthite, a calcium-rich silicate

anticyclonic atmospheric circulation having a sense of rotation about the local vertical opposite to that of the planet's rotation

antitail the part of a comet's dust tail which, due to geometric projection, appears on the sunward side of the nucleus

aphelion the point in a heliocentric orbit farthest from the Sun

Apollo a class of asteroids having semimajor axes greater than 1.0 AU and perihelion distances less than 1.017 AU

asthenosphere a weak spherical shell located between the lithosphere and upper mantle

astrometry the branch of astronomy concerned with measuring the positions of celestial bodies in the sky

astronomical unit (AU) 149,597,870 km, which is approximately the average Earth-Sun distance

Aten a class of asteroids having semimajor axes less than 1.0 AU and aphelion distances greater than 0.983 AU

aurora a glow produced by atoms and ions in a planetary ionosphere

bar a unit of pressure equal to 0.987 atmosphere

baroclinic instability an unstable energy distribution in an atmosphere, because of which winds arise to move excess energy from one region to another

barycenter the center of mass of a system of celestial bodies, such as the Earth-Moon system

basalt igneous rock, produced by the cooling of lava, which consists primarily of silicon, oxygen, iron, aluminum, and magnesium

bow shock a thin current layer around an interplanetary obstacle, across which the solar wind's velocity drops and its plasma becomes heated, compressed, and deflected

breccia rock composed of broken rock fragments that was fused together by finer-grained material during an impact

brown dwarf an object with a mass somewhere between that of a star and a planet, but not massive enough for fusion to occur at its core

C-type asteroid one of the two most common classes of asteroid, located primarily in the outer part of the main belt

caldera a volcanic crater, often resulting from the partial collapse of the summit of a shield volcano

Centaurs small bodies in the outer solar system that have been scattered inward from the Kuiper belt by gravitational interactions

central peak the mountain or group of mountains at the center of a large impact crater or impact basin

chaotic terrain jumbled depressions and isolated hills found on Mars, possibly produced by collapse after the removal of subsurface groundwater or ice

chondrites a class of stony meteorites that usually contain chondrules; *carbonaceous chondrites* also include carbon compounds, while those of the *Type I* or *C1* variety contain no chondrules

chondrules small spherical grains, usually composed of iron, aluminum, or magnesium silicates, found in abundance in primitive stony meteorites

chromosphere the region of the solar atmosphere lying between the photosphere and corona

clathrate a solid mineral in which one molecular component is trapped in cavities in the crystalline lattice of the host compound

coesite a dense form of silica found in impact craters that is synthesized by subjecting quartz to very high pressure

coma the spherical gaseous envelope, typically about 150,000 km in diameter, that surrounds the nucleus of a comet

continental drift the gradual motion of the Earth's continents due to plate tectonism

convection vertical atmospheric or fluid motions resulting in transport and mixing

co-orbital satellites two satellites that share almost exactly the same orbit but which avoid collision through gravitational interaction

Coriolis force the force that operates on anything traveling from the pole of a rotating body to its equator; produces cyclonic and anticyclonic weather patterns on Earth

corona (1) the Sun's hot, highly ionized, and luminous atmosphere; (2) a circular formation of ridges enclosing a central area of jumbled relief on Venus

co-rotation the rotation of plasma in a planetary magnetosphere turning at the same angular rate as the planet's magnetic field

cosmic rays high-energy atomic nuclei (mostly protons) that enter the solar system from interstellar space

crater density a measure of impact-crater crowding on a planetary surface, usually given as the number of craters of a given size per unit area

crater ray a streak of ejected material extending radially beyond a crater's rim

crust the outer solid layer of a planet

cryovolcanism the resurfacing of an icy surface by flowing ice or ice-liquid mixtures

current sheet the two-dimensional surface within a magnetosphere that separates magnetic fields of opposite polarities

dendritic denoting a river system with a branching (treelike) pattern

deuterium an isotope of hydrogen containing a neutron

differentiation the processes by which planets and satellites develop layers or zones of different chemical and mineralogical composition

eccentricity a value that defines the shape of an ellipse; the ratio of the center-to-focus distance to the semimajor axis

ecliptic plane the plane defined by Earth's orbit; the *ecliptic* is the Sun's apparent path on the celestial sphere

eclogite a dense granular rock composed essentially of garnet and pyroxene

ejecta blanket the material surrounding an impact crater that was thrown out during its formation

eucrites a class of basaltic meteorites believed to have originated on the asteroid Vesta

extremophiles microbes that exist in environmentally extreme regions, such as sea ice, deep-sea vents, hot springs, and salt and soda lakes

feldspar one of several aluminum-rich silicate minerals found in meteorites and other rocks

fluvial produced by flowing water

forward scattering the tendency of very small particles to redirect light at large phase angles (away from the illumination source)

fusion crust the glassy veneer of melted minerals that forms on the surface of a meteorite during its atmospheric descent

geoid the equipotential surface ("mean sea level") of the Earth's gravitational field

geosynchronous an orbit whose period is equal to that of the rotation period of the primary body

graben a surface depression bounded on its long sides by normal faults

granulation a mottling effect on the Sun's photosphere caused by columns of gas rising from convection

greenhouse effect the trapping of infrared radiation by gases in a planet's atmosphere, producing an elevated surface temperature

Hadley cell circulation in a planet's atmosphere in which air heated by the Sun rises and moves poleward, then descends and returns toward the equator

heliopause the edge of the heliosphere, where the pressure of the solar wind equals that of the interstellar medium

helioseismology the study of the Sun's interior by analyzing oscillations in the solar convective zone

heliosphere the region in space permeated by the Sun's gases and magnetic field

Hirayama family a group of asteroids with similar orbital elements, indicating a probable common origin due to a collision sometime in the past

hydrostatic equilibrium the state of balance in a fluid between gravity and pressure

hydroxyl the chemical radical OH, consisting of an oxygen and hydrogen atom bound together

inclination the angle between an object's orbital plane and a reference plane, usually the ecliptic (for heliocentric orbits) or a planet's equator (for satellites)

interstellar grains small solid particles that permeate interstellar space; they may also be a component of comets and meteorites

ionopause the boundary level of the ionosphere

ionosphere the upper region of a planet's atmosphere in which many atoms are ionized

iridium a rare-earth element that is scarce in Earth's crust but relatively abundant in meteorites

irregular satellite a satellite whose orbit is highly inclined (often retrograde), highly eccentric, or both

isostasy the state in which an object is subject to equal pressure from all sides; in geophysics, isostasy means that the gravitational attraction on topography is balanced by its buoyancy with respect to dense underlying material

kamacite an alloy of iron and a small amount of nickel, found in some meteorites

kerogen insoluble organic material found in rocks

Kirkwood gaps voids in the asteroid belt where the orbital period of the asteroids are simple fractions of the orbital period of Jupiter

KREEP lunar basaltic material rich in radioactive elements; the acronym is from *K* for potassium, *REE* for rare-earth elements, *P* for phosphorus

Kuiper belt a disk-shaped region beyond the orbit of Neptune that contains countless icy objects; the source region for short-period comets

Lagrangian points the five equilibrium points in the restricted three-body problem; two Lagrangian points (L_4 and L_5), located at the vertices of equilateral triangles formed by the two primaries, are stable; the other three are unstable and lie on the line connecting the two primaries

Laplace plane the plane in which a satellite's or ring's orbit is forced to lie because of perturbations by planetary oblateness, the Sun, and nearby satellites

libration a small oscillation, or apparent rocking, exhibited by a synchronously rotating satellite in a slightly eccentric orbit

lithosphere the stiff upper layer of a planetary body, including the crust and part of the upper mantle

Lorentz force the force on a charged particle in the presence of an electric field due to its motion across a magnetic field

luminosity a measure of the total amount of radiation emitted by a glowing object

M-type asteroids asteroids composed primarily of metal

mafic denoting minerals that have a high magnesium and iron content

magma mobile or fluid rock material (lava), or any material that behaves like silicate magma in the Earth

magnetopause the outer boundary of a planetary magnetosphere

magnetosheath the region between a planetary bow shock and magnetopause in which the solar wind plasma flows around the magnetosphere

magnetosphere the region of space surrounding a planet in which the planet's magnetic field dominates that of the solar wind

magnetotail the portion of a planetary magnetosphere pulled downstream by the solar wind

mantle the part of a planet between its crust and core

mare an area on the Moon or Mars that appears darker and smoother than its surroundings; lunar maria are basalt flows

mascon subsurface *mass concentrations* that cause large-scale gravity anomalies on the Moon and other bodies

megaregolith the regolith structure thought to exist throughout an asteroid

mesosphere an atmospheric region, between the stratosphere and the thermosphere, characterized by a broad temperature maximum

meteor shower the enhancement of meteors caused by Earth's passage (usually at the same time each year) through the particles distributed along a comet's orbit

micrometeorite a very small meteorite or meteoritic particle less than a millimeter in diameter

mixing ratio the ratio of the mass of a given atmospheric gas to that of the atmosphere as a whole

Mohorovičić discontinuity the boundary between the Earth's crust and upper mantle

neutrino a neutral relativistic particle of very small (presumably zero) rest mass

noble gases the rare and inert gases helium, neon, argon, krypton, xenon, and radon

nucleus (comet) the core of a comet, typically a few kilometers in diameter, consisting of a mixture of ices and solid silicate and carbonaceous grains

oblateness the centrifugal distortion of an otherwise spherical planet, defined as the difference between the equatorial and polar radii, divided by the equatorial radius

obliquity the angle between an object's axis of rotation and the pole of its orbit

occultation the passage of one object in front of another of smaller apparent angular size, such as when an asteroid briefly blocks a star from view

olivine a metal-rich silicate mineral found in meteorites and other rocks

Oort cloud a roughly spherical volume, extending more than 100,000 AU from the Sun, in which up to a trillion small icy bodies are thought to reside; the source region for "new" and long-period comets

opacity the ability of a medium to block light or other radiation

opposition effect a nonlinear surge in reflected light from an object when it is observed at a small phase angle

optical depth a measure of the radiation absorbed as it passes through a medium; the medium is termed *transparent* when the value is zero, *optically thin* when less than 1, and *optically thick* when greater than 1

organic a compound containing carbon and hydrogen, though not necessarily those compounds associated with life

outgas to release gases from the interior of a planet, often in association with volcanic activity

P wave a seismic disturbance whose oscillations are in the direction of the wave's propagation

paleomagnetism the ancient magnetic fields recorded in rocks

palimpsest a roughly circular spot on icy satellites thought to identify a former crater and rim

Pangea a hypothetical supercontinent that included all the land masses of the Earth before the Triassic period (225 million years ago)

perihelion the point in a heliocentric orbit closest to the Sun

phase angle the angle on a surface between the direction of illumination (usually the Sun) and that of the observer or observing instrument

photodissociation the breakdown of molecules due to the absorption of light, especially ultraviolet sunlight

photometry the branch of astronomy concerned with measuring the brightness of a celestial object at various wavelengths

photosphere the intensely bright surface of the Sun

plagioclase a common rock-forming silicate mineral

planetesimal a primordial body of intermediate size, up to perhaps 1 km across, which accreted into planets or asteroids

planetology the study of the physical and chemical properties of planets

plasma completely ionized gas, consisting of free electrons and atomic nuclei, in which the temperature is too high for neutral atoms to exist

plasma tail a narrow cometary tail stretching directly away from the Sun consisting of plasma swept back by the solar wind; also called *ion tail*

plate tectonism the movement of segments of the lithosphere under the influence of convection in the mantle

polarimetry the branch of astronomy concerned with measuring the orientation and extent of the polarization in light (or other radiation) coming from a celestial object

polarization the process of affecting radiation, especially light, such that the electromagnetic vibrations are not randomly oriented

Poynting-Robertson effect the drag on a small orbiting particle due to the asymmetrical re-radiation of absorbed light

Precambrian the earliest era in geologic time, encompassing all Earth history before the beginning of the Cambrian era, 570 million years ago

precession a slow, periodic conical motion of the rotation axis of a spinning body

primordial existing at or near the very beginning of the solar system, about 4.6 billion years ago

prograde orbiting or rotating in the same direction as the prevailing direction of motion

protoplanet an early accumulation of material from the solar nebula that lead to the formation of a planet

pyroxene rock-forming silicates having 1:1 ratios of metal oxides (magnesium, iron, or calcium) to silicon oxide

radiogenic heat heat produced by the radioactive decay of potassium, uranium, thorium, and other isotopes

radionuclide a radioactive isotope of an element

rarefaction a decrease in density and pressure due to the passage of a sound wave or other compression wave

refractory denoting an element that vaporizes at high temperatures; examples are uranium, calcium, and aluminum

regolith the fragmented rocky debris, produced by meteoritic impact, that forms the uppermost surface on planets, satellites, and asteroids

regular satellite a prograde-moving satellite whose orbit has low eccentricity and inclination

remote sensing any technique for measuring an object's characteristics at a distance

resonance a state in which one orbiting object is subject to periodic gravitational perturbations by another

retrograde orbiting or rotating opposite to the prevailing direction of motion

rille a trenchlike valley, up to several hundred km long and 1 to 2 km wide, on the surface of the Moon or other satellites

Roche limit the critical distance inside which an idealized satellite with no tensile strength would shatter due to tidal forces exerted by its parent planet

S-type asteroid one of the two most common classes of asteroid, located primarily in the inner part of the main belt

S wave a seismic disturbance whose oscillations are perpendicular to the direction of the wave's propagation

scale height the altitude over which atmospheric density decreases by the factor e (about 2.7183)

scarp a cliff produced by faulting or erosion

secondary crater an impact crater produced by ejecta from a primary impact

seismic tomography a technique used to produce three-dimensional maps of the Earth's interior by analyzing seismic waves

semimajor axis half of the longest axis (diameter) of an ellipse, used to define the mean orbital distance of an object from its primary

shepherd satellite a satellite that sustains the structure of a planetary ring through its close gravitational influence

siderophile denoting an element soluble in molten iron and often occurring in the native state (iron, cobalt, nickel and gold)

shield volcano a broad volcano built up through repeated, non-explosive eruptions of highly fluid basalts

silicate a rock or mineral whose crystalline structure is dominated by bonded silicon and oxygen atoms

SNC meteorites three small classes of basaltic meteorites (shergottites, nakhlites, and chassignites) thought to have been ejected from Mars's surface during one or more impacts millions of years ago

solar constant the total amount of solar energy irradiating a given surface area in a given interval of time; at 1 AU, this value is 1,367 watts per square meter

solar flare a sudden release of energy in or near the Sun's photosphere that accelerates charged particles into space

solar nebula the disk-shaped cloud of gas and dust where the solar system formed (also known as *protosolar nebula, protoplanetary nebula*)

solar wind the high-speed outflow of energetic charged particles and entrained magnetic field lines from the solar corona

spectroscopy the study of the light emitted from or reflected by a body (its spectrum)

spherule a round particle of rock and iron oxide formed as molten material flows off a meteorite during its passage through Earth's atmosphere

spokes dark radial streaks in Saturn's ring system

sputtering a process in which atomic particles strike a surface at high speed, altering its chemistry and ejecting atoms or molecular fragments

stereoisomers two compounds of the same composition in which the atoms are arranged as mirror images of each other; an equal mixture of right- and left-handed stereoisomers is termed *racemic*

stishovite an extremely dense form of silica produced by the very high pressure of a meteorite impact

stratigraphy a sequence of rock layers that provides information on the geologic history of a region

stratosphere the layer of a planet's atmosphere in which the temperature remains roughly constant with altitude; on Earth, it is above the troposphere and below the ionosphere

strewn field the area of a meteorite fall, with pieces of meteorite lying in a pattern tracing the original body's fall trajectory

subduction the process by which by one crustal plate is forced under another

sunspot a relatively cool, dark area on the solar photosphere

superrotation the characteristic of an atmosphere that rotates faster than the planet itself

synchrotron radiation electromagnetic energy emitted when a charged particle is accelerated about a line of magnetic field

tectonism associated with the forces acting in a planet's crust

tektite a small, rounded, piece of silicate glass formed during terrestrial impacts

tholin a complex hydrocarbon created by the irradiation of a mixture of volatile compounds containing carbon, hydrogen, nitrogen, and oxygen

tidal heating the heating of a satellite's interior due to the tidal friction induced by the strong gravitational field of its planet

thermosphere the atmospheric region where the temperature rises due to ionospheric heating

Titius-Bode law a numerical series derived in the 18th century used to approximate the average distances of the planets from the Sun

transient cavity the initial, temporary crater formed during an impact event

Trojans asteroids located in the two stable Lagrangian points of Jupiter's orbit (60° preceding and following the planet)

tropopause the boundary between the troposphere and the stratosphere in Earth's atmosphere

troposphere the lowest level of the Earth's atmosphere, where most weather takes place, and the convection-dominated region of other planetary atmospheres

viscous relaxation the tendency of high or low topographic features to assume the local mean elevation through gradual deformation over geologically long periods of time

volatiles elements or molecules with low melting temperatures; examples are potassium, sodium, water, and ammonia

zonal jets high-speed winds confined to a limited range of latitude in an atmosphere

zodiacal light a faint glow in the night sky caused by sunlight scattering off interplanetary dust near the plane of the ecliptic

Authors, Suggested Readings, and Illustration Credits

PREFACE

J. KELLY BEATTY is the senior editor of *Sky & Telescope* magazine in Cambridge, Massachusetts, where for more than 20 years he has reported on developments in planetary science. While earning a B.S. in geology at the California Institute of Technology, Beatty worked on imaging results from the Mariner 9 and 10 interplanetary missions. He also holds a M.S. in science journalism. His versatility as a writer has won a large following within scientific circles as well as the general public, and on the strength of that reputation he was among the first Western journalists to gain firsthand access to the Soviet space program. Beatty also contributes to a wide range of other magazines, newspapers, and encyclopedias. Asteroid 2925 Beatty was named on the occasion of his marriage in 1983, and in 1986 he was one of the 100 semifinalists for NASA's Journalist in Space program.

CAROLYN COLLINS PETERSEN is an award-winning science writer whose articles have appeared in a wide range of newspapers and magazines since 1980. She holds degrees in education (B.S., 1978) and in journalism (M.A., 1996) from the University of Colorado. For eight years she was a member of the HST Goddard High Resolution Spectrograph team, during which time she coordinated observations for the Ulysses Comet Watch Network. Petersen co-authored *Hubble Vision* with GHRS principal investigator John C. Brandt. She has also written planetarium programs seen in hundreds of facilities around the world, and she maintains the Henrietta Leavitt Flat Screen Space Theater, an online planetarium show and space image gallery. In 1996 Petersen joined Sky Publishing Corporation as editor of its books and products division.

ANDREW CHAIKIN is a Boston-based science writer with lifelong interest in planetary science. He holds a Bachelor's degree in geology from Brown University, and he has served as a research geologist at the Smithsonian Institution's Center for Earth and Planetary Studies in Washington. In 1980 Chaikin joined the editorial staff of *Sky & Telescope*, where he covered space-exploration missions for six years. Chaikin wrote *A Man on the Moon*, the widely acclaimed chronicle of the Apollo lunar astronauts and their experiences, and more recently *Air & Space*. He served as a technical consultant for the HBO miniseries *From the Earth to the Moon*. Now a contributing editor for *Popular Science*, he also writes frequently for *Air & Space/Smithsonian, Discover,* and other publications.

DON DAVIS is widely recognized as one of the finest astronomical artists in the world. His meticulous attention to detail and accuracy has won admiration from artists and scientists alike. In recent years he has both mastered and advanced a host of innovative computerized illustration techniques. Davis won an Emmy award for his groundbreaking work in the classic PBS series *Cosmos,* and his many collaborations with the late Carl Sagan included the dramatic opening sequence for the film *Contact.* Davis's stunningly realistic portrayals can be appreciated in such publications as Time-Life Books, *Parade, Smithsonian,* and *Natural History.*

Greeley, R., and R. Batson, *The NASA Atlas of the Solar System* (Cambridge University Press, 1997)

Lewis, J. S., *Physics and Chemistry of the Solar System* (Academic Press, 1996)

McSween, H. Y., *Stardust to Planets: A Geological Tour of the Universe* (St. Martin's Griffin, 1995)

Morrison, D., and T. Owen, *The Planetary System* (Addison-Wesley, 1995)

Shirley, J. H., and R. W. Fairbridge, eds., *Encyclopedia of Planetary Sciences* (Chapman & Hall, 1997)

Weissman, P. R., and others, *Encyclopedia of the Solar System* (Academic Press, 1998)

Wood, J. A., *The Solar System* (Prentice Hall, 1979)

Front: NASA/JPL

CHAPTER 1: Exploring the Solar System

DAVID MORRISON is the Director of Space at NASA's Ames Research Center, where he manages basic and applied research programs in the space, life, and Earth sciences, with emphasis on astrobiology. Prior to this position he was a professor of astronomy at the University of Hawai'i and director of NASA's 3-meter Infrared Telescope Facility atop Mauna Kea. His research involves asteroids and the satellites of the outer planets. Morrison was a member of the Voyager imaging-science team and serves as an interdisciplinary scientist on the Galileo mission. He also chaired NASA's Solar System Exploration Committee, and in 1981 he served at NASA Headquarters as acting deputy associate administrator for space science.

Cowen, R., "Scooping Up a Chunk of Mars" (*Science News, 153,* 265–267, 25 April 1998)

Fimmel, R. O., J. Van Allen, and E. Burgess, *Pioneer: First to Jupiter, Saturn, and Beyond* (National Aeronautics and Space Administration, SP-446, 1980)

McDougall, W. A., *...The Heavens and the Earth: A Political History of the Space Age* (Basic Books, 1985)

Morrison, D., *Exploring Planetary Worlds* (Scientific American Library, 1993)

Murrill, M. B., "The Grandest Tour Voyager" (*Mercury, 22,* 66–77, May-June 1993)

Sobel, D., "Among Planets" (*New Yorker, 72,* 84–90, 9 December 1996)

Washburn, M., *Distant Encounters: The Exploration of Jupiter and Saturn* (Harcourt Brace Jovanovich, 1983)

Front: NASA/JPL *1:* NASA/JPL *2:* Patricia Jacobberger (NASM) and Gerald Jellison *3a:* Frank B. Hitchens *3b:* Reta Beebe (New Mexico State Univ.), NASA *3c:* NASA/JPL *3d:* John Clarke (Univ. Michigan), NASA *3e:* Glenn Orton (JPL) *3f:* Imke De Pater (Univ. California, Berkeley) *4:* Dan Woods (NASA Headquarters) *5-7:* NASA/JPL *8:* NASA *9:* NASA, Malin Space Science Systems *10:* NASA/JPL *Table 1:* Sky Publishing *Table 2:* Sky Publishing

CHAPTER 2: Origin of the Solar System

JOHN A. WOOD studied geology at Virginia Polytechnic Institute (B.S., 1954) and the Massachusetts Institute of Technology (Ph.D., 1958). While at MIT, and later at the Enrico Fermi Institute, University of Chicago, he explored the geologic properties of meteorites. Since then Wood has pursued this subject, and the larger question of the origin of the planets, at the Smithsonian Astrophysical Observatory in Cambridge, Massachusetts. Wood participated in the analysis of Apollo lunar samples and was an investigator on the Magellan mission to Venus. He is the author of *Meteorites and the Origin of Planets* (McGraw-Hill, 1968) and *The Solar System* (Prentice-Hall, 1979). In 1999 he will chair the National Academy of Sciences' Committee on Space Exploration of the Space Studies Board.

Artymowicz, P., "Beta Pictoris: An Early Solar System?" (*Annual Reviews of Earth and Planetary Science, 25,* 175–219, 1997)

Davies, J., "Frozen in Time" (*New Scientist, 152,* 36–39, 13 April 1996)

Killian, A. M., "Playing Dice with the Solar System" (*Sky & Telescope, 78,* 136–140, August 1989)

Lada, C., "Deciphering the Mysteries of Stellar Origins" (*Sky & Telescope, 93,* 18-24, May 1993)

Lissauer, J. J., "Planet Formation" (*Annual Reviews of Astronomy and Astrophysics, 31,* 129–174, 1993)

O'Dell, C. R., and S. V. W. Beckwith, "Young Stars and Their Surroundings" (*Science, 276,* 1355–1359, 30 May 1997)

Podosek, F. A., and P. Cassen, "Theoretical, Observational, and Isotopic Estimates of the Lifetime of the Solar Nebula" (*Meteoritics, 29,* 6–25, 1994)

Wetherill, G. W., "Formation of the Earth" (*Annual Reviews of Earth and Planetary Science, 18,* 205-256, 1990)

Wood, J. A., "Chondritic Meteorites and the Solar Nebula" (*Annual Reviews of Earth and Planetary Science, 16,* 53–72, 1988)

Front: Robert O'Dell and S. Wong (Rice Univ.), NASA *1:* Robert O'Dell and S. Wong (Rice Univ.), NASA *2:* Mark McCaughrean (MPIA), Robert O'Dell, NASA *3:* John Wood (Harvard Univ.), Frank Shu (Univ. California, Santa Cruz) *4:* Christopher Burrows (STScI/ESA), NASA *5:* John Wood *6:* Donald Brownlee (Univ. Washington) *7:* John Wood *8:* Don Davis *9:* John Wood *10:* Don Davis, David Hughes (Univ. Sheffield) *11:* Peter McGregor and Mark Allen (MSSSO/Australian National University) *12:* John Wood *13:* John Wood *14:* NASA/JPL

CHAPTER 3: The Sun

KENNETH LANG has been a professor at Tufts University since 1974. From 1969 to 1973 Lang held postdoctoral positions at Cambridge University, Cornell University, and the California Institute of Technology, where he pioneered radio interferometric investigations of the Sun. From 1990 to 1992 he served at NASA Headquarters as a visiting senior scientist. The Solar Physics Division of the American Astronomical Society awarded Lang its 1997 popular-writing award for his article "Unsolved Mysteries of the Sun — Part I," published in *Sky & Telescope* in August 1996. He is the author of *Astrophysical Formulae* (now in its third edition) and *Sun, Earth and Sky.*

Acton, L. W., and others, "The Yohkoh Mission for High-Energy Solar Physics" (*Science, 258,* 618–625, 1992)

Bahcall, J. N., "Progress and Prospects in Neutrino Astrophysics" (*Nature, 375,* 29–34, 1995)

Golub, L., and J. M. Pasachoff, *The Solar Corona* (Cambridge University Press, 1997)

Harvey, J., "Helioseismology" (*Physics Today, 48,* 32–38, October 1995)

Kennedy, J. R., "GONG: Probing the Sun's Hidden Heart" (*Sky & Telescope, 92,* 20–25, October 1996)

Kippenhahn, R., *Discovering the Secrets of the Sun* (John Wiley and Sons, 1994)

Lang, K. R., *Sun, Earth and Sky* (Springer-Verlag, 1995)

Lang, K. R., "SOHO Reveals the Secrets of the Sun" (*Scientific American, 276,* 32–39, March 1997)

Marsden, R. G., and E. J. Smith, "Ulysses: Solar Sojourner" (*Sky & Telescope, 91,* 24–30, March 1996)

Solomey, N., *The Elusive Neutrino: A Subatomic Detective Story* (W. H. Freeman, 1997)

Front: Carl Foley. (University College London) *1:* Don Davis *2:* Kenneth Lang (Tufts Univ.) *3:* Yoji Totsuka (University Tokyo) *4:* Fred Espenak (NASA/Goddard) *5:* Carl Foley (University College London) *6:* SOHO/EIT Consortium *7:* Harold Zirin (Caltech) *8:* David Hathaway (NASA/Marshall) *9:* William Livingston (National Solar Obs.) *10:* Don Davis *11:* Guenter Brueckner (NASA/GODDARD), SOHO/LASCO Consortium *12:* Loren Acton (Montana State Univ.) *13:* James Kennedy (National Solar Obs.) *14:* Philip Scherrer (Stanford Univ.) *15:* SOHO/SOI/MDI Consortium, Alexander Kosovichev *16:* SOHO/SOI/MDI Consortium *17:* SOHO/SOI/MDI consortium *18:* Neal Hurlburt (Stanford-Lockheed Inst. for Space Research) *19:* Alan Title (Stanford), NASA *20:* SOHO/EIS Consortium, Bernhard Fleck (NASA/Goddard) *21:* Edward Smith (JPL) *22:* Richard Willson (JPL) *23:* Greg Slater and Gary Linford (Lockheed Palo Alto Research Lab.) *24:* I.-Julianna Sackmann (Caltech) *Table 1:* Kenneth Lang (Tufts Univ.)

CHAPTER 4: Planetary Magnetospheres and the Interplanetary Medium

JAMES A. VAN ALLEN is one of the pioneers of space physics, having initiated scientific work with high-altitude rockets in 1946. He has been a principal investigator on 24 Earth-orbiting, interplanetary, and planetary spacecraft, including Pioneers 10 and 11. In 1958, he discovered the radiation belts of Earth with instrumentation on Explorer 1, the first U.S. satellite. Van Allen's research since then has emphasized planetary magnetospheric physics. He was a professor of physics and head of the Department of Physics and Astronomy at the University of Iowa from 1951 until his retirement in 1985.

FRANCES BAGENAL is an associate professor of astrophysical and planetary sciences at the University of Colorado at Boulder. She received her Ph.D. from the Massachusetts Institute of Technology in 1981. Her main research interests are the magnetospheres of the Jovian planets and the interactions between space plasmas and smaller bodies such as the Galilean satellites and Pluto. Bagenal has participated on the scientific teams of the Voyager, Galileo, and Deep Space 1 missions.

Bagenal, F., "Giant Planet Magnetospheres" (*Annual Reviews of Earth and Planetary Science, 20,* 289–328, 1992)

Connerney, J. E. P., "Magnetic Fields of the Outer Planets" (*Journal of Geophysical Research, 98,* 18,659–18,679, 25 October 1993)

Cronin, J. W., and others, "Cosmic Rays at the Energy Frontier" in *Magnificent Cosmos* (*Scientific American,* 62–67, spring 1998)

Friedlander, M., *Cosmic Rays* (Harvard Univ. Press, 1989)

Jokipii, J. R., and F. B. McDonald, "Quest for the Limits of the Heliosphere" (*Scientific American, 272,* 58–63, April 1995)

Kivelson, M. G., *Introduction to Space Physics* (Cambridge University Press, 1995)

Lanzerotti, L. J., and C. Uberoi, "Earth's Magnetic Environment" (*Sky & Telescope, 76,* 360–362, October 1988)

Lanzerotti, L. J., and C. Uberoi, "The Planets' Magnetic Environments" (*Sky & Telescope, 77,* 149–152, February 1989)

Van Allen, J. A., "Electrons, Protons, and Planets" (*Icarus, 122,* 209–232, 1996)

Front: Don Davis *1:* Frances Bagenal (Univ. Colorado) *2:* Peter Riley (LANL) *3:* Frances Bagenal *4:* John Connerney (NASA/Goddard) *5:* Frances Bagenal *6:* Frances Bagenal *7:* Don Davis *8:* Margaret Kivelson and Christopher Russell (Univ. California, Los Angeles) *9:* Frances Bagenal *10:* Margaret Kivelson and Christopher Russell *11:* Louis Frank (Univ. Iowa) *12:* Imke de Pater (Univ. California, Berkeley) *13a:* Frances Bagenal *13b:* John Spencer (Lowell Obs.) *14:* Nicholas Schneider (Univ. Colorado) and John Trauger (JPL) *15:* Frances Bagenal *16:* John Clarke (Univ. Michigan), NASA *17:* John Trauger (JPL), NASA *18:* Frances Bagenal *19:* Stamatios Krimigis (Johns Hopkins Univ.) *20:* Frances Bagenal *21:* Frances Bagenal *22:* Janet Luhmann (Univ. California, Berkeley) *23:* Applied Physics Lab., Johns Hopkins Univ. *24:* Marcia Neugebauer (JPL) *Table 1:* Frances Bagenal

CHAPTER 5: Cometary Reservoirs

PAUL WEISSMAN is a senior research scientist at the Jet Propulsion Laboratory in Pasadena, California. His prime area of study is the physics and dynamics of small bodies in the solar system, especially comets. He pioneered Monte Carlo dynamical studies of the Oort cloud and of the evolution of long-period comets. Weissman is a Galileo co-investigator and an interdisciplinary scientist on the Rosetta mission to Comet Wirtanen. He also serves as project scientist for the Deep Space 4/Champollion mission to Comet Tempel 1. Weissman received his Ph.D. in planetary and space physics from the University of California at Los Angeles in 1978.

Duncan, M. J., and H. F. Levison, "A Disk of Scattered Icy Objects and the Origin of Jupiter-Family Comets" (*Science, 276,* 1670–1672, 13 June 1997)

Duncan, M., T. Quinn, and S. Tremaine, "The Origin of Short-Period Comets" (*Astrophysical Journal Letters, 328,* 69–73, 15 May 1988)

Flamsteed, S., and others, "Where Comets Come From" (*Discover, 16,* 79-87, November 1995)

Luu, J. X., and D. C. Jewitt, "The Kuiper Belt" (*Scientific American, 274,* 46-52, May 1996)

Weissman, P., "The Oort Cloud" (*Nature, 344,* 825–830, 26 April 1990)

Weissman, P. R., and H. F. Levison, "The Population of the Trans-neptunian Region" in *Pluto and Charon* (S. A. Stern and D. J. Tholen, eds., University of Arizona Press, Tucson, 559–604, 1997)

Front: Don Davis *1:* Paul Weissman (JPL) *2a:* Shigemi Numazawa (Japan Planetarium Laboratory) *2b:* Dennis di Cicco (Sky Publishing) *3:* Paul Weissman *4:* Don Davis *5:* Sky Publishing *6:* Martin Duncan (Queens Univ.) *7:* Sky Publishing *8:* Bradford Smith (Univ. Hawaii) and Richard Terrile (JPL) *9:* David Jewitt (Univ. Hawaii) and Jane Luu (Harvard Univ.) *10:* Martin Duncan and Harold Levison (Southwest Research Inst.) *11:* Paul Weissman and Harold Levison *12:* Martin Duncan

CHAPTER 6: The Role of Collisions

EUGENE M. SHOEMAKER was awarded a Ph.D. from Princeton in 1960. He served as a geologist with the U.S. Geological Survey (1948–1997), professor at the California Institute of Technology (1969–1985), and on the staff of Lowell Observatory (1993–1997). His contributions to geology, astronomy, and planetary science include creating a consistent theory for impact cratering based on his study of the Barringer Meteor Crater and the application of this knowledge to the Ranger, Surveyor, Apollo, and Voyager space missions. Shoemaker developed methods of lunar geologic mapping and established the lunar geologic time scale. He is best known as the "Father of Astrogeology" and is also remembered for his co-discovery of Comet Shoemaker-Levy 9. He was killed in an automobile collision on a remote track in Australia's outback on 18 July 1997.

CAROLYN S. SHOEMAKER has been a staff astronomer at Lowell Observatory since 1993, a research professor of astronomy at Northern Arizona University, and a volunteer at the U.S. Geological Survey in Flagstaff, Arizona, since 1982. With Eugene Shoemaker, she collaborated in the Palomar Asteroid and Comet Survey from 1982–1994. Her discovery of 32 comets within 11 years makes her the most successful comet hunter of all time. Among these discoveries was Comet Shoemaker-Levy 9, which collided with Jupiter in July 1994. She has also discovered 41 Earth-approaching asteroids, 67 Mars-crossers, 47 Jupiter Trojans, and many other asteroids of unusual motion. She received her B.A. in 1949 and her M.A. in 1950 from Chico State College.

Alvarez, W., *T. rex and the Crater of Doom* (Princeton University Press, 1997)

Beatty, J. K., and D. H. Levy, "Crashes to Ashes: A Comet's Demise" (*Sky & Telescope, 90,* 18–26)

Chapman, C. R., and D. Morrison, *Cosmic Catastrophies* (Plenum, 1989)

Gallant, R. A., "Journey to Tunguska" (*Sky & Telescope, 87,* 38–43, June 1994)

Grieve, R. A. F., and E. M. Shoemaker, "The Record of Past Impacts on Earth" in *Hazards Due to Comets and Asteroids*, T. Gehrels, ed. (University of Arizona Press, 417–462, 1994)

Hoyt, W. G., *Coon Mountain Controversies: Meteor Crater and the Development of Impact Theory* (University of Arizona Press, 1987)

Lewis, J. S., *Rain of Iron and Ice: The Very Real Threat of Comet and Asteroid Bombardment* (Addison-Wesley, 1996)

Melosh, H. J., *Impact Cratering: A Geological Process* (Oxford University Press, 1989)

Shoemaker, E. M., "Asteroid and Comet Bombardment of the Earth" (*Annual Reviews of Earth and Planetary Science, 11*, 461–494, 1983)

Spencer, J. R., and J. Mitton, eds., *The Great Comet Crash* (Cambridge University Press, 1995)

Spudis, P. D., *The Geology of Multi-Ring Impact Basins: The Moon and Other Planets* (Cambridge University Press, Cambridge Planetary Science Series, 1993)

Front: NASA/Langley *1:* Hubble Space Telescope Comet Team *2:* Don Davis *3:* Don Davis *4:* David Roddy and Karl Zeller (U.S. Geological Survey) *5:* Richard Grieve (Natural Resources Canada) *6:* Richard Grieve *7:* Alan Hildebrand and Mark Pilkington (Geological Survey Canada) *8:* Don Davis *9:* NASA/Johnson *10:* Lick Observatory *11:* Walter Kiefer (Lunar and Planetary Inst.) *12:* Eugene Shoemaker *13:* Robert Strom (Univ. Arizona) *14:* NASA/JPL *15:* David Jewitt (Univ. Hawaii) and Jane Luu (Harvard Obs.) *16:* Sky Publishing *17:* Philip Nicholson (Cornell Univ.) *18:* Paul Schenk (Lunar and Planetary Inst.) *19:* Clark Chapman (Southwest Research Inst.) and David Morrison (NASA/Ames) *20:* Eugene Shoemaker *Table 1:* Eugene Shoemaker *Table 2:* Eugene Shoemaker *Table 3:* IAU Minor Planet Center

CHAPTER 7: Mercury

FAITH VILAS received her Ph.D. in planetary sciences from the University of Arizona in 1984. There she specialized in the spectral characteristics of asteroids in the outer solar system. Currently, she is a space scientist at NASA's Johnson Space Center. Her duties include designing space-based experiments to monitor artificial space debris orbiting Earth. She continues to study the mineralogical composition of asteroids, especially the alteration of their surfaces by liquid water. Vilas is the principal editor of the seminal reference volume *Mercury*. In 1996–97 she chaired the Division for Planetary Sciences of the American Astronomical Society. In 1988, the International Astronomical Union honored her by naming asteroid 3507 with her surname.

Harmon, J. K., "Mercury Radar Studies and Lunar Comparisons" (*Advances in Space Research, 19*, 1487-1496, 1997)

Nelson, R. M., "Mercury: The Forgotten Planet" (*Scientific American, 277*, 56–63, November 1997)

Robinson, M. S., and P. G. Lucey, "Recalibrated Mariner 10 Color Mosaics: Implications for Mercurian Volcanism" (*Science, 275*, 197–200, 10 January 1997)

Sprague, A. L., and others, "Mercury's Feldspar Connection: Mid-IR Measurements Suggest Plagioclase" (*Advances in Space Research, 19*, 1507–1510, 1997)

Strom, R. G., *Mercury: The Elusive Planet* (Smithsonian Institution Press, 1987)

Vilas, F., C. R. Chapman, and M. S. Matthews, eds., *Mercury* (University of Arizona Press, 1988)

Front: Calvin Hamilton (LANL) *1:* Andrew Potter (Lunar and Planetary Inst.) *2:* Robert Strom (Univ. Arizona) *3:* David Mitchell (Univ. California, Berkeley) *4:* Mark Robinson (Northwestern Univ.) *5:* Mark Robinson *6:* Mark Robinson *7:* Mark Robinson and Paul Lucey (Univ. Hawaii) *8:* Mark Robinson and Paul Lucey *9:* Calvin Hamilton (LANL) *10:* Mark Robinson *11:* Don Davis *12:* Mark Robinson *13:* Don Davis *14:* Andrew Potter (Lunar and Planetary Inst.) *15:* Ann Sprague (Univ. Arizona) *16:* Martin Slade (JPL) and Duane Muhleman (Caltech) *17:* John Harmon (National Atmospheric and Ionospheric Center)

CHAPTER 8: Venus

R. STEPHEN SAUNDERS is a senior research scientist at the Jet Propulsion Laboratory. He received his Ph.D. from Brown University in 1970. He has studied the geological processes of the Moon, Venus, and Mars extensively. He is Director of JPL's Regional Planetary Image Facility and is project scientist for the Planetary Data System. Currently, he is also project scientist of the Mars Surveyor Program 2001 mission, the first of a series of missions that will lead to a return of the first sample from Mars. Saunders was a former member of the U.S. Peace Corps and served as a geologist in Ghana from 1963 to 1965.

Barsukov, V. L., and others, eds., *Venus Geology, Geochemistry, and Geophysics* (Univ. of Arizona Press, 1992)

Basilevskiy, A. T., "The Planet Next Door" (*Sky & Telescope, 77*, 360–368, April 1989)

Bougher, S. W., D. M. Hunten, and R. S. Phillips, eds., *Venus II* (University of Arizona Press, 1997)

Cattermole, P., *Venus: The Geological Story* (Johns Hopkins University Press, 1994)

Grinspoon, D. H., *Venus Revealed: A New Look Below the Clouds of Our Mysterious Twin Planet* (Addison-Wesley, 1997)

Price, M., and J. Suppe, "Mean Age of Rifting and Volcanism on Venus Deduced from Impact Crater Densitites" (*Nature, 372*, 756–759, 22/29 December 1994)

Solomon, S. C., "The Geophysics of Venus" (*Physics Today, 46*, 49–55, July 1993)

Stofan, E. R., "The New Face of Venus" (*Sky & Telescope, 86*, 22–31, August 1993)

Front: NASA/JPL *1:* NASA/JPL *2:* NASA/JPL *3:* NASA/JPL *4:* NASA/JPL *5:* NASA/JPL *6:* NASA/JPL *7:* NASA/JPL *8:* Carlé Pieters (Brown Univ.), Russian Academy of Sciences *9:* Peter Ford (MIT), NASA/JPL *10:* NASA/JPL *11:* NASA/JPL *12:* NASA/JPL *13:* NASA/JPL *14:* NASA/JPL *15:* NASA/JPL *16:* NASA/JPL *17:* Peter Ford and Gordon Pettengill (MIT) *18:* NASA/JPL *19:* NASA/JPL *20:* NASA/JPL *21:* NASA/JPL *22:* NASA/JPL

CHAPTER 9: Planet Earth

DON L. ANDERSON is director of the Seismological Laboratory and a professor of geophysics at the California Institute of Technology, where he received his Ph.D. in geophysics and mathematics in 1962. His interests lie in the origin, evolution, structure, and composition of Earth and other planets. Anderson served with Chevron Oil Company, the Air Force Cambridge Research Laboratory, and the Arctic Institute of North America from 1955 to 1958. He was elected to the National Academy of Sciences in 1982, and he served as president of the American Geophysical Union from 1988 to 1991. Anderson was awarded the Crafoord prize in 1998.

Anderson, D. L., "The Earth as a Planet: Paradigms and Paradoxes" (*Science, 223*, 347–535, 27 January 1984)

Cox, A., and R. B. Hart, *Plate Tectonics: How It Works* (Blackwell Scientific, 1986)

Fowler, C. M. R., *The Solid Earth* (Cambridge University Press, 1990)

Schneider, D., "A Spinning Crystal Ball" (*Scientific American, 275*, 28–30, October 1996)

Taylor, S. K., and S. M. McLennan, "The Evolution of Continental Crust" (*Scientific American, 274*, 76–81, January 1996)

Front: NASA/Johnson *1:* Thomas Hunt, Kalmbach Publishing *2:* Don Anderson (Caltech) *3:* Guy Masters (Scripps Inst. of Oceanography) *4:* Don Anderson *5:* Dixon Rohr (Lamont-Doherty Earth Observatory) *6:* Adam Dziewonski (Harvard Univ.) *7:* NASA *8:* Walter Smith (NOAA) and David Sandwell (Scripps Inst. of Oceanography) *9: Scientific American 10:* Don Davis *11:* Stephen Grand (Univ. Texas), Rob van der Hilst (MIT) *12:* Lisa Gahagan (Plates Project, Univ. Texas) *13:* Frank Lemoine and James Frawley (NASA/Goddard) *Tables 1 and 2:* Don Anderson

CHAPTER 10: The Moon

PAUL D. SPUDIS is a staff scientist at the Lunar and Planetary Institute in Houston, Texas. He received his Ph.D. in geology from Arizona State University in 1982. Spudis specializes in the impact and volcanic processes operating on the terrestrial planets, with particular focus on the geologic history and evolution of the Moon. He is a principal investigator in NASA's Planetary Geology Program and was deputy leader of the science team for the Department of Defense's Clementine mission. His current research includes analysis of lunar remote-sensing data and various studies leading to the establishment of a permanent base on the Moon.

Chaikin A., *A Man on the Moon.* (Viking Press, New York, 1994)

Hartmann, W. K., and others, eds., *Origin of the Moon* (Lunar and Planetary Institute, 1986)

Heiken G. H., and others, eds., *The Lunar Sourcebook: A User's Guide to the Moon.* (Cambridge University Press, 1991)

Ryder, G., "Apollo's Gift: The Moon." (*Astronomy, 22,* 40–45, July, 1994)

Spudis, P. D., *The Once and Future Moon* (Smithsonian Institution Press, 1996)

Taylor, G. J., "The Scientific Legacy of Apollo." (*Scientific American, 271,* 26–33, July, 1994)

Taylor, S. R., "The Origin of the Moon." (*American Scientist, 75,* 469–477, September-October 1987)

Wilhelms, D. E., *The Geologic History of the Moon* (U.S. Geological Survey Professional Paper 1348, 1987)

Wilhelms, D. E., *To a Rocky Moon: A Geologist's History of Lunar Exploration* (University of Arizona Press, 1993)

Front: Paul Spudis, NASA *1:* Lick Observatory *2:* Paul Spudis (Lunar and Planetary Inst.) *3a:* Paul Spudis *3b:* Nicolas Stacy and Donald Campbell (Cornell Univ.) *4:* NASA/Johnson *5a:* NASA/Johnson *5b:* Paul Spudis *6:* NASA/Johnson *7a:* NASA/Johnson *7b:* Paul Spudis *8:* Paul Spudis *9:* NASA/Johnson *10a:* NASA/Johnson *10b:* Paul Spudis *11:* Paul Spudis *12a:* NASA/Johnson *12b:* Paul Spudis *13:* Paul Spudis, NASA *14:* Maria Zuber (MIT), NASA *15:* Paul Spudis *16:* Susan Pullan (Geological Survey Canada) *17:* Sky Publishing *18:* Don Davis *19:* Alastair Cameron (Center for Astrophysics) *20:* Paul Spudis *21:* Don Wilhelms *22:* NASA/Ames *Table 1:* Paul Spudis *Table 2:* Ross Taylor

CHAPTER 11: Mars

MICHAEL H. CARR received his Ph.D. from Yale University in 1960. He has been on the scientific staff of the U.S. Geological Survey since 1962, serving as the chief of its Branch of Astrogeologic Studies in Flagstaff, Arizona, from 1974 to 1978, and is currently engaged in research in Menlo Park, California. His interests include modeling the various roles volcanism, water, and ice have played throughout Martian history. Carr has served as the leader of the Viking Orbiter imaging team and on the scientific teams for Lunar Orbiter, Mariner 9, Voyager, and Galileo. He is currently a member of the Mars Global Surveyor imaging team, studying the function of water in the Martian past.

Baker, V. R., *The Channels of Mars* (University of Texas Press, 1982)

Carr, M. H., *The Surface of Mars* (Yale University Press, 1981)

McSween, H. S., "Nor any Drop to Drink" (*Sky & Telescope, 90,* 18–23, December 1995)

Scientific Results of the Viking Project (*Journal of Geophysical Research, 82,* 3959–4681, 1977)

Sheehan, W., *The Planet Mars: A History of Observation and Discovery* (University of Arizona Press, 1996)

Special issue of Mars Pathfinder Results (*Science, 278,* 1734–1774, 5 December 1997)

Viking Orbiter Views of Mars (U.S. National Aeronautics and Space Administration, SP-441, 1980)

Front: NASA/JPL *1:* David Crisp (JPL), WFPC2 Science Team *2:* Don Davis, NASA/JPL *3:* NASA/JPL *4:* R. Rieder (Max Planck Inst. Chemie), NASA *5:* Don Davis, NASA/JPL *6:* Edward Guinness and Thomas Stein (Washington Univ.) *7:* MOLA Science Team, NASA *8:* NASA/JPL *9:* Crofton Farmer and Peter Doms *10:* Michael Carr (USGS, Menlo Park), NASA *11:* Dale Schneeberger (JPL) *12:* Annie Allison and Alfred McEwen (USGS, Flagstaff) *13:* Don Davis *14:* Calvin Hamilton (LANL) *15:* Alfred McEwen *16:* Michael Malin (MSSS), NASA *17:* NASA/JPL *18:* NASA/JPL *19:* Michael Carr *20:* NASA/JPL *21:* Michael Malin (MSSS), NASA *22:* NASA/JPL *23:* Alfred McEwen *24:* NASA/JPL *25:* NASA/JPL *Table 1:* R. Rieder, NASA

CHAPTER 12: Interiors of the Terrestrial Planets

JAMES W. HEAD, III, is a professor of geological sciences at Brown University, where he received his Ph.D. in 1969. From 1968 to 1972, he participated in selecting the lunar landing sites and astronaut training for the Apollo program, and he continues to be involved with astronaut training. Head's research centers on the study of the processes that form and modify planetary surfaces and crusts, and how these processes vary with time and interact to produce the historical record preserved on the planets. He has served on the science teams for NASA's Magellan, Mars Surveyor, and Galileo missions, and on the Soviet Union's Venera 15/16 and Phobos missions.

Cattermole, P., *Planetary Volcanism: A study of Volcanic Activity in the Solar System* (John Wiley & Sons, 1996)

Chapman, C. R., *Planets of Rock and Ice* (Scribner's, 1982)

Greeley, R., *Planetary Landscapes* (Chapman & Hall, 1993)

Hartmann, W. K., *Moons and Planets* (Wadsworth, 1993)

Head, J. W., and S. C. Solomon, "Tectonic Evolution of the Terrestrial Planets" (*Science, 213,* 62–76, 3 July 1981)

Murray, B. C., and others, *Earthlike Planets* (W. H. Freeman, 1981)

Mursky, G., *Introduction to Planetary Volcanism* (Prentice-Hall, 1996)

The Near Planets, part of the "Voyage Through the Universe" series (Time-Life, 1989)

Front: Don Davis *1:* James Head (Brown Univ.) *2:* James Head *3:* James Head *4:* James Head *5:* James Head *6:* James Head *7:* Don Davis *8:* James Head *9:* NASA/Johnson *10:* James Head *11:* NASA/JPL *12:* Calvin Hamilton (LANL), NASA *13:* Calvin Hamilton *14:* NASA/JPL *15:* NASA/JPL *16:* Sky Publishing *17:* NASA/JPL *18:* James Head *19:* NASA/JPL *20:* James Head

CHAPTER 13: Atmospheres of the Terrestrial Planets

BRUCE M. JAKOSKY is a professor of geology and a member of the Laboratory for Atmospheric and Space Physics at the University of Colorado at Boulder. Jakosky received his Ph.D. in planetary science from the California Institute of Technology in 1982. His research focuses on the geology and remote sensing of plane-

AUTHORS AND SUGGESTED READINGS

tary surfaces, the evolution of the Martian atmosphere and climate, and planetary exobiology. He has participated in the Viking, Solar Mesosphere Explorer, Clementine, Mars Observer, and Mars Global Surveyor missions. He has recently served as editor of the *Journal of Geophysical Research (Planets)* and written a book on the possible existence of extraterrestrial life.

Albee, A. L., F. D. Palluconi, and R. E. Arvidson, "Mars Global Surveyor Mission: Overview and Status" (*Science, 279,* 1671–1672, 13 March 1998)

Atreya, S. K., J. B. Pollack, and M. S. Matthews, eds., *Origin and Evolution of Planetary and Satellite Atmospheres* (University of Arizona Press, 1989)

Bougher, S. W., D. M. Hunten, and R. S. Phillips, *Venus II* (University of Arizona Press, 1997)

Grinspoon, D. H., *Venus Revealed* (Addison-Wesley, 1997)

Hunten, D. M., and A. L. Sprague, "Origin and Character of the Lunar and Mercurian Atmospheres" (*Advances in Space Research, 19,* 1551–1560, 1997)

Jakosky, B. M., *The Search for Life on Other Planets* (Cambridge University Press, 1998)

Kargel, J. S., and R. G. Strom, "Global Climatic Change on Mars" (*Scientific American, 275,* 80–88, November 1996)

Kieffer, H. H., and others, eds., *Mars* (University of Arizona Press, 1992)

CHAPTER 14: Interiors of the Giant Planets

WILLIAM B. HUBBARD received a Ph.D. in astronomy from the University of California at Berkeley in 1967, for which he explored stellar structure and evolution of condensed objects. Prior to joining the University of Arizona in 1973, Hubbard conducted research at the California Institute of Technology, the University of Texas at Austin, and the O. Yu. Schmidt Institute in Moscow. At the University of Arizona, Hubbard has been director of its Lunar and Planetary Laboratory and is currently professor of planetary sciences. His research is devoted to giant-planet interiors, brown dwarfs, and extrasolar giant planets, as well as the occultation of stars by solar-system objects.

Bergstralh, J. T., E. D. Miner, and M. S. Matthews, eds., *Uranus* (University of Arizona Press, 1991)

Cruikshank, D. P., *Neptune and Triton* (University of Arizona Press, 1995)

Gehrels, T., and M. S. Matthews, *Saturn* (University of Arizona Press, 1984)

Hartmann, W. K., *Moons and Planets* (Wadsworth Publ. Co., 1993)

Hubbard, W. B., *Planetary Interiors* (Van Nostrand Reinhold, 1984)

Kafatos, M., R. S. Harrington, and S. P. Maran, *Astrophysics of Brown Dwarfs* (Cambridge University Press, 1985)

Marcy, G., and Butler, P., "The Diversity of Planetary Systems" (*Sky & Telescope, 95,* 30–37, March 1998)

CHAPTER 15: Atmospheres of the Giant Planets

ANDREW P. INGERSOLL's interests in oceans and atmospheres was nurtured at Harvard University, where he was awarded his Ph.D. in 1966. He has been a professor of planetary science at the California Institute of Technology since 1976. Ingersoll is interested in how planetary atmospheres and oceans work — how they redistribute heat, why winds and currents flow as they do, and how their climates change. He has served on scientific teams for the Pioneer Venus, Voyager, Pioneer Jupiter, Pioneer Saturn, and Galileo missions.

Alexander, A. F. O'D., *The Planet Saturn* (Dover, 1980)

Beebe, R., *Jupiter: The Giant Planet* (Smithsonian Institution Press, 1997)

The Far Planets, part of the "Voyage Through the Universe" series (Time-Life, 1988)

Ingersoll, A. P., "Uranus" (*Scientific American, 256,* 38–45, January 1987)

Lindal, G. F., D. N. Sweetnam, and V. R. Eshleman, "The Atmosphere of Saturn: An Analysis of the Voyager Radio Occultation Measurements" (*Astronomical Journal, 90,* 1136–1146, 1985)

CHAPTER 16: Planetary Rings

JOSEPH A. BURNS is a professor of engineering and astronomy at Cornell University, where he received his Ph.D. in 1966. His current research concerns planetary rings, small bodies of the solar system, orbital evolution, and tides, in addition to the rotational dynamics and strength of planets, satellites, and asteroids. Burns is a member of the imaging team on the Cassini mission to Saturn and an associate of the imaging investigation team for the Galileo mission to Jupiter. He was the editor of *Icarus,* the principal journal of planetary science, for nearly 20 years. Until recently, he chaired COMPLEX, the National Research Council's advisory body for solar system exploration.

Beatty, J. K., "Rings of Revelation" (*Sky & Telescope, 92,* 30–33, August 1996)

Cuzzi, J. N., and L. W. Esposito, "The Rings of Uranus" (*Scientific American, 257,* 52–66, July 1987)

Elliot, J., and R. Kerr, *Rings* (MIT Press, 1984)

French, R., and others, "Dynamics and Structure of the Uranian Rings" in *Uranus,* J. T. Bergstralh and others, eds. (University of Arizona Press, 327–409, 1991)

Greenberg, R., and A. Brahic, eds., *Planetary Rings* (Univ. of Arizona Press, 1984)

Lissauer, J. J., "Shepherding Model for Neptune's Arc Ring" (*Nature, 318*, 544–545, 12 December 1985)

Nicholson, P. D., and others, "Observations of Saturn's Ring-Plane Crossings in August and November 1995" (*Science, 272*, 509–515, 26 April 1996)

Porco, C., and others, "Neptune's Ring System" in *Neptune and Triton*, D. P. Cruikshank, ed. (University of Arizona Press, 703–804, 1995)

Moons and Rings, part of the "Voyage Through the Universe" series (Time-Life, 1991)

Front: Don Davis 1: F. O'D. Alexander, *The Planet Saturn 2:* Sky Publishing *3:* Heikki Salo (UNAM, Mexico) *4:* Sky Publishing *5:* Jeffrey Cuzzi (NASA/Ames) *6:* Frank Shu (Univ. California, Berkeley) *7:* Don Davis *8:* NASA/JPL *9:* Don Davis *10a:* NASA/JPL *10b:* NASA/JPL *11:* Imke de Pater (Univ. California, Berkeley) *12:* James Elliot (MIT) *13:* Erich Karkoschka (Univ. Arizona), NASA *14:* NASA/JPL *15:* NASA/JPL *16:* NASA/JPL *17:* NASA/JPL *18:* NASA/JPL *19:* NASA/JPL *20:* Lonne Lane (JPL) *21:* Imke de Pater *22:* NASA/JPL *23:* Erich Karkoschka, NASA *24:* Jeffrey Cuzzi and Paul Estrada (NASA/Ames) *25:* NASA/JPL *26:* NASA/JPL *27:* Mark Showalter (NASA/Ames) *28:* Philip Nicholson (Cornell) *29:* Mark Showalter *30:* NASA/JPL *Table 1:* Joseph Burns (Cornell Univ.)

CHAPTER 17: Io

TORRENCE V. JOHNSON, the project scientist for the Galileo mission, has long been interested in the Galilean satellites of Jupiter. He earned a Ph.D. in planetary science at the California Institute of Technology in 1970. His research involves remote sensing of the Galilean satellites, the Moon, the terrestrial planets, large planetary satellites, and asteroids. He is a senior research scientist at the Jet Propulsion Laboratory. Johnson was as a member of the Voyager imaging team.

Johnson, T. V., and others, "Io: Evidence for Silicate Volcanism in 1986" (*Science, 242*, 1280–1283, 2 December 1988)

McGrath, M. A., "Io and the Plasma Torus" (*Science, 278*, 237–238, 10 October 1997)

Morrison, D., ed., *Satellites of Jupiter* (University of Arizona Press, 1982)

Peale, S. J., P. Cassen, and R. T. Reynolds, "Melting of Io by Tidal Dissipation" (*Science, 203*, 892–894, 2 March 1979)

Smith, B. A., and others, "The Jupiter System Through the Eyes of Voyager 1" (*Science 204*, 951–972, 1 June 1979)

Front: NASA/JPL *1:* NASA/JPL *2:* Sky Publishing *3:* Zareh Gorjian and Eric DeJong (JPL) *4:* Roger Clark (USGS, Denver) *5:* Charles Yoder (JPL *6a:* NASA/JPL *6b:* John Spencer (Lowell Obs.), NASA *7:* NASA/JPL *8:* NASA/JPL *9:* NASA/JPL *10:* NASA/JPL *11:* Torrence Johnson (JPL) *12:* NASA/JPL *13:* Ashley Davies (JPL) *14:* Don Davis *15:* NASA/JPL *16:* NASA/JPL *17a:* NASA/JPL *17b:* Nicholas Schneider (Univ. Colorado) *18:* Alfred McEwen (Univ. Arizona) *19:* Don Davis *Table 1:* Torrence Johnson (JPL) *Table 2:* Rosaly Lopes-Gautier (JPL)

CHAPTER 18: Europa

RONALD GREELEY received his Ph.D. from the University of Missouri at Rolla in 1966. Before becoming a professor of geology at Arizona State University in 1977, Greeley worked on the Apollo Lunar program and analyzed planetary data at NASA's Ames Research Center for 11 years. He was also on the Viking science team from 1976 to 1980. His research includes the study of impact craters, wind processes on terrestrial planets, field stud-

ies of basaltic volcanism, and geologic mapping of Mars, Venus, Io, Europa, and Callisto. Greeley serves on editorial panels for several journals and is associate editor for the planetary science series of Cambridge University Press.

Anderson, J. D., and others, "Europa's differentiated internal structure: Inferences from two Galileo encounters" (*Science, 276*, 1236–1239, 23 May 1997)

Belton, M. J. S., and others, "Galileo's first images of Jupiter and the Galilean satellites" (*Science, 274*, 377–385, 18 October 1996)

Cassen, P. M., S. J. Peale, and R. T. Reynolds, "Structure and thermal evolution of the Galilean Satellites" in *Satellites of Jupiter*, D. Morrison, ed. (University of Arizona Press, 93–128, 1982)

Greeley, R., and others, "Europa: Initial Galileo Geological Observations" (*Icarus, 135*, September 1998)

Lucchitta, B. K. and L. A. Soderblom, "Geology of Europa" in *Satellites of Jupiter*, D. Morrison, ed. (University of Arizona Press, 521–555, 1982)

Malin, M. C., and D. C. Pieri, "Europa" in *Satellites*, J. A. Burns and M. S. Matthews, eds. (University of Arizona Press, 689–716, 1986)

Milstein, M., "Diving into Europa's Ocean" (*Astronomy, 25*, 38–43, October 1997)

Schubert, G., T. Spohn, and R. T. Reynolds. "Thermal Histories, Compositions, and Internal Structures of the Moons of the Solar System" in *Satellites*, J. A. Burns and M. S. Matthews, eds. (University of Arizona Press, 224–292, 1986)

Squyres, S. W., and others, "Liquid Water and Active Resurfacing on Europa" (*Nature, 301*, 225–226, 20 January 1983)

Front: NASA/JPL *1:* NASA/JPL *2:* Eric DeJong and Zareh Gorjian (JPL) *3:* NASA/Ames *4:* Ronald Greeley (Arizona State Univ.), NASA *5:* Ronald Greeley *6:* Ronald Greeley *7:* Don Davis *8:* Robert Sullivan (Cornell Univ.) *9:* Ronald Greeley *10:* Cynthia Phillips (Univ. Arizona) *11:* Ronald Greeley *12:* Ronald Greeley *13:* Ronald Greeley *14:* NASA/JPL *15:* Ronald Greeley *16:* Ronald Greeley *17:* Ronald Greeley

CHAPTER 19: Ganymede and Callisto

ROBERT T. PAPPALARDO received his B.A. in geological science from Cornell University in 1986 and his Ph.D. from Arizona State University in 1994. He is currently a senior research associate at Brown University. He has investigated the geological histories of icy outer planet satellites, including the large interior upwellings that have shaped the surface of Miranda. He has worked with the Galileo imaging team to plan many of the Galileo observations of Europa and Ganymede, and his present research concentrates on the analysis of these Galileo images.

Belton, M. J. S., and others, "Galileo's First Images from Jupiter's Orbit" (*Science, 274*, 377–385, 18 October 1996)

McCord, T. B., and others, "Organics and Other Molecules in the Surfaces of Callisto and Ganymede" (*Science, 278*, 271–275, 10 October 1997)

McKinnon, W. B., and E. M. Parmentier, "Ganymede and Callisto" in *Satellites* (J. A. Burns and M. S. Matthews, eds., University of Arizona Press, 718–763, 1986)

Schenk, P. M., "The geology of Callisto" (*Journal of Geophysical Research, 100*, 19,023–19,040, 1995)

Shoemaker, E. M., and others, "The Geology of Ganymede," in *Satellites of Jupiter*, D. Morrison, ed. (University of Arizona Press, 435–520, 1982)

Spencer, J., "Jupiter's Odd Bunch" (*New Scientist, 154*, 42–45, 5 May 1997)

Front: NASA/JPL *1:* Tilmann Denk (DLR, Germany), NASA *2:* Wendy Calvin (Lowell Observatory) *3:* NASA/JPL *4:* Don Davis, Jay Melosh (Univ. Arizona) and William McKinnon (Washington Univ.) *5:* NASA/JPL *6:* Bernd Giese (DLR) *7:* NASA/JPL *8:* Robert Pappalardo (Brown Univ.) *9:* Marc Parmentier (Brown Univ.) *10:* NASA/JPL *11:* NASA/JPL *12:* NASA/JPL *13:* Paul Schenk (Lunar and Planetary Inst.) *14:* NASA/JPL *15:* NASA/JPL *16:* NASA/JPL *17:* James Klemaszewski (Arizona State Univ.), Marcia Segura (JPL), and Jan Yosimizu *18:* NASA/JPL *19:* NASA/JPL *20:* NASA/JPL *21:* NASA/JPL *22:* Zareh Gorjian and Eric DeJong (JPL) *Table 1:* Robert Pappalardo

CHAPTER 20: Titan

TOBIAS OWEN received his Ph.D. from the University of Arizona in 1965. He is currently a professor of astronomy at the Institute for Astronomy, University of Hawai'i. He was a member of the Viking lander science team and is presently an interdisciplinary scientist on both the Galileo mission to Jupiter and the Cassini-Huygens mission to Saturn and Titan. Owen is pursuing a systematic study of the planetary atmospheres. His efforts yielded the first detection of carbon monoxide on Titan, molecular nitrogen on Pluto, and the discovery of deuterium enrichment on Mars.

Hunten, D. M., and others, "Titan" in *Saturn*, T. Gehrels and M. S. Matthews, eds. (University of Arizona Press, 671–759, 1984)

Lebreton, J. P., and D. Matson, eds., "The Huygens Probe: Science, Payload, and Mission Overview" (*ESA Bulletin*, No. 92, November 1997)

Lindal, G. F., and others, "The Atmosphere of Titan: An Analysis of the Voyager 1 Radio Occultation Measurments" (*Icarus, 53,* 348–363, February 1983)

Lunine, J. I., "Does Titan Have Oceans?" (*American Scientist, 82,* 134-143, March-April 1994)

Morrison, D., T. Owen, and L. A. Soderblom, "The Satellites of Saturn" in *Satellites*, J. A. Burns and M. S. Matthews, eds. (Univ. of Arizona Press, 764–801, 1986)

Owen, T., "Titan" (*Scientific American, 246,* 98–109, February 1982)

Smith, P. H., and others, "Titan's Surface, Revealed by HST Imaging" (*Icarus, 119,* 336–349, February 1996)

Toon, O. B., and others, "Methane Rain on Titan" (*Icarus, 75,* 255–284, August 1988)

Front: Don Davis *1:* Erich Karkoschka (Univ. Arizona), NASA *2:* Robert Danehy, Tobias Owen (Univ. Hawaii) *3:* NASA/JPL *4:* Virgil Kunde (NASA/Goddard) *5:* Tobias Owen *6:* Peter Smith and Mark Lemmon (Univ. Arizona), NASA *7:* Ralph Lorenz (Univ. Arizona) *8:* Don Davis *Table 1:* Tobias Owen

CHAPTER 21: Triton, Pluto, and Charon

DALE P. CRUIKSHANK spent the years following his undergraduate studies working for the late Gerard P. Kuiper, first at Yerkes Observatory and later at the University of Arizona, where he received his Ph.D. in 1968. Cruikshank spent almost 18 years at the Institute for Astronomy, University of Hawai'i, observing the surfaces and atmospheres of planets, satellites, asteroids and comets. In 1988, he transferred to NASA's Ames Research Center, where he studies Io, Iapetus, Pluto, Triton, and other unusual small bodies. He serves as an associate editor for *Icarus*, the international planetary science journal.

Crosswell, K., "The Titan/Triton Connection" (*Astronomy, 21,* 26–36, April 1993)

Cruikshank, D. P., ed., *Neptune and Triton* (University of Arizona Press, 1995)

Hoyt, W. G., *Planets X and Pluto* (University of Arizona Press, 1980)

Littmann, M., *Planets Beyond: Discovering the Outer Solar System* (John Wiley & Sons, 1990)

Stern, A., and J. Mitton, *Pluto and Charon: Ice Worlds on the Ragged Edge of the Solar System* (John Wiley and Sons, 1998)

Stern, S. A., and D. J. Tholen, eds., *Pluto and Charon* (University of Arizona Press, 1997)

Tombaugh, C. W., and P. Moore, *Out of the Darkness: The Planet Pluto* (Stackpole Books, 1980)

Front: Alfred McEwen, NASA/JPL *1:* Alfred McEwen, USGS *2:* Alfred McEwen *3:* NASA/JPL *4:* Alfred McEwen *5:* NASA/JPL *6:* Dale Cruikshank (NASA/Ames); Eric Quirico (CNRS) *7:* NASA/JPL *8:* Alan Harris (JPL) *9:* Sky Publishing *10:* Randall Kirk (USGS, Flagstaff) *11:* Lowell Observatory *12:* David Tholen (Univ. Hawaii) and Edward Tedesco (JPL) *13:* David Tholen *14:* Alan Stern (Southwest Research Inst.) *15:* Tobias Owen (Univ. Hawaii) *16:* Marc Buie (Lowell Observatory) *17:* Alan Stern and Marc Buie *18:* Sky Publishing *19:* Don Davis

CHAPTER 22: The Icy Satellites

WILLIAM B. McKINNON has been fascinated with the outer solar system since boyhood. After studying physics and geology as an undergraduate, he attended the California Institute of Technology and in 1980 defended a Ph.D thesis on impact cratering mechanics. He was at the University of Arizona before joining the faculty at Washington University in St. Louis, where he is now a professor. His current research interests include cratering, Pluto and the Kuiper belt, and the origin and evolution of icy satellites, especially Europa. McKinnon helped write *The Integrated Strategy for the Planetary Sciences, 1995–2010,* for the National Research Council, and he agitates as a member of the "Pluto Underground."

Jankowski, D. G., and S. W. Squyres, "Solid-state Ice Volcanism on the Satellites of Uranus" (*Science, 241,* 1322–1325, 9 September 1988)

Littmann, M., *Planets Beyond: Discovering the Outer Solar System* (Wiley & Sons, 1990)

McKinnon, W. B., "Sublime Solar System Ices" (*Nature, 375,* 535–536, 15 June 1995)

Morrison, D., and others, "The Satellites of Saturn" in *Satellites,* J. Burns and M. S. Matthews, eds. (University of Arizona Press, 764–801, 1986)

Noll, K. S., and others, "Detection of Ozone on Saturn's Satellites Rhea and Dione" (*Nature, 388,* 45–47, 3 July 1997)

Rothery, D. A., *Satellites of the Outer Planets: Worlds in Their Own Right* (Clarendon Press, 1992)

Wilson, P. D., and C. Sagan, "Spectrophotometry and Organic Matter on Iapetus" (*Icarus, 122,* 92–106, July 1996)

Front: Calvin Hamilton, NASA *1:* Paul Schenk (Lunar and Planetary Inst.), NASA *2:* Don Davis, William McKinnon (Washington Univ.) *3:* Paul Schenk *4:* USGS, NASA *5:* Paul Schenk *6:* Calvin Hamilton (LANL) *7:* NASA/JPL *8:* William McKinnon *9:* Paul Schenk *10:* Calvin Hamilton *11:* NASA/JPL *12:* Sky Publishing *13a:* NASA/JPL *13b:* Calvin Hamilton *14:* Paul Schenk *15:* Paul Schenk *16:* Paul Schenk *17:* Paul Schenk *18:* NASA/JPL *Table 1:* William McKinnon

CHAPTER 23: Small Bodies: Patterns and Relationships

WILLIAM K. HARTMANN is a senior scientist at the Planetary Science Institute, a division of the San Juan Capistrano Research Institute. He received his Ph.D. in astronomy from the University of Arizona in 1966, and his research focuses on collisions, accretion, and the early cratering history of the planets. Hartmann served as a co-investigator on the Mariner 9 imaging team, and his telescopic observations include co-discovery of cometary activity on the unusual object 2060 Chiron. Hartmann has written several astronomy textbooks and popular illustrated books, and he is a noted astronomical artist.

Burns, J. A., and M. S. Matthews, eds., *Satellites* (University of Arizona Press, 1986)

Chapman, C., "Two Shades Beyond Neptune" (*Nature, 392,* 16–17, 5 March 1998)

Hartmann, W. K., *Moons and Planets* (Wadsworth, 1993)

Stern, S., and H. Campins, "Chiron and the Centaurs: Escapees from the Kuiper Belt" (*Nature, 382,* 8 August 1996)

Front: NASA/JPL *1a:* Eleanor Helin (JPL) *1b:* European Southern Observatory *2:* Karen Meech (Univ. Hawaii) *3:* Sky Publishing *4:* William Hartmann (Planetary Science Inst.) *5:* William Hartmann *6:* William Hartmann *7:* Charles Wood (Univ. North Dakota) *8:* NASA/JPL *9:* Rob Johnson *10:* NASA/JPL *11:* Calvin Hamilton (LANL) *12:* NASA/JPL *Table 1:* Sky Publishing *Table 2:* Sky Publishing

CHAPTER 24: Comets

JOHN C. BRANDT retired in 1998 from the University of Colorado, Boulder, where he was a senior research associate in the Laboratory for Atmospheric and Space Physics and a professor in the department of astrophysics and planetary sciences. He received his Ph.D. in astronomy and astrophysics from the University of Chicago in 1960. He served on the science teams of the Solar Maximum Mission and the International Cometary Explorer. He was the principle investigator on the Goddard High Resolution Spectrograph, one of the original instruments on the Hubble Space Telescope. His research includes the studies of comets, solar physics, and planetary atmospheres.

Battrick, B., E. J. Rolfe, and R. Reinhard, eds., *20th ESLAB Symposium on the Exploration of Halley's Comet* (European Space Agency, SP-250, 1986)

Brandt, J. C., and R. D. Chapman, *Introduction to Comets* (Cambridge University Press, 1981)

Brandt, J. C., and M. B. Niedner, Jr., "The Structure of Comet Tails" (*Scientific American, 254,* 48–56, January 1986)

Delsemme, A. H., "Whence Come Comets?" (*Sky & Telescope, 77,* 260–264, March 1989)

Grewing, M., F. Praderie, and R. Reinhard, eds., *Exploration of Halley's Comet* (Springer-Verlag, 1988)

Mumma, M. J., "Hyakutake's Interstellar Ices" (*Nature, 383,* 581–582, 17 October 1996)

Newcott, W. R., "The Age of Comets" (*National Geographic, 192,* 94–109, December 1997)

Whipple, F. W., *The Mystery of Comets* (Smithsonian Institution, 1985)

Wilkening, L. L., ed., *Comets* (University of Arizona Press, 1982)

Front: Akira Fujii *1:* Dennis Young *2:* Lowell Observatory *3:* Kelly Beatty (Sky Publishing) *4:* William Liller *5:* Halley Multicolor Camera team *6:* Michael Belton (NOAO) *7:* Don Davis *8:* Akira Fujii *9:* Armand Delsemme (Univ. Toledo) *10:* Guido Pizarro (European Southern Obs.) *11:* Gabriele Cremonese (Padua Astronomical Obs.) *12:* Johns Hopkins Univ., Applied Physics Lab. *13:* Paul Feldman (Johns Hopkins Univ.) *14:* Susan Wyckoff and Peter Wehinger (Arizona State Univ.) *15:* Harold Weaver (Johns Hopkins Univ.), NASA *16:* Don Davis *17:* Harold Weaver, NASA *18:* Mayo Greenberg (Leiden Univ.) *19:* Donald Brownlee (Univ. Washington) *20:* Sky Publishing *21:* Shigemi Numazawa (Japan Planetarium Lab.) *22:* E. Kolmhofer and H. Raab (Johannes Kepler Univ.) *23:* Carey Lisse (NASA/Goddard) *24:* John Brandt (Univ. Colorado) *Table 1:* John Brandt (Univ. Colorado) *Table 2:* Neil Bone (British Astronomical Assoc.)

CHAPTER 25: Asteroids

CLARK R. CHAPMAN is an institute scientist at the Boulder, Colorado, office of the Southwest Research Institute. Chapman received his Ph.D. in planetary science from the Massachusetts Institute of Technology in 1972. He studies planetary cratering and small bodies of the solar system, having served on the science teams for the Galileo and Near Earth Asteroid Rendezvous missions. Chapman has chaired the Division for Planetary Sciences of the American Astronomical Society, and he was president of Commission 15 (Physical Properties of Asteroids and Comets) of the International Astronomical Union. Chapman is a widely sought speaker and author on asteroids and their role in Earth history.

Asphaug, E., "New Views of Asteroids" (*Science, 278,* 2070–2071, 19 December 1997)

Binzel, R. P., T. Gehrels, and M. S. Matthews, eds., *Asteroids II* (University of Arizona Press, 1989)

Binzel, R. P., M. A. Barucci, and M. Fulchignoni, "The Origins of the Asteroids" (*Scientific American, 265,* 88–94, October 1991)

Cunningham, C. J., *Introduction to Asteroids* (Willmann-Bell, 1988)

Gehrels, T., ed., *Asteroids* (University of Arizona Press, 1979)

Lewis, J. S., *Mining the Sky: Untold Riches from the Asteroids, Comets, and Planets* (Addison-Wesley, 1996)

Front: NASA/JPL *1:* Edward Bowell (Lowell Obs.) *2:* Marc Buie (Lowell Obs.) *3:* Don Davis *4:* Mark Sykes (Univ. Arizona) *5:* Scott Hudson (Washington State Univ.), Steven Ostro (JPL) *6:* Steven Ostro, Scott Hudson *7a:* Don Davis *7b:* Clark Chapman (Southwest Research Inst.) *8:* Clark Chapman *9:* Don Davis *10:* Peter Thomas (Cornell Univ.), NASA *11:* Jeffrey Bell (Univ. Hawaii) *12:* Joseph Veverka (Cornell Univ.) *13:* NASA/JPL *14:* NASA/JPL *15:* NASA/JPL *16:* Applied Physics Lab., Johns Hopkins Univ. *Table 1:* Clark Chapman *Table 2:* IAU Minor Planet Center

CHAPTER 26: Meteorites

HARRY Y. MCSWEEN, JR., is a professor of geology at the University of Tennessee in Knoxville. Since receiving a Ph.D. at Harvard in 1977, he has studied the petrology and chemistry of extraterrestrial materials. McSween has worked extensively on Martian meteorites, particularly on their implications for understanding the geology of Mars, and on the thermal alteration histories of various kinds of chondrites. He has served as president of the Meteoritical Society. McSween is also a participating scientist for the Mars Pathfinder and Mars Global Surveyor missions. He has written a geochemistry textbook and three popular books on meteorites and planetary science.

Allegre, C., *From Stone to Star, A View of Modern Geology* (Harvard University Press, 1992)

Bone, N., *Meteors* (Sky Publishing, 1993)

Dodd, R. T., *Thunderstones and Shooting Stars* (Harvard University Press, 1986)

Hutchinson, R., *The Search for Our Beginning: An Enquiry Based on Meteorite Research* (Oxford University Press, 1983)

Kerridge, J. F., and M. S. Matthews, eds., *Meteorites and the Early Solar System* (University of Arizona Press, 1988)

McSween, H. Y., *Meteorites and Their Parent Planets* (Cambridge University Press, 1987)

Norton, O. R., *Rocks from Space* (Mountain Press Publishing, 1998)

Wasson, J. T., *Meteorites: Their Record of Early Solar System History* (W. H. Freeman, 1985)

Front: Jeff Smith *1:* NASA/Johnson *2:* NASA/Johnson *3:* NASA/Johnson *4:* Inst. of Meteoritics, Univ. New Mexico *5:* Richard Norton, *Rocks From Space* *6:* Harry McSween (Univ. Tennessee) *7:* Natural History Museum, London *8:* Ursula Marvin (Smithsonian Astrophysical Obs.) *9:* Jeff Smith *10:* Daniel Ball (Arizona State Univ.) *11:* Smithsonian Institution *12:* Donald Brownlee (Univ. Washington) *13:* Robert Clayton (Univ. Chicago) *14:* Sky Publishing *15:* Oliver Farrington, *Meteorites, Their Structure, Composition, and Terrestrial Relations* *16:* Roy Lewis (Univ. Chicago) *17:* Harry McSween *18:* Harry McSween *19:* NASA/Johnson *20:* NASA/Johnson *21:* NASA/Johnson *22:* T. McCoy (Smithsonian Inst.) *23a:* John Valley (Univ. Wisconsin) *23b:* Allan Treiman (Lunar and Planetary Inst.) *Table 1:* Frank Wlotzka (Max Planck Inst. für Chemie) *Table 2:* Sky Publishing

CHAPTER 27: Life in the Solar System

GERALD A. SOFFEN is the director of university programs at NASA's Goddard Space Flight Center. He received his Ph.D. in biology in 1961 from Princeton. Prior to his appointment at Goddard, Soffen served as director of life sciences at NASA Headquarters. In that capacity he oversaw the agency's biomedical, space-biology, and exobiology programs. As project scientist for Viking missions to Mars, Soffen coordinated the mission's scientific investigations of Mars from orbit, during entry, and on the Martian surface. He recently coordinated the creation and staffing of NASA's Astrobiology Institute.

Cooper, H. S. F., Jr., *The Search for Life on Mars* (Harper and Row, 1980)

Fredrickson, J. K., and T. C. Onstott, "Microbes Deep Inside the Earth" (*Scientific American, 275,* 68–73, October 1996)

Gibson, E. K., and others, "The Case for Relic Life on Mars" (*Scientific American, 277,* 58–63, December 1997)

Goldsmith, D., *The Hunt for Life on Mars* (Penguin, 1997)

Horgan, J., "The Sinister Cosmos" (*Scientific American, 276,* 18–20, May 1997)

Horowitz, N. H., *To Utopia and Back* (W. H. Freeman, 1986)

Jakosky, B. M., "Searching for Life in Our Solar System" in *The Magnificent Cosmos* (*Scientific American,* Spring 1998)

Lemonick, M. D., *Other Worlds: The Search for Life in the Universe* (Simon & Schuster, 1998)

Madigan, M. T., and B. L. Marrs, "Extremophiles" (*Scientific American, 276,* 82–87, April 1997)

McSween, H. Y., *Fanfare for Earth: The Origins of Our Planet and Life* (St. Martin's Press, 1997)

Front: Don Davis *1:* Don Davis *2:* Dudley Foster (Woods Hole Oceanographic Inst.) *3:* Sky Publishing *4:* Charles Fisher (Penn State University)

5: NASA/JPL *6:* NASA/JPL *7:* NASA/JPL *8:* NASA/JSC *9:* NASA/JSC *10:* *Science,* NASA *11a:* Hojatollah Vali (McGill Univ.) *11b:* Susan Wentworth (NASA/Johnson) *12:* NASA/JPL *13:* Don Davis *14:* Michael Kaiser (NASA/Goddard) *15:* Rudolf Hanel (NASA/Goddard)

CHAPTER 28: Other Planetary Systems

R. PAUL BUTLER is a staff astronomer at the Anglo-Australian Observatory in Sydney, Australia. He received a Ph.D. in astronomy from the University of Maryland in 1993. Along with longtime collaborator Geoffrey Marcy, Butler has developed the most precise astronomical Doppler measurement technique in the world and discovered six of the first eight known extrasolar planets. Beginning in 1996 Butler and Marcy extended their planet survey from 100 stars at Lick Observatory in California to 700 stars using the Lick, Keck, and the Anglo-Australian Telescopes. These are the only active surveys currently capable of detecting planets around nearby stars resembling those in our own solar system.

Angel, J. R. P., and N. J. Woolfe, "Searching for Life on Other Planets" (*Scientific American, 274,* 60–66, April 1996)

Beckwith, S. V. W., and A. I. Sargent, "Circumstellar Disks and the Search for Neighbouring Planetary Systems" (*Nature, 383,* 139–144, 12 September 1996)

Black, D. C., "Other Suns, Other Planets" (*Sky & Telescope, 92,* 20–27, August 1996)

Black, D. C., and M. S. Matthews, *Protostars and Planets II* (University of Arizona Press, 1985)

Croswell, K., *Planet Quest: The Epic Discovery of Alien Solar Systems* (Free Press, 1997)

Goldsmith, D., *Worlds Unnumbered: The Search for Extrasolar Planets* (University Science Books, 1997)

Marcy, G., and R. P. Butler, "Detection of Extrasolar Planets" (*Annual Review of Astronomy and Astrophysics, 36,* 1998)

Sargent, A. I., and S. V. W. Beckwith, "The Search for Forming Planetary Systems" (*Physics Today, 46,* 22–29, April 1993)

Front: Wilhelm Kley and Pawel Artymowicz *1:* Sky Publishing *2:* Roger Sinnott (Sky Publishing) *3:* Sky Publishing *4:* Sky Publishing *5:* Geoffrey Marcy (Univ. California, Berkeley) *6:* Geoffrey Marcy and Paul Butler (Anglo-Australian Obs.) *7:* Sky Publishing *8:* Sky Publishing *9:* Robert O'Dell (Rice Univ.), NASA *10:* Stuart Weidenschilling and F. Marzari (Planetary Science Inst.) *11:* Christopher Burrows (Space Telescope Science Inst.), NASA *12:* Steven Simpson (Sky Publishing) *13:* Wilhelm Kley (Max Planck Inst.) and Pawel Artymowicz (Stockholm Obs.) *14:* Sky Publishing *15:* William Borucki (NASA/Ames) *16:* NASA/JPL *17:* Neville Woolf and Roger Angel (Univ. Arizona) *Table 1:* Sky Publishing

Index